# 16th International Conference on Electron Dynamics in Semiconductors, Optoelectronics and Nanostructures 2009

# (EDISON 16)



**Journal of Physics: Conference Series Volume 193**

## Montpellier, France
## 24-28 August 2009

**Editors:**

**Luca Varani**
**Gérald Bastard**
**Christophe Palermo**

ISBN: 978-1-61738-283-3

**Printed from e-media with permission by:**

Curran Associates, Inc.
57 Morehouse Lane
Red Hook, NY 12571
www.proceedings.com

**Some format issues inherent in the e-media version may also appear in this print version.**



Printed by Curran Associates, Inc. (2010)



Institute of Physics
Dirac House, Temple Back
Bristol BS1 6BE UK

Phone: 44 1 17 929 7481
Fax: 44 1 17 920 0979

techtracking@iop.org

# TABLE OF CONTENTS

## NONEQUILIBRIUM CARRIER TRANSPORT IN BULK MATERIALS, HETEROSTRUCTURES AND NANOSTRUCTURES

## INTERSUBBAND TRANSITIONS AND QUANTUM CASCADE LASERS



## TERAHERTZ PHENOMENA IN SEMICONDUCTOR MATERIALS AND DEVICES

## SEMICONDUCTOR-BASED SPINTRONICS



## TRANSPORT IN ORGANIC MATERIALS, GRAPHENE AND CARBON NANOTUBES

## QUANTUM CARRIER DYNAMICS IN ULTRASMALL STRUCTURES, NANOSTRUCTURES AND DEVICES

Journal of Physics: Conference Series **193** (2009) 011001 doi:10.1088/1742-6596/193/1/011001

# Edison16
Montpellier 2009

On behalf of the organizing committee we welcome all participants to the **16th International Conference on Electron Dynamics In Semiconductors, Optoelectronics and Nanostructures (EDISON)**.

Since the first meeting, organized in 1973 in Modena (Italy), this international conference has been regularly organized mostly each two years, and after 28 years EDISON is back to Montpellier for its 16th edition.

This conference was previously known as *"Hot Carriers In Semiconductors"* (HCIS). For some years the name was changed into *"Nonequilibrium Carrier Dynamics in Semiconductors"* (keeping the acronym HCIS). Then, during the last meeting of the international advisory committee in Tokyo, the decision was taken to find a new denomination that should be more adapted to the latest evolutions in the domains covered by the conference. After several discussions and a vote within the committee, the new acronym of EDISON was chosen. Therefore EDISON 16 can be considered as the last of a long series of conferences and, at the same time, the first of a new cycle of conferences.

The primary focus of EDISON is to provide an international meeting place to discuss the latest progress in the field of nonequilibrium classical and quantum carrier dynamics, from semiconductor bulk materials up to advanced structures and nanometric devices. This includes: classical incoherent transport, as well as coherent transport, usually analyzed by means of ultrafast electrical and optical excitations. To accompany the newest developments, TeraHertz phenomena, with a special attention to quantum cascade lasers, have been added to the list of topics. More recently new subjects have been included such as spintronics, transport in organic materials, graphene and carbon nanotubes. This list of topics is regularly modified and updated at each edition of the conference to represent but also anticipate the latest scientific evolutions.

Journal of Physics: Conference Series **193** (2009) 011001 doi:10.1088/1742-6596/193/1/011001

The distinctive features of EDISON are:

- A long-lasting history of 36 years.
- A fruitful interplay between fundamental and applied science due to a balanced participation of physicists and engineers.
- Emphasis is put on the physics of future devices rather than on the optimization of existing devices.
- A balance between theoretical and experimental contributions characterizes the scientific program.
- The scientific and advisory committees are composed by well-recognized scientists.

This year the scientific program was composed by more than 200 contributions divided into 60 oral presentations and 2 poster sessions. To keep a high scientific level, two independent referees have reviewed all submitted abstracts. We are happy to have drawn attendance from more than 30 different countries.

We hope that EDISON 16 has been an excellent opportunity for you to meet with colleagues from all over the world in an informal atmosphere, create an opportunity for networking, sharing of scientific and technical information and the building of trust relationships internationally.

Luca Varani
Christophe Palermo
Gérald Bastard
EDITORS

All papers published in this volume of *Journal of Physics: Conference Series* have been peer reviewed through processes administered by the proceedings Editors. Reviews were conducted by expert referees to the professional and scientific standards expected of a proceedings journal published by IOP Publishing.

EDISON 16 IOP Publishing
Journal of Physics: Conference Series **193** (2009) 012001 doi:10.1088/1742-6596/193/1/012001

# Space-charge-effect-induced large magnetoresistance in silicon

**M P Delmo[1], S Kasai[2], K Kobayashi[1], and T Ono[1]**

[1] Institute for Chemical Research, Kyoto University, Uji 611-0011, Kyoto, Japan,
[2] National Institute for Material Science,1-2-1 Sengen, Tsukuba, Ibaraki, Japan

Email: delmo@scc1.kuicr.kyoto-u.ac.jp

**Abstract**. We report that silicon shows large positive magnetoresistance (MR) between 0 and 3 T more than 1,000 per cent at 300 K and 3,000 per cent at 25 K. The experiment on lightly doped silicon reveals that when the carrier density decreases below $10^{13}$ cm$^{-3}$ the MR becomes pronounced due to the space-charge effect. We propose that the quasi-neutrality breaking of the space-charge effect can introduce electric field inhomogeneity, similar to the situation in other semiconductors where the non-saturating positive MR was observed. We have realized a magnetoresistive device that works at room temperature with simpler structure in a way different from other known metal-semiconductor hybrid devices.

## 1. Introduction

Consider a piece of material that has non-zero conductivity and has two ohmic contacts at both sides. An electric current is injected into the material from one of the contacts and extracted from the other as a finite voltage difference is maintained between them. Because there are usually many positive and negative charges that are equal in number in the material, the injected small amount of charge is immediately compensated by the background charges and by the charges from the extraction contact to achieve quasi-neutrality, which results in the uniform electric field in the material [1]. The conventional free electron model describes that the electrons have constant drift velocity and the external magnetic field perpendicular to the current does not affect the two-terminal resistance between the contacts [1]. Now what happens when more electrons are injected to break the quasi-neutrality? The accumulation of the injected electrons, namely the space-charge effect [2-4], occurs to generate an electric field unevenly distributed inside the material. The motion of the electrons is no longer trivial; the absence of the compensating charges makes the electrons correlate with each other via the unscreened Coulomb interaction such that their velocity becomes dependent on the local electric field and hence on their trajectory. The Hall field that is generated by the carrier motion in the magnetic field in the broken quasi-neutrality is expected to be very different from the simple Hall field that the conventional free electron model describes. Although it is naively expected that in this regime the modification of the electron trajectory by the magnetic field may result in an unusual electron transport such a possibility has never been explored, where we have found large magnetoresistance in silicon.

## 2. Experimental

In this experiment we used n-type silicon substrate (*n*-Si) that is lightly doped with phosphorus, and intrinsic silicon substrate (*i*-Si) supplied by Furuuchi Chemical Corporation. The nominal resistivities of *n*-Si and *i*-Si substrates are 10-20 Ωcm and > 1,000 Ωcm at room temperature, respectively. The carrier density of *n*-Si at 25 K and *i*-Si substrates at 300 K are $n = 2.73 \times 10^{11}$ cm$^{-3}$ and $n = 1.78 \times 10^{12}$ cm$^{-3}$, respectively. The mobilities ($\mu$) of *n*-Si at 25 K and *i*-Si at 300 K are $\mu = 2 \times 10^4$ cm$^2$V$^{-1}$s$^{-1}$ and $\mu$ = 1,300 cm$^2$V$^{-1}$s$^{-1}$, respectively. We fabricated the ohmic In/*n*-Si/In and In/*i*-Si/In devices by soldering and alloying indium directly to the silicon substrate at 350°C for five minutes, and measured its magnetotransport properties in current-in-plane configuration via two-terminal method. The magnetic field is applied perpendicular to the current direction. The device structure and measurement setup is schematically shown in the figure 1. For the detailed description of the experiment see reference [5].

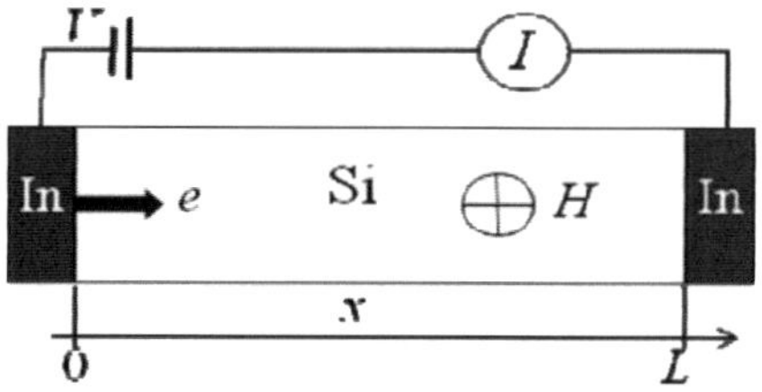


**Figure 1.** Schematic illustration of the device structure and measurement setup. Device length ($L$) is typically $L$ = 0.1 mm ~ 4 mm.

### 3. Results and Discussion

Figure 2a shows the typical current-voltage (*I-V*) characteristics of In/*n*-Si/In device as a function of magnetic field at 25 K. The *I-V* curve for $H$ = 0 T (solid line) shows three different transport regimes; linear ($|V| < 1$ V), nonlinear (3 V< $|V| < 50$ V), and breakdown ($|V| > 64$ V) regimes. The crossover from the linear to the nonlinear regime at larger values of $V$ indicates that the device is in the space-charge regime. As $V$ increases, the injected carrier density starts to exceed the original carrier density of the silicon to cause the electrons to accumulate around the injection contact [2-4]. Instead of Ohm's law, the electric field induced by the space-charge determines the current so as to satisfy the current continuity and Poisson equations. The high mobility and low carrier density favor this phenomenon [5]. Actually, the *I-V* curve at 3 V < $|V|$ < 50 V obeys the Mott-Gurney law, given by $J = 9\varepsilon\mu V^2/8L^3$, which holds for unipolar space-charge-limited transport [2-4] ($J$ is the current density, $\varepsilon$ is the dielectric constant of silicon, $\mu$ is the mobility and $L$ is the distance between the contacts (see figure 1)). Above the Mott-Gurney regime the *I-V* curve shows breakdown at ~ 64 V as the electric field created by the bias voltage and the accumulated space-charge exceeds the threshold that the device can sustain. The application of magnetic field perpendicular to the device decreases the current at a fixed bias voltage as shown in figure 2a. This results in the positive magnetoresistance ratio MR($H$) = $[R(H) - R(H = 0)]/R(H = 0) = \Delta R/R$, where $R(H)$ is the two-terminal resistance ($V/I$) at magnetic field $H$. Surprisingly, the MR in the Mott-Gurney regime is larger than that in the Ohmic regime. As shown in figure 2b, in the Mott-Gurney regime MR is ~ 700 % at 3 T at $V$ = 20 V, which is almost four times larger than that in the Ohmic regime at $V$ = 0.2 V. The magnetoresistance is much reduced when the magnetic field is applied parallel to the current direction, indicating that the present effect results from the orbital magnetoresistance [5]. Furthermore, as the magnetic field increases the breakdown voltage is suppressed by the magnetic field making the device more resistive, and the breakdown voltage is shifted to higher voltages. As a result, MR ≈ 3,000 % for $V$ = 63.5 V at 3 T just before the breakdown. Figure 2b shows the MR as a function of $H$ for various bias voltages at 25 K. The MR($H$) is increasing with increasing $V$, which indicates that the space-charge effect induces the MR.

The observed results have been confirmed as the effect of electron transport through the silicon by measuring the internal voltage differences between voltage probes in multi-probe silicon device [5]. The measurement reveals that in the space-charge transport regime the voltage differences is proportional to $x^{3/2}$ in agreement with the Mott-Gurney theory [2-4] that the electric field is spatially distributed in proportion to $x^{1/2}$ owing to the electrons around the injection contact. Moreover, the magnetoresistance measurement of the multi-probe silicon device reveals that the MR increases linearly and in a non-saturating behavior against $H$ in the space-charge regime [5]. This indicates that inhomogeneity induced in the Mott-Gurney regime by the broken quasi-neutrality is responsible for the large MR effect [5]. It is known that inhomogeneous semiconductors exhibit non-saturating and linear positive MR effect caused by the current distortion in the magnetic field due to the inhomogeneity [6-9, 11-13]. Recently, large MR effect was reported to occur in boron-doped silicon with Schottky contacts at low temperature [14], for which the impact ionization at the Schottky barrier is responsible as proposed in reference [15]. The mechanism we proposed to explain our results based on the space-charge effect is different.

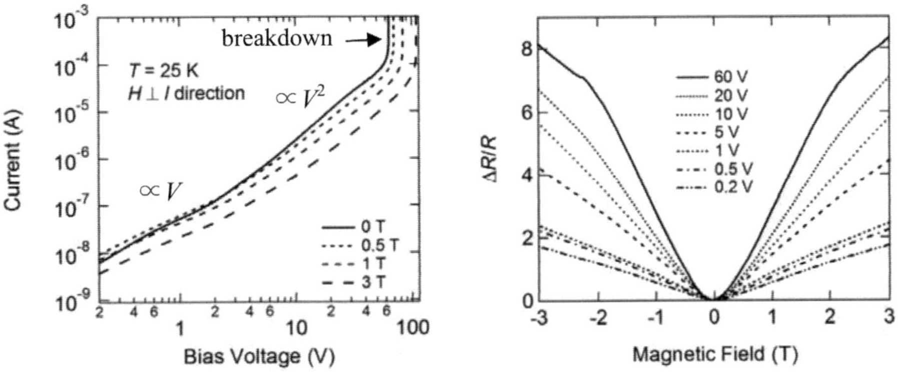


**Figure 2.** (a) $I$-$V$ characteristics of the In/$n$-Si/In device at 25 K for various applied magnetic fields. (b) Magnetoresistance ($\Delta R/R$) as a function of magnetic field at 25 K for various bias voltages. $\Delta R/R$ is in ohmic regime for $V < 1$ V and space-charge regime for $V > 5$ V.

Figure 3 shows the resistance as a function of temperature, $R(T)$, for $H = 0$ and 3 T. To address the wide temperature range, the data was obtained for the constant current condition (0.2 mA). Above 40 K, the $R(T)$ under $H = 0$ and 3 T shows no significant difference. On the other hand, below 40 K the two resistance curves separate into two distinct curves and significant MR effect is measured in this regime. The two $R(T)$ curves only separate when the value of the carrier density of the silicon (right-hand scale) is below ~ $10^{13}$ cm$^{-3}$. This result corroborates the proposition that space-charge effect induces the MR effect because space-charge effect favors materials with low carrier density.

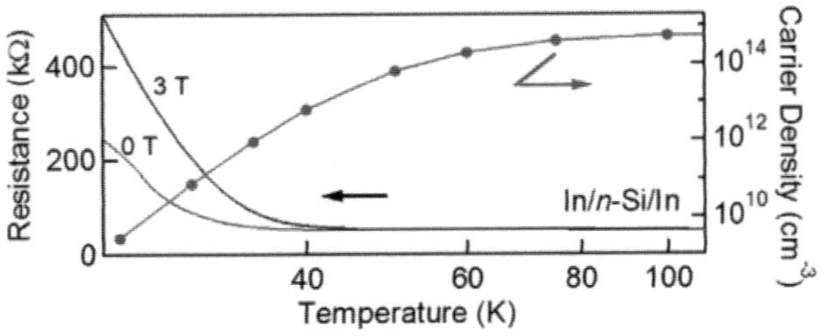


**Figure 3.** Resistance of In/$n$-Si/In as a function of temperature at $H = 0$ and 3 T. The line with dots is the temperature dependence of the carrier density of $n$-Si substrate.

On the basis of the above arguments, a similar effect is expected to occur at higher temperature if the carrier density is low and the mobility is high. Figure 4 shows the $I$-$V$ characteristics of In/$i$-Si/In device at 300 K for various applied magnetic fields. At $H = 0$ T, the device shows ohmic behavior at low bias voltages but the behavior changes for $V > 20$ V, indicating space-charge-effect. In this regime the current is significantly suppressed by the magnetic field, similar to that in the In/$n$-Si/In device. Further increase in $V$ induces breakdown in the device and the breakdown voltage shows significant magnetic field dependence that it shifts with increasing $H$. The MR exceeds ~ 1000 % at 3 T just before the breakdown at 71.2 V. The MR ratio curve around the breakdown voltage is not linear in $H$ and is very sensitive to the bias voltage that determines the internal electric field.

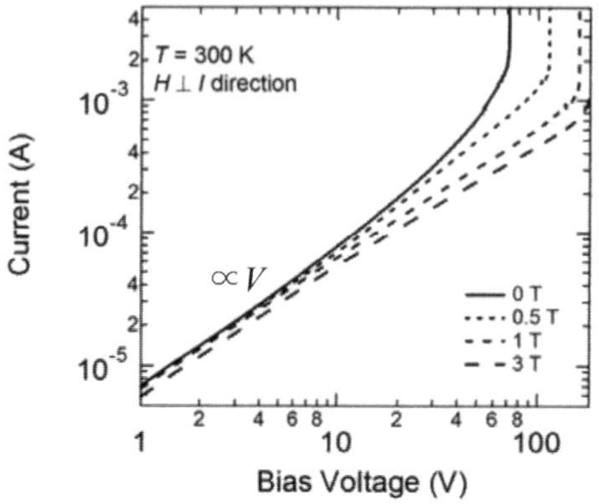


**Figure 4.** *I-V* characteristics of In/*i*-Si/In device at 300 K for various applied magnetic fields.

## 4. Conclusion

We have shown that space-charge effect can induce large positive magnetoresistance in silicon. The size of the effect is very large that the MR is in the order of 1000 % for different temperature ranges. The magnetic field dependence of the magnetoresistance shows that it is linear and non-saturating, which suggests that inhomogeneity is important in understanding the present MR effect in silicon. Technologically, this unusually large MR effect in silicon can be utilized to develop advanced silicon-based magnetoelectronics technologies.

## Acknowledgement

We appreciate discussions with T Shinjo, H Akinaga, H Sakakima, Y Iye, J Ohe, S Takahashi, and T Susaki. This is partly supported by KAKENHI, ICR Grants for Young Scientists, the Asahi Glass Foundation, the Sumitomo Foundation, and JSPS Research Fellowships for Young Scientists.

## References

[1] Ashcroft N W and Mermin N D 1976 *Solid State Physics* (Orlando: Holt, Rinehart, and Winston)
[2] Sze S M and Kwok K 2007 P*hysics of Semiconductors 3rd Edition* (New Jersey: John Wiley & Sons Inc.)
[3] Rose A 1955 *Phys. Rev.* **97** 1538
[4] Lampert M A 1956 *Phys. Rev.* **103** 1648
[5] Delmo M P, Yamamoto S, Kasai S, Ono T, and Kobayashi K 2009 *Nature* **457** 1112
[6] Xu R, Husmann A, Rosenbaum T F, Saboungi M –L, Enderby J E, and Littlewood P B 1997 *Nature* **390** 57
[7] Parish M M and Littlewood P B 2003 *Nature* **426** 162
[8] Parish M M and Littlewood P B 2005 *Phys. Rev. B* **72** 094417
[9] Zhang T, Harris J J, Branford W R, Bugoslavsky Y V, Clowes S K, and Cohen L F 2006 *Appl. Phys. Lett.* **88** 012110
[10] Solin S A, Thio T, Hines D R, and Heremans J. J. 2000 *Science* **289** 1530
[11] Herring C 1960 *J. Appl. Phys.* **31** 1939
[12] Sampsell J B and Garland J C 1976 *Phys. Rev. B* **13** 583
[13] Stroud D and Pan F P 1976 *Phys. Rev. B* **13** 1434
[14] Schoonus J J H M, Bloom F L, Wagemans W, Swagten H J M, and Koopmans B 2008 *Phys. Rev. Lett.* **100** 127202
[15] Sun Z G, Mizuguchi M, Manago T, and Akinaga H 2004 *Appl. Phys. Lett.* **85** 5643

Journal of Physics: Conference Series **193** (2009) 012002 doi:10.1088/1742-6596/193/1/012002

# Electrothermal Monte Carlo validation of a hydrodynamic model for sub-micron silicon devices

**O Muscato and V Di Stefano**

Dipartimento di Matematica e Informatica, Università degli Studi di Catania, Viale A. Doria 6, 95125 Catania – Italy

E-mail: muscato@dmi.unict.it

**Abstract.** An Extended hydrodynamic model for electron and phonon transport has been formulated, on the basis of the Maximum Entropy Principle, in order to describe off-equilibrium phenomena in sub-micron silicon devices. With this procedure one obtains a closed system, with an energy phonon transport equation which is consistent with the thermodynamic principles. To validate this model, an Electrothermal Monte Carlo simulator has been developed. Simulation results for a 1D $n^+ - n - n^+$ silicon diode are presented.

## 1. Introduction

The very large scale integration technology has allowed the shrinkage of a single device to sub-micron dimensions. The volumetric heat generation rate inside such small semiconductor devices can be expected to be very high since the device operating power cannot be reduced below a certain level. The resulting increases in the device local temperature can significantly affect the device electrical performance and reliability. Thus, it is necessary to understand the mechanisms of heat generation and dissipation inside the device as well as their effects on electrical performance and reliability. From the physical point of view, self-heating starts when the nearly-free conduction band electrons in a semiconductor are accelerated by the electric field. The electrons gain energy from the field, then lose it by inelastically scattering with boundaries and phonons, heating up the lattice through the mechanism known as Joule heating. High energy electrons ($\simeq$ 50 meV) scatter more effectively with optical phonons, with a time scale of about 0.1 ps. Optical phonons do not migrate since their group velocity can be neglected. They decay to acoustic phonons through phonon-phonon interactions with a time scale $\simeq$ 10 ps. Acoustic phonons propagate at speed of sound and dominate heat conduction in silicon. The development of Monte Carlo (MC) tools that deal with the lattice self-heating are getting larger attention within the scientific community [1],[2]. A sophisticated Electrothermal Monte Carlo (ETMC) tool which takes into account the previous phonon scattering mechanisms has been developed [3]: the MC code for the carrier Boltzmann equation solution has been self-consistently coupled to fluid-like energy balance equations for optical and acoustic phonons. But the MC simulation suffers of some disadvantages as the large computational requirements and limitations inherent in the finite number of simulated particles, which pose serious limitations for industrial applications.

A kinetic description of the electron-phonon system can be obtained by means of the Bloch-Boltzmann-Peierls (BBP) equations [4], coupled with the Poisson equation. The fundamental

Journal of Physics: Conference Series **193** (2009) 012002 doi:10.1088/1742-6596/193/1/012002

approach is to solve the BBP, but this is a daunting task, also from the numerical viewpoint. An alternative is to obtain from the BBP hydrodynamic models, which are a good engineering-oriented approach to simulate sub-micron devices. To have a simple example of application, in the following we have adopted parabolic-band approximation in order to avoid the complicacies of a more realistic form of the band, moreover linear dispersion relation for the acoustic modes, and constant dispersion relation for the optical modes have been considered.

## 2. Extended Hydrodynamic model

Hydrodynamic models, for the macroscopic quantities associated to the electron-phonon transport, can be obtained by multiplying the BBP equations by appropriate weight functions, and integrating over the momentum space. In a such way a set of balance equations for the macroscopic quantities called moments, are obtained. The two main problems one must tackle are 1. how to "close" the resulting hierarchy of equations, and 2. how to model the production terms (i.e., the moments over the collisional operator). In [5] a hydrodynamic model for the electrons (based on the physical moments electron density $n$, velocity $V^i$, and energy $W$) and a two-step energy conservation model for the optical energy density $W_{op}$, and the acoustic phonon energy density $W_{ac}$ were introduced. But this model is unsatisfactory because the Fourier law for the electron heat flow has been used, which is not a valid approximation anywhere in the device as shown by MC simulation results [6]. One way to tackle this problem is to introduce an extended hydrodynamic model, in which high order moments are taken into account. By introducing the higher order moments electron energy flux $S^i$, and the flux of acoustic phonon energy density $Q^i$, from the BBP, one obtains the following hydrodynamic-like transport equations [7],[8]

$$\frac{\partial n}{\partial t} + \frac{\partial (nV^i)}{\partial x^i} = 0 \tag{1}$$

$$\frac{\partial (nm^\star V^i)}{\partial t} + \frac{\partial (nU^{ij})}{\partial x^j} + neE^i = -\frac{nm^\star V^i}{\tau_p} \tag{2}$$

$$\frac{\partial (nW)}{\partial t} + \frac{\partial (nS^i)}{\partial x^i} + neV_iE^i = -n\frac{W - W_{op}^\star}{\tau_W} \tag{3}$$

$$\frac{\partial (nS^i)}{\partial t} + \frac{\partial (nF^{ij})}{\partial x^j} + neE_jG^{ij} = -\frac{nS^i}{\tau_S} \tag{4}$$

$$\frac{\partial W_{ac}}{\partial t} + \frac{\partial Q^i}{\partial x^i} = \frac{W_{op} - W_{ac}}{\tau_{op}} \quad , Q^i = -\kappa\frac{\partial T_{ac}}{\partial x^i} + \frac{\tau_{ac}}{\tau_p}v_s^2 nm^\star V^i \tag{5}$$

$$\frac{\partial W_{op}}{\partial t} = n\frac{W - W_{op}^\star}{\tau_W} - \frac{W_{op} - W_{ac}}{\tau_{op}} \tag{6}$$

In this model the constitutive equations (or state equations) for the higher order moments $F^{ij}$,$G^{ij}$,$U^{ij}$ have been obtained by means of the Maximum Entropy Principle (MEP hereafter) on the basis of information theory [9],[10]. The MEP offers a definite procedure to provide an approximation of the electron and phonon distribution functions, in terms of a finite number of moments. Now the Fourier law, for the electron heat flow, is not invoked because the energy flux $S^i$ is considered a new variable. The closure for $Q^i$, given in eq.$(5)_2$, has been obtained by using the so called *Maxwellian iteration*, in which the phonon heat flux equation is linearized under the hypothesis of small temperature gradients. The production terms (appearing in the right-hand-side), as a first guess, have been approximated with relaxation terms, consistently with the first principle of thermodynamics. The numerical solution, in the stationary 1D case, have been obtained by discretizing the above system using central differences and the one-dimensional box method. The resulting system has been solved via Newton's method. Finally we observe that in this model the "lattice temperature" is taken to be the acoustic phonon temperature $T_{ac}$, because this is the mode responsible for diffusion.

## 3. ETMC validation

Once we have established a consistent hydrodynamic model, it is mandatory to check its results by comparisons with MC or experimental data, in order to clarify the limit of validity of the model. To this purpose, we have developed an ETMC simulator where the particle-based electron transport is coupled with the lattice heat diffusion equation

$$C\frac{\partial T_L}{\partial t} + \frac{\partial Q^i}{\partial x^i} = \mathcal{H} \quad , Q^i = -\kappa \frac{\partial T_L}{\partial x^i} \,. \tag{7}$$

The lattice heating generation rate $\mathcal{H}$ has been calculated during the ETMC simulation by accounting the exchange of phonons between the electrons and the lattice. Since the lattice temperature $T_L$ is a new variable, we have added this variable in the scattering table following the scheme introduced in [3]. We have considered a 1D $n^+ - n - n^+$ silicon diode where the cathode and the anode have an extension of 100 nm, doped to a density of $10^{19} cm^{-3}$, while the channel is 100 nm-long doped to a density of $10^{16} cm^{-3}$, and the applied bias is 1.2 V.

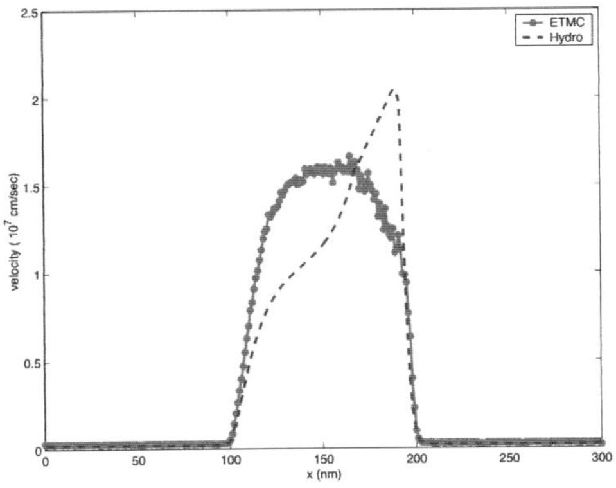


**Figure 1.** Average electron velocity profile under steady state operation.

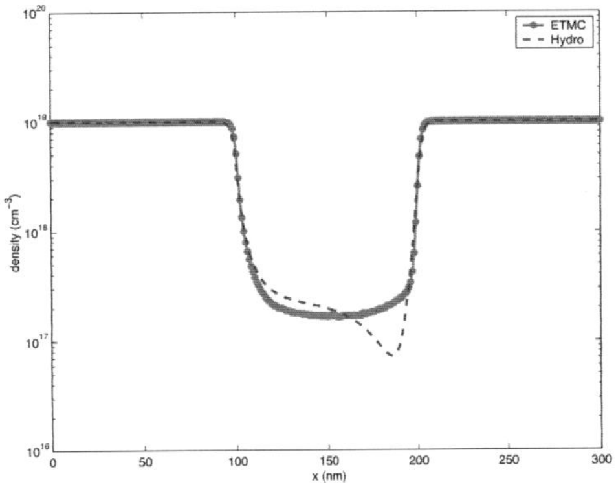


**Figure 2.** Electron density profile under steady state operation.

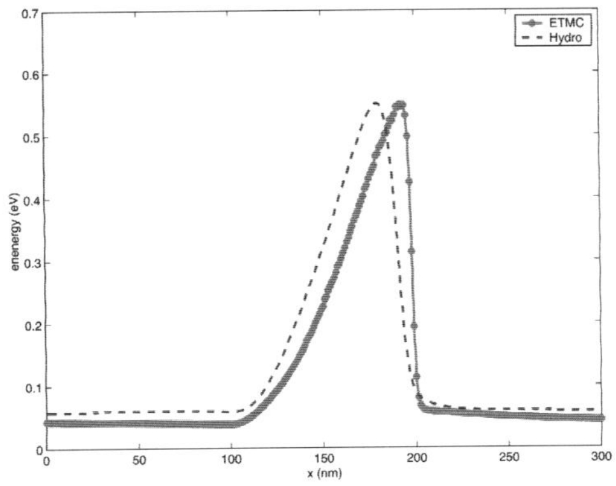


**Figure 3.** Electron energy profile under steady state operation.

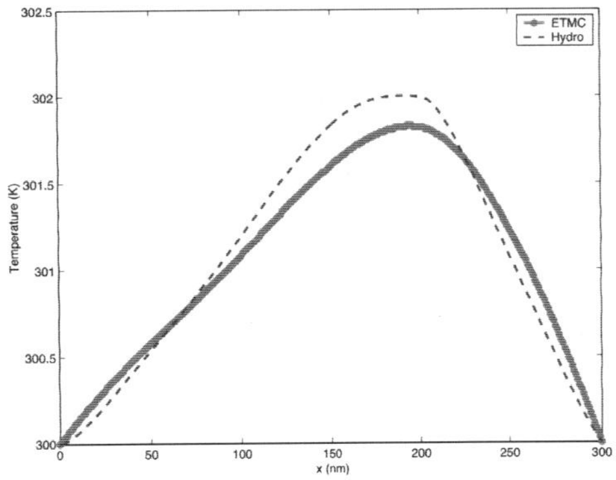


**Figure 4.** Lattice temperature profile under steady state operation.

In figures 1, 2 we plot respectively the electron velocity and density along the device obtained with the hydrodynamic and ETMC models. We can observed that the hydrodynamic model,

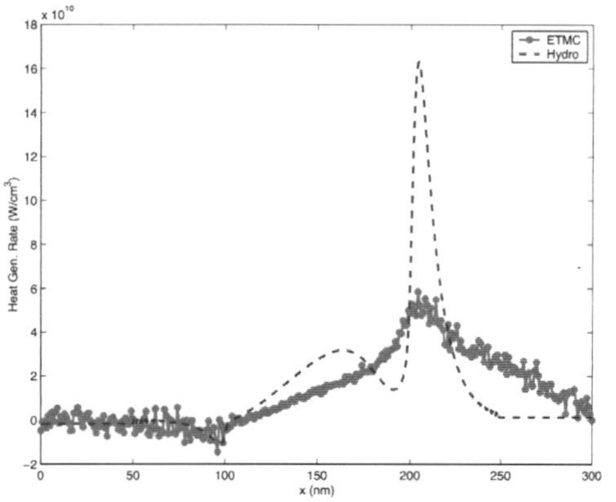


**Figure 5.** The heat generation rate under steady state operation.

produces a spurious spike near to the second junction, which is an effect typical of each hydrodynamic code [11]. A refinement of the spatial mesh shows already this spike, which is an indication of the occurrence of some sort of singularity in the stationary solution, which we shall not study anymore in this paper. The figure 3 shows the electron energy for both models. In figure 4 we plot the acoustic phonon temperature $T_{ac}$ obtained with the hydrodynamic model, and the lattice temperature $T_L$ obtained with the ETMC model through eq.(7). In figure 5 we show the heat generation rate produced in the device where $\mathcal{H}$, in hydrodynamic model, is given by the right-hand-side of equation $(5)_1$. We notice that the heat generation rate is more pronounced in hydrodynamic case, which is consistent with the higher hydrodynamic temperature shown in figure 4. We guess that this effect is due to the modeling of the production term. In conclusion, the general behaviour of the extended hydrodynamic model is confirmed by ETMC simulations, except at some singular point. We guess that a more sophisticated modeling of the production terms as well as an ETMC code self-consistently coupled to fluid-like energy balance equations for optical and acoustic phonons, could improve the results. Work along this path is in progress and will be published in the next future.

**Acknowledgments**
This work has been supported by "Progetti di ricerca di Ateneo", Università degli Studi di Catania.

**References**
[1] Pop E, Sinha S and Goodson K E 2006 *Proc. IEEE* **94** 1587
[2] Sadi T, Kelsall R W and Pilgrim N J 2007 *IEEE Trans. Electr. Devices* **54** 332
[3] Raleva K, Vasileska D, Goodnick S M and Dzekov T 2008 *J. Comput. Electron.* **7** 226
[4] Ferry D K 1991 *Semiconductors* (New York N.Y.: Maxwell Macmillan Editions) p 354
[5] Lai J and Majumdar A 1996 *J. Appl. Phys.* **79** 7353
[6] Stettler M A, Alam M A and Lundstrom M S 1993 *IEEE Trans. Electr. Devices* **40** 733
[7] Muscato O, Di Stefano V and Milazzo C 2008 *J. Comput. Electron.* **7** 142
[8] Muscato O and Di Stefano V 2008 *J. Appl. Phys.* 12 **104** 124501
[9] Müller I and Ruggeri T 1998 *Rational Extended Thermodynamics* (Berlin: Springer-Verlag) p 393
[10] Jou D, Casas-Vázquez J and Lebon G 2001 *Extended irreversible thermodynamics* (Berlin: Springer-Verlag) p 517
[11] Anile A M, Junk M, Romano V and Russo G 2000 *M3AS* **10** 833

Journal of Physics: Conference Series **193** (2009) 012003 doi:10.1088/1742-6596/193/1/012003

# Hydrodynamic modelling of electron transport in submicron $Hg_{0.8}Cd_{0.2}Te$ diodes

**M Daoudi[1], A Belghachi[1], C Palermo[2], H Marinchio[2] and L Varani[2]**

[1] Laboratory of semiconductors devices physics, University of Béchar, P.O. Box 417 Béchar, Algeria.
[2] Institut d'Electronique du Sud (UMR CNRS 5214), Université Montpellier II, Place E. Bataillon, 34095 Montpellier Cedex 5, France.
E-mail: daoudimadz@yahoo.fr

**Abstract.** We simulate electron transport in ultra small mercury-cadmium-telluride $n^+$-$n$-$n^+$ diodes using a hydrodynamic approach. A numerical staggered solution is employed to treat the coupled hydrodynamic and Poisson equations, where the spatial profiles of the main transport parameters within the diodes are analyzed including the Auger generation-recombination processes. Our numerical results show that, even for low applied voltages, impact ionization processes are activated and affect dramatically the current-voltage characteristics of the $Hg_{0.8}Cd_{0.2}Te$ diode.

## 1. Introduction

Carrier transport in advanced submicron devices is not always described with sufficient accuracy by the conventional drift-diffusion (DD) model. As a matter of fact, the DD model does not describe velocity overshoot, diffusion associated with carrier temperature gradients, or the dependence of impact ionization (II) rates on the carrier energy distributions. The limitations of the DD model indicate the need for more general transport models. Two of the more significant classes of such transport models are Monte Carlo (MC) and hydrodynamic (HD). The most accurate kinetic description is given by the MC method because it can take into account explicitly both the energy band-structure and the various scattering phenomena characteristics of the studied material. Therefore it enables a direct calculation of important transport quantities (distribution function, carrier density, drift velocity, mean energy, etc) but at a cost of long computation times and stochastic noise in the output data [1]. On the other hand, the results obtained from the MC simulations permit us also to calculate transport coefficients that can be used as input parameters for more simplified HD models. In this sense, the HD description of electron transport has been extensively applied to the analysis and design of semiconductor devices because it provides a useful compromise between computational simplicity and physical accuracy [2].

In our previous work [3], we have demonstrated the robustness of the HD model to simulate the electronic transport in bulk HgCdTe (Mercury-Cadmium-Telluride, MCT) at 77 K. The majority of MCT-based devices contain a cadmium fraction $x = 0.2$ which allows, at 77 K, the detection in the 8-14 $\mu$m spectral region and which is thus a widely used alloy for infrared optoelectronics applications. The consequence of this alloy proportion is a narrow semiconductor band-gap of about 0.1 eV: in particular, degeneracy and impact ionization processes are activated from low electric fields of the order of 100 V/cm [4]. Despite of his wide interest, few results are available in the literature concerning a physical modelling of transport in MCT-based devices. In this paper, we present the simulation of one-dimensional $Hg_{0.8}Cd_{0.2}Te$ $n^+$-$n$-$n^+$ diodes using the HD approach coupled to the

Journal of Physics: Conference Series **193** (2009) 012003 doi:10.1088/1742-6596/193/1/012003

Poisson equation. The main goal is to investigate the diode electrical behaviour under far-from-equilibrium conditions including the impact ionisation processes.

## 2. Hydrodynamic transport model

The considered HD model consists of a set of equations expressing the conservation of charge, momentum and energy for each species of carriers. These equations are derived by taking the first three moments of the Boltzmann transport equation. For the one-dimensional case (taking into account the electron-hole generation-recombination processes) these equations take the form [5]:

$$\frac{\partial n}{\partial t}+\frac{\partial (nv)}{\partial x}=g_n n, \tag{1}$$

$$\frac{\partial v}{\partial t}+v\frac{\partial v}{\partial x}+\frac{1}{n}\frac{\partial (nQ_v)}{\partial x}+\frac{q\mathrm{E}}{m^*}=-\frac{v}{\tau_v}, \tag{2}$$

$$\frac{\partial \varepsilon}{\partial t}+v\frac{\partial \varepsilon}{\partial x}+\frac{1}{n}\frac{\partial (nQ_\varepsilon)}{\partial x}+q\mathrm{E}v=-\frac{\varepsilon-\varepsilon_0}{\tau_\varepsilon}; \tag{3}$$

where $q$ is the absolute value of the electric charge, $n$ the electron density. The quantities $v$, $\varepsilon$ represent the average velocity and mean energy, respectively, $E$ the electric field, $Q_v \equiv \langle \delta v^2 \rangle = \langle v^2 \rangle - \langle v \rangle^2$ the variance of velocity fluctuations and $Q_\varepsilon \equiv \langle \delta v \delta \varepsilon \rangle = \langle v\varepsilon \rangle - \langle v \rangle \langle \varepsilon \rangle$ the covariance of velocity-energy fluctuations. Still, $m^*$ is the electron effective mass, $g_n$ the carrier generation term, and $\tau_v$ and $\tau_\varepsilon$ are the momentum and energy relaxation times. For the modelling of voltage driven operation, that is where a constant voltage is applied between the structure terminals, equations (1), (2) and (3) are coupled to the Poisson equation:

$$\frac{\partial E}{\partial x}=\frac{q}{\varepsilon_{sc}}(n-N_d) \tag{4}$$

where $N_d$ is the ionized donor density and $\varepsilon_{sc}$ the lattice dielectric constant.

This system of equations contains four unknowns, namely the electron density, the average velocity, mean energy and electric field. Moreover, it contains several input parameters: the effective mass, the velocity and energy relaxation times, the velocity variance and velocity-energy covariance. We note that the parameters of equations (1)-(3) are, by definition, functions of the mean energy and they have been obtained by a previously developed MC simulation of the bulk material under stationary conditions [6].

## 3. Simulation results

We consider an $Hg_{0.8}Cd_{0.2}Te$ $n^+$-$n$-$n^+$ diode in which the low-doped diode base $n$ of length $L$ is sandwiched between two heavily doped contacts $n+$ of length $d$. The parameters of the simulated structure are: the $n^+$ regions are each 0.2 $\mu$m thick with a doping density of $5.4\times10^{15}$ cm$^{-3}$, the $n$ central region is 0.2 $\mu$m thick with a doping density of $10^{14}$ cm$^{-3}$. Initially, the electrons are in thermal equilibrium with the lattice at $T = 77$ K. The carriers densities are assumed to be equal to the doping densities (fully ionized impurities). We present in Figure 1 the carrier density, mean energy, drift velocity and electric field versus position. Three different applied voltages are considered: 12, 15, 21 and 27 mV in order to describe electron behavior from under non-equilibrium conditions.

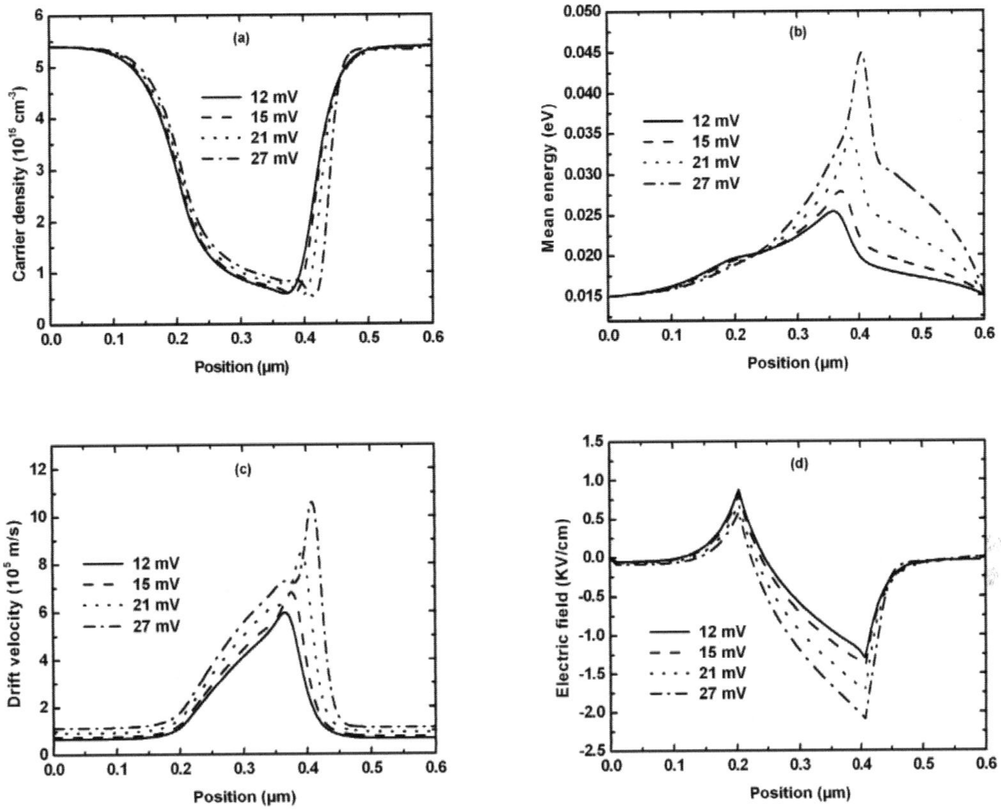


**Figure 1.** Stationary profiles accounting for impact ionization of the: (a) electron density, (b) mean energy, (c) drift velocity, (d) electric field as functions of position for a $Hg_{0.8}Cd_{0.2}Te$ $n^+$-$n$-$n^+$ diode of length 0.6 $\mu$m for different applied bias at 77 K.

From the simulation results we observe that a significant electron injection is present in the central region due to the diffusion from the $n^+$ regions in conjunction with the quasi-linear variation of the electric field [see Figures 1(a) and 1(d)]. When the voltage applied to the structure increases, the potential barrier in the first $n^+/n$ homojunction decreases, and the potential mostly drops inside the $n$ region. As a consequence of the fall in the energy barrier, the positive value of the electric field in the first $n^+/n$ homojunction decreases, and a high negative electric field is found in the active layer ($n$ region), which reaches an extremum of about −2.1 kV/cm in the second $n/n^+$ homojunction. Furthermore, in the $n$ region a significant number of high-energy electrons penetrate in the active layer [see Figure 1(b)]. Therefore, as it can be seen in figure 1(c), the average electron velocity exhibits a significant overshoot reaching a value of $1.1\times10^6$ m/s near the high-field region (the saturation velocity in MCT is of the order of $5\times10^5$ m/s at 77 K [6]).

The current-voltage characteristics with and without including impact ionization processes for the same structure are shown in Figure 2.

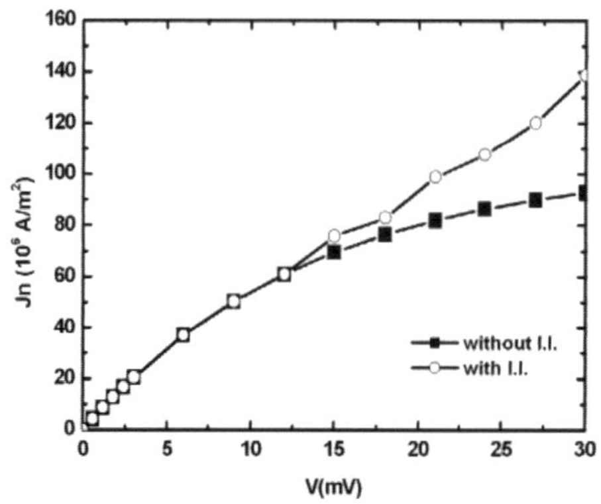


**Figure 2.** Current-voltage characteristics of $Hg_{0.8}Cd_{0.2}Te$ $n^+$-$n$-$n^+$ diode at 77K with (○) and without (■) taking into account impact ionization (I.I.).

We observe that for the highest considered biases, that is above 21 mV, the impact ionization processes become important and the current density increases rapidly. This behavior is related to the high sensitivity to the bias of $Hg_{0.8}Cd_{0.2}Te$ material in which generation processes are activated for electric fields much smaller that 1 kV/cm.

## 4. Conclusion

Device modelling is playing a very important role in the development of semiconductor device design. In this work, a semiconductor device simulator based on the hydrodynamic balance equations has been developed to simulate the electronic transport in $Hg_{0.8}Cd_{0.2}Te$ $n^+$-$n$-$n^+$ submicron diodes. The numerical results show the presence of hot carrier effects such as the spatial velocity overshoot appearing even for low applied voltages due to the small effective mass characteristic of this material. Moreover, the onset of impact ionization processes, which affects dramatically the current-voltage characteristics, has been put in evidence for relatively small bias voltages with respect to similar devices fabricated with standard semiconductor materials.

**Acknowledgements:** this work is partly supported by Algerian ministry of higher education and research (CNPRU) project N° D03820070006.

## References

[1] Blotekjaer K 1970 *IEEE Trans. Electron Devices* **17** 38
[2] Jacoboni C and Reggiani L 1983 *Rev. Mod. Phys.* **55** 645
[3] Daoudi M, Belghachi A, Varani L, and Palermo C 2008 *Eur. Phys. J. B* **62** 15
[4] Gelmont B, Lund B, Kim K, Jensen G U, Shur M, Fjeldly T A 1992 *J. Appl. Phys.* **71**, 4977
[5] Gružinskis V, Starikov E, Shiktorov P, Reggiani L, Saraniti M, Varani L 1993 *Semicond. Sci. Technol.* **8** 1283
[6] Palermo C, Varani L, Vaissière J C, Starikov E, Shiktorov P, Gružinskis V, Azaïs B 2009 *Solid State Electron.* **53** 1316

Journal of Physics: Conference Series **193** (2009) 012004 doi:10.1088/1742-6596/193/1/012004

# Time dependent transport in 1D micro- and nanostructures: solving the Boltzmann and Wigner–Boltzmann equations

**W. Magnus**[1,2], **F. Brosens**[1] **and B. Sorée**[2]

[1] Universiteit Antwerpen, Physics Department, Groenenborgerlaan 171, B-2020 Antwerpen, Belgium

[2] Interuniversity Microelectronics Centre (IMEC), Kapeldreef 75, B-3001 Leuven, Belgium

E-mail: wim.magnus@ua.ac.be, fons.brosens@ua.ac.be, bart.soree@imec.be

**Abstract.** For many decades the Boltzmann distribution function has been used to calculate the non-equilibrium properties of mobile particles undergoing the combined action of various scattering mechanisms and externally applied force fields. When the latter give rise to the occurrence of inhomogeneous potential profiles across the region through which the particles are moving, the numerical solution of the Boltzmann equation becomes a highly complicated task. In this work we highlight a particular algorithm that can be used to solve the time dependent Boltzmann equation as well as its quantum mechanical extension, the Wigner–Boltzmann equation. As an illustration, we show the calculated distribution function describing electrons propagating under the action of both a uniform and a pronouncedly non-uniform electric field.

## 1. Introduction

In this paper we propose the method of characteristic curves as an alternative approach to solve the time dependent Boltzmann equation and other kinetic equations with a similar kernel structure, such as the Wigner–Boltzmann equation. In section 2 we briefly outline the corresponding algorithm which is thoroughly described in [1]. In section 3, we illustrate and discuss how the algorithm can be exploited most conveniently to calculate the Boltzmann distribution function for an electron gas moving in two different force field profiles, whereas section 4 provides the conclusions.

## 2. Characteristic curves of the Boltzmann equation

Consider an ensemble of mobile, charged particles with band mass $m$ moving through a force field $\mathbf{F}(\mathbf{r})$. Providing a classical tool to quantify the probability of finding a particle with momentum $\mathbf{p}$ at a position $\mathbf{r}$ at an arbitrary time instant $t > 0$, the Boltzmann distribution function $f(\mathbf{r}, \mathbf{p}, t)$ satisfies the time-dependent Boltzmann transport equation (BTE)

$$\frac{\partial f(\mathbf{r},\mathbf{p},t)}{\partial t} + \frac{\mathbf{p}}{m} \cdot \boldsymbol{\nabla} f(\mathbf{r},\mathbf{p},t) + \mathbf{F}(\mathbf{r}) \cdot \boldsymbol{\nabla}_{\mathbf{p}} f(\mathbf{r},\mathbf{p},t) = Q_{\mathrm{B}}[f(\mathbf{r},\mathbf{p},t)]. \tag{1}$$

Appearing as a functional of $f(\mathbf{r}, \mathbf{p}, t)$, the Boltzmann collision integral $Q_{\mathrm{B}}[f(\mathbf{r}, \mathbf{p}, t)]$ incorporates the transition probabilities associated with all relevant, microscopic scattering mechanisms that may affect the particle flow. A straightforward algorithm providing an iterative, numerical solution of Eq. (1) can be constructed by observing that the classical trajectories $\mathbf{R}(s)$ and $\mathbf{P}(s)$, traversed by a particle in phase space between the time instants $s = -t$ and $s = 0$ are in fact the characteristic curves of the BTE. Indeed, inserting Newton's equations

$$\frac{\mathrm{d}\mathbf{R}(s)}{\mathrm{d}s} = \frac{\mathbf{P}(s)}{m}, \quad \frac{\mathrm{d}\mathbf{P}(s)}{\mathrm{d}s} = \mathbf{F}[\mathbf{R}(s)], \tag{2}$$

together with the function $T(s) = t + s$ into Eq. (1), we may turn the left-hand side of the latter into a complete differential, reduce the BTE to an ordinary differential equation and integrate it formally while interpreting $\mathbf{r}$ and $\mathbf{p}$ as "initial" position and momentum,

$$f(\mathbf{r},\mathbf{p},t) = f_0(\mathbf{R}(-t),\mathbf{P}(-t)) + \int_0^t \mathrm{d}s\, Q_{\mathrm{B}}[f(\mathbf{R}(s-t),\mathbf{P}(s-t),s)], \tag{3}$$

where we need to specify the initial distribution function $f_0(\mathbf{r},\mathbf{p})$. Requiring merely the history of the distribution function during the time interval $[-t, 0]$, Eq. (3) can easily be recast in a forward step algorithm

$$f(\mathbf{r},\mathbf{p},t_n) = f(\mathbf{R}(-\Delta t),\mathbf{P}(-\Delta t),t_{n-1}) + \Delta t\, Q_{\mathrm{B}}\big[f(\mathbf{R}(-\Delta t),\mathbf{P}(-\Delta t),t_{n-1})\big], \tag{4}$$

by choosing an equidistant time grid $t_n = n\Delta t$, $n = 0, 1, 2, \ldots, N$. The time step $\Delta t$ should be small compared with the system's characteristic times such as dwell times, scattering times etc., whereas $t_N$ represents the numerical onset of the steady state (if any), i.e. the time at which $f(\mathbf{r},\mathbf{p},t_N)$ does not appreciably differ from its previous values.

Though being devised originally to tackle the BTE, one may as well use the algorithm defined in Eq. (4) to investigate the Wigner–Boltzmann distribution function $f_{\mathrm{W}}(\mathbf{R},\mathbf{p},t)$ obeying the Wigner–Boltzmann transport equation (WBTE) [1, 5, 6, 7]

$$\frac{\partial f_{\mathrm{W}}(\mathbf{R},\mathbf{p},t)}{\partial t} + \frac{\mathbf{p}}{m}\cdot\boldsymbol{\nabla}_{\mathbf{R}} f_{\mathrm{W}}(\mathbf{R},\mathbf{p},t) = Q_{\mathrm{B}}[f_{\mathrm{W}}(\mathbf{R},\mathbf{p},t)] + Q_{\mathrm{W}}[f_{\mathrm{W}}(\mathbf{R},\mathbf{p},t)] \tag{5}$$

that may be considered a pragmatic quantum mechanical extension of the BTE where $\mathbf{r}$ is replaced with $\mathbf{R}$, the so-called "center-of-mass" coordinate. As such, it borrows the Boltzmann collision integral $Q_{\mathrm{B}}$ from the BTE to include decoherent scattering processes while non-locality effects are described by the Wigner kernel

$$W(\mathbf{R},\mathbf{p}) = -\frac{\mathrm{i}}{\hbar}\int \mathrm{d}^3x \left[U\left(\mathbf{R}+\frac{\mathbf{x}}{2}\right) - U\left(\mathbf{R}-\frac{\mathbf{x}}{2}\right)\right] \exp\left(-\frac{\mathrm{i}\mathbf{p}\cdot\mathbf{x}}{\hbar}\right). \tag{6}$$

Entering the Wigner collision term

$$Q_{\mathrm{W}}[f_{\mathrm{W}}(\mathbf{R},\mathbf{p},t)] = \int \frac{\mathrm{d}^3p'}{(2\pi\hbar)^3} f(\mathbf{R},\mathbf{p}',t) W(\mathbf{R},\mathbf{p}-\mathbf{p}'), \tag{7}$$

the kernel $W(\mathbf{R},\mathbf{p})$ accumulates all non-local contributions arising from the potential energy function $U(\mathbf{r})$ which is related to the force field by $\mathbf{F}(\mathbf{r}) = -\boldsymbol{\nabla} U(\mathbf{r})$. In contrast to earlier works [5, 7] relying as well on classical paths to solve the WBTE, we follow an alternative approach and rewrite the WBTE such that its left-hand side is identical to that of the BTE. Accordingly, the right-hand side consists of an effective collision term, given by

$$Q[f_{\mathrm{W}}(\mathbf{R},\mathbf{p},t)] = Q_{\mathrm{B}}[f_{\mathrm{W}}(\mathbf{R},\mathbf{p},t)] + Q_{\mathrm{W}}[f_{\mathrm{W}}(\mathbf{R},\mathbf{p},t)] + \mathbf{F}(\mathbf{R})\cdot\boldsymbol{\nabla}_{\mathbf{p}} f_{\mathrm{W}}(\mathbf{R},\mathbf{p},t), \tag{8}$$

with $\mathbf{F}(\mathbf{R}) = -\boldsymbol{\nabla}_{\mathbf{R}} U(\mathbf{R})$. In spite of its internal complexity $Q_{\mathrm{W}}[f_{\mathrm{W}}]$ shares at least one feature with its classical counterpart $Q_{\mathrm{B}}[f_{\mathrm{W}}]$, namely the mere dependence on times prior to the time instant $t$ appearing in the left-hand side of the WBTE. Consequently, the forward step algorithm introduced in Eq. (4) can be usefully invoked to solve the time dependent WBTE as well.

## 3. Examples

As an illustration, we calculate the distribution function of electrons moving through a simple semiconductor device.

### *3.1. n-i-n diode*

First, we consider an intrinsic semiconductor layer being sandwiched between two $n^+$-doped reservoirs to form a n-i-n diode. For the sake of simplicity, we treat this structure as a 1D device with density and field profiles merely depending on the $z$ coordinate and we discard any inelastic scattering events ($Q_\mathrm{B} = 0$). On the other hand, as the force field $F(z)$ turns out to be strongly localized inside the intrinsic region and the adjacent junction areas, it proves convenient to rephrase the Wigner collision integral in terms of the Fourier transform $\tilde{F}(k)$ of the force field:

$$Q_\mathrm{W}(Z,p,t) = \frac{1}{\hbar}\int_{-\infty}^{+\infty} \mathrm{d}k\, \tilde{F}(2k)\, \mathrm{e}^{2\mathrm{i}kZ} \,\cdot\, \frac{f(Z,p-\hbar k,t) - f(Z,p+\hbar k,t)}{k}. \tag{9}$$

Starting from an initial distribution with an empty intrinsic region, we reach a steady state after about 300 time steps. The corresponding distribution function revealing the particle diffusion into the central layer is shown in Fig. (1).

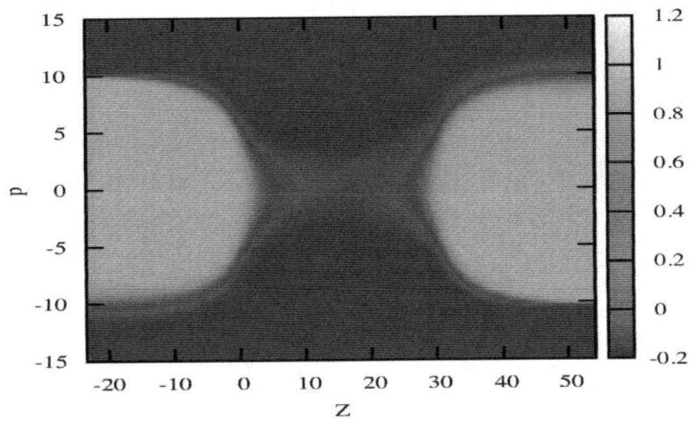


**Figure 1.** Wigner–Boltzmann distribution for a n–i–n diode after 300 time steps. The intrinsic region extends along the interval $0 \leqslant Z \leqslant 31.1$. All quantities are expressed in atomic units.

### *3.2. Polaron scattering*

Next, we consider an electron gas moving in a bulk, polar semiconductor and being driven by a uniform electric field $F$ applied along the $z$-direction. Consequently, the Wigner kernel reduces to the Boltzmann force term with $f(\mathbf{r},\mathbf{p},t) = f(p_z, p_\perp, t)$ and $p_\perp \equiv \sqrt{p_x^2 + p_y^2}$, while $Q_\mathrm{B}$ describes the interaction of the electrons with longitudinal optical (LO) phonons having a dispersion free energy $\hbar\omega_\mathrm{LO}$ [9] or an equivalent temperature $\Theta = \hbar\omega_\mathrm{LO}/k_\mathrm{B}$. Given the lattice temperature $T$, the corresponding transition probabilities read

$$\Pi(\mathbf{p} \to \mathbf{p}') = \frac{1}{|\mathbf{p} - \mathbf{p}'|^2}\left[A_\mathrm{e}\,\delta\!\left(\frac{p^2}{2} - \frac{p'^2}{2} - 1\right) + A_\mathrm{a}\,\delta\!\left(\frac{p^2}{2} - \frac{p'^2}{2} + 1\right)\right], \tag{10}$$

where the absorption(emission) coefficients $A_\mathrm{a}(A_\mathrm{e})$ are proportional to the Fröhlich coupling constant $\alpha$,

$$A_\mathrm{e} = \frac{\alpha}{\pi\sqrt{2}}(1+\nu), \quad A_\mathrm{a} = \frac{\alpha}{\pi\sqrt{2}}\nu, \quad \nu = \frac{1}{\mathrm{e}^{\Theta/T} - 1}, \tag{11}$$

and all quantities are expressed in polaronic units ($\hbar = m = \omega_\mathrm{LO} = 1$). Fig. (2) compares the

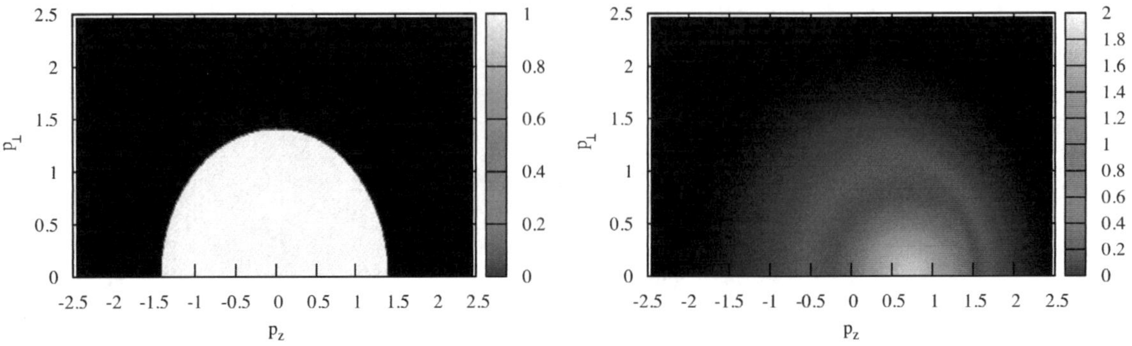


**Figure 2.** Polaron distribution function in equilibrium (left) and in the presence of a uniform electric field after 500 time steps (right). The following parameters have been used: $A_{\rm a} = 0.2$, $A_{\rm e} = 1.0$ and $F = 1.0$. All quantities are expressed in polaronic units ($\hbar = m = \omega_{\rm LO} = 1$).

initial distribution $f_0(\mathbf{p})$ with the steady-state distribution attained after 500 time steps. The former describes a configuration with all electrons gathering inside the sphere $|\mathbf{p}| = \sqrt{2}$ while the latter develops an sharp maximum shifting to positive $p_z$-values due to the action of the uniform electric field. It should be noted that the same steady state distribution is found when $f_0(\mathbf{p})$ is chosen otherwise, e.g. a simple Maxwellian.

## 4. Conclusion

The characteristic curves method provides a natural computational framework to evaluate the time-dependent Wigner and Wigner–Boltzmann distribution functions. Accordingly, only an initial distribution function needs to be specified and no artificial boundary conditions directly impinging on the steady state (such as fixing the chemical potentials of the contacts) need to be imposed.

[1] Brosens F and Magnus W 2009 Carrier transport in nanodevices: revisiting the Boltzmann and Wigner distribution functions *physica status solidi (b)* B **267** 1656
[2] Geurts B J 1991 Modelling transport in submicron structures using the relaxation time Boltzmann equation *J. Phys.: Condens. Matter* **3** 9447
[3] Cáceres M J, Carrillo J A and Goudon T 2003 Equilibrium rate for the linear inhomogeneous relaxation-time Boltzmann equation for charged particles *Communications in PDE* **28** 969
[4] Cáceres M J, Carrillo J A, Gamba I, Majorana A and Shu C W 2006 DSMC versus WENO-BTE: A double gate MOSFET example *J. Comput. Electron.* **5** 471
[5] Barker J R and Murray S 1983 A quasi-classical formulation of the Wigner function approach to quantum ballistic transport *Phys. Lett.* **A 93** 271
[6] Bordone P, Pascoli M, Brunetti R, Bertoni A, Jacoboni C and Abramo A 1999 Quantum transport of electrons in open nanostructures with the Wigner-function formalism *Phys. Rev.* **B59** 3060
[7] M. Nedjalkov, H. Kosina, S. Selberherr, C. Ringhofer and D. K. Ferry, *Unified particle approach to Wigner–Boltzmann transport in small semiconductor devices*, Phys. Rev. **B 70**, 115319 (2004) – and references therein.
[8] Rhew J H and Lundstrom M S 2002 Drift-diffusion equation for ballistic transport in nanoscale metal-oxide-semiconductor field effect transistors *J. Appl. Phys.* **92** 5196
[9] Devreese J T and Evrard R 1976 On the momentum distribution of electrons in polar semiconductors for high electric field *phys. stat. sol. (b)* **78** 85

Journal of Physics: Conference Series **193** (2009) 012005 doi:10.1088/1742-6596/193/1/012005

# Electron Transport Properties of Gallium Nitride for Microscopic Power Device Modelling

**B. Benbakhti[1,⋆], M. Rousseau[2], A. Soltani[2], J-C. De Jaeger[2]**

[1]University of Glasgow, Glasgow, Scotland G12 8LT - UK

[2]IEMN, Avenue Poincar, Villeneuve dAscq 59652 - France

E-mail: ⋆brahim@elec.gla.ac.uk Tel +44 141 330 2964 Fax +44 141 330 4907

**Abstract.** The design of power GaN devices has to take into account the impact of temperature on device materials due to highly dissipated power and a consequent large self-heating. The accurate knowledge of transport properties as a function of the lattice temperature is essential in order to make a good thermal management to optimise the device performance. In this paper, accurate expressions describing the main transport properties as function of temperature and electric field for wurtzite GaN have been extracted starting from Monte Carlo simulations and then using a genetic algorithm. In particular, these expressions take into account the abrupt change in electron velocity slope at a low electric field (∼20 kV/cm). Using the same methodology, we have determined the temperature dependence of other physical parameters such as the low field mobility, saturation velocity, critical electric field and the corresponding peak velocity in a temperature range of 300 K - 700 K. The results show a very good agreement between the theoretical and experimental values.

## 1. Introduction

Semiconductor devices based on III-Nitrides have been widely accepted as a good vehicle for high frequency and high power applications [1][2]. In particular, gallium nitride presents remarkable electrical and thermal properties such as a high bandgap (3.4 eV), a high saturation velocity ($v_{sat} > 1\times10^7 cm/s$) and a high breakdown electric field (3 MV/cm) associated to an excellent thermal conductivity (1.6 W/cm.K) [3][4].

III-Nitrides based devices can take the wider bias ranges without inducing breakdown of the device. At very high biases, a self-heating occurs in the device and considerably increases the lattice temperature which negatively affects transport properties [4][5]. Therefore, the simulation of such devices has to account for thermal effects in order to obtain more accurate results [6][7]. The numerical investigations of such devices need an accurate knowledge of the temperature dependence of the transport properties, mainly the carrier velocity.

In this paper, the transport properties of GaN are determined from bulk Monte Carlo data. Starting from Monte Carlo results and using Genetic Algorithm (GA), it is possible to extract an analytical function, which describes the electron drift velocity of wurtzite-GaN versus the lattice temperature and electric field. The GA is used to determine the lattice temperature dependence of other physical parameters such as the low field mobility, saturation velocity, critical electric field and the corresponding peak velocity.

## 2. Monte Carlo calculations

The transport properties of GaN are obtained from stationary Monte Carlo (MC) simulations for different lattice temperatures (in a range of 300 K - 700 K by step of 50 K) and for ionised impurity concentrations of $10^{15}/cm^3$ and $10^{17}/cm^3$, respectively [8]. The used MC code is based on a three isotropic and non parabolic valley model. The scattering mechanisms included in the simulations are optical acoustic phonons, piezoelectric, intervalley scattering and ionized impurities. The valley parameters are deduced from band structure calculations [9], and the coupling constants involved in scattering rates are taken from [10].

Figure 1 shows that the average drift velocity drops as the lattice temperature increases at a fixed electric field. This is due to the total scattering rate increasing as the lattice temperature increases [12]. Indeed, the higher temperature results in the increase of the phonon scattering rate due to a larger phonon occupation number [13]. The dominant scattering mechanisms, at low electric fields, are polar optical phonon absorption, an interaction with ionised impurities and acoustic phonon scattering. As the electric field increases, a substantial number of electrons get sufficient energy and consequently the polar optical phonon emission becomes possible.

The slope of the electron velocity at low electric field decreases abruptly at approximately 20 kV/cm due to the onset of polar optical phonon emission [14]. However, this phenomenon vanishes at high temperatures. At high electric fields, close to 200 kV/cm, the electrons gain enough energy for a transfer into the satellite valleys by the intervalley phonon scattering and therefore a peak velocity is observed. In addition, this increased relaxation of electron energy and momentum results in a higher threshold electric field for heating of electrons. This increases their probability to scatter into the satellite valleys [15]. Thus, if we assume the constant separation energy for satellite valleys (we neglect a band gap renormalization due to the change in temperature), the threshold electric field increases with temperature while the peak velocity decreases. The Monte Carlo results show that the electron effective masses and the energy relaxation times are nearly identical for all considered lattice temperatures. Thus, mobility is the only quantity that strongly depends on lattice temperature.

## 3. Results and discussion

Starting from Monte Carlo data, the GA was implemented in order to determine a new expression of drift electron velocity for GaN versus electric field in respect to a temperature range of 300 K - 700 K:

$$\begin{aligned} V(E,T) &= \frac{\frac{\mu_n(T)}{1+E/E_3(T)}\,E + \frac{v_0+v_{sat}(T)[E/E_2(T)]^4}{1+[E/E_2(T)]^4}[E/E_C(T)]^\alpha}{1+[E/E_C(T)]^\alpha} \qquad (1)\\ v_0 &= \left[2V_{peak}(T) - \frac{\mu_n(T)E_C(T)}{1+E_C(T)/E_3(T)}\right]\left[1+(E_C(T)/E_2(T))^4\right] - v_{sat}(T)\,(E_C(T)/E_2(T))^4 \end{aligned}$$

where $T$ is the lattice temperature in Kelvin, $\mu_n$ is the low field mobility, $E_C$ is the critical electric field, $v_{peak}$ is the peak velocity and $v_{sat}$ is the saturation velocity. These parameters are determined as function of the temperatures. A good agreement is observed between Monte Carlo results and the analytical determinations for the two ionized impurity concentrations (figure 1).

Based on GA optimisation, the extracted analytical expressions describing the low field mobility $\mu_n[cm^2/V.s]$ as well as saturation $v_{sat}[cm/s]$ velocity at high electric fields and for two ionized impurity concentration as a function of the lattice temperature are:

| | $N_D$=$10^{15}/cm^3$ | $N_D$=$10^{17}/cm^3$ |
|---|---|---|
| $\mu_n(T)$ | $634-204\,\frac{T}{300}+80.9\times10^3\,exp\left(-4\,\frac{T}{300}\right)$ | $120+4.9\times10^3\,exp\left(-1.88\,\frac{T}{300}\right)$ |
| $v_{sat}(T)$ | $2.05\times10^7-78.9\times10^2\,T$ | $3.56\times10^7-17\times10^3\,T$ |

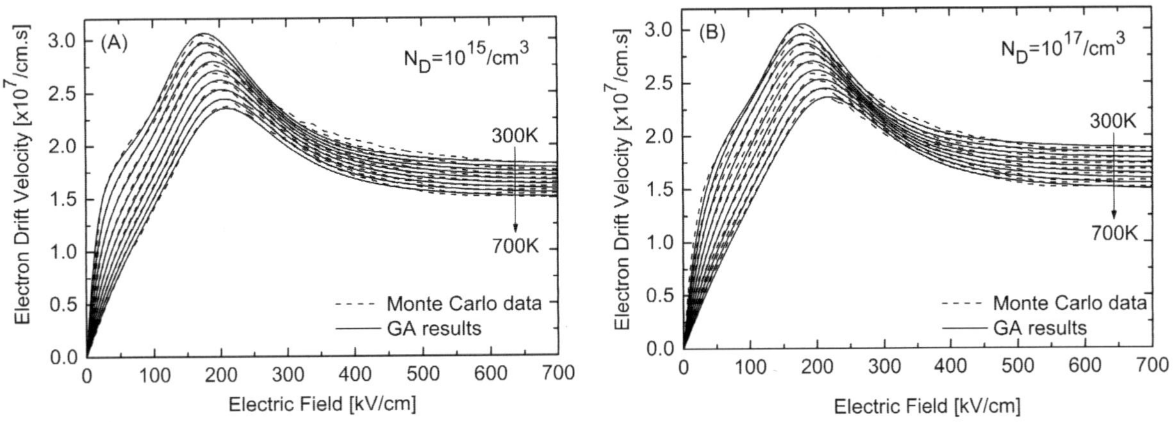


**Figure 1.** Electron drift velocity versus electric field of GaN for different lattice temperatures (in a range of 300 K - 700 K by step of 50 K): (A) $N_D$=$10^{15}/cm^3$ and (B) $N_D$=$10^{17}/cm^3$. Monte Carlo data (dash line) are compared to the genetic algorithm results (solid line).

Figure 2 shows the measured and analytical determination of the low field mobility at $N_D$=$10^{15}/cm^3$ as well as comparison between the analytical determination and results obtained from Ref.[16] for $N_D$=$10^{17}/cm^3$. The electron mobility saturation is observed at very high temperature primarily limited by longitudinal optical (LO) phonon scattering. The measurements of low field mobility have been carried out in the temperature range of 300 K to 500 K using Hall effect. We observe the large drop of the electron mobility between $N_D$=$10^{15}/cm^3$ and $N_D$=$10^{17}/cm^3$. The theoretical predictions are compared to experimental results [16] observing a very good agreement.

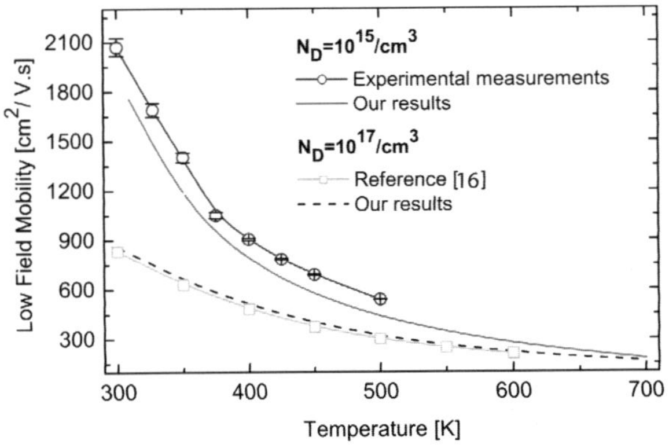


**Figure 2.** Low field mobility of GaN versus lattice temperature. Measured (circles) and analytical determination (solid line) at $N_D$=$10^{15}/cm^3$ are compared to the analytical determination (dash line) and results of Ref. [16] (squares) at $N_D$=$10^{17}/cm^3$.

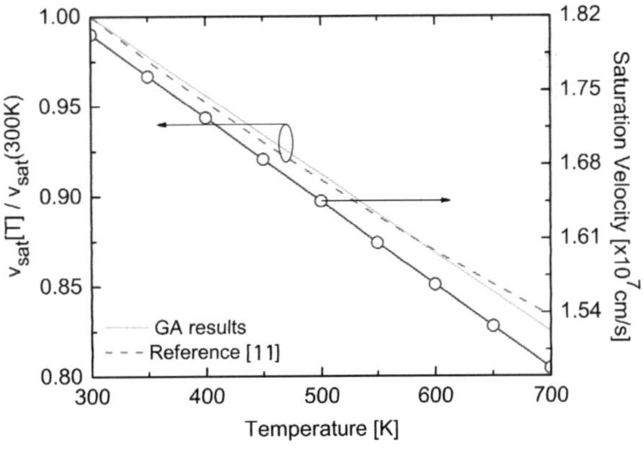


**Figure 3.** Saturation velocity of GaN versus lattice temperature for $N_D$=$10^{15}/cm^3$. Comparison between the results of Ref. [11] (dash line) and the GA expression (solid line).

Figure 3 show that the saturation velocity decreases significantly when the temperature increase. The velocity versus electric field evolution is due to the larger electron effective mass increase in the satellite valleys. A very good agreement is also observed for $N_D$=$10^{15}/cm^3$ when our results

are compared to results of Ref. [11].
For the critical electric field and the associated peak velocity, we have obtained the following expressions for the temperature dependence:

| | $N_D$=$10^{15}/cm^3$ | $N_D$=$10^{17}/cm^3$ |
|---|---|---|
| $E_{peak}(T)$ | $135 \times 10^3 + 10^2\ T$ | $132.5 \times 10^3 + 1.25 \times 10^2\ T$ |
| $v_{peak}(T)$ | $3.6 \times 10^7 - 175.3 \times 10^2\ T$ | $3.56 \times 10^7 - 170 \times 10^2\ T$ |

The drop in the peak velocity with temperature is less than those currently observed in other semiconductors such as GaAs [11]. When the lattice temperature varies from 300 K to 700 K, the associated electric field increases by about 25 % and the peak velocity decreases by about 23 % (figure 3).
In the 300 K - 700 K temperature range, the evolutions of the other parameters used in the drift velocity expression as a function of the temperature are given by the following equations:

| | $N_D$=$10^{15}/cm^3$ | $N_D$=$10^{17}/cm^3$ |
|---|---|---|
| $E_2(T)$ | $9.3 \times 10^5 \left(\frac{T}{300}\right)^{0.1166}$ | $19.3 \times 10^4 \left(\frac{T}{300}\right)^{0.1818}$ |
| $E_3(T)$ | $52.5 \times 10^2\ exp\left(1.65\ \frac{T}{300}\right) - 150.2 \times 10^2$ | $11.16 \times 10^3\ exp\left(1.57\ \frac{T}{300}\right) - 183.6 \times 10^2$ |
| $\alpha(T)$ | $4.15 + 74.6 \times 10^{-3}\ exp\left(1.1\ \frac{T}{300}\right)$ | $5.15 + 9.68\ exp\left(-2.12\ \frac{T}{300}\right)$ |

## 4. Conclusion

In this paper, we have extracted an analytical expression describing the dependence of the electron velocity as function of lattice temperature and electric field. This extraction is carried out starting from Monte Carlo simulations and then using GA. The same methodology is used to determine the lattice temperature dependent of low field mobility, saturation velocity, critical electric field and the corresponding peak velocity. The obtained analytical expressions describe the main transport properties of GaN in a temperature range of 300 K - 700 K. In particular, these expressions take into account the abrupt change in the electron velocity at low electric field. The results show a very good agreement between the theoretical and experimental results. This accurate analytical determination is absolutely necessary for simulation of power GaN devices which need an accurate information about electron velocity. The temperature dependence of electron velocity must be taken into account due to the highly dissipated power and the consequent large device self-heating. This constitutes an essential tool for the prediction of thermal management in GaN based power devices.

## References

[1] E. Alekseev, D. Pavlidis, Solid State Electronics 44 (2000) 941-947.
[2] Y. Chang, Y. Zhang, Y. Zhang, K.-Y. Tong, Journal of Applied Physics 99(17) (2006) 044501.
[3] B.-M. Green et al., IEEE Transactions on Microwave Theory and Techniques 49(12) (2001) 2486-2493.
[4] J. Park, M.-W. Shin, C.-C. Lee, IEEE Electron Devices Letters 24(7) (2003) 424-426.
[5] L.-F. Eastman et al., IEEE Transactions on Electron Devices 48(3) (2001) 479-485.
[6] Y.-F. Wu et al., IEEE Electron Device Letters 25(3) (2004) 117-119.
[7] T. Palacios el al., IEEE Electron Device Letters 26(11) (2005) 781-783.
[8] F. Dessenne et al., Materials Science and Engineering B 50(3) (1997) 315-318.
[9] J. Kolnik, I.H. Oguzman, K.F. Brennan, R. Wang, P.P. Ruden, W. Wang, J. Appl. Phys. 78 (1995) 1033.
[10] B. Gelmont, K. Kim, M.S. Shur, J. Appl. Phys. 74 (1993) 1818.
[11] U.-V. Bhapkara, M.-S. Shur, Journal of Applied Physics 82(4) (1997) 1649-1655.
[12] R.-P. Joshia, S. Viswanadha, P. Shah, R.-D. Rosario, Journal of Applied Physics 93(8) (2003) 4836-4842.
[13] A. Reklaitis, L. Reggiania, Journal of Applied Physics 95(12) (2004) 7925-7935.
[14] N. Mansour, K.-W. Kim, N.-A. Bannov, M.-A. Littlejohn, Journal of Applied Physics 81(6) (1997) 2901-2903.
[15] M. Goanoa et al., Journal of Applied Physics 88(11) (2000) 6467-6475.
[16] J.-D. Albrecht el al., Journal of Applied Physics 83(9) (1998) 4777-4781.

Journal of Physics: Conference Series **193** (2009) 012006 doi:10.1088/1742-6596/193/1/012006

# Alternating-current and direct-current electrical transport properties of diamond diodes with diamond surfaces of the different morphology

**A. M. Rodrigues**

Departamento de Física, FCT, Universidade do Algarve, Campus de Gambelas, 8000-139 Faro, Portugal

E-mail:arodrig@ualg.pt

**Abstract**. The alternating current (ac) conductance ($f$=200 Hz) was measured as a function of applied voltage at ambient temperature for two diodes with aluminium/diamond rectifying contacts on chemical vapour-deposited diamond films of 5 μm in thickness with different morphology, nucleation and growth surfaces. Comparison of direct current (dc) conductance curves (deduced from experimental current–voltage curves) with ac conductance curves for the diodes revealed different transport mechanisms for two interfaces. The current–voltage characteristic of gold/diamond/gold structures shows that the two interfaces have different resistance values. The results demonstrate that the nucleation surface has a greater surface density of states than the growth surface, and this contributes to electrical transport through the potential barrier. Cole–Cole curves show a discrepancy between the Debye semicircle and experimental results and illustrate the fundamental differences between the two diodes. These results provide additional proof that the principal source of dispersion admittance is the surface states. The current–voltage characteristic of diamond diodes with diamond films of 1 μm in thickness shows two distinct resistance states (high and low resistance). By applying an appropriate bias, resistance switching from one state to the other state can be achieved. The experimental results are discussed on the basis of filamentary conduction

## 1. Introduction

Carbon-based materials such as nanocrystalline diamond, polycrystalline diamond, amorphous carbon and carbon nanotubes are of immense scientific and technological interest in different fields such as electronics, electrochemistry, thermal applications, optics, and biology. The quality of the polycrystalline diamond films used in the present study is comparable to that of heteroepitaxial films reported in the literature [1]. Analysis of the diamond surfaces revealed that the growth surface was very rough, whereas the nucleation surface had a more uniform texture. Two types of structures were produced using diamond film of 5 μm in thickness. One sample (designated S1) was prepared by depositing a rectifying aluminium (Al) contact on the diamond growth surface and the other (designated S2) by depositing a rectifying Al contact on the nucleation surface. As previously reported [2], the electrical properties of the devices are influenced by the surface morphology and microstructure. Investigation of the I–V characteristics revealed that the diode with a rectifying contact on the nucleation surface had better diode properties and lower conductivity. In the present study, complementary comparative studies of the alternating current (ac) and direct current (dc) conductance for the two different devices confirm our previous results [2] regarding charge transport. The impedance was measured and plotted in Cole–Cole representation for the growth surface and nucleation surface. The I–V characteristics of a similar device fabricated with a thinner diamond film (~1 μm) are also described.

## 2. Experimental

The diamond film used in this study was deposited on a *n*-type silicon substrate by MPCVD using a mixture of methane (1%) and hydrogen (99%). Details of the deposition conditions, substrate preparation, and quality of the diamond films can be found elsewhere [3, 4]. Diamond film of 1–5 μm in thickness was obtained by removing the substrate with HF solution. The morphology of the two free surfaces was analysed by atomic force microscopy. This confirmed that the growth surface was relatively rough (~2% of the total film thickness) and formed by microcrystals with a common <111> orientation distributed homogeneously on the surface. Micro-Raman measurements showed no significant spectral differences for different regions of the surface. The nucleation surface was exposed after the substrate was removed and corresponds to the first layer of diamond deposition. This surface had an extremely flat morphology with average roughness ~4 nm, and a compact texture of very small, round grains. To produce the diodes, Au and Al contacts were deposited by thermal evaporation on opposite surfaces of the diamond film. Two types of diodes were prepared, with the rectifying Al contact on either the growth surface (S1 and S4) or the nucleation surface (S2). A symmetric structure, denoted S3, was prepared by depositing Au contacts on both diamond surfaces. Current–voltage measurements were performed using a Keithley 487 picoammeter/voltage source. Capacitance and conductance measurements were carried out in a vacuum chamber at a pressure of approximately $10^{-5}$ mbar using a Fluke PM6306 RLC meter with a resolution of 0.1 pF in the frequency range between 100 Hz and 1 MHz. Device fabrication and all electrical measurements were performed in the Centre for Electronics, Optoelectronics and Telecommunications, Algarve University.

## 3. Results and discussion

Analysis of the *I–V* curves for symmetric device S3 showed that the two diamond surfaces had different morphology and established electrical interfaces with the same metal (Au). The contact on the growth surface (rough surface) had lower resistance than the contact on the nucleation surface (flat surface) and this dissimilarity in resistance was responsible for the very small rectification. This observation prompted our first study of devices with rectifying Al contacts on diamond with two different surfaces [2]. To analyse the electrical transport mechanism in more detail and to complement the previous study, the dc and ac conductance of these devices was investigated in the present study. The *I–V* curves were derivative [2] to determine the dc conductance ($\sigma_{dc}$) as a function of the applied voltage for devices S1 and S2. Figures 1 and 2 compare these results with the ac conductance ($\sigma_{ac}$) as a function of the applied voltage, which was measured experimentally at a test frequency of 200 Hz at ambient temperature for each device. The curves for forward bias show that in general $\sigma_{ac}$ is greater than $\sigma_{dc}$ for both devices. For S1 the $\sigma_{dc}$ and $\sigma_{ac}$ curves tend to converge with increasing voltage, whereas for S2 the difference between the $\sigma_{dc}$ and $\sigma_{ac}$ curves increases with the voltage. These results indicate that the dominant electrical transport mechanism is different for each interface. In the S1 interface charge transport in the stationary regime predominates over charge transport in the alternate regime, in S2 the opposite is the case. This indicates that electronic states exist in the diamond gap that actively contribute to charge transport through the Schottky barrier. It is known that conduction in the ac regime is favoured by the presence of electronic states in the interface when the test frequency is sufficiently low so that the states can be filled and emptied during one cycle. In the stationary dc regime and for forward bias, all the electronic states are filled and consequently do not participate in electrical transport. The transport regime ac prevalence with Fermi energy dislocation in the direction of the conduction band indicates that the concentration of electronic states that participate in this

Journal of Physics: Conference Series **193** (2009) 012006 doi:10.1088/1742-6596/193/1/012006

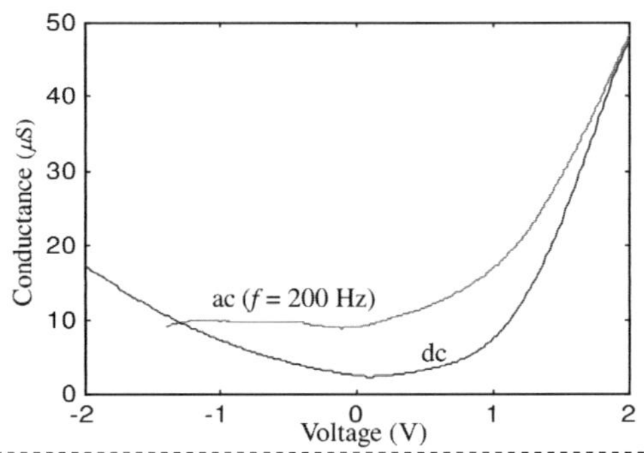


**Figure 1.** Conductance for diode S1.

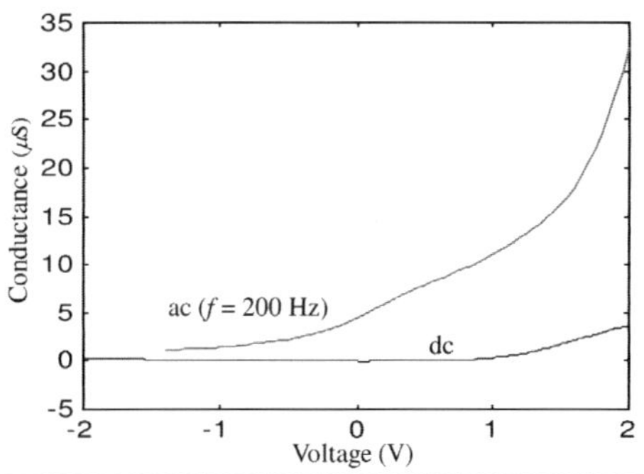


**Figure 2.** Conductance for diode S2.

process increases inside the forbidden band in the direction of the valence band of the diamond. The acceptor impurity concentrations obtained from a Mott-Schottky plot [2] for the growth and nucleation surfaces correspond at $6\times10^{17}$ cm$^{-3}$ for S1 and $8\times10^{16}$ cm$^{-3}$ for S2. The conclusion is that although the nucleation surface has a greater number of surface density states, the number of acceptor impurities is lower than for the growth surface.

Understanding of the physical origin of impurities is important when considering the growth process of the films. According to this model, the growth of crystals on the nucleation surface initially involves accumulation of diamond-like-carbon (DLC), followed by diamond growth in a planar region and finally by epitaxial growth of diamond normal to the surface of the substrate plane. This results in many small diamond grains with small inter-grain spaces on the nucleation surface. The growth surface has larger microcrystals and inter-grain spaces and a smaller number of microcrystals. Based on this growth model of the diamond films, it can be inferred that the nucleation surface is formed essentially of DLC whereas the growth surface is composed of diamond microcrystals and grain boundaries. Assuming that this model is adequate, it can be concluded that the carbon form found in the nucleation region is more easily compacted and thus accepts few impurities, but more surfaces states are generated.

Admittance measurements provided more information about the nature of the two rectifying contacts and the role of morphology in the interface quality [2, 4]. Both devices were described by a simple equivalent circuit model (inset in Figure 3) for which the components $C_d$, $R_d$, $C_b$ and $R_b$ correspond to the capacitance and resistance of the depletion and bulk regions, respectively. A graphic representation of capacitance and conductance in the form of Cole–Cole plots for devices S1 and S2 is presented in Fig. 3. The fit parameters [2] are $C_d$=5 nF, $R_d$=0.4 MΩ, $C_b$=0.2 nF and $R_b$=1 kΩ for sample S1 and $C_d$=1 nF, $R_d$=0.5 MΩ, $C_b$=0.3 nF and $R_b$=0.3 kΩ for S2, in accordance with the electrical characteristics of the samples. For both devices the semicircle is flatter than a typical Debye curve, which is a perfect semicircle centred on the $x$-axis. Comparison of the curves for the two devices reveals greater compression for S2 than for S1, indicating that the former has greater admittance dispersion, which can be attributed to the surface states. Figure 3 also suggests another incomplete semicircle for low frequencies. This possibility exists because, as previously demonstrated [4], polycrystalline diamond can make two different contributions to admittance: due to grains at low frequencies and to grain boundaries at high frequencies. In this case a more complete circuit model that contains explicit grain and grain boundary characteristics is required.

Thinner films of approximately 1 μm exhibited characteristics of filamentary conduction (device S4). Figure 4 shows a typical $I$–$V$ curve for this device. Over a complete cycle originating at 0 V and traversing the voltage interval between −8 V and 8 V, a pronounced hysteresis is evident that is a consequence of the abrupt change from less conductor state to a more conductor state. These transitions occur at positive and negative voltage. The current density is relatively high and is two orders of magnitude greater than for the devices S1 and S2, and the noise is also higher. This high

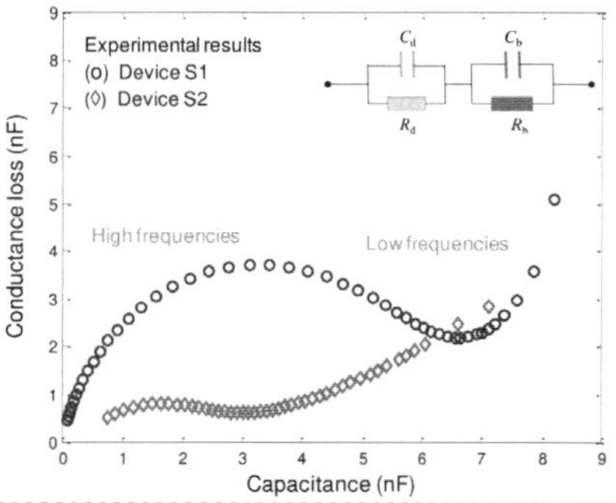


**Figure 3.** Cole-Cole plot for diodes S1 and S2.

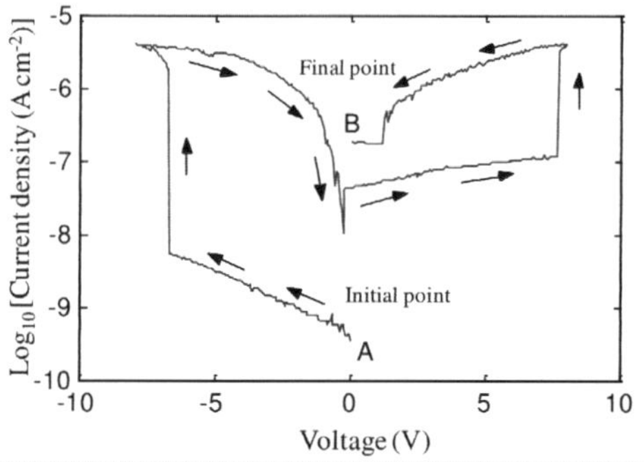


**Figure 4.** Typical filamentary conduction for devices of 1 $\mu$m in thickness.

current density due to the change in states possibly corresponds to a mixture of electronic and chemical processes. This behavior is characteristic of filamentary conduction, which has also been observed for different amorphous inorganic semiconductors. Another feature of the samples was that no relaxation peak was observed in the ac admittance spectrum.

## 4. Conclusions

The *I–V* characteristic of symmetric devices revealed that the rough surface had lower resistance than the flat surface, causing extremely little rectification. Comparison of ac and dc conductance for the different aluminium/diamond interfaces revealed that the nucleation surface had a higher surface density that contributed to transport through the barrier. The dependence of this contribution on voltage suggests that this density increases of the interior of the forbidden band to valence band. This result is according with the literature [2] that considers that the nucleation surface has more defects. The surfaces states seem to have physical origin different of the accept impurities that contribute for bulk transport. The Cole-Cole plot shows a grade of the dispersion superior in interface of the nucleation surface indicating that the admittance dispersion apparently is due to surface states [2]. The fit parameters reflect the electrical conditions of the samples. The possibility of an incomplete semicircle at low frequencies suggests the possibility of two relaxation processes, one related to grains and the other to grain boundary regions of the diamond.

Current–voltage curves for diamond diodes with diamond films of 1 μm in thickness revealed that application of an adequate voltage leads to resistance switching between two distinct resistance states. These samples present characteristics of filamentary currents that have been observed in other semiconductors. Moreover, no relaxation peak was observed in the admittance curve for this device.

## Acknowledgement

The author would like to thank Dr. H. Gomes of the Centre for Electronics, Optoelectronics and Telecommunications, Algarve University.

## References

[1] Willander M, Friesel M, Wahab Q and Straumal B 2006 *J. Mater. Sci. Mater. Electr.* **17** 1.
[2] Rodrigues A M, Gomes H L, Rees J A, Pereira L and Pereira E 1999 *Phys. Status Solidi* (a) **174** 165
[3] Rodrigues A M 2008 *J. Appl. Phys.* **103** 083708
[4] Rodrigues A M 2008 *Diamond Relat. Mater.* **17** 1264
[5] Rodrigues A M 2007 *Appl. Surf. Sci* **253** 5992
[6] Kao K C and Hwang W 1981 *International series on the science of the solid state* **vol 14** (Oxford New-York: Pergamon Press)

Journal of Physics: Conference Series **193** (2009) 012007 doi:10.1088/1742-6596/193/1/012007

# Analysis of the resonant tunneling diode with the stepped pre-barrier

**Roman Yatskiv[1], Jan Voves[2]**
[1]Institute of Photonics and Electronics, Academy of Sciences of the Czech Republic, Prague, Czech Republic
[2] Department of Microelectronics, CTU FEE Prague, Prague, Czech Republic

E-mail: yatskiv@ufe.cz

**Abstract.** Resonant-tunnelling diodes (RTD) incorporating an emitter stepped pre-barrier are studied both theoretically and experimentally. The simulation of I-V characteristics of modified AlAs-GaAs double barrier RTD grown by molecular beam epitaxy with the stepped pre-barrier is presented. An 1D quantum transport simulator Wingreen based on the nonequilibrium Green functions (NEGF) is used in our case. Our result show that the coupling between energy levels in the emitter quantum-well and the main quantum well leads to the plateau behaviour of the I-V curves. The two plateau regions on the I-V characteristics have been observed in experimental and simulation results.

## 1. Introduction

In recent years, the resonant tunneling diodes (RTD) have attracted the attention of scientists due to their high frequency performance up to the terahertz region. It is possible to realize oscillators, mixers, switches and various kinds of functional circuits with RTD structures. It is well known that the current-voltage (I-V) curves of the resonant tunneling diodes (RTD) exhibit a characteristic plateau-like behavior and hysteresis. A proper and complete interpretation of the I-V curves is still a controversial issue.

Some groups [1-4] utilized self-consistent solutions of the time-dependent Wigner function equation (WFE) and Poisson equation (PE) for a GaAs-AlGaAs based double-barrier RTD. These simulations revealed intrinsic high-frequency oscillations in the tunneling current. Plateau-like behavior in the I-V curves was obtained from the time-average of the current oscillations. The results showed that the formation of the emitter quantum well (EQW) and the coupling of its quasi-bound state to the state in the main quantum well (MQW) is the physical mechanism responsible for both the hysteresis and the plateau-like behavior.

Other approach is based on the non-equilibrium Green's functions (NEGF). Gardner, Klimeck and Ringhofer [5-6] performed numerical calculations by the program NEMO based on the NEGF approach. They found that scattering in the emitter reduces hysteresis by broadening the quasi-bound state formed in the emitter as well. Nevertheless, the plateau region did not appear in their results. In our previous work, we presented simulations performed by the NEGF based simulator Wingreen. The plateau region of the same AlAs-GaAs RTD in both experimental I-V curves and those obtained by simulations was portrayed [7].

In this work, we present the simulation of the modified AlAs-GaAs double barrier RTD grown by molecular beam epitaxy with the stepped pre-barrier in the emitter region. The simulation results have been compared with experimental measurements. Creation of the two plateau regions in negative differential resistance (NDR) region of the I-V curve is explained by coupling between energy levels in the emitter quantum-well and the main quantum well.

## 2. Experiment and simulation

The AlAs-GaAs double barrier RTD with the stepped pre-barrier in the emitter region were grown by molecular beam epitaxy on semi-insulating GaAs oriented in the [100] direction. The growth starts with a buffer layer of 250 nm of undoped GaAs. The double barrier quantum well (DBQW) structures consist of the pair identical 3.4 nm AlAs barriers sandwiching a 5.9 nm GaAs well. The stepped pre-barrier consist of three 4 nm $Al_xGa_xAs$ layers with different values of x. 10 nm u-GaAs spacer layers and 100 nm GaAs:Si supply layers were grown on the both sides of the DBQW. The $n^+$ top and bottom contact layers were 100 nm and 500 nm thick respectively. Ti/Au ohmic contacts were fabricated on the anode and cathode layers. Mesa structures were formed by photolithography and etching. The active areas of diodes were 20x20 $\mu m^2$. Schematic cross-sections RTD with the stepped pre-barrier and the band edge diagram are shown in figure 1(a) and 1(b) respectively.

The DC I-V characteristics were measured by Hewlett Packard picoampermeter/DC voltage source 4140B. The measurements were carried out for different temperatures, that have been changed from 77 to 300K for increasing and decreasing votage slopes in both directions [8]. The theoretical calculations were performed by the Wingreen simulator [9] based on the NEGF approach.

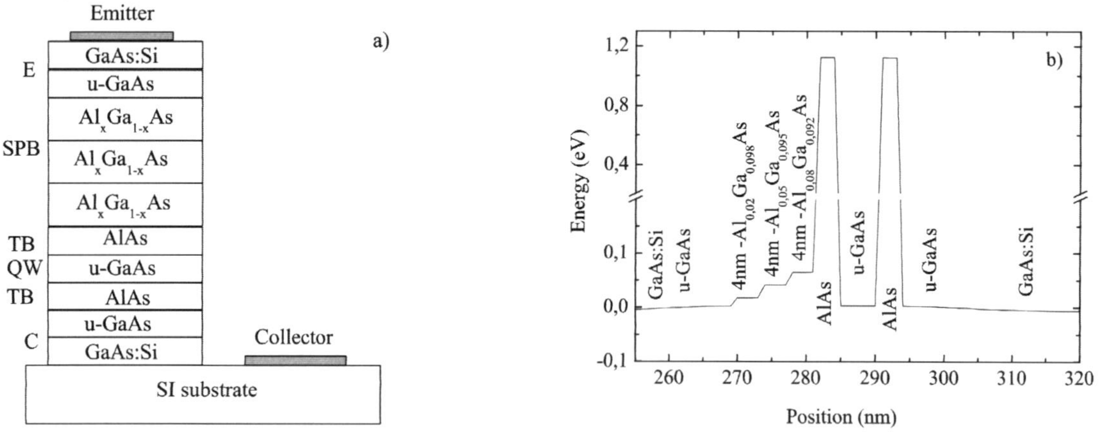


**Figure 1.** Schematic cross section (a) and conduction band profile (b) of the resonant tunnelling diodes with the stepped pre-barrier. SPB-stepped pre-barrier, QW-quantum well, TB-tunnelling barrier.

In figure 2 we show comparison of experimental and calculated current-voltage characteristics at 77K. Both experimental and simulation curves show two plateau regions in the NDR region. The plateau behaviour is restricted to bias voltage between 0.42 and 0.62 V for experimental curve and between 0.24 and 0.73 V for simulated curve. Since the behaviour of I-V curves outside this voltage region is well understood, attention is now focused on the behaviour of the curves inside the bistability region.

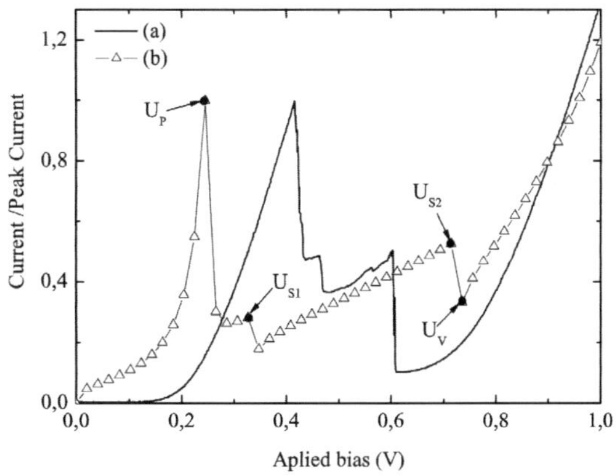


**Figure 2.** I-V characteristics RTD structure with the stepped pre-barrier in the emitter measured (a) and simulated (b) at 77 K

Figure 3 shows the self-consistent potential and the electron density versus bias for the important bias points. These data can be used for the explanation of the plateau region creation. Due to a depletion of electrons in front of the emitter barrier a wide quantum well is formed in the emitter region.With an increase of the bias, energy level in the main quantum well (QW) couples with energy level in the first triangular well in the emitter and the first plateau is observed. The second plateau is formed, when the further triangular well in emitter region is populated.

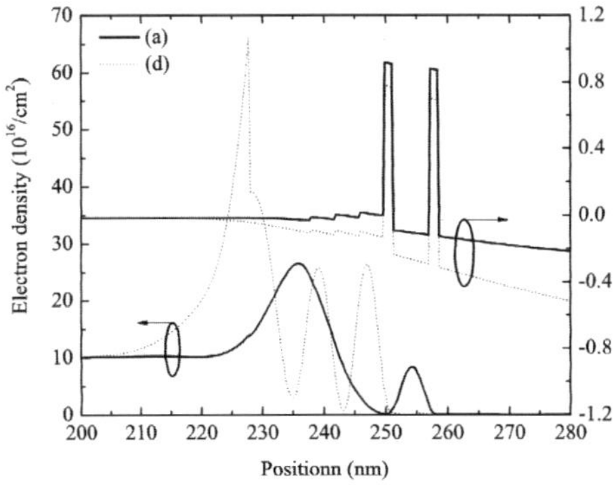


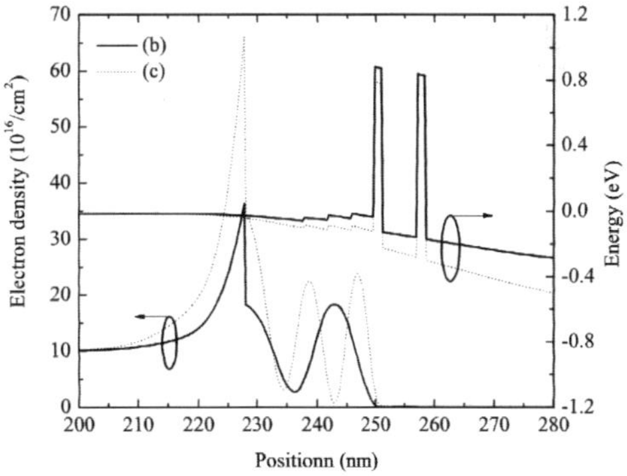


**Figure 3.** Electron densities and conduction band profiles of the modified RTD structure with the stepped pre-barrier in the emitter simulated at the bias of the peak current $U_P$= 0.24V (a), of the first and second plateau maximum $U_{S1}$=0.32V (b), $U_{S2}$=0.71V (c) and of the valley current $U_V$=0.73V (d).

## 3. Conclusions

We presented both theoretical and experimental studies ofresonant tunnelling diodes incorporating an emitter stepped pre-barrier. The results showed that the formation of the emitter quantum well and the coupling of its quasi-bound state to the state in the main quantum well is the physical mechanism responsible for plateau-like behaviour. In a forward bias sweep there is a decrease of electron density in the emitter due to the interference between the injected and the reflected electron waves, and a deep and wide quantum well is formed in the emitter. With an increase of the bias, energy level in the main QW couples with that in EQW. This coupling leads to the plateau-like structure.

## 4. Acknowledgment

We would like to thank to Dr. M. Cukr and to Dr. Z Vyborny from Institute of Physics, The Czech Academy of Science for the MBE growth of the RTD samples. The work has been financially supported by Academy of Sciences of the Czech Republic, grants KJB200670901, KAN401220801 and KAN400100652 and by the grant No. MSM 6840770014 The Ministry of Education CR.

## References

[1] Biegel B A, Plummer J D 1997 *IEEE Trans. El. Dev.* **44** 733
[2] Zhao P, Cui H L, Woolard D L, Jensen K J, Buot F A 2000 *Int. J. Mod. Phys. B* **14** 411
[3] Zhao P, Cui H L, Woolard D L, Jensen K J, Buot F A 2000 *J. of Appl. Phys.* **87** 1337
[4] Zhao P, Woolard D L, Gelmont B L, Cui H L 2003 *J. of Appl. Phys.* **94** 1833
[5] Klimeck G 2003 *J. of Comp. Electronics* **2** 177
[6] Gardner G L, Klimeck G, Ringhofer C 2004 *J. of Comp. Electronics* **3** 95
[7] Voves J, Třebicky T, Jackiv R 2007 *J. of Comp. Electronics* **1-3** 255
[8] Jackiv R, Třebický T, Voves J, Výborný Z, Cukr M, Jurka V 2004 *Proc. Int. Conf. on Microelectronics* (Nis, SERBIA) vol. 1 p. 359
[9] Indlekofer K M, Malindretos J http://www.fz-juelich.de/isg/mbe/software.html

Journal of Physics: Conference Series **193** (2009) 012008 doi:10.1088/1742-6596/193/1/012008

# Quasi-ballistic electron transport through silicon nanocrystals

**Nobuya Mori[1], Hideki Minari[1], Shigeyasu Uno[2], Hiroshi Mizuta[3], and Nobuyuki Koshida[4]**

[1]Graduate School of Engineering, Osaka University, Suita, Osaka 565-0871, Japan
[2]Graduate School of Engineering, Nagoya University, Nagoya, Aichi 464-8603, Japan
[3]School of Electronics and Computer Science, University of Southampton Southampton SO17 1BJ, United Kingdom
[4]Graduate School of Engineering, Tokyo University of Agriculture and Technology Koganei, Tokyo 184-8588, Japan

E-mail: nobuya.mori@eei.eng.osaka-u.ac.jp

**Abstract.** Monte Carlo simulation of electron transport through a silicon nanocrystal array is performed to clarify the mechanism of high-energy electron emission from a porous silicon diode. The electronic states are calculated within an empirical tight-binding approximation. In the Monte Carlo simulation, optical-phonon emission and elastic tunneling processes are taken into account. It is found that initial electron acceleration, which is enhanced due to the discrete nature of lower energy levels, plays an essential role in generating high-energy electron emission.

## 1. Introduction

Recently, electron transport properties of nanocrystalline Si (nc-Si) have attracted considerable attention along with the remarkable progress of nc-Si material control technologies [1]. It is widely known that porous Si (PS) comprises tree-like network of many nanometer size nc-Si. When high voltage ($\sim 10-20\,\mathrm{V}$) is applied to a PS region, high-energy electrons are emitted from the surface [2] (see Fig. 1). This unique property has been exploited for ballistic electron emitters and other various applications. Essential features of the observed emission energy distribution are summarized in Ref. [3]; which include the fact that the energy difference between the Fermi level at the Si substrate and the peak energy of the emitted distribution, $E_{\mathrm{loss}}$ (see Fig. 2), is nearly independent on the applied voltage $V_{\mathrm{a}}$. Although high-energy electron emission from a PS diode is clearly observed, the detailed mechanism has not been fully understood and explained. In the preset study, we have performed Monte Carlo simulation of electron transport through Si quantum dots (QDs) to clarify the underlying physics of the high-energy electron emission.

## 2. Analysis of the emission energy distribution

### *2.1. Energy levels*

We use an sp$^3$s$^*$d$^5$ empirical tight-binding method [4] to calculate electron energy levels $E_i$ ($i = 1, 2, \cdots$) in a Si QD. Figure 3 shows a level separation,$\Delta\ E_i = E_{i+1} - E_i$, as a function of the electron energy for a Si QD with a diameter of $d = 4.1\,\mathrm{nm}$. The quantum confinement makes the level separation larger for lower energy region $E \lesssim 1.5\,\mathrm{eV}$. However, the level separation

Journal of Physics: Conference Series **193** (2009) 012008 doi:10.1088/1742-6596/193/1/012008

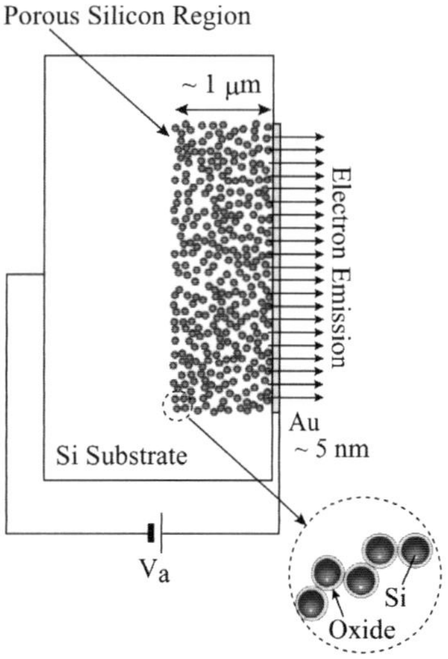


**Figure 1.** Schematic illustration of the porous silicon diode. The silicon substrate is electrochemically treated to form $\sim 1\,\mu$m thick porous silicon region. When positive bias is applied to the Au surface electrode, electrons are emitted from the surface.

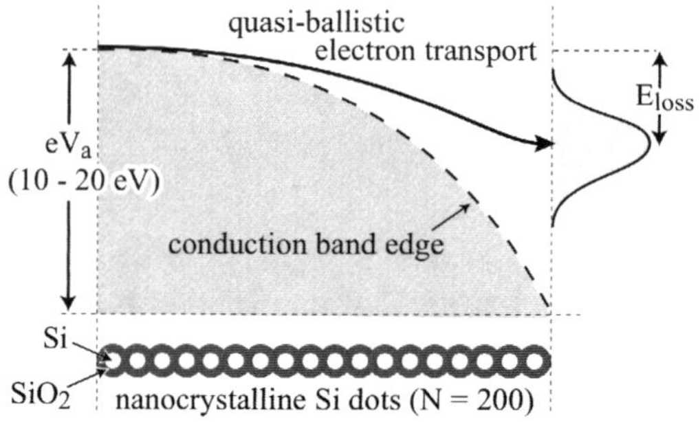


**Figure 2.** Schematic illustration of the conduction-band profile under positive bias $V_{\mathrm{a}}$. Major voltage drop occurs across the porous silicon region due to its high resistivity.

becomes quite small ($\Delta E_i < 10\,\mathrm{meV}$) for higher energy region. As will be shown later, this emergence of two regions of low-energy discrete levels and high-energy quasi-continuous levels plays an important role in generating the electron emission from a PS diode.

### 2.2. Optical-phonon emission rate

Within the Fermi's golden rule, the optical-phonon emission rate, $W_i$, from an initial energy level $E_i$ can be written as

$$W_i = \frac{2\pi}{\hbar} \sum_j \sum_{\boldsymbol{q}} |\langle \Psi_j(\boldsymbol{x})|U(\boldsymbol{x})|\Psi_i(\boldsymbol{x})\rangle|^2 \delta(E_i - E_j - \hbar\omega_0), \tag{1}$$

where $j$ is a final state index,$\Psi_{\ i}(\boldsymbol{x})$ is the wavefunction associated with the $i$-th energy level, $\hbar\omega_0$ is the optical-phonon energy, $U(\boldsymbol{x})$ is a Fourier component of the electron–optical-phonon interaction potential

$$U(\boldsymbol{x}) = D_{\mathrm{op}} \left( \frac{\hbar}{2\rho V \omega_{\ 0}} \right)^{1/2} \mathrm{e}^{\mathrm{i}\boldsymbol{q}\cdot\boldsymbol{x}}, \tag{2}$$

$D_{\mathrm{op}}$ is the optical deformation potential, $\rho$ is the mass density, and $V$ is the sample volume. Note that we set the phonon occupation factor $N_0 = [\exp(\hbar\omega_0/kT) - 1]^{-1}$ ($kT$ is the thermal energy) to 0 and consider the emission process only, which should introduce a negligible error in the final results because we consider high-energy electrons with $E \sim 10\,\mathrm{eV}$. From Eqs. (1) and (2), we obtain

$$W_i = \frac{\pi D_{\mathrm{op}}^2}{2M\omega_0} g(E_i - \hbar\omega_0), \tag{3}$$

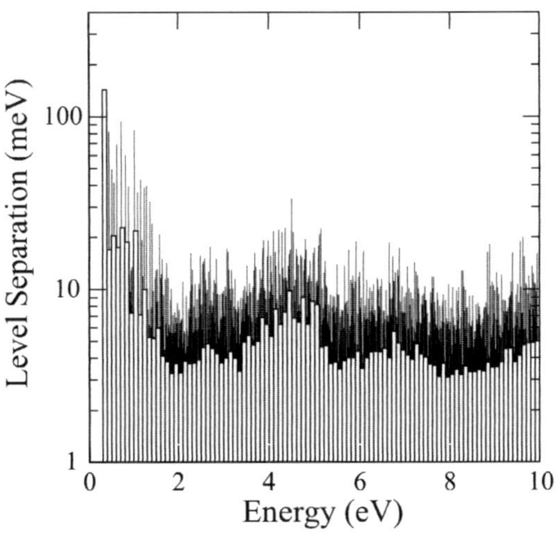


**Figure 3.** Level separation as a function of electron energy for a Si quantum dot with a diameter of $d = 4.1\,\mathrm{nm}$ (vertical solid line). Box shows an average over a 10 meV interval.

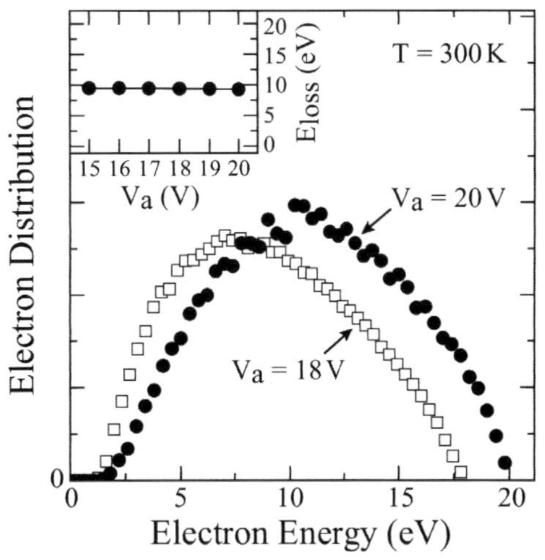


**Figure 4.** Energy distribution of the emitted electrons for applied voltage $V_\mathrm{a} = 18\,\mathrm{V}$ (□) and 20 V (● ) at $T = 300\,\mathrm{K}$. Inset shows applied-voltage dependence of the energy difference between the Fermi level at the Si substrate and the peak energy of the emitted distribution.

where $M$ is the mass of a silicon atom and the density-of-states $g(E)$ is given by

$$g(E) = \frac{2}{N}\sum_j \delta(E - E_j), \tag{4}$$

with $N$ being the number of silicon atoms in the QD. Here use has been made of $F_{ij} = \Omega_\mathrm{D}^{-1}$ ($\Omega_\mathrm{D}$ is the QD volume) for

$$F_{ij} = \int |\Psi_i(\boldsymbol{x})|^2 |\Psi_j(\boldsymbol{x})|^2 \mathrm{d}\boldsymbol{x}. \tag{5}$$

The value of the optical deformation potential $D_\mathrm{op}$ for high-energy electrons should be different from that for electrons near the conduction band edge. In the present study, we estimated the value from the energy shift caused by the lattice displacement.

We consider acoustic-phonon scattering as a scattering process which causes collisional broadening of the density of states. The level broadening $\Gamma$ can be calculated in a similar way as in the optical-phonon emission rate and we have

$$\Gamma = D_\mathrm{ac}\left(\frac{kT}{\rho s_l^2 \Omega_\mathrm{D}}\right)^{1/2}, \tag{6}$$

where $D_\mathrm{ac}$ is the acoustic deformation potential and $s_l$ is the longitudinal sound velocity. The level broadening is introduced by replacing $\delta(E)$ in Eq. (4) with $\exp[-(E/\Gamma)^2]/(\sqrt{\pi}\,\Gamma)$.

*2.3. Monte Carlo simulation*

We use a Monte Carlo method to simulate the dynamics of electrons inside a PS diode. An electron in a QD has a probability to emit an optical phonon and a probability to escape from the QD through elastic tunneling. The optical phonon emission rate is given by Eq. (3). The elastic tunneling time, $\tau_t$, is related to the miniband width $\Delta$ as $\hbar/\tau_t = \Delta/4$. We use a simple one-dimensional Kronig-Penny model [5] to calculate$\Delta$.

Figure 4 shows energy distribution of the emitted electrons for applied voltage $V_a = 18\,\mathrm{V}$ and $20\,\mathrm{V}$. We used the following parameters: the QD diameter $d = 4.1\,\mathrm{nm}$, the $SiO_2$ thickness $t_{ox} = 1\,\mathrm{nm}$, the optical phonon energy $\hbar\omega_0 = 60\,\mathrm{meV}$, the acoustic deformation potential $D_{ac} = 9.5\,\mathrm{eV}$, and the longitudinal sound velocity $s_l = 9.04 \times 10^5\,\mathrm{cm/s}$. The data in Fig. 4 is the energy distribution of electrons just before the emission from the PS region and dose not include effects of the workfunction of the electrode. We assume that the potential profile from the cathode to the anode electrodes is of the form $V(x) \propto x^{3/2}$ considering the space-charge-limited operation. Coulomb interaction, which will affect the energy level distribution and the potential profile, is consider to be weak as PS diodes usually operate at room temperature under high applied bias. Figure 4 clearly shows that electrons are quasi-ballistically emitted from a PS diode. There is good agreement between the calculated energy distribution and the experimental results [1]. We find that initial electron acceleration, which is enhanced in QDs due to the discrete nature of lower energy levels [6], plays an essential role in generating high-energy electron emission. Once an electron is accelerated to the higher energy state ($E \gtrsim 1.5\,\mathrm{eV}$), it can ballistically travel through the quasi-continuous states and is emitted from the anode electrode with high energy. We also find that the simulation results are consistent with the experimental finding that $E_{loss}$ is nearly independent on $V_a$ (see inset of Fig. 4).

## 3. Summary

We have performed Monte Carlo simulation of electron transport through Si QDs to clarify the underlying physics of high-energy electron emission from a PS diode. We used an empirical tight-binding method to calculate electron energy levels in a Si QD. Using the energy levels, we calculated the optical-phonon emission rate. We then performed Monte Carlo simulation of electron motion through Si QDs. The results clearly show that electrons are quasi-ballistically emitted from a PS diode. We find that initial electron acceleration, which is enhanced in QDs due to the discrete nature of lower energy levels, plays an essential role in generating high-energy electron emission.

## Acknowledgments

This work was supported in part by a Global COE Program at Osaka University "Center for Electronic Devices Innovation" and a Grant-in-Aid for Scientific Research on Priority Areas "Post-Scaling Technology" from the Ministry of Education, Culture, Sports, Science and Technology.

[1] Komoda T and Koshida N 2009 *Device Applications of Silicon Nanocrystals and Nanostructures*, Koshida N, New York, Springer-Verlag, 251-291
[2] Koshida N, Ozaki T, Cheng X and Koyama K 1995 *Jpn. J. Appl. Phys.* **34** L705
[3] Uno S, Nakazato K, Yamaguchi S, Kojima A, Koshida N and Mizuta H 2003 *IEEE Trans. Nanotechonol.* **2** 301
[4] Jancu J M, Scholz R, Beltram F and Bassani F 1998 *Phys. Rev. B* **57** 6493
[5] Mizuta H, Uno S, Mori N, Oda S and Koshida N 2009 *Device Applications of Silicon Nanocrystals and Nanostructures*, Koshida N, New York, Springer-Verlag, 197-221
[6] Bockelmann U and Bastard G 1990 *Phys. Rev.* B **42** 8947

Journal of Physics: Conference Series **193** (2009) 012009 doi:10.1088/1742-6596/193/1/012009

# Green Function Treatment of Electronic Transport in Narrow Rough Semiconductor Conduction Channels

**Pierre N. Martin**[1,2] **and Umberto Ravaioli**[1,2]
[1]Department of Electrical and Computer Engineering,
[2]Beckman Institute for Advanced Technology and Science,
University of Illinois, Urbana-Champaign, Urbana, IL 61801, USA
E-mail: `pmartin7@illinois.edu, ravaioli@illinois.edu`

**Abstract.** We explore the effect of geometrical fluctuations on the electronic transport in rough Si nanowire (NW) thermoelectric devices of diameter $D < 10$ nm. At this scale, the quantum nature of transport is accounted in the computation of energy dependent transmission coefficients through a recursive green function algorithm. The rough 3D NW geometry is used as a direct input to simulations through the roughness height $\Delta$ and autocovariance length $L$. Using a non parabolic band structure, the channel conductance above 0.1 eV is drastically reduced in such NW with high $D/\Delta$ ratio. In addition, the roughness induced resistivity is only increased by 6% on the first energy level of 10 nm Si channels with $\Delta = 7.7$ Å, showing possible application for high thermoelectric figures of merit $ZT$.

While semiconductor nanowires (NW) have found viable applications in field effect transistors [1], interconnects [2], and thermoelectrics [3], scattering of charge carriers with the surface nevertheless limits transport in these structures, owing to their high surface-to-volume aspect ratio. Recent experimental and theoretical work on such NW [3, 4] showed that, through careful nano-engineering of intentionally rough edges, it is possible to decouple thermal and electrical transport, making such devices extremely appealing for novel applications in thermoelectric energy conversion. In particular, it is believed than in thin NW of diameter $D <15$ nm, where electrons are pushed away from the surface, phonons may be scattered more efficiently than electrons by asperities at the NW surface [5], leading to high thermoelectric figures of merit $ZT$. Conventional formalism which accounts for electron – surface roughness scattering is based on the derivation of a perturbed hamiltonian of the system due to geometrical fluctuations [6, 7] and is adapted to a 2D electron gas (2DEG) conducting near an interface. However, at low NW scales, this approach does not account for the quantum nature of electron transport, where charge carriers theoretically impinge on a series of potential barriers resulting from geometrical constrictions in the transport direction [8]. In this study we compute an energy dependent electrical quantum conductance in thin Si NW from a Recursive Green Function (RGF) approach [9], including evanescent modes, non-parabolic bands, a 3D description of NW, and a model for arbitrary rough surfaces through the root mean square (RMS) roughness height $\Delta$ and auto-correlation length $L$. In the absence of other scattering mechanisms, we show that surface roughness contributes to an increased resistivity for high energy carriers in Si NW of diameter $D < 10$ nm, while having little effect on the lower energy levels, thus building good hopes for application of rough thin NW in thermoelectrics.

The quality of thermoelectric devices is measured by the dimensionless figure of merit $ZT = (S^2\sigma/\kappa)T$, where $S$ is the Seebeck coefficient, $T$ the temperature, $\sigma$ the electrical conductivity and $\kappa$ the thermal conductivity. With this respect, a standard approach to achieve performance is to design a "phonon glass - electron crystal", with high electron mobility and poor thermal conductivity. Recently, thermal conductivity in Si NW with etched rough edges was experimentally reduced by a factor of about 100 in comparison to bulk crystalline Si [3, 4], to nearly the value of amorphous Si. In order to provide supplementary information for the design of efficient thermoelectric devices, the present model relates the transmission of charge carriers in a rough Si NW conduction channels to the NW diameter, the experimental roughness parameters $\Delta$and $L$, and the electron energy $E$. While considering geometrical fluctuations as the sole source of resistivity in NW, the transmission probability $T(E)$ of carriers with energy $E$ is directly linked to the electrical current $I$ flowing between two Büttiker probes held at Fermi energies $E_{F,1}$, $E_{F,2}$ through Landauer's formula $I = 2e/h \int_{E_{F,1}}^{E_{F,2}} T(E)\, dE$. Hence, in the ideally smooth case, the channel conductance varies in steps of integer multiples of the fundamental quantum conductance $e^2/h$.

The 3D conduction channel is modeled by a 1D tight binding (TB) chain where each site $q$ represents a cross section of the whole 3D TB domain (Fig. 1). The NW is assumed in contact with two semi-infinite leads of perfectly smooth NW of the same diameter. The quantum nature of NW transport is accounted by solving the tight binding Schrödinger equation (TBSE) in a series of 2D cross sections along the transport direction, and recursively computing the transmission between adjacent sections through the overlap of the wavefunctions [9] $(\mathbf{V}_{q,q+1})_{nm} = V \sum_j \psi_n^*(j)\psi_m(j)$, where $\psi_n(j)$ and $\psi_m(j)$ are the transverse wavefunctions at the $j^{th}$ cross-sectional mesh point at sites $q$ and $q+1$ respectively, and $V = \hbar^2/(2m_x^* dx^2)$ is the hopping potential between adjacent sites. The domain is discretized on a 3D mesh of spacing $dx = dy = dz$ where $x$ is the transport direction. The NW boundaries are defined by a potential step equal to the Si work function$\Phi_{Si} = 4.8$ eV, while the interior of the wire is kept at 0 eV, hence modeling a Si NW suspended in vacuum. It is noticeable that the simulation domain is larger than the actual NW, thus allowing eigenfunctions to diffuse out of the NW boundaries (Fig. 2). The transmission probability from eigen-state $m$ at the leftmost site $l$ to state $n$ at the rightmost site $r$ is

$$t_{mn} = -2iV(\sin\theta_m\theta_n)^{1/2} e^{i(\theta_m l - \theta_n r)} \langle n|\mathbf{G}_{rl}|m\rangle \quad (1)$$

where $\theta_n \equiv k_{x,n}dx$, $k_{x,n}$ is the wavevector available for longitudinal motion, and $\mathbf{G}_{rl}$ is the Green Function (GF) propagator matrix between $r$ and $l$. The total transmission at energy $E$ is $T(E) = \sum_{m,n} |t_{mn}(E)|^2$. The NW resistivity due to geometrical constrictions is then obtained from $\mathbf{G}_{rl}$. Since the propagator $\mathbf{G}$ is known for a perfectly smooth semi-infinite lead, the algorithm functions in a recursive way. Once the right lead is attached, at the step $q$ of the recursion, the propagator $\mathbf{G}^R_{r,q+1}$ between site $r$ and $q+1$ is computed. The site $q$ is then

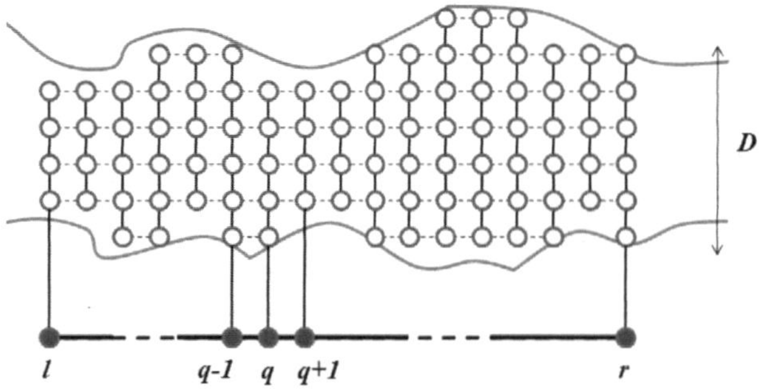


**Figure 1.** 2D Model of the RGF approach in a rough NW. Each transverse chain is represented by a single site in the 1D system, and the propagators **G** are computed to connect the energy modes between neighbor sites.

attached in order to determine the new propagators $\mathbf{G}^{L+R}_{r,q}$ and $\mathbf{G}^{L+R}_{q,q}$

$$\mathbf{G}^{L+R}_{r,q} = \mathbf{G}^{R}_{r,q+1}\mathbf{V}_{q+1,q}\left(\mathbf{I}-\mathbf{G}^{L}_{q,q}\mathbf{V}_{q,q+1}\mathbf{G}^{R}_{q+1,q+1}\mathbf{V}_{q+1,q}\right)^{-1}\mathbf{G}^{L}_{q,q} \tag{2}$$

$$\mathbf{G}^{L+R}_{q,q} = \left(\mathbf{I}-\mathbf{G}^{L}_{q,q}\mathbf{V}_{q,q+1}\mathbf{G}^{R}_{q+1,q+1}\mathbf{V}_{q+1,q}\right)^{-1}\mathbf{G}^{L}_{q,q} \tag{3}$$

Non-parabolicity is included in the computation of the longitudinal wavevectors [10, 11]

$$E_{y,z}\left[1+2\alpha(E_x+E_{y,z})\right] = E - E_x \tag{4}$$

where $\alpha$ is the coefficient of non-parabolicity, $E_{y,z}$ is the TBSE cross section energy, and $E_x = \hbar^2 k_x^2/(2m_x^*)$ is the effective energy available for longitudinal transport. As a result of non-parabolicity, transverse energy levels appear shifted down (Fig. reffig:Fig3). Conduction is assumed along the $\mathbf{\Gamma}-\mathbf{X}$ direction with $m_x^* = 0.19m_0$ and $\alpha = 0.5$. Non-parabolic bands allow better accuracy in the computation of roughness limited $T(E)$ at $E > 0.1$ eV.

We model transport in 100 nm long NW of average width $D < 10$ nm. Surface roughness alleviates the degeneracy of the cross section eigen-modes (Fig. 2). Subsequently, the computation of $T(E)$ is averaged over 500 randomly generated NW. The effect of RMS height$\Delta$ is compared at different diameters and varying energy (Fig. 3). Increasing $\Delta$ results in a series of higher potential steps encountered along the x-direction, thus reducing the average transmission on rough NW. The quantum features of conduction disappear at high $\Delta$ where the step-like behavior is smoothened as a consequence of carrier back-scattering from surface asperities. In addition, the resistivity incured at similar RMS height is stronger in NW of smaller diameter.

Noticeably, surface roughness has limited impact on the transmission of the first energy level, particularly at $D = 10$ nm, in accordance with Wang et al. [8]. Thus, geometrical fluctuations are expected to have only a minimal influence on the low field NW mobility. In this regime of operation, where thermal conductivity varies as $(D/\Delta)^2$ [4], it is expected to achieve high $ZT > 1$. The RMS height $\Delta$ measures the average height of roughness peaks or constrictions. This statistical picture is consistent with the case of chemically grown NW, where one may control the amplitude of geometrical variations rather than the actual shape of the edge. In

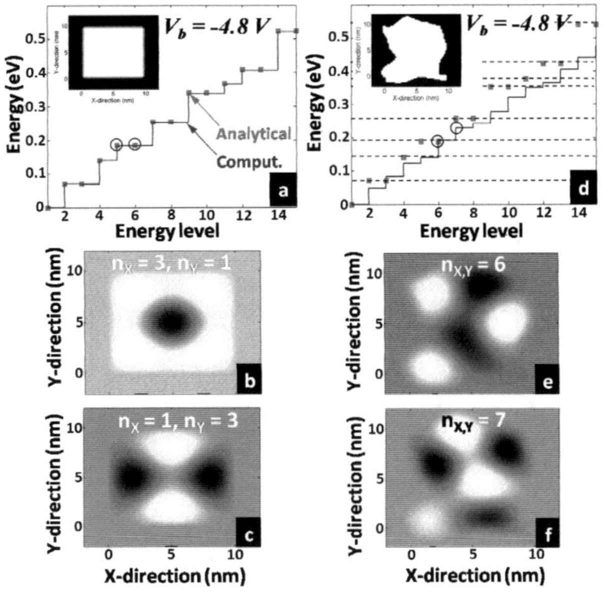


**Figure 2.** **(a)** Energy levels in a prefectly smooth NW. **(b)**, **(c)** Examples of cross-section wavefunctions in a perfectly smooth NW. **(d)** Energy levels in a rough NW. **(e)**, **(cf)** cross-section wavefunctions in a rough NW. $D =$ 10 nm, $\Delta = 5$ Å and $L$ =22 Å throughout.

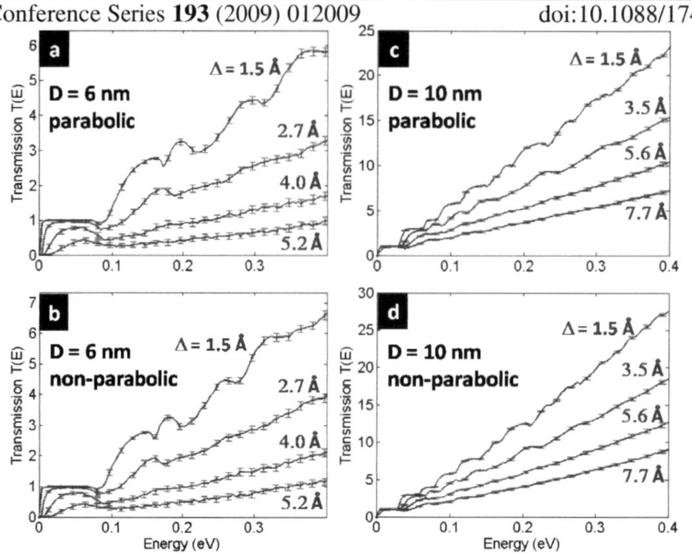


**Figure 3.** Effect of $\Delta$ on the quantum transmission in rough Si NW of diameter 6 nm (**a**), (**b**), and 10 nm (**c**), (**d**). $L = 22$ Å throughout.

10 nm NW with $\Delta$ = 7.7 Å, we observed constrictions of effective diameter down to 6 nm. Hence, for high$\Delta$ /$D$ ratio, transmission does not occur until sufficiently high energy is reached to tunnel through constriction barriers. Nevertheless, the surface is generally comparable to a random noise around the average diameter. As a result, higher variance of NW behavior is observed when $E$ is near the transition between energy levels. The standard deviation of NW transmission is also increased at high$\Delta$ /$D$ ratio, but remains small at the center of energy steps.

As a conclusion, we explored the effect of surface roughness on electronic transport in thin Si NW of diameter $D < 10$ nm based on a recursive green function algorithm. This approach allows a 3D representation of the NW geometry through the statistical parameters $\Delta$and $L$, and further includes the effect of non-parabolicity of the Si band structure. As the quantum nature of charge carriers is accounted for, an energy dependent transmission is computed for such devices. With geometrical fluctuations as the sole electronic decay process, conductance at energies $E > 0.1$ eV is drastically decreased in such NW with high$\Delta$ /$D$ ratio. Nevertheless, surface asperities have little effect on low energy transport, leading to possibly high thermoelectric figure of merit $ZT$ in this regime of operation. To test such assumption, it is necessary to additionally model thermal transport in channels with dimensions below the average phonon mean free path.

This work was supported by the DARPA through the IMPACT Center for Advancement of MEMS/NEMS VLSI grant n. HR0011-06-1-0046.

## References

[1] Singhan N, Agarwal A, Bera L, et al. 2006 *IEEE Electron Dev. Lett.* **27** 383.
[2] Steinhogl W 2005 *Appl. Phys. Lett.* **97** 023706.
[3] Hochbaum A, Chen R, Delgado R, et al. 2008 *Nature* **451** 163.
[4] Martin P, Aksamija Z, Pop E and Ravaioli U 2009 *Phys. Rev Lett.* **102** 125503.
[5] Ramayya E, Vasileska D, Goodnick S, et al. 2008 *8th IEEE Conference on Nanotechnology* 339 (2008).
[6] Goodnick S et al. 1983 *J. Vac. Sci. Technol. B* **1** 803.
[7] Yamakawa S et al. 1996 *J. Appl. Phys.* **79** 911.
[8] Wang J, Polizzi E, Ghosh A et al 2005 *Appl. Phys. Lett.* **87** 043101.
[9] Sols F, Macucci M, Ravaioli U and Hess K 1989 *Journ. Appl. Phys.* **66** 3892.
[10] Godoy A, Yang Z, Ravaioli U et al. 2005 *J. Appl. Phys.* **98** 013702.
[11] Lopez-Villanueva J, Melchor I, Cartujo P, et al 1993 *Phys. Rev. B* **48** 1626.

Journal of Physics: Conference Series **193** (2009) 012010 doi:10.1088/1742-6596/193/1/012010

# Prediction of Reduced Thermal Conductivity in Nano-Engineered Rough Semiconductor Nanowires

**Pierre N. Martin[1,2], Zlatan Aksamija[1,2], Eric Pop[1,2,3] and Umberto Ravaioli[1,2]**

[1]Department of Electrical and Computer Engineering, [2]Beckman Institute for Advanced Technology and Science, [3]Micro- and Nano-Technology Laboratory, University of Illinois, Urbana-Champaign, Urbana, IL 61801, USA

E-mail: pmartin7@illinois.edu, epop@illinois.edu

**Abstract.** We explore phonon decay processes necessary to the design of efficient rough semiconductor nanowire (NW) thermoelectric devices. A novel approach to surface roughness-limited thermal conductivity of Si, Ge, and GaAs NW with diameter $D < 500$ nm is presented. In particular, a frequency-dependent phonon scattering rate is computed from perturbation theory and related to a description of the surface through the root-mean-square roughness height $\Delta$ and autocovariance length $L$. Using a full phonon dispersion relation, the thermal conductivity varies quadratically with diameter and roughness as $(D/\Delta)^2$. Computed results are in agreement with experimental data, and predict remarkably low thermal conductivity below 1 W/m/K in rough-etched 56 nm Ge and GaAs NW at room temperature.

Semiconductor nanowires (NWs) have drawn much attention for their potential applications in field effect transistors [1], interconnects [2], thermoelectrics [3, 4, 5], and heterostructures [6]. Given their high surface-to-volume aspect ratio, the most prominent size effect limiting transport originates from scattering of carriers with the surface. Recent experimental and theoretical work [4, 7] suggested that potentially high thermoelectric figures of merit $ZT$ could be achieved by scattering phonons more effectively than electron at the interfaces of rough NW. In particular, thermal conductivity in Si NW with etched rough edges was experimentally reduced by a factor of about 100 in comparison to bulk crystalline Si, to nearly the value of amorphous Si. Indeed, it is expected that vibrations of the crystal lattice interfere with the spatial fluctuations of the NW boundaries. Hence, it is theoretically relevant to adopt a picture where the interference process is stronger in thin NW for a given roughness root-mean-square height (RMS)$\Delta$ , and vary with the wavelength of incident phonons and the temperature of the crystal lattice. Nevertheless, most of the work in phonon transport only accounts for surface roughness scattering by a constant fitting parameter which reflects an average probability of diffuse scattering at the NW surface. We recently proposed [7] a perturbative approach to phonon – surface roughness scattering, where a matrix element is derived for such interactions. This model has shown excellent agreement to experimental observations on rough Si NW below 50 nm where the effect of surface roughness is the strongest, yielding $ZT \approx 0.6$. In order to assist the design of efficient thermoelectric NW devices, the present work explores the trade-offs between different scattering mechanisms in Si, Ge, and GaAs necessary to achieve high figures of merit $ZT > 1$.

The roughness of a surface is generally quantified by its RMS height $\Delta$ and auto-correlation length $L$. The present model accounts for the variation of a NW thermal conductivity with the experimentally observable$\Delta$, $L$, and the phonon angular frequency $\omega$; this approach

Journal of Physics: Conference Series **193** (2009) 012010 doi:10.1088/1742-6596/193/1/012010

accurately predicts phonon-surface roughness scattering to contribute to the reduction of thermal conductivity by a factor $(D/\Delta)^2\omega^{-2}$ where $D$ is the NW diameter. In addition, the quality of thermoelectric devices is measured by the dimensionless figure of merit $ZT = (S^2\sigma/\kappa)T$, where $S$ is the Seebeck coefficient, $T$ the temperature, $\sigma$ the electrical conductivity and $\kappa$ the thermal conductivity. With this respect, a standard approach to achieve performance is to design a "phonon glass - electron crystal", with high electron mobility and poor thermal conductivity. Among those thermally resistive mechanism in a semiconductor crystal, scattering of phonons by isotopes or in three-phonon anharmonic decays is also of considerable importance.

Our model assumes that no phonon is emitted in the surrounding environment. This condition reflects the case where NW are wrapped in a medium of considerably different $\kappa$, as it is for semiconductor NW wrapped in oxide or suspended in vacuum. While propagating along wires of diameter $D < 500$ nm, phonons scatter from a series of constriction along the transport direction which result from the surface asperities and are reflected by a perturbed Hamiltonian $H'$ of the system. The interface roughness is considered as a space varying dilatation $\delta(r)$ of the wire, which auto-correlation function is assumed to be exponential$\Delta(r) = \Delta\exp(-r/L)$. Following the derivation of Klemens [8], the matrix element for a perturbation due to a space varying dilatation reflects the transitions of phonons from a state of momentum $\mathbf{k}$ to $\mathbf{k}'$ such as

$$|\langle \mathbf{k}|H'|\mathbf{k}\rangle|^2 = \frac{4\gamma^2}{3V_{ol}}\omega^2(\langle n'\rangle + 1)\Delta(\mathbf{k} - \mathbf{k}') \quad (1)$$

where $V_{ol}$ is the volume of the device, $\gamma$ is Grüneisen's parameter, $\omega$ is the angular frequency of the incident phonon, and$\Delta(\mathbf{q}) = \pi\Delta^2L^2\exp(-L^2\mathbf{q}^2/4)$ is the roughness power spectrum associated to$\Delta(r)$ Close to equilibrium, the occupation of the destination phonon branch $\langle n'\rangle$ is given by the Bose-Einstein distribution and reflects the temperature dependence of the model $\langle n'\rangle = (\exp(\hbar\omega/k_BT)-1)^{-1}$. The probability of transitions from $\mathbf{k}$ to $\mathbf{k}'$ due to surface roughness is directly proportional to the matrix element derived above. As a consequence of the $\omega^2$ factor, low frequency phonons experience little thermal resistance from surface asperities. In addition, the dependence on the autocorrelation length $L$, which represents the average width of roughness peaks, shows that smoothly varying surfaces favor scattering events of the specular type.

We use a Gilat-Raubenheimer (GR) [9] scheme to compute the surface roughness scattering

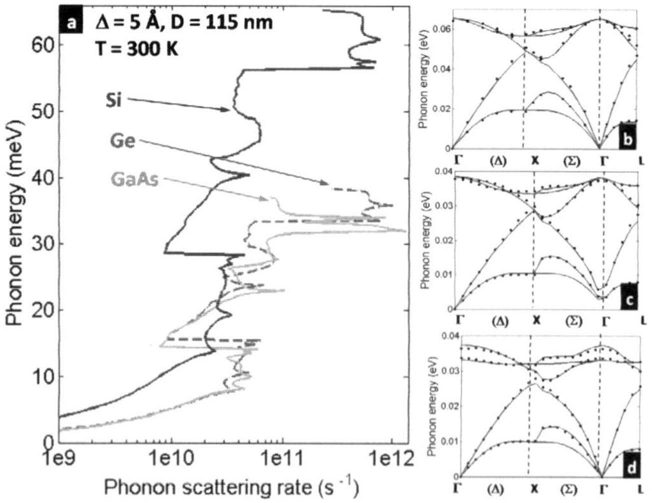


**Figure 1.** **(a)** Phonon – surface roughness scattering rates computed at room temperature in Si, Ge, GaAs. **(b)**,**(c)**,**(d)** Computed full phonon branches in the FBZ of Si, Ge, and GaAs respectively. Dots are extracted from experimental observations [10, 11] for comparison. $L = 22$ Å throughout.

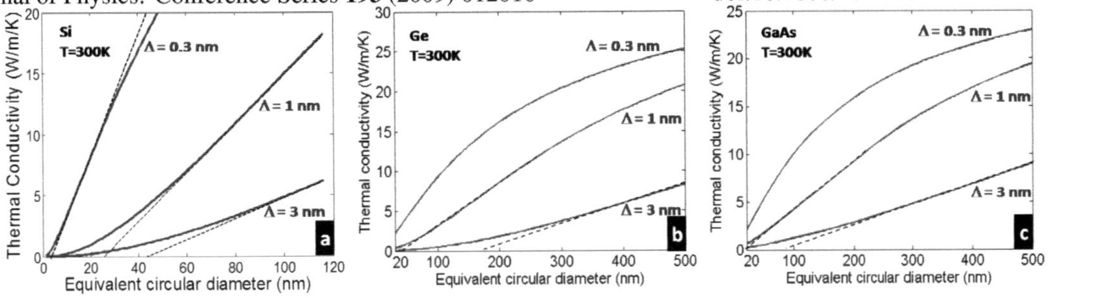


**Figure 2.** Effect of NW diameter on thermal conductivity for varying roughness RMS in **(a)** Si [7], **(b)** Ge , and **(b)** GaAs. $L = 22$ Å throughout.

rate integrals in Si, Ge, and GaAs (see Fig. 1.a)

$$\tau_{i,j}^{-1}(E) = \frac{2\pi}{\hbar N_i(E)} \int_{E'=E_i} \frac{|\langle \mathbf{k}|H'|\mathbf{k}'\rangle|^2}{\nabla_{\mathbf{k}'} E'(\mathbf{k}')} dS \tag{2}$$

where $\tau_{i,j}^{-1}$ is the transition rate from phonon branch $i$ to $j$ due to surface roughness and $E'(\mathbf{k})$ is the phonon dispersion relation on the destination branch $j$. In order to account for the frequency dependence of phonon-surface roughness interference process, full dispersion relations are used, which are obtained from an adiabatic bond charge model, and approximate the dispersion relations observed in Si, Ge and GaAs [10, 11] with a 1-2.5% error (Fig. 1.b, c, d).

The latter surface roughness scattering rate is added to the total scattering rate by Mathiessen's rule. Subsequently, we compute thermal conductivity in Holland's formalism [12]

$$\kappa = \frac{\hbar^2}{6\pi^2 k_B T^2} \sum_i \int \omega_i^4(q) \tau_i^{tot}(q) \frac{\exp(\hbar\omega_i/k_B T)}{(\exp(\hbar\omega_i/k_B T) - 1)^2} dq \tag{3}$$

where $i$ goes over all incident transverse and longitudinal branches, and the wave vector $q$ models conduction in the NW along the $\mathbf{\Gamma}$-($\Delta$)-$\mathbf{X}$ direction. Computations of $\kappa$ include boundary, isotope, 3 phonon Umklapp, and normal decay processes. Parametric laws are used for the latter rates as described in [7], which are fitted to best reflect experimental bulk $\kappa$ values [12, 13]. Isotope concentrations of group IV semiconductors match their natural occurrence, with 92.2 % of $^{28}$Ge, and noticeably only 36.5 % of $^{74}$Ge [13]. As a result of its heavier atomic mass and high natural isotope concentration, Ge shows lower $\kappa$ throughout all simulations (Fig. 2 and 3).

In Fig. 2 and 3, NW thermal conductivity is predicted to decrease in rough nanowires (high RMS$\Delta$ ). The $(D/\Delta)^2$ behavior is apparent in thin NW of diameter $D < 115$ nm, showing that phonon–surface roughness scattering is the dominant resistive process in this case. For intermediate diameters ($D/\Delta \approx 100$ for Ge and GaAs, 50 for Si, current experiments [4] report $D = 22 - 115$ nm and $\Delta = 0.3 - 5$ nm), the linear dependence in the diameter suggests that phonons dominantly decay from classical boundary scattering $\tau_B^{-1}(q) = v_s/D$, with $v_s$ the average speed of sound in each branch. For wider $D$, $\kappa$ converges to its bulk value.

Low NW thermal conductivity results from the combination of all concurrent interference processes mentioned above. In particular, the effect of artificially high $\Delta$ and low $D$ decreases thermal conductivity below 1 W/m/K in Ge and GaAs NW of diameter $D = 56$ nm. At this scale, fabrication of rough etched NW remains feasible [4], and the perturbative approach holds validity as the ratio$\Delta$ /$D$ is only of a few %, showing good agreement with experimental results (Fig. 3.b). Noticeably, the best fit for the natural Ge sample shows strong isotope and weak Umklapp scattering. As a result, Ge experiences the starkest influence from surface roughness

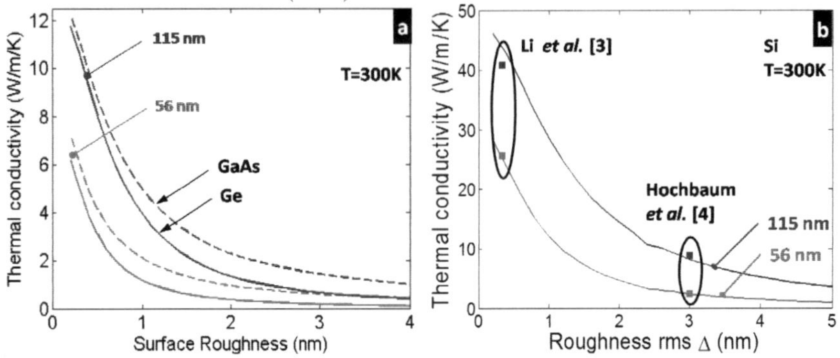


**Figure 3.** Effect of roughness RMS on NW thermal conductivity in **(a,** ——**)** Ge and **(a,** - - - -**)** GaAs, and **(b)** Si [7] at room temperature. $L = 22$ Å throughout.

at high temperatures, making it a suitable candidate for design of rough NW thermoelectric devices. Our model further predicts weak interference between low energy phonons and surface asperities. Consequently, we noticed very little variation of thermal conductivity with respect to the roughness RMS at temperatures below 100 K.

In summary, we modeled the thermal properties of artificially rough Si, Ge and GaAs NW for thermoelectric applications based on a perturbative treatment of the interaction between lattice vibrations and surface asperities. This approach, based on a full phonon dispersion, accurately accounts for the frequency dependence of phonon scattering processes resulting from surface roughness, isotope, boundary, and anharmonic decays. Our model predicts $\kappa \propto (D/\Delta)^2$ for thin NW, yielding remarkably low thermal conductivity in rough etched Ge and GaAs NW of diameter $D < 115$ nm, thus revealing those materials as potential candidates for efficient thermoelectric devices. Similar scattering mechanisms nevertheless limit electron mobility, making it necessary to model charge transport in order to find optimal thermoelectric figures of merit $ZT$. In this scope, it is noticeable that isotopes have no influence on electronic conduction, and surface roughness may only have limited impact at low $D$ where electrons tend to be pushed away from the interface. Hence, those two factors are expected to be major design axes in the engineering of efficient semiconductor NW thermoelectric devices.

This work was supported in part by the DARPA through the IMPACT Center for Advancement of MEMS/NEMS VLSI grant n. HR0011-06-1-0046 (PM), the DOE Computational Science Graduate Program of the OSNNSA under contract DE-FG02-97ER25308 (ZA), and the Nanoelectronics Research Initiative (NRI) SWAN center (EP).

## References

[1] Singhan N, Agarwal A, Bera L, et al. 2006 *IEEE Electron Dev. Lett.* **27** 383.
[2] Steinhogl W 2005 *Appl. Phys. Lett.* **97** 023706.
[3] Li D et al 2003 *Appl. Phys. Lett.* **83** 2934.
[4] Hochbaum A, Chen R, Delgado R, et al. 2008 *Nature* **451** 163.
[5] Ramayya E, Vasileska D, Goodnick S, et al. 2008 *8th IEEE Conference on Nanotechnology* 339 (2008).
[6] Dames C and Chen G 2004 *J. Appl. Phys.* **95** 682.
[7] Martin P, Aksamija Z, Pop E and Ravaioli U 2009 *Phys. Rev Lett.* **102** 125503.
[8] Klemens P 1958 *Solid State Phys.* **7** 1.
[9] Gilat G and Raubenheimer L 1966 *Phys. Rev.* **144** 390.
[10] Strausch D and Dorner B 1990 *J. Phys. Cond. Mat.* **2** 1457.
[11] Nilsson G and Nelin G 1972 *Phys. Rev. B* **6** 3777.
[12] Holland M 1964 *Phys. Rev.* **134** 471.
[13] Asen-Palmer M et al. 1997 *Phys. Rev. B.* **56** 9431.

Journal of Physics: Conference Series **193** (2009) 012011 doi:10.1088/1742-6596/193/1/012011

# Controlled Breaking of Phase Symmetry in a "Which-Path?" Interferometer

**Vadim I. Puller**[1], **Yigal Meir**[1,2]

[1]Department of Physics, Ben-Gurion University of the Negev, Beer Sheva 84105 Israel
[2] The Ilse Katz Center for Meso- and Nano-scale Science and Technology, Ben-Gurion University, Beer Sheva 84105, Israel

E-mail: `puller@bgu.ac.il`

**Abstract.** Linear response conductance of a two terminal Aharonov-Bohm (AB) interferometer is an even function of magnetic field, as dictated by Onsager-Büttiker relations. We discuss an experimental setup in which this phase symmetry can be broken in a controlled way. To this end we consider a "Which Path?" detector, which consists of an interferometer with quantum dots (QDs) in each of its arms, one of the QDs being capacitively coupled to a nearby quantum point contact (QPC). This coupling results in breaking of the phase symmetry of the AB oscillations, given that a finite voltage bias is applied to the QPC. Sweeping in the same time the level in the other QD across the Fermi level allows one to observe smooth change of the phase from 0 to $\pi$, rather than an abrupt jump observed in isolated AB interferometers. We explore the possibility of using this setup in order to measure the transmission phase through a QD.

## 1. Introduction

Linear response conductance of a two-terminal Aharonov-Bohm (AB) interferometer is an even function of magnetic flux, as required by Onsager-Büttiker relations [1, 2]. This means that the phase of AB oscillations can take only values 0 or $\pi$. If a quantum dot (QD) is placed in one of the arms of the interferometer, the phase may jump abruptly between 0 or $\pi$ when the Fermi level passes through the level in the QD [3, 4]. This jump is extremely sharp due to the fact that the lowest harmonic of AB oscillations takes zero value, the oscillations being completely dominated by higher order harmonics [5].

Such *phase symmetry* precludes measuring directly the transmission phase via a QD [3]. This obstacle may be overcome by breaking the phase symmetry, e.g., by using a multiterminal interferometer [4, 6].

Other ways to break the phase symmetry include studying the interferometer in the regime of nonlinear transport [7, 8, 9, 10, 11] or coupling it to a non-equilibrium environment [12]. We address the latter case by considering a "Which Path?" detector [13, 14], which consists of an AB interferometer containing a QD in each of its arms, and a quantum point contact (QPC) capacitively coupled to one of the QDs, as shown in Fig. 1a. The QPC plays the role of an environment and causes breaking of the phase symmetry, but only when a finite bias voltage is applied to it [12]. Thus, instead of the phase jump in the linear response conductance, expected when the level in the QD, not coupled to the QPC, is swept across the Fermi level, the AB phase smoothly flows between values 0 and $\pi$.

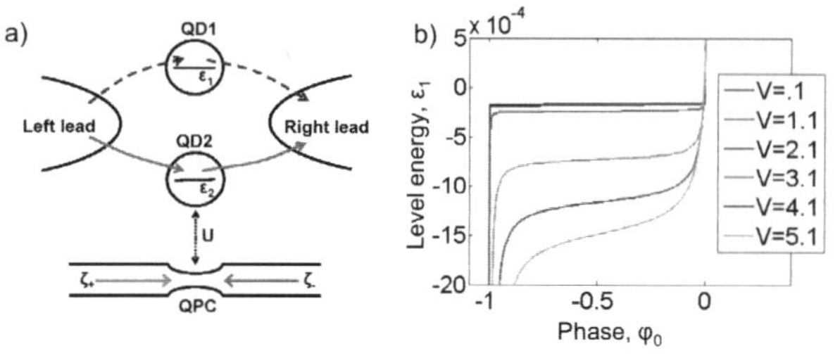


**Figure 1.** a) Schematics of "Which path?" interferometer studied in this paper. b) The phase of AB oscillations vs the energy of the level in the reference arm, $\epsilon_1$, for different values of QPC bias.

## 2. Formulation of the problem

We describe the system in Fig. 1a by Hamiltonian $H = H_I + H_{QPC} + H_C$. $H_I = \sum_\alpha \epsilon_\alpha d_\alpha^+ d_\alpha + \sum_{k,\mu} \epsilon_{k\mu} c_{k\mu}^+ c_{k\mu} + \sum_{\alpha,k,\mu} (t_{\alpha\mu} d_\alpha^+ c_{k\mu} + h.c.)$ which includes the terms describing the QDs, the leads, and tunneling between them. Operator $d_\alpha$ destroys an electron in QD $\alpha = 1, 2$, $c_{k\mu}$ destroys an electron in state $k$ of lead $\mu = L, R$. The strength of coupling between the QD $\alpha$ and lead $\mu$ is characterized by$\Gamma_\alpha^\mu = 2\pi N_\mu |t_{\alpha\mu}|^2$, where $N_\mu$ is the density-of-states in lead $\mu$. Magnetic field enters via the complex phases of tunneling matrix elements $t_{\alpha\mu}$: $t_{1L}^* t_{1R} t_{2R}^* t_{2L} = \pm |t_{1L}^* t_{1R} t_{2R}^* t_{2L}| e^{i\varphi}$, where $\varphi = 2\pi\Phi/\Phi_0$, $\Phi$ is the magnetic flux threading the interferometer,$\Phi_0 = hc/e$ is the flux quantum, and sign $\pm$ depends on parities of QD states.

$H_{QPC} = \sum_{p,\nu} E_{p\nu} a_{p\nu}^+ a_{p\nu}$ describes the QPC, the states in which are right/left-movers ($\nu = \pm$) characterized by quantum numbers $p, p'$; $a_{p\nu}$ is the corresponding destruction operator.

$H_C = d_2^+ d_2 \sum_{p,\nu;p'\nu'} U_{\nu\nu'} a_{p\nu}^+ a_{p'\nu'}$ is the capacitive coupling between the QPC and QD2.

We calculated the current through the interferometer using the Green's function approach [16], taking account for the coupling to the QPC as a self-energy correction in electron Green's function up to the first order in coupling strength $g_{\nu\nu'} = g_{\nu'\nu} = 2\pi\rho_\nu\rho_{\nu'} |U_{\nu\nu'}|^2$ [15] ($\rho_\nu$ is the density-of-states in the QPC). The current consists of three distinct contributions: (i) the one resembling the current via an interferometer uncoupled from the QPC, the coupling having effect of suppressing the visibility of AB oscillations; (ii) and (iii) the two contributions proportional to coupling strength $g_{\nu\nu'}$ which are respectively even and odd in magnetic field. The latter two contributions generally do not vanish at zero source-drain bias, $V_{sd}$, given that the bias applied to the QPC, $V_{QPC}$, is not zero. Below we describe (a) the nature of the current component odd in magnetic field, and (b) the reasons why the current does not vanish at zero source-drain bias. The detailed derivation and full mathematical expressions will be given elsewhere.

## 3. Processes contributing to the current odd in magnetic field

The probability of a process in which electron is transferred between the leads can be written as (up to the third order in lead-to-lead tunneling amplitude $V_{nm}$)

$$2\pi \left\{ |V_{fi}|^2 + 2\Re \left[ V_{fi}^* \frac{V_{fm} V_{mi}}{E_m - E_i + i0^+} \right] \right\} \delta(E_f - E_i), \tag{1}$$

where indices $i, f$, and $m$ refer to the initial, final and intermediate states. When we sum over all of these states that are available the imaginary part of the denominator, treated as $1/(E + i0^+) = 1/E - i\pi\delta(E)$, projects out the states where initial, final and intermediate energies lay on the same energy shell, $E_i = E_f = E_m$ [11, 17]. The phase of the resulting contribution to AB oscillations is shifted by $\pi/2$.

One can show [11] that the only such processes that will give non-zero contribution to the odd AB oscillations are either elastic processes like the ones in Figs. 2c,e, where electron first absorbs energy from the QPC and then re-emits it, or the inelastic processes like those in Figs. 2a,b,d,f.

There are pairs of processes, such as those in Figs. 2c,d, whose contributions mutually cancel out: the processes in such a pair are composed of identical matrix elements, but effectively describe electron motion around the interferometer in opposite directions.

Non-cancelling processes include the two processes in Figs. 2e,f which add up (the process in Fig. 2e describes tunneling of a hole, which implies an additional minus sign), as well as the inelastic processes shown in Figs. 2a,b (which do not have pairs). All such processes begin with creation of an electron-hole pair in one of the leads, followed by tunneling to the other lead [11].

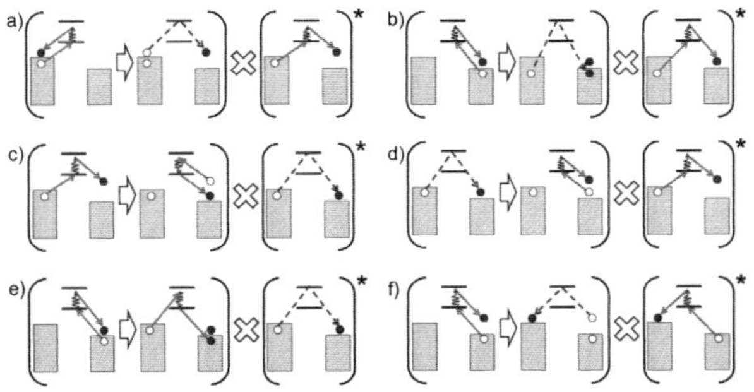


**Figure 2.** Various transport processes that may contribute to odd AB oscillations: a) and b) two inelastic processes; c) and d) two processes that mutually cancel out; e) and f) two processes that add together.

## 4. Aharonov-Bohm conductance

The upper left panel of Fig. 3 shows full linear response AB conductance when the bias applied to the QPC is $V_{QPC} = 0.1$. (The other parameters are (in relative units)$\Gamma_1^L = \Gamma_2^L = .02$, $\Gamma_1^R = \Gamma_2^R = .05$, $g_{\nu\nu'} = 0.05$, $\epsilon_2 = 2$, $\epsilon_F = 0$.) The AB oscillations are nearly even in magnetic flux, their odd component (upper right of Fig. 3) being many orders of magnitude smaller than the even one. The phase jump occurs around $\epsilon_1 \approx 0$, where the oscillations change from having a minimum at zero magnetic field to having a maximum. Near the jump the lowest harmonic of the oscillations, $\sim \cos(\varphi)$, vanishes and they are dominated by the second harmonic, $\sim \cos(2\varphi)$.

The bottom panels of Fig. 3 show the situation at high bias, $V_{QPC} = 4.1$. The AB oscillations are now asymmetric in magnetic flux. Although close to the phase jump they are still dominated by the 2nd harmonic, the lowest harmonic does not disappear, but enters with a phase different from either 0 or $\pi$ ($\sim \cos(\varphi - \varphi_0)$, $\varphi_0 \neq 0, \pi$). The amplitude of the odd oscillations (lower right panel of Fig. 3) is now comparable to that of the even ones.

Fig. 1b shows AB phase $\varphi_0$ as a function of energy $\epsilon_1$ for different values of $V_{QPC}$. At low bias the phase jumps abruptly between 0, and $\pi$, but it changes smoothly between these two values as the bias increases.

Our calculation shows that in the limit$\Gamma_{22}, \Gamma_{11} \ll |\epsilon_2 - \epsilon_F|$ the part of the differential conductance odd in magnetic flux is given by $G^{odd}(\varphi, \epsilon_F, \epsilon_1) = \sin(\varphi) \times A(\epsilon_F)\Re[G_1^r(\epsilon_F)]$, where $G_1(\epsilon_F)$ is the Green's function of QD1, calculated without account for QD2 and the QPC, while factor $A(\epsilon_F)$ is independent on energy $\epsilon_1$. Transmission via QD1, when QD2 is disconnected, is given by $G_{QD1}(\epsilon_1) = G_0\left(\Gamma_{11}^L\Gamma_{11}^R/\Gamma_{11}\right)\Im[G_1^r(\epsilon_F)]$. Thus, the transmission phase via QD1, defined by $G_1^r(\omega) = |G_1^r(\omega)|\, e^{i\varphi_{QD1}(\omega)}$, as a function of $\epsilon_1$ can be deduced from $\tan[\varphi_{QD1}(\epsilon_F)] \sim G^{odd}(\varphi, \epsilon_F, \epsilon_1)/G_{QD1}(\epsilon_1)$.

## 5. Photon-assisted tunneling

Elastic processes are forbidden at zero bias, but electrons can still be transferred between the leads by means of processes shown Figs. 2a,b,e, absorbing energy from the QPC. However, in order to have non-zero net current, the device should possess some kind of left/right asymmetry: either due to non-symmetric coupling to the leads or due to the presence of magnetic field. While the former case results in a current even in magnetic flux, the letter leads to a current odd in magnetic flux, since changing sign of the magnetic field is equivalent to interchanging the leads.

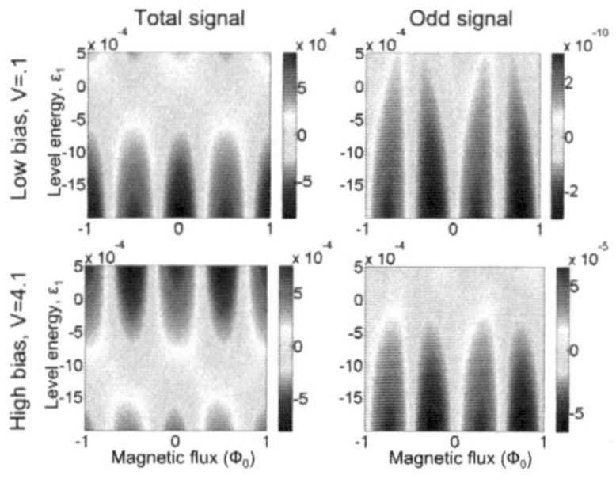


**Figure 3.** Linear response conductance as a function of magnetic flux (horizontal axis) and the energy of the level in the reference arm, $\epsilon_1$ (vertical axis). The upper (lower) panels correspond to the QPC bias $V_{QPC} = .1$ ($V_{QPC} = 4.1$); left (right) panels show total (odd) AB conductance.

## 6. Conclusion

We studied the symmetry of AB oscillations in a "Which Path?" detector, which consists of an AB interferometer with QDs in each of its arms, with one of the QDs capacitively coupled to a QPC. When the QPC is biased, the AB oscillations in response conductance of the ABI loose their symmetry in respect to the sign of the magnetic field. The effect is pronounced in the regime of phase jump, when the energy level in the other QD is swept across the Fermi level, and can level, and can be used to measure the transmission phase via a QD.

In addition, the current through the interferometer does not disappear at zero source-drain bias due to the possibility of energy transfer from the biased QPC. Existence of such a current requires asymmetry introduced either by unequal coupling to the leads or by the magnetic field, and the current itself is asymmetric in the magnetic field.

## References

[1] Gefen Y, Imry Y and Azbel M Ya, 1984 *Surf. Sci.* **142** 203
Büttiker M, Imry Y and Azbel M Ya *Phys. Rev. A* **30** 1982
Levy Yeyati A and Büttiker M 1995 *Phys. Rev. B* **52** R14360
Yacoby A, Schuster R, and Heiblum M 1996 *Phys. Rev. B* **53** 9583
[2] Büttiker M 1986 *Phys. Rev. Lett.* **57** 1761
[3] Yacoby A, Heiblum M, Mahalu D and Shtrikman H 1995 *Phys. Rev. Lett.* **74** 4047
[4] Schuster R, Buks E, Heiblum M, Mahalu D, Umansky D and Shtrikman H 1997 *Nature* **385** 427
[5] Yacoby A, Schuster R and Heiblum M 1996 *Phys. Rev. B* **53** 9583
[6] Aharony A, Entin-Wohlman O, Halperin B I, Imry Y 2002 *Phys. Rev. B* **66** 115311
[7] Bruder C, Fazio R and Schoeller H 1996 *Phys. Rev. Lett* **76** 114
[8] Sánchez D and Büttiker M 2004 *Phys. Rev. Lett.* **93** 106802
[9] Spivak B and Zyuzin A 2004 *Phys. Rev. B* **93** 226801
[10] Marlow C A, Taylor R P, Fairbanks M, Shurubalko I and Linke H 2006 *Phys. Rev. Lett.* **96** 116801
Zumbühl D M, Marcus C M, Hanson M P and Gossard A C 2006 *Phys. Rev. Lett.* **96** 206802
van der Wiel W G, Nazarov Yu V, De Francheski C, Fujisawa T, Elzerman J M, Huizeling E W G M, Tarucha S and Kouwenhoven L P 2003 *Phys. Rev. B* **67** 033307
Leturcq R, Sánchez D, Götz G, Ihn T, Ensslin K, Driscoll D C, Gossard A C 2006 *Phys. Rev. Lett.* **96** 126801
Angers L, Zakka-Bajjani E, Deblock R, Guéron S, Bouchiat H, Cavanna A, Gennser U, Polianski M 2007 *Phys. Rev. B* **75** 115309
[11] Puller V, Meir Y, Sigrist M, Ensslin K and Ihn Th 2009 *Phys. Rev. B* **80** 035416
[12] Khym G L and Kang K 2006 *Phys. Rev. B* **74** 153309
Sánchez D and Kang K 2008 *Phys. Rev. Lett.* **100** 036806
[13] Buks E, Schuster R, Heiblum M, Mahalu D and Umansky V 1998 *Nature* **391** 871
[14] Aleiner I L, Wingreen N S and Meir Y 1997 *Phys. Rev. Lett.* **79** 3740
[15] Silva A and Levit S 2001 *Phys. Rev. B* **63** 201309(R)
[16] Meir Y and Wingreen N S 1992 *Phys. Rev. Lett.* **68** 2512
Jauho A-P, Wingreen N S and Meir Y 1994 *Phys. Rev. B* **50** 5528
[17] Holstein T 1961 *Phys. Rev.* **124** 1329
Entin-Wohlman O, Imry Y and Aharony A 2004 *Phys. Rev. B* **70** 075301

Journal of Physics: Conference Series **193** (2009) 012012 doi:10.1088/1742-6596/193/1/012012

# Percolation transport in an AlGaN/GaN heterostructure

**N. Sawaki[1,3], XX. Han[2], Y.Honda[2], and M.Yamaguchi[2]**
[1] Dept. Electrical Engg., Aichi Instit.of Technology, *Yakusa, Toyota, 470-0392 Japan*
[2] Dept. Electronics, Nagoya University, *Chikusa-ku, Nagoya 464-8603 Japan*
[3] Nagoya Industrial Science Research Institute, *Chikusa-ku, Nagoya 464-0819 Japan*

**Abstract.** Magneto-transport in an AlGaN/GaN heterostructure has been studied at 4.2K and it was found that the magneto-resistance is negative at low fields. The negative components disappeared by inserting a thin AlN film between the AlGaN barrier layer and the GaN layer. The negative component is attributed to the presence of alloy disorder at the hetero-interface. It increases linearly followed by saturation, which is characteristic to the percolation transport in random potential.

## 1. Introduction

AlGaN/GaN hetero-structure attracts much attention in the progress of high frequency/high power transistors. Since the III-nitrides have large energy band gaps, hetero-structures made of III-nitrides shows large band offsets. As the results, we might achieve an electron device with superior performances. But, an alloy such as AlGaN or GaInN has large potential fluctuations due to random distribution of the elements, which in turn affects the transport properties at hetero-interfaces. This fact is in contrast to the case of AlGaAs/GaAs heterostructure where the alloy scattering has rather a fractional role in the electron mobility [1]. In spite of a long history in the studies of AlGaN/GaN, only a few studies have been reported to the magneto-transport phenomena.

In this study, we have investigated the magneto-resistance in an AlGaN/GaN heterostructure and found that the magneto-resistance at low fields is negative, while it is positive in an AlN/GaN hetero-structure. And the appearance of the negative components is attributed to the presence of percolation transport in the random distribution of scattering potential.

## 2. Samples

The samples were made by metal-organic vapour phase epitaxy (MOVPE) on a (0001) sapphire substrate [2]. Following the deposition of a 30nm thick low temperature GaN buffer layer, a 1.0μm thick nominally non-doped GaN layer followed by a 20 ~25 nm thick $Al_xGa_{1-x}N$ ( x=0.15~0.30)

Table 1. 2DEG density and mobility in $Al_xGa_{1-x}N$/AlN/GaN hetero-structure at 17K

| Sample | AlN thickness (nm) | Al composition (x) | 2DEG density ($10^{11}cm^{-2}$) | Mobility ($cm^2/Vs$) |
|---|---|---|---|---|
| **#01** | 0 | 0.15 | 5.5 | 6,640 |
| **#02** | 0 | 0.20 | 8.4 | 5,690 |
| **#03** | 0 | 0.30 | 12.3 | 3,810 |
| **#11** | 0 | 0.24 | 10.0 | 6,680 |
| **#12** | 1 | 0.24 | 9.6 | 26,100 |
| **#13** | 2 | 0.24 | 11.1 | 13,400 |

Journal of Physics: Conference Series **193** (2009) 012012 doi:10.1088/1742-6596/193/1/012012

barrier layer was grown at 1080 $^0$C. In some samples, a thin AlN layer was grown between the GaN layer and the AlGaN barrier layer. In order to characterize the electronic properties, the 2DEG density and mobility were investigated with van der Pauw configuration at low temperatures. Typical results are summarized in Table 1. The electron density was 5~12 x $10^{12}$ cm$^{-2}$ depending on the Al composition in the barrier layer, that is attributed to different confinement efficiency due to different strength of the piezo-electric filed induced at the hetero-interface. The electron mobility was investigated as a function of the temperature. We found that the mobility is determined by alloy disorder at the hetero-interface [2].

## 3. Magneto-transport measurements

Magneto-transport at 4.2K was analyzed under magnetic field applied normal to the hetero-interface up to 7 Tesla. Typical results for samples #01~#03 are depicted in Fig.1(a). It is obvious that the magneto-resistance shows Shubnikov-de Haas (SdH) oscillations at high magnetic fields. In a sample with high Al composition exhibits weak SdH oscillation reflecting the low electron mobility in the sample. In all samples under study, we have negative magneto-resistance components at low fields. This is the main interest of the present study.

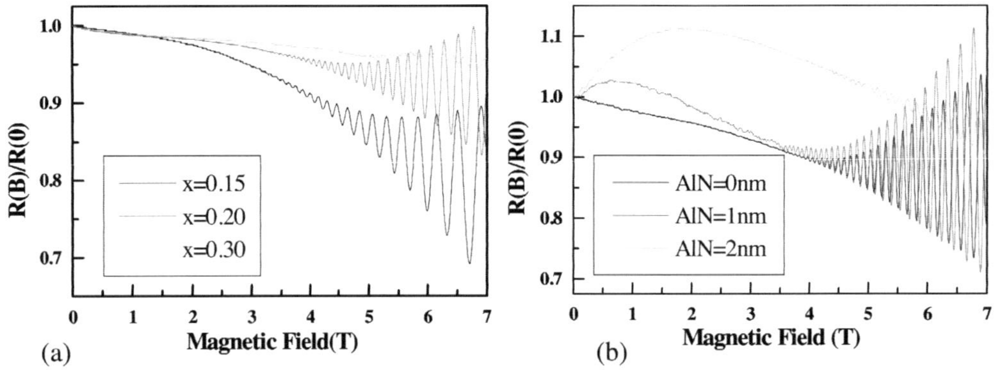


Fig.1. Normalized resistance R as a function of the magnetic field in AlGaN/GaN heterostructure with (a)different Al composition (Sample #01~#03) and (b)different AlN thickness (Sample #11~#13).

Similar measurements were performed for a series of samples with an AlN intermediate layer (sample #11~#13) and the results are displayed in Fig.1(b). We can see that the negative magneto-resistance component at low fields disappears by the insertion of a thin AlN layer. In case of the sample with a 2nm thick AlN layer, the AlN layer would act as an actual potential barrier to the 2DEG at the hetero-interface. Thus we might conclude that the negative component is attributed to the AlGaN alloy used as the barrier layer or the presence of alloy disorder at the hetero-interface.

In order to confirm the fact, the nature of the scattering events was investigated. By analyzing the amplitude of the SdH oscillation as a function of the magnetic field (Dingle plots)[2], the quantum scattering time in each sample was determined as summarized in Table 2, where the transport scattering times were also depicted, determined by the corresponding electron mobility at 17K

As shown in Table 2, by inserting a thin AlN intermediate layer, the transport scattering time or the electron mobility was increased by a factor. The scattering time deduced from the SdH measurements (quantum scattering time), on the other hand, was not changed much. As the result, the ratio of the transport scattering time against the quantum scattering time became as large as 13~14 in a sample with 1~2 nm thick AlN intermediate layer. This fact suggests that short range scattering events such as interface roughness and/or alloy disorder predominates the scattering events in samples without an AlN intermediate layer, while a long range scattering (presumably non-uniform polarization fields) predominates in a sample with AlN layer. In conclusion, the negative magneto-resistance is associated with the short range scattering centres due to interface roughness and/or alloy disorder.

Table 2. Transport and quantum scattering time and their ratio

| Sample | Transport Scattering Time. (ps) | Quantum Scattering Time (ps) | Ratio |
|---|---|---|---|
| **#11** | 0.84 | 0.22 | 3.8 |
| **#12** | 3.26 | 0.23 | 14.2 |
| **#13** | 1.68 | 0.13 | 13.0 |

In Fig.1, the magneto-resistance at high fields is modified by the SdH effect which oscillates along a base line. As was shown by Katsuno et al [3], the base line is well expressed by $\alpha B^2$ at low fields ($\alpha$ is an adjustable parameter). Adopting the method, the behaviour of the positive/negative components at low fields were extracted and the results are depicted in Fig.2. The plots shown in Fig.2 suggest that the positive/negative component increases at low field followed by saturation. Careful investigation of the positive components for samples #12 and #13 suggests that we have a normal magneto-resistance components expressed by $\delta R(B) \sim 1/\{\sigma + \beta B^2\}$, where $\sigma$ is the conductivity at zero magnetic field and $\beta$ is a parameter representing the magneto-resistance [4]. This normal component should show quadratic increase as $B^2$ followed by saturation in agreement with the experimental observation. While, in samples without an AlN interlayer (Sample #01~#03, and #11) the negative component increase linearly followed by saturation. This is the point of the present investigation.

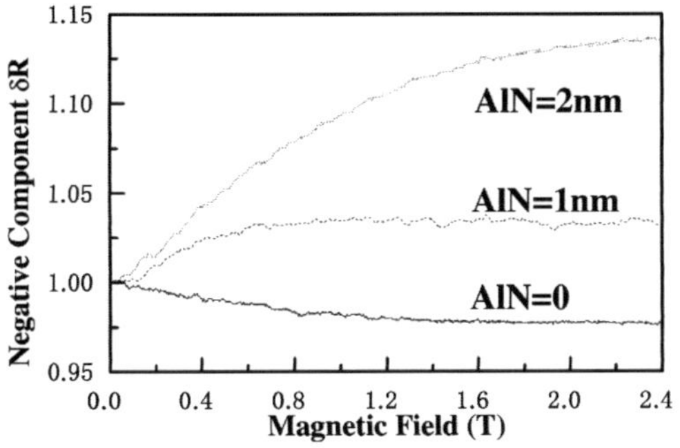


Fig.2. Negative/Positive component at low magnetic fields

**4. Origin of the negative magneto-resistance: Percolation transport in random potentials**

In the present samples, the SdH effect appears at fields higher than 3 T. This suggests that the negative/positive components at low fields as shown in Fig.2 should be accounted for neglecting the formation of Landau levels. According to the semi-classical theory for the magneto-transport, the magneto-resistance is positive irrespective to the nature of scattering events and we may expect the component described as $\delta R(B) \sim 1/\{\sigma + \beta B^2\}$, as described earlier [4]. This is just the case observed for samples with a thin AlN layer. The negative components found in samples without AlN layer, on the other hand, must be attributed to a different mechanism.

Almost three decades ago, negative magneto-resistance was found in various materials/structures, which was attributed to quantum interference effect [5] or weak localization [6]. But the interference effect may not be applicable to the present case, because the electron mobility/transport scattering time is not so long at 4.2K. The positive magneto-conductance (or the negative magneto-resistance) due to weak localization is represented by [7]

$$\delta G = \{We^2/\pi\hbar L\}[\Psi(1/2 + z) - \ln(z)] \quad (1)$$

where

$$1/z = 8\pi eBL_\phi^2/h \qquad (2)$$

In equations (1) and (2), $W$ and $L$ are the width and the length of the sample, respectively, $\Psi(z)$ the di-gamma function, $L_\phi$ is the coherence length. Equation (1) shows that the $\delta G/G$ should be proportional to $B^2$ at low magnetic fields, which is not the present case. Therefore the WL should be omitted in our observation.

Recent studies on the transport in random networks often find negative magneto-resistance at low fields [8,9]. Similar phenomenon was found in a delta-doped GaAs sample [3] or an antidote lattice [10], where the 2DEG transport is violated by random distribution of scattering potentials. The characteristic behaviour of the negative component is such that it increases linearly at low fields followed by saturation, which is just the case observed in Fig.2. Katsuno et al [3] studied the case in an Si delta doped GaAs, where the transport is due to the 2DEG in the delta doped layer. Because of the random distribution of the Si donors, the 2DEG is subject to non-uniform potential distribution or the non-uniform Fermi energy in the 2DEG plane. According to Katsuno et al, the negative component is attributed to the random nature of the transport in 2DEG, or the transport is due to percolation like network. More recently, a numerical simulation on the random lattice has been performed to support the phenomenon [11].

Recently, negative magnetoresistance was found in an AlGaN/GaN hetero-structure[12], where the negative component persisted up to 15 Tesla along with a weak SdH oscillation. They tried to express the results with a model formula for a percolation transport. Bu the analytical results depends strongly on parameters used and discussion is rather ambiguous. Thus, the negative-magneto-resistance at high fields is the subject of further studies. They did not find clear evidence of the negative components as shown in our present work.

In conclusion, the magneto-transport in AlGaN/AlN/GaN heterostructure has been investigated as a function of the alloy composition in the barrier layer and the thickness of the AlN thin layer. Negative magneto-resistance component was found in low magnetic fields in samples without a thin AlN layer. This showed that the negative component is attributed to the presence of alloy disorder in the hetero-structure. Comparing the negative component found in a 2DEG transport in an Si delta doped GaAs sample, the phenomenon is attributed to the percolation transport in random media.

**Acknowledgments**

Use of the facilities of Nagoya University Akasaki Research Center is acknowledged. Part of this work has been performed with the Grant in Aid for scientific research from the Japanese Society for the Promotion of Sciences (JSPS) under contract #16106001.

[1] Ando T, Fowler AB, and Stern F 1982 *Review of Modern Physics* **54** 437
[2] Han XX, Honda Y, Narita T, Yamaguchi M, and Sawaki N 2007 *J.Phys.:Condens. Matter* **19** 046204
[3] Katsuno M, and Sawaki N 1999 *Semicond. Sci. & Technol.*, **14** 549
[4] Seeger K 1999 *Semiconductor Physics,* Chap.4
[5] Thornton T J, Pepper M, Ahmed H, Andrews D and Davies G J 1986 *Phys.Rev.Lett.* **56** 1198
[6] Nambu T, Kawaji S and Kuboki K 1986 *J.Phys.Soc.Jpn.* **53** 682
[7] Feng Y, Thornton T J, Harris J J and Williams D *Appl.Phys.Lett.* **60** 94
[8] Dunford JL, Dhirani AA, and Statt BW 2006 *Phys. Rev.* **B74** 115417
[9] Wang J, Ma XC, Qui Y, Fu YS, Ji SH, Lu L, Jia JF, and Xue QK 2007 *Appl. Phys. Lett.* **90** 113109
[10] Polyakov DG, Evers F, Mirlin AD, and Wolfle P 2001 *Phys. Rev.* **B64** 205306
[11] Ju S, Cai TY, and Li ZY 2005 *Appl. Phys. Lett.* **87** 172504
[12] Cho HI, Gusev GM, Kvon ZD, Renard VT, Lee JH, and Portal JC 2005 *Phys. Rev.* **B71** 245323

Journal of Physics: Conference Series **193** (2009) 012013 doi:10.1088/1742-6596/193/1/012013

# Quantum Maximum Entropy Principle and the Moments of the Generalized Wigner Function

**M. Trovato**

Dipartimento di Matematica, Università di Catania, Viale A. Doria, 95125 Catania, Italy

**L. Reggiani**

Dipartimento di Ingegneria dell' Innovazione and CNISM, Università del Salento, Via Arnesano s/n, 73100 Lecce, Italy

E-mail: trovato@dmi.unict.it – lino.reggiani@unile.it

**Abstract.** By introducing a quantum entropy functional of the reduced density matrix, we construct a rigorous scheme to develop quantum hydrodynamic models. The principle of quantum maximum entropy permits to solve the closure problem for a quantum hydrodynamic set of balance equations corresponding to an arbitrary number of moments in the framework of extended thermodynamics. Quantum contributions are expressed in powers of $\hbar^2$.

## 1. Introduction

Hydrodynamical (HD) models are essential for a physical-mathematical description of the space-time evolution of any kind of fluids. For a rigorous derivation of an HD model the main difficulty is identified in the closure problem associated with the constraint that to solve a finite set of moment equations it is necessary the knowledge of higher order moments [1]. In classical mechanics, the introduction of a maximum entropy principle (MEP) has proven to be very fruitful in solving the closure problem to any degree of approximation [2]. In the quantum MEP (QMEP), the main issue is to define a proper quantum entropy for the explicit incorporation of statistics into problems involving a system of identical particles. The aim of this work is to address this issue and construct a complete set of quantum balance equations (QHD) which are rigorously closed within a global QMEP. Four main results are pointed out in the present paper.

## 2. The generalized Wigner equation, the QHD models and the QMEP approach

We consider a fixed number $N$ of identical particles introducing the general Hamiltonian $H = \int d^3r \Psi^\dagger(\mathbf{r}) \left[ -\frac{\hbar^2}{2m}\nabla^2 + V(\mathbf{r}) \right] \Psi(\mathbf{r}) + \sum_{k=2}^{L} 1/k! \int d^3r_1 \cdots \int d^3r_k \Psi^\dagger(\mathbf{r_1}) \cdots \Psi^\dagger(\mathbf{r_k})\ V(\mathbf{r_1}, \ldots, \mathbf{r_k})$ $\Psi(\mathbf{r_k}) \cdots \Psi(\mathbf{r_1})$ in the Fock space, with many-body interactions and, the statistical density matrix $\rho$, being $Tr(\rho) = 1$ and $\Psi$ the particle field operator [3]. Analogously, in the coordinate space representation we can define the reduced density matrix [4] of single particle, $\langle \mathbf{r}|\widehat{\varrho}|\mathbf{r}'\rangle = \langle \Psi^\dagger(\mathbf{r}')\Psi(\mathbf{r})\rangle = Tr(\rho\Psi^\dagger(\mathbf{r}')\Psi(\mathbf{r}))$ that in an arbitrary representation will take the form $\langle \nu|\widehat{\varrho}|\nu'\rangle = \langle a^\dagger_{\nu'} a_\nu\rangle = Tr(\rho\, a^\dagger_{\nu'} a_\nu)$ being $\nu$, $\nu'$ single particle states, $a_\nu$, $a^\dagger_{\nu'}$ the annihilation and creation operators for these states, $\langle \cdots \rangle$ the statistical mean value with $Tr(\widehat{\varrho}) = N$. Thus, we can define

[4] the *reduced* Wigner function

$$\mathcal{F}_{\mathcal{W}} = (2\pi\hbar)^{-3} \int d^3\tau e^{-\frac{i}{\hbar}\tau\cdot\mathbf{p}} \langle \Psi^\dagger(\mathbf{r}-\tau/2)\Psi(\mathbf{r}+\tau/2)\rangle \tag{1}$$

Accordingly, we look for a function $\widetilde{\mathcal{M}}(\mathbf{r},\mathbf{p})$ in phase space that *corresponds* unambiguously to a single particle operator $\widehat{\mathcal{M}}(\widehat{\mathbf{r}},\widehat{\mathbf{p}})$, introducing the Weyl-Wigner transform $\mathcal{W}(\widehat{\mathcal{M}}) = \widetilde{\mathcal{M}}$ and the inverse Weyl-Wigner transform $\mathcal{W}^{-1}(\widetilde{\mathcal{M}}) = \langle \mathbf{r}|\widehat{\mathcal{M}}|\mathbf{r}'\rangle$. Following an usual script [4, 5], we obtain the equation of motion for the *reduced Wigner function*

$$i\hbar\frac{\partial}{\partial t}\mathcal{F}_{\mathcal{W}}(\mathbf{r},\mathbf{p}) = \int D\xi \left[\widetilde{\mathcal{H}}\left(\mathbf{r}'+\tau/2,\,\mathbf{p}'+\phi/2\right) - \widetilde{\mathcal{H}}\left(\mathbf{r}'-\tau/2,\,\mathbf{p}'-\phi/2\right)\right]\mathcal{F}_{\mathcal{W}}(\mathbf{r}',\mathbf{p}') \tag{2}$$

where $D\xi = d^3r'd^3p'd^3\tau d^3\phi\, e^{\frac{i}{\hbar}[\tau\cdot(\mathbf{p}'-\mathbf{p})+\phi\cdot(\mathbf{r}-\mathbf{r}')]}$ and $\widetilde{\mathcal{H}}$ the phase function of the single particle operator $\widehat{\mathcal{H}} = \langle\mathcal{H}\rangle$, being $\mathcal{H} = -\hbar^2/2m\nabla^2 + V(\mathbf{r}) + \sum_{k=1}^{L-1}(1/k!)\int d^3r_1\cdots\int d^3r_k\ \Psi^\dagger(\mathbf{r_1})\cdots\Psi^\dagger(\mathbf{r_k})$ $V(\mathbf{r},\mathbf{r_1},\ldots,\mathbf{r_k})\Psi(\mathbf{r_k})\cdots\Psi(\mathbf{r_1})$. Then, by expanding the integrand of (2) around $\tau = 0$ and using the Fourier integral theorem, we obtain the formal full expansion, to all orders in $\hbar$, of the equation of motion for the *reduced Wigner function* in the generalized Hartree approximation [6]

$$\frac{\partial\mathcal{F}_{\mathcal{W}}}{\partial t} + \frac{p_k}{m}\frac{\partial\mathcal{F}_{\mathcal{W}}}{\partial x_k} = \sum_{l=0}^{\infty}\frac{(i\hbar/2)^{2l}}{(2l+1)!}\left[\frac{\partial^{2l+1}V_{eff}}{\partial x_{k_1}\cdots\partial x_{k_{2l+1}}}\right]\left[\frac{\partial^{2l+1}\mathcal{F}_{\mathcal{W}}}{\partial p_{k_1}\cdots\partial p_{k_{2l+1}}}\right] \tag{3}$$

where Einstein convention is used on the saturated indices, and the effects of interactions are entirely contained in the effective potential $V_{eff}(\mathbf{r})$. We remark that, if we do not describe in explicit way the many body interactions, then the expansion (3) will take only a formal value. As relevant application of this approach we consider a Bose gas with many-body contact interactions [7] and set $V(\mathbf{r_1},\mathbf{r_2},\ldots,\mathbf{r_k}) = c_{k-1}\prod_{i=1}^{k-1}\delta(\mathbf{r_i}-\mathbf{r_{i+1}})$ for $\forall\ k \geq 2$. In this case we have

$$V_{eff}(\mathbf{r}) = V(\mathbf{r}) + \sum_{k=1}^{L-1}\frac{c_k}{k!}g^{(k)}(\mathbf{r})\,[n(\mathbf{r})]^k \qquad \text{with} \qquad g^{(k)}(\mathbf{r}) = \frac{\langle[\Psi^\dagger(\mathbf{r})]^k[\Psi(\mathbf{r})]^k\rangle}{[n(\mathbf{r})]^k} \tag{4}$$

where $g^{(k)}(\mathbf{r})$ is the k-order correlation function [8]. We stress that, by considering explicitly the relation (4), then the kinetic equation (3) loses its formal value and, consequently, all non-linear phenomena imputable to weak interactions between bosons can be expressed in terms of increasing powers of density. The advantage of this approach will be evident in the corresponding QHD system. In this sense, a theory based on Eq. (3) supplemented by Eq. (4) is a first major result of the work, because all closure relations imputable to the contact interactions, are *explicitly determined* as known polynomial functions of the macroscopic field variable $n(\mathbf{r})$ [9]. We remark that these results can be generalized by including explicitly the spin degrees of freedom. Equations (2),(3) can be supplemented by others interaction terms to describe a variety of physical systems, including Fermi liquids [5], non-ideal gases and plasma [10].

Below we develop the extended three-dimensional QHD model associated with (3). By considering the operator $\widehat{\mathcal{M}}(\widehat{\mathbf{r}},\widehat{\mathbf{p}})$ and the corresponding phase function $\widetilde{\mathcal{M}}(\mathbf{r},\mathbf{p})$ we define the macroscopic *local moment* $M(\mathbf{r},t) = \int d^3p\ \widetilde{\mathcal{M}}(\mathbf{r},\mathbf{p})\ \mathcal{F}_{\mathcal{W}}(\mathbf{r},\mathbf{p},t)$ of $\widehat{\mathcal{M}}$. As in classic extended thermodynamics [1], by introducing the group velocity $u_i = p_i/m$, we define the mean velocity $v_i = n^{-1}\int d^3p\ u_i\ \mathcal{F}_{\mathcal{W}}$, the peculiar velocity $\widetilde{u}_i = u_i - v_i$, and the quantity $\widetilde{\varepsilon} = m\widetilde{u}^2/2$. Thus, we consider the set of traceless kinetic fields [11] $\widetilde{\mathcal{M}}_A = \{\widetilde{\varepsilon}^s,\ \widetilde{\varepsilon}^s\widetilde{u}_{i_1},\ \ldots,\ \widetilde{\varepsilon}^s\widetilde{u}_{\langle i_1}\widetilde{u}_{i_2}\cdots\widetilde{u}_{i_r\rangle}\}$ and the corresponding set of *central moments* $M_A(\mathbf{r},t) = \{M_{(s)},\ M_{(s)|i_1},\ \ldots,\ M_{(s)|\langle i_1\cdots i_r\rangle}\}$ where, by construction, it is $M_{(0)|i_1} = 0$, and $M_{(s)|\langle i_1 i_2\cdots i_r\rangle} = \int\ d^3p\ \widetilde{\varepsilon}^s\,\widetilde{u}_{\langle i_1}\widetilde{u}_{i_2}\cdots\widetilde{u}_{i_r\rangle}\mathcal{F}_{\mathcal{W}}$ with $s = 0,1,\ldots\mathcal{N}$ and $r = 1,2,\ldots$ M. In particular, by using a finite but arbitrary number of scalar and vectorial kinetic fields $\widetilde{\mathcal{M}}_A = \{\widetilde{\varepsilon}^s,\ \widetilde{\varepsilon}^s\widetilde{u}_i\}$ we obtain in correspondence the set of scalar and vectorial central moments $M_A = \{M_{(s)}, M_{(s)|i}\}$ with $s = 0,\cdots\mathcal{N}$. Accordingly, for $\mathcal{N} = 0$, as set of macroscopic variables we get the *numerical density* $n = M_{(0)}$ and the *velocity* $v_i$. For $\mathcal{N} = 1$ we get in addition $M_{(1)}$ and $M_{(1)|i}$, which admit a direct physical interpretation being $M_{(1)} = 3/2\,P$ and $M_{(1)|i} = q_i$ respectively, the *internal energy density* (with $P$ the pressure) and the *heat flux density*. By contrast, for $\mathcal{N} > 1$, as macroscopic variables, we consider also some scalar and vectorial moments of higher order. Multiplying (3) by $\widetilde{\mathcal{M}}_A$, integrating over $\mathbf{p}$ we exactly determine a theory that is consistent up to terms of order $\hbar^2$. Thus, the moments $\{v_i, M_A\}$ must satisfy the following extended QHD$_2$ system

$$\dot{n} + n\,\frac{\partial v_k}{\partial x_k} = 0\,, \qquad \dot{v}_i + \frac{1}{n}\frac{\partial M_{(0)|ik}}{\partial x_k} + \frac{1}{m}\frac{\partial V_{eff}}{\partial x_i} = 0, \tag{5}$$

$$\dot{M}_{(s)} + M_{(s)}\,\frac{\partial v_k}{\partial x_k} + \frac{\partial M_{(s)|k}}{\partial x_k} + s\,m\,M_{(s-1)|ik}\,\frac{\partial v_i}{\partial x_k} - s\frac{m}{n}\,M_{(s-1)|i}\,\frac{\partial M_{(0)|ik}}{\partial x_k} =$$
$$\frac{\hbar^2}{24}\,s(s-1)\left\{(s-2)\,\frac{\partial^3 V_{eff}}{\partial x_{\langle i}\partial x_j \partial x_{k\rangle}}\,M_{(s-3)|\langle ijk\rangle} + \frac{3}{5}\frac{(1+2s)}{m}\,\frac{\partial^3 V_{eff}}{\partial x_r \partial x_r \partial x_k}\,M_{(s-2)|k}\right\}, \tag{6}$$

$$\dot{M}_{(s)|i} + M_{(s)|i}\frac{\partial v_k}{\partial x_k} + \frac{\partial M_{(s)|ik}}{\partial x_k} + smM_{(s-1)|ijk}\frac{\partial v_j}{\partial x_k} + M_{(s)|k}\frac{\partial v_i}{\partial x_k} - \frac{M_{(s)}}{n}\frac{\partial M_{(0)|ik}}{\partial x_k} -$$
$$sm\frac{M_{(s-1)|ij}}{n}\frac{\partial M_{(0)|jk}}{\partial x_k} = \frac{\hbar^2}{24}\,s\left\{(s-1)(s-2)\,\frac{\partial^3 V_{eff}}{\partial x_r \partial x_j \partial x_k}\,M_{(s-3)|\langle rjki\rangle} + \right.$$
$$\frac{3(s-1)}{m}\frac{(3+2s)}{7}\left(\frac{\partial^3 V_{eff}}{\partial x_r \partial x_r \partial x_k}\,M_{(s-2)|\langle ki\rangle} + \frac{\partial^3 V_{eff}}{\partial x_{\langle k}\partial x_{r\rangle}\partial x_i}\,M_{(s-2)|\langle kr\rangle}\right) +$$
$$\left.\frac{(4s^2+8s+3)}{5\,m^2}\,\frac{\partial^3 V_{eff}}{\partial x_r \partial x_r \partial x_i}\,M_{(s-1)}\right\} \quad \text{with} \quad s = 1, \cdots \mathcal{N}\,. \tag{7}$$

The set (5) - (7) is a second major result of the work: in particular, for $\hbar \to 0$, it recovers the classic form of extended thermodynamics [1]. The set contains unknown constitutive functions that are represented by the central moments of higher order $H_A = \{M_{\mathcal{N}+1},\, M_{(l)|\langle ij\rangle},\, M_{(r)|\langle ijk\rangle},\, M_{(s)|\langle ijqk\rangle}\}$ with $l = 0, \cdots \mathcal{N}$; $r = 0, \cdots \mathcal{N}-1$; and $s = 0, \cdots \mathcal{N}-3$.

In general, the closure problem of a set of balance equations is tackled using the QMEP formalism. Accordingly, to take into account *ab initio* the Bose and Fermi statistics, we follow the Landau strategy [12]. Thus, for a non-interacting system in non-equilibrium conditions, the quantum entropy can be determined in terms of the reduced density matrix $\widehat{\varrho}$ [13]. Then, by generalizing existing definitions [14], as quantum entropy we take the *global* quantity

$$S(\widehat{\varrho}) = -k_B Tr(\widehat{\Phi}(\widehat{\varrho})) \qquad \text{with} \qquad \widehat{\Phi}(\widehat{\varrho}) = \widehat{\varrho}\left\{\ln\left(\frac{\widehat{\varrho}}{y}\right) \pm y\,\widehat{\varrho}^{-1}\left(\widehat{I} \mp \frac{\widehat{\varrho}}{y}\right)\ln\left(\widehat{I} \mp \frac{\widehat{\varrho}}{y}\right)\right\} \tag{8}$$

where $k_B$ is the Boltzmann constant, $\widehat{I}$ is the identity operator, $y = (2\tilde{s}+1)$ is the spin degeneration and, the $\pm$ signs refer to fermions and bosons, respectively. By using the QMEP, we search the extremal value of entropy subject to the constraint that the information on the physical system is described by the $M_A(\mathbf{r}, t)$. To this purpose, we consider the new *global* functional [15]

$$\widetilde{S} = S - \int d^3r \left\{\sum_{A=1}^{\mathrm{N}} \widetilde{\lambda}_A \left[\int d^3p\, \widetilde{\mathcal{M}}_A\, \mathcal{F}_W - M_A\right]\right\} \tag{9}$$

being $\widetilde{\lambda}_A(\mathbf{r}, t)$ the *local Lagrange multipliers* to be determined. One can show that $\delta\widetilde{S} = 0$ implies

$$\widehat{\varrho} = y\left\{\exp\left[\mathcal{W}^{-1}\left(\sum_{A=1}^{\mathrm{N}} \lambda_A(\mathbf{r}, t)\,\widetilde{\mathcal{M}}_A\right)\right] \pm \widehat{I}\right\}^{-1} \tag{10}$$

with $\lambda_A = \widetilde{\lambda}_A / k_B$. The (10) is a third major result of the work. We remark that, by using the (8), Bose or Fermi statistics are implicitly taken into account and, from relation (10), we obtain

$$\mathcal{F}_W = (2\pi\hbar)^{-3}\,\mathcal{W}\left(\widehat{\varrho}[\lambda_A(\mathbf{r}, t), \widetilde{\mathcal{M}}_A]\right)\,. \tag{11}$$

From the functional form (10)-(11) of the reduced Wigner function, we use the (3) to obtain a set of equations for the constraints. This set completely represents the closed $\mathrm{QHD}_2$ model (5)-(7) in which all the constitutive functions are determined starting from their kinetic expressions. Thus, for a given number of moments $M_A$, we can consider a consistent expansion around $\hbar$ of the Wigner function. In this way we separate classical from quantum dynamics, and obtain order by order corrections terms. In particular, by using the Moyal formalism, one can prove [16] that the Wigner function, and hence the central moments, can be expanded in even power of $\hbar$ being $\mathcal{F}_{\mathcal{W}} = \sum_{k=0}^{\infty} \hbar^{2k}\, \mathcal{F}_{\mathcal{W}}^{(2k)}$ and $M_A = \sum_{k=0}^{\infty} \hbar^{2k}\, M_A^{(2k)}$. With this approach, the dynamics of the system is described by resolving, order by order, a closed QHD set of balance equations for the moments $\{M_A(\mathbf{r},t)\}$. To this end, the Lagrange multipliers $\lambda_A$ are determined by inverting, order by order, the constraints $M_A = (2\pi\hbar)^{-3} \int d^3p\; \widetilde{\mathcal{M}}_A\, \mathcal{W}\left(\widehat{\varrho}[\lambda_B(\mathbf{r},\,t), \widetilde{\mathcal{M}}_B]\right)$ and the inversion problem can be solved only by assuming that also the Lagrange multipliers admit for an expansion in even powers of $\hbar$, i.e. $\lambda_A = \lambda_A^{(0)} + \sum_{k=1}^{\infty} \hbar^{2k}\, \lambda_A^{(2k)}$. With these assumptions, and using (10) and (11), we succeed in determining the following expression for $\mathcal{F}_{\mathcal{W}}$

$$\mathcal{F}_{\mathcal{W}} = \frac{\widetilde{y}}{e^{\Pi} \pm 1}\left\{1 + \sum_{r=1}^{\infty} \hbar^{2r} P_{2r}^{\pm}\right\}, \tag{12}$$

where $\widetilde{y} = y/(2\pi\hbar)^3$, $\Pi = \sum \lambda_A\, \widetilde{\mathcal{M}}_A$ and the terms $P_{2r}^{\pm}$ are expressed by recursive formulas.

The (12) is a fourth major result of the work. By considering terms up to first order in the quantum correction, the Lagrange multipliers are obtained as solutions of constraints $M_A + \mathcal{O}(\hbar^4) = \int d^3p\; \widetilde{\mathcal{M}}_A\, \mathcal{F}_{\mathcal{W}}$. In this case, $M_A$ must satisfy the $\mathrm{QHD}_2$ system (5)-(7). From the knowledge of the Lagrange multipliers, both the Wigner function and the constitutive functions $H_A$ can be determined explicitly. We remark that: (i) Using the general definition (8) we incorporate the indistinguishability principle of a system of identical particles, by developing a nonlocal theory that contains implicitly the Fermi and Bose statistics, a result which was left open since the Wigner seminal papers [16]. (ii) When $\hbar \to 0$ we recover the expressions $\lambda_A^{(0)} = \lambda_A^{(0)}(M_B^{(0)})$ and $\mathcal{F}_{\mathcal{W}}^{(0)}$ obtained in the framework of classic MEP approach [1] for a Fermions or Bosons system.

[1] Müller I and Ruggeri T 1998 *Rational Extended Thermodynamics* Vol. **37**, Springer-Verlag New York.
[2] Trovato M and Reggiani L 2006 *Phys. Rev. B.* **73** 245209 and references therein.
[3] We neglect both, the symbol $\widehat{\ }$ to refer to operators acting in the Fock space, and the spin degrees of freedom.
[4] Klimontovich Yu 1958 *Soviet Phys. JETP.* **6** 753. Brittin W and Chappell W 1962 *Rev. Mod. Phys.* **34** 620
[5] Silin V P (1958), *Soviet Phys. JETP* **6**, 387. (1958).
[6] Ploszajczak M and Rhoades-Brown M J, (1985) *Phys. Rev. lett.* **55**, 147; (1986) *Phys. Rev. D* **33**, 333686.
[7] Fersino E, Mussardo G and Trombettoni A 2008 *Phys. Rev. A.* **7** 053608
[8] Parkins A S and Walls D F 1998 *Phys. Reports.* **303** 1, p.71 §7.6 and references therein.
[9] By using the (4), for $L = 2$ we have $V_{eff}(\mathbf{r}) = V(\mathbf{r}) + c_1\, n(\mathbf{r})$. Analogously for $L > 2$, a reasonable description of the low-energy dynamics, can be given assuming values approximatively constant for the remaining correlation functions (see §7.6, in [8]). Thus, we obtain $V_{eff}(\mathbf{r}) = V(\mathbf{r}) + \sum_{s=1}^{L-1} \mathrm{u}_s\, [n(\mathbf{r})]^s$ where the coefficients $\mathrm{u}_s$ describe the many body interactions in the mean-field approximation.
[10] Klimontovich Yu 1982 *Kinetic theory of nonideal gases & nonideal plasmas*, Pergamon, and references therein.
[11] $A_{\langle i_1 \cdots i_n \rangle}$ is the traceless symmetric part of tensor $A_{i_1 \cdots i_n}$ respect to indices contained between $\langle\,\rangle$. Thus, for example, the quantity $A_{\langle kr \rangle i}$ must be intended as it follows $A_{\langle kr \rangle i} = (A_{kri} + A_{rki})/2 - 1/3\, \delta_{kr}\, A_{lli}$.
[12] Landau L D and Lifshitz E M 1959 *Statistical Physics*, Vol. **5**, Chap. V §54, Pergamon Press, London-Paris.
[13] In non-equilibrium conditions, the quantum entropy can be determined in terms of occupation numbers[12] $S = -k_B \sum_\nu y[\langle \overline{N_\nu} \rangle \ln \langle \overline{N_\nu} \rangle \pm (1 \mp \langle \overline{N_\nu} \rangle) \ln (1 \mp \langle \overline{N_\nu} \rangle)]$ being $\langle \overline{N_\nu} \rangle = \langle a_\nu^\dagger a_\nu \rangle / y$. If we consider the Schrodinger equation of single particle $[\widehat{\mathcal{H}}(\mathbf{r}) - E_\nu]\varphi_\nu(\mathbf{r}) = 0$ then, in stationary conditions, both the reduced density matrix $\widehat{\varrho}$ and the operator $(8)_2$ $\widehat{\Phi}(\widehat{\varrho})$ are diagonal in the base $\varphi_\nu$ being $\langle \nu | \widehat{\varrho} | \nu' \rangle = \langle a_\nu^\dagger a_\nu \rangle\, \delta_{\nu\nu'}$ and $\langle \nu | \widehat{\Phi}(\widehat{\varrho}) | \nu' \rangle = y[\langle \overline{N_\nu} \rangle \ln \langle \overline{N_\nu} \rangle \pm (1 \mp \langle \overline{N_\nu} \rangle) \ln (1 \mp \langle \overline{N_\nu} \rangle)]\, \delta_{\nu\nu'}$. Consequently, we obtain the $(8)_1$.
[14] Von Neumann J 1955 *The Mathematical Foundations of Quantum Mechanics* Princeton Press, Princeton.
[15] E.T. Jaynes E T (1957) *Phys. Rev.* **108**, 171. Degond P and Ringhofer C (2003) *J. Stat. Phys.* **112**, 587.
[16] Wigner E 1932 *Phys. Rev.* **40** 749. Hillery M, O'Conell R, Scully M and Wigner E 1984 *Phys. Rep.* **106** 121.

# Experimental study of transport in InAs Quantum Hot Electron Transistor

**Thibaut Daoud, Guilhem Boissier, Jan Devenson, Alexei N. Baranov and Roland Teissier**

Institut d'´Electronique du Sud (UMR 5214, CNRS/UM2)
Université Montpellier 2, Place E. Bataillon, cc067
34095 Montpellier, France

roland.teissier@ies.univ-montp2.fr

**Abstract**. We study a Quantum Hot Electron Transistor, an original transistor made of InAs/AlSb heterostructures with high speed intrinsic transport. It is a unipolar vertical transport device, based on the concept of hot electron transistor. Devices using different heterostructure designs have been grown, processed and characterized. Both resonant and parasitic transport of electrons is studied by mean of the static output characteristics of the transistors. We demonstrated that a key parameter is the design and thickness of the collector barrier. For 15nm-thin barriers, we obtained static current gain of 5 associated to a base transit time of 140 fs.

## 1. Introduction

Shifting towards high indium content alloys, in order to benefit from the increase in electron mobility and high field velocity of these materials, is a general trend of high speed III-V microelectronics in recent years. Significantly improved performances, with frequency $f_t$=760 GHz, have been obtained with a Heterojunction Bipolar Transitor (HBT) using indium rich InGaAs material [1].

The work presented here aims to exploit the attractive properties of InAs to further extend the limits of vertical electron transport. Different attempts have been made to design HBTs with InAs or GaSb-based materials (also known as 6.1 Å material family) [2]. Our approach which is based on the concept of hot electron transistor [3] is radically different. The Quantum Hot Electron Transistor (QHET) that we developed [4] is a unipolar vertical transport device which uses a resonant transport through the exited states of a quantum well (QW).

We will present in this paper studies of electron transport in QHET through the analysis of static characterizations of devices made of different heterostructures. In particular, we will focus on the role of the collector barrier that controls base-collector leakage currents and also transit times through the base layer.

## 2. Device structure

### *2.1. Principle of operation*

The QHET consists of a base layer made of a N-doped ~10nm thick InAs quantum well, an emitter and collector barrier made of graded InAs/AlSb superlattices, and a 100nm-thick InAs collector.

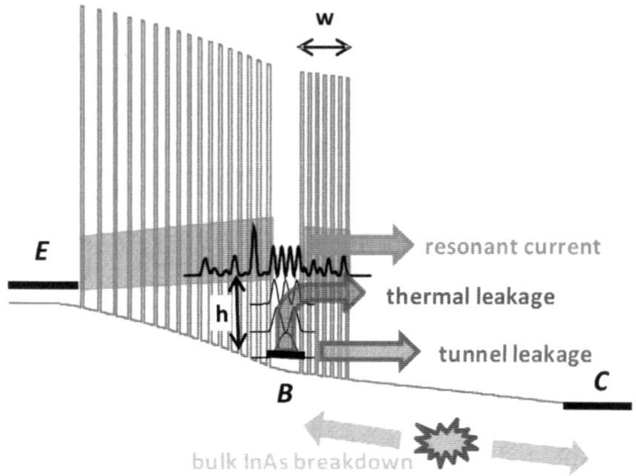


**Figure 1.** Conduction band profile of the InAs QHET under bias (in grey). The thick black lines represent the Fermi level in the three electrodes; the wavefunctions of relevant electron levels in the base layer are drawn ; the light grey areas show schematically the electron minibands of the InAs/AlSb superlattices forming the emitter-base junction and the collector barrier. The main resonant and parasitic (thermal and tunnel leakages) transport channels are also indicated.

Majority electrons in the base are confined in the QW ground state thanks to the collector barrier, while minority hot electrons transit from emitter to collector through coherent transport via a QW excited state which is resonant with the emitter and collector barrier minibands (Fig.1).

Its two main advantages over HBTs are the low base sheet resistance accessible with moderate n-type doping levels (typically $10^{18}$ $cm^{-3}$) and very high speed electron transport in the InAs collector.

*2.2. Fabrication*

The structures were grown by molecular beam epitaxy on InAs substrates, using a Riber Compact 21 solid source machine equipped with As and Sb valved cracking cells. N-type InAs layers were doped with Silicon. Thanks to the expertise of IES in the growth of state of the art InAs/AlSb heterostructures for quantum cascade lasers [5], high quality structures were available for this work.

Large double mesa devices with 10x10 $\mu m^2$ emitter and 20x30$\mu m^2$ base have been fabricated, using standard UV lithography. Ohmic contacts on N-InAs were made of non alloyed Cr/Au metallization. In this device fabrication, a key technological step is the accurate contacting of the thin base. A high precision layer-by-layer etching process has been developed for that purpose [4].

## 3. Experimental study

*3.1. Device characteristics*

Devices using different heterostructure designs, have been tested. The resonant transport of electrons was studied by means of the static output characteristics of the transistors (Fig.2).

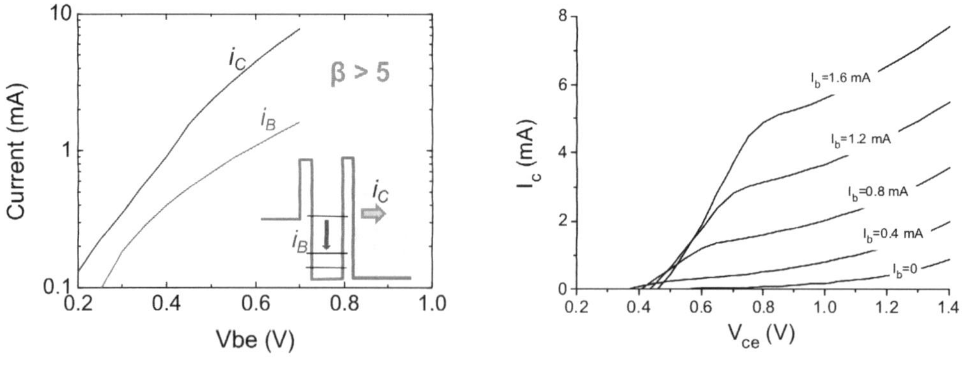


**Figure 2.** Gummel plot (left) and common emitter static output characteristics (right) of a 10x10 $\mu m^2$ device with reduced parasitic currents. A current gain of 5 and collector breakdown voltage of 1.4V are obtained.

Because of the n-type of the base, leakage current from base to collector and voltage dynamic of the collector junction are critical issues. We report in the following on the experimental optimization of the collector barrier that controls collector parasitic currents. The main parasitic mechanisms are illustrated on Fig.1: majority electrons from the doped base layer can escape via thermal activation to the collector barrier miniband or tunnel through the collector barrier. In addition a breakdown of the bulk InAs layer in the collector junction can arise under high electric fields.

### *3.2. Thermal leakage*

Thermally activated escape from the base strongly depends on the position of the collector barrier miniband (*h* parameter on Fig.1). We studied the base-collector junction characteristics for two structures, with *h* equal to 0.3 and 0.45eV respectively. The experimental curves are shown on Fig.3, indicating that thermal leakage becomes negligible with the higher barrier for Vbc<1.5V. For reduced barrier height, the current is much higher and scales according to a simple activation energy model assuming thermal equilibrium.

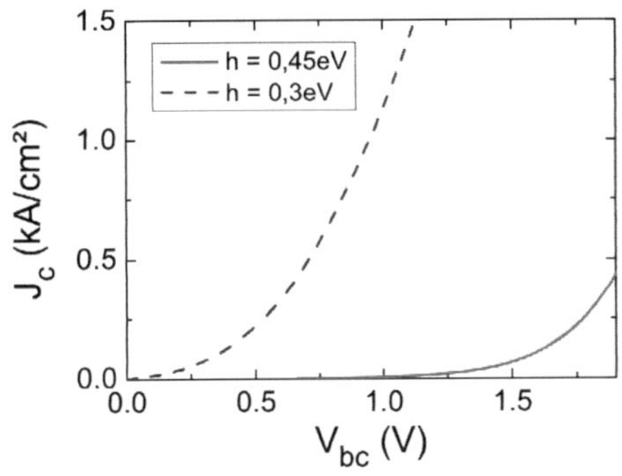


**Figure 3.** Base-collector leakage for different collector barrier heights, defined as the difference between the base Fermi energy and the bottom of collector barrier miniband.

### *3.3. Tunnel leakage*

Another possible leakage current from the base is tunnelling to the collector. This mechanism, involving transfer from a localized state to a continuum of states, is non resonant, hence weakly dependent on the bias of the base-collector junction. In addition it must be proportional to the electron density in the base. We compared the leakage current at low bias of two structures with different base doping levels (Fig.4). Up to 0.8V, the current is very small and independent on doping level. It demonstrates that for the used barrier thickness of 18nm (6 QWs), tunnelling is negligible.

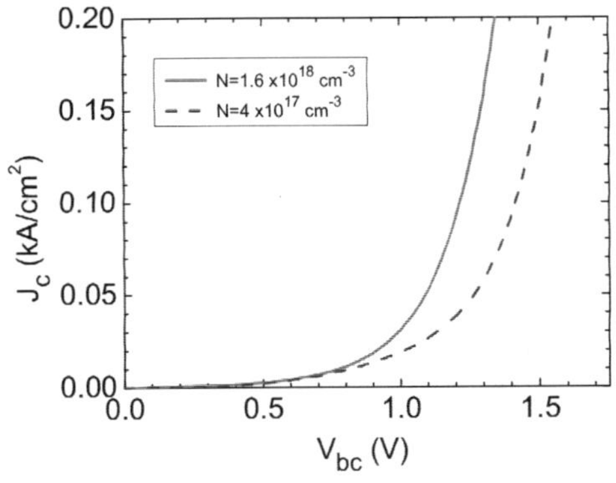


**Figure 4.** Base-collector leakage for different base doping levels

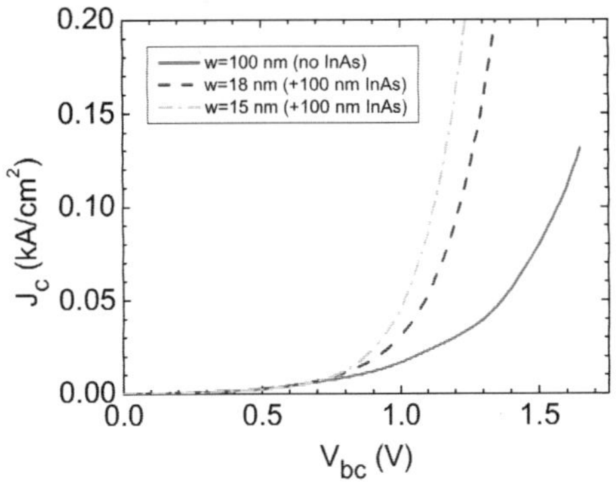


**Figure 5.** Base-collector leakage for different collector barrier widths.

### *3.4. Bulk collector breakdown*

At higher bias, the collector current increases rapidly. In order to understand the origin of this current and see the importance of breakdown in the bulk InAs of the collector junction, we studied a device in which we replaced InAs by a InAs/AlSb superlattice with the same composition as the collector barrier or, in other words, we used a 100nm-thick collector barrier. The comparison of this structure with the standard one (18 nm-thick barrier) is shown on Fig.5. For the device with thick barrier, the current is only due to thermal activation to the superlattice miniband. This thermal current must be of the same intensity for the thinner barriers whose miniband is at the same energy (same $h$). Hence the difference between the two characteristics comes from current generated in the InAs collector due to the onset of impact ionization in the high field region of this small bandgap material. We also studied a device with thinner barrier of 15nm (5 QWs). We observe a similar breakdown voltage, but no signs of increased parasitic tunnel current, thus confirming that this latter effect is still negligible.

### *3.5. Base transit time*

The collector barrier thickness has also an important role on the electron transit time in the base, which is the time necessary to transfer electrons from the base exited state to the collector high field region. This time can be estimated from the static current gain which is, in a simple picture (see inset on Fig.2), the ratio of the electron relaxation time ($\tau_{ISB}$) in the base QW to the base transit time ($\tau_t$). The value of $\tau_{ISB}$ can be calculated to 0.7 ps, assuming that the main relaxation mechanism is LO-phonon emission. For the different devices presented in Fig. 5 with barrier widths of 100, 18 and 15 nm we measured current gains of 2, 3 and 5 respectively. This shows that transit time is significantly reduced when the collector barrier is thinned, with a value of $\tau_t$=0.14 ps for the 15 nm-thick barrier.

## 4. Conclusions

Through the experimental study of different structures, we explored the ability of the collector barrier superlattice to control the transport in InAs QHET. The height of the miniband controls the thermal leakage of electron from the base. We demonstrated that the tunnel current is negligible even for barriers as thin as 15 nm. The reduction of barrier thickness leads to an increase of the static current gain thanks to a reduction of base transit time. As a main conclusion this study demonstrated that thinner barriers are feasible and would be a way to further increase the gain.

## Acknowledgements

The authors would like to thank the support of French ANR under the programs TACHUS (ANR-08-BLANC-0299).

## References

[1] W. Snodgrass, W. Hafez, N. Harff, and M. Feng, *2006 International Electron Device Meeting technical digest*, pp. 338-341, (2006).
[2] Bennett BR, Magno R, Boos JB, Kruppa W, Ancona MG, *Solid-State Electr.* **49** (12), 1875-1895, (2005); Averett KL, Wu X, Koch MW, et al., *J. of Crystal Growth* **251**, 852-857, (2003); P. D. Moran, D. Chow, A. Hunter, and T. F. Kuech, *Appl. Phys. Lett.* **78**, 2232 (2001).
[3] M. Heiblum, Tunneling hot electron transfer amplifiers, Solid State Electron. 24, 343, (1981); A.F.J. Levi, T.H. Chiu, *Appl. Phys. Lett.* **51**, 984, (1987).
[4] T. Daoud, J. Devenson, A.N. Baranov, R. Teissier, *Springer Proceedings in Physics*, **119**, 163-165, (2008); T. Daoud, G. Boissier, J. Devenson, G. Sabatini, L. Varani, A.N. Baranov, R. Teissier, Conference Proceedings – Indium Phosphide and related Materials, 381-384, (2008).
[5] J. Devenson, O. Cathabard, R. Teissier, A.N. Baranov, *Appl. Phys. Lett.* **91**, 141106 (2007).

# Electrical Excitation of Surface Phonon-Polaritons in III-V Heterostructures: a Monte Carlo study

**F Mazzamuto[1], J Saint-Martin[1], A Bournel[1], P Dollfus[1], A Archambault[2], F Marquier[2] and J-J Greffet[2]**
[1]Institut d'Electronique Fondamentale; Univ. Paris-Sud, CNRS; 91405 Orsay, France
[2]Institut d'Optique, CNRS, Univ Paris-Sud, Laboratoire Charles Fabry, Campus Polytechnique, RD128, 91127 Palaiseau cedex-France

E-mail: fulvio.mazzamuto@u-psud.fr

**Abstract.** The surface phonon polaritons (SPP) excitation by hot electrons of an appropriate two-dimensional gas could make possible to implement an opto-electronic device emitting around 10 THz. To quantitatively assess the SPP emission by inter-subband electron relaxation in a GaAs/AlGaAs quantum well, accurate electron/SPP scattering rates have been included in a particle Multi-Subband Monte Carlo (MSMC) simulator. Our results suggest some possible optimizations to enhance the SPP emission in order to design new Tera-Hertz sources.

## 1. Introduction

Surface plasmons polaritons have been widely investigated and now offer the possibility of designing optoelectronic devices operating in the THz frequency range [1]. The same potential is expected for phonon polaritons in semiconductor structures and devices. Recently, phonon-polariton waves have been experimentally studied on the surface of SiC crystal [2], and infra-red emission of thermally excited SPPs has been observed in polar materials, by grating-induced diffraction [3]. This paper investigates an SPP excitation method by hot electrons inter-subband transitions in a 2DEG. An AlGaAs/GaAs quantum well heterostructure and the surface wave over the air/GaAs interface are studied. Within the Fermi golden rule, analytical expressions of SPP scattering rates are first derived using electromagnetic surface wave quantization and simplified SPP dispersion relation. A Multi-Subband Monte Carlo (MSMC) simulator [4] has been modified appropriately to consider these electron/SPP scattering rates. The most important scattering phenomena governing the transport properties have been considered, i.e. interaction with SPP, bulk polar optical phonons (POP) and acoustic phonons (AP). Our simulations provide an accurate evaluation of the number of SPP emitted per unit of time as well as their wave vector distribution.

## 2. SPP scattering rates

The surface phonon scattering model developed here for an air/semiconductor interface, as presented in Fig. 1, will be used as a first approximation in device simulations of Section 3. Consider an electron of wave vector $K$ in the $m$th subband of a 2DEG. The transition from this state to the state $K'$ in the subband $m'$ via electron-phonon interaction is described by the coupling matrix $M^{\sigma}$ ($\sigma = -1$ for phonon emission, +1 for absorption) whose square element can be expressed in the first order perturbation theory as [5]

$$\left|M^{\sigma}_{mK,m'K'}\right|^2 = \frac{e^2}{m^{*2}}\frac{\hbar^3\omega}{2AG(q_{//})}(n_{q_{//}}+1/2-\sigma/2)\left|Z_1 u.K+\sigma q_{//}Z_2 u_z.n\right|^2\delta(K+\sigma q_{//}\quad K') \tag{1}$$

where A is the surface area of the 2DEG. The term $G(q_{//})$ in (1) is obtained via the quantization of the electromagnetic surface wave associated with an SPP and is expressed by

$$G(q_{//}) = \frac{\varepsilon_0\omega^2}{2}\left\{\frac{1}{2\kappa_{1\perp}}\left[\left(1+\frac{q_{//}{}^2}{\kappa_{1\perp}{}^2}\right)\frac{d}{d\omega}(\omega\varepsilon(\omega))-\left(1+\frac{q_{//}{}^2}{\kappa_{1\perp}{}^2}\right)\varepsilon(\omega)\right]\quad\frac{q_{//}{}^2}{\kappa_{2\perp}{}^3}\right\} \tag{2}$$

The SPP is defined by its angular frequency $\omega$, its in-plane wave vector $q_{//}$ and its normal absorption components $\kappa_{1(2)\perp}$ characterizing the vanishing of the electric field in the first (second) material. The two considered materials are air (labelled by subscript 2) and semiconductor (subscript 1), thus only the GaAs dielectric function $\varepsilon(\omega)$ is involved [6]. In (1) $u$ and $u_z$ are the unit vector in SPP wave direction and the perpendicular unitary vector, respectively (see Fig. 2). The energy distribution of SPPs is described by the Bose-Einstein statistics $n_{q_{//}}$. The overlap integrals $Z_1$ and $Z_2$ are given by

$$Z_1 = \int e^{-\kappa_{1\perp}|z|}\xi^*_{m'}(z)\xi_m(z)dz \qquad Z_2 = \frac{1}{\kappa_\perp}\int e^{-\kappa_{1\perp}|z|}\xi^*_{m'}(z)\frac{d\xi_m}{dz}dz \tag{3}$$

where $\xi_m(Z)$ is the electron envelope function for the subband $m$

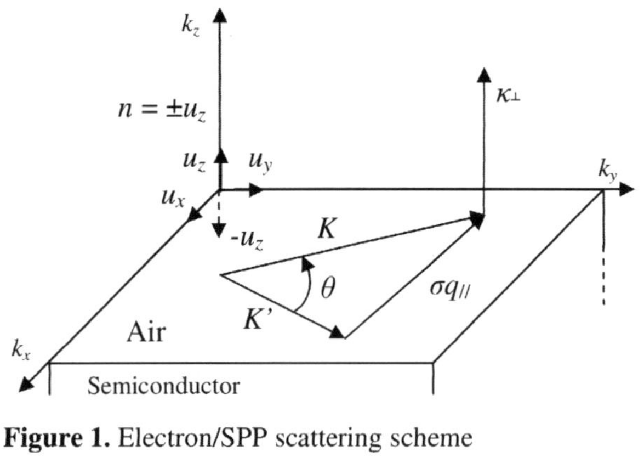


**Figure 1.** Electron/SPP scattering scheme

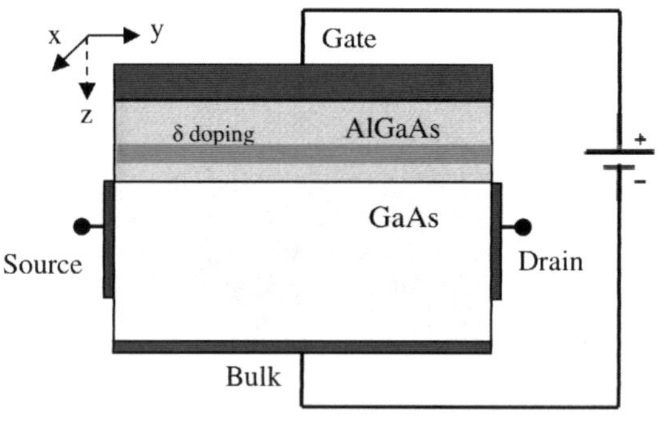


**Figure 2** Structure of simulated device

The SPP dispersion is approximated by the step function $\omega(q_{//}) = \omega_{SPP}U(q-q_0)$, where $\omega_{SPP}$ is chosen as the asymptotic value of the SPP dispersion ($\hbar\omega_{SPP} = 36\,\text{meV}$) and $q_0$ is the wave vector limit value satisfying the surface resonance condition ($q_0 = 1.61\times10^6\ \text{m}^{-1}$). Integrating $M^\sigma$ over all possible final states gives the electron/SPP scattering rate $\lambda(E,m)$ for an electron of kinetic energy $E$ in subband $m$:

$$\lambda(E,m) = C_{SPP}\cdot\omega_{SPP}\int d\theta\cdot n_{q_{//}+1/2-\sigma/2}\left(1+2\alpha(E+\sigma\hbar\omega_{SPP})\right)\frac{\left|Z_1u.K+\sigma q_{//}Z_2u_z.n\right|^2}{G(q_{//})} \tag{4}$$

where $C_{SPP} = e^2/4\pi m^*$, $\theta$ is the deviation angle, $e$ is the electron charge and $m^*$ is the electron effective mass corrected by a non-parabolicity coefficient $\alpha$. Only the $\Gamma$ valley of GaAs is considered here.

The resulting inter-subband and intra-subband SPP emission rates for the 1st and 2nd subbands in an AlGaAs/GaAs quantum well heterostructure at room temperature are plotted in figure 3. Figure 4 compares the most frequent SPP emission rate for electrons in the first subband with that of the most frequent emission of bulk polar optical phonons (POP) and acoustic phonons (AP). It is worth noting that the emission rate corresponding to inter-subband transition $SPP_{2-1}$ from the second to the first subband which generates an SPP with short wave vector is the only scattering rate that decreases when initial electron energy increases (fig. 3). It is also shown that for low energy electrons, SPP emission rates values are about one decade less than that of POP emission which is the dominant scattering mechanism in III-V materials. However, these two emission rates become similar at high energy.

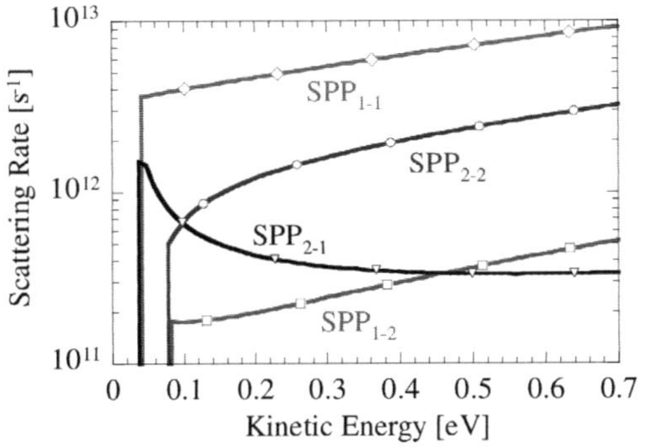


**Figure 3** SPP/electron intra-subband and inter-subband emission rates.

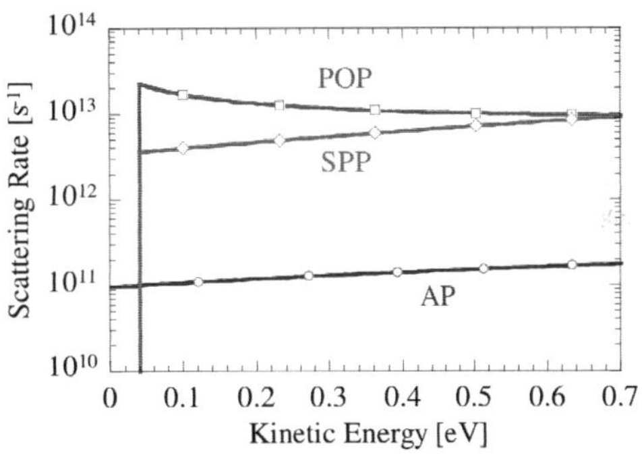


**Figure 4.** Scattering rate of most frequent AP, POP and SPP processes in the 1st subband.

## 3. MSMC simulation results

To analyse the electronic transport and to assess the potentiality of hot electron to enhance the SPP emission, a Multi-Subband Monte Carlo simulator including SPP scattering rates was applied for the device presented in fig. 2. First, a 1D Schrodinger/Poisson simulation was performed along the $z$ axis to provide quantum energy levels and electron wave functions for the biased structure based on a typical AlGaAs/GaAs HEMT. These wave functions are used to calculate the SPP scattering rates within the model developed in Section 2. Frozen-field MSMC simulations were then performed. Such type of simulation solves the Boltzmann equation without any assumption on the distribution function, which actually allows a qualitative and quantitative analysis of electron/SPP interaction properties.

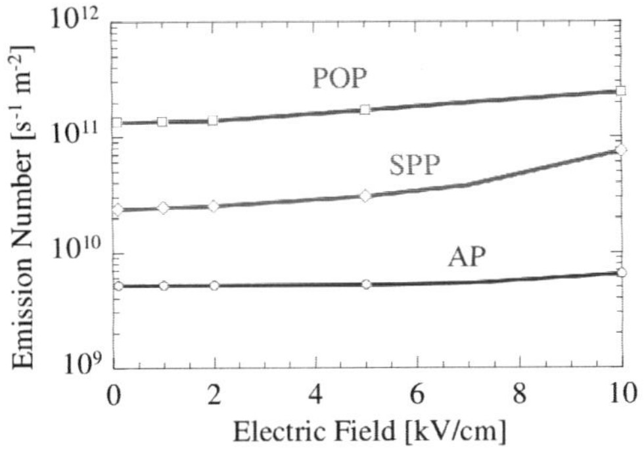


**Figure 5.** Number of SPP, POP and AP emitted per second and per surface area as function of electric field $E_y$. $\Delta E_{12}$ is fixed at $\hbar\omega_{SPP}$.

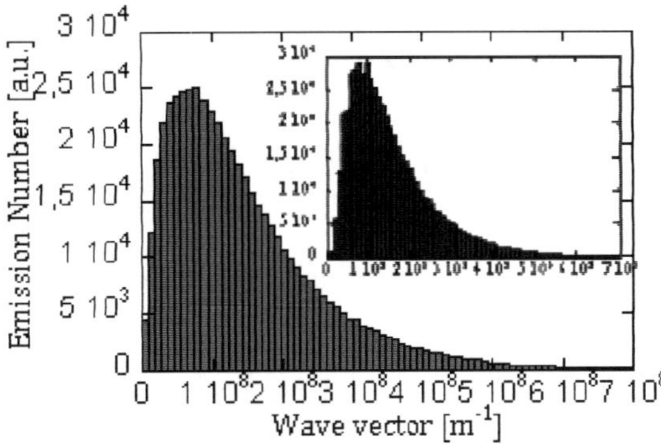


**Figure 6.** Spectrum of emitted SPP wave vector for $E_y = 1$kV/cm and $\Delta E_{12} = \hbar\omega_{SPP}$. In inset: same spectrum for $\Delta E_{12} = 30$ meV

The numbers of emitted phonons (SPP, POP and AP) per second and per unit area of surface as a function of the electric field in the transport direction are plotted in figure 5. Consistently with scattering rates of Fig. 4, the number of POP emissions is always less than one decade smaller than that of SPP emissions. However, because of decreasing POP emission rate at high energy, increasing the electric field is more favourable to the SPP emission than to the POP one.
To produce THz radiations, short wave vector SSPs may be collected by a grating structure at the top of the device. To evaluate this possibility the SPP wave vector distribution has been investigated together with its dependence on the energy difference $\Delta E_{12}$ between the 1st and 2nd subbands. Maximum of short wave vector SPP emission, useful for THz emission, is obtained for $\Delta E_{12}$ just equal to the SPP energy $\hbar\omega_{SPP} = 36\,\text{meV}$ (fig. 6). For other values of $\Delta E_{12}$ (e.g. as in inset of fig. 6, for $\Delta E_{12} = 30$ meV), the most frequent scattering mechanism (not shown) is the intra subband band SPP emission ($SPP_{1\text{-}1}$ in figure 3). This transition generates SPPs with wave vector on the order of magnitude of electron wave vectors i.e. equal to about $10^8$ m$^{-1}$. SPPs with short wave vector are more efficiently generated by electron transitions from 2nd to 1st subband. This scattering mechanism is enhanced for $\Delta E_{12} = \hbar\omega_{SPP}$. In this case, high electron kinetic energy is not necessary to emit SPPs because of decreasing $SPP_{2\text{-}1}$ scattering rate (see Figure 3). This low wave vector phonon emission depends on $\Delta E_{12}$ which is controlled by the gate voltage. To make possible the THz radiation from SPPs of wave vector as close as possible to $10^8$ m$^{-1}$ the grating structure should be optimized with a period smaller than 50 nm. In this case, one can benefit from a strong intra-subband emission of SPPs by high energy electrons of the first subband.

## 4. Conclusion

Theoretical estimation of the emission of SPPs via intra- and inter-subband interactions in an AlGaAs/GaAs two-dimensional electron gas has been carried out using appropriate multi-subband Monte Carlo simulation. Our calculations show that SPP emissions are not negligible compared to other mechanisms of phonon emission, which supports the idea that electrically-induced SPPs have the potential to generate THz radiations. Our results also suggest that a device appropriately biased to favour the non-stationary transport of hot electrons may be helpful to generate a large amount of SPPs and thus to design an efficient THz source. The other key point is then the design of the diffraction grating structure which has to couple efficiently SPPs with photons. Further theoretical works are needed to model accurately the effects on SPP emission of the gate and of the material stack, including the presence of the additional grating structure. Next studies will focus on alternative materials and multi-interface structures to explore more deeply the potentialities of electronic SPP emission.

**Acknowledgments**
This work was partially supported by the French ANR through project ANR-07-NANO-043 LAPSUS.

**References**
[1] MacDonald K F, Samson Z L, Stockman M I, and Zheludev N I 2008 Ultrafast active plasmonic *Nature Photonics* **3** 55-58
[2] Kazantsev D 2006 Phonon-polariton waves on the surface of SiC crystal *JETP Letters* **83** 323-6
[3] Greffet J-J, Carminati R, Joulain K, Mulet J-P, Mainguy S and Chen Y 2002 Coherent emission of light by thermal sources *Nature* **416** 61-4
[4] Saint-Martin J, Bournel A, Monsef F, Chassat C and Dollfus P 2006 Multi sub-band Monte Carlo simulation of an ultra-thin double gate MOSFET with 2D electron gas *Semiconductor Science and Technology* **21** L29-L31
[5] Archambault A 2008 Éléments de matrice du couplage entre un puit quantique et les plasmons de surface d'une interface plane (*Internal report)*
[6] Blakemore J S 1982 Semiconducting and other major properties of gallium arsenide *Journal of Applied Physics* **53** R123-R81

# Experimental demonstration of resonant-tunneling-diode operation beyond quasibound-state-lifetime limit

**M. N. Feiginov**
Technische Universität Darmstad, 64283 Darmstadt, Germany
E-mail: feiginov@ont.tu-darmstadt.de

**D. Roy Chowdhury**
Universitaet Duisburg-Essen, 47057 Duisburg, Germany

**Abstract.** We show, first, that the charge relaxation (response) time of resonant-tunneling diode (RTD) can be significantly shorter or longer than the resonant-state lifetime, depending on RTD operating point and RTD parameters. Coulomb interaction of electrons is responsible for the effect. Second, we demonstrate that the operating frequencies of RTDs are limited neither by resonant-state lifetime, nor by relaxation time; particularly in the RTDs with heavily doped collector, the differential conductance can stay negative at the frequencies far beyond the limits imposed by both time constants. We provide experimental evidences for both effects.

## Introduction

Resonant-tunneling diodes (RTDs, see Fig. 1) are the most simple structures with resonant tunneling. Such structures were intensively studied in the last several decades, they are used for studies of fundamental aspects of tunneling and they are also attractive for practical applications. E.g., RTDs belong to the fastest [1] operating devices nowadays and their negative differential conductance (NDC) is useful for high-frequency oscillators [2, 3]. The question that arises in connection with any electronic device: what is limiting its response time and its operating frequencies? The response of RTD is connected to population and depopulation of the resonant states in the quantum well (QW) of the diode. Therefore, it is typically accepted [4, 5] that the inherent response-time limitation of such structures should be determined by the resonant-state lifetime ($\tau$). Also it is intuitively expectable [5] that the current and conductance of the structure should tend to zero, when $\omega\tau \gg 1$.

One of us has shown theoretically previously that the intuitive picture is not generally correct [6, 7, 8]. The objective of the present work is to demonstrate that experimentally. We show that the response time or, equivalently, the time constant of charge-relaxation processes ($\tau_{rel}$) can be significantly shorter than $\tau$. Such behavior is the consequence of the Coulomb interaction between the charge carriers in the structures [6, 7]. We also show that, under certain conditions, the Coulomb interaction can lead to the opposite effect, when $\tau_{rel}$ becomes longer than $\tau$. Additionally, we demonstrate that, as predicted [8], the resonant-tunneling conductance can stay large and it can stay even negative in RTDs, when $\omega\tau \gg 1$ and $\omega\tau_{rel} \gg 1$. Such effects have not been observed till recently [9].

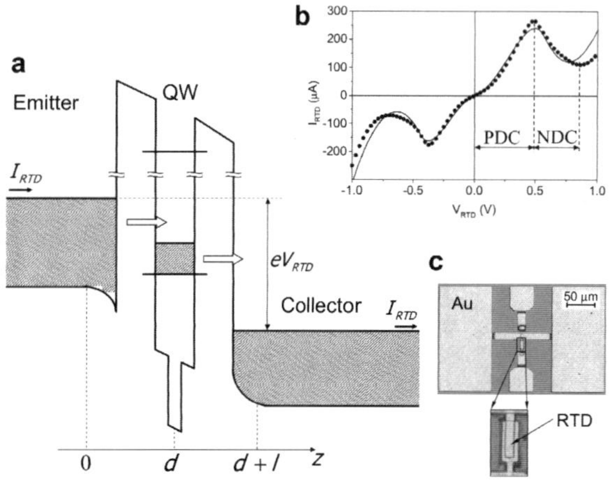


**Figure 1.** **a**, Schematic band diagram of the fabricated RTD. Panel **b** shows measured (dots) and simulated (solid line) room-temperature I-V curves of $3 \times 15$ $\mu m^2$ RTD with the following layer sequence: QW with 1.4 $nm$ $In_{0.53}Ga_{0.47}As$ / 3.0 $nm$ $In_{0.68}Ga_{0.32}As$ / 1.4 $nm$ $In_{0.53}Ga_{0.47}As$ layers is sandwiched between $\approx 3.5$ $nm$ (nominal value) $AlAs$ barriers and $In_{0.53}Ga_{0.47}As$ emitter and collector regions n-doped at the level of $10^{18}$ $cm^{-3}$ with 1.5 $nm$ spacers close to the barriers. **c**, The photograph of fabricated RTDs with connecting circuitry for microwave-probe measurements.

## 1. Experimental

For the experimental demonstration of the above effects, we were studying InGaAs/AlAs RTDs with heavily doped collector specially designed (see subscript to Fig. 1) to have approximately equal effective thicknesses (including the screening and depletion lengthes) of the emitter ($d$) and collector ($l$) barriers. $l \approx d$ is a necessary condition to achieve NDC when $\omega\tau_{rel} \gg 1$ and $\omega\tau \gg 1$ [9]. Without loss of generality of the obtained results and for the sake of ease of measurements only, the barriers of RTDs were made thick ($\approx 3.5$ $nm$), so that the value of $\tau$ is $\approx 100$ $ps$. For such diodes, the conditions $\omega\tau \gg 1$ and $\omega\tau_{rel} \gg 1$ are already well satisfied at the frequencies below 10 $GHz$ and the effects we are looking for could be demonstrated with standard microwave measurement techniques.

The layout of the fabricated RTD structures is shown in Fig. 1c. RTD is connected via air bridges and other metal lines to the contact pads for the standard microwave-probe measurements. The admittance of RTD with the surrounding parasitic circuitry has been measured with a conventional vector network analyzer. To determine the contribution of the parasitic circuitry, the additional short and open structures were fabricated and measured, where RTD is replaced by a short (for the measurement of the inductive contribution) or open (for the capacitive contribution) circuits. The parasitic parameters measured in such a way are supported by simulation results (with CST Microwave Studio). When the parasitic contributions to the measured admittance are excluded, we get the resultant "pure" RTD admittance, see Fig. 2. The plots of conductance in the NDC region clearly show that RTD conductance stay negative in the whole measurement range of 12 $GHz$, where the conditions $\omega\tau_{rel} \gg 1$ and $\omega\tau \gg 1$ are satisfied.

Further, we simulate the static and dynamic characteristics of RTDs. The sequential-tunneling approximation is used in our self-consistent static simulations, we take into account the nonparabolicity in the barriers, finite (room) temperature, thermionic emission over the barriers, tunneling through the second subband in QW. The simulated curves are in very good agreement with the measured ones, they describe well the peak and valley regions of the I-V curves both

Journal of Physics: Conference Series **193** (2009) 012016 doi:10.1088/1742-6596/193/1/012016

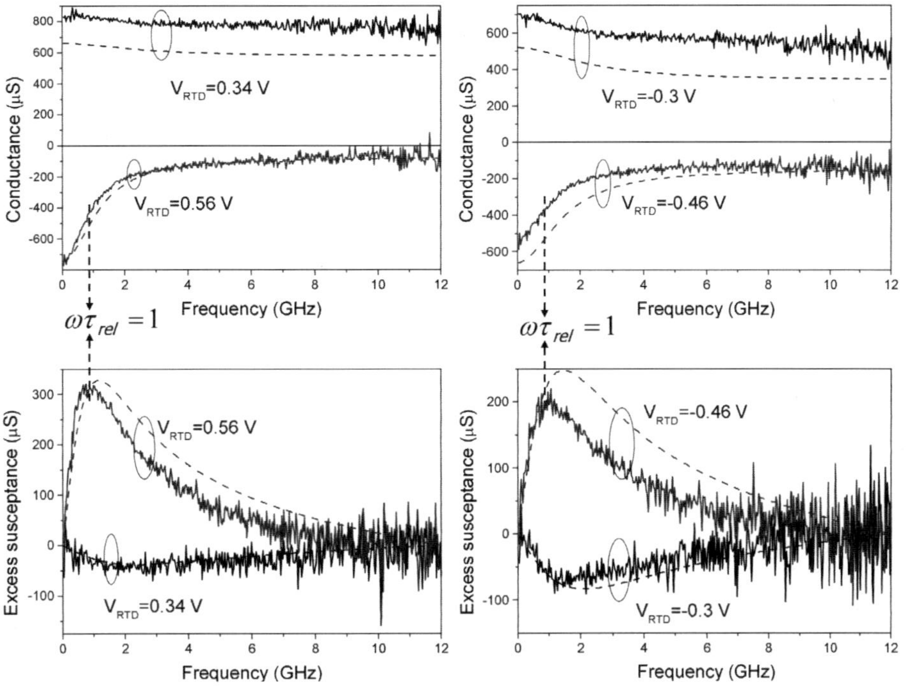


**Figure 2.** The plots of measured (solid lines) and simulated (dashed lines) RTD conductances ($Re(G)$) and excess susceptances ($Im(G) - \omega C_{RTD}$) for forward (0.34 $V$ and 0.56 $V$) and reverse ($-0.3$ $V$ and $-0.46$ $V$) biases, $C_{RTD}$ is the geometrical RTD capacitance. The biases of 0.34 $V$ and $-0.3$ $V$ are in the PDC region of the I-V curve (see Fig. 1b) and those of 0.56 $V$ and $-0.46$ $V$ are in NDC region. The RTD was designed in such a way that the values of $1/2\pi\tau$ and $1/2\pi\tau_{rel}$ in NDC region are close to 1 $GHz$. The plots of RTD conductance at 0.56 $V$ and $-0.46$ $V$ show that the conductance stays negative, even when $\omega\tau_{rel} \gg 1$ and $\omega\tau \gg 1$: $\omega\tau_{rel} \approx 12$ and $\omega\tau \approx 8$ at 12 $GHz$.

for the forward and reverse biases applied to RTD (see Fig. 1b). The RTD parameters used in the simulations (the barrier thicknesses are 3.3 $nm$ and 3.2 $nm$) are in good agreement with the nominal MBE growth parameters and the parameters of the layers determined by X-ray diffraction on the calibration test structures (the barriers were 3.6 $nm$ and 3.5 $nm$). Further, the AC characteristics of RTD have been calculated relying on our previously developed theory [7, 8]. The simulated and measured AC admittances are in good agreement with each other both in the positive differential conductance (PDC) and NDC regions and also for the forward and reverse biases, see Fig. 2.

Knowing the RTD parameters from the simulations and comparison with experimental data above, we calculate $\tau$ and $\tau_{rel}$ for our diodes, Fig. 3. The values of $\tau_{rel}$ can be also determined from the admittance plots (Fig. 2). As it has been shown by one of us theoretically [7, 8], the characteristic conductance roll-off frequency and the peak in the excess susceptance (see Fig. 2) correspond to the condition $\omega\tau_{rel} = 1$. The experimental values of $\tau_{rel}$ determined in such a way are also plotted in Fig. 3. The figure shows that $\tau_{rel}$ determines the roll-off of the AC characteristics of RTD, while $\tau$ has even qualitatively different behavior. The value of $\tau$ is decreasing with bias as a consequence of the decrease of the collector-barrier height, whereas

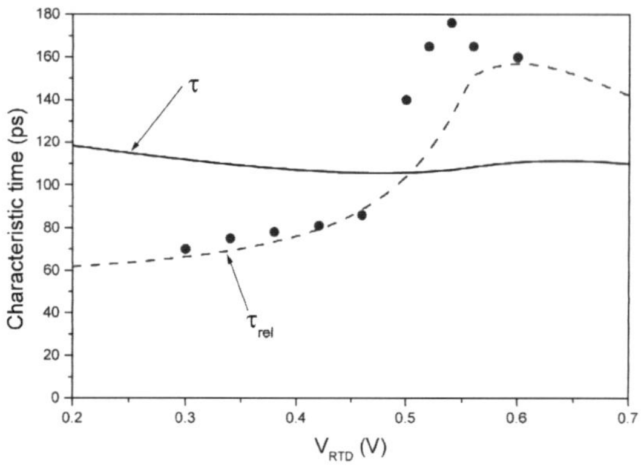


**Figure 3.** The plots for simulated values of $\tau$ (solid line) and $\tau_{rel}$ (dashed line) for forward bias. The values of $\tau_{rel}$ determined from the experimental RTD conductance and susceptance data (see Fig. 2) are marked with dots.

$\tau_{rel}$ is increasing, as a consequence of the Coulomb-interaction effects [6, 7], when one goes from PDC to NDC region of the I-V curve. The experimental and theoretical values of $\tau_{rel}$ are in quantitative agreement with each other. Both the simulations and measurements do indeed show that $\tau_{rel}$ is by a factor of $\approx 2$ shorter (longer) than $\tau$ in the PDC (NDC) regions.

Although the measurements presented here are done at relatively low frequencies, the same effects should be observable also at much higher frequencies. E.g., the same RTDs with 1.5 $nm$ barriers would have $\tau \approx 0.5$ $ps$ and $1/2\pi\tau \approx 300$ $GHz$. One of the practical consequences of the presented results is that the frequency range, where the differential conductance of RTD stays negative and realization of RTD oscillators is possible, should be limited neither by $\tau$ nor by $\tau_{rel}$ in RTDs with heavily doped collectors. Such diodes should allow one to overcome the upper frequency of $\approx 0.7$ $THz$ achieved [2, 3] with RTD oscillators till now. Additionally, although we present in this work the data on the most simple resonant-tunneling structure and at relatively low frequencies, the observed effects (Coulomb shortening of the relaxation time and conductance beyond the resonant-state-lifetime limit) are also expected in the more complicated structures (multi-barrier structures, single-electron-transistor-like structures, quantum-cascade lasers, etc.) and at the frequencies much higher than those covered in the present work and even in the THz frequency range.

The financial support of DFG and the Alexander von Humboldt Foundation are gratefully acknowledged.

[1] Scott J S, Kaminski J P, Wanke M, Allen S J, Chow D H, Lui M and Liu T Y 1994 *Appl. Phys. Lett.* **64** 1995–1997
[2] Brown E R, Söderström J R, Parker C D, Mahoney L J, Molvar K M and McGill T C 1991 *Appl. Phys. Lett.* **58** 2291–2293
[3] Asada M, Suzuki S and Kishimoto N 2008 *Jpn. J. Appl. Phys.* **47** 4375–4384
[4] Sollner T C L G, Brown E R, Goodhue W D and Le H Q 1990 *Physics of Quantum Electron Devices* (*Springer Series in Electronics and Photonics* vol 28) ed Capasso F (New York: Springer-Verlag)
[5] Brown E R, Parker C D and Sollner T C L G 1989 *Applied Physics Letters* **54** 934–936
[6] Feiginov M N 2000 *Applied Physics Letters* **76** 2904–2906
[7] Feiginov M N 2000 *Nanotechnology* **11** 359–364
[8] Feiginov M N 2001 *Applied Physics Letters* **78** 3301–3303
[9] Feiginov M N and Chowdhury D R 2007 *Applied Physics Letters* **91** 203501 (pages 3)

Journal of Physics: Conference Series **193** (2009) 012017 doi:10.1088/1742-6596/193/1/012017

# Study of the high-frequency performance of III-As nanojunctions using a three-dimensional ensemble Monte Carlo model

**Toufik Sadi and Jean-Luc Thobel**

Institut d'Electronique, de Microélectronique et de Nanotechnologie (IEMN), UMR-CNRS 8520, Université Lille 1, 59652 Villeneuve d'Ascq Cédex, France

E-mail: Toufik.Sadi@IEMN.Univ-Lille1.fr

**Abstract.** We apply a well-calibrated three-dimensional Monte Carlo simulator using finite-element meshing to study the intrinsic high-frequency (HF) behaviour of three-terminal nanojunctions based on InGaAs/InAlAs heterostructures. To obtain a reliable prediction of device performance, we use a realistic model for surface charge effects at high-frequency conditions. In this work, we perform an analysis of the dynamics of electron transport in the devices and present a prediction of their intrinsic HF performance. The results demonstrate the suitability of these nanostructures for application in the terahertz regime, and illustrate the influence of surface charge effects in this frequency range.

## 1. Introduction

It is widely accepted that future improvement in device performance should not be expected from size reduction but rather from profound change in geometry. In this context, the development of reliable three-dimensional (3D) simulation tools are essential to help designers to determine the best structures, and provide insight into their physical behaviour which is still not completely understood especially at high-frequencies. The use of nanojunctions has been suggested as one possible solution to the scaling problem. Such structures have attracted substantial interest in recent years, due to their non-linear properties [1]. We report results from the study of the intrinsic high-frequency (HF) behaviour of T-branch junctions (TBJs) based on InGaAs/InAlAs heterostructures, using a 3D Monte Carlo simulator. TBJs incorporating InGaAs channels are promising for terahertz applications due to the high 2DEG mobility in these structures. These nanodevices function as voltage rectifiers, under a push-pull biasing scheme, due to quasi-ballistic transport and surface charge effects in these devices. TBJs can also be used in logic circuits as Multiplexer/Demultiplexer (MUX/DEMUX) structures, as discussed below. The simulated TBJs are shown in Fig. 1. Details of the InGaAs/InAlAs heterostructure are given in [1].

## 2. Simulation Model, Results and Discussion

The simulator self-consistently couples 3D non-equilibrium electron dynamics with a solution of the 3D Poisson's equation using the finite-element method [2]. As demonstrated in [1], the correct inclusion of the real device geometry in the 3D space is necessary for the accurate analysis of electron transport, which is important to understand the physics of these devices. The model

Journal of Physics: Conference Series **193** (2009) 012017 doi:10.1088/1742-6596/193/1/012017

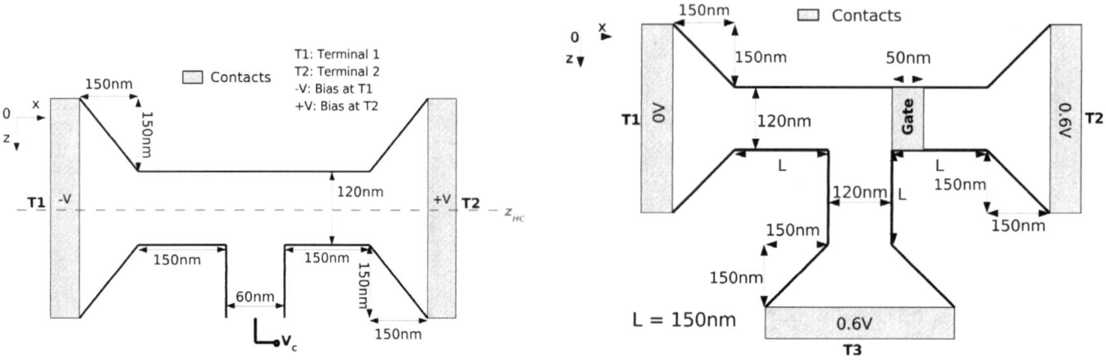


**Figure 1.** (left figure) TBJ structure simulated by applying a push-pull biasing scheme to study rectification, by observing the output potential $V_C$. (right figure) The TBJ structure with a gate contact used as a MUX/DEMUX structure. See [1] for more details about these structures.

minimises significantly the need for parameter fitting to match experimental results, and leads to more realistic macroscopic and microscopic characteristics. The simulator incorporates a realistic surface charge model designed specifically for HF simulations. This is necessary since surface charge effects play an important role in the performance of these devices characterised by a relatively high surface-to-volume ratio. The newly-developed charge model is partially based on the model used in our previous work for DC analysis [1], and allows the adjustment of surface charge density according to electron concentrations near the free surfaces. To reduce the computational cost, the applied potentials at terminals $T_1$ ($V_L$) and $T_2$ ($V_R$) are assumed to follow a squarewave form, with the same frequency $f$. In general, the biasing scheme chosen such that push-pull polarisation is maintained at every instant of the simulation. In the first half-period (case 1) $V_L$ is set to $-V_1$ and $V_R$ is set to $+V_1$, while in the second half-period (case 2) $V_L$ is set to $+V_2$ and $V_R$ is set to $-V_2$. The HF charge model is applied as follows. First, we perform DC simulations to determine the fixed charge distribution in each surface of the device, for both biasing cases (we use $\sigma_1$ to define the charge density at a given surface for case 1, and $\sigma_2$ to define the charge density for case 2). Since little is known about the fixed charge at the surfaces we suggest three models for considering surface effects at HF conditions. The first model (Model I) allows instantaneous change of $\sigma$ (from $\sigma_1$ to $\sigma_2$ or vice-versa) with the applied potential. The second model (Model II) allows a time-dependent change of $\sigma$: if a change in the applied potential occurs from case $m$ to case $n$ ($m = 1, 2$, $n = 1, 2$), $\sigma$ is given by

$$\sigma = \sigma_m + [(\sigma_n - \sigma_m)][(1 - \exp\{-t/\tau\} + \exp\{-(T/2)/\tau\}/2)]. \tag{1}$$

$t$ is the time elapsed since the last potential change, $\tau$ is the time constant and $T$ is the time period. The third model (Model III) sets the value of $\sigma$ to the average value $(\sigma_1 + \sigma_2)/2$. Clearly, Models I and III represent special (opposite) cases of Model II, corresponding to $\tau \sim 0$ and $\tau \sim \infty$, respectively. The simulations are run assuming room temperature (300 K).

For the analysis of the simple TBJ structure (without a gate contact) shown in Fig. 1, we generate results at 1THz. The biasing scheme is chosen such that push-pull polarisation ($V_L = -V$ and $V_R = V$) is satisfied in the first half-period. In the second half-period, both $V_L$ and $V_R$ are set to zero. In the case of Model II, $\tau$ is set to $T$ for the sake of illustration. Fig. 2 shows the variation of the average $V_C$ (representing the central potential at the bottom of the vertical branch) with bias for the three charge models, for 1THz as well as for the DC case. In Fig. 2, we also show the time dependence of $V_C$, at a high bias. The results demonstrate how the choice of the charge model influences considerably the predicted results. Model I gives a

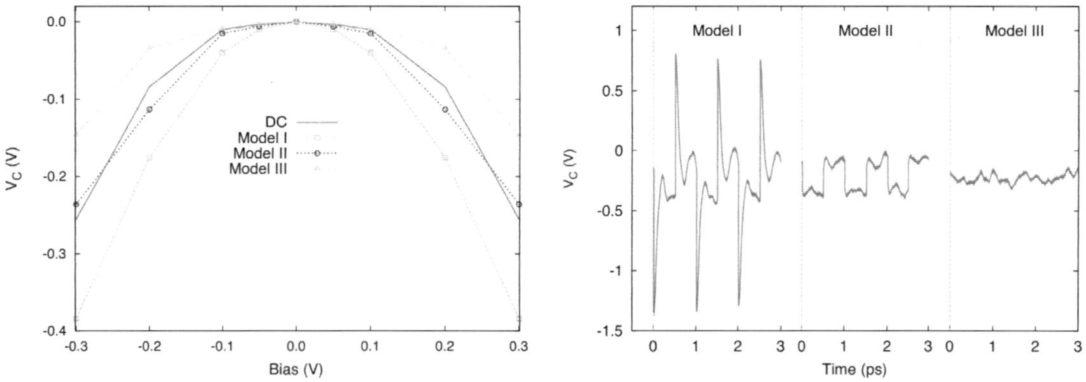


**Figure 2.** (left figure) Variations of the average $V_C$ with $V$. (right figure) Variations of $V_C$ with time at $V = 0.3$V. The results are obtained from the three surface charge models, for a frequency of 1THz. Results for the DC case are also shown in the $V_C$ vs $V$ figure.

strong time-dependence of $V_C$. In this case, the electron population does not settle to a steady state, leading to unrealistic average $V_C$ values. Model III gives a much less time-dependence of $V_C$. While both models represent extreme cases of the influence of surface charge, model III may be more suitable at high frequencies, since the surface charge is expected to respond relatively slowly to bias changes. At 1THz, this model is shown to give results that are similar in both half-periods, suggesting that surface charge effects may be an important factor in determining the HF behaviour of TBJs. Such possibility leads to an important conclusion: For an improved performance at high-frequencies, future designs should incorporate a reduced junction length to minimise the influence of surface charge effects, and enhance the role of ballistic transport on the switching behaviour of TBJs. In general, realistic results are expected to lie between these two extremes, by choosing, for example, a time constant value $\tau$ (see Fig. 2) from calibration with HF experimental data, which are not available at the moment.

For a deeper understanding of the physics governing device operation at high frequencies, we show in Fig. 3 the variation of the mean electron velocity and energy along the channel. The results are taken at the $z$ coordinate ($z_{HC}$) corresponding to the middle of the device (see Fig. 1). These results are shown for the first-half period ($V_L = -0.3$, $V_R = +0.3$). As discussed above, model I leads to unrealistic results. In this case, microscopic results show a significant reduction in the average velocity, which is verified to be due to an increase in electron transitions to the upper valleys. Analysis of the microscopic results obtained from models II and III show the expected asymmetry in velocity and energy. The average (and peak) velocities and energies are reduced at higher frequencies. The energy profiles, showing a constant increase in electron energy in the junction, are maintained at 1THz. These profiles, which are at the origin of the rectifying behaviour of TBJs, suggest that such devices maintain their non-linear behaviour at high frequencies, and therefore are suitable for terahertz applications.

TBJs can be used as MUX/DEMUXs for logic applications [3], by introducing a Schottky gate contact in the structure, as explained in [1]. We apply the simulator to demonstrate the device operation at high frequencies but also the influence of surface charge effects. Fig. 1 shows this structure and the biasing scheme used for this study. The applied gate potential $V_G$ is assumed to follow a squarewave form with a frequency of 1THz. The magnitudes are +0.75V in the first half-period and -0.75V in the second half-period. When $V_G$ is positive most of the current flows through terminal T2 ($I_2$), and when $V_G$ is negative most of the current flows through terminal T3 ($I_3$). This explains its characteristic as a MUX/DEMUX. Fig. 4 shows the variations of the current with time, when using models I and II ($\tau = T$). While

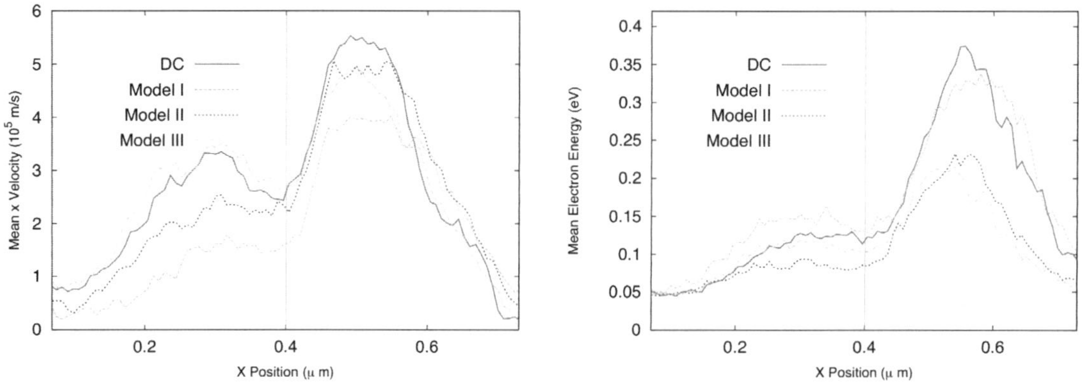


**Figure 3.** Variations of (left figure) the mean electron velocity in the $x$ direction and (right figure) kinetic energy in the channel at $z = 0.21\mu$m and $V = 0.3$V. The results are obtained from the three surface charge models, for a frequency of 1THz. DC results are also shown. The vertical line shows the centre of the device in the $x$ direction.

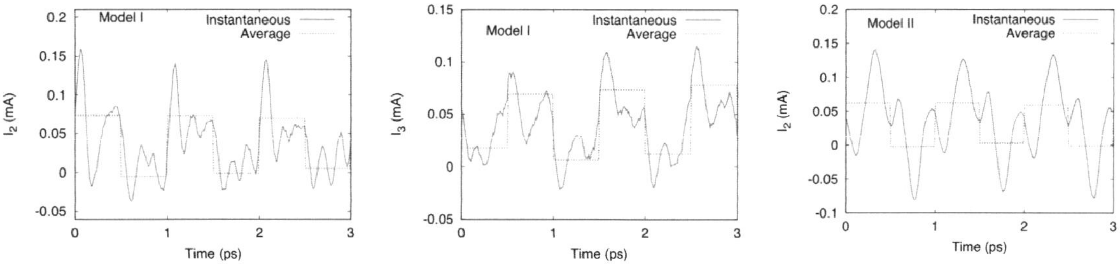


**Figure 4.** Time-variations of (left figure) $I_2$ and (middle figure) $I_3$, for a 1THz frequency and surface charge model I. The figure on the right shows the variation of $I_2$, for surface charge model II, for the same frequency.

the MUX/DEMUX function is maintained at 1THz, surface charge models give considerably different average current values, demonstrating the importance of understanding surface charge effects for a complete understanding of the behaviour of nanojunctions at high-frequencies.

## 3. Conclusions

This work presents an analysis of the intrinsic high-frequency behaviour of III-As TBJs, using a 3D Monte Carlo simulator incorporating a finite-element Poisson solver. Results demonstrate how the behaviour of the devices as switches is sustained at the terahertz frequency range, and most importantly illustrate the role played by surface charge effects in this regime.

## Acknowledgments

The authors would like to thank F. Dessenne from the IEMN for valuable discussions.

## References

[1] Sadi T, Dessenne F and Thobel J L 2009 *J. Appl. Phys.* **105** 053707
[2] Gerstmayr J and Schöberl J 2006 *Multibody. Syst. Dyn.* **15** 309
[3] Mateos J, Vasallo B G, Pardo D, Gonźales T, Galloo J S, Roelens Y *et al 2003 Nanotechnology* **14** *117*

# Monte Carlo analysis of Gunn oscillations in narrow and wide band-gap asymmetric nanodiodes

**T. González[1], I. Iñiguez-de-la Torre[1], D. Pardo[1], J. Mateos[1] and A. M. Song[2]**

[1] Departamento de Física Aplicada, Universidad de Salamanca, Plaza de la Merced s/n, 37008 Salamanca, Spain
[2] School of Electrical and Electronic Engineering, University of Manchester, Manchester M60 1QD, United Kingdom

E-mail: tomasg@usal.es

**Abstract.** By means of Monte Carlo simulations we show the feasibility of asymmetric nonlinear planar nanodiodes for the development of Gunn oscillations. For channel lengths about 1 μm, oscillation frequencies around 100 GHz are predicted in InGaAs diodes, being significantly higher, around 400 GHz, in the case of GaN structures. The DC to AC conversion efficiency is found to be higher than 1% for the fundamental and second harmonic frequencies in GaN diodes.

## 1. Introduction

Gunn oscillations (GOs) are typically used to produce microwave emitters. Increasing the oscillation frequency to enter the THz range is a challenging issue. Materials exhibiting negative differential mobility and high saturation velocity are appropriate to this end. GaN, with a saturation velocity about $2 \times 10^7$ cm/s and a short energy relaxation time about 0.15 ps, is a good candidate. However, despite big efforts spent in the last years [1,2], mainly focused on vertical structures, only very recently promising indications of GOs were experimentally observed in GaN diodes [3]. In the case of planar structures, GOs in GaAs/AlGaAs heterodiodes were also recently measured [4].

The asymmetry in the channel of a self-switching diode (SSD), a lately developed planar non-linear nanodevice with the topology shown in figure 1 [5], is especially appropriate for the development of GOs, as demonstrated numerically in [6] for the case of a diode with an InGaAs channel. Higher frequencies are expected in SSDs with GaN channels by virtue of its superior saturation velocity and the reduction of the dead space needed for the onset of the oscillations. Moreover, because of the planar design of SSDs, heat dissipation, a technological challenge in power devices, could be managed in an efficient way by a correct design of the channels and their separation.

Monte Carlo (MC) simulations have proved to be quite useful for the analysis of the static, dynamic and noise behavior of SSDs [7,8], and have been commonly used to study GOs in vertical GaN diodes [9]. In this work we demonstrate, by means of MC simulations, the feasibility of SSDs for the development of GOs, both for narrow (InGaAs) and wide (GaN) band-gap channel materials.

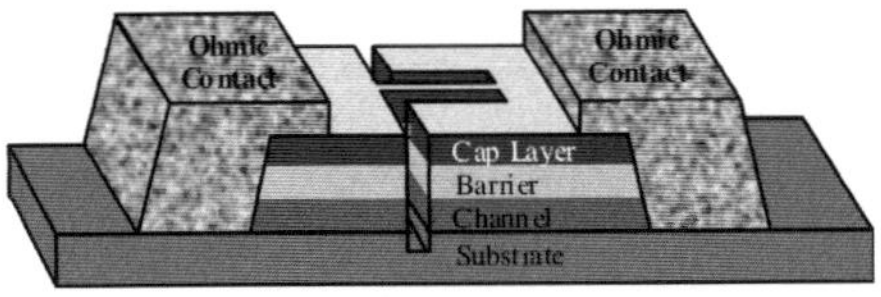


**Figure 1.** Three-dimensional geometry of the SSD. The SSD is fabricated with just one lithographic step, by simply etching L-shaped insulating grooves onto a semiconductor layer to define a narrow channel with broken symmetry.

## 2. Monte Carlo model

A semiclassical MC simulator self-consistently coupled with a 2D Poisson solver is used for the analysis. To account for the 3D nature of the diodes, a *background doping* $N_{Db}$ is considered when solving Poisson equation and a negative charge density $\sigma$ is placed at the semiconductor-dielectric boundaries of the insulating trenches. Details of the model can be found in [10].

In the case of the InGaAs structure, we have considered $N_{Db}=10^{17}$ cm$^{-3}$ and $\sigma$=-0.2x10$^{12}$ cm$^{-2}$, while for the GaN one $N_{Db}$=2x10$^{17}$ cm$^{-3}$ and $\sigma$=-0.2x10$^{12}$ cm$^{-2}$. The non-simulated dimension Z, which allows to determine the value of the current provided by a single SSD, is estimated as $Z=n_s/N_{Db}$, taking values of 2x10$^{-5}$ and 4x10$^{-5}$ cm in the InGaAs and GaN SSDs, respectively (for $n_s$=2x10$^{12}$ and 8x10$^{12}$ cm$^{-2}$, typical values of sheet electron density in InGaAs and GaN channels, respectively). All the simulations are performed at room temperature.

## 3. Results and discussion

### *3.1. InGaAs SSDs*

Initially, we analyze the case of an SSD with InGaAs channel. Figure 2(a) shows the geometry of the diode with the values of the main parameters. The channel length is $L_C$=1425 nm and the width is $W_C$=60 nm. It has been found that to ease the development of GOs a wide vertical trench should be used, $W_V$=200 nm in our case. The length of the anode access region [right contact in figure 2(a)] has been chosen to be long enough (625 nm) to allow electrons thermalize before reaching the contact, process which is rather slow and takes a long distance in InGaAs because of the low density of states in the Γ valley.

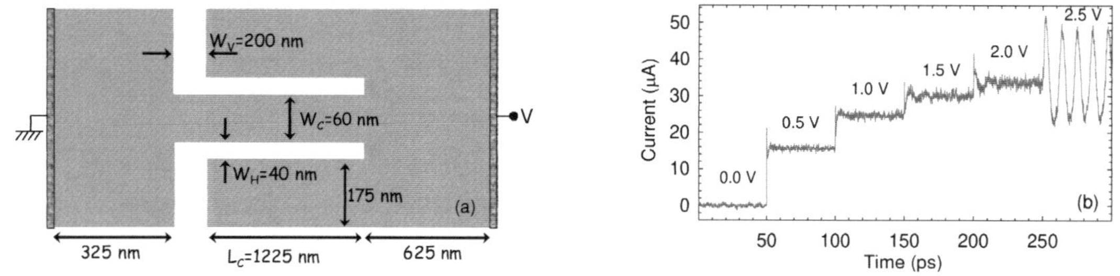


**Figure 2.** (a) Geometry of the InGaAs SSD. (b) Current sequence for $\Delta V$=0.5 V applied every 50 ps.

Figure 2(b) shows the current sequence obtained when the voltage applied to the SSD is incremented in steps of 0.5 V every 50 ps. Oscillations arise for voltages higher than 2.0 V. In particular, clear self-sustained oscillations are observed for 2.5 V, with a frequency of about 100 GHz, similar to that reported in [6]. The origin of the oscillations is the formation of domains of U-valley electrons near the entrance of the channel (where the field is higher due to the asymmetry of the diode) and their subsequent displacement along it, eventually evolving into dipoles. The oscillation frequency basically corresponds to the saturation velocity of InGaAs (about 10$^7$ cm/s) divided by the distance covered by the domains (about 1 μm).

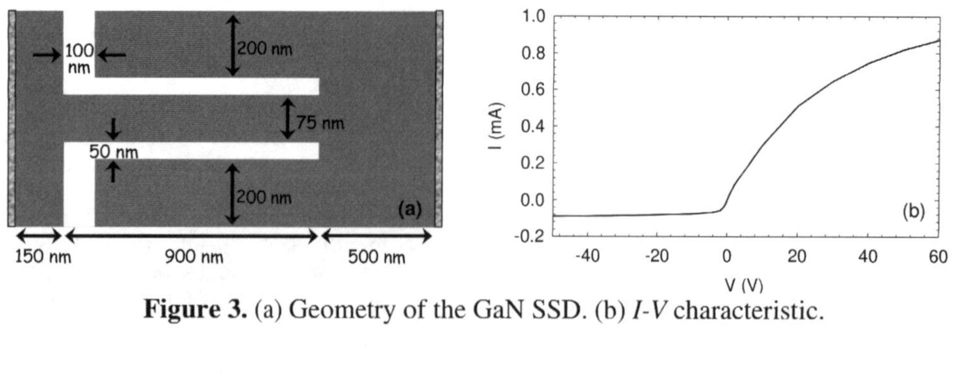


**Figure 3.** (a) Geometry of the GaN SSD. (b) *I-V* characteristic.

### *3.2. GaN SSDs*

In the case of GaN SSDs, we have analyzed the structure depicted in figure 3(a): a diode with channel length $L_C$=900 nm and width $W_C$=75 nm. Figure 3(b) shows the *I-V* curve obtained for this diode. The rectifying behaviour is due to electrostatic effects caused by the asymmetry of the structure [5,7]. Current sequences at several applied voltages are plotted in figure 4(a). As observed, when the applied voltage exceeds 30 V, current oscillations appear. The main frequency of such oscillations is represented in figure 4(b). Values around 400 GHz are obtained (much higher than in the InGaAs SSD), decreasing with the increase of the applied voltage due to a lower drift velocity of the domains.

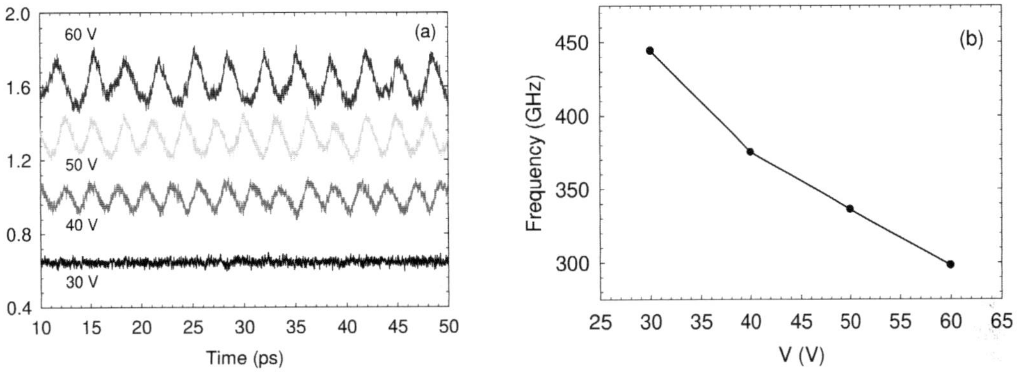


**Figure 4.** (a) Current sequences calculated in the diode of figure 3(a) for several applied voltages. Each sequence is displaced 0.25 mA with respect to the previous one in order to be clearly distinguished. (b) Frequency of the current oscillations vs. applied voltage.

To evaluate the DC to AC conversion efficiency ($\eta$=$P_{AC}$/$P_{DC}$) a standard procedure is used [9]: a single-tone sinusoidal potential with amplitude $V_{AC}$ is superimposed to a DC bias $V_{DC}$ (50 V in our case). Under such conditions, the dissipated DC power $P_{DC}$ and the time-average AC power $P_{AC}$ are evaluated. Positive values of $\eta$ correspond to resistive behavior of the diode, while negative values of $\eta$ indicate AC generation from DC. Figure 5(a) shows $\eta$ as a function of the frequency of the AC excitation for three different amplitudes. Negative values higher than 1% are achieved around 340 GHz (frequency of the oscillation taking place when only DC bias is present) and, interestingly, even larger efficiencies (about 2%) are obtained at around half such a frequency. As expected, by increasing the amplitude of the AC signal, the efficiency is enhanced. In figure 5(b), the phase shift between the current response and the AC excitation leading to negative values of $\eta$ at 340 GHz can be observed.

To identify the origin of the oscillations, figure 6 shows profiles of several microscopic quantities along the center of the channel obtained at different time moments during one period of the AC signal for the case $V_{DC}$=50 V, $V_{AC}$=10 V and *f*=340 GHz. As observed, high-field domains shift along the

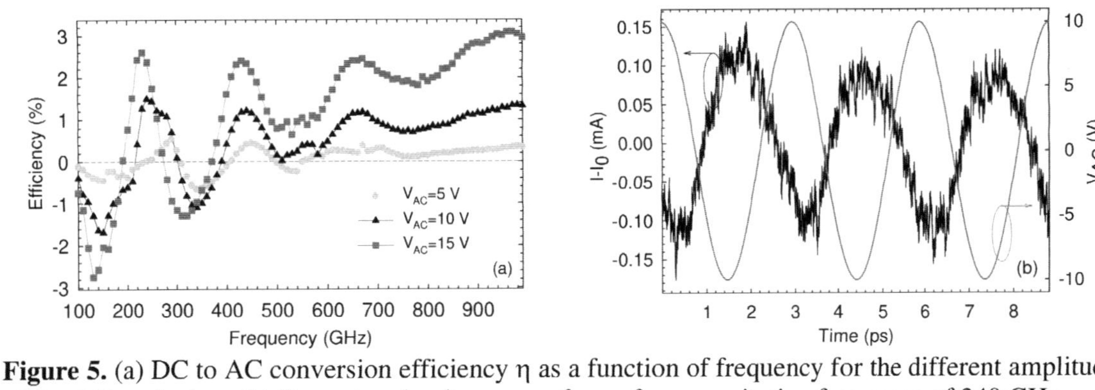


**Figure 5.** (a) DC to AC conversion efficiency η as a function of frequency for the different amplitudes of the AC excitation. (b) Current and voltage waveforms for an excitation frequency of 340 GHz.

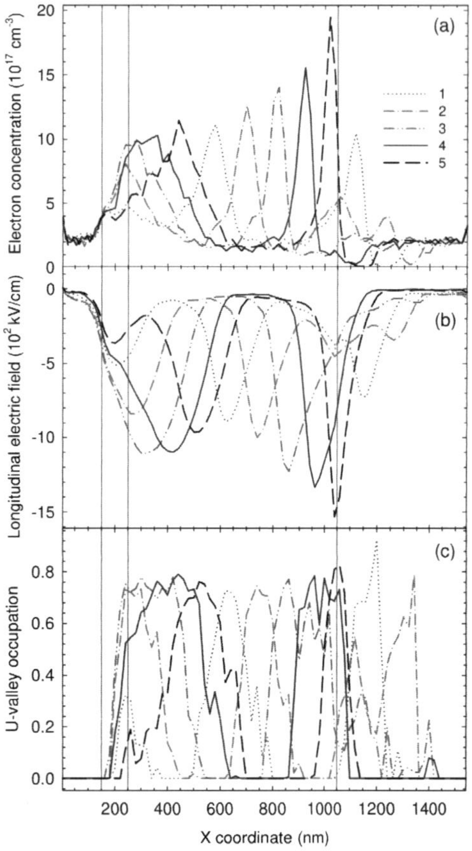


**Figure 6.** Profiles of (a) carrier concentration, (b) electric field and (c) U-valley occupation along the center of the channel for $V_{DC}$=50 V, $V_{AC}$=10 V and $f$=340 GHz at five equidistant time moments during one period of the excitation. Vertical lines indicate the position of the vertical trench and the end of the channel.

channel, coinciding with accumulations of slow electrons in the upper U-valley. The accumulations are originated at the entrance of the channel, where the electric field takes high values due to the presence of the vertical trench. This feature of the SSD geometry is especially appropriate for the development of the GOs. At this particular frequency, two domains coexist in the structure, which indicates that the oscillations observed under DC bias, taking place at 340 GHz for 50 V (figure 4), correspond to the second harmonic of the fundamental frequency of the diode. Indeed, the high efficiency observed in figure 5(a) at around 150 GHz corresponds to the presence of just one domain in the diode (fundamental frequency), which, in addition to the DC bias, requires an external AC component to be excited.

## 4. Conclusions

By means of MC simulations, the suitability of planar asymmetric nanodiodes for the development of GOs both in narrow and wide band-gap semiconductors has been demonstrated. In the case of GaN SSDs, oscillations at frequencies of several hundreds of GHz can be achieved. Arrays of SSDs in parallel can provide levels of current high enough for practical applications.

## Acknowledgments

This work has been partially supported by the Dirección General de Investigación (MEC) and FEDER (project TEC2007-61259/MIC and Acción Integrada HF2007-0014) and by the Consejería de Educación, Junta de Castilla y León (projects SA019A08 and GR270).

## References:

[1] Mutamba K, Yilmazoglua O, Sydloa C, Mira C, Hubbardb S, Zhaob G, Daumillerc I and Pavlidis D 2006, *Superlatt. Microstruct.* **40** 363
[2] Yilmazoglu O, Mutamba K, Pavlidis D and Karaduman T 2007 *Electronics Lett.* **43** 480
[3] Yilmazoglu O, Mutamba K, Pavlidis D and Karaduman T 2008 *IEEE Trans. Electron Devices* **55** 1563
[4] Pilgrim N J, Khalid A, Dunn G M and Cumming D R S 2008, *Semicond. Sci. Technol.* **23** 075013
[5] Song A M, Missous M, Omling P, Peaker A R, Samuelson L and Seifert W 2003 *Appl. Phys. Lett.* **83** 1881
[6] Xu K Y, Wang G and Song A M 2008 *Appl. Phys. Lett.* **93** 233506
[7] Mateos J, Vasallo B G, Pardo D and González T 2004 *Appl. Phys. Lett.* **86** 212103
[8] Iñiguez-de-la-Torre I, Mateos J, Pardo D and González T 2008 *J. Appl. Phys.* **103** 024502
[9] Sevik C and Bulutay C 2004 *J. Appl. Phys.* **85** 3908
[10] Mateos J, Vasallo B G, Pardo D, González T, Galloo J S, Roelens Y, Bollaert S and Cappy A, 2003 *Nanotechnology* **14** 117

Journal of Physics: Conference Series **193** (2009) 012019 doi:10.1088/1742-6596/193/1/012019

# Conductance crossovers in coherent surface transport on Y nanojunctions

**Andrea Bertoni[1], Giampaolo Cuoghi[1,2], Giulio Ferrari[3] and Guido Goldoni[1,2]**

[1] S3 National Research Center, INFM-CNR, 41125 Modena, Italy
[2] Dipartimento di Fisica, Università di Modena e Reggio Emilia, Modena, Italy
[3] CNISM Unità di Ricerca di Modena, Modena, Italy

E-mail: `andrea.bertoni@unimore.it`

**Abstract.** Conductance characteristics of a nonplanar two-dimensional electron gas (2DEG) can expose the role of its bending on the 2DEG electronic states. In particular, the presence of an effective geometric potential can be revealed. Here, we present a numerical study of the coherent electron transport on Y nanojunctions of three cylindrical 2DEGs, including a proposal for the experimental detection of the geometric potential. We describe the analytical approach leading to the reduction of the problem dimensionality from 3D to 2D and sketch our simulation scheme.

## 1. Introduction

The quantum dynamics of a particle constrained on a thin semiconductor layer can be effectively represented by a two-dimensional (2D) equation of motion, as far as the layer thickness is so small that the dynamics in the directions parallel to the layer can be decoupled from the orthogonal one, where only the ground state of the confinement potential is occupied. While for planar layers, a trivial 2D version of the effective-mass Schrödinger equation can be sufficient to get the main electronic-states features, a 2D system that is curved within the 3D space, requires a different model. In fact, the system curvature must be included into the 2D dynamics through an effective geometric potential (GP) that attracts a carrier towards the regions with stronger curvature. Furthermore, the equation of motion should be modified in order to account for the transformation between the 2D and 3D coordinate system, as we will illustrate in the next section.

Before proceeding, we stress that the topic introduced above has a relevance that goes beyond theoretical physics. In fact, a number of nanostructures have been recently realized where carriers are confined on a surface wrapped around a core semiconductor nanowire via epitaxial overgrown[1], and the controlled growth of branched nanowires has been clearly demonstrated[2]. The fabrication of Y junctions of cylindrical 2DEGs is possibly the next step of these experimental activities. We will show in Section 3 that transport characteristics of the latter Y junctions are able in principle to expose the presence of the GP. Indeed, although the GP has been conjectured long time ago[3] and it is commonly included in the modeling of curved 2D systems[4], its direct experimental evidence has not been achieved so far.

Finally, in Section 4 we briefly analyze the dependence of the junction coherent transmission on the angle between two of the branches.

## 2. Equation of motion on a curved surface

Let us consider a curved surface whose parametric equation is $\mathbf{r} = [x(q_1, q_2), y(q_1, q_2), z(q_1, q_2)]$ and a quantum particle in the 3D space. Following the limiting procedure proposed in Refs. [5] and [6], we take a confining potential that is zero in close proximity of the surface and infinite elsewhere. The particle dynamics is separable in the directions orthogonal and parallel to the surface. In particular, the former direction can be neglected, while in the latter direction, the steady-states of the quantum particle can be obtained through the following modified Schrödinger equation[7]

$$-\frac{\hbar^2}{2m}\sum_{i,j=1}^{2}\frac{1}{\sqrt{g}}\frac{\partial}{\partial q_i}\left[\sqrt{g}\left(\mathbf{g}^{-1}\right)_{ij}\frac{\partial}{\partial q_j}\right]\psi - \frac{\hbar^2}{2m}\left[M^2 - K\right]\psi + V\psi = E\psi(q_1, q_2), \qquad (1)$$

where $\mathbf{g}(q_1, q_2)$ is the metric tensor on the surface, $g(q_1, q_2)$ its determinant, $M(q_1, q_2)$ and $K(q_1, q_2)$ its local mean and Gaussian curvature, respectively, and $V(q_1, q_2)$ a possible external potential. It is straightforward to note that the term containing the two curvatures acts as an additional potential term. Furthermore, since $M^2 \geq K$, this potential is always attractive. We note here that an external magnetic field can be also included in the above equation[8], with its effect being separable from that of the surface curvature[9].

We are interested in the scattering states of the 2D structure. To be specific, we consider a Y junction of three cylinders, as depicted in Fig. 1. The system has three open boundaries, namely the three circumferences $A$, $B$ and $C$ at the edges of the three branches, with radii $R_A$, $R_B$ and $R_C$, respectively. We take the surface coordinate $q_1$ along the circumferences and $q_2$ orthogonal

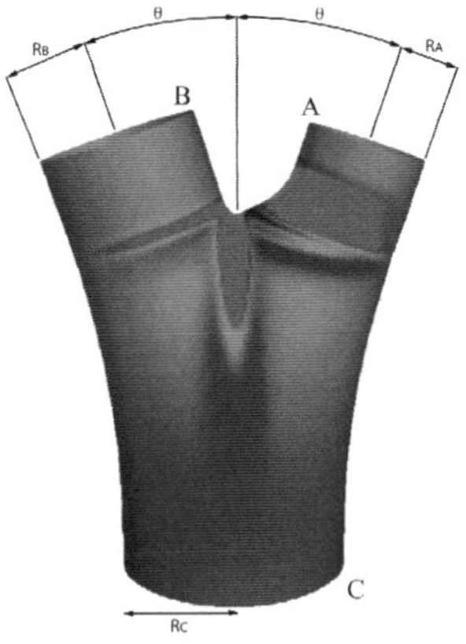


**Figure 1.** Curved surface forming a Y junction, where the two uppermost branches form an angle $\theta$ with the lower cylinder. In the simulation presented we take $R_A = 15$ nm $R_B = 20$ nm $R_C = 30$ nm. The negative value of the GP term $-\frac{\hbar^2}{2m}\left[M^2 - K\right]$ is reported, in colorcode (red=$-30$ meV, blue=0) on the surface.

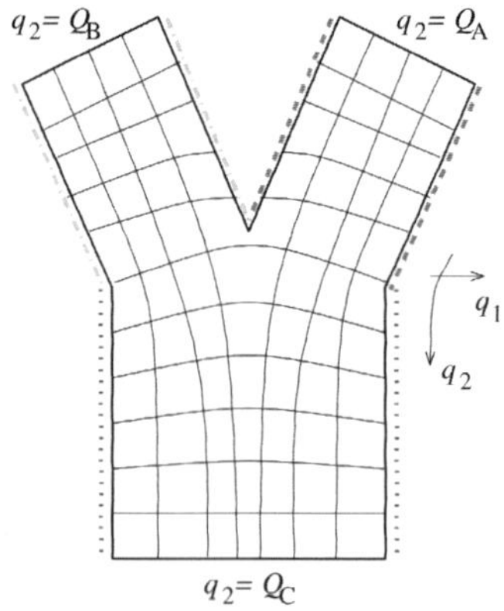


**Figure 2.** Sketch of the simulation domain with the 2D coordinate system $[q_1, q_2]$. The lateral edges with the same dashed/dotted line are to be considered mutually connected to form periodic conditions.

to them, as shown in Fig. 2, where the $[q_1, q_2]$ simulation domain is sketched, together with its coordinate system. The boundary $\alpha = A, B, C$ has coordinates $(q_1 \in [0, 2\pi R_\alpha], q_2 = Q_\alpha)$. On the cylindrical lead connected to $A$ (not shown in Fig. 1) the wave function $\psi$ for a given energy $E$ can be written as [10]

$$\psi(q_1, q_2 < Q_A) = \sum_{m=0}^{M^A} a_m^A \chi_m^A(q_1) e^{ik_m^A q_2} + \sum_{m=0}^{M^A} b_m^A \chi_m^A(q_1) e^{-ik_m^A q_2} + \sum_{m=M^A+1}^{\infty} b_m^A \chi_m^A(q_1) e^{k_m^A q_2}, \quad (2)$$

and it has a similar expression in the other leads, non reported for brevity. The physical meaning of Eq. (2) is simple. The first term represents the incoming component as a combination of the different orthogonal modes of the lead $\chi_m^A(q_1)$, weighted with the injection coefficients $a_m^A$. The second term represents the combination of the outgoing traveling component, weighted with the outgoing coefficients $b_m^A$ (they are the reflection coefficients in case no injection occurs from the other leads). The third term represents the evanescent waves. If we define $\epsilon_m^\alpha$ as the energy of the transverse mode $\chi_m^\alpha$, $m = M^\alpha$ represents the last mode with $\epsilon_m^\alpha < E$ and the evanescent waves have $\epsilon_m^\alpha > E$. On the other hand, for the traveling waves, $\frac{(\hbar k_m^\alpha)^2}{2m} = (E - \epsilon_m^\alpha)$ is the kinetic part of the particle energy.

In order to obtain the scattering states of the structure, once we fix an energy $E$, we solve numerically Eq. (1) on the internal points of the domain (Fig. 2), together with Eq. (2) on the boundaries. We stress that the outgoing coefficients $b_m^\alpha$ are unknowns of the problem. The solution of the two coupled equations is essentially the *quantum transmitting boundary method*[10].

In our simulations, we obtain the transmission coefficients by injecting the carriers only in the ground mode of lead $A$, i.e. $a_0^A \neq 0$ and zero otherwise. On the other hand, for the conductance calculation we populate the incoming modes with a Fermi distribution following the Landauer-Büttiker approach.

## 3. Effect of the GP on the conductance

Here we show how the presence of the GP term $\frac{-\hbar^2}{2m}\left[M^2 - K\right]$ in Eq. (1) produces a remarkable effect on the conductance of a GaAs-based Y junction. In Fig. 3 we report the conductance between leads $A$ and $C$ (solid line) and between leads $A$ and $B$ (dashed line), calculated through a Landauer-Büttiker model at 2 K. Leads $B$ and $C$ are always kept at the same potential. We take a Fermi energy of 0.6 meV from the bottom of the conduction band and do not consider charging effects inside the device. Left (right) panel shows the results with (without) the GP term included in Eq. (1).

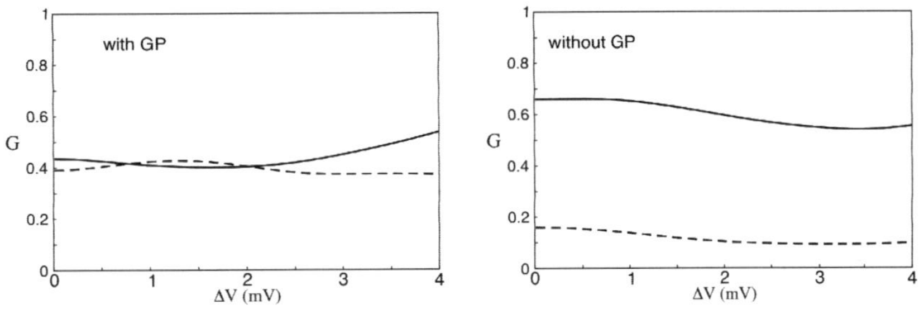


**Figure 3.** Conductance at a temperature of 2 K between two couples of cylindrical leads of the Y junction shown in Fig. 1. The solid (dashed) line represents the conductance between $A$ and $C$ ($A$ and $B$). The inclusion of the GP leads to crossovers of the two curves.

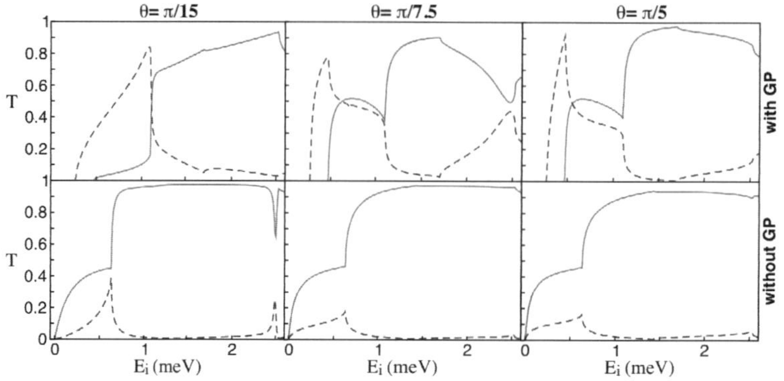


**Figure 4.** Transmission probability as a function of the injection energy with (a) and without (b) the GP at three different angles $\theta$ (see Fig. 1). The carriers enter the Y junction from the ground transverse mode of $A$ and are transmitted in $B$ (dashed line) and $C$ (solid line). The thresholds of the transverse modes in $B$ and $C$ emerge as peaks or discontinuities in the curves.

A double crossing between the two curves occurs if the GP is included, while they remain well separated otherwise. The reason of this behavior can be ascribed to the different surface curvatures in the cylindrical leads, that produce, through the GP term, different energy shifts on the fundamental transverse modes $\chi_0^\alpha$, as detailed in Ref. [11]. On the other hand, if the GP is not included, the ground-state energies $\epsilon_0^\alpha$ are independent on the radii $R_\alpha$, since the 1D lines of the boundaries have periodic boundary conditions.

## 4. Dependence of the coherent transmission on the branches angle

In order to understand the robustness of the effect described above, we calculated the transmission coefficient for different values of the angle $\theta$ between the two uppermost branches and the lowermost one (see Fig. 1). When a very small angle ($\theta \approx \pi/15$, left panels of Fig. 4) is considered, the coherent transmission in $B$ is in general reduced with respect to larger $\theta$. However, for angles comparable to experimental Y-branched nanowires ($\pi/10 < \theta < \pi/4$) we find that the transmission spectrum is essentially unchanged, as can be gathered from the central and right panels of Fig. 4. A similar behavior can be expected for the conductance, this confirming the robustness of the presence of the conductance crossings as a signature of the GP.

## References

We are pleased to thank Fiorenzo Bastianelli and Massimo Rudan for helpful discussions and suggestions.

[1] Fontcuberta i Morral A, Spirkoska D, Arbiol J, Heigoldt M, Morante J R and Abstreiter G 2008 *Small* **4** 899
[2] Wang D, Qian F, Yang C, Zhong Z and Lieber C M 2004 *Nano Letters* **4** 871
[3] DeWitt B S 1957 *Rev. Mod. Phys.* **29** 377
[4] see e.g. Duclos P and Exner P 1995 *Rev. Math. Phys.* **7** 73
Aoki H, Koshino M, Takeda D, Morise H and Kuroki K 2001 *Phys. Rev. B* **65** 35102
Meyer G J, Dias N L, Blick R H and Knezevic I 2007 *IEEE Trans. on Nanotech.* **6** 446
Ferrari G, Goldoni G, Bertoni A, Cuoghi G and Molinari E 2009 *Nano Lett.* **9** 1631
[5] Jensen H and Koppe H 1971 *Ann. Phys.* **63** 586
[6] da Costa R C T 1981 *Phys. Rev. A* **23** 1982
[7] Marchi A, Reggiani S, Rudan M and Bertoni A 2005 *Phys. Rev. B* **72** 035403
[8] Ferrari G, Bertoni A, Goldoni G and Molinari E 2008 *Phys. Rev. B* **78** 115326
[9] Ferrari G and Cuoghi G 2008 *Phys. Rev. Lett.* **100** 230403
[10] Lent C S and Kirkner D J 1990 *J. Appl. Phys.* **67** 6353
[11] Cuoghi G, Ferrari G and Bertoni A 2009 *Phys. Rev. B* **79** 073410

Journal of Physics: Conference Series **193** (2009) 012020 doi:10.1088/1742-6596/193/1/012020

# Mechanisms of a negative differential conductivity in a short-channel $In_{0.52}Ga_{0.48}As/In_{0.53}Al_{0.47}As$ HEMT

**M L Orlov**

Institute for Physics of Microstructures of the Russian Academy of Sciences (IPM RAS) 603950 Nizhny Novgorod, Russia

E-mail: orlovm@ipm.sci-nnov.ru

**Abstract**. The effect of negative resistivity observed at anomalously low voltages in output characteristics of modulation-doped $In_{0.53}Ga_{0.47}As/In_{0.52}Al_{0.48}As$ field-effect transistors is considered. In the discussed experiments, the threshold for appearance of negative resistivity depends not only on the gate length, which was mentioned before, but also on the gate–drain voltage. It is shown that the negative differential resistivity observed at output characteristics of the short-channel $In_{0.53}Ga_{0.47}As/In_{0.52}Al_{0.48}As$ field-effect transistor at anomalously low threshold voltages is related to the formation of the second transport channel. This effect is a result of resonance transition of hot electrons from upper levels of the quantum well to the above-barrier layer via the states of ionized donor impurity.

## 1. Introduction

One of the mechanisms most actively discussed in recent years is the one related to the formation of a descending section in the transistor's output characteristic; this effect can be caused, by example, by interlayer transport of hot electrons in the heterostructures with two-dimensional (2D) electron gas [1]. The most interesting features of this effect are observed in transistors fabricated on the basis of the $In_{0.53}Ga_{0.47}As/In_{0.52}Al_{0.48}As$ heteropair. The distinctive feature of such transistors in comparison with a traditional GaAs/GaAlAs high electron mobility transistor (HEMT) is the deeper quantum well ($U_{QW} \approx 0.5eV$) (that forms the transport channel in the InGaAs layer), and the absence of side valleys within the depth of the quantum well (the energy distance between the valleys $\Gamma$ and $L$ is $E_{\Gamma L} \geq 0.55\text{eV}$); these valleys make possible the manifestation of effects related to the transport of electrons between the valleys. In these structures, the effects related to the interlayer electron transport manifest themselves most clearly, which was shown by Mensz et al. [2]. Moreover, they found for the first time that, in the case of the gate length $L_G < 1\mu\text{m}$, one observes a decrease in the value of the threshold voltage $U_D$ for the formation of the descending section in output characteristics of a transistor from $U_D = U_{QW} = 0.5\text{V}$ at $L_G \geq 1\mu\text{m}$ to the value of $U_D \approx 0.35\text{V}$ at $L_G \approx 0.6\mu\text{m}$. Even smaller values of the threshold voltage, $U_D \leq 0.1\text{V}$, were observed for field effect transistors (FETs) [3]. However, possible mechanisms of formation of a section with negative differential resistivity (NDR) in output characteristics of a transistor at low voltages applied between the source and drain of the transistor were not discussed in available publications so far.

Journal of Physics: Conference Series **193** (2009) 012020 doi:10.1088/1742-6596/193/1/012020

## 2. Effect of a negative differential conductivity in a short-channel transistor

In this study, we analyze the results of measurements of output characteristics of a short-channel $In_{0.53}Ga_{0.47}As/In_{0.52}Al_{0.48}As$ transistor fabricated using the traditional technological scheme [3] with high-mobility 2D electrons in the transport channel. Conducted experiments show that it is possible to observe the NDR section in the output characteristics of transistors at anomalously low threshold voltages, $U_D < 0.1V$, at a lowered temperature [3] for negative and positive voltages $U_G$ at the gate. The conducted measurements show not only that, as the length of the transistor gate is decreased, the threshold voltage decreases but also that this voltage depends directly on the gate–drain voltage. In what follows, we consider possible causes of appearance of features in experimental curves.

In figure 1, we show the output characteristics $J_D$ ($U_D$) of the transistor that belongs to the same series of structures that were used in the studies reported in [3]. Measurements were carried out at temperature $T = 4K$ for several voltages applied to the transistor electrodes. The presented results and ones obtained by other groups show that the low-field NDR ($U_D < U_{QW}$) is observed in a wide range of temperatures and applied voltages; however, the minimum voltage at the gate $U_G$, at which the NDR section appears, is significantly different for different structures. The section of negative resistivity at output characteristics of the transistor represents the source of noise in the structure. Therefore, an analysis of possible causes of the NDR formation in the transistor channel and the effect of NDR on the emission spectrum of a transistor require careful consideration.

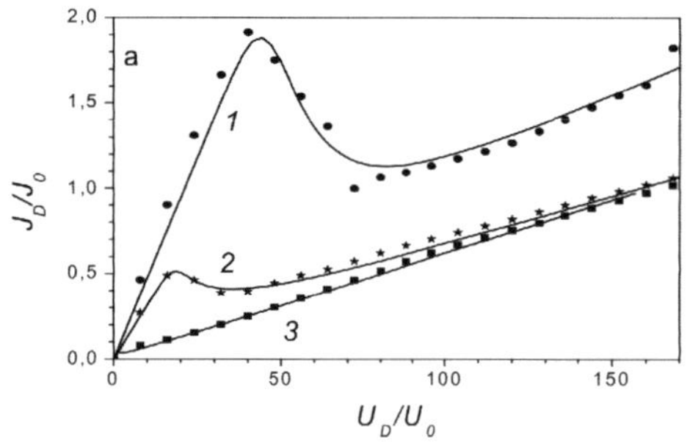


**Figure 1.** Experimental (points) output characteristics of a short-channel $In_{0.53}Ga_{0.47}As/In_{0.52}Al_{0.48}As$ transistor at $T = 4K$ and for the voltage at the transistor gate $U_G = 0.1V(1), 0.05V(2), 0V(3)$. Lines represent the theoretical dependences obtained in the context of two-temperature transport model.

The specific feature of the transistor manifests itself in the dependence of the threshold voltage on the gate voltage in the transistor (figure 1) and was not taken into account in previous studies. The threshold voltage in the short-channel structure increases as the gate potential is increased at a constant gate length; i.e., an anomalously low threshold for appearance of NDR in the *I–V* characteristic of a short-channel transistor structure is a consequence of a decrease not only in the effective gate length but also in the transverse electric field applied to the structure. It is also worth noting that, as the gate voltage $U_G$ is varied, the voltage $U_{DG}$ between the drain and gate remains unchanged, which indicates that the resonance mechanism of formation of low-threshold NDR can manifest itself in the sample.

## 3. Transfer mechanism of the hot electrons in a short-channel FET

We now discuss the mechanisms that can be responsible for experimentally observed phenomena. In accordance with numerous previous studies [2] where it was shown in a conclusive way that the formation of NDR in output characteristics of $In_{0.53}Ga_{0.47}As/In_{0.52}Al_{0.48}As$ FETs with a long channel ($L_G > 1\mu m$) is related to the effect of interlayer transport of hot electrons in real space. In order to assess the NDR effect in the structures considered here, we assume that there is the transport of hot electrons between two spatially separated channels in the system; these channels differ in transport properties of electrons. Possible causes of the appearance of additional transport channels in the samples will be considered separately.

In order to analyze the observed current dependences, we use the two-temperature model of electron transport; this model showed itself to be an advantage in practice [4]. In figure 1 with the solid lines we show the results of calculations of the current $J_D$ flowing through the structure in relation to the voltage $U_D$ between the drain and source of the transistor. The final theoretical dependences obtained as a result of the considered calculation procedure are in excellent agreement with experimental data shown in figure 1. Although the appearance of two transport channels in the real space and the redistribution of electrons between these channels ensuring the descending portion in the *I–V* characteristic of the structure are established beyond doubt, the position of the second channel and causes bringing about its appearance remain the subject of discussions.

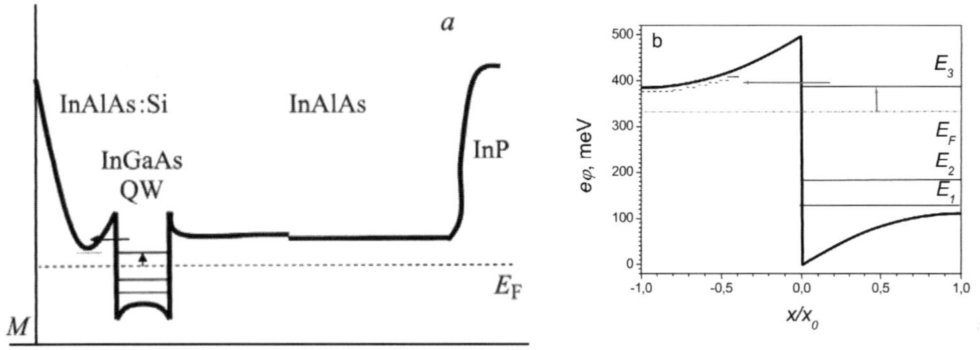


**Figure 2.** (a) The structure of the conduction-band bottom and (b) the distribution of potential in the vicinity of heteroboundary of the quantum well (QW) with the upper barrier layer doped at a level of $N_D = 2\times10^{18}\text{cm}^{-3}$; $T = 5\text{K}$. The spectrum of energy levels in the quantum well, the position of the Fermi level $E_F$, and positions of the levels of shallow donor centers are shown (dashed lines). $x_0 = 22.5$nm is the Debye screening radius.

A possible cause of the appearance of NDR in the *I–V* characteristic can be related to transitions of electrons heated by an electric field from lower subbands of quantum confinement $E_{21} \approx eU_D$ to upper subbands. If electrons remain within the quantum well as a result of interlevel transitions, then, due to the fact that the electrons' effective masses are the same in different subbands of quantum confinement, the only possibility for appearance of NDR in the structure is related to a significant difference in characteristic times of scattering of electrons at the upper and lower levels. The electrons that reside at the lower level of dimensional quantization and are localized in the center of the quantum well are scattered only slightly by the potential of impurity atoms in barrier layers. The maxima of the wave function of the electrons at the second level of dimensional quantization are located near the boundaries of barrier layers; as a result, these electrons experience more pronounced scattering and their mobility is much lower compared with the mobility of electrons at the first level of the quantum well. However, according to the model under consideration, formation of NDR in the *I–V* characteristic is impossible if we have only a reduction in the electron mobility as a result of the transition of electrons to the second level of the quantum well (compare curves *1* and *3* in figure 1). The required variation in the density of states $N_{1(2)}$ can be attained only as a result of transport of 2D electrons in the real space from the quantum well to the states in the barrier layer. On the other hand, estimations of the position of the Fermi level in the structure with the quantum-well thickness of 20 nm and the electron concentration higher than $8\times10^{11}\text{cm}^{-2}$ show that the two lower subbands of dimensional quantization are filled. In the case under consideration, the electron concentration in the transistor channel was as high as $n \approx 4\times10^{12}\text{cm}^{-2}$ and the Fermi energy equalled $E_F = \pi n\hbar^2/m_e^* \approx 220\text{meV}$; i.e., the Fermi energy vastly exceeded the energy of the dimensional-quantization level $E_2$ ($E_2 - E_1 \approx 60\text{meV}$). Thus, the considered mechanism of formation of the desscending section in the *I–V* characteristic is possible only in the region of low electron concentrations in the transistor channel.

In the case of experiments under consideration (figure 1), it is more reasonable to assume that, as a result of heating of the charge carrier by an electric field, electrons from the Fermi level transfer to the third energy level $E_3$ in the quantum well ( $E_3 - E_1 = 300\text{meV}$ , $E_3 - E_F \approx 80\text{meV}$ ). Numerical analysis of the structure of the energy band in the system (figure 2) on the basis of the Poisson equation shows that only electrons from the third energy level in the quantum well can transfer to the neighboring barrier layer. Calculations show that, at the doping level $N_D \approx 10^{18}\text{cm}^{-3}$ of the barrier layer with the thickness $d_D$, the potential arising in the structure as a result of redistribution of charges due to their transfer from the impurity centers to the quantum well has the amplitude ~200 mV. The value of the potential φ modulating the bottom of the quantum well can be estimated on the basis of the following considerations. Ionized donors and electrons in the structure are spatially separated and their charges are equal by magnitude. In addition, as a result of quantization of electrons in the potential well of the InGaAs layer, their charge is almost uniformly distributed over the layer's thickness. It is quite reasonable to replace the charge of electrons in the right-hand side of the Poisson equation by an equivalent charge of ionized acceptors ($N_A = N_D$). In this case, a solution of the Poisson equation at the half-period for a structure with a completely compensated charge of free carriers can be written as $\varphi^*_{\max} - \varphi^*_{\min} = (1/8)N^*_D d^{*2}_D$ , where $\varphi^* = e\varphi / kT$ , $d^*_D = d_D / x_0$, $x_0 = (\varepsilon_0 kT / 4\pi n_i e^2)^{1/2}$ , and $N^*_D = N_D / n_i$ . For the values of the parameters $N_D = 2 \times 10^{18} cm^{-3}$, $d_D = d_{ch} = 2 \times 10^{-6} cm$, and $\varepsilon_0 = 14$, we correspondingly obtain $e\varphi_{\max} - e\varphi_{\min} = 260\text{meV}$ , which is quite consistent with the results of a more precise numerical analysis (figure 1). This value of the potential is quite sufficient to ensure the possibility of resonance transitions of electrons from the third level of the quantum well to ionized shallow impurity centers in the upper barrier layer of the structure. These electrons pass to the conduction band as a result of breakdown in the pulling electric field. The resonance model of electron transitions between the layers is supported by the experimentally observed (see figure 1) dependence of the threshold voltage on the gate voltage and existence of a threshold voltage $U_{DG}$ defined by the difference between potentials of the gate and drain ( $U_{GD} = 50\text{mV}$ for a transistor with the *I–V* characteristic shown in figure 1). The above field gives rise to a threshold for tunneling of electrons between the states in the main and additional channels. An increase in the voltage at the transistor gate by $\Delta U_G$ at a specified voltage at the transistor drain brings about the disappearance of the NDR in the *I–V* characteristic. However, as the current through the structure is increased, i.e., as the voltage at the drain electrode is increased, the system again transfers to the resonance state if the condition $\Delta U_D = \Delta U_G$ is satisfied (see figure 1). Differences in the geometry of the structural elements of specific devices also give rise correspondingly to different values of $U_{GD}$ and $\Delta U_D$ for different transistors. The probability of formation of resonance electronic states in the potential well of the δ-doped InGaAs:Si layer is low as a result of scattering of electrons by the impurities' potential (the free path $l_p = N_D^{1/3} \approx 10\text{nm} < d_{QW} \approx 20\text{nm}$ ). As a consequence of the multivalley character of the electron spectrum in the InAlAs layer and different electron masses in 2D (InGaAs) and 3D (InAlAs) channels in the system, the condition $R = N_2 / N_1 >> 1$ required for observation of NDR in output characteristics of the transistor is easily satisfied.

So, the new resonance mechanism of the hot electron transfer in a short-channel FET, that explains the peculiarities of the observed NDR effect, is proposed.

**References**

[1] Gribnikov Z S, Hess K and Kosinovsky G A 1995 *J. Appl. Phys.* **77** 1337
[2] Mensz P M, Garbinski P A, Cho A Y, Sivco D L and Lurie S 1990 *Appl. Phys. Lett.* **57** 2558
[3] Knap W, Lusakowski J, Parenty T, Bollaert S, Cappy A, Popov V V and Shur M S 2004 *Appl. Phys. Lett.* **84** 2331
[4] Orlov M L and Orlov L K 2009 *Semiconductors* **43** 652

# RF doubling and rectification in three-terminal junctions: experimental characterization and Monte Carlo analysis

**I. Iñiguez-de-la-Torre[1], T. González[1], D. Pardo[1], C. Gardès[2], Y. Roelens[2], S. Bollaert[2], A. Curutchet[3], C. Gaquiere[2] and J. Mateos[1]**

[1]Departamento de Física Aplicada, Universidad de Salamanca,
Plaza de la Merced s/n, 37008 Salamanca, Spain
[2]Institut d'Electronique, de Microélectronique et de Nanotechnologie (IEMN),
Université de Lille, UMR CNRS 8520, BP 60069, 59652 Villeneuve d'Ascq, France
[3]Laboratoire IMS, Université Bordeaux 1,
351 Cours de la Libération, 33405 Talence, France

E-mail: indy@usal.es

**Abstract**. The feasibility of three-terminal junction devices for frequency doubling and rectification reaching the terahertz range is studied both experimentally and numerically (by means of Monte Carlo simulations). Experiments at room temperature up to 94 GHz for detection and to 4 GHz for doubling and phase detection are shown. The influence of the width of the branches and the shape of the junction (T-shaped vs. Y-shaped) on the frequency response of the devices has been studied. Cut-off frequencies found for frequency doubling are lower than for the rectifying operation of the devices. This difference is attributed to the slow characteristic time of carrier penetration into the stem.

## 1. Introduction

Nanometer sized Three-Terminal Junctions (TTJs) [1] based on high mobility semiconductors are emerging as potential building blocks for analog and digital integrated circuits. Ballistic transport of electrons inside these structures results in an attractive nonlinear response [2]. In virtue of the parabolic behavior of the stem voltage when biasing the left and right branches in push-pull fashion, several applications like rectification, frequency doubling and mixing, and operation as digital logic gates are possible [3]. Although their static behavior is already well understood, not much attention has been paid till now to the study of their RF response. Owing to its small size (and intrinsic high impedance), the cutoff frequency of these devices is mainly limited by its RC time constant. The aim of this work is to analyze the dependence of the cutoff frequency of TTJs on their geometry: width of the horizontal and vertical branches and angle between the left and right branches (T-shaped vs. Y-shaped junctions). To fully understand the device operation and predict its intrinsic cutoff frequencies, a Monte Carlo (MC) simulator is used. Experimental results of TTJs working as detectors at 94 GHz and as frequency doublers and phase detectors at 4 GHz are also shown.

## 2. Monte Carlo simulations

TTJs are fabricated by defining the boundaries of three connecting semiconductor channels on the top of a heterolayer. In this case we use a high In content InGaAs channel to get maximum mobility. Our simulation tool is a Monte Carlo code employed in previous works to understand the physics of TTJs, successfully reproduce the experimental measurements and optimize the design of several kinds of ballistic nanodevices [2, 5-8]. The results shown in this paper have been obtained with a surface charge profile calculated at equilibrium conditions by means of the self-consistent model presented in [6-7], then used when a voltage (DC or AC) is applied to the different branches. By using this "frozen" equilibrium surface charge profile we assume that the capture/emission times of the surface states (typically of about 1 μs) is several orders of magnitude longer than the maximum period used for exciting the TBJ (1 ns). The geometry of the simulated devices is sketched in figure 1. The bias is applied in push-pull fashion to the left and right branches and the potential at the bottom of the vertical branch, called $V_C$, is measured (or simulated) in open circuit conditions. For the T-shaped junctions we consider two parameters to analyze their influence on the dynamic response: the width of the horizontal branches $W_{HOR}$ and the width of the vertical stem $W_{VER}$.

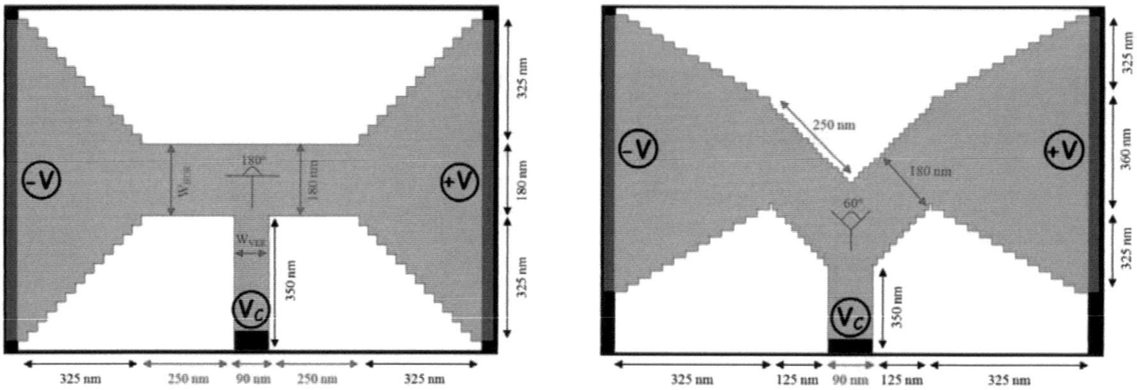


**Figure 1.** Topology of the simulated structures. (a) T-shaped junction and (b) Y-shaped junction

Figure 2 shows the DC value of $V_C$ and the current flowing through the horizontal branches for several junctions. When the width of the vertical branch is reduced, the parabolicity in the bell-shape of the $V_C$-$V$ curves is enhanced, while providing a similar current (as expected since $W_{HOR}$ is the same [6]). On the other hand, the stem voltage values are more negative in the Y-shaped junction than in the T-shaped one because of the enhanced electron penetration into the vertical branch. A narrower horizontal branch leads to less negative values of $V_C$ (and obviously to lower current). This is not consistent with experimental measurements (that show an enhancement of the bell shape of $V_C$ vs. $V$) due to the bias dependence of the surface charge profile (explanation can be found in Ref. 7). The "frozen" equilibrium surface charge profile is used here in order to simplify the analysis of the results.

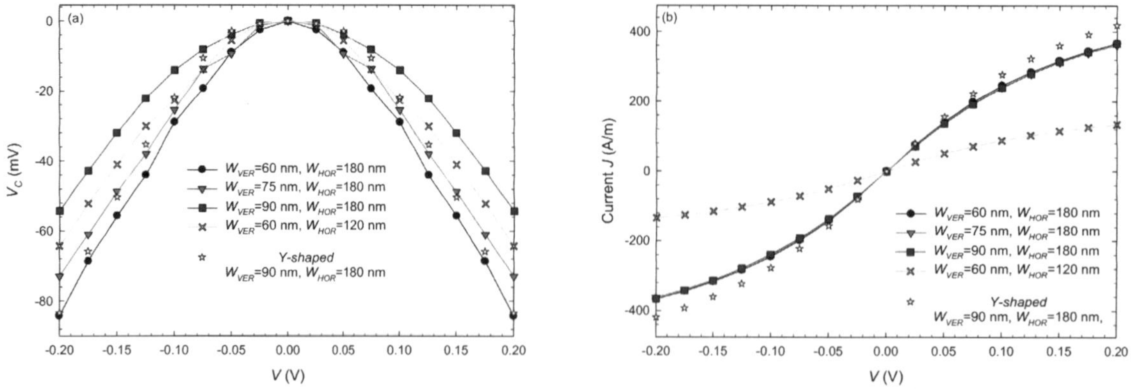


**Figure 2.** MC values of (a) bottom potential $V_C$ and (b) current as a function of the push-pull bias calculated in T-shaped junctions with different $W_{VER}$ and $W_{HOR}$ and in a Y-shaped junction.

Journal of Physics: Conference Series **193** (2009) 012021 doi:10.1088/1742-6596/193/1/012021

For the dynamic study of the TTJs, sinusoidal signals of 0.1 V of amplitude are applied to the left and right branches, while monitoring the value of $V_C$. The amplitude of its response informs about the capability of operating the TTJs as frequency doublers, while the average DC value of $V_C$ provides the rectification capability. Both quantities are plotted in dB in figure 3 for the different TTJs studied (once subtracted the low frequency value). As a general feature, the cut-off for the amplitude of the output [figure 3(b)] appears at much lower frequencies than for the mean value [figure 3(a), with cut-off frequencies around 1 THz]. Different mechanisms are involved in the cut-off of both quantities. The DC value of $V_C$ (used for rectification and thus detection) is mainly related to the electron horizontal transport, which is very fast. As a result, its cut-off is hardly influenced by the width of the vertical branch. On the other hand, the AC amplitude of $V_C$ (and thus frequency doubling) is controlled by the penetration of carriers in the stem, so that its width ($W_{VER}$) clearly changes the cut-off frequency (higher frequencies for wider stems, in which carriers enter more easily). Since the electric field that forces electrons to enter/leave the vertical branch following the excitation signal is small, the characteristic time of such a process is much longer than the transit time associated to horizontal transport, so that the cut-off frequency for the amplitude of the output AC signal is much lower than in the case of the DC value, the difference being more significant the narrower the stem is. Nevertheless, wider stems provide less negative values of $V_C$ at low frequency, as observed in figure 2(a). As a consequence, for an optimized response $W_{VER}$ must be chosen depending on the required type of operation and frequency. On the other hand when reducing the width of the horizontal branch ($W_{HOR}$), a slight increase of the cut-off frequency is observed in both quantities; however, the matching to the typical 50 Ω lines would be worse due to the higher impedance of the TTJ. Concerning the shape, the Y geometry much improves the global performance of the device as a result of a stronger injection of carriers into the stem (having a more pronounced influence on the cut-off for frequency doubling). A better performance is expected when the angle of the junction is decreased.

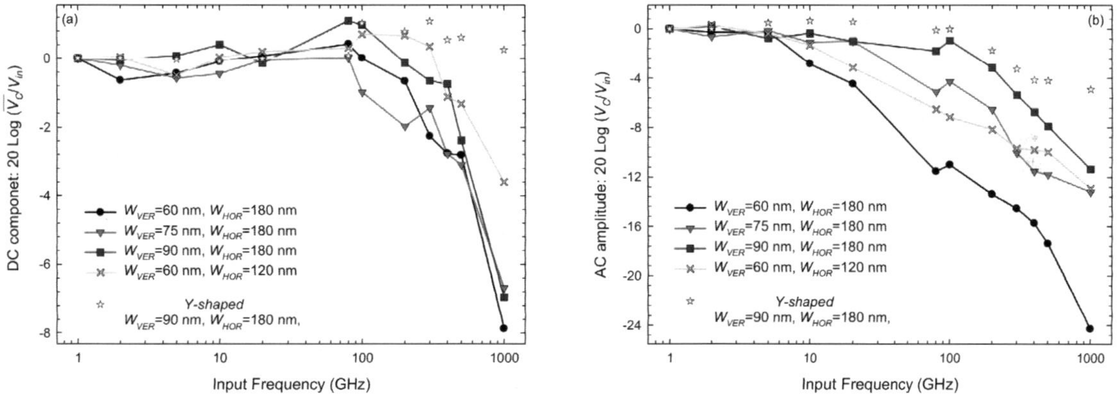


**Figure 3.** Average DC value and AC amplitude of $V_C/V_{in}$ (in dB) as a function of the frequency of a sinusoidal push-pull input excitation of amplitude of $V_{in}$=0.1 V. For comparison, the low frequency value has been subtracted for each curve.

## 3. Experimental measurements

We report here experimental results illustrating the functionalities of TTJs previously analyzed by MC simulations. The geometry of a double Y-shaped junction has been defined by means of reactive ion etching on a typical InP-HEMT epilayer. Details of the fabrication process can be found in [4,7]. The complete frequency characterization of the structure between 3 and 40 GHz was presented in Ref. 8. In figure 4(a) it can be observed that the junction is still perfectly operative as detector at 94 GHz, with a measured sensitivity of 0.076 mV/μW (without DC bias). Concerning the devices acting as frequency doublers and phase detectors, we provide here new results obtained using a setup based on a Large Signal Network Analyzer (LSNA). After adequate calibration steps, the injected and reflected waves (phase and magnitude) at the input and output of the device are used to determine the currents and voltages in the frequency or time domain. In figure 4(b) an excellent frequency doubling for an input signal of 4 GHz can be observed: a quasi sinusoidal signal at 8 GHz is clearly obtained at the output. In order to check the possibility of phase detection, signals have been applied to the right and left branches with three different phase-shifts. The results of these measurements are shown in figure 4(c). They are in a very good agreement with the simulations. The main improvement with respect to previous works [4] is the optimization of the ohmic contacts and coplanar waveguide design in order to minimize extrinsic crosstalk capacitances without excessively increasing access resistances (their high impedance is one of the main problems of these ballistic nanodevices).

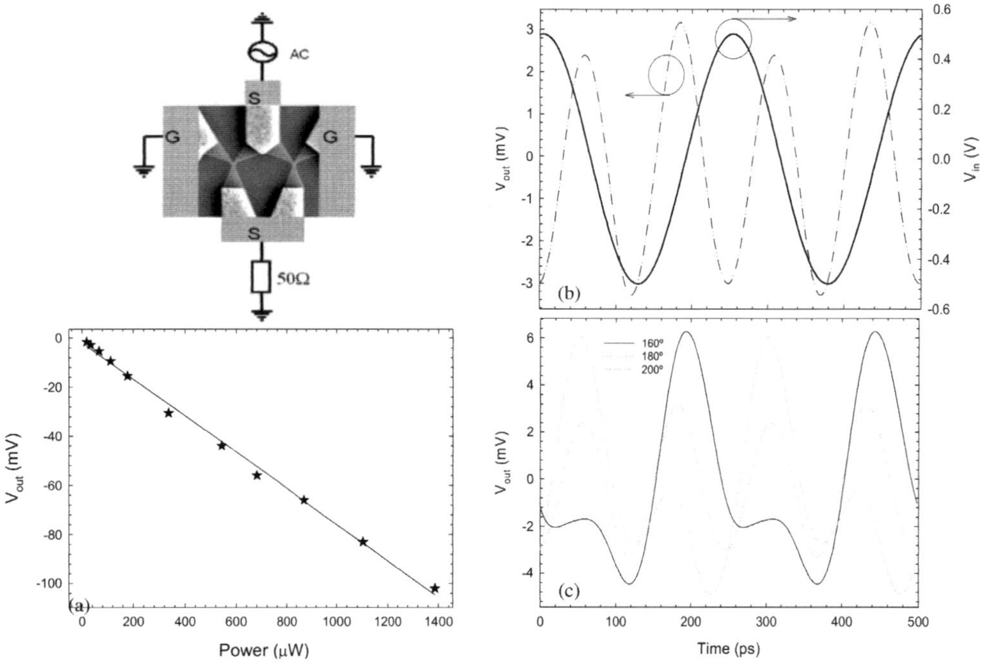


**Figure 4.** (a) $V_C$ vs. microwave (94 GHz) power injected to the double Y-shaped junction shown above with $V_{bias}$=0. LSNA characterization at 4 GHz working as (b) frequency doubler with a -5dBm input signal and (c) phase detector (output for signals with different phase-shift between inputs).

## 4. Conclusions

By using MC simulations, the influence of the shape and branch sizes of TTJs on their dynamic behavior has been studied. Signal detection up to 94 GHz and frequency doubling and phase detection up to 4 GHz have been experimentally confirmed in a double Y-shaped junction.

This work has been partially supported by the Dirección General de Investigación (MEC) and FEDER (project TEC2007-61259/MIC and Acción Integrada HF2007-0014) and by the Consejería de Educación, Junta de Castilla y León (projects SA019A08 and GR270).

## References

[1] Shorubalko I *et. al.* 2001 *Appl. Phys. Lett.* **79** 1384
[2] Mateos J *et. al.* 2003 *IEEE Trans. Electron Devices.* **50** 1897
[3] Sun J *et. al.* 2007 *Nanotechnology* **18** 195205
[4] Gardès C *et. al.* 2008 *Int. J. Nanotech.* **5** 796
[5] Mateos J *et. al.* 2003 *IEEE Trans. Electron Devices* **50** 1897
[6] Iñiguez-de-la-Torre I *et. al.* 2007 *Semicond. Sci. Technol.* **22** 663
[7] Iñiguez-de-la-Torre I *et. al.* 2009 *J. Appl. Phys.* **105** 094504
[8] Bednarz L *et. al.* 2006 *IEEE Trans Nanotech.* **5** 750

Journal of Physics: Conference Series **193** (2009) 012022 doi:10.1088/1742-6596/193/1/012022

# Monte Carlo simulation of charge transport in amorphous chalcogenides

**E Piccinini[1,2], F Buscemi[1,2], M Rudan[1,3], R Brunetti[2,4] and C Jacoboni[2,4]**

[1] ARCES, Alma Mater Studiorum, University of Bologna, Via Toffano 2/2, I-40125 Bologna, Italy

[2] S3 Research Center, CNR-INFM, Via Campi 213/A, Modena I-41125, Italy

[3] Department of Electronics, Computer Science and Systems, University of Bologna, Viale Risorgimento 2, I-40136 Bologna, Italy

[4] Physics Department, University of Modena and Reggio Emilia, Via Campi 213/A, I-41025 Modena, Italy

E-mail: enrico.piccinini@unimore.it

**Abstract.** The $I(V)$ characteristics of amorphous GST devices show a peculiar S-shape behavior, that is a swift rise of the current along with a voltage snap-back. This type of characteristics led to a growing research interest in view of the future application of such materials to the manufacturing of phase-change memory devices. In this work we adopt a generalization of the variable-range hopping theory to simulate charge transport in a layer of amorphous $Ge_2Sb_2Te_5$ sandwiched between two planar metallic electrodes. The numerical implementation of a current-driven Monte Carlo code allows one both to provide a complete microscopic particle picture of electrical conduction in the device and to better analyze the mechanisms governing the snap-back effect.

## 1. Introduction

Chalcogenide materials are nowadays the focus of many research efforts owing to their application as phase-change materials in non-volatile memories. Particular attention has been devoted to the $Ge_2Sb_2Te_5$ (GST) compound that has been recognized as the most suitable for industrial applications. When the GST is in its amorphous phase, it exhibits a threshold switching in the conduction characteristic. The latter consists in a transition from a low- to a high-conductive state once the applied bias reaches a critical threshold voltage and an S-shaped negative differential resistance behavior in the I(V) curve is found [1]. The theoretical understanding of such a mechanism is crucial in view of the fabrication of innovative nonvolatile memories.

The transport properties of amorphous chalcogenides have been studied with different schemes relying either on a trap-controlled conduction [2, 3] or on the generation-recombination and impact ionization mechanisms via localized states [1]. The large concentration of localized states arising from structural defects and acting as donor- or acceptor-like traps has been shown by experimental and theoretical investigations [4, 5, 6]. This seems to indicate the essential role of trapped-carrier transitions in the electrical behavior of amorphous GST ($a$-GST).

## 2. Physical model and results

We have considered a device consisting of a 3D region of $a$-GST, with a cross section $\sigma = 270$ nm$^2$ and length $\ell = 27$ nm, sandwiched between two planar contacts, in which a number $N_t$ of randomly-positioned donor-like traps exist. Carriers can hop among traps by tunneling. An energy-level, randomly chosen within a narrow band of width$\Delta$ $E$ centered at the Fermi level, is attributed to each trap. In order to mimic contacts, two infinite "reservoirs" of both carriers and empty states at the equilibrium Fermi energy have also been considered. The reservoirs can at any time inject electrons into the traps of the GST or host electrons coming from the GST. The electrical neutrality is ensured by fixed negative charges not involved in the transport process. The transition rate $S_{ij}$ for a carrier's hopping from an occupied site $i$ to an empty site $j$

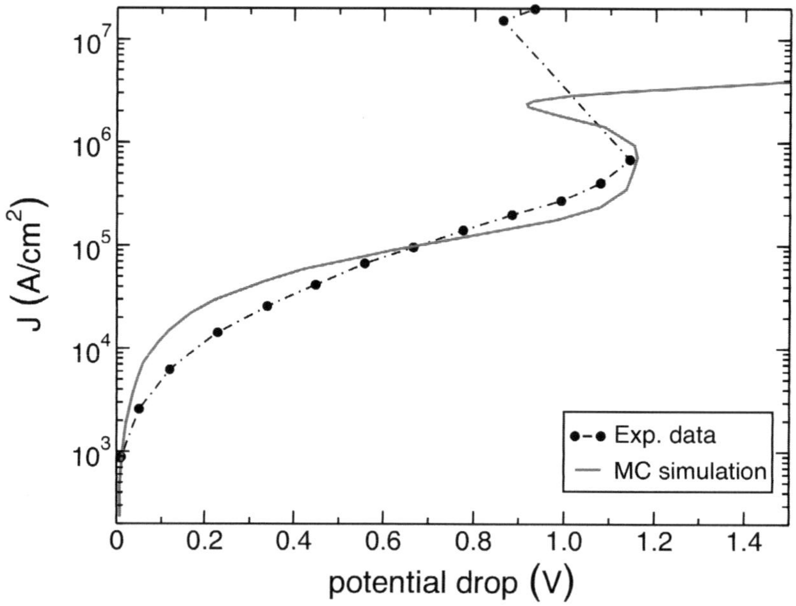


**Figure 1.** Measured and simulated current density vs voltage. The parameters used in the Monte Carlo simulations are: $\beta = 1/20$, $\nu_0 = 1.35 \cdot 10^{13}$ s$^{-1}$, $N_t = 1.48 \cdot 10^{19}$ cm$^{-3}$, and $\alpha_0 = 4.54 \cdot 10^6$ cm$^{-1}$. The experimental data are taken from Ref. [2]

is evaluated according to the variable-range hopping theory [7]:

$$S_{ij} = \begin{cases} \nu_0 T_{ij} \exp\left[-\frac{\Delta\epsilon}{k_B T}\right] & \text{if}\Delta \quad \epsilon > 0 \\ \nu_0 T_{ij} & \text{if}\Delta \quad \epsilon \le 0 \end{cases} . \tag{1}$$

Here, $\nu_0$ is the attempt-to-escape frequency, $T_{ij}$ is the transmission coefficient through the potential barrier between sites $i$ and $j$, separated by the distance $R_{ij}$; $\Delta\epsilon = \epsilon_j + e\varphi_j - \epsilon_i - e\varphi_i$ indicates the energy difference between the initial and final states, $\epsilon_i$ and $\epsilon_j$ being the intrinsic energy levels of traps $i$ and $j$, $\varphi_i$ and $\varphi_j$ being the values of the electric potential at the two considered sites. In the original formulation [8] the transmission coefficient $T_{ij}$ does not depend upon electric field. However such an assumption is not realistic for a non-negligible shift of the energy barrier between the two traps due to electrostatic potential, as in the case at hand. Taking into account the effect of the electric field on the barrier [9], $T_{ij}$ has been assumed to be

$$T_{ij} \propto e^{-2\alpha R_{ij}}, \qquad \alpha^2 = \alpha_0^2 - \frac{m_0 e\beta}{\hbar^2}|\phi_i - \phi_j|. \tag{2}$$

In Eq. (2) $\alpha_0$ is the inverse of a characteristic tunnelling distance, $m_0$ the free-electron mass and $\beta$ a phenomenological parameter. We report in Fig. 1 the $J(V)$ curve ($J = I/\sigma$ being the current

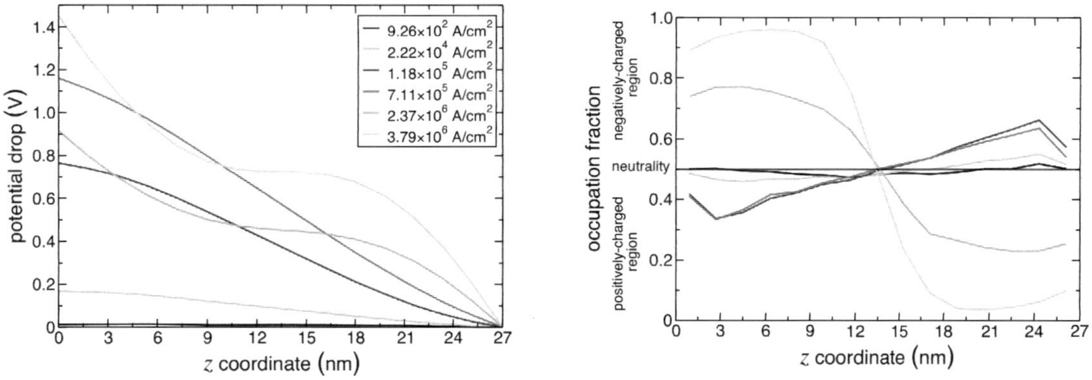


**Figure 2.** Left: potential profile inside the device. Each line refers to a different applied current density indicated in the legend box. Electrons enter the device from the right contact. Right: trap occupation fraction with respect to the axial coordinate of the device.

density) of the simulated device, obtained by tuning the parameters on experimental data [2]. The voltage values have been calculated as averages of the final output over 192 independent simulations. The typical features of the electric conduction in $a$-GST are found: a subthreshold region characterized by an Ohmic part at the lowest currents, a subsequent exponential regime, and a negative differential-resistance region. where the potential difference between contacts is reduced at increasing currents. The behavior exhibited by the upper part of the curve indicates the existence of a limiting current for the simulated model and suggests also that a different transport process like band conduction, not considered in the present analysis, must set in to sustain the prescribed current.

The outcome of the model can be interpreted from a microscopic viewpoint in terms of the trap-occupation fraction along the device and the internal electric potential profile (Fig. 2). In the subthreshold region the occupation fraction is almost constant and close to the neutrality value of 0.5, and the potential drop between electrodes is linear. For current densities larger than $10^5$ A/cm$^2$ the hopping process is not efficient enough in transferring carriers from GST region to the collecting contact and charges tend to accumulate in the region close to the drain contact. This implies that a counterfield adds to the field generated by the two contacts. A further increase of the current enhances such effect and produces the feed-back effect necessary for the S-shape of the $J(V)$ characteristics. In fact, the traps close to the emitting contact are emptied with high rates and carriers are transferred close to the collecting contact, giving this way origin to a step-shaped profile for the trap-occupation fraction. This behavior can also be demonstrated analytically and is of fundamental importance for the occurrence of the snap-back effect.

Fig. 4 shows the probability distribution of the number of hops needed to transfer a carrier from the emitting to collecting contact, for the $I(V)$ characteristics displayed in Fig. 3. It is worth noting how at low currents the carriers emitted from one contact must hop many times to reach the drain, whilst at higher currents only a small number of hops is required. Such a behavior is related to the increase of the internal electric field that lowers the energy barriers between the traps. As a net effect, carriers are transferred from one contact to the other one by means of few long-range transitions, which are responsible for the snap-back effect. As the threshold-switching region is approached, carriers can tunnel across half the device (or more), thus enhancing the formation of opposite charge domains close to the contacts. Furthemore, the microscopic analysis of the transport process does not show evidences of carrier preferred pathways.

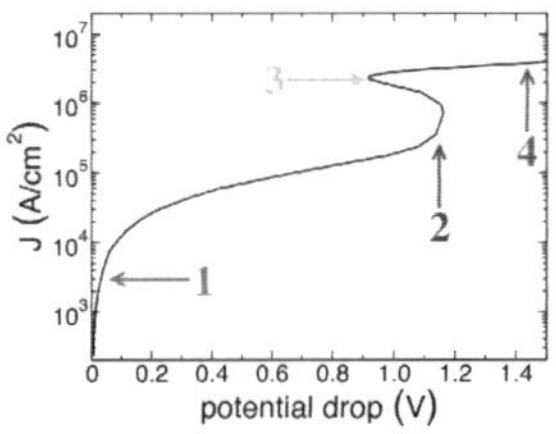


**Figure 3.** The simulated $J(V)$ characteristic already shown in Fig. 1. The arrows and the numbers indicate the points where the trajectory statistics has been performed.

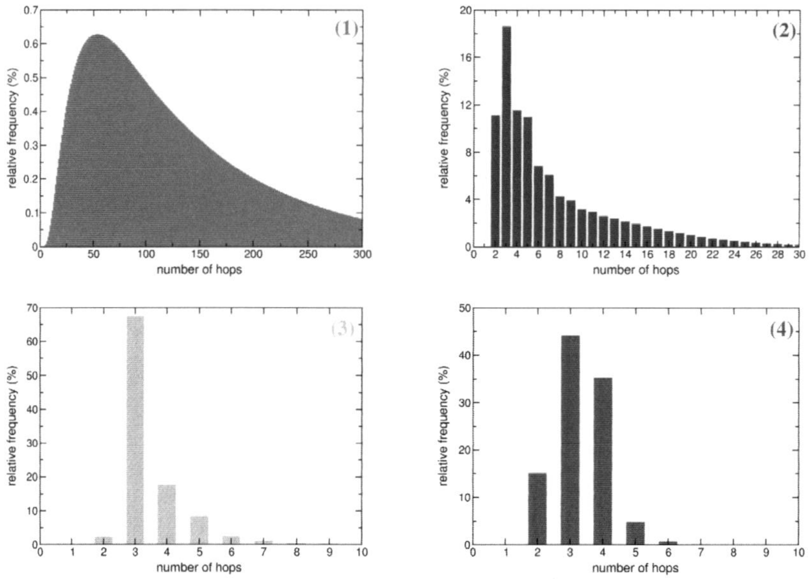


**Figure 4.** Number of hops needed to transfer a carrier from one contact to the other. Each graph refers to a different applied current density $J$, indicated by the arrows and the numbers in Fig. 3. The simulated device has a cross section of 900 $nm^2$ and a lenght of 30 nm.

## 3. Conclusions

We have investigated the charge transport across a a device made of a nanometric layer of *a*-GST sandwiched between two metal contacts by means of a microscopic particle description. The mechanisms ruling the threshold switching have been investigated. Our result shows that the snap-back effect is related to the formation of domains of charges that modify the potential profile within the device, allowing, in turn, for higher currents to be sustained with a smaller potential drop between contacts.

## Acknowledgments

This work has been carried out under the Intel Corporation contract 34524/2007, whose support is gratefully acknowledged.

## References

[1] Pirovano A, Lacaita A L, Benvenuti A, Pellizzer F, and Bez R 2004 *IEEE Trans. Electron. Devices* **51** 452
[2] Ielmini D, and Zhang Y 2007 *J. Appl. Phys.* **102** 054517
[3] Ielmini D 2008 *Phys. Rev. B* **78** 03538
[4] Thomas C B 1976 *J. Phys. D: Appl. Phys.* **9** 1976
[5] Lee B S, Abelson J R, Bishop S G, Kang D, Cheong B, and Kim K 2005 *J. Appl. Phys.* ,**97** 093509.
[6] Caravati S, Bernasconi M, Kühne T D, Krack M, and Parrinello M 2009 *J. Phys.: Condens. Matter* **21** 255501
[7] Mott N F and Davis E A 1961 *Electronic processes in noncrystalline materials*, (Claredon Press, England: Oxford)
[8] Miller A and Abrahams E 1960 *Phys. Rev.* **120** 745
[9] Buscemi F, Piccinini E, Brunetti R, Rudan M, and Jacoboni C 2009 submitted to *J. Appl. Phys.*

Journal of Physics: Conference Series **193** (2009) 012023 doi:10.1088/1742-6596/193/1/012023

# Current oscillations excited by optical phonon emission in GaN $n^+nn^+$ diodes: Monte Carlo simulations

**A. Iñiguez-de-la-Torre, T. González and J. Mateos**

Departamento de Física Aplicada, Universidad de Salamanca,
Plaza de la Merced s/n, 37008 Salamanca, Spain

E-mail: javierm@usal.es

**Abstract**. Under certain conditions, usually at low temperature, when optical phonon emission is the dominant scattering mechanism, plasma instabilities, driven by the streaming motion of carriers, can appear in $n^+nn^+$ diodes and may originate current oscillations. In this work, by means of Monte Carlo simulations, we investigate the optimum conditions for the onset of such mechanism in GaN $n^+nn^+$ diodes: applied bias, temperature, doping and length of the active *n* region. Simulations show that current oscillations at frequencies as high as 400-600 GHz can be achieved for temperatures below 40 K, with some degree of tunability by means of the applied voltage and length of the *n* region. Oscillations are enhanced for bias conditions providing a characteristic frequency close to the plasma frequency of the *n* region.

## 1. Introduction

Recently, the analysis of electron transport in $n^+nn^+$ structures based on semiconductors with high-energy optical phonons is attracting increasing attention due to the possibility of THz signal generation [1]. When operating at small bias and low temperature, in a regime where optical phonon emission (OPE) is the dominant scattering mechanism, plasma instabilities [2], driven by the streaming motion of carriers, may originate current oscillations [3]. The frequency of such oscillations corresponds to the inverse of the transit time between two OPEs. These phenomena have been analyzed so far, by means of Monte Carlo (MC) simulations, for the case of InN [3] and InP [1,4] $n^+nn^+$ structures. However, in the case of GaN, a wide band-gap semiconductor of growing interest, only ideal bulk material properties have been studied [5].

The aim of this work is to investigate current oscillations excited by OPE in GaN $n^+nn^+$ diodes by using a realistic MC model including the influence of dislocations. We analyze the dependence of the onset of the oscillations, their frequency and amplitude on the bias conditions, temperature, doping and length of the active *n*-region.

## 2. Monte Carlo model and simulated structures

We have developed an ensemble MC simulator self-consistently coupled with a 2D Poisson solver for the analysis of GaN $n^+nn^+$ diodes. The conduction band of GaN is modelled by three non-parabolic spherical valleys ($\Gamma_1$, U and $\Gamma_3$). The scattering mechanisms considered in the simulations are: ionized impurities, acoustic, optical and intervalley phonons, piezoelectric and dislocation scattering. Special attention has been paid to the implementation of dislocation scattering, which can be crucial to correctly reproduce electron mobility in bulk GaN or in the channel of GaN based HEMTs. However, since $n^+nn^+$ diodes are vertical structures, dislocations are essentially parallel to electron transport, so that dislocation scattering does not have any influence on the results. Ohmic boundary conditions are considered at the contacts. In order to easily perform a frequency analysis of electrical fluctuations and detect the presence of current oscillations, the time-domain current sequences obtained from the MC simulations are Fourier transformed (once subtracted the average value) into the frequency domain to determine the current noise spectra.

The simulated GaN $n^+nn^+$ diodes [sketched in the inset of figure 1(a)] are just a low doped $n$ region of length $L$ (the effect of both doping level and length will be studied in next section), sandwiched between two 50 nm $n^+$ layers, with a higher doping of $10^{17}$ cm$^{-3}$, at which end ohmic contacts maintain charge neutrality.

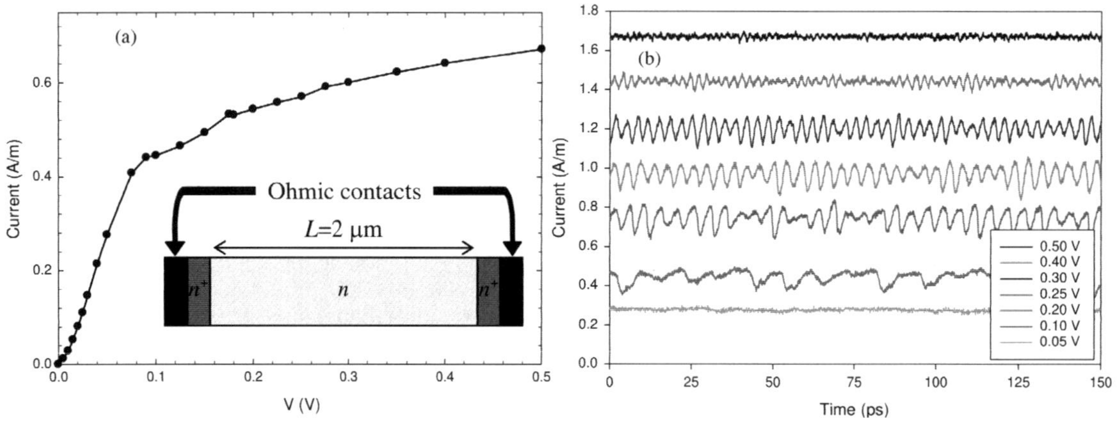


**Figure 1.** (a) *I-V* curve and (b) time evolution of the current at 15 K in the GaN $n^+nn^+$ diode sketched as an inset. The doping/length of the $n^+$ and $n$ regions are $10^{17}$ cm$^{-3}$/50 nm and $10^{15}$ cm$^{-3}$/2 μm, respectively. In order to improve the readability, in (b) the curves are vertically shifted by 0.2 A/m each, starting from the one corresponding to $V$=0.2 V.

## 3. Optical phonon transit time resonance (OPTTR)

Figure 1(a) shows the *I-V* curve of a GaN $n^+nn^+$ diode with an $n$ region of length $L$=2 μm and doping $10^{15}$ cm$^{-3}$ calculated at 15 K. A kink is evident when the applied voltage reaches the threshold for OPE (the optical phonon energy $\varepsilon_p$ is 0.091 eV), and also at twice and three times such a value. Interestingly, associated with the onset of OPE, oscillations emerge in the current, as shown in figure 1(b). However, when the bias is much increased (V=0.5 V), current oscillations disappear. The mechanism at the origin of current oscillations is the optical phonon transit time resonance, OPTTR [1]. Ideally, when only OPE scattering is present, and under the action of a constant electric field $E$, an electron moves ballistically until reaching the energy for emitting an optical phonon, thus returning back into the initial low-energy state and starting the next cycle. The time needed for such a process is given by [1]

$$\tau = \frac{\sqrt{2m^* m_0 \varepsilon_p}}{eE}, \qquad (1)$$

where $e$ is the electron charge and $m^*$ and $m_0$ the effective and free electron masses, respectively. In the case of our $n^+nn^+$ diodes, the electric field is nearly zero in the $n^+$ regions, and therefore OPTTR takes place only in the $n$ region, where the electric potential is self consistently adapted to electron concentration, so that $E$ is not constant and exhibits interesting spatial and temporal features. In order to understand the phenomenon, profiles of OPE rate, electron concentration, electric field, electron energy and velocity along the previous GaN $n^+nn^+$ diode are shown in figure 2 for different applied voltages. The acceleration/stopping cycles are clearly observed in the figure, with a straight coincidence between high OPE rates and low electron velocity and energy zones. Interestingly, electron accumulations take place at those regions, enhancing the inhomogeneity of the electric field, which becomes spatially periodic. As the applied bias is raised, the number of OPE mechanisms undergone by electrons while crossing the *n* region increases, each one producing an electron accumulation.

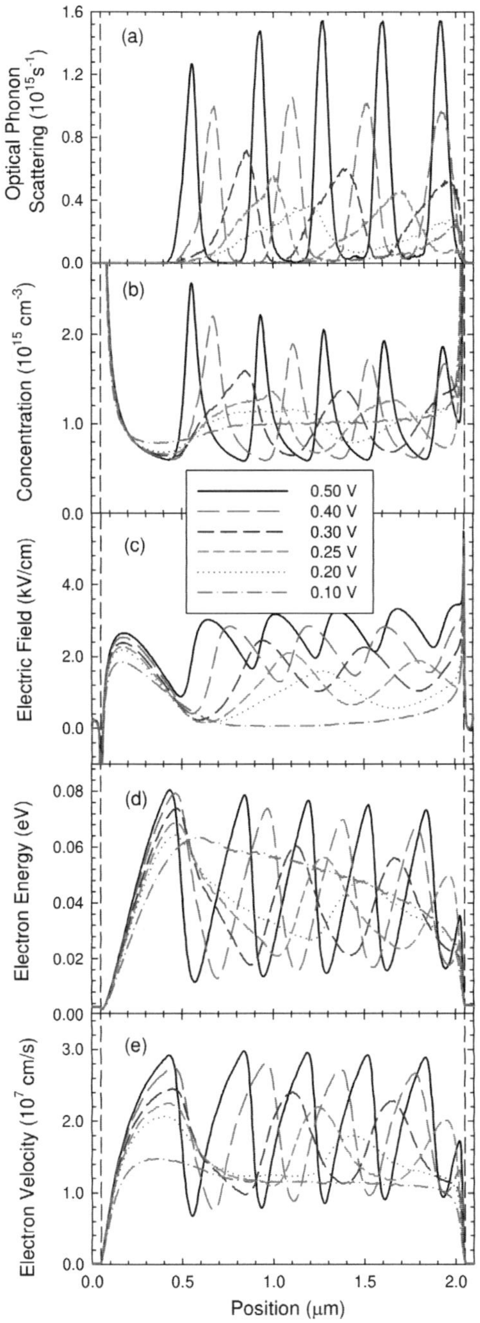


**Figure 2.** Profiles of (a) OPE rate (b) electron concentration, (c) electric field, (d) electron energy and (e) electron velocity along the GaN $n^+nn^+$ diode of figure 1 for different *V* at 15 K.

It is remarkable that for V=0.5 V, when almost no oscillation is observed in the current, electrons flowing through the *n* region still show well defined an spatially periodic OPE profile. This fact indicates that OPTTR is not enough for generating current oscillations; a feedback mechanism is additionally necessary for the time synchronization of the OPE events. This feedback is provided by the self-consistent electric field, which, modulated by the electron concentration, imposes a correlation between the OPEs taking place at different positions and time moments. When an OPE occurs at a given position, the local increase of electron concentration is self-consistently coupled to the electric field, which leads further OPEs to be delayed in time and shifted in space, thus originating the observed current oscillations. When the bias is too high, the spatial modulation of the electric field is much weaker due to very short distance between consecutive OPEs [figure 2(c)], what degrades the synchronization effect until oscillations disappear.

The current spectrum corresponding to *V*=0.25 V shows a clear peak at around 250 GHz, figure 3(a) (and also higher harmonics). In figure 3(b) the frequency of the oscillations is plotted vs. the applied bias, showing a good agreement with the ideal behavior of equation (1) (with 25 GHz error bars and using $E=V/L$, the mean electric field in the *n* region), even if the real electric field is strongly non-homogeneous. Regarding the amplitude of the oscillations, it takes the maximum value [figure 3(b), left axis] for an intermediate voltage, *V*=0.25 V, when the electric-field-related OPE modulation is optimum.

When the temperature is increased to 40 K, other scattering mechanisms compete with OPE (mainly optical phonon absorption), and thus the electron motion between consecutive OPEs is not completely ballistic, and the coherence of the OPTTR is degraded. This reflects in the amplitude of current oscillations [figure 3(b)], whose maximum appears at lower *V* and is one order of magnitude smaller than for 15 K, while their frequency still follows the theoretical value given by equation (1). If the temperature is further increased to 77 K, the OPTTR disappears and no current oscillation is observed.

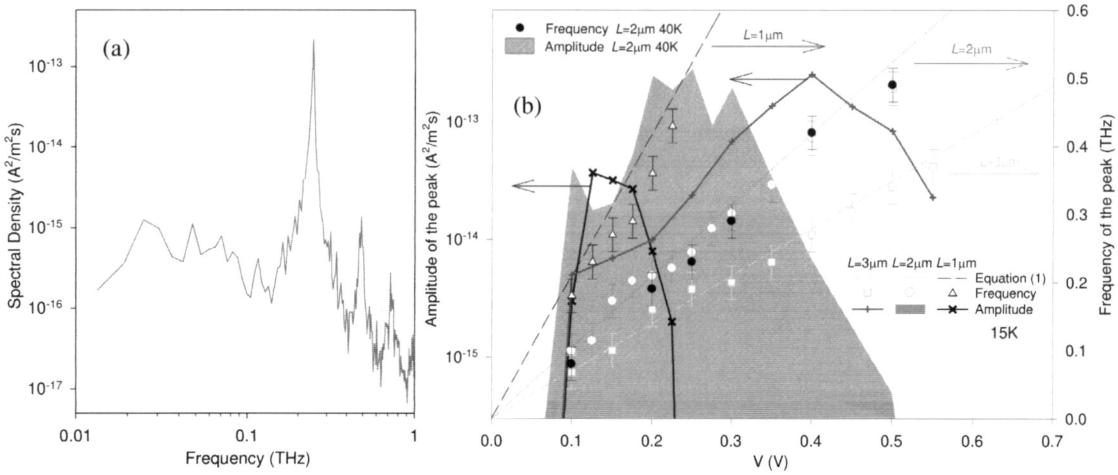


**Figure 3.** (a) Spectral density of the fluctuations of the current flowing through the GaN $n^+nn^+$ diode of figure 1 for $V$=0.25 V and 15 K. (b) Frequency of the oscillations (with 25 GHz error bars) and amplitude of the corresponding peak in the spectral density (right and left axes, respectively) as a function of the applied voltage for different lengths of the $n$ region. For $L$=2.0 μm results for 15 and 40 K are given. In (b) the theoretical oscillation frequency given by equation (1) is also plotted.

If the length of the $n$ region is decreased, the bias range for which the current oscillations come into view is reduced (and also their amplitude) since, as already explained, by shortening the distance between consecutive OPEs, the feedback provided by the electric field weakens. For $L$=1.0 μm, oscillations take place only in the range $V$=0.1-0.25 V, figure 3(b), with frequencies slightly lower than predicted by equation (1). If $L$ is further reduced to 0.5 μm, current oscillations do not appear at all. On the other hand, for $L$=3.0 μm, current oscillations are present in a much wider bias range (0.1-0.6 V). Interestingly, for all the values of $L$, the frequency of the oscillations providing maximum amplitude is around 200-250 GHz, in the range of the plasma frequency of the $n$ region, as expected by the strong relationship between OPTTR oscillations and plasma instabilities [2].

When the doping of the $n$ region is increased (results not shown), space charge effects are enhanced, and so does the feedback (now effective in shorter distances) associated to the electric field, thus providing oscillations with much higher amplitude that persist up to higher voltages. The calculated oscillation frequencies are larger than those predicted by equation (1). The significant value of the electric field in the $n^+$ regions and the higher plasma frequency of the $n$ region may cause that discrepancy.

## 4. Acknowledgements

This work has been partially supported by the Dirección General de Investigación (MEC) and FEDER (project TEC2007-61259/MIC and Acción Integrada HF2007-0014) and by the Consejería de Educación, Junta de Castilla y León (projects SA019A08 and GR270).

## References

[1] Starikov E, Shiktorov P, Gruzinskis V, Varani L, Palermo C, Sabatini G, Marinchio H, Laurent T and Reggiani L 2009 *J. Stat. Mech.* **2009** P01035
[2] Bakshi P and Kempa K 1995 *Superlatt. Microstruct.* **17** 363
[3] Gruzinskis V, Shiktorov P, Starikov E, Reggiani L, Varani L and Vaissiere J C 2004 *Semicond. Sci. Technol.* **19** S173
[4] Gruzinskis V, Shiktorov P and Starikov E 2008 *Acta Phys. Polon. A* **113** 943
[5] Starikov E, Shiktorov P, Gruzinskis V, Reggiani L, Varani L, Vaissiere J C and Zhao J H 2001 *IEEE Trans. Electron Devices* **48** 438

# Energy relaxation in IR laser excited $Hg_{1-x}Cd_xTe$

**A K Storebo[1], T Brudevoll[1], O Olsen[2], O C Norum[2] and M Breivik[3]**

[1]FFI (*Norwegian Defence Research Establishment*), P.O. Box 25, NO-2027 Kjeller, Norway
NTNU (*Norwegian University of Science and Technology*)
[2]*Department of Physics and* [3]*Department of Electronics and Telecommunications*
NO-7491 Trondheim, Norway

E-mail: asta-katrine.storebo@ffi.no

**Abstract.** IR laser excitation of $Hg_{1-x}Cd_xTe$ by low-fluence femtosecond and high fluence microsecond pulses was explored for the technologically important alloy fractions $x \sim 0.2$ and $x \sim 0.28$. We have used first principles (LAPW) electronic structure calculations and finite element modelling, supported by Monte Carlo simulation for the description of femtosecond pulse carrier relaxation and the transport parameters. Laser wavelengths considered were 6.4 - 10.6 μm for $x \sim 0.2$ and 3.8 - 4.8 μm for $x \sim 0.28$, with an incident 1 microsecond pulse fluence of 2 J/cm$^2$. Many energy transfer mechanisms are invoked due to the long timescales of the microsecond pulses, and a main challenge is therefore to elucidate how these interplay in situations away from thermal equilibrium. Mechanisms studied include one- and two-photon absorption (OPA and TPA) across the band gap, inter-valence band absorption (IVA) between light- and heavy hole bands, electron-hole recombination/impact ionization, band gap renormalisation, intra-band free carrier absorption (FCA), excess carrier temperatures, non-equilibrium phonon generation, and refractive index changes. In the high fluence case, lattice temperatures evolve considerably during the laser pulse in response to the heated carriers. The chosen photon energies lie just above the band gap at the starting lattice temperature of 77 K, and nonlinear effects therefore dominate as the material heats up and the band gap begins to exceed the photon energy. Because of the low photon energy we must rely on Auger recombination, inter-valence band absorption and free carrier absorption to heat the carrier plasma. Although some $Hg_{1-x}Cd_xTe$ material parameters are now relatively well known, existing data for many of the processes are inadequate for cases far away from thermal equilibrium. Furthermore, the role of Auger recombination in relation to non-intrinsic recombination has been a matter of debate lately. In this respect, information from experiments can only be interpreted correctly if the intrinsic behaviour of the material is well known, and we have therefore assumed intrinsic $Hg_{1-x}Cd_xTe$ for the present study.

## 1. Introduction

Excitation by low-fluence, ultrashort laser pulses is a central technique for extracting information about non-equilibrium carrier dynamics in semiconductors. Using photon energies far exceeding the fundamental band gap, such experiments can provide a very direct insight into carrier relaxation and hot phonon generation phenomena by isolating these processes from subsequent effects that occur on longer timescales. Hot, non-equilibrium optical phonons are then usually created due to the sudden, simultaneous relaxation of a large amount of photoexcited carriers entering the conduction band far above the conduction band edge. After the ultrashort laser pulse has passed and the short lived hot optical phonons have decayed, the overall lattice temperature is left almost unchanged. Energy transfer due to long, high fluence laser pulses on the other hand, increases the lattice temperature and involves many additional physical effects. In particular, if the photon energy of the incident radiation just exceeds the band gap, absorption can be highly nonlinear and depend strongly on wavelength, radiation intensity, carrier temperatures, and lattice temperatures. Using a long photon wavelength would also allow intra-band free carrier (FCA) and inter-valence band (IVA) absorption between the light and heavy hole bands to be studied. The high fluence in such experiments can be sufficient to raise the overall lattice temperature by several hundred K. Although the total number of carriers generated in the conduction band can be comparable to that of a short pulse experiment, the majority of carriers are well thermalized at carrier temperatures which lie much closer to the overall lattice temperature. The difference between carrier and lattice temperatures will often be smaller than 100 K. Nevertheless, such small carrier excess temperatures can have a major impact on processes occurring near the band gap. For example, a slightly elevated carrier temperature can strongly reduce the saturation or "bleaching" of one-photon absorption across the band gap, or affect other long-timescale parameters, like carrier lifetimes and impact ionization rates.

## 2. Simulation Results and Discussion

Short pulse Monte Carlo (MC) simulation results (electrons only) shown in figure 1 are for the material $Hg_{0.72}Cd_{0.28}Te$ excited at $\lambda = 1064$ nm assuming an initial lattice temperature of 300 K. Due to the small effective mass of the conduction band, the electrons receive most of the photonic energy. They enter the conduction band with an initial energy corresponding to ~ 6000 K. Delayed cooling at higher carrier densities is clearly seen, and this effect is attributed to the creation of hot phonons. The experiments carried out by Pelouch et al. [1] apparently did not show any strong hot phonon effect, but the presence of multiple relaxation time constants was seen. We plan to analyze this experiment more thoroughly in a future study, when further upgrades to our MC simulator have been introduced.

In the long pulse length case, Auger lifetimes are important for calculating carrier temperatures, since this process conserves the energy of the recombination within the carrier system. Interestingly, these lifetimes have become a matter of debate lately. Some of the measured lifetimes in $Hg_{1-x}Cd_xTe$ and similar materials are very short [2] while others are very long [3]. Theoretical predictions using simplified band models suggest relatively short Auger lifetimes [4], whereas results from more elaborate models [5] predict much longer lifetimes. For an update on possible causes for this (and e g the suspected contribution from non-intrinsic recombination mechanisms in the lifetime measurements) we recommend references [6] and [7]. Here we have taken an average between the conventional short estimates used in a previous work on $Hg_{0.72}Cd_{0.28}Te$ [8] and the longer estimates of Krishnamurty [5]. Effectively, this increases the Auger lifetime by a factor of 5-6 compared to [8]. Uncertainties of this magnitude are usually considered 'normal'. We further generalized the expressions to accommodate different carrier and lattice temperatures. High fluence simulations were carried out for two undoped 10 μm thick films of $Hg_{0.72}Cd_{0.28}Te$ and $Hg_{0.8}Cd_{0.2}Te$, in ideal thermal contact with a highly conductive underlying substrate with a temperature of 77 K before illumination. The samples were subjected to an incident radiation intensity of 2 $MW/cm^2$ with a laser pulse length of 1 microsecond. As expected, the introduced changes resulted in much smaller initial excess carrier temperatures than in the previous work [8], but more importantly, the physics now becomes more diversified. Obviously, any extraction of data from non-equilibrium laser experiments must be performed with utmost care, since controlling the state variables of the material is essential for an accurate measurement. With lower carrier temperatures, carrier migration to higher valleys becomes less important. Still, a much needed paper displaying the energy differences between e g the Γ valley and the L valley has recently been published [9]. In figure 1 we show our corresponding gap-corrected results using first principles LAPW [10] electronic structure calculations. Figure 2 shows the temporal evolution of physical variables at the surface during the 1 μs laser pulse, for two wavelengths in each material. Initially the photon energy exceeds the band gap by approximately 0.1 eV and 0.05 eV for the respective short and long wavelengths incident on each material. As the band gaps increase with

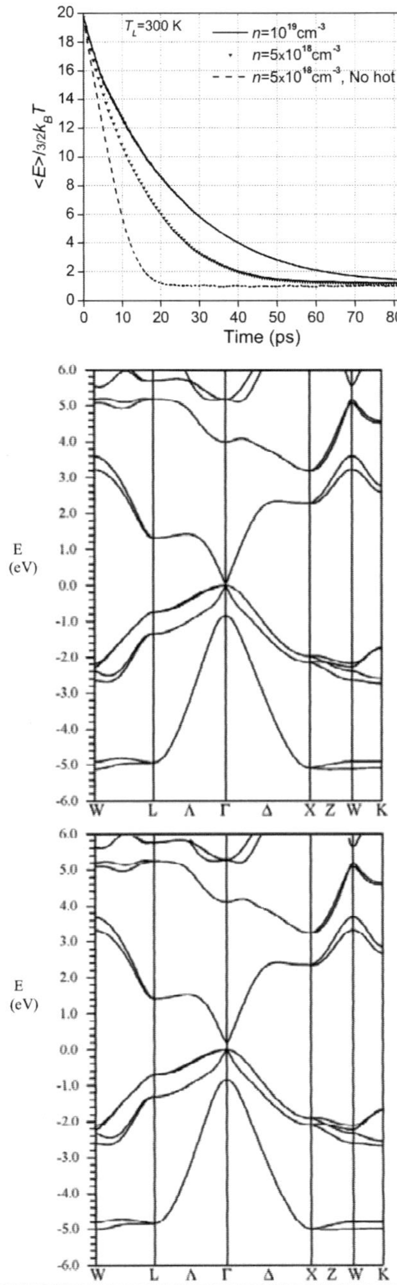


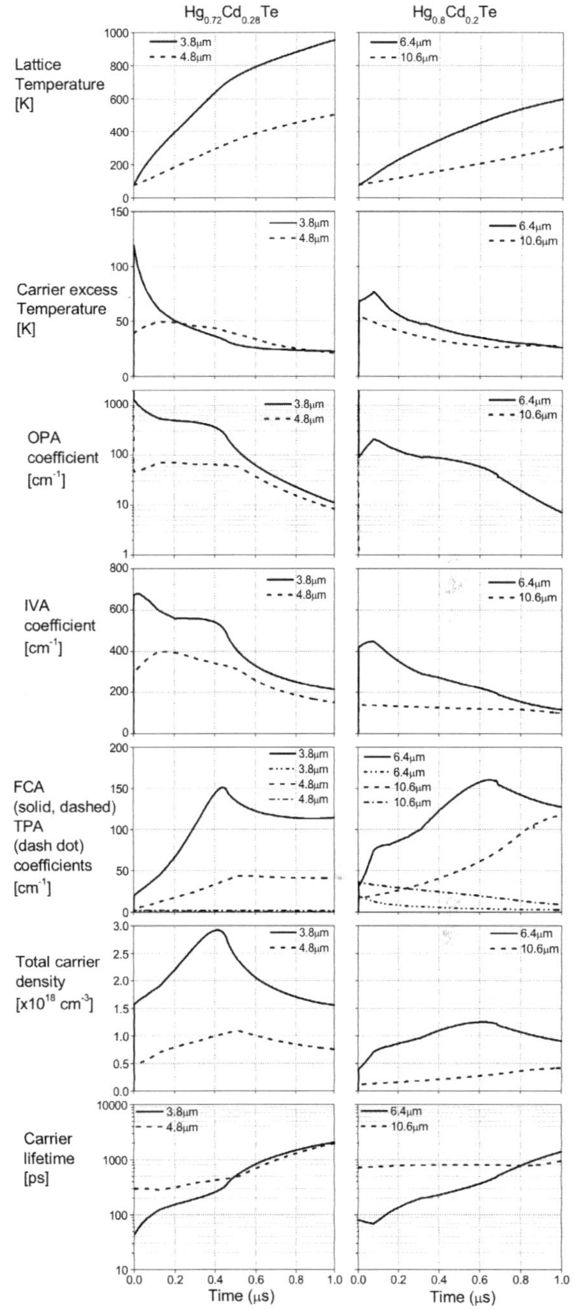


**Figure 1.** Energy relaxation after $\lambda$ = 1064 nm femtosecond laser pulse excitation, x = 0.28 (upper graph), and calculated band structures for $Hg_{0.8}Cd_{0.2}Te$ and $Hg_{0.72}Cd_{0.28}Te$ respectively, at 77 K (lower graphs).

**Figure 2.** Temporal evolution of some key variables at the surface during a 1 μs laser pulse. Two laser wavelengths for each material, $Hg_{0.72}Cd_{0.28}Te$ and $Hg_{0.8}Cd_{0.2}Te$. The melting point of $Hg_{1-x}Cd_xTe$ is 1000 K.

temperature, one-photon absorption (OPA) is weakened and the other energy transfer processes begin to dominate. In fact, only in $Hg_{0.72}Cd_{0.28}Te$ at a laser wavelength of 3.8 μm does OPA have a very prominent role. Here, carrier excess temperatures are high for a short period at the beginning of the pulse, and are driven by OPA combined with Auger recombination. After termination of regular OPA the lattice temperature increases more slowly, resulting in a 'kink' in the lattice temperature curve. For the other cases, OPA is severely bleached, and in $Hg_{0.8}Cd_{0.2}Te$, with its narrow conduction band, carriers generated by two-photon absorption (TPA) slightly reverses OPA at the 10.6 μm laser wavelength, resulting in an OPA absorption coefficient fluctuating around zero (not shown in figure 2).

## 3. Conclusions

Energy relaxation in two representative, intrinsic $Hg_{1-x}Cd_xTe$ alloys has been investigated. Specifically, we have investigated the low gap material $Hg_{0.8}Cd_{0.2}Te$ and the medium gap material $Hg_{0.72}Cd_{0.28}Te$. MC simulation of short pulse electronic relaxation shows reduced cooling due to the generation of hot phonons, but final conclusions have to await a more comprehensive study. For the long pulse length case, we have demonstrated a need for non-equilibrium rather than equilibrium parameters for the absorption and recombination mechanisms. This is necessary in order to determine excess carrier temperatures more accurately as a function of the applied laser intensity. In one case, high initial carrier excess temperatures obviously played an important role by dispersing the carriers over a large range of energy states, thereby reducing saturation (bleaching) of one-photon absorption across the band gap. Furthermore, we emphasized the importance of knowing the true intrinsic $Hg_{1-x}Cd_xTe$ behaviour when interpreting experimental results, and presented band plots displaying the energy differences between conduction band valleys.

## References

[1] Pelouch W S, Schlie, L A 1996 Ultrafast carrier dynamics and saturable absorption in HgCdTe *Appl. Phys.Lett* **68** (10), 1389-1391

[2] Chazapis V, Blom H A, Vodopyanov K L, Norman A G, Phillips C C 1995 Midinfrared picosecond spectroscopy studies of Auger recombination in InSb *Phys. Rev. B* **52** (4) 2516-2521

[3] Ciesla C M, Murdin B N, Phillips T J, White A M, Beattie A R, Langerak C J G M, Elliot C T, Pidgeon C R 1997 Auger Recombination Dynamics in Highly Excited HgCdTe *Phys. Stat. Sol.(b)* **204** 121-124

[4] Kinch M A, Aqariden F, Chandra D, Liao P-K, Schaake H F, Shih H D 2005 Minority Carrier Lifetime in p-HgCdTe *J. Electron. Mater.* **34** (6) 880-884

[5] Krishnamurty S, Yu Zhi Gang, Gonzalez L P, Guha S 2007 Accurate evaluation of nonlinear absorption coefficients in InAs, InSb, and HgCdTe alloys *J. Appl. Phys.* **101**, 113104

[6] Dubikovskiy V, Hagan D J, Stryland E W Van 2008 Large nonlinear refraction in InSb at 10 μm and the effects of Auger recombination *J. Opt. Soc. Am. B* **25** (2) 223-235

[7] Grein C H, Flatte M E, Chang Y 2008 Modeling of Recombination in HgCdTe *J. Electron. Mater.* **37**, 1415-1419

[8] Storeboe A K, Brudevoll T, Stenersen K 2008 Calculated temperature rise in midinfrared laser irradiated $Hg_{0.72}Cd_{0.28}Te$ *J. Appl. Phys.* **103**, 053108

[9] Penna M, Marnetto A, Bertazzi F, Bellotti E, Goano M 2009 Empirical Pseudopotential and Full-Brillouin-Zone **k·p** Electronic Structure of CdTe, HgTe, and $Hg_{1-x}Cd_xTe$ *J. Electron. Mater.* Published online 30. April 2009

[10] Blaha P, Schwarz K, Madsen G K H., Kvasnicka D, and Luitz J 2001 An Augmented Plane Wave+Local Orbitals Program for Calculating Crystal Properties, Vienna University of Technology

Journal of Physics: Conference Series **193** (2009) 012025 doi:10.1088/1742-6596/193/1/012025

# Evidence of non-local impact ionization in CNT and HgCdTe

**J S Marsland[1] and A G Atanu**

Department of Electrical Engineering and Electronics, University of Liverpool, Brownlow Hill, Liverpool, L69 3GJ, UK.

marsland@liv.ac.uk

**Abstract**. Two aspects of the non-local nature of impact ionization, dead space and resonance, are investigated. The very small excess noise factor measured for mercury cadmium telluride photodiodes can only be explained if the hole to electron ionization coefficient ratio, $k$, is very small and the impact ionization dead space is also considered. A maximum value of $k$ for HgCdTe is estimated in this paper. In addition, recent measurements of the reverse photocurrent in single wall carbon nanotubes show a well defined flat region at a multiplication of 1.6. This is argued to be evidence for resonant behaviour in impact ionization for carbon nanotubes.

## 1. Introduction

The classic 1966 theory of McIntyre [1] gives the following relationship for the excess noise factor, $F(M)$, as a function of the multiplication, $M$, and the ionization coefficient ratio, $k$.

$$F(M) = kM + \left(2 - \tfrac{1}{M}\right)(1 - k) \quad (1)$$

For diodes where the injected carriers are electrons, $k$ is the ratio of the hole ionization coefficient, β, to the electron ionization coefficient, α. McIntyre [1] assumed that the ionization coefficients are a function of the local electric field only. However, the probability of impact ionization depends upon kinetic energy rather than field, so that, although ionization coefficients will tend to an asymptotic value for a given field, there is a region where carriers can not impact ionize because they have insufficient energy. This 'dead space' region is a non-local effect and was first investigated by van Vliet et al [2] in 1979, then independently in 1990 by Marsland [3] and Saleh et al [4] and by numerous others since. These papers all demonstrated that the dead space acts to reduce the excess noise factor so that equation (1) is an upper limit which applies in cases where the dead space effect can be neglected. HgCdTe photodiodes provide the clearest experimental evidence [5,6] of the dead space effect. The very small excess noise factor at high multiplications requires the combination of a significant dead space effect and $k$ to be very small. However an upper limit for $k$ in HgCdTe has not been established and that is addressed in this paper.

Marsland [7-9] has predicted that there is a resonance effect in addition to the dead space effect. Although resonance effects have been demonstrated by Monte Carlo simulation for HgCdTe [10], there has been no experimental evidence due to the very high electric field required. However recently Gabor et al [11] have reported reverse photocurrent measurements in single wall carbon nanotubes

[1] To whom any correspondence should be addressed.

which show a distinct flat region at a multiplication of 1.6. It is argued in section 3 that this is the first experimental evidence for the resonance effect.

## 2. Non-local impact ionization in mercury cadmium telluride

Multiplication of up to 5,300 has been measured in HgCdTe photodiodes with excess noise factors very close to unity [5,6]. McIntyre's noise equation predicts an excess noise factor $F(M) \approx 2 + 5300k$ for $M = 5300$. An excess noise factor significantly less than 2 can only be explained if the non-local nature of impact ionization is considered as has been demonstrated by [2-4]. The Monte Carlo study reported by Ma et al [10] showed that a resonant behaviour was observed in the excess noise factor as shown by the solid line in figure 1. Very similar resonant behaviour has been predicted by Marsland [8,9] for $k = 0$ and that is shown by a dashed line on figure 1. There is excellent agreement between the two results for $M \leq 2$. It should be noted that if each injected electron results in a multiplication of either 1 or 2, and there is no possibility of a multiplication greater than 2, then the excess noise factor is given by equation (2) for a multiplication (average over all carriers) in the interval 1 to 2.

$$F(M) = (3M - 2)/M^2 \qquad 1 \leq M \leq 2 \qquad (2)$$

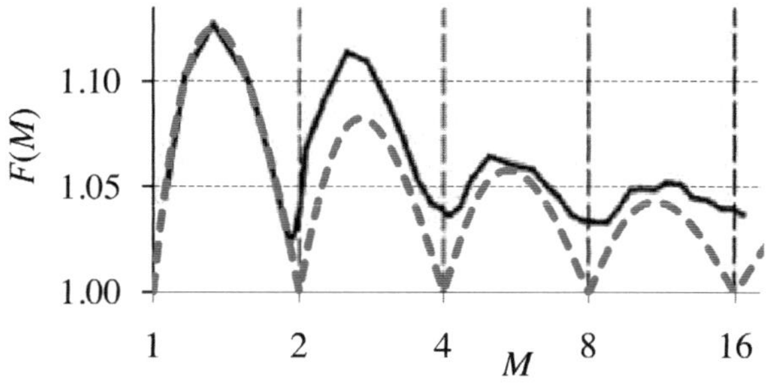


**Figure 1.** Excess noise factor versus multiplication; solid line Monte Carlo [10], dashed line [8,9] $\sigma$=1 $\lambda$=0.00333 $l$=0.0415

The results diverge for $M > 2$ although they both show the same resonant minima at 4, 8 and 16. Exact agreement is not expected because the results from [8,9] assume a parameter set ($\sigma$=1, $\lambda$=0.00333, $l$=0.0415) appropriate for an extremely high electric field. Also there may be some sampling uncertainty in the Monte Carlo simulation and the curve shown in figure 1 from [8,9] assumes $k = 0$. More recent Monte Carlo simulations [12] do not reproduce the resonance effect when more realistic alloy scattering and impact ionization rates are used and this may be why the resonance effect has not been observed experimentally.

### 2.1. Excess noise factor for $M = 1000$ and $k \neq 0$.

Following the methodology described in [8] the excess noise factor is calculated for $M$ up to 1000 and non zero values of $k$. Whereas $k$ can be unambiguously defined when $\alpha$ and $\beta$ are history independent, in the nonlocal case this is not as straightforward because $\alpha$ and $\beta$ are dependent on the distance, $z$, travelled from a point ($z = 0$) where the carriers have no kinetic energy. In this paper $k$ is taken to be the ratio of the asymptotic values of the ionization coefficients given by the following equation.

$$k = \frac{\beta(z\to\infty)}{\alpha(z\to\infty)} \qquad (3)$$

The ionization coefficients are parameterized using three coefficients; $l$ the dead space length and $\sigma$ and $\lambda$ which determine the slope of the rise and fall of the ionization pathlength p.d.f. as described in [9]. The asymptotic ionization coefficients are given by equation (4).

$$\alpha, \beta(z \to \infty) = (\lambda(\sigma + \sigma^{-1}) + l)^{-1} \qquad (4)$$

In the results shown in figure 2, two sets of parameters have been used for $\alpha$; for the dotted line and the squares the parameters are $\sigma$=$\lambda$=0.3162 and $l$=0.15 representing a high field with an overshoot factor of 0.97 and for the dashed line and circles, $\sigma$=1, $\lambda$=0.00333, $l$=0.0415, representing an extremely high field with an overshoot factor of 5.36. The parameters for $\beta$ are selected to achieve either $k$ = 0.01 or 0.001 with $l$ constant for both carriers and either $\sigma$ constant (open symbols) or the $\sigma\lambda$ product constant (closed symbols). The dashed and dotted lines show the $k$ = 0 case and the solid line shows McIntyre's equation (1) for comparison.

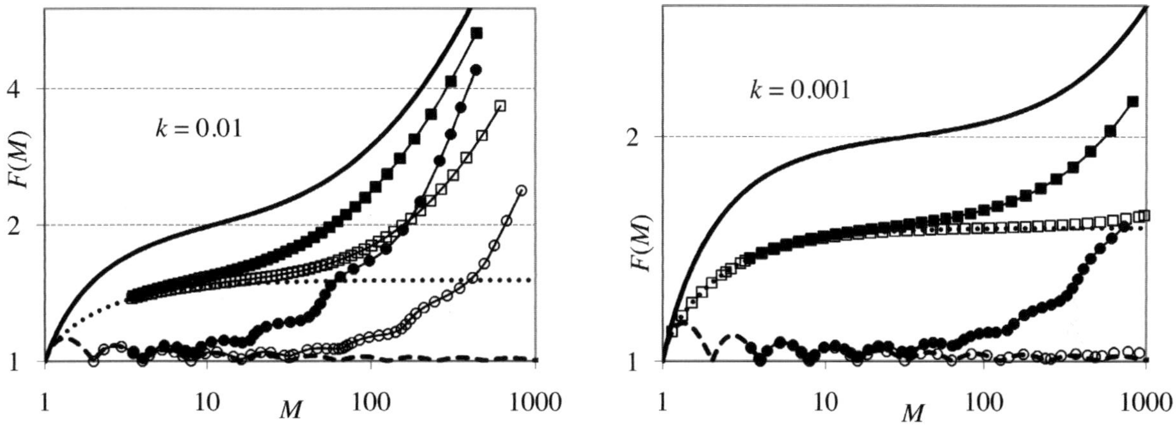


**Figure 2.** Excess noise factor versus multiplication. Solid line McIntyre's equation (1), dotted line $k$ = 0 overshoot = 0.97, dashed line $k$ = 0 overshoot = 5.36, squares overshoot = 0.97, circles overshoot = 5.36, closed symbols $\sigma\lambda$ constant, open symbols $\sigma$ constant.

Considering the graph for $k$ = 0.01, all four sets of results give $F(M)$ greater than 2 at $M$ less than 1000 so it can be concluded that $k$ is less than 0.01 for HgCdTe. The open symbols represent the best case (i.e. least noise) but they are probably unrealistic because the constant $\sigma$ assumption results in a nonlocal ionization coefficient for holes that has a very 'soft' turn on; $\beta$ only achieves 53% (squares) and 28% (circles) of its asymptotic value within the multiplication region.

As expected $F(M)$ is less for $k$ = 0.001. Indeed using the constant $\sigma$ approach (open symbols) there appears to be very little deviation from the $k$ = 0 case. However this is misleading because the hole ionization has an extremely soft turn on so that the nonlocal ionization coefficient for a hole that traverses the complete multiplication region is only 9% (squares) and 4% (circles) of $\beta(z\rightarrow\infty)$. So $k$ is meaningless in these cases and the closed symbols are more realistic for $k$ = 0.001. In conclusion the experimental results for $F(M)$ in HgCdTe implies that $k$ is less than 0.001 in that material.

## 3. Evidence for the resonance effect in single wall carbon nanotubes

Avalanche multiplication and breakdown has been observed in semiconducting single wall carbon nanotubes (CNT) [13,14]. Recently CNT pin diodes, fabricated and measured at Cornell University and University of Alberta [11], have shown a stepped response in the photocurrent as shown in figure 3(a). These measurements were made at 4K using CNT with a bandgap of 0.28eV. The reverse bias data from the three curves in figure 3(a) with the greatest photon flux is plotted in figure 3(b) normalized to a multiplication, $M$, of 1 at zero applied bias.

The flattening in the multiplication curve at $M$ ~ 1.6 and possible further flat regions at higher multiplications may be the first experimental observation of the resonance effect predicted in [8]. However the first predicted level region occurs at $M$ = 2. The lower multiplication value of 1.6 can be explained by considering phonon interactions in CNT. Unlike conventional semiconductors, the optical phonon m.f.p. and energy is a significant fraction of the m.f.p. and threshold energy for impact ionization. Under these conditions, the ionization pathlength p.d.f. can split into two (or more) peaks; a peak at shorter pathlengths, of population $N$, for quasi-ballistic carriers which do not emit an optical phonon and further peak (or peaks), of population $(1 - N)$, for carriers that have one or more optical phonon interactions. The first flat region in the multiplication curve would correspond to $M = (1 + N)$. An excess noise factor equal to 1 has been predicted [8] for level regions in the multiplication curve at $M = 2, 4, 8$, etc. However the same cannot be predicted for a level region at $M = 1.6$ where $F(M) \approx 1.1$ using equation (2).

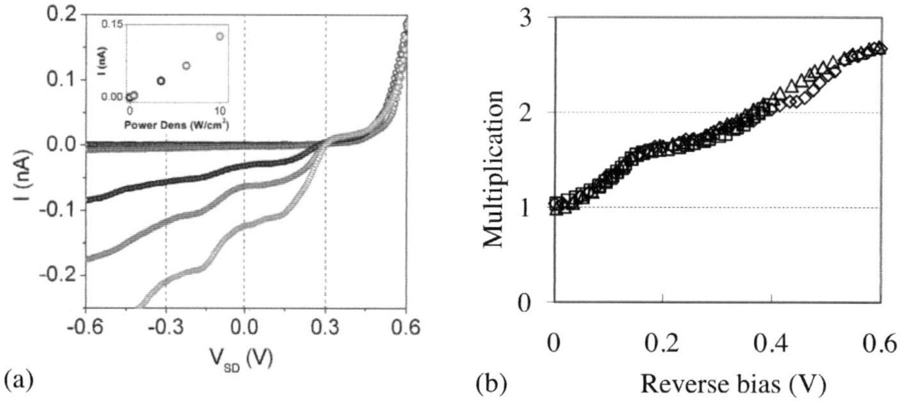


**Figure 3.** (a) Photocurrent versus applied voltage from [11] and (b) multiplication versus reverse bias.

## 4. Conclusion

Calculations of the excess noise factor for small, non-zero $k$ values using non-local ionization coefficients indicate that $k$ is less than 0.001 in HgCdTe. In photocurrent measurements made on carbon nanotubes, a distinct flat region at a multiplication of 1.6 is argued to be the first experimental evidence for the resonance effect.

## References

[1] McIntyre R 1966 *IEEE Trans Electron Dev.* **ED-13** 164
[2] Van Vliet KM, Friedmann A and Rucker LM 1979 *IEEE Trans Electron Dev.* **ED-26** 752
[3] Marsland JS 1990 *J. Appl. Phys.* **67** 1929
[4] Saleh BEA, Hayat MM and Teich MC 1990 *IEEE Trans Electron Dev.* **ED-37** 1976
[5] Beck J, Wan C, Kinch M, Robinson J, Mitra P, Scritchfield R, Ma F and Campbell J 2006 *J. Electron. Mater.* **35** 1166
[6] Perrais G, Gravrand O, Baylet J, Destefanis G and Rothman J 2007 *J. Electron. Mater.* **36** 963
[7] Marsland JS 2002 *Electron. Lett.* **38** 55
[8] Marsland JS 2009 *J. Mater. Sci: Mater. Electron.* **20** S514
[9] Marsland JS 2009 *Phys. Stat. Sol. A* **206** 874
[10] Ma F, Li X, Campbell JC, Beck JD, Wan C and Kinch MA 2003 *Appl. Phys. Lett.* **83** 785
[11] Gabor NM, Zhong Z, Bosnick K, Park J and McEuen PL 2009 "Extremely efficient multiple electron-hole pair generation in carbon nanotube photodiodes", accepted for publication in *Science*
[12] Derelle S, Bernhardt S, Haidar R, Primot J, Deschamps J and Rothman J 2009 *IEEE Trans Electron Dev.* **ED-56** 569
[13] Bosnick K, Gabor N and McEuen P 2006 *Appl. Phys. Lett.* **89** 163121
[14] Liao A, Zhao Y and Pop E 2008 *Phys. Rev. Lett.* **101** 256804

# Carrier transport in THz quantum cascade lasers: Are Green's functions necessary?

**A Mátyás[1,2], T Kubis[3], P Lugli[2] and C Jirauschek[1,2]**

[1] Emmy Noether Research Group "Modeling of Quantum Cascade Devices", TU München, D-80333 München, Germany
[2] Institute of Nanoelectronics, TU München, D-80333 München, Germany
[3] Walter Schottky Institute, TU München, D-85748 Garching, Germany

E-mail: alparmat@mytum.de

**Abstract.** We have applied two different simulation models for the stationary carrier transport and optical gain analysis in resonant phonon depopulation THz Quantum Cascade Lasers (QCLs), based on the semiclassical ensemble Monte Carlo (EMC) and fully quantum mechanical non-equilibrium Green's functions (NEGF) method, respectively. We find in the incoherent regime near and above the threshold current a qualitative and quantitative agreement of both methods. Therefore, we show that THz-QCLs can be successfully optimized utilizing the numerically efficient EMC method.

## 1. Introduction

Accurate transport models are required to identify the detrimental effects [1, 2, 3, 4] that decrease the occupation inversion and the optical gain in state of the art THz quantum cascade lasers. Two of the most widely used models are the non-equilibrium Greens function method (NEGF) and the Ensemble Monte Carlo method (EMC). The NEGF method treats incoherent scattering such as energy dissipation and momentum relaxation as well as coherent effects such as multibarrier tunneling and correlation effects on an equal footing [5]. However exact inclusion of higher order many particle interactions such as the inelastic electron-electron scattering is typically unfeasible. In particular the numerical load of NEGF calculations usually prohibits a systematic optimization of QCL structure parameters. The EMC method is based on the semi-classical solution of the Boltzmann equation and provides a straightforward way of taking two-body processes such as electron-electron scattering into account. The most important advantage of EMC, however, is the comparably low numerical complexity which allows the systematic study of all QCL parameters. Nevertheless, EMC does not consider non-diagonal scattering [6] and misses an exact and consistent treatment of the broadening of the laser states. We show in this paper that the EMC method reproduces NEGF results in regimes where coherent transport plays a minor role. Since the threshold current of typical resonant phonon depopulation structures lies in the incoherent regime [7], the EMC method is a well suited method for modeling transport around the operation bias. In particular, the numerical efficiency of EMC allows for the systematic improvement of future THz-QCL designs.

Journal of Physics: Conference Series **193** (2009) 012026 doi:10.1088/1742-6596/193/1/012026

## 2. Results

We apply the EMC and NEGF method on stationary transport in the THz-QCL structure of [8] at a lattice temperature of 40 K and a sheet doping density of 1.95 $\times 10^{10}$ $cm^{-2}$. A single period of this QCL consists of a sequence of GaAs and $Al_{0.15}Ga_{0.85}As$ layers of the widths (30) 92 (55) 80 (27) 66 (41) 155 Å, with the $Al_{0.15}Ga_{0.85}As$ barriers given in parenthesis. Please find details of our implementation of the EMC method in [1, 9] and the NEGF method in [5]. In both methods, we take all scattering mechanisms relevant for this QCL, i.e. electron-phonon, interface roughness [10] scattering and electron-electron interaction in the Hartree approximation into account. In addition, we also implement inelastic electron-electron scattering in the EMC method.

In order to illustrate the typical confinement energies and resonant laser states, figure 1 shows the magnitude squared of the wavefunctions used in the EMC method for the specified QCL at the peak gain bias of 54.16 mV per period. Upper and lower laser levels are marked by solid lines.

States of adjacent periods are separated by the energy of an LO-phonon (36 meV). Thus, electrons traverse between the states of adjacent periods by the resonant emission of an LO-phonon. Thereby, the lower laser level gets emptied [11, 12].

We want to emphasize that the EMC method models electronic transport with eigenfunctions of the Hermitian Hamiltonian of the QCL structure. Incoherent scattering mechanisms that limit the electronic lifetime are not included in the determination of the QCL states. Thus, the electronic states in figure 1 are given as discrete levels. In EMC, elastic and inelastic scattering is only included in the calculation of the state occupancy. In contrast, the resonant energies, the state linewidths and state occupancies are self-consistently determined in the NEGF method.

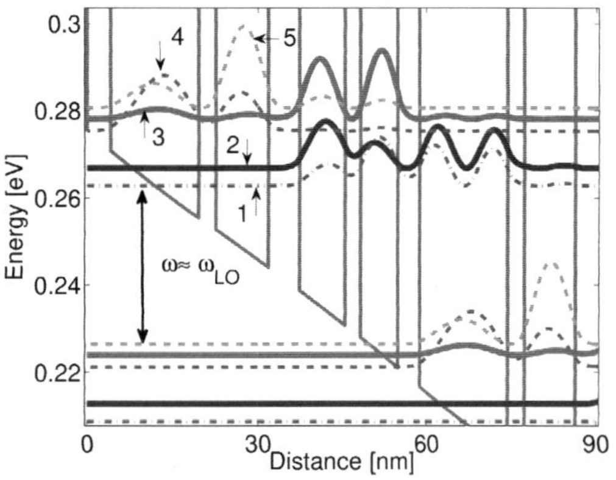


**Figure 1.** Energy levels and squared wavefunctions of the QCL at the peak-gain bias of 54.16 mV per period given by the EMC method.

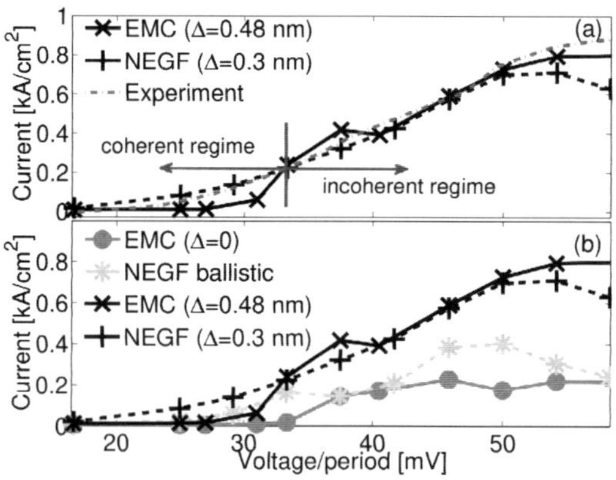


**Figure 2.** (a) Comparison of the simulated currents to experimental results. (b) Different cases of transport, which are shown to illustrate the contribution of the different scattering and transport mechanisms.

In spite of this discrepancy, both models reproduce the experimental current-voltage (IV) characteristics for bias voltages above 30 mV per period, as shown in figure 2 (a). For bias voltages below 30 mV per period, the EMC method yields a very small current density. In this voltage regime, the energy separation of the electronic states is too small to allow for the emission of LO-phonons. Here, the dominant transport mechanism is the coherent multi-barrier tunneling [5]. This is illustrated in more detail in figure 2 (b), as it shows the IV-characteristics resulting from a ballistic NEGF calculation when all incoherent scattering mechanisms are neglected (grey-dashed). At low bias voltages, results of the ballistic calculation agree with results of NEGF calculations including incoherent scattering (black-dashed). This shows that the transport is coherent in this regime. Since coherent transport is neglected in EMC (solid in figure 2(a)), the EMC method underestimates the current density in this regime. However, we want to point out that the current density of the EMC method remains very small in exactly that bias range which the NEGF formalism identifies as the coherent regime.

For voltages above 30 mV per period, the incoherent regime sets in. Here, incoherent scattering significantly enhances the current density in both methods. This is illustrated in figure 2 (b) for the example of interface roughness scattering. Calculations ignoring rough interfaces yield in the NEGF formalism (green dashed) and in the EMC method (orange solid) a significantly smaller current density than the respective calculations including this effect (black dashed for NEGF, black solid for EMC). Although the two methods require different interface roughness parameters (slightly larger interface roughness step height) to fit the experimental current, EMC can reproduce results of fully quantum mechanical calculations in the incoherent regime.

A well known issue comes about when the discrete sharp energy levels in the EMC method get aligned [13]. Any anticrossing of discrete states yield highly delocalized wavefunctions. Thus, the EMC method tends to overestimate the form factor and accordingly the scattering rate when discrete states are aligned. In this situation, the overestimation of scattering in the EMC method causes artificial current spikes [1] and often unreliable gain. For this reason, bias voltages that correspond to aligned discrete states are commonly avoided in EMC calculations [13].

In contrast to the IV-characteristics, the optical gain is sensitive to the linewidth of the resonant states for every bias voltage. Therefore, we have augmented in [9] the EMC method by a procedure that generates a spectrum of resonant QCL states with finite lifetime. In this procedure, the discrete energy spectrum in figure 1 is multiplied by Lorentzian functions of widths that correspond to the calculated out-scattering rates. In order to show that this procedure gives reasonable results, we compare in figure 3 the local density of states (LDOS) of the EMC (a) and the NEGF method (b).

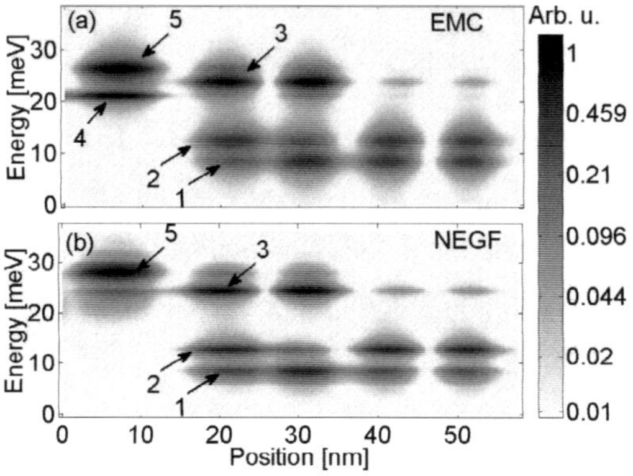


**Figure 3.** (a) EMC and (b) NEGF LDOS of the simulated QCL at the peak-gain bias of 54.16 mV per period. We observe similar linewidths and the missing level #4 in case of NEGF.

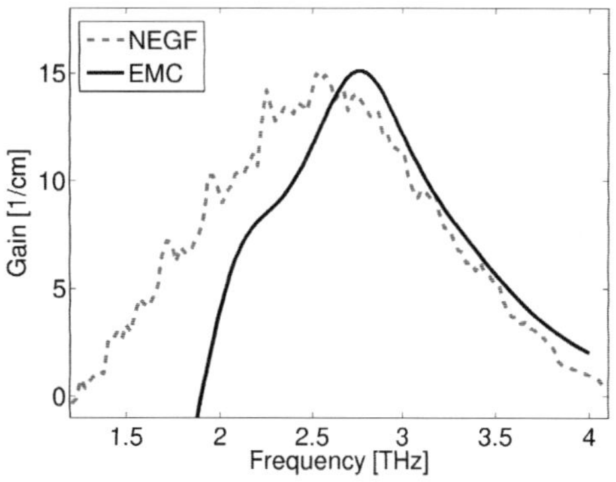


**Figure 4.** The optical gain in EMC (solid) and NEGF (dashed) at a bias of 54.16 mV per period. We get quantitative agreement. The slight assymetry in EMC is explained in the text.

We see a reasonable agreement between the resulting LDOS of both methods. The only significant difference is the missing resonance #4 in the NEGF calculation. The resonance #4 in the EMC method originates from an anticrossing of state #5 with a state of the preceding QCL period (see figure 1 and 3 (a)). We model in our NEGF implementation adjacent QCL periods with field free leads [14]. Therefore, the states of the preceding period lie at slightly different energies and this state anticrossing with state #5 is absent in the NEGF model (compare figure 3 (b) with (a)).

For the calculation of gain in the EMC method we use the same procedure as for the LDOS, but here the Lorentzians are weighted with the occupation inversion [9]. Optical gain in the NEGF method is calculated in linear response [5] taking into account the occupation inversion and the finite linewidths. The qualitative agreement in the LDOS of both models (see figure 3) explains the good agreement of the predicted gain spectra shown in figure 4. We find the small discrepancy for photon frequencies below approximately 2 THz to originate from the missing resonance #4 in the NEGF calculation, see figure 3 (a).

## 3. Conclusion

We have compared results of the semi-classical EMC and the quantum mechanical NEGF method for stationary charge transport in a recently fabricated THz-QCL. We find in the incoherent transport regime that the semi-classical transport model nicely agrees in the predicted current density and optical gain with results of the NEGF method. In particular, both models reproduce experimental results. For bias voltages significantly below threshold, coherent multi-barrier tunneling dominates the transport. Since EMC does not include this type of transport, it underestimates the current density in this regime. Nevertheless, the optimization of QCL designs requires reliable predictions close to and above threshold. Therefore, we have shown that for the purpose of design optimization, the numerical load of a detailed NEGF calculation can be avoided and the numerically efficient EMC method is appropriate.

## 4. Acknowledgments

C Jirauschek and A Mátyás acknowledge support the Emmy Noether program of the German Research Foundation (DFG, JI115/1-1), T Kubis acknowledges support from the Deutsche Forschungsgemeinschaft (SFB 631), the Österreichische Fonds zur Förderung der Wissenschaft (SFB IRON) and the Nano Initiative Munich.

## References

[1] Jirauschek C, Scarpa G, Lugli P, Vitiello M S and Scamarcio G 2007 *Journal of Applied Physics* **101** 086109
[2] Belkin M A, Fan J A, Hormoz S, Capasso F, Khanna S P, Lachab M, Davies A G and Linfield E H 2008 *Optics Express* **16** 3242
[3] Jirauschek C 2009 *IEEE Journal of Quantum Electronics* **45** (in print)
[4] Kumar S, Hu Q and Reno J L 2009 *Applied Physics Letters* **94** 131105
[5] Kubis T, Yeh C, Vogl P, Benz A, Fasching G and Deutsch C 2009 *Physical Review B* **79** 195323
[6] Lee S C, Banit F, Woerner M and Wacker A 2006 *Physical Review B* **73** 245320
[7] Kubis T, Yeh C and Vogl P 2007 *Physica Status Solidi C* **5** 233–235
[8] Benz A, Fasching G, Andrews A M, Martl M, Unterrainer K, Roch T, Schrenk W, Golka S and Strasser G 2007 *Applied Physics Letters* **90** 101107
[9] Jirauschek C and Lugli P 2009 *Journal of Applied Physics* **105** 123102
[10] Leuliet A, Vasanelli A, Wade A, Fedorov G, Smirnov D, Bastard G and Sirtori C 2006 *Physical Review B* **73** 085311
[11] Williams B S, Callebaut H, Kumar S, Hu Q and Reno J L 2003 *Applied Physics Letters* **82** 1015
[12] Kumar S, Williams B S, Kohen S, Hu Q and Reno J L 2004 *Applied Physics Letters* **84** 2494
[13] Callebaut H, Kumar S, Williams B S, Hu Q and Reno J L 2003 *Applied Physics Letters* **83** 207
[14] Kubis T, Yeh C and Vogl P 2007 *Journal of Computational Electronics* **7** 432–435

# Electronic and magnetic states in core multishell nanowires: Edge localization, Landau levels and Aharonov-Bohm oscillations

**G Ferrari[1], G Cuoghi[2], A Bertoni[3], G Goldoni[2,3] and E Molinari[2,3]**

[1] CNISM Unità di Ricerca di Modena, [2] Dipartimento di Fisica, Università di Modena e Reggio Emilia, [3] CNR-INFM S3 Research Center, Via Campi 213/A, 41100 Modena, IT

E-mail: giulio.ferrari@unimore.it

**Abstract.** We study the electronic states of hexagonal core multishell semiconductor nanowires, including the effect of magnetic fields. We find that the two dimensional electron states formed at the interface between different layers are mostly localized at the six edges of the hexagonal prism, and behave as a set of quasi-1D quantum channels. They can be manipulated by magnetic fields either parallel or perpendicular to the wire axis. These results can be rationalized in terms of Aharonov-Bohm oscillations or Landau level formation. We also show that inter-channel coupling and magnetic behavior is influenced by the geometric details of the nanowires.

## 1. Introduction

New inorganic semiconductor systems are emerging where carriers are confined in non-planar two dimensional electron states (2DESs). Among these nano-structures, we study the properties of 2DESs formed in multishell semiconductor nanowires (NWs) with hexagonal cross-section. Samples of this type have been recently realized experimentally [1, 2]. The 2DES is obtained in coaxial structures which are fabricated similarly to standard layered heterostructures, but using a hexagonal substrate, provided by a free standing NW, rather than the usual planar substrate, for multilayer overgrowth of lattice matched materials. The resulting 2DESs have a diameter in the tens of nm range and a cross-section shape determined by the NW used as a substrate.

The application of a magnetic field in the axial direction induces Aharonov-Bohm (AB) oscillations of the energy of the system, due to the enclosing of the magnetic flux through the shells. Since the semiconductor NWs have relative large diameters, the measure of such effect is quite accessible [3, 4]. The magneto-transport properties of similar systems, with a cylindrical cross-section, have been also investigated by applying a magnetic field perpendicular to the tube axis. Peculiar magneto-resistance properties emerged, that have no correspondence in similar planar structures [5].

## 2. Edge localization

In order to properly describe 2DESs with hexagonal cross-sections, the 2D quantum equation of motion for carriers on curved surfaces has been adopted. In fact, it can be derived by a limiting procedure starting from the full 3D Schrödinger equation [6, 7]. This technique allows

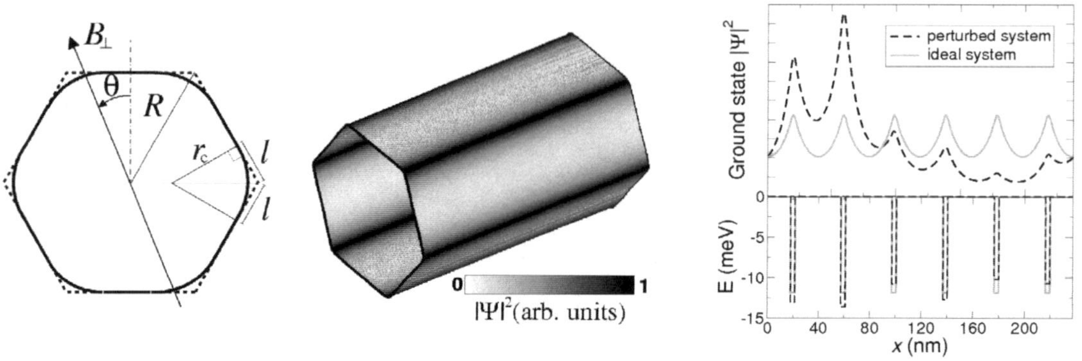


**Figure 1.** Left: the cross-section of an hexagonal NW is modeled from a regular hexagon (dashed line). The radius of the circumscribed circle is $R$. Each edge is rounded for a length $l$ and with a constant radius of curvature $r_c$. A transverse magnetic field $B_\perp$ forms an angle $\theta$ with a line normal to a facet. The azimuthal position on the rounded polygon (solid line) is given by an $x$ coordinate. Center: charge density of the ground state of a carrier in a core multishell NW (gray scale), showing preferential localization along the edges (darker regions). The state shown here is for a GaAs-based nanostructure. Right: confining potential $V_{geo}(x)$ (bottom panel) and charge density $|\Psi|^2$ of the ground state, shown both for the ideal case (solid line) with all the wells of equal depth, and for a perturbed system (dashed line) with values for each well of chosen randomly between $\pm 15\%$.

to model the effective attractive potential arising in the curved regions of 2D nanostructures [2]. The left panel of Fig. 1 shows the model of the system, whose cross-section results to be a rounded-hexagon. Electrons are strictly confined to the perimeter of the hexagon. In order to simulate the round-off of the edges of realistic devices, the full length of the side of the hexagon is $R$, but a piece of length $l$ is removed at every side of each edge, and replaced by a round corner with a constant radius s of curvature $r_c = \sqrt{3}l$. This finite curvature gives rise to an attractive potential [6, 7], whose intensity is given by

$$V_{geo} = -\frac{\hbar^2}{8m^* r_c^2}, \tag{1}$$

resulting, for the present model, in a set of six square wells located at the edges, as shown in the bottom of the right panel of Fig. 1. In the above formula, $m^*$ is the effective mass, that has been taken to be that of a GaAs shell. The parameters for the rounded-hexagonal cross-section have been taken from Ref.[2], that is $R = 40$ nm and $l = 2$ nm, so that the resulting wells of Fig. 1 are 11.9 meV deep and 4 nm wide. This description corresponds to an ideal case, where the growth of the NW and the overgrowth of the shells is perfectly homogeneous all around the wire. The magnetic states of this ideal case have been studied in detail in Ref.[8]. In real cases, it is most likely that fluctuations may occur in the growth, so that the corner sharpness or the shell thickness might be inhomogeneous, breaking the 6-fold symmetry of the system.

In order to study the effect of the above symmetry breaking, we have randomly varied the depth of the wells, and kept the width unaltered (as shown in the right panel of Fig. 1), and then studied the effect on the electronic and magnetic states. The upper graph of the right panel of Fig. 1 shows the square moduli of the eigen-functions corresponding to the ground states. Due to the presence of the wells at the corners, the probability-density of the ground state shows maxima in the edges, this leading to the formation of six 1D channels in the hexagonal 2DES with higher probability density in the edges. We note how the random perturbation of the well

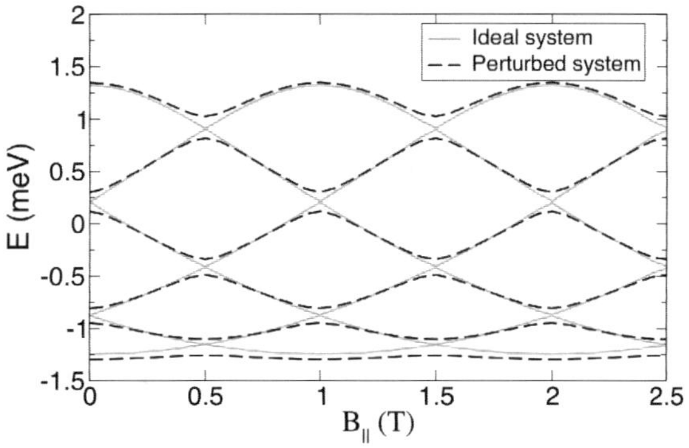


**Figure 2.** Magnetic levels as a function of an axial magnetic field. In the ideal case (solid line) the AB oscillations affect groups of six levels that cross each other while oscillating, while in the perturbed case (dashed line) the energy degeneracy occurring at $\Phi = (n + 1/2)\Phi_0$ $(n = 0, 1, 2...)$ is removed and each level oscillates with the AB period$\Phi_0$.

depths changes the relative probabilities of finding the carrier in each edge, but still preserves overall the edges as regions of preferential localization. The same edge-localization has been demonstrated also for other materials and shapes of the NWs [8].

## 3. Axial field B$_{||}$

When a magnetic field is introduced, the expression of the Schrödinger equation on a curved surface should be derived with attention. In fact, a proper choice of the gauge allows to decouple the dynamics on the surface from the perpendicular one and to avoid couplings between the curvature and the field in the dynamical equation [9]. Solving the proper surface Schrödinger equation, we observe that, by applying an axial magnetic field, the energies show AB oscillations with respect to the field intensity as expected, and that the effect of the hexagonal cross-section is still present and important. Figure 2 shows the first six energy levels as a function of the applied field $B_{||}$, both in the ideal and perturbed cases, so that we can recognize the different behaviors depending on the symmetry of the shell. In the ideal case, the levels oscillate in bunches of six levels crossing each other [8], while, in the perturbed case, the introduction of the random potential breaks the grouping, and the level crossing is split, so that each level oscillates without crossings. In both cases, however, the period of the oscillation is the AB period$\Phi_0$. Note that only the first six levels are shown in Fig. 2; for an analysis of the higher energy levels see Ref.[8].

## 4. Transverse field B$_\perp$

The analysis of the effect of the magnetic field can be repeated for a field B$_\perp$ perpendicular to the NW axis. We showed in a previous work [10] that in cylindrical systems a strong transverse magnetic field induces the formation of Landau levels in the regions of the 2DES orthogonal to the direction of the field. Therefore, these regions behave as 1D channels where the carriers tend to localize. This mechanism is competing with the edge-localization described in Section 2, hence, depending on the orientation $\theta$ of the facets with respect to the field direction: the perpendicular magnetic field can be used as a tool to enhance the localization in the edges (Landau levels for $\theta = \pi/6$) or to extract carriers from the wells (for $\theta = 0$), as shown in Fig. 3. In the central panel of the latter figure, we show the density of states (DOS) for the ideal GaAs shell described in the previous sections, when B$_\perp$ is applied with an orientation $\theta$ with respect to the facets. If the field is sufficiently strong to induce Landau localization [10], the edge-localized

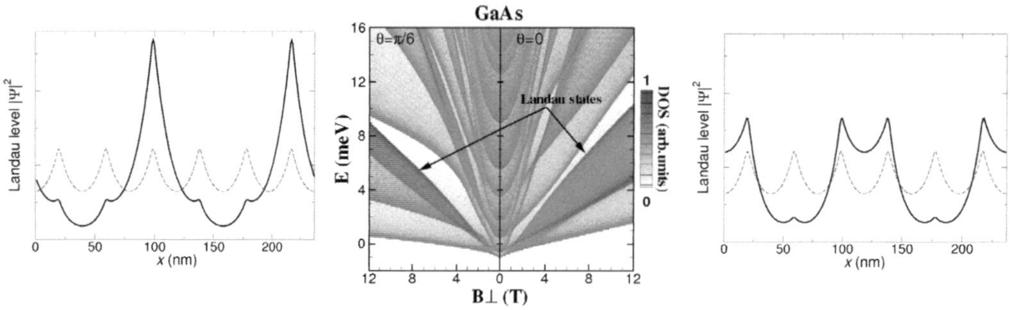


**Figure 3.** Results are shown for the ideal case with an unperturbed 6-fold symmetry. Center: DOS as a function of a perpendicular magnetic field and energy, shown for two different orientations $\theta$ of the facets with respect to the field direction. Darker color correspond to higher DOS. The states corresponding to the Landau levels are indicated by the arrows. Left: carrier probability density of a Landau state for $\theta = \pi/6$, showing a localization increase in two of the six edges. Right: carrier probability density of a Landau state for $\theta = 0$, showing a depletion of the edge-localized states. The zero-field ground state (dashed line) is shown as a comparison.

states can be depleted or further filled, depending on $\theta$, as shown in the side panels of Fig. 3 for $\theta = \pi/6$ (left panel, enhancement) and $\theta = 0$ (right panel, depletion).

## 5. Conclusions

We have calculated the eigen-functions, energies and DOS of carriers in a shell of a core-multishell NW with rounded-hexagonal cross-section. We have shown that the edges are regions of preferential localization: this effect is persistent also against fluctuations of the edge sharpness and shell width. When an axial magnetic field is applied, the energy of the system oscillates with the AB period, corresponding to the quantum flux, and the crossing of the levels reflects the symmetry of the NW and the defects in the shell homogeneity. On the other hand, a magnetic field applied perpendicularly to the NW axis can be used to control the edge localization, allowing to strengthen or weaken it.

This work has been partially supported by project FIRB-RBIN04EY74 and Cineca Calcolo Parallelo 2008.

## References

[1] Mohan P, Motohisa J and Fukui T 2006 *Appl. Phys. Lett.* **88** 133105
[2] Fontcuberta i Morral A, Spirkoska D, Arbiol J, Heigoldt M, Morante J R and Abstreiter G 2008 *Small* **4** 899
[3] Tsumura K, Nomura S, Mohan P, Motohisa J and Fukui T 2007 *Jpn. J. Appl. Phys.* **46** L440
[4] Richter T, Blömers C, Lüth H, Calarco R, Indlekofer M, Marso M and Schäpers T 2008 *Nano Lett.* **8** 2834
[5] Vorob'ev A B, Friedland K-J, Kostial H, Hey R, Jahn U, Wiebicke E, Yukecheva J S and Prinz V Y 2007 *Phys. Rev. B* **75** 205309
[6] da Costa R C T 1981 *Phys. Rev. A* **23** 1982
[7] Cuoghi G, Ferrari G and Bertoni A 2009 *Phys. Rev. B* **79** 073410
[8] Ferrari G, Goldoni G, Bertoni A, Cuoghi G and Molinari E 2009 *Nano Lett.* **9** 1631
[9] Ferrari G and Cuoghi G 2008 *Phys. Rev. Lett.* **100** 230403
[10] Ferrari G, Bertoni A, Goldoni G and Molinari E 2008 *Phys. Rev. B* **78** 115326

# Monte Carlo investigation of electron gas cooling in electric field in n-type InSb crystals

**S Ašmontas, and R. Raguotis**[1]

Semiconductor Physics Institute, A. Goštauto 11, Vilnius, Lithuania

E-mail: asmontas@pfi.lt

**Abstract.** The behaviour of the electron ensemble in the steady-state and response to the microwave frequency electric field was investigated in n-type InSb crystals by the Monte Carlo method. The effect of electron energy cooling below the lattice equilibrium thermodynamic value in highly compensated n-InSb is obtained at lattice temperature 78 and 50 K. Cooling effect vanishes with the increase of the electron density.

## 1. Introduction

The effect of electron gas cooling below the thermodynamic equilibrium temperature in the electric field in a semiconductor has been predicted analytically in [1]. This effect reveals in highly compensated semiconductor in the case of energy scattering dominated by the optical phonon emission, while momentum is governed by the charge impurity scattering. Electrons accelerated by the electric field get optical phonon energy and quickly emitted it. And due to the intensive impurity scattering electrons localize at this condition quite long. It leads to the case that their total energy can be smaller than the equilibrium one. When electron density increases electron-electron (e-e) scattering removes electron cooling due to the mutual mixing between passive and active areas divided by the optical phonon energy. Experimentally electron gas cooling in electric field was firstly observed in [2,3] in highly compensated n-InSb. Numerically the effect of electron gas cooling in n-InSb after strep-like switchover steady-state electric field was firstly obtained by the Monte Carlo (MC) method in [4]. In the presented work we continue the corresponding MC treatment. The results of the simulation of the main kinetic electrons characteristics, such as drift velocity, mean energy, etc. were performed in quite low electric field in compensated n-type InSb crystals. Lattice temperature was taken equal to 78 K and 50K. The behaviour of electron ensemble in the steady-state electric field as response to the microwave frequency field was investigated.

Ensemble (EMC) MC method [5] and the modified combined scattering rate (CSR) MC method [6] have been used. CSR modulation is based on the concept of ensemble scattering events. CSR technique avoids the short-time-step procedure inherent to conventional EMC methods. It is especially convenient when electron-electron (e-e) scattering is included. The n-InSb parameters were typical and similar to those used in [4,7].

## 2.Results and discussion

Electron-impurity scattering mechanism in MC simulation was taken in well-known Brooks-Herring (BH) relaxation time approximation [8]. Figure 1 shows the dependencies upon energy of the electron scattering probabilities for the lattice temperature T=78 K. Impurity scattering rate decreases with the increase of the screening [8], which increases with the electron density. This rate essentially exceeds the one when electron density increased to $n_e = 1\times10^{15}$ cm$^{-3}$ (closed circles) although impurity densities stay the same, i.e., $N_I = 2\times10^{15}$ cm$^{-3}$. Open circles correspond to the experimental situation [2,3], where electron cooling was obtained. For this case e-e scattering has not been included into MC calculation due to the small electron density. When e-e scattering was included ($n_e = 1\times10^{15}$ cm$^{-3}$) e-e scattering removes electron cooling as it will be shown latter. The dependence for e-e scattering is plotted as a function on the energy corresponding to the wave-vectors difference of the colliding electrons – $|\mathbf{k}_1 - \mathbf{k}_2|$. It follows that e-e scattering becomes comparable with the impurity scattering rate for the given conditions.

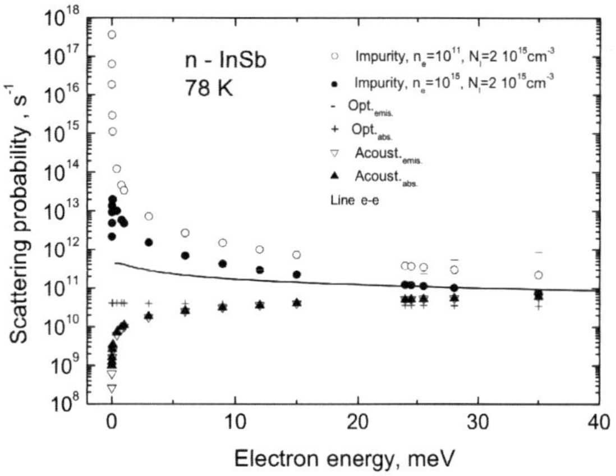


**Figure 1.** Electron scattering probability in n-type InSb crystal at tattice temperature T=78 K.

The calculated dependencies of electron mean energy on electric field are shown in figure 2 for lattice temperature T=78K.

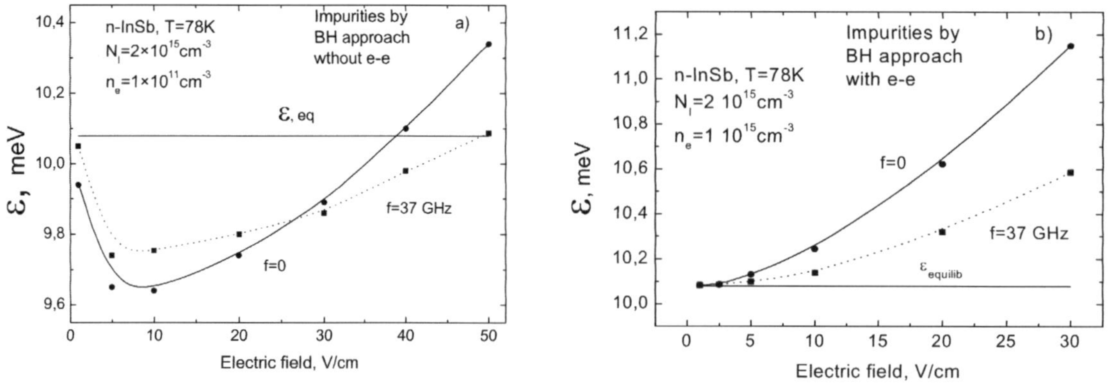


**Figure 2.** Mean electron energy dependence on electric field in n type InSb at T=78K. a) – strongly compensated n-InSb without e-e. b)- electron density increases to $10^{15}$ cm$^{-3}$, with e-e

It is seen from figure 2a that electron energy falls below the equilibrium value in highly compensated n-InSb in low electric field region. It is evidently that some factors influence these dependencies. They are – the time which electrons spend in active optical scattering region, the inertia, at al.. Electrons in steady field more effectively penetrate into active optical phonon region and their energy increases over equilibrium at smaller electric field values (figure 2a). At f=37 GHz due to inertia electrons more weakly emit optical phonons. Therefore, and the cooling effect is decreased.

The cooling effect of electrons has been predicted in [1] for the condition when the time of electron free flight needed for the optical phonon absorption is greater than the time needed for the phonon emission. In the typical case (E=10V/cm, figure 2a) the absorption and emission rates obtained in calculation are comparable ~ near 3.9 $10^{10}$ $s^{-1}$. The rate of phonon emission is slightly larger (about 1%). So, the results obtained in [1] can not be directly applied. The determination of electron cooling including their thermo transfer in active region (equation (53') in [1]) is also unfulfilled. Nevertheless, at such marginal conditions MC simulation obtains electron cooling although the cooling effect is small ~ of the order of a few percent.

Seeking to the wider elucidation of the cooling effect MC simulation was performed at smaller lattice temperature - T= 50K. The corresponding results are plotted in figure 3.

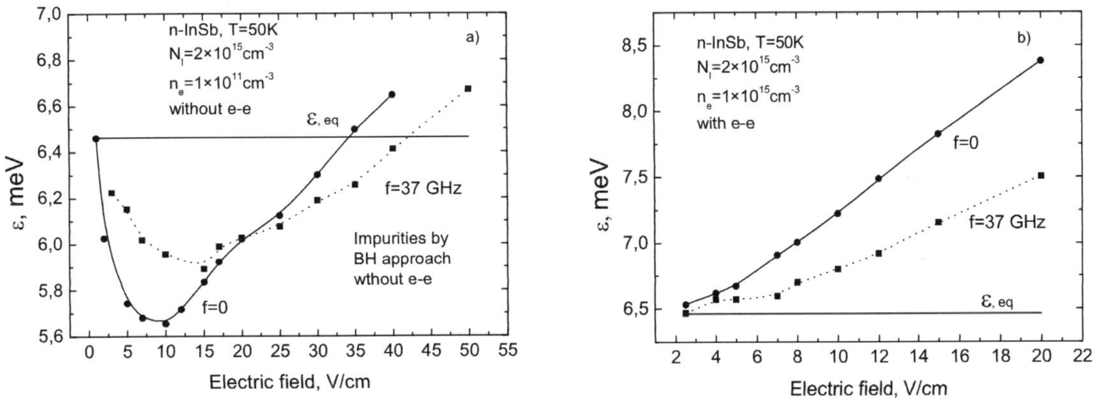


Figure 3. Mean electron energy dependence on electric field in n type InSb at T=50K Marking the same as in Figure 2.

Now, the cooling effect reveals more evidently. Electron energy at 10 V/cm diminishes to ~ 12%. The rates of phonon emission and absorption are correspondingly equal to ~ 5.2 and 4.8 $10^{10}$ $s^{-1}$, and differ about ~ 8 %.

It was shown in [9] that when $n_e$ increases e-e scattering results in electrons interchanging their energy due to the mutual mixing between passive and active areas divided by the optical phonon energy. Electron cooling vanishes with the electron density. We performed corresponding simulation with increased electron density to $n_e = 1\times10^{15}\,cm^{-3}$ and with the same impurity density $N_I = 2\times10^{15}\,cm^{-3}$. E-e scattering was modelled in Brooks Herring approach. Results are shown in Fig. 2b, 3b. It is seen that e-e scattering removes electron cooling. This result coincides with obtained experimentally [2,3] and theoretically [1,4].

Journal of Physics: Conference Series **193** (2009) 012028 doi:10.1088/1742-6596/193/1/012028

## 2. Conclusions

The results obtained by numerical MC calculations show the important role of impurity scattering on electron transport in n-InSb and prove the possibility of electron gas cooling in low electric field and low lattice temperatures below the equilibrium energy value.

We note that analytical theory [1] and experiment [2,3] show more strong cooling effect. Though, as it was estimated in our paper, the requirements admitted in [1] does not hold for the presently simulated case. Here we used well known semi-classical Brooks-Herring approach for the electron-impurity scattering description. The results obtained with Conwell-Weisskopf approach gives similar energy behaviour, however cooling is smaller [10]. It follows that electron energy evidently and velocity remarkably depends on the impurity scattering description.

A number of refinements to the impurity scattering theory have been made. Ridley has shown reconciliation of these theories. More works have explored the effects of dielectric screening, multiple screening, compensation, correlation from donor-acceptors pairs, central cell scattering, corrections due to higher Born approximations and so on [11-13]. The consistent deployment of the conductivity theory of highly impure semiconductors at low temperatures is performed in [14]. Therefore the results obtained in this report require further investigation seeking to obtain better agreement between experiment [2,3] and theory [1].

## References

[1] Gribnikov Z S, Kochelap V A 1970 *ZETF* **58** 1046
[2] Ašmontas S P, Požela Yu K and Subačius L E 1981 *JETF Lett.,* **33** 580
[3] Ašmontas S, Požela J, Subačius L, Valušis L G 1988 *Solid-St. Electron.* **31** 701
[4] Ašmontas S and Raguotis R 2008 *Acta Phys. Polonica* 929
[5] Jacoboni C and Lugli P 1988 *The Monte Carlo method for Semiconductor device Simulation* (Springler, New York)
[6] Matulionis A, Raguotis R and Katilius R 1977 *Phys. Rev.* **B 56** 2052
[7] Akarsu M, Özbaş Ö 2002 *Turk. J. Phys.* **26** 283
[8] Ridley B K 1982 *Quantum process in semiconductors* (Clarendon Press Oxford)
[9] Levinson Y B, Mažuolytė G E 1966 *ŽETF* **50** 1048-1054
[10] Ašmontas S and Raguotis R 2008 *29th International Conference on the Physics of Semiconductors* Rio de Janeiro.
[11] Chattopadhyay D and Quesser H J 1981 *Rev. Mod. Phys*. 745
[12] Rimbey P R, Mahan G D 1985 *J. Appl. Phys.* **57** 2812
[13] Lancefield D, Adams A R and Fisher M A 1987 *J. Appl. Phys.* **62** 2342
[14] Shklovskii B I, Efros A L 1979 *Electrical properties of heavily doped semiconductors* (M.: Nauka) [in Russian].

**Acknowledgment.** This work was supported in part by Lithuanian State Science and Studies Foundation under the contracts No. P-01/2009 and No. V-21/2009.

# Novel Planar Gunn Diode Operating in Fundamental Mode up to 158 GHz

**Chong Li[1], A Khalid[1], N Pilgrim[2], M C Holland[1], G Dunn[2] and D S R Cumming[1]**

[1]Department of Electronic and Electrical Engineering, University of Glasgow, Glasgow G12 8LT, United Kingdom

[2]School of Engineering and Physical Sciences, University of Aberdeen, Aberdeen AB24 SFX, United Kingdom

E-mail: lichong@elec.gla.ac.uk

**Abstract**. We show the experimental realisation of fundamental mode operation of planar Gunn diode structures fabricated in GaAs/AlGaAs quantum wells. The electron density in the active channel is enhanced by positioning double delta-doping layers on either side. Small signal measurement shows that a typical device exhibits negative resistance up to 158 GHz. Using this device structure we have demonstrated a planar Gunn oscillator working at 115.5 GHz with an output power of -28 dBm.

## 1. Introduction

Gunn diodes have been studied as a source of stable and low-noise solid-state microwave radiation since the early 1960s. Most research work on Gunn diodes has concentrated on "vertical" structures, but relatively few researchers have explored "planar" architectures. Recent research indicates vertical Gunn diodes can work up to sub-Terahertz frequency range. Among these, InP-based vertical Gunn diodes have shown second harmonic oscillation at a frequency of 425 GHz [1] and simulations suggest GaN-based vertical Gunn diodes may possibly produce even higher frequency fundamental-mode oscillation [2]. However, almost all vertical Gunn diodes are embedded in bulky cavities and the RF power is extracted from regular rectangular waveguide, leading to higher cost and difficulty with integration to modern MMIC technology. Although some researchers have reported "planar" Gunn devices [3, 4], they are in fact still "vertical", because current flow is perpendicular to the epitaxial layers in the devices. Recently, a "true" planar GaAs-based Gunn diode has successfully been demonstrated at 108 GHz [5]. The device has a HEMT-like structure but without a gate. This novel technology also has the potential to permit the fabrication of different devices with different oscillating frequencies on the same wafer, because the frequency of the diodes is determined by lithographically controlled anode and cathode separation. This opens up new possibilities for complex monolithic circuit designs.

In this paper, we present recent work on planar Gunn diodes. The devices are fabricated on to an electron rich inversion layer. The electron density in the inversion layer places a limit on the maximum achievable operating frequency since there is anode-cathode region must be long enough for Gunn domains to form. In this work we present devices into which an additional delta-doping layer on each side of the active channel has been implemented. As these additional delta-doping layers increase the electron density within the active layer, it becomes possible to form satisfactory Gunn domains within a shorter transit region. This design has made it possible to experimentally observe improved performance. Typical devices exhibit negative resistance up to 158 GHz and output power of -28 dBm has been extracted at a frequency of 115.5 GHz. This frequency is the highest recorded from GaAs-based planar Gunn diodes for the fundamental mode of oscillation.

## 2. Fabrication and measurements of planar Gunn diode

The semiconductor material was grown by molecular beam epitaxy to match layer properties specified by device modeling [6]. Figure 1a is a schematic view of the device. Metal alloys (Pd/Ge/Au/Pd/Au) were deposited on the top of multiple graded layers of GaAs/InGaAs in order to aid the formation of low resistance contacts (approximately 0.15 Ω·mm). Beneath the contact layers there a 15 nm layer of highly doped n-GaAs was incorporated. The thickness of this layer was optimized by Monte Carlo simulation [6]. Below it is a 20 nm barrier layer that is un-doped $Al_{0.23}Ga_{0.77}As$. Two δ-doping ($8 \times 10^{11}$ $cm^{-2}$) layers were inserted evenly in this layer during growth. The channel layer was made of 50 nm un-doped GaAs. Another 20 nm $Al_{0.23}Ga_{0.77}As$ layer with two δ-doping ($8 \times 10^{11}$ $cm^{-2}$) barrier was grown under the channel layer. The channel layer was estimated to have an electron charge density of the order of approximately $10^{17}$ $cm^{-3}$ [8, 9].

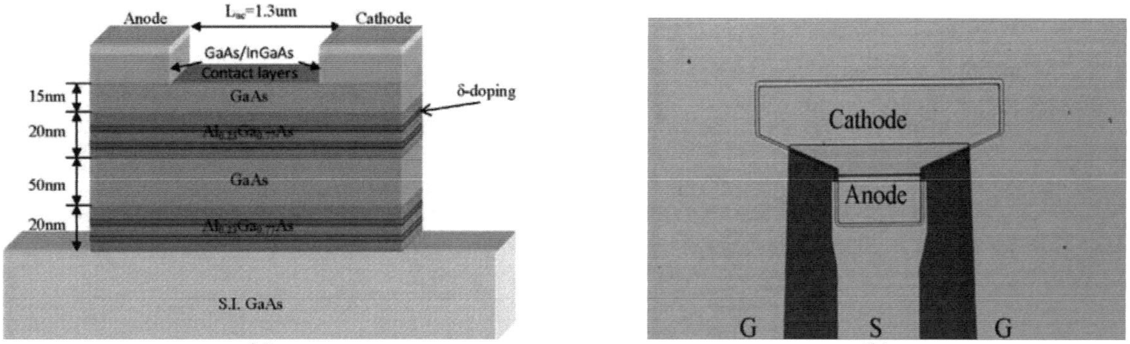


(a) (b)

**Figure 1.** (a) Schematic view of epitaxy layers of planar Gunn diode. (b) Optical image of an actual planar Gunn diode embedded in CPW. The CPW has characteristic impedance of 50 ohm. Width of the CPW signal conductor is 60 um and ground-ground separation is 140 um.

Figure 1b shows the optical image of a planar Gunn diode embedded in a co-planar waveguide (CPW) in order to allow for RF measurements. Small-signal measurements on individual planar Gunn diodes were carried out using similar techniques to those that have previously been used to characterise vertical Gunn diodes [3] and other similar one-port active devices e.g. IMPATT diodes [7]. The measurement setup was based on two sets of PNAs (Agilent) covering the frequency range from DC to 110 GHz and 140 to 220 GHz. Both sets of PNAs used GSG 100 μm-pitch probes that were supplied by GGB industries. The calibration substrates (Microtech technology) were 109-102B for DC-110 GHz and CS-15 for 140-220 GHz, respectively.

Figure 2 shows the small-signal measurement results after the diode was de-embedded from CPW measuring pads over the frequency range of 140-160 GHz. It can be seen from Figure 2a that the reflection coefficient ($S_{11}$) of the planar Gunn diode is over 0 dB up to 158 GHz when the bias is 2.8 V; it means the power reflected from the diode is greater than the power incident on it. This demonstrates that the diode generates RF power up to 158 GHz. This directly leads to a negative resistance for the diode, as seen in Figure 2b, according to the following equation

$$Z_{11} = \frac{1 - S_{11}}{1 + S_{11}} \bullet Z_0 \tag{1}$$

where $Z_0 = 50\Omega$. With appropriate design of external circuits [7], an oscillator working over 150 GHz may be achieved.

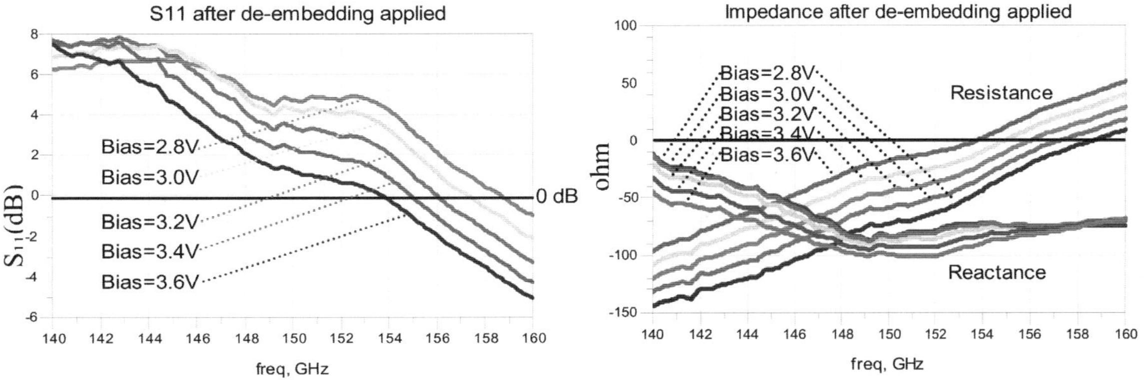


(a) (b)

**Figure 2.** Small-signal measurement results (140-160 GHz) of a planar Gunn diode as external bias varied from 2.8 to 3.6 V by a step of 0.2 V. (a) Reflection coefficients ($S_{11}$). (b) Impedance of the diode derived from $S_{11}$.

## 3. Planar Gunn Oscillator

On the same wafer we constructed a planar Gunn oscillator. Since most Gunn diodes are embedded in resonant circuits in order to improve their frequency stability and tunability [1, 4]. We also designed a simple CPW resonator to operate at the frequency at which the planar Gunn diode may oscillate. The circuit layout is shown in Figure 3a. It was drawn in L-edit (Tanner EDA) for smooth connection with e-beam fabrication process. The circuit was optimised by commercial FEM (Finite Element Method)-based software HFSS (Ansoft) and the simulation result is shown in Figure 3b.

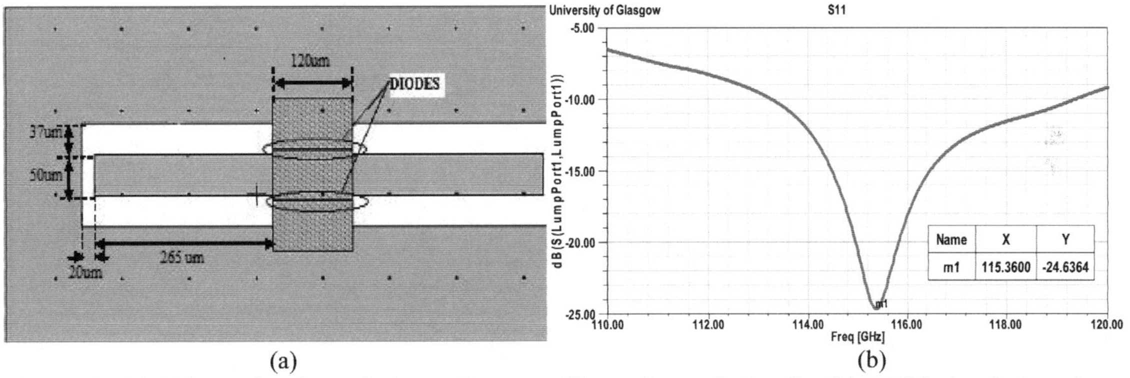


(a) (b)

**Figure 3.** (a) Schematic view of planar Gunn oscillator drawn in L-edit. (b) HFSS simulation shows the circuit resonates at 115.36 GHz and its Q factor is about 150.

The oscillator contains two 1.3 um (anode-cathode distance) planar Gunn diodes and an open-end coplanar waveguide. The two diodes were placed back-to-back in parallel in order to make it a two-port device and to easily integrate with the CPW resonator. The left end of CPW signal line was left open to permit the application of a DC bias to the diodes. However, the gap had to be kept small enough to make the circuit behave as a short-circuited at the frequency of interest so that the diodes could sit where the maximum electric field appeared on the waveguide. The right end was left fully open to enable on-wafer probing. By using HFSS, the optimal geometry of the circuit was finalized as shown in Figure 3a and the resonant frequency of this circuit is 115.36 GHz as shown in Figure 3b.

Once the device was fabricated, its output power and oscillation frequency were measured. The measurement setup consisted of 100 μm-pitch W-band GSG probe, W-band harmonic mixer (WHMP-10 from Farran Technology) and Spectrum Analyzer (E8364B from Agilent). The output power has been corrected from conversion loss induced by the mixer and other systemic losses from cables and probe. Figure 4 illustrates the oscillator produced -28 dBm at 115.536 GHz which is very close to the resonant frequency of the circuit.

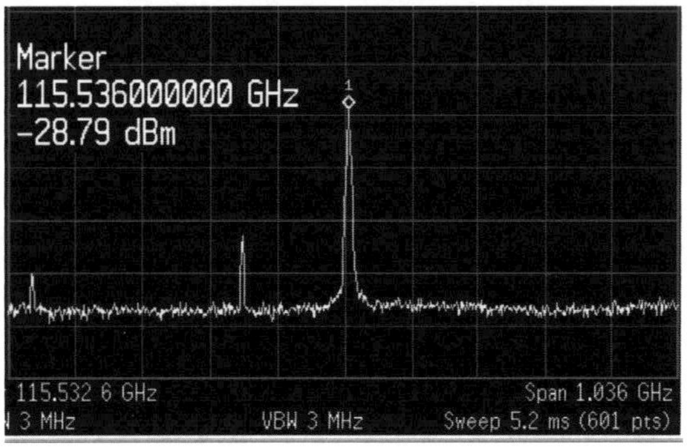


**Figure 4.** The oscillator produces -28 dBm output power at 115.5 GHz after corrections. The mixer has conversion loss of 28 dB from manufacture's datasheet. Other losses from cables and probe contribute about 2 dB. This signal is confirmed to be the fundamental mode of oscillation. Note: the central peak is the real signal and other peaks are spurious.

## 4. Conclusions

We have demonstrated a novel structure of planar Gunn diode. With additional delta-doping layers, electron density within the active channel was enhanced. Higher electron density led to better domain formation that resulted in better performance of the Gunn diodes. Small-signal measurement indicated the device exhibited negative resistance up to 158 GHz, which may be used to construct an oscillator working in the region of 150 GHz with appropriate circuit design. To date an oscillator has been constructed on the same wafer that produced a signal with power of -28 dBm at 115.5 GHz. This is the highest record of oscillation frequency from true planar Gunn devices. Further work will be carried out to optimize oscillator design to achieve higher frequency and higher output power.

## Acknowledgement

The authors would like to thank JWNC for providing fabrication facilities and EDC for supplementing measurement equipment. This project is in part sponsored by EPSRC and e2v technology.

## References

[1] Eisele H 2006 InP Gunn devices for 400-425 GHz *Electronics Letters* **42** p385-6
[2] Alekseev E, Pavlidis D, Sutton W E, Piner E and Redwing J 2001 GaN-Based Gunn Diodes: Their Frequency and Power Performance and Experimental Considerations. *IEICE Transactions on Electronics E Series C* **E84-C** p1462-7
[3] Kim M R, Lee S D, Chae Y S and Rhee J K 2007 Design and Fabrication of Planar GaAs Gunn Diodes. *IEICE Transactions on Electronics* **E91-C** p693-8
[4] Forster A, Lepsa M I, Freundt D, Stock J and Montanari S 2007 Hot electron injector Gunn diode for advanced driver assistance systems *Applied Physics A: Materials Science & Processing* **87** p545-58
[5] Khalid A, Pilgrim N J, Dunn G M, Holland M C, Stanley C R, Thayne I G and Cumming D R S 2007 A Planar Gunn Diode Operating Above 100 GHz *IEEE Electron Device Letters* **28** p849-51
[6] Pilgrim N J, Khalid A, Dunn G M and Cumming D R S 2008 Gunn Oscillations in Planar Heterostructure Diodes *Semiconductor Science and Technology* **23** 075013
[7] Schoellhorn C J, Morschbach M, Xu H, Zhao W and Kasper E 2004 S-Parameter Measurements of the Impedance of mm-Wave IMPATT Diodes in Dependency on the Current Density *Jounal of Microwaves and Optoelectronics* **3** p81-96
[8] Gomes V M S, Chaves A S and Leite J R 1987 Self-consistent Calculations of the Two-dimensional Electron Density in Modulation-doped Superlattices *Physical Review* ***B*** **35** p3984-3989
[9] Singh B N, Deb S, Mishra G P and Sarkar S K 2009 Effects of Double Delta Doping on Millimeter and Sub-millimeter Wave Response of Two-Dimensional Hot Electrons in GaAs Nanostructures *International Journal of Electonics, Circuits and Systems* **3** p32-39

# Transport measurements on InAs/GaSb superlattice structures for mid-infrared photodiode

**C. Cervera, J.P. Perez, R. Chaghi, J.B. Rodriguez, P. Christol**

Institut d'Electronique du Sud, UMR-CNRS 5214, Université Montpellier 2
34 095 Montpellier, France

**L. Konczewicz, S. Contreras**

Groupe d'Etude des Semiconducteurs, UMR-CNRS 5650, Université Montpellier 2
34 095 Montpellier, France

E-mail: cervera@ies.univ-montp2.fr

**Abstract**. In this communication, we report on electrical transport measurements of non-intentionally doped InAs/GaSb Superlattice structures grown by Molecular Beam Epitaxy. Resistivity and Hall Effect measurements were performed on two samples, corresponding to the same SL structure that has been grown on two different substrates: one on semi-insulating GaAs substrate, another on n-type GaSb substrate. To carry out the electrical measurements, the conducting GaSb substrate of the second sample has been removed. The study were performed in the temperature range 77-300K, for magnetic fields of 0.38 T.. The both samples exhibited a change in type of conductivity from p-type at low temperature to n-type near room temperature.

## 1. Introduction

Thanks to recent results obtained in the mid-infrared domain (MWIR 3-5$\mu$m), InAs/GaSb superlattice (SL) is now considered as a new material system[1] for the fabrication of high performance infrared (IR) p-i-n photodiodes suitable for thermal imaging camera[2]. To enhance performances and/or temperature operation, improvements of device technology as well as a better knowledge on fundamental properties of the SL photodiodes are still necessary. In particular, one needs to have a better understanding of carrier transport in this material where a reproducible change in type of conductivity as a function of temperature has been observed by different groups[3,4].

Hall Effect measurement is an efficient way to determine the effective carrier concentration in quantum structure. For the practical applications the InAs/GaSb SL photodiode structures are grown on lattice matched GaSb substrate and in this case there is no insulating substrate allowing accurate Hall measurements. The structures can be grown on semi-insulating GaAs substrates[5], but it is well-known that the important density of defects inducing by the lattice mismatch between the SL structure and the GaAs substrate can lead to the experimental conclusions which cannot be directly applied into the lattice matched structures grown on GaSb substrate. Consequently, it is necessary to develop an experimental procedure to measure transport properties of high quality SL structure without influence of the substrate.

In this paper, we report Hall Effect measurements on a MWIR SL sample after the GaSb substrate has been mechanically and chemically removed. In this case, only the carriers from the SL materials

Journal of Physics: Conference Series **193** (2009) 012030 doi:10.1088/1742-6596/193/1/012030

take part in electrical conduction process and we compare results with those obtained for the same SL structure grown on semi-insulating GaAs substrate.

## 2. Experimental

The non-intentionally-doped (nid) InAs/GaSb SL structures were grown by molecular beam epitaxy (MBE) on n-type (001) GaSb substrate (called: GaSb-based sample) and on semi-insulating (001) GaAs substrate (called: GaAs-based sample), mounted on In-free molybdenum holders. The both structures are identical and composed of 600 periods of 8 monolayers (MLs) InAs / 8 MLs GaSb for a total thickness of about 3 $\mu$m. In the case of SL structure grown on GaSb substrate, an InAsSb layer was inserted between the SL and the substrate as a stop-layer to allow removing the conducting GaSb substrate by an appropriate technological process.

5x5 $mm^2$ piece of epilayers of GaSb-based sample was first cleaved and frontside-bonded on an $Al_2O_3$ host substrate using liquid epoxy adhesive. The GaSb substrate was then mechanically thinned up to 80/100 $\mu$m in a polishing machine utilizing various roughness abrasive pads. The last GaSb substrate micrometers were selectively wet etched by using a diluted $Cr0_3$/HF based solution with an etch rate of about 5-6 $\mu$m/min. Next, to remove the InAsSb stop-layer, a standard citric acid ($C_6H_8O_7$) based solution with a 100:1 InAsSb:GaSb selectivity was used[6]. Further details on the technological procedure will be the subject of a forthcoming paper.

Concerning GaAs-based sample (semi-insulating substrate), the measurements have been done on two type of samples : with and without the GaAs substrate. To remove the GaAs substrate, the same technological procedure was employed with a chemical wet etching using a $C_6H_8O_7/H_2O/H_2O_2$ based solution. To perform transport measurements, the prepared samples (5x5 $mm^2$) were placed inside the cryostat installed between the electromagnet pole pieces. The contacts were made of indium. The Hall coefficient and the resisitivity were measured in Van der Pauw configuration by a standard DC method. For Hall Effect measurement the magnetic field of 0.38 T was applied. The temperature inside the cell was measured using Pt100 resistor and stabilized within 0.1 K accuracy.

## 3. Results

First of all, we wanted to verify the influence of the substrate on Hall measurements. The Figure 1 shows the results obtained from measurements on two pieces of the sample grown on GaAs substrate: first one with the substrate and second one with substrate removed.

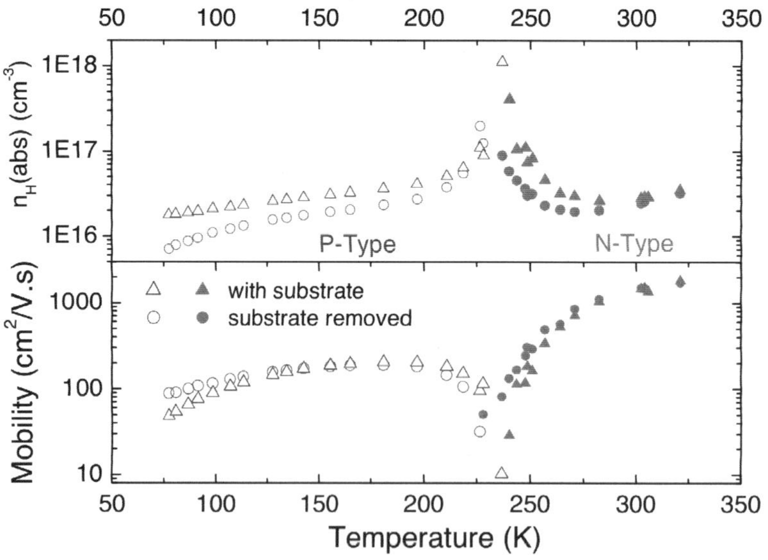


**Figure 1.** Hall carrier concentration and Hall mobility versus temperature measured on the nid SL GaAs-based sample. The carrier type switches from p-type (open symbol) to n-type (full symbol) at 235°C.

As it can be seen from this figure an overall good agreement was obtained between the two measurements, confirming that the method used to remove the substrate didn't modify the carrier transport properties. The carrier density variation ($n_H$) with the temperature shows values in the range of 1.8 x$10^{16}$ $cm^{-3}$ – 3.5 x$10^{16}$ $cm^{-3}$ between 77 and 300 K. One can notice the carrier-type switch from p-type to n-type occurring at 235°C, that reveals the compensated nature of the SL.

Mobility in the range of 48 $cm^2/V.s$ at 77 K (p-type material), and 1720 $cm^2/V.s$ at 300 K (n-type material) was measured. Those fairly low values can be attributed to the high density of threading dislocations due to the large lattice-mismatch existing between the SL structure and the GaAs substrate.

Similar measurements were made on GaSb-based sample without substrate. Temperature dependence of the effective carrier concentration $n_H$ and Hall coefficient $R_H$ are presented in Figure 2. A reproducible change of sign of Hall coefficient $R_H$ is observed. It means that the change of type of conductivity of the nid SL occurs around $T_{crit} \sim 120$ K. At low temperature the sample is p-type with carrier density $p_H\,(100K) = 2.8\times10^{15}\ cm^{-3}$. At room temperature, the SL is n-type with $n_H\,(300K) = 1.5\times10^{16}\ cm^{-3}$. We can remark that at low temperature, the carrier density value is one order of magnitude lower than the value measured on the sample grown on GaAs (Figure 1).

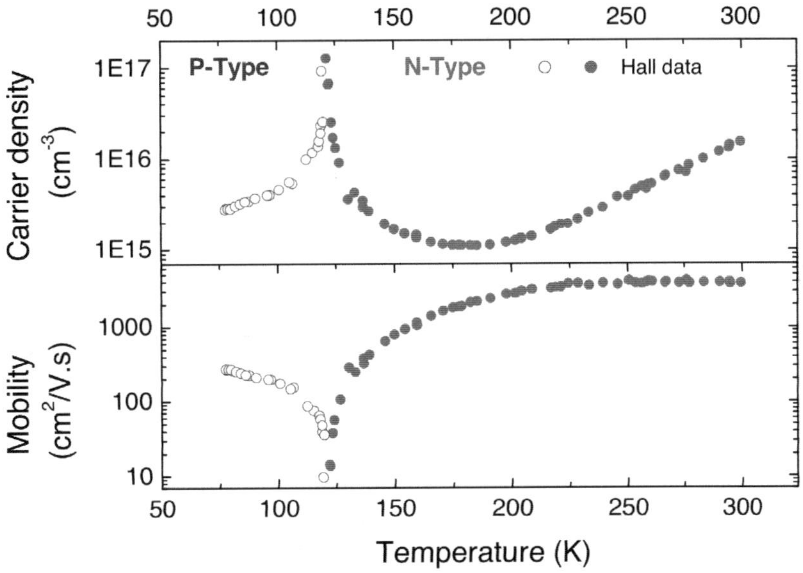


**Figure 2.** Measured apparent Hall carrier concentration $n_H$ and Hall mobility versus temperature, results for the nid SL grown on GaSb substrate. The carrier type switches from p-type (open symbol) to n-type (full symbol) at 120°C.

Mobilities measured on GaSb-based sample were also notably higher compared to the GaAs-based sample. The values measured reached respectively: 270 $cm^2/V$ at 77K (p-type material) and 3850 $cm^2/V.s$ at 300 K (n-type material).

Comparison between GaAs-based and GaSb-based samples with substrates removed is presented on Figure 3. It can be seen that the character of temperature variation of apparent Hall concentration for the both samples is very similar. At low temperature range (p-type conduction) they are characterised by the same thermally activated process. The analogue situation takes place at room temperatures, in n-type conduction regime. Of course, due to the lower hole concentration the switch

Journal of Physics: Conference Series **193** (2009) 012030 doi:10.1088/1742-6596/193/1/012030

from p-type to n-type conduction in GaSb-based sample occurs in lower temperature than in GaAs-based one. This difference in hole concentration manifests itself in mobility vs temperature variation. In case of GaSb-based sample the low temperature mobility (p-type conduction) is much higher and decreases with increasing temperature. In GaAs-based sample the increase of hole mobility with temperature indicate that the ionized impurity scattering mode is predominant in this material.

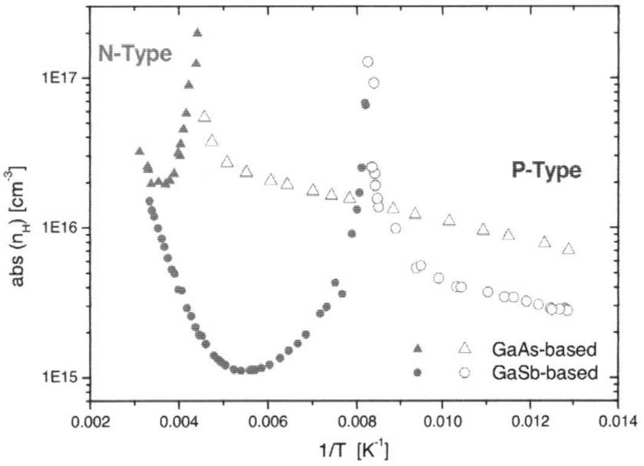


**Figure 3a.** Apparent Hall carrier concentration $n_H$ versus temperature, results for the nid SL grown on GaSb and GaAs substrate respectively.

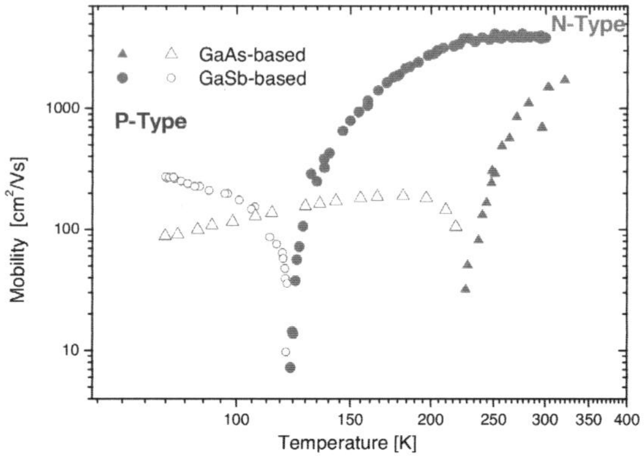


**Figure 3b.** Hall mobilty versus temperature, results for the nid SL grown on GaSb and GaAs based samples respectively, without substrates (log to log).

## 4. Conclusion

Up till now the only way to study the electrical transport in InAs/GaSb SL for MWIR photodiodes applications was to grown the structures on semi-insulating GaAs substrates. However the results of such a study could be deform by important density of defects inducing by the lattice mismatch between the SL structure and the GaAs substrate. In the presented paper, we have demonstrated that the resistivity and Hall Effect can be measured on a structure in which the substrate has been mechanically and chemically removed. It allows studying electrical properties of InAs/GaSb SL grown on a lattice-matched GaSb substrate. From the Hall data, a change of conductance from p-type below 120 K to n-type for higher temperatures was observed. A hole concentration and a mobility of $2.8x10^{15}$ $cm^{-3}$ and 270 cm²/Vs were measured at 77 K.

This reliable method for Hall measurements of SL carrier transport properties could prove very helpful for the sake of material improvement, photodiode design and simulation.

## References

[1] Rogalski A 2007 *Infrared Phys. Technol.* **50** 240.

[2] Rehm R, Walther M, Schmitz J, Fleiβner J, Fuchs F, Ziegler J, Cabanski W 2006 *Opto-Electron. Rev.* **14** 19

[3] Szmulowicz F, Elhamri S, Haugan J, Brown G.J, Mitchel W.C 2007 *J. Appl. Phys.* **101**, 043706

[4] Konczewicz L, Contreras S, Aït-Kaci H, Cuminal Y, Rodriguez J.B, Christol P 2009 *Phys. Stat. Sol B* **246,** 643

[5] Hoffman C.A, Meyer J.R, Youngdale E.R, Bartoli F.J, Miles R.H 1993 *Appl. Phys. Let.* **63**, 2210

[6] Dier O, Lin C, Grau M, Aman M-C 2004 *Semicond. Sci. Technol.* **19** 1250

Journal of Physics: Conference Series **193** (2009) 012031 doi:10.1088/1742-6596/193/1/012031

# Electron drift velocity in high electric fields in zincblende, wurtzite and rocksalt GaN

R Brazis, D Nausewicz, and R Raguotis

Semiconductor Physics Institute, A. Goštauto 11, Vilnius, Lithuania

E-mail: brazis@pfi.lt

**Abstract.** The influence of GaN crystal symmetry (zincblende, wurtzite, and rocksalt) on the drift velocity of electrons in high electric fields is studied employing Monte Carlo method. The highest drift velocity as well as the most pronounced negative differential mobility is predicted in the zincblende phase. Wurtzite-phase crystals turn out to be inferior in these respects due to the complications of electron and phonon dispersion brought by the lowering symmetry. Experimental data on wurtzite phase reported by Wraback et al. are in excellent agreement with the data of modelling taking into account intense (low-energy phonon-mediated) scattering of electrons between the lowest- and the closely-located upper gamma-valleys. Barker et al. revealed experimentally the velocity-field dependence in cubic GaN to be higher than that in wurtzite phase, however, the experimental drift velocity turned out to be lower than the theoretical one. This might be the signature of coexisting different phases in experimental samples. The rocksalt GaN (still expecting experimental verification) is predicted to manifest the lowest drift velocity of electrons and no negative differential mobility, as a result of the fast exchange of electrons between the lowest-energy X-valleys preventing electron runaway to the upper gamma-valley.

## 1. Introduction

Gallium nitride crystals are in the focus of recent material science and technology seeking for applications in high-power electronics and ultrafast photonics. Remarkable differences in the electron and phonon dispersion in zincblende (ZB), wurtzite (WZ) and rocksalt (RS) phases present a challenge for the research of kinetic characteristics of these phases. Experimental data on electron drift velocity in high electric field have been reported for the WZ and ZB phases, leaving the RS phase unexplored. Monte Carlo (MC) modelling has been presented primarily without a detailed discrimination of parameters for the WZ and ZB phases, leaving the RS phase unexplored again. In the presented paper the results of MC simulation are reported for the three phases of GaN crystals, and compared with the available experimental data.

## 2. Model and parameters

Calculations have been performed with standard MC code [1] using the set of parameters given in Table 1. The set is tentative, as many of the GaN electronic parameters are not yet well established.

ZB GaN parameters have been selected seeking to obtain better coincidence of the calculated drift velocity with the available experimental data.

The set of WZ parameters has been selected in our previous work [2].

The parameters for RS phase were determined from the electron and phonon dispersion branches [3, 4]. Other parameters for RS were taken by the analogy with ZB and WZ phases. The RS GaN has indirect band gap with the lowest conduction-band valley in the X point. The highest valleys lie quite atop. They do not manifest in the kinetic phenomena in the treated electric field range.

Non-elastic optical and acoustic scattering mechanisms were taken into account [1]. Impurity and piezo-acoustic scattering were employed in accordance with Ref. 5. Impurity scattering has been treated using the third-body exclusion method [5], which encompasses well-known Brooks-Herring and Conwell-Weisskopf approximations. Lattice temperature was equal to $T = 300$ K.

**Table 1. GaN parameters**

| | ZB | WZ[2b] | RS[3,4] |
|---|---|---|---|
| Valley separation, eV | 0.8[6] (Γ-X) | 0.4 ($\Gamma_1$–$\Gamma_3$) | 2 (X-Γ) |
| | 2.2[6] (Γ-L) | 2.1 ($\Gamma_1$-LM) | 4 (X-L) |
| Electron effective mass | 0.193[7] (Γ) | 0.2 ($\Gamma_1$) | 0.47 (X) |
| | 0.4[8] (X) | 0.22 ($\Gamma_3$) | 0.4 (Γ) |
| | 0.6[8] (L) | 0.6 (LM) | 1 (L) |
| Nonparabolicity coefficient, $eV^{-1}$ | 0.21[8] (Γ) | 0.183 ($\Gamma_1$) | 0 (X) |
| | 0.065[8] (X) | 0.065 ($\Gamma_3$) | 0 (Γ) |
| | 0.029[8] (L) | 0.029 (LM) | 0 (L) |
| Polar optical phonon energy, meV | 92[9,10] | 92 and 26 | 77 |
| Equivalent-valley phonon energy, meV | | | 37 (X) |
| Non-equivalent intervalley phonon energy, meV | 65.84[11] | 2 | 19 ($X_{ij}$) |
| Acoustic deformation potential, eV | 12[9] | 8.3 | 7.2 |
| Intervalley coupling constant, $10^9$eV/cm | 1.2 | 1 | 1 |
| Mass density, g/cm$^3$ | 6.1[10] | 6.15 | 7.83 |
| Sound velocity, km/s | 4.57[10] | 5.94 | 6.15 |
| Piezoelectric const., C/m$^2$ | 0.5[12] | 0.375 | 0.375 |
| Static dielectric const. | 10.4[13] | 9.95 | 57 |
| Optic dielectric const. | 5.47[13] | 5.35 | 7 |

## 3. Results

The results obtained by the MC simulation are presented in Figure 1 (lines) together with the available experimental data. The MC results in the case of ZB phase are plotted for compensated semiconductor with electron concentration equal to $n_e = 1\times10^{17}$ cm$^{-3}$ and impurity density $N_I = 1\times10^{18}$ cm$^{-3}$. The calculated drift velocity for ZB case gives somewhat larger values than measured ones. The calculated mobility of electrons in the low-field limit (~ 50 V/cm) in ZB case is about 500 cm$^2$/Vs. This value is nearly two times larger than experimental one [14]. The authors [14] performed full-band MC simulation and obtained good coincidence with experimental results up to the fields of 120 kV/cm. However, at higher fields their MC values of the drift velocity rolled over too soon.

MC calculations for WZ phase GaN were performed using wurtzite model proposed in Ref. [2]; this model is suggested accounting for the inelastic scattering of electrons on the additional low-energy phonon modes (~26 meV), and the satellite valley location close (400 meV) to the conduction band bottom allowing for the fast (acoustic-phonon mediated) exchange of electrons between both the gamma-valleys. The impurity and electron densities were taken equal to $1\times10^{17}$. As it is seen, the good agreement is obtained with the time of flight experiment [15].

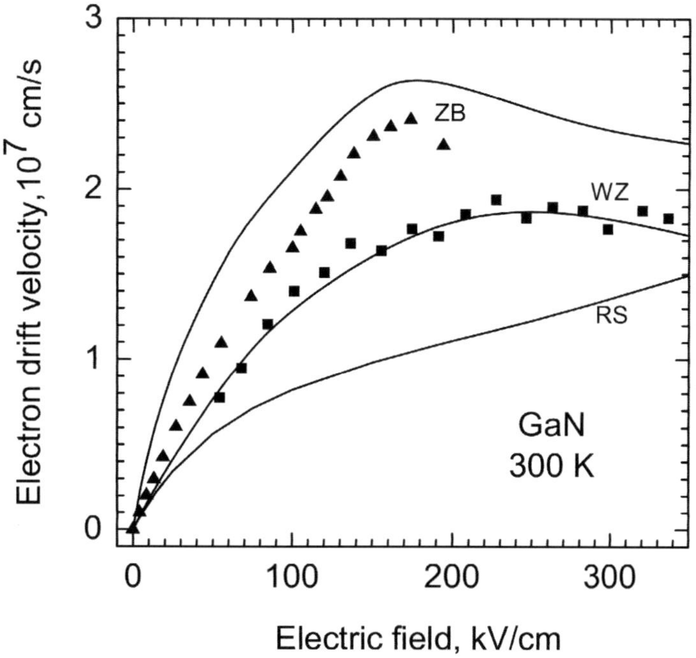


**Figure 1.** Drift velocity versus electric field in three phases of GaN crystal. Close triangles – experiment [14], closed squares – experiment [15]. Lines – present MC.

Rocksalt GaN characteristics are quite different from those of ZB and WZ phases. The lowest valley is at X point and the next nearest Γ valley is located at ~ 2 eV higher. Therefore, the highest-valley electron population does not exceed 0.1 %. The main scattering mechanism is polar optical one. The scattering intensity on deformation optical vibrations is about one order of magnitude smaller. The calculated mobility at low fields is of the order of 200 $cm^2/Vs$.

## 4. Conclusions

The Monte Carlo simulation of electron drift velocity has been performed for zincblende, wurtzite and rocksalt phase of GaN crystals at $T$ = 300 K.

Using the updated parameters (Tab.1), and including impurity compensation, we obtain significantly lower electron drift velocity and low-field mobility in ZB phase, as compared to our previous work [2b]. However, the velocity is seen to remain higher than experimental one (Fig.1). This may be due to the still non-precisely determined ZB electronic parameters, or due to the presence of inclusions of the other phases that could reduce the spatially-average electron mobility in the experimental samples. The possibility of an admixture of ZB and WZ phases has been actually mentioned in [14]; RS phase inclusions seem to be feasible, as well. We leave the topic for separate investigation.

The strong point for WZ GaN applications is the availability of versatile technology. As the WZ phase grows most easily, it is likely to be relatively free of other phase inclusions, allowing for quite adequate MC modelling of electron drift velocity. The weak point of this phase is its relatively low peak drift velocity of electrons.

Modelling of electron transport in the RS phase is performed to our best knowledge for the first time. The rocksalt GaN reveals the lowest drift velocity of electrons and no negative differential mobility.

This is due to the fast exchange of electrons between the lowest-energy X-valleys preventing electron runaway to the high-located upper gamma-valley.

Such a remarkable difference in the behaviour of electrons in ZB, WZ and RS phases presents a challenge for further experimental and theoretical research aiming at manufacturing heterostructures comprising of chemically identical but different-symmetry compounds.

**Acknowledgements**

The research has been conducted in the framework of the COST Action MP0805 under the auspices of the Agency of International Science and Technology Development Programmes in Lithuania.

**References:**

[1] Jacoboni C, Lugli P 1989 *The Monte Carlo Method for Semiconductor Devise Simulation* (Wien: Springler-Werlag)
[2] Brazis R and Raguotis R 2004 *J. Appl. Phys.* **85** 609 (a); 2006 *Opt. & Quant. Electr.* **38,** 339 (b)
[3] Chen Z V, Lv M Y, Li L X, Wang Q, Zhang X Y, Liu X P 2006 *Thin Solid Film* 2433
[4] Duman S, Bağcı S, Tütüncü H M, Uğur G, Srivastava G P 2006 *Diamond & Related Materials* **15** 1175–1178
[5] Ridley B K 1982 *Quantum processes in semiconductor* (Oxford: Clarendon Press)
[6] Berrah S, Boukortt P and Abid H 2008 *Sem. Phys., Quant. Electr. & Optoelectr.* **11** 59
[7] Rinke P, Winkelnkemper, M, Qteish A, Bimberg D, Neugebauer J and Scheffler M 2006 *Appl. Phys. Lett.* **89** 161919
[8] Foutz B E, Eastman L E 1997 *Appl. Phys. Lett.* **70** 2849
[9] Weber M 2005 *Analysis of Zincblende-Phase GaN* Thesis (Georgia institute of Technology)
[10] Kolnik J, Oguzman J, Brennan K F, Wang R and Ruden P P, Wang J 1995 *J. Appl. Phys.* 1033
[11] Zi J, Wan J X, Wei G, Zhang K, Xie X 1996 *J. Phys Cond. Matter* **8** 6323
[12] Aidoğu S, Akarsu M and Özbas Ö 2005 *Rom. Journ. Phys* **50** 1047
[13] Akarsu M, Aidoğu S, Özbas Ö 2005 *Rom. Journ. Phys.* **50** 869
[14] Barker J M, Akis R, Thornton T J, Ferry D K and Goodnick S M 2002 *Phys. Stat. Sol.* **190** 263
[15] Wraback M, Shen H, Curano J C, Li T, Campbell J C, Schurman M J and Ferguson I T 2000 *Appl. Phys. Lett.* **76** 1155

# Tunnelling transport in nonuniform barriers as a possible way for current control

**V.A. Kozlov and V.A. Verbus**

Institute for Physics of Microstructures, Russian Academy of Science, 603950 Nizhny Novgorod, GSP-105, Russia

kozlov@ipm.sci-nnov.ru

**Abstract**. This paper presents the results of computer simulation of the under barrier tunnelling transport through the nonuniform heterostructure barriers containing spherical or cylindrical nano objects. Embedding of nano objects in the barrier enables one to construct the semiconductor barriers with predefined properties. Application of external potential to these objects exponentially changes the current through the barrier and it can be used for current control and amplification of signals.

## 1. Introduction

The scale reduction of semiconductor devices requires the decrease of barrier dimensions. In the short part of nanometer range the semiconductor barriers produced by the bandgap engineering are nonuniform in plane. This can lead to new features of the hot electron transport in device. The aim of this paper is the simple quantum description of the tunneling transport [1] in the semiconductor barriers containing nano objects of spherical or cylindrical shapes. It is very useful for construction of new semiconductor devices with the external control of tunneling current.

In a conventional type FET, the electron transport in a channel along $z$ direction is described by propagating Bloch waves $a_k \exp(ikz)$, where $k$ is the wave vector and $a_k$ is the amplitude. To change the current we need to change the width of the channel by gate potential. This corresponds to the change in the number of propagating modes. It leads to linearly dependence of the current on a channel width. In contrast to this description, the tunneling current through the barrier is described by the exponential damping wave functions $a_k \exp(-\eta z) + b_k \exp(+\eta z)$ , where $a_k$ and $b_k$ are the amplitudes of forward and backward damping waves ($k = i\eta$). In this case a change in the barrier height leads to the correspondent change of the exponent power $\eta z$ and the current control becomes more effective. So the question is: how to change the barrier potential by applying external voltage.

To realize this control we considered the semiconductor barrier with the built-in nanoscale objects. Two shapes of objects were considered: spherical and cylindrical. The potential of the spherical object which is embedded in a channel may be changed, for example, by the gate potential. The potential of cylindrical one may be changed by applying voltage to the end of the cylinder. The influence of spherical objects on the current was described by the application of the Mie theory [2] for the scattering of electromagnetic waves by sphere. The cylindrical objects were described similarly. To calculate the current in the barrier we took into account both the forward and the backward waves.

## 2. Gamow Peak and Gamow Window

It is well known that tunneling processes in quantum mechanics were firstly introduced by George Gamow [3]. Now in nuclear and stellar physics the terms Gamow peak (Gp) and Gamow window (Gw) are frequently used for the description of penetration of the ensemble of carriers through the barrier. The term Gp denotes the carrier energy corresponding to maximum of tunneling penetration probability, and Gw is the energy interval corresponding to half maximum of this probability. The carriers in the ensemble are in thermal equilibrium with temperature T. The Gp arises due to the multiplication of two factors. The first factor is the tunneling transmission coefficient which increases with the increase of carrier energy. The second one is the Maxwell-Boltzmann or Fermi distribution function which decreases with the decrease of energy. It means that the Gp isn't a resonance type and at adverse conditions it can be rather wide or absent. It is interesting to compare the conditions of existing Gp in stars and heterostructures. For example the temperature of the internal silicon shell of iron core in the supernova (type II) star is approximately equal to several $T_9$ (where $T_9 = 10^9$ K ). If we estimate a typical nuclear potential $V_{nuc}$ to be several MeV, and a typical bandoffset in heterostructures $V_{hs}$ to be 100 meV, then we obtain the realistic temperature of heterostructure: $T_{hs} = T_9 (V_{hs}/V_{nuc}) = T_2 = 100$ K. This demonstrates common features in the different branches of hot carrier physics.

Consider the plane heterostructure barrier of thickness $d$ with the flat barrier potential equal to $V_b$. The wave function of carrier inside the barrier consists of the sum of two damping components. The first component exponentially increases in forward direction, but the second increases in the opposite direction. It corresponds to the distribution of minority carriers inside the base of bipolar transistor and one may hope to obtain the amplification of signal if it will be possible to change the barrier potential. Note that for calculation of the current it is necessary to take into account both forward and backward waves. To describe the tunneling transmission, it is useful to introduce the thickness potential $V_d$ by the formula: $V_d = \hbar^2 / (2m^* d^2)$ so that $V_d$ depends only on thickness $d$ and effective mass $m^*$. The tunneling transmission coefficient $T_{tun}$ for carrier with energy $E$ through the barrier (for small transmission) can be written in the form:

$$T_{tun}(E) = 16\,\frac{E}{V_b}\left(1-\frac{E}{V_b}\right)\exp\left(-2\sqrt{\frac{V_b - E}{V_d}}\right). \tag{1}$$

Thus the dependence of penetration probability for ensemble of carriers with temperature $T$ versus energy is proportional to $P(E) = T_{tun}(E)\exp(-E/V_t)$, parameter $V_t = k_B T$ is the so called thermal potential, $k_B$ is the Boltzmann constant. The penetration probabilities for two heterostructure barriers with the same thickness ($d$ = 10 nm) and effective mass ($m^* = 0.1\,m_0$) for temperature ($T = 100\,K$), are shown in figures 1 and 2. The only difference is the value of barrier potential: $V_b$ = 100 meV for figure 1 and $V_b$ = 200 meV for figure 2. These figures are very much alike and the main difference is in the penetration rates. The fall-off in penetration is equal to 160 times when the barrier potential increase is equal to 2 times. So it is possible to get the signal amplification. Especially, effective current control will be achieved if we build-in resonance levels into the barrier and the Gamow peak and resonance level coalesce together. The penetration of carriers through Gw produces the inverted distribution of transmitted carriers and a high frequency negative differential conductivity may arise.

## 3. Under barrier tunnelling transport with built-in spherical or cylindrical nano objects.

To describe the tunnelling transport through the barrier with embedded spherical (or cylindrical) object (considering as scatterer) we solved the Mie problem [2] in the single scattering approximation.

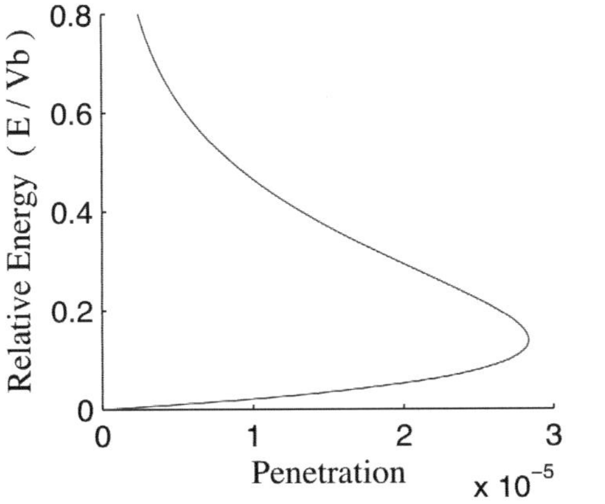


**Figure 1**. Penetration vs. relative energy for barrier with the potential $V_b$ = 100 meV.

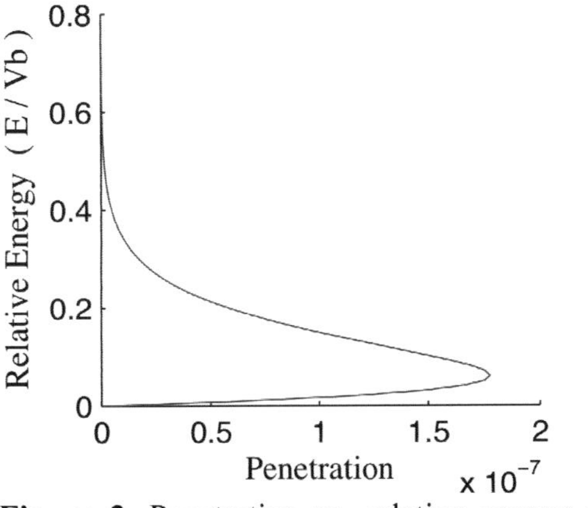


**Figure 2**. Penetration vs. relative energy for barrier with the potential $V_b$ = 200 meV.

In our approach the scatterer contained the radial dependent potential which was approximated by the step function with constant potential inside each shell layer. The wave function $\psi_j(r,\vartheta)$ inside the shell layer $j$ was taken in the following form:

$$\psi_j(r,\vartheta)=\sum_{l=0}^{\infty}(c_{jl}^{(1)}(\vartheta)h_l^{(1)}(k_j r)+c_{jl}^{(2)}(\vartheta)h_l^{(2)}(k_j r)) \tag{2}$$

where $r$ is a radius from the centre of the scatter in a spherical coordinate system, $\vartheta$ is the angle, $l$ is the index of spherical harmonic, $c_{jl}^{(1)}(\vartheta)$ and $c_{jl}^{(2)}(\vartheta)$ are the coefficients containing Legendre polynoms $P_l(cos\vartheta)$. The functions $h_l^{(1)}(k_j r)$ and $h_l^{(2)}(k_j r)$ are the spherical Bessel functions of the third kind describing diverge and converge spherical waves. The scatterer consists of several spherical layers and $j$ is the index of the layer, $k_j=(2m_j(E-V_j)/\hbar^2)^{1/2}$ is the wave vector in the $j$ layer. Outside the scatterer we added to the expression (2) the actuate plain waves: $A\exp(ikr\cos\vartheta)+B\exp(-ikr\cos\vartheta)$. The coefficients $A$ and $B$ are the amplitudes of forward and backwards plain waves, propagating along direction $z=r\cos\vartheta$ . We fulfilled the Sommerfield boundary conditions at infinity together with the boundary conditions $\psi_j(r_j)=\psi_{j+1}(r_j)$ and $(\partial\psi_j(r_j)/\partial r)/m_j=(\partial\psi_{j+1}(r_j)/\partial r)/m_{j+1}$. This gave us the possibility to calculate the coefficients $c_{jl}^{(1)}(\vartheta)$ and $c_{jl}^{(2)}(\vartheta)$ for finding the wave function by formula (2).

From the wave function we found the tunnelling current density inside the barrier including scatterer. The spatial distribution of $j_z$ and $j_y$ components of the current which flow from left side of the barrier ( $z=z_{left}=-0.02\mu$ ) to the right side ( $z=z_{right}=+0.02\mu$ ) are shown in figures 3-6. These figures demonstrate the nonuniform character of the current density. The scatterer is located in the position $z=0$ and $y=0$. The figures 3, 4 correspond to the attractive case: $V_s<V_b$ where $V_s$ is the potential of scatterer and $V_b$ is the barrier potential. In this case the current flows mainly through the scatterer and its surrounding region. Thus the scatterer forms an additional channel for current. The figures 5, 6 correspond to the repulsive case: $V_s>V_b$. In this case the current flows mainly far away from the scatterer and penetration of the barrier decreases. It also means that the leakage current to the scatterer is small enough and it is possible to control the total current through barrier by the change of scatterer potential with small power.

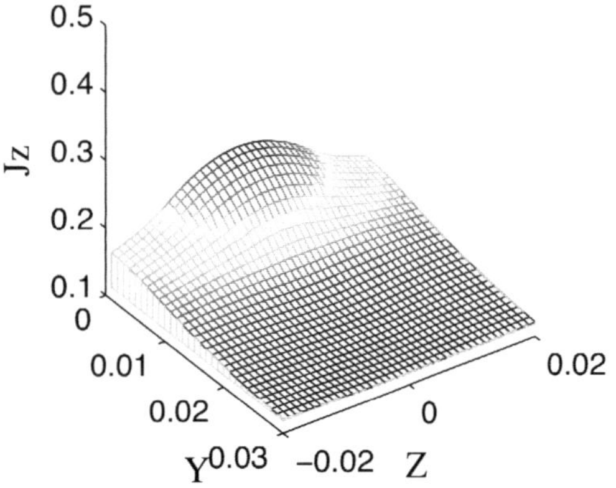


**Figure 3**. Distribution of the current $J_z$ in the barrier when $V_s$ is less than $V_b$.

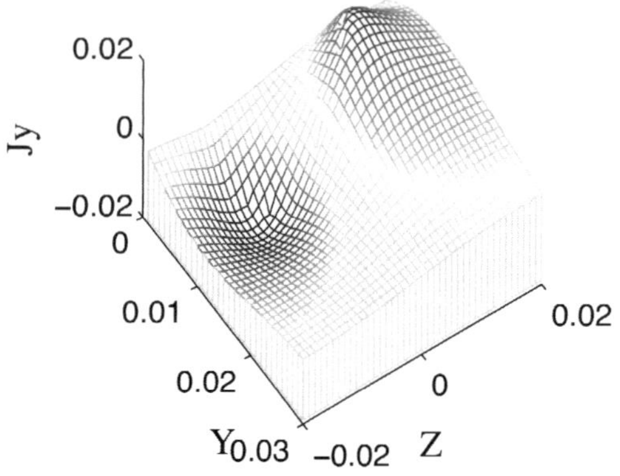


**Figure 4**. Distribution of the current $J_y$ in the barrier when $V_s$ is less than $V_b$.

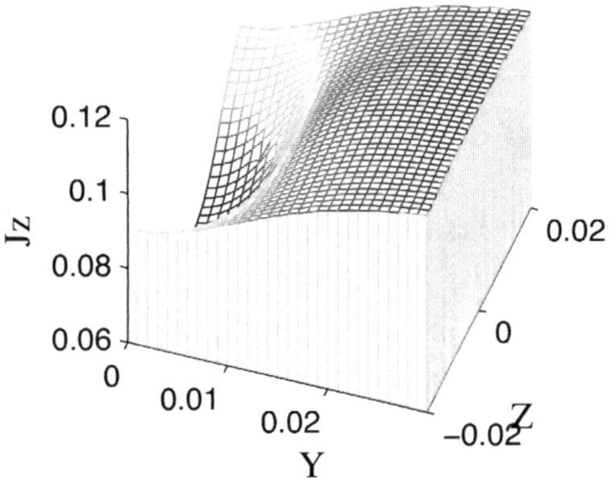


**Figure 5**. Distribution of the current $J_z$ in the barrier when $V_s$ exceed $V_b$.

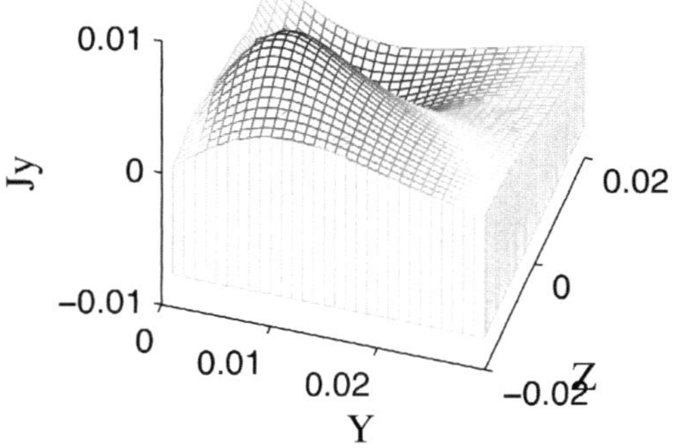


**Figure 6**. Distribution of the current $J_y$ in the barrier when $V_s$ exceed $V_b$.

In summary, we have developed the algorithm for computer simulation of under barrier transport of electrons through the nonuniform semiconductor barriers containing nanometer scatterers. The scatterers may have spherical or cylindrical shapes with radial dependent potential. The results of simulation of the tunneling current density in the non uniform barrier demonstrate the possibility of the current control by the external potential applied to embedded scatterers, especially for cylindrical shape. These results hold promise for the development of new types of active devices with embedded quantum wires into the barrier, like a grid in the vacuum triode. It is also shown that the tunneling through Gamow window produce the population inversion of transmitted carriers.

## References

[1] Price P.J. 2004 Semiconductor Science and Technology **19** S24
[2] Ishimaru A. 1978 *Wave Propagation and Scattering in Random Media* (NY: Academic Press)
[3] Gamow G. 1928 *Z. Phys.* **51** 204

Journal of Physics: Conference Series **193** (2009) 012033 doi:10.1088/1742-6596/193/1/012033

# Anharmonic decay of non-equilibrium intervalley phonons in silicon

**Zlatan Aksamija and Umberto Ravaioli**

Electrical and Computer Engineering Department and Beckman Institute

University of Illinois at Urbana/Champain, Urbana, IL. 61801

aksamija@illinois.edu

**Abstract**. We study phonons produced by transitions between the equivalent X valleys in silicon. We use the Monte Carlo method first to select stochastically the time between phonon collisions, and then to select a final-state pair of phonons from the probability distribution for anharmonic decay. Our results show that intervalley phonons decay into one near-equilibrium transverse acoustic phonon and another intermediate longitudinal phonon either on the acoustic or optical branch. This second phonon has energies between 40 and 50 meV and undergoes another decay before turning into a pair of near-equilibrium transverse acoustic phonons, presenting an additional potential bottleneck.

## 1. Introduction

Silicon has been the primary material of CMOS technology for decades due to its ability to withstand very high electric fields present in the channels of nanoscale devices. Fields exceeding 100kV/cm are commonly found in the current deep sub-micron channel devices. When a high field is applied to a silicon device, such as in the drain end of the channel of a nanoscale MOSFET transistor, electrons are accelerated by the field and they reach sufficient energies to begin making transitions between the six equivalent valleys in the crystallographic **X** direction [1]. Such high fields accelerate conduction electrons to energies above 50meV, sufficient to produce optical phonons. Due to the indirect band-gap structure of silicon, such transitions occur primarily between the X valleys, which are located around the point <0.85 0 0>. Two types of transitions between the valleys are possible. One is a back-scattering type of transition, where an electron transitions to a valley in the direction opposite from the valley from which it originated. This transition is called g-type and it has, in general, the effect of reversing the electron's momentum vector. The phonons produced by this transition are also referred to as g-type phonons [2].

The other significant inter-valley transition is between valleys in directions orthogonal to each other, or nearest-neighbor valleys. This transition, and the phonons produced from it, are called f-type. Both f- and g-type inter-valley transitions typically involve long wavelenth optical and longitudinal phonons, which are also termed f-type and g-type phonons according to the type of transition from which these phonons originated. Since the six valleys are located at **k**=<0.85 0 0>, the phonons inlvolved in the intervalley transitions can be found by momentum conservation by taking the difference between the locations of the initial and final valleys of the electrons. Since most of the energy emitted by electrons are intervalley phonons, they can accumulate and impact performance.

Journal of Physics: Conference Series **193** (2009) 012033 doi:10.1088/1742-6596/193/1/012033

## 2. Anharmonic decay of intervalley phonons in silicon

In this study, we are interested in the behavior of intervalley phonons after they are generated by the electron transitions. The phonon produced by electrons scattering from a valley to any of its four neighbor valleys is called f-type, while the phonon arising from scattering to the opposite valley is called g-type. The momentum of the g-type phonon can be computed from momentum conservation by taking the difference between initial and final electron states. This gives a phonon momentum q=<1.7 0 0> which, when reduced to the first Brillouin zone, is <0.3 0 0>. Similarly, f-type phonons have momenta of q=<0.85 0.85 0>, which again reduced to the first Brillouin zone become q=<1 0.15 0.15>. Based on this observation, we will assume f-type phonons have momenta centered around the point **q**=<1.0 0.15 0.15>, while g-type phonons have momenta centered around **q**=<0.3 0 0> . Once emitted by the electron-phonon scattering process, these intervalley phonons will move slowly due to their low group velocity [1], which is especially prominent in the optical phonon branches. Phonons can decay due to the anharmonicity of the atomic potentials. Anharmonic decay is caused by the cubic and higher terms in the crystal potential. The strongest of these terms is cubic and it gives rise to the three-phonon process where a phonon breaks up, or decays, into two phonons, possibly on different branches, while at the same time conserving total crystal momentum and energy. This implies that the strongest decay path is through the three-phonon process which allows the phonons generated by the electron-phonon coupling to decay into pairs of phonons, each with smaller energy, but with higher group velocities.

The probability of every such decay can be obtained from perturbation theory, which says that the probability of the three-phonon decay is given by the product of the cubic matrix element and a resonance factor which ensures energy conservation. The resonance factor is given by the sinc function and, in the limit of infinite collision time, this distribution approaches the Dirac delta function which gives perfect energy conservation. Based on this derivation, lifetimes of acoustic phonons were computed in the early works of Klemens [3], Callaway [4], and Holland [5]. Lifetimes of optical phonons were also first considered by Klemens [6] based on a very simple model of the lattice consisting of a linear chain of atoms. This assumption produces very simple forms for both the phonon dispersion relationship, as well as for the anharmonic coupling between the optical and acoustic modes. Klemens also assumed that the optical phonons decay into a pair of acoustic phonons on the same branch but with opposite momenta, which does not apply to the present case. A more detailed study, based on DFT calculations [7], focused on the decay of the Raman-active zone center optical mode, and showed that, in contrast to Klemens' work, optical phonons in silicon decay into pairs of acoustic phonons mostly involving one longitudinal and one transverse branch. Those results were all brought together, tabulated, and summarized in [8]. In this paper, Rowlette and Goodson show that the decay rate for the intervalley phonons relevant to transport in silicon can be assumed to be independent of phonon energy, and depends linearly on temperature.

## 3. Final state selection after anharmonic decay

The problem of selecting final-state pairs for the three-phonon decay process can be approached by the Monte Carlo method. In this work, the Monte Carlo [9] algorithm is used to select the length of free flight for each phonon in the simulation and to select the momenta of the pair of phonons generated by 3-phonon decay. The first of these two steps starts when a phonon is generated by the electron scattering. The average life-time of optical phonons is known from previous studies and measurements [8], and can be estimated to be around 2ps for optical phonons at room temperature. The probability of a particle to stay in its present state is described by the Poisson process so the time of phonon decay can be chosen stochastically by using the Monte Carlo method [9]. This gives the time of phonon decay as $t_{decay} = -\tau_{anh.} \ln(r)$ where $\tau_{anh.}$ is the lifetime of the phonon due to anharmonic decay, and r is a random number uniformly distributed on the unit interval. Once the time of the decay process is found, then we must search for a final state. This second step is accomplished using the rejection algorithm on the probability distribution for anharmonic decay including the matrix element. The probability distribution of anharmonic decay was obtained by Klemens [3] from quantum-mechanical perturbation theory as the product of the anharmonic matrix element for the three-phonon process and the time-dependent resonance factor.

(1.1) $P(\vec{q},\vec{q}') = \langle \vec{q} | H' | \vec{q}' \rangle^2 \frac{1-\cos\Delta\omega t}{\Delta\omega^2 t}$

This factor is used as a probability kernel in the Monte Carlo method to select final states of one of the two phonons involved in the anharmonic decay, while the second phonon's final state is then chosen based on momentum conservation. The perturbation Hamiltonian due to cubic anharmonicities is given by:

(1.2) $H' = c_3(\vec{q},\vec{q}',\vec{q}'')\, a*(\vec{q})\, a*(\vec{q}')\, a(\vec{q}''),$

where the coefficients a and a* are phonon creation and annihilation operators, and the anharmonicity coefficient can be derived from a Gruneisen model [3] as:

(1.3) $c(\vec{q},\vec{q}',\vec{q}'') = \frac{2\gamma M v^2}{\sqrt{3G}} q q' q''.$

Placing all the above elements together and expanding the creation and annihilation operators produces the final form of the scattering kernel. The expression was simplified assuming that the modes represented by $\vec{q}'$ and $\vec{q}''$ are in equilibrium:

(1.4) $P(\vec{q},\vec{q}') = 2c(\vec{q},\vec{q}',\vec{q}'')^2 \frac{\hbar}{M^3 \omega\omega'\omega''} \frac{1-\cos\Delta\omega t}{\Delta\omega^2 t} \left( N_{eq}' - N_{eq}'' \right).$

The first of the three phonons involved in the anharmonic three-phonon decay is given by the initial phonon and its momentum $\vec{q}$. This momentum depends on the type of transition we are interested in (g-type or f-type). Then the second phonon momentum ( $\vec{q}'$ ) is chosen uniformly at random from the entire first Brillouin zone. Finally, the third phonon is produced by momentum conservation as the difference between the initial and the randomly chosen momenta $\vec{q}'' = \vec{q} - \vec{q}' + \vec{G}$, where $\vec{G}$ is a basis vector in reciprocal space ensuring that the resulting $\vec{q}''$ falls within the first Brillouin zone. For processes where all three vectors are contained in the first Brillouin zone, the value of $\vec{G}$ is zero. These processes are called *Normal*. In contrast, those processes where one of the final momentum vectors, in the present case $\vec{q}''$, lands outside of the first Brillouin zone have non-zero values of G and are called *Umklapp* [3]. Such processes play a large role in thermal resistance, and participate in the decay of intravalley optical phonons. Therefore, both types of processes are considered as candidates for scattering. The energy $\hbar\omega(\vec{q})$ of all three phonons involved is obtained from the full phonon dispersion relationship computed numerically from the Adiabatic Bond Charge model of Weber [10] by diagonalizing the dynamical matrix for each value of the momentum vector. This procedure is repeated on a fine grid in 3-dimensional momentum space and stored so that the value of the phonon energy can be obtained at any point in the first Brillouin zone with great accuracy by linear interpolation.

## 4. Results

Results show that most of the Longitudinal Optical (LO) g-type phonons decay into one Longitudinal Acoustic (LA) and one Transverse Acoustic (TA) phonon, in agreement with MD [2] and DFT calculations [7], and that the LA phonons produced by the anharmonic decay are non-equilibrium and have energies around 40meV. This can lead to reabsorption of non-equilibrium phonons by the electrons, especially the LA branch which couples strongly with electrons. This effect can have a large impact on the hot electron effect and reliability. Increasing the absorption of LA phonons imparts additional energy on the electron population, and can cause more electrons in the tail of the distribution, producing an increase in the number of hot electrons. This increase can then lead to more hot electron effects, such as oxide degradation, thereby impacting both reliability and performance.

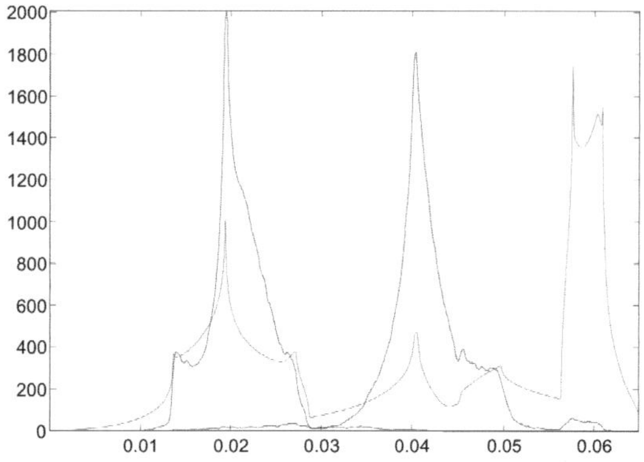


Figure 1: Energy distribution of phonons produced by the anharmonic decay of g-type Longitudinal Optical (LO) phonons. Most of the 100,000 simulated phonons break up into one low energy acoustic and one higher energy longitudinal phonon. The phonon DOS (blue) is plotted for reference.

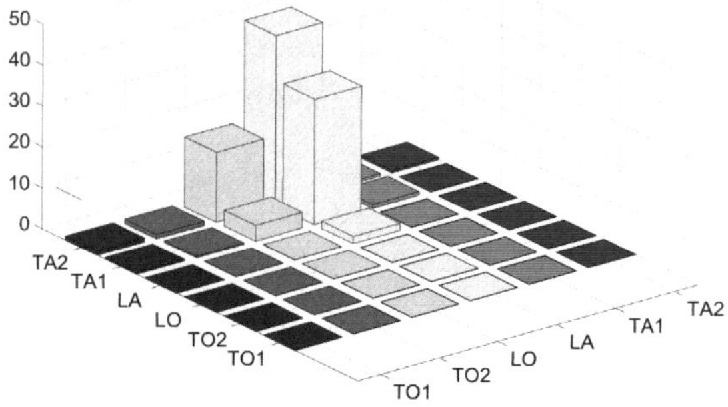


Figure 2: Bar plot of the percentage of phonons on each of the six phonon branches that are produced by the anharmonic decay of g-type LO phonons. Most gLO phonons decay into combinations of TA+LA and also TA+LO phonons, leading to an increase in both LO and LA phonons.

**Acknowledgements:** This work was supported by the National Science Foundation ITR grant SBC PU 01-1045-01. One of the authors (ZA) acknowledges support from the Department of Energy Computational Science Graduate Fellowship Program of the Office of Science and National Nuclear Security Administration in the Department of Energy under contract DE-FG02-97ER25308.

**Referenes**

[1] E. Pop, R. W. Dutton, and K. E. Goodson, "Analytic band Monte Carlo model for electron transport in Si including acoustic and optical phonon dispersion," *J. Appl. Phys.*, vol. 96, no. 9, pp. 4998–5005, Nov. 2004.

[2] S. Sinha, P. K. Schelling, S. R. Phillpot, and K. E. Goodson, "Scattering of g-process longitudinal optical phonons at hotspots in silicon," *J. Appl. Phys.*, vol. 97, no. 2, p. 023702-023709, Jan. 2005.

[3] P. G. Klemens, "Thermal Conductivity and Lattice Vibrational Modes", in Solid State Physics, edited by F. Seitz and D. Turnbull, vol. 7, pp. 1-98, 1958.

[4] J. Callaway, "Model for Lattice Thermal Conductivity at Low Temperatures", *Phys. Rev.*, 113, pp. 1046-1051, 1959.

[5] M. G. Holland, "Analysis of lattice thermal conductivity," *Phys. Rev.*, vol. 132, no. 6, pp. 2461–2471, Dec. 1963.

[6] P. G. Klemens, "Anharmonic decay of optical phonons," *Phys. Rev.*, vol. 148, no. 2, p. 845, Aug. 1966.

[7] A. Debernardi, S. Baroni, and E. Molinari, "Anharmonic phonon lifetimes in semiconductors from density-functional perturbation theory," *Phys. Rev. Lett.*, vol. 75, no. 9, pp. 1819–1822, Aug. 1995.

[8] J. Rowlette and K. E. Goodson, "Fully Coupled Nonequilibrium Electron–Phonon Transport in Nanometer-Scale Silicon FETs," *IEEE Trans. Electron Devices*, vol. 55, p. 220 (2008).

[9] C. Jacoboni and L. Reggiani, "The Monte Carlo method for the solution of charge transport in semiconductors with applications to covalent materials," *Rev. Mod. Phys.*, vol. 55, no. 3, pp. 645–705, Jul. 1983.

[10] W. Weber, "Adiabatic Bond Charge Model for the Phonons in Diamond, Si, Ge, and Alpha-Sn", *Phys. Rev. B*, vol. 16, pp. 4789-4803, 1977.

# Kinetics of shallow acceptor impact ionization and recombination in p-Ge

**M.S. Kagan, I.V. Altukhov, V.P. Sinis, S.K. Paprotskiy**

Institute of Radio Engineering and Electronics of Russian Ac. Sci.
11 Mokhovaya, 125009 Moscow - Russia

kagan@cplire.ru

**Abstract**. The time evolution of current in p-Ge doped with shallow acceptors was studied. The dependences of impact ionization and recombination times on electric field and uniaxial pressure were found. The role of Poole-Frenkel ionization of shallow acceptors was analyzed.

## 1. Introduction

Stimulated THz emission has been observed in uniaxially strained p-Ge with an optical resonator [1-3]. The origin of the THz emission was attributed to intra-center optical transitions between resonant and localized acceptor levels. A population inversion in the p-Ge occurs for states of shallow acceptors split under pressure. The amplitude-rise time (and possible frequency band of modulation) of the THz emission is determined by relaxation times of conductivity that is the relaxation times of carrier distribution function and of non-uniform distributions of current (current filaments caused by impurity breakdown) and field (electric domains formed at negative differential), which can arise in strained p-Ge. In this report we present studies of transient currents in Ga-doped Ge at different uniaxial pressures, which allowed us to find directly the characteristic times of impact ionization and recombination for shallow acceptors in Ge.

First, the impact ionization of shallow impurities in Ge have been observed in 1953 [4] as a sharp rise of conductivity at some threshold field. The intensive studies of this effect mainly at dc conditions and its first theoretical treatments [5 - 10] have been performed in mid-50s to 70s (see, e.g., [11] and references within). Kinetics of field ionization of shallow donors in Si and Ge has been studied in 80s (see [12] and references within). The studies of carrier capture on attractive Coulomb centers have been done mainly during this period. The first explanation of high values of the capture cross sections have been done in 1960 by M. Lax [13] introducing a cascade scheme of the capture (see [14] for references). In this report, the kinetics of field ionization and recombination in p-Ge with shallow acceptors was studied at different uniaxial pressures.

## 2. Experimental

Ga-doped p-Ge crystals with Ga concentration of $2\times10^{12}$ to $10^{14}$ $cm^{-3}$ were studied at the liquid He temperature. Two kinds of samples were exploited: with intermediate compensation (10-30%) and highly (>90 %) compensated material. The match-like samples of a square cross section of 0.5 to 1 $mm^2$ and 6 to 10 mm long were cut in the [111] crystallographic direction. The uniaxial pressure $P$ and electric field $E$ were applied along this direction. Voltage pulses of 0.2 to 1 μs duration were applied to two contacts deposited on the long (lateral) facet of the sample and spaced by 4 to 9 mm.

To study the field ionization, we recorded the time dependence of current at the leading front of voltage pulse. To find the field dependence of recombination time, we used the step-like voltage pulse (see figure 4) keeping higher voltage fixed and varying the lower voltage. The current decay on the back front should reflect the recombination of carriers created by electric field ionization of acceptors.

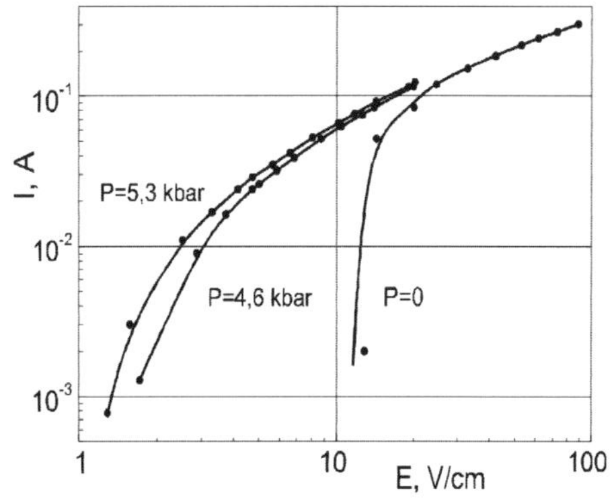


Figure 1. Current-voltage characteristics. Ge:Ga, $N_A = 2*10^{14}$ cm$^{-3}$, $N_D \sim 7*10^{13}$ cm$^{-3}$

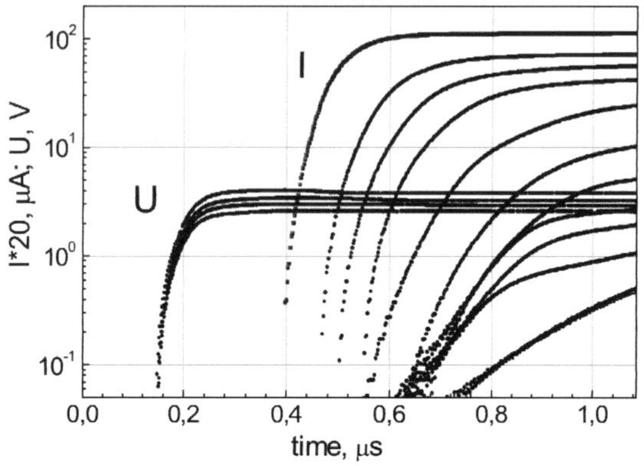


Figure 2. Current and voltage pulse shapes at initial stage of breakdown. For I(t): U=2.8 to 4.1 V

### 3. Results and discussion

Current-voltage characteristics at different pressures for the sample with $N_A = 2\times10^{14}$ cm$^{-3}$, $N_D \sim 7\times10^{13}$ cm$^{-3}$ ($N_A$ and $N_D$ are acceptor and compensating donor concentrations) are shown in figure 1. They have the vertical parts caused by current filament formation. The breakdown threshold shifts to lower voltage with increasing pressure due to a decrease in acceptor binding energy. The leading fronts of voltage and current pulses just above the threshold are presented in figure 2 in semi-log scale.

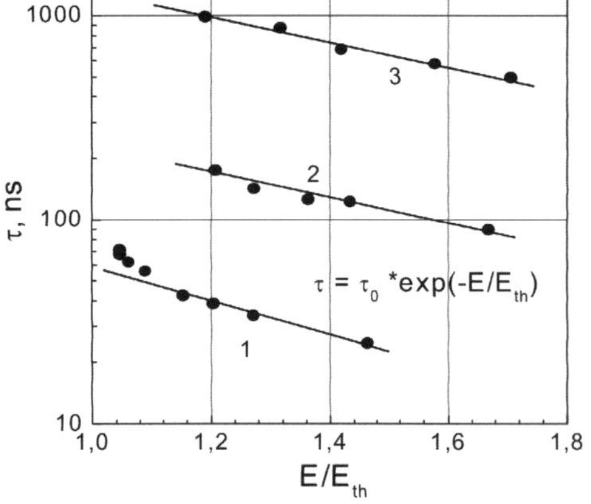


Figure 3. Electric field dependences of impact ionization times for samples with different compensations.
1: $N_A = 2*10^{14}$ cm$^{-3}$, $N_D \sim 2*10^{13}$ cm$^{-3}$;
2: $N_A = 2*10^{14}$ cm$^{-3}$, $N_D \sim 7*10^{13}$ cm$^{-3}$;
3: $N_A = 1*10^{13}$ cm$^{-3}$, $N_D \sim 9.6*10^{12}$ cm$^{-3}$.
$E_{th}$ is the threshold field of breakdown.

The current growth at impurity breakdown obeys the exponential law with a characteristic time decreasing with applied voltage. The field dependences of the times found from these data are presented in figure 3. The generation-recombination kinetics can be expressed as $dp/dt = \beta(E)p(N_A-N_D) - \alpha(E)pN_D$ ,where $p$ is the free hole concentration, $\beta$ and $\alpha$ are impact ionization and capture coefficients. At breakdown conditions $\beta >> \alpha$ and $p(t) = p_{st}\exp(t/\tau(E))$, where $p_{st}$ is stationary value of $p$ and $\tau^{-1} = \beta(E)(N_A-N_D)$. The dependence of impact ionization coefficient on electric field found from the current rise time turns out to be the exponential function of field, $\beta(E) \propto \exp(E/E_{th})$.

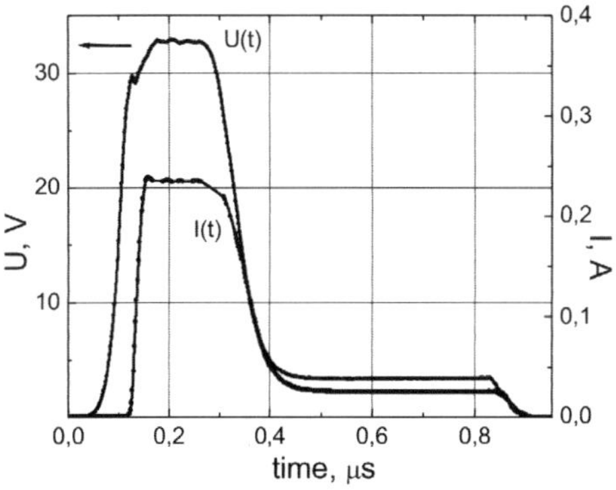


Figure 4. Voltage and current pulse shapes.

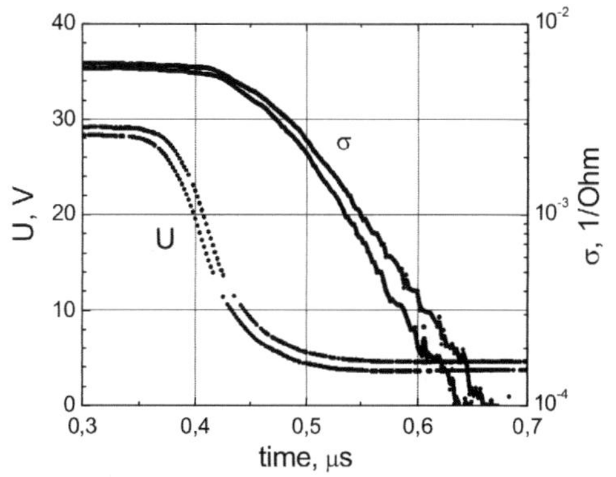


Figure 5. Conductivity decay at decreasing voltage.

Using the voltage pulse shaped as shown in figure 4, we could determine hole recombination times. The conductivity decay at the voltage step is shown in figure 5. It is due to capture of holes, which are created by field ionization, at shallow acceptors. The decay is of the exponential character. To find the field dependence of this time we changed the voltage of the second plateau. The recombination times $\tau_R$ in dependence on electric field for different pressures are presented in figure 6. The initial parts of these dependences seem natural, because the probability of capture at attractive impurities should decrease with increasing carrier energy. Then the recombination time falls down and the maxima in these dependences coincide exactly with the threshold of impurity breakdown. The reason for this decreasing of recombination times is that the ionization rate at the breakdown is much larger than the capture rate. Simple calculations show that the reciprocal capture time is the sum of the reciprocal times of generation and capture. So, above the breakdown threshold, the relaxation time for generation-recombination balance is determined by impact ionization rate.

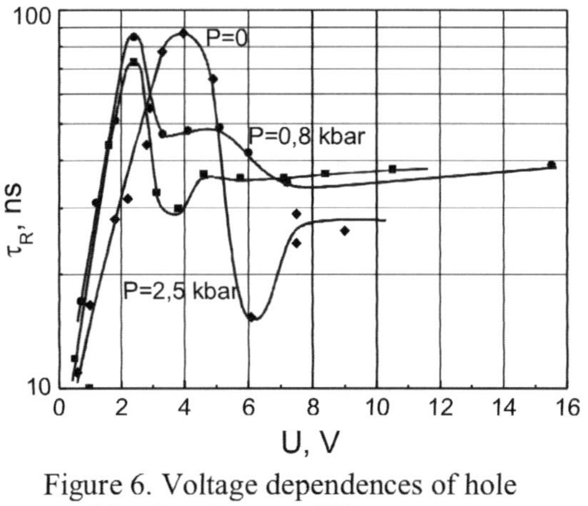


Figure 6. Voltage dependences of hole recombination times at different pressures.

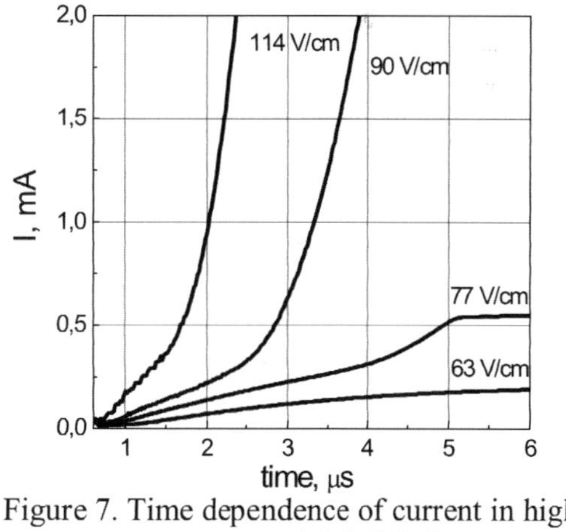


Figure 7. Time dependence of current in highly compensated (~96%) Ge.

Transient currents in highly compensated material are quite different from those for samples with low compensation. Initially, the current grows linearly with time and then, after reaching some critical current, an exponential growth begins (figure 7). Note that the high compensation leads to essentially larger voltages for the impact ionization than in previous samples. The linear growth of current is the result of increased photo-ionization rate (with background radiation) caused by a decrease of impurity binding energy due to Poole-Frenkel effect that is lowering the potential barrier of impurity center in the external field. The I-V curves presented in figure 8 confirm this conclusion: they have no vertical parts and obey the exponential law with the square root of field in the exponent. To determine the impact ionization times, we used the exponential parts of I-V curves. The dependence τ(E) for this case shown in figure 3 (curve 3) has the same character as for previous samples. However, the field dependences of recombination times shown in figure 9 are quite different from those obtained for samples with low compensation (figure 6). The dependences consist of two parts. Both of them can be described by exponential function with squared voltage in the exponent. We connect the kinks in these curves with the change in character of field heating of carriers from diffusion heating to ballistic one (streaming motion).

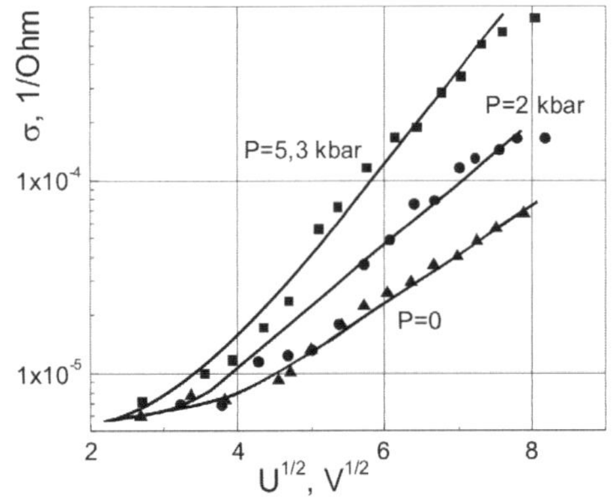


Figure 8. Conductivity vs $U^{1/2}$ plot demonstrated Poole-Frenkel effect.

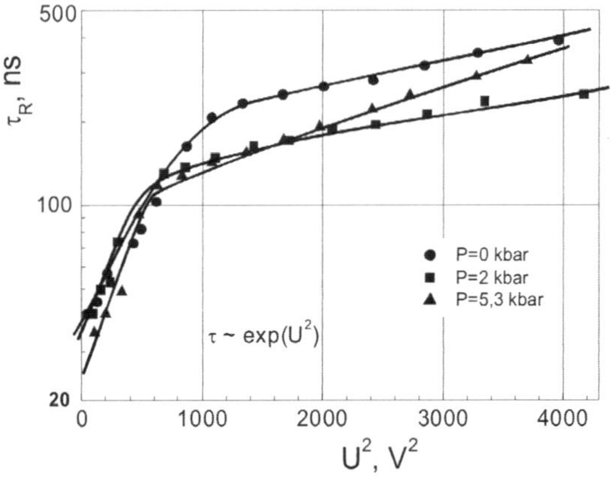


Figure 9. Recombination times in log - $U^2$ scale at different pressures. High compensation.

In summary, studies of transient currents at impurity breakdown give the possibility for direct determination of carrier capture and field ionization times. The transient processes turn out to be rather complicated and strongly depended on doping level and compensation.

**Acknowledgments**
The work is supported by RFBR grants 07-02-00857, 08-02-01464, 08-02-91323 and RAS programs "Quantum nanostructures" and "Problems of radio physics".

**References**

[1] Kagan M S, in: *Towards the first silicon laser*, *NATO Science Series*, eds L Pavesi, S Gaponenko, and L Dal Negro (Kluwer Academic Publishers), **359** (2003).
[2] Kagan M S, Altukhov I V, Sinis V P, Chirkova E G, Yassievich I N, and Kolodzey J, *Journal of Communications Technology and Electronics*, **48**, 1047 (2003).
[3] Kagan M S, Altukhov I V, Sinis V P, Chirkova E G, Paprotskiy S K, Yassievich I N , Odnoblyudov M A, Prokofiev A A, and Kolodzey J, *ECS Transactions*, **3**, 745 (2006).
[4] Sclar N, Burstein E, Tunner W J, Davisson J W, *Phys. Rev.*, **91**, 216; **92**, 858 (1953).
[5] Ryder E J, Ross J M, Kleiman D A, Phys. Rev., **85**, 1342 (1954).
[6] Keldysh L V, *ZhETP* (*Sov. Phys. JETP*) **37**, 713 1959.
[7] Vul B M, Zavaritskaya E I, Keldysh L V, *Proc. USSR Ac. Sci.*, **135**, 1361 (1960).
[8] Ascarelly G, Brown S, *Phys. Rev.*, **120**, 1615 (1960).
[9] Yamashita J, *J. Phys. Soc. Japan*, **16**, 720 (1961).
[10] Zylbersztein A, *Phys. Rev.*, **127**, 744 (1962).
[11] Schoell E, *Nonequilibrium Phase Transitions in Semiconductors*, Springer (1987).
[12] Dargis A., Zhurauskas S., Zhurauskene, Phys. Lett. A, **157**, 419 (1991)
[13] Lax M, *Phys. Rev.*, **119**, 1502 (1960).
[14] Abakumov V N, Perel V I, Yassievich I N, *Nonradiative Recombination in Semiconductors,* in *Modern Problems in Condensed Matter Sciences*, **33**, edited by V M Agranovich and A A Maradudin (North-Holland, Amsterdam, 1991).

Journal of Physics: Conference Series **193** (2009) 012035 doi:10.1088/1742-6596/193/1/012035

# Monte Carlo simulation of ballistic transport in high-mobility channels

**G. Sabatini, H. Marinchio, C. Palermo, L. Varani, T. Daoud and R. Teissier**

Institut d'Electronique du Sud (CNRS UMR 5214) - Université Montpellier II (France)

**H. Rodilla, T. González, and J. Mateos**

Departamento de Fisica Aplicada - Universidad de Salamanca (Spain)

E-mail: `sabatini@ies.univ-montp2.fr`

**Abstract.** By means of Monte Carlo simulations coupled with a two-dimensional Poisson solver, we evaluate directly the possibility to use high mobility materials in ultra fast devices exploiting ballistic transport. To this purpose, we have calculated specific physical quantities such as the transit time, the transit velocity, the free flight time and the mean free path as functions of applied voltage in InAs channels with different lengths, from 2000 nm down to 50 nm. In this way the transition from diffusive to ballistic transport is carefully described. We remark a high value of the mean transit velocity with a maximum of $14\times10^5$ m/s for a 50 nm-long channel and a transit time shorter than 0.1 ps, corresponding to a cutoff frequency in the terahertz domain. The percentage of ballistic electrons and the number of scatterings as functions of distance are also reported, showing the strong influence of quasi-ballistic transport in the shorter channels.

## 1. Introduction

Modern technology allows the reduction of the dimensions of electronic devices deeply into the nanometric scale, thus reducing the electron transit time inside the structures. The increase of the operation speed obtained by downsizing the device dimensions can be boosted by the use of very high mobility materials such as InGaAs and, more recently, InAs. For instance, recent studies showed that High Electron-Mobility Transistors (HEMT) with InGaAs channels can be used as emitters or detectors in the THz domain [1]. On the other hand, despite its small gap, InAs has attracted recent interest for advanced electronic applications because it presents a higher mobility. The so-called Quantum Hot Electron Transistor (QHET) is an example of an innovative high-speed transistor using the attractive properties of InAs [2]. By means of Monte Carlo (MC) simulations coupled with a two-dimensional Poisson solver, we can evaluate the possibility to use this semiconductor as the active zone of ultra fast devices exploiting ballistic transport. In particular, we have calculated specific microscopic physical quantities such as the transit time, the transit velocity, the free flight time and the mean free path as functions of applied voltage in InAs channels with different lengths, from 2000 nm down to 50 nm. The percentage of ballistic electrons and the number of scatterings as functions of length are also reported. In this way the transition from diffusive to ballistic transport is carefully described.

## 2. Model description

We consider different InAs semiconductor channels at room temperature with two contacts at both ends. The width $W$ is fixed to 200 nm and the length $L$ can be shorter (longer) than the carrier mean free path so that ballistic (diffusive) regimes can be analysed. The carrier concentration $N_D$ is equal to $10^{16}$ cm$^{-3}$. Electrons are injected in the channel at a constant rate from contacts treated as thermal reservoirs according to the model detailed in Ref. [3], which is well adapted to the case of mesoscopic conductors.

The MC model here used to simulate transport at room temperature has been previously validated by comparison with experimental data [4]. For the conduction band we have used three non-parabolic spherical valleys. The scattering mechanisms which are included are the collisions with ionized impurities, the transitions due to absorption and emission of polar and nonpolar optical phonons, the collisions with acoustic (elastic) phonons, and the intervalley scatterings. Carrier-carrier interaction is neglected. Impact ionization is not taken into account even if, because of the small InAs gap, it could influence the results mainly for the ballistic channels and when the applied voltage is above 0.4 V.

## 3. Transport analysis

To evaluate the ballistic or diffusive character of transport and the possibility for this type of devices to work at THz frequencies we have simulated different channels of various lengths, from 2000 nm to 50 nm. First of all we have calculated two basic parameters: the average number of collisions $S$ and the average transit time of an electron which moves from cathode to anode $T_t$. In our model we consider the exit of an electron from the device as a scattering event.

We remark that, for shorter channels, $S$ decreases (Fig. 1(a)). For low voltages $S$ is practically constant and varies from 1 for $L$ = 50 nm to 30 for $L$ = 2000 nm. Therefore for 50 and 100 nm-long devices under ohmic conditions the majority of electrons are completely ballistic. On the other hand, when the voltage reaches a value of approximately 1.0 V, intervalley transfer is activated and the number of scatterings increases very fast. A similar behavior is found for $T_t$ (Fig. 1(b)) where we notice a first region (quasi-ballistic region) where $T_t$ decreases with voltage, typical of a ballistic behavior, and a second region, after intervalley onset (diffusive region), where it increases with voltage. In the case impact ionization mechanisms were considered, a similar effect (but less significant) would be found for lower applied voltages (around 0.4 V).

Starting from the previous parameter we have extracted the longitudinal transit velocity $V_t = L/T_t$. As we can see in Fig. 1(c) $V_t$ at equilibrium is different from zero, because it corresponds to the mean velocity of electrons which from the cathode arrive to the anode. The shorter the channel the higher the velocity because of the more ballistic character of transport. As long as the applied voltage is low, the number of the scatterings undergone by an electron decreases and so the longitudinal transit velocity increases. We also remark that the shape of velocity shows a maximum just before the onset of intervalley transfers. In the diffusive region, contrary to the ballistic one, the velocity decreases with voltage. If we take into account impact ionization mechanisms, by avoiding intervalley scattering, the electron velocity should increase [5]. We have also reported in Fig. 1(d) the free flight time $T_f = T_t/S$. At equilibrium $T_f$ tends towards the transit time for the ballistic channels , while when $L$ increase, and the transport becomes more diffusive, $T_f$ tend toward the same value. When biasing the channels, we first observe a slow decrease of $T_f$ and then a sharp drop around V = 1V, again due to the onset of intervalley mechanisms. To better understand the behavior of electrons in these channels we have also extracted the longitudinal mean free path $L_p = L/S$ reported on Fig. 1(e). At equilibrium and in the quasi-ballistic channels $L_p$ increases with $L$. As expected for the ballistic channels, $L_p$ coincides with $L$ since there is practically no scattering. Moreover, at low voltages it is practically constant because the increase of $V_t$ is balanced by the reduction of $T_f$. For higher voltages intervalley transfer leads to a strong reduction of $L_p$.

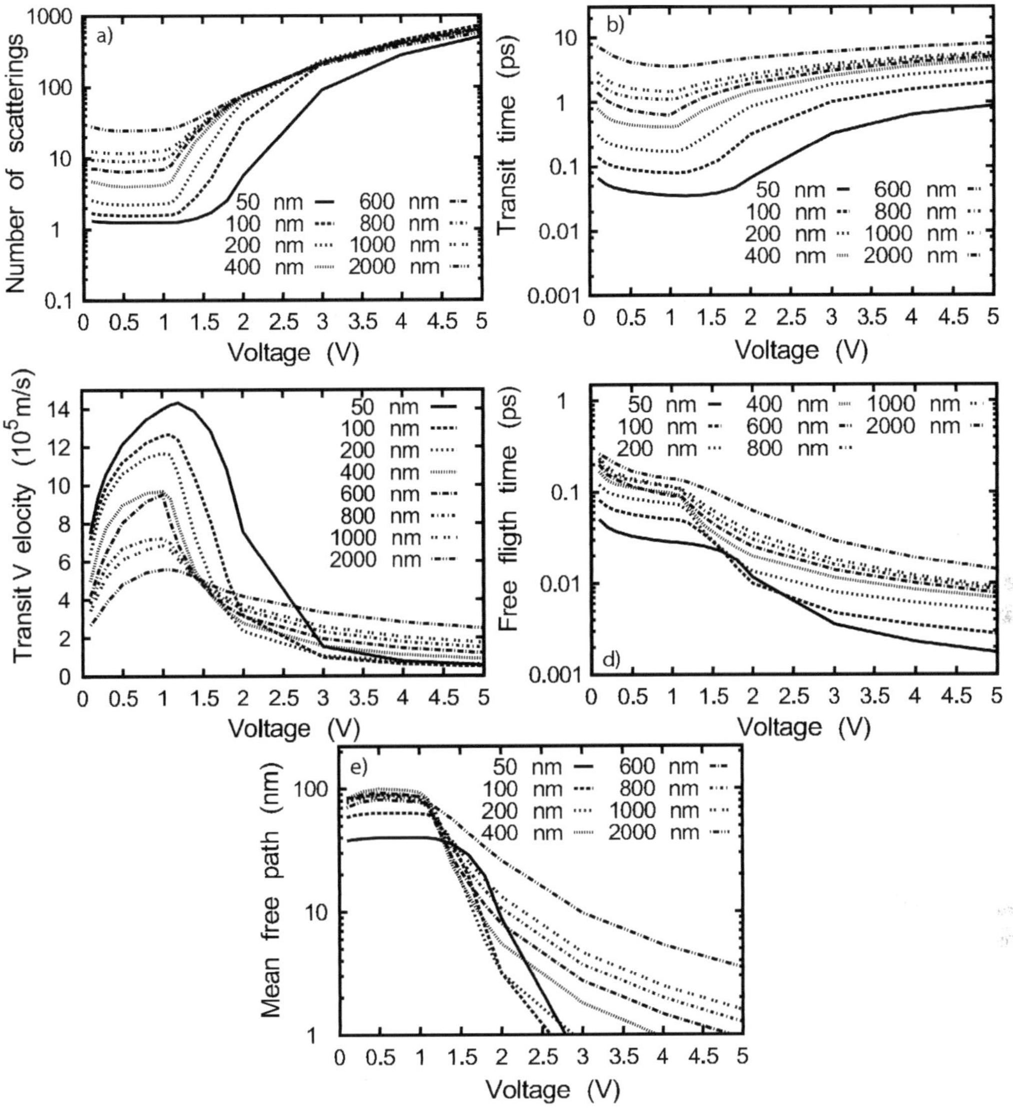


**Figure 1.** Number of scatterings from cathode to anode $S$ (a), transit time $T_t$ (b), longitudinal transit velocity $V_t$ (c), free flight time $T_f$ (d) and longitudinal mean free path $L_p$ (e) as functions of applied voltage in InAs channels of different lengths $L$ from 50 nm to 2000 nm.

## 4. Evolution of ballistic electrons

We investigate now the evolution of the number of interactions undergone by electrons on a channel of length $L$ with an applied average electric field of 1 kV/cm. Here the exit from the channel is not considered as a collision. Figure 2(a) shows the percentage of carriers as a function of the scatterings undergone from cathode to anode for several channel lengths. In the case of a long channel (beyond 400 nm) the curve exhibits a bell shape. This is typical of a stationary diffusive transport where the maximum is centered on a non-zero average number of scatterings. The shorter the channel is the more the maximum of the distribution approaches to zero. The distribution is tightened and decreases exponentially, which is typical of non-stationary transport where the majority of carriers undergoes a maximum of one interaction, or

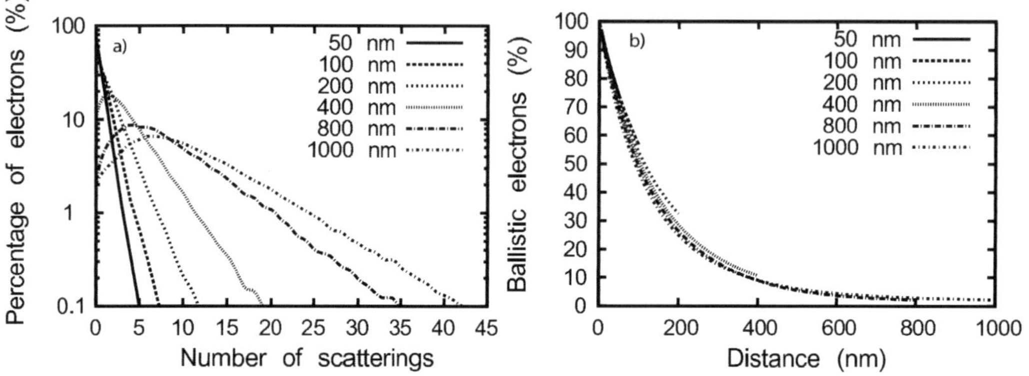


**Figure 2.** Percentage of electrons going from cathode to anode as a function of the number of scatterings (a) and percentage of ballistic electrons going from cathode to anode as a function of the distance from the cathode (b) for an average electric field of 1 kV/cm, in channels of different lengths $L$ from 50 nm to 2000 nm.

are completely ballistic. For $L$ = 100 nm, we have 54% of ballistic electrons and 73% for a 50 nm-long channel. In Fig. 2 (b) we show the number of strictly ballistic electrons, as a function of the distance from the cathode for the same average electric field of 1 kV/cm. We remark the high quantity of ballistic electrons in the shorter structures, equal, for example, to 35% at the end of a 200 nm-channel, sign of non-stationary quasi-ballistic transport. For the shorter lengths ($L \leq 400$ nm) the percentage of ballistic electrons depends slightly on the length of the channel while for the longer lengths the different curves coincide. For the longer channels transport is pratically stationary and there is only a negligible quantity of ballistic electrons at the end of the channel.

## 5. Conclusions

To investigate the diffusive or ballistic character of transport and to evaluate the possibility of InAs channels to work at THz frequencies, we have calculated different microscopic parameters. The transit time, in the case of 50 nm or 100 nm channels can be equal or lower than 0.1 ps, that in frequency corresponds to 10 or more terahertz. In fact for these lengths we remark that already at low voltages there is a great number of ballistic electrons equal, for example, to 73% for a 50 nm-long channel. The limit of length to have a cut off frequency in the THz range (transit time lower than 1 ps) is between 600 nm and 800 nm. We remark also that the mean longitudinal transit velocity is very high with a maximum value of $14\times10^5$ m/s for the shortest channel. However, to have the best performances the applied voltage must be kept under the limit imposed by intervalley scattering. In the case impact ionization mechanisms were considered, avoiding intervalley scattering, the electron velocity should further increase [5].

## 6. References

[1] Lusakowski J, Knap V, Dyakonova N, Varani L, Mateos J, Gonzalez T, Roelens Y, Bollaert S, Cappy A and Karpierz K 2005 *J. Appl. Phys.* **97** 064307
[2] Daoud T, Boissier G, Dvenson J, Sabatini G, Varani L, Baranov A, Teissier R 2008 *20th International Conference on Indium Phosphide and Related Materials, IPRM 2008* 1
[3] González T, Mateos J, Pardo D, Bulashenko O M and Reggiani L 1998 *Semicond. Sci. Technol.*, **13**, pp. 714
[4] Mateos J, González T, Pardo D, Hoel V, and Cappy A 2000 *IEEE Transactions on Electron Devices* **47** 250
[5] Rodilla H, González T, Pardo D, and Mateos J 2009 *J. Appl. Phys.*, **105**, p. 113705

Journal of Physics: Conference Series **193** (2009) 012036 doi:10.1088/1742-6596/193/1/012036

# Modeling Self-Heating Effects in Nanoscale SOI Devices

**D. Vasileska[1], S. M. Goodnick[1] and K. Raleva[2]**

[1]Department of ECEE, Arizona State University, Tempe, AZ 85287-5706, USA

[2]Faculty of Electrical Enginnering and ITs, University Sts. Cyril and Methodi, Skopje, Republic of Macedonia

Corresponding Author: vasileska@asu.edu

**Abstract**. We have developed electro-thermal device simulator that at present self-consistently solves the Boltzmann Transport Equation (BTE) for the electrons and the energy balance equations for acoustic and optical phonons. The simulator has been successfully used in investigation of self-heating effects in different technology nodes of nano-scale fully-depleted (FD) silicon on insulator (SOI) devices and dual gate (DG) device structures. Our simulation results suggest that self-heating has less degrading effect on the on-current in smaller devices in which non-stationary transport and velocity overshoot effect dominate the carrier transport. We also find that the choice of the boundary conditions on the gate (Dirichlet vs. Neumann) has significant impact on the current degradation. The role of Neumann and leaky Dirichlet boundary conditions at the source and drain contact is also examined and estimates on the overall current degradation are presented. Finally we illustrate that the use of anisotropic and temperature dependent thermal conductivity is needed to properly determine the temperature of the hot spot which is important parameter regarding device reliability.

## 1. Introduction

There are basically two reasons why self-heating in silicon on insulator (SOI) devices is very important: (1) the underlying buried oxide layer (BOX) that is typically made of $SiO_2$ to reduce the fabrication cost and also get better bottom $Si/SiO_2$ interface has thermal conductivity that is 100 times smaller than the value of bulk Si material, thus representing significant thermal barrier for the heat flow into the substrate. Second, since the active silicon film has thickness between 10 nm and 70 nm, depending on the technology generation, phonon boundary scattering significantly reduces the thermal conductivity of thin Si films, as it is shown in Figure 1 below where we present the simulation results for the thermal conductivity by solving the Phonon Boltzmann Transport Equation in the relaxation time approximation, experimental data of Asheghi and co-workers [1,2] and our fit to the Monte Carlo results for which we have followed the approach of Sondheimer [3], that takes into account phonon boundary scattering by assuming it to be purely diffusive. Namely, the thermal conductivity of a semiconductor film of a thickness $a$, under the assumption that the $z$-axis is perpendicular to the plane of the film, the surfaces of the film being at $z$=0 and $z$=$a$, is given by:

$$\kappa(z) = \kappa_0(T) \int_0^{\pi/2} \sin^3\theta \left\{ 1 - \exp\left( -\frac{a}{2\lambda(T)\cos\theta} \right) \cosh\left( \frac{a-2z}{2\lambda(T)\cos\theta} \right) \right\} d\theta \qquad (1)$$

where $\lambda(T)$ is the mean free path expressed as $\lambda(T) = \lambda_0(300/T)$ nm where room temperature mean free path of bulk phonons is taken to be $\lambda_0 = 290$ nm.

Selberherr [4,5] has parameterized the temperature dependence [6] of the bulk thermal conductivity in the temperature range between 250K and 1000K. In our case we find that the appropriate expression is:

$$\kappa_0(T) = \frac{135}{a + bT + cT^2} \quad \text{W/m/K} \tag{2}$$

where $a = 0.03$, $b = 1.56\times10^{-3}$, and $c = 1.65\times10^{-6}$. As can be seen from the results presented in Figure 1 Equations (1) and (2) give excellent agreement with the experimental and the theoretical data reported in a later Asheghi paper [2].

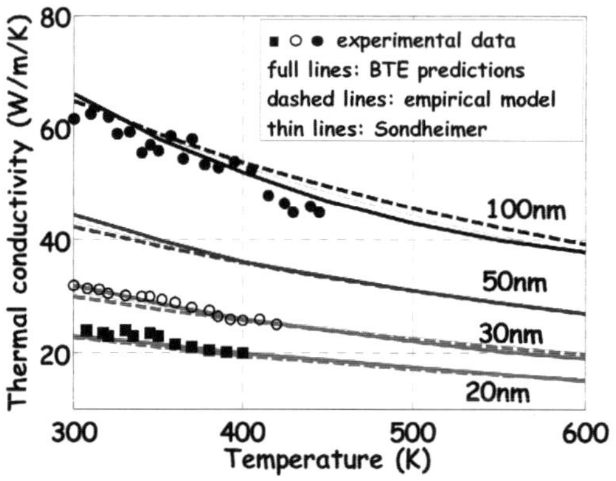


Figure 1. Silicon film thickness dependence of the average thermal conductivity at T=300 K vs. active silicon layer thickness. Experimental data are taken from the work of Asheghi and co-workers [1,2].

In the two sections that follow we first highlight the importance of the inclusion of the silicon-film thickness dependence of the thermal conductivity (Section 2) and in Section 3 we will discuss the inclusion of proper boundary conditions and the importance/unimportance of simulation of a larger domain from the device active region.

## 2. Importance of the inclusion of the anisotropic thermal conductivity regarding reliability concerns

We compare electro-thermal simulation results for various models for two different gate length devices, 25 nm and 180 nm gate length devices in Figure 2. Here Dirichlet boundary conditions are assumed on the gate and back contact (300K), while the other boundaries are treated as Neumann boundaries (no heat flow), as discussed elsewhere [7]. As seen from the results presented in Figure 2, with the inclusion of the full thickness and temperature dependence of the thermal conductivity, there is a significant temperature rise along the channel. This temperature increase can have significant implications on the device reliability and its lifetime. Thus, the results presented in Figure 2 suggest that if proper determination of the peak temperature of the hot-spot is needed then the complete model that we propose (given in Eqs. (1) and (2)) must be used for the thermal conductivity in the energy balance equation for acoustic phonons.

## 3. Boundary Conditions and proper choice of the device simulation domain

In the analysis that we have performed in obtaining the results presented in Figure 2 we have assumed Dirichlet boundary conditions at the bottom of the BOX for reasons that will be explained below, Dirichlet boundary conditions at the gate electrode and Neumann boundary conditions at the source and drain contact, and at the side boundaries.

In what follows we first illustrate that modelling of the substrate underneath the BOX is not necessary, and then we show that the same conclusion is valid for the gate electrode if it is not near a hot spot. Let us first concentrate on the substrate. In Table 1 we summarize the simulation results for 25 nm FD SOI device for the case when the substrate is not modelled and when the substrate is modelled. It is evident from the results presented in Table 1 that the boundary condition for the lattice temperature at the bottom of the substrate very well maps as a boundary condition at the bottom of the BOX. Thus, modelling of the substrate would just result in an unnecessary increase in the required CPU time.

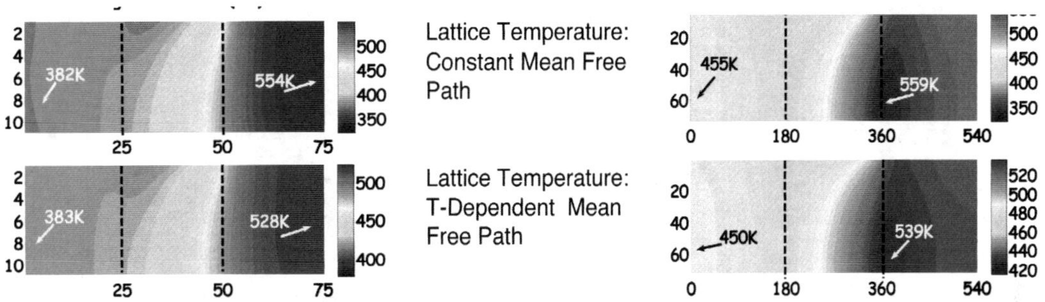


Figure 2. Lattice temperature contour plots for a 25 nm channel length device (left) and a 180 nm gate-length device (right). X-axis is device length from source to drain in nm and y-axis is silicon film thickness in nm. Parameter in this figure is the variation of the mean free path for phonons. The Si film layer thickness is 10 nm for the 25 nm channel length device and 72 nm for the 180 nm channel length device. Bias conditions are: $V_{GS}=V_{DS}=1.2$ V for the 25 nm channel length device and $V_{GS}=V_{DS}=1.8$ V for the 180 nm channel length device.

Next issue that we investigate is whether 300 K is an appropriate boundary condition at the gate electrode and whether Neumann boundary conditions are the appropriate choice of boundary conditions for the source and drain electrode. To test this conjecture we extend the length of the metal gate (Figure 3 – top left panel) and source and drain electrodes and assume the best case scenario, namely we neglect the Joule heating. At the end of the three electrodes we set Dirichlet boundary conditions for the lattice temperature of 300 K. The results of these simulations are shown in Figure 3. We see that if the gate is not near a hot spot then the 300K boundary condition on the gate is the correct choice as copper metal gate has very large thermal conductivity ($\kappa_{th}$=400W/m/K). Regarding the source and drain contacts, for the case when we neglect the Joule heating (we solve the heat conduction equation in this region), which is the best case scenario, we get about 3% reduction in the current degradation. This also is not very large value and for larger devices will represent even less significant correction. Thus, everything depends whether the device is near a hot-spot or not. If it is then Neumann boundary conditions are appropriate and if it is not then some sort of leaky Dirichlet boundary conditions at the source and drain electrode are appropriate.

Table 1. The role of the substrate on the current degradation due to self-heating in 25 nm FD SOI device ($V_{GS}$=1.2V, $V_{DS}$=1.2V).

| Type of simulation | Gate Electrode Temperature (K) | Substrate Electrode Temperature (K) | 25 nm FD-SOI nMOSFET substrate region is not modeled | | 25 nm FD-SOI nMOSFET with substrate region (250nm) | |
|---|---|---|---|---|---|---|
| | | | Current (mA/um) | Current Decrease (%) | Current (mA/um) | Current Decrease (%) |
| isothermal | 300 | 300 | 1.771 | \ | 1.768 | \ |
| thermal | 300 | 300 | 1.646 | 6.7 | 1.639 | 6.9 |

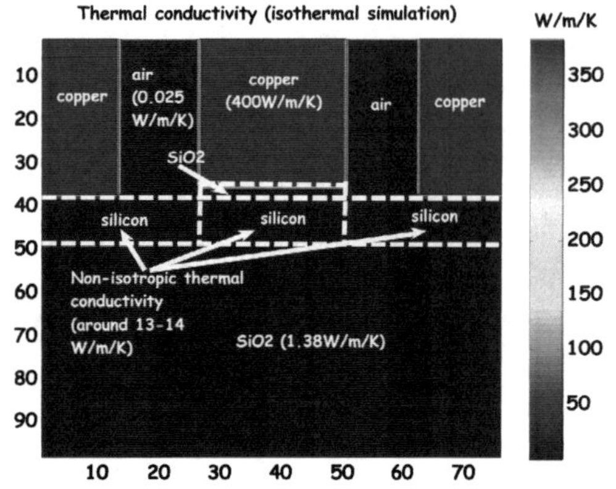


| | Tgate | Tsource | Tdrain | Tsides | Current Degradation |
|---|---|---|---|---|---|
| 1. | 300K | Neumann | Neumann | Neumann | 6.7% |
| 2. | 300K | Neumann | Neumann | 300K | 3.8% |
| 3. | Neumann | Tsource | Neumann | 300K | 11.6% |
| 4. | Neumann | Tsource | Neumann | Neumann | 38.7% |
| 5. | 300K | 300K | 300K | Neumann | 3.96% |

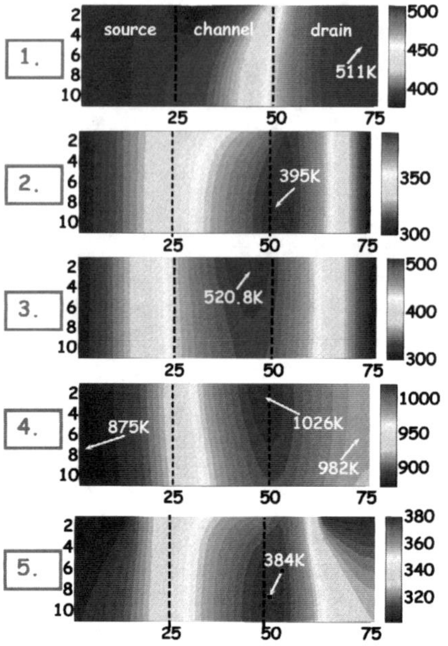


Figure 3. Top left – Device structure being simulated. Bottom left table - Current degradation due to self-heating effects in 25nm FD-SOI nMOSFET ($V_{GS}$ = 1.2V, $V_{DS}$ = 1.2V) for different boundary conditions. Simulation results #1-4 are obtained when metal electrode regions are not included in the simulation domain. Simulation result #5 is obtained when metal electrode regions are included in the simulation domain. Right panel - Lattice temperature profile in the active Si-layer for different boundary conditions as given in the table. In the 2D plots, x-axis is device length and y-axis is device width, both in nm.

## Summary

In summary, our investigations suggest that the choice of the proper boundary conditions very much depends on the operation regime of the device and whether the device is near a hot spot or not. We also find that proper determination of the temperature of the hot spot requires usage of the thickness and temperature dependent thermal conductivity for thin silicon films. Therefore, the results presented here can significantly help in better design of SOI devices from the thermal perspective.

## References

[1] M. Asheghi,., M. N. Touzelbaev., K. E. Goodson, Y. K. Leung, and S. S. Wong., ASME Journal of Heat Transfer, Vol.120, pp. 30-33, 1998
[2] W. Liu and M. Asheghi, J. Appl. Phys., Vol. 98, 123523-1 (2005).
[3] E. H. Sondheimer, Advances in Physics, Volume 1, Issue 1 January 1952 , pages 1 – 42.
[4] V. Palankovski and S. Selberherr, Journal Microsystem Technologies, Vol. 7, pp. 183-187 November, 2001.
[5] Silvaco Manual (www.silvaco.com).
[6] M. Asheghi,., M. N. Touzelbaev., K. E. Goodson, Y. K. Leung, and S. S. Wong., ASME Journal of Heat Transfer, Vol.120, pp. 30-33, 1998.
[7] K. Raleva, D. Vasileska and S. M. Goodnick, IEEE Trans. Electron Dev. Volume 55, Issue 6, June 2008 Page(s):1306 – 1316.

Journal of Physics: Conference Series **193** (2009) 012037 doi:10.1088/1742-6596/193/1/012037

# Monte Carlo simulation of high-energy transport of electrons and holes in bulk SiGeC alloys

**M Michaillat[1], D Rideau[1], F Aniel[2], C Tavernier[1] and H Jaouen[1]**

[1] STMicroelectronics, 850, rue Jean Monnet, 38926 Crolles, France
[2] IEF - CNRS UMR 8622, Bât. 220, Université Paris-Sud, 91045 Orsay, France

E-mail: `marc.michaillat@st.com`

**Abstract.** High-energy transport properties of charge carriers in ternary random $Si_{1-x-y}Ge_xC_y$ alloys are investigated using Full-Band Monte Carlo simulations. Models for scattering mechanisms include phonon scattering, impact ionization and alloy scattering. Phonon scattering rates are wave-vector dependent and calculated consistently with the Full-Band structure. Impact ionization rates are modeled using analytical Keldysh formulas fitted to previously reported *ab initio* results. We derive a model for alloy scattering rate specific to ternary random alloys. It involves 2 effective alloy potentials which are independently calibrated on experimental mobility measurements. Presented Monte Carlo simulation results are shown to be in very good agreement with a variety of high-energy transport measurements, including drift velocities, impact ionization coefficients and quantum yields. Effects of alloy composition on the electrical characteristics of $Si_{1-x-y}Ge_xC_y$ alloys are investigated.

## 1. Introduction

Performance optimization of advanced heterojunction bipolar transistors has led to the introduction of germanium and carbon in the device base, yielding SiGeC base layers biaxially strained on Si. TCAD simulation of the electrical operation of such SiGeC/Si HBTs should include transport parameter models relevant to ternary SiGeC alloys. The aim of this work is to investigate high-energy carrier transport in $Si_{1-x-y}Ge_xC_y$ materials within the alloy composition domain defined by $0 < x < 100\%$ and $0 < y < 2\%$, by means of Full-Band Monte Carlo (MC) method. The physical models for scattering mechanisms are described and compared with litterature in section 2. In section 3, simulation results for high-field transport properties in $Si_{1-x-y}Ge_xC_y$ materials are reported.

## 2. Scattering models

Our MC simulations rely on SiGe band structures calculated with Tight-Binding method [1]. In order to simulate high-energy transport phenomena, it is necessary to accurately describe the dispersion relation up to 5 eV above the band minima, consequently 5 conduction bands and 3 valence bands are accounted for.

Phonon scattering rates are wave-vector dependent and are calculated consistently with the Full-Band structure [2]. Phonon modes in SiGeC alloys are modeled by 4 Si-Si modes and 4 Ge-Ge modes. Alloy modes such as C-C modes or mixed Si-Ge modes are neglected. The probability for a carrier of wave-vector $k$ in band $\nu$ to interact with a phonon mode $\eta$ is given

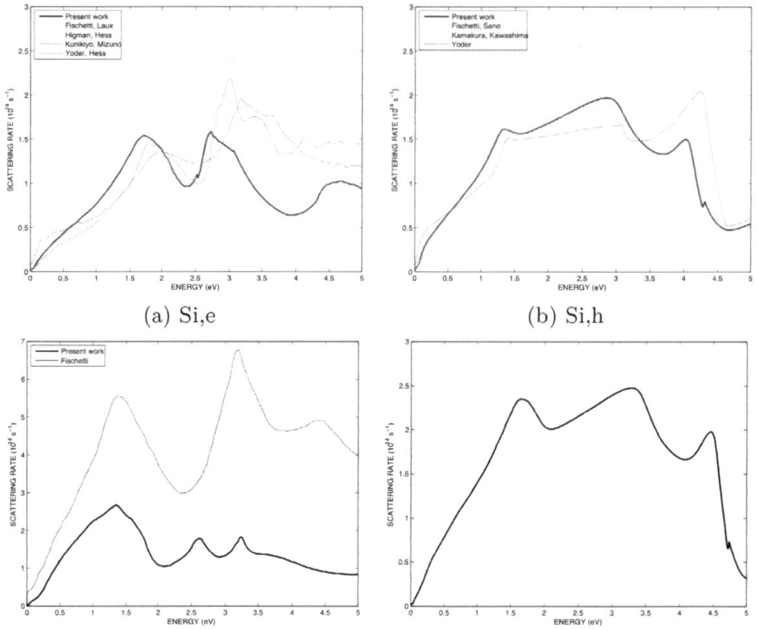


(a) Si,e (b) Si,h

(c) Ge,e (d) Ge,h

Figure 1: Electron- and hole-phonon scattering rates as a function of carrier energy in bulk Si and Ge at 300 K. Our results are compared with Full-Band results exposed in Refs.[12, 6] and [7] for Si and Ge, respectively.

by:

$$P_{\eta,\nu}(k) = \sum_{\nu',k'} \frac{\pi}{\rho\omega_\eta(q)} \Delta_\eta^2 \times \mathcal{I}(k,k')^2 \times \delta\left(E_\nu - E'_{\nu'} \mp \hbar\omega_\eta(q)\right)\left(n_\eta(q) + 1/2 \pm 1/2\right) \qquad (1)$$

where upper and lower signs respectively refer to phonon emission and absorption. In (1), phonon dispersions $\omega_\eta(q)$ are approximated by isotropic quadratic expressions, as in Ref.[3]. Electron overlap integrals $\mathcal{I}(k,k')$ are modeled within the Nordheim approximation [4]. Hole overlap integrals are described by Wiley's angular expressions [5]. Deformation potentials$\Delta_\eta$ are treated as empirical parameters. As such they are assumed to depend on the square root of the initial carrier energy, in a similar way as in Ref.[6]. The total carrier-phonon scattering rates in bulk Si and Ge are plotted in Fig.1. In the case of Si, it can be seen that our calibrated rates are in excellent agreement with the consensus established by the modeling community. However in the case of Ge, although several studies have applied Full-Band techniques to the simulation of charge carrier transport [7, 8, 9, 4, 10], only the early work of Fischetti was found to have reported the total electron-phonon scattering rate magnitude. It appears that our results are in strong disagreement with Fischetti's, however it must be stressed that our scattering rate was calibrated together with a realistic impact ionization rate derived from *ab initio* calculations, whereas the early work of [7] relied on an arbitrary Keldysh formula. In SiGeC alloys, the total phonon rate is expressed as a weighted sum of the scattering rates due to Si-Si and Ge-Ge modes [11]:

$$P_{\eta,\nu}(x,y,k) = (1-x) \cdot P_{\eta,\nu}^{Si}(k) + x \cdot P_{\eta,\nu}^{Ge}(k) \qquad (2)$$

Impact ionization rates in bulk Si and Ge are modeled with analytical energy-dependent Keldysh formulas:

$$P_{II}(E) = P_0 \times (E - E_0)^{\alpha_0} \qquad (3)$$

Table 1: Keldysh formula parameters for the impact ionization rate in bulk Si and Ge.

| Material | Carrier | $P_0$ ($\mathrm{eV}^{-\alpha_0}\ \mathrm{s}^{-1}$) | $\alpha_0$ | $E_0$ (eV) | Ref. |
|---|---|---|---|---|---|
| Si | e | $1.0 \times 10^{11}$ | 4.6 | 1.1 | [13] |
| | h | $1.14 \times 10^{12}$ | 3.4 | 1.49 | [14] |
| Ge | e | $4.0 \times 10^{11}$ | 4.8 | 0.8 | [8] |
| | h | $1.2 \times 10^{12}$ | 3.8 | 1.0 | [15] |

The parameters $P_0$, $E_0$ and $\alpha_0$ are adjusted so that the empirical rate in (3) fits *ab initio* calculated ionization rates. Parameter values together with references of the *ab initio* studies from which they were determined are indicated in table 1. In SiGe alloys, Keldysh parameters are linearly interpolated between their respective values in Si and Ge. Since no information was found concerning impact ionization in bulk carbon or in carbon alloys, C mole fraction $y$ is assumed to have no influence on impact ionization rates.

Once models for phonon scattering and impact ionization have been specified, simulations of carrier transport in pure Si and Ge materials can be carried out. MC results are compared to a variety of experimental data, including drift velocity characteristics (cf. Fig.2a), impact ionization coefficients (cf. Fig.2b), and quantum yield measurements.

Simulation of charge carrier transport in random alloys requires the inclusion of an additional scattering mechanism, namely alloy scattering. Our model for alloy scattering extends the commonly used model specific to binary random alloys (see e.g. Refs.[11, 16]) to the more general case of ternary random alloys. Alloy scattering rate in ternary $Si_{1-x-y}Ge_xC_y$ is expressed as:

$$P_{alloy}(x,y,E) = \frac{2\pi}{\hbar}\Omega_c \cdot \mathcal{D}(E) \cdot \left[x(1-x)\cdot U^2_{Si-Ge} + y(1-y)\cdot U^2_{Si-C} - 2xy \cdot U_{Si-Ge}\cdot U_{Si-C}\right] \quad (4)$$

where $\mathcal{D}(E)$ is the density of states per spin. In (4), $U_{Si-Ge}$ and $U_{Si-C}$ are effective alloy potential parameters, which are independently calibrated to reproduce low-field mobility measurements in binary $Si_{1-x}Ge_x$ [17] and $Si_{1-y}C_y$ [18] alloys. When the particle undergoes an alloy scattering event, the MC particle post-scattering state $(k', \nu')$ is randomized on the energy-conserving surface.

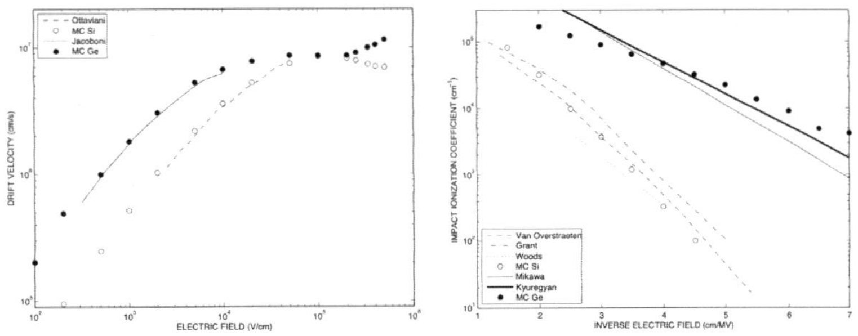


(a) drift velocity (b) impact ionization coefficient

Figure 2: Comparison between MC simulations and measurements for the drift velocity and impact ionization coefficient of holes in bulk Si and Ge.

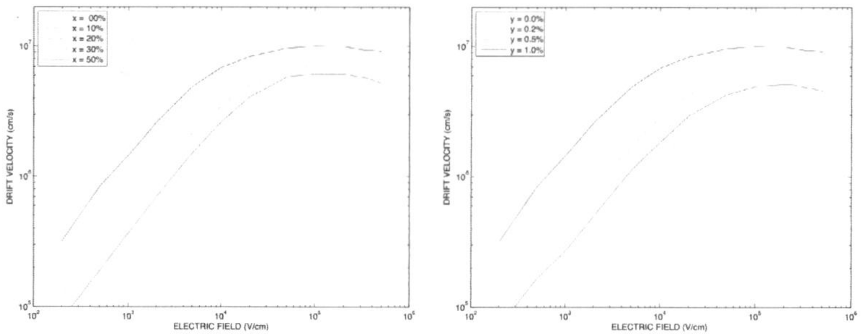


(a) $Si_{1-x}Ge_x$ (b) $Si_{1-y}C_y$

Figure 3: Electron drift velocity as a function of electric field in bulk binary SiGe and SiC alloys. MC simulation results are shown for several alloy compositions.

## 3. Simulation results

Based on the bandstructure and scattering models exposed in section 2, high-field properties of $Si_{1-x-y}Ge_xC_y$ alloys can be investigated by constant-field single-particle MC simulation. In order to study how the substitutional Ge and C concentrations affect SiGeC electrical properties, we plot in figures 3a and 3b the electron velocity-field characteristics as obtained from MC simulations of binary SiGe and SiC alloys, respectively. It can be seen that both Ge- and C-alloying result in a strong degradation of the electron mobility and saturation velocity. However equivalent results are obtained with 10% Ge and 0.2% C, emphasizing the higher scattering strength of substitutional C atoms as compared to Ge atoms.

## 4. Conclusion

In this paper, the Monte Carlo technique has been applied to the simulation of high-field properties of ternary SiGeC materials. All the relevant scattering mechanisms have been physically accounted for, and specifically adressed to both electrons and holes. Results for high-energy transport in bulk Si and Ge have been extensively compared to measurements. The models should yield a robust basis for the study of charge carrier transport in SiGeC alloys.

## References

[1] Rideau D, Feraille M and Ciampolini L 2006 *Phys. Rev.* B **74** 195208
[2] Fischetti M V and Laux S E 1988 *Phys. Rev.* B **38** 9721–45
[3] Pop E, Dutton R W and Goodson K E 2004 *J. Appl. Phys.* **96** 4998–5005
[4] Nguyen P H, Hofmann K R and Paasch G 2003 *J. Appl. Phys.* **94** 375–86
[5] Wiley J D 1971 *Phys. Rev.* B **4** 2485–93
[6] Kamakura Y, Kawashima I and Degushi K 1999 *Proc. IEDM (Washington)* pp 727–30
[7] Fischetti M V 1991 *IEEE Trans. Electron Devices* **38** 634–49
[8] Fischetti M V, Sano N and Laux S E 1996 *Proc. SISPAD (Tokyo)* pp 43–4
[9] Nguyen P H, Hofmann K R and Paasch G 2002 *J. Appl. Phys.* **92** 5359–70
[10] Ghosh B, Wang X and Fan X F 2005 *IEEE Trans. Electron Devices* **52** 547–53
[11] Bufler F M and Meinerzhagen B 1998 *J. Appl. Phys.* **84** 5597–602
[12] Abramo A *et al.* 1994 *IEEE Trans. Electron Devices* **41** 1646–54
[13] Kamakura Y, Mizuno H and Yamaji M 1994 *J. Appl. Phys.* **75** 3500–6
[14] Kunikiyo T, Takenaka M and Morifuji M 1996 *J. Appl. Phys.* **79** 7718–25
[15] Sano N, Fischetti M V and Laux S E 1998 *Proc.* $6^{th}$ *IWCE (Osaka)* pp 198–201
[16] Fischetti M V and Laux S E 1996 *J. Appl. Phys.* **80** 2234–52
[17] Gaworzewski P, Tittelbach-Helmrich K and Penner U 1998 *J. Appl. Phys.* **83** 5258–63
[18] Osten H J and Gaworzewski P 1997 *J. Appl. Phys.* **82** 4977–81

Journal of Physics: Conference Series **193** (2009) 012038 doi:10.1088/1742-6596/193/1/012038

# A determinist method solving the Wigner transport equation for quantum transport in silicon nanowires

**Sylvain Barraud and Emmanuelle Sarrazin**
CEA-LETI-Minatec, 17 rue des martyrs, 38054 Grenoble - cedex 09, France
E-mail: sylvain.barraud@cea.fr

**Abstract.** A determinist method is presented for solving the steady-state Wigner transport equation in nanoscale MOSFET devices. The three-dimensional quantum transport is computed by solving the coupled 3D Poisson and 2D Schrödinger equations (by a mode-space approach) with the 1D Wigner transport equation along the source-drain direction. Numerical simulations are performed to demonstrate the ability of the Wigner function formalism to correctly reproduce quantum transport properties in *gate-all-around* silicon nanowire MOSFETs.

## 1. Introduction

One-dimensional nanostructures such as nanowires are attractive building block to develop novel nanoscale electronic devices. Advances have been made in the fabrication of semiconductor nanowire transistors in which the quantum nature of carrier dynamics cannot longer be neglected [1]. Among the modeling techniques of these devices, the Nonequilibrium Green function (NEGF) formalism is often used [2–4]. Although scattering mechanisms remain a difficult problem to adress, this formalism is especially suitable for quantum transport in mesoscopic devices. Another approach capable of handling quantum coherent propagation and dissipative scattering effects is based on the Wigner function formalism [5, 6]. An advantage of the Wigner equation as compared to NEGF method is the strong analogy with the Boltzmann equation since it is a phase space description similar to the classical Boltzmann distribution function. This analogy will allow to easily highlight the impact of expected quantum effects like tunneling and interference phenomena on performances of nano-scale electronic devices.

In this work, we present a 3D quantum simulation of *gate-all-around* silicon nanowire using a direct solution of Wigner transport equation (WTE). The 3D ballistic quantum transport is computed by solving the coupled 3D Poisson - 2D Schrödinger equations by a mode-space approach with the 1D Wigner transport equation along the source-drain direction. Our motivation is to demonstrate the possibility to use an alternative to the NEGF method combining both a correct description of quantum transport properties and a strong analogy with the semi-classical Boltzmann transport equation usually used.

## 2. Wigner's function formalism for quantum transport

Let us consider the 3D geometry of silicon nanowire shown in figure 1. The $x$ and $y$ axes represent the quantization directions, and $z$ axis represents the transport direction. The length of source/drain (S/D) extensions is $16\,nm$ and the square cross-section is equal to $3\,nm \times 3\,nm$. The N-type S/D extensions are doped $5 \times 10^{25}\ m^{-3}$ and the silicon channel is assumed undoped.

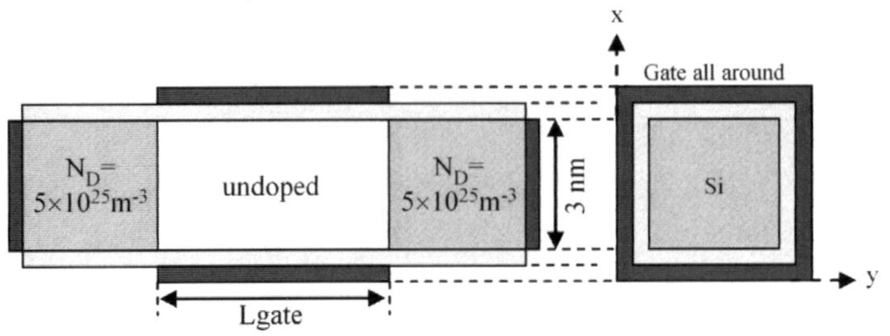


**Figure 1.** *Gate-all-around* silicon nanowire simulated in this work. $x$ and $y$ axes represent the quantization direction and $z$ axis represents the transport direction.

The effective mass Hamiltonian of the device satisfies the 3D Schrödinger equation,

$$H^{3D}\Psi(x, y, z) = E\Psi(x, y, z), \tag{1}$$

with the Hamiltonian of the system defined as

$$H^{3D} = -\frac{\hbar^2}{2}\frac{\partial}{\partial x}\left(\frac{1}{m^x_{eff}}\frac{\partial}{\partial x}\right) - \frac{\hbar^2}{2}\frac{\partial}{\partial y}\left(\frac{1}{m^y_{eff}}\frac{\partial}{\partial y}\right) - \frac{\hbar^2}{2}\frac{\partial}{\partial z}\left(\frac{1}{m^z_{eff}}\frac{\partial}{\partial z}\right) + V(x, y, z), \tag{2}$$

where $m^x_{eff}$, $m^y_{eff}$ and $m^z_{eff}$ are the effective masses along the $x$, $y$ and $z$ axes, respectively, and $V(x, y, z)$ is the conduction band-edge profile. In order to significantly reduce the computational burden associated with a full 3D real-space solution, the so-called mode-space approximation is used. Then, the nanowire can be subdivided into a large number of slice along the transport direction $z$. At every discretized real-space point $z$, the energy subbands $E_\nu$ and the corresponding wavefunctions are obtained from 2D Schrödinger's equation. Then, the Wigner transport equation is solved for each energy subband. By analogy with the Boltzmann transport equation, the Wigner transport equation is computed in order to determine the distribution function dependent on both the position $z$ and the wavevector $k$. Performing the Fourier transformation of the Liouville-von Neumann equation, the Wigner transport equation can be written as [7],

$$\frac{\partial f_w(z, k, t)}{\partial t} + \frac{\hbar k}{m_{eff}}\frac{\partial f_w(z, k, t)}{\partial z} + \int_{-\infty}^{+\infty} dk' f_w(z, k', t)V_w(z, k - k') = \left(\frac{\partial f_w}{\partial t}\right)_{coll} \tag{3}$$

Here $f_w(z, k, t)$ is the 1D Wigner function at position $z$, wavevector $k$ and time $t$. The second term of the left-hand side is the kinetic term. It is exactly the same as the corresponding term of the Boltzmann equation. Quantum-interference effects are induced by the non-local potential term represented by the third term of the left-hand side. The non-local Wigner potential $V_w(z, k - k')$ is calculated from the potential energy $E_\nu$. It is defined as [7],

$$V_w(z, k - k') = \frac{2}{\pi\hbar}\int_0^{+\infty} dz' \sin[2(k - k')z'][E_\nu(z + z') - E_\nu(z - z')] \tag{4}$$

As can be seen, the Wigner function in $k$ and $z$ depends in a non-local manner on all other momentum points $k'$ and also on the potential at all other locations $z \pm z'$. Due to the non-local potential, we will show in the next section that the steady state Wigner function $f_w(z, k)$ may be negative in some points of the phase space.

The Wigner function being determined, we can readily calculated the 1D carrier density defined as,

$$n(z) = \int_{-\infty}^{+\infty} \frac{dk}{2\pi} f_w(z,k) \tag{5}$$

as well as the 3D quantum charge density used for solving the Poisson equation. Then, the potential is updated and 2D Schrödinger's equation in each slice of the device can be solved again. Once the self-consistency between Poisson-Schrödinger-Wigner equations is obtained, the current density $j(z)$ can be calculated.

$$j(z) = \int_{-\infty}^{+\infty} \frac{dk}{2\pi} \frac{\hbar k}{m_{eff}^z} f_w(z,k) \tag{6}$$

## 3. Results and discussion

The validity of our Wigner function-based simulator is evaluated by computing current-voltage electrical characteristics of surround-gated silicon nanowire shown in figure 1.

The two lowest energy subbands of the six ellipsoidal $\Delta$ valleys of the silicon conduction band are plotted in figure 2. The corresponding 1D electron density calculated for each energy subband is shown in figure 3. It appears that due to the strong confinement in silicon nanowire, most of electrons are located in the first energy subband. The Wigner function corresponding to the first energy subband resulting from $V_{gs} = 0.2\,V$ and $V_{ds} = 0.4\,V$ is shown in figure 4. We can see that oscillations appear at the end of the channel due to quantum interferences. In the highly-doped S/D extensions, the Wigner function is similar to a Maxwell-Boltzmann distribution. However, in the channel, electron's velocity increases and a peak of ballistic distribution into the positive range of wavevector is observed. Moreover, due to the non-local potential, the Wigner function is negative in some points of the phase-space. Oscillations of Wigner's function in the channel, not observed for the Boltzmann distribution function, clearly demonstrate the ability of the Wigner function to account for quantum-interference effects.

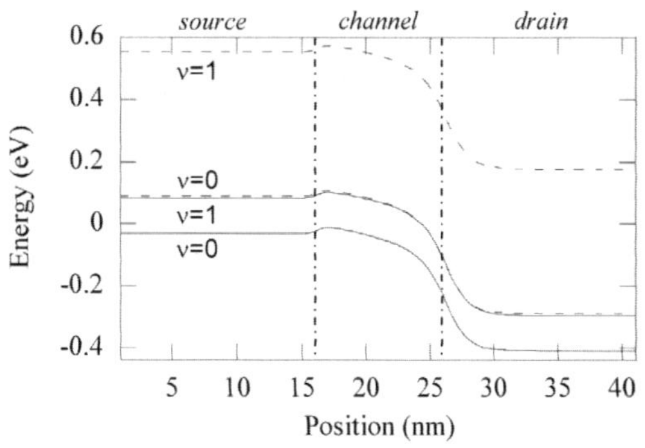


**Figure 2.** Profile of energy subbands along the transport direction for a gate bias $V_{gs} = 0.3\,V$ and a drain bias $V_{ds} = 0.4\,V$. The six ellipsoidal $\Delta$ valleys of silicon conduction band are represented.

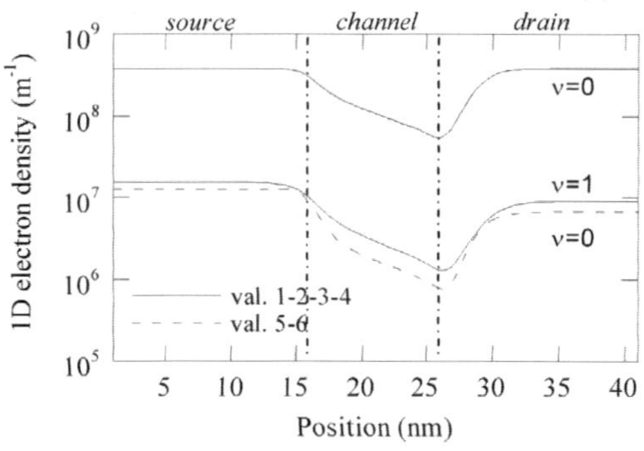


**Figure 3.** 1D electron density along the transport direction for a gate bias $V_{gs} = 0.3\,V$ and a drain bias $V_{ds} = 0.4\,V$. The six ellipsoidal $\Delta$ valleys of silicon conduction band are represented.

The electrical characteristics $I_{ds} - V_{gs}$ resulting from solution of Wigner transport equation for $V_{ds} = 0.4\,V$ are plotted on figure 5. In order to validate the Wigner-function based simulator, the results are compared with the NEGF method. A good agreement is observed between both approaches. Three channel lengths varying from $L_c = 10\,nm$ down to $L_c = 4\,nm$ are simulated. For a channel length $L_c = 10\,nm$, a subthreshold slope of around $60\,mV/dec$ is recovered. On the other hand, when the channel length is reduced down to $4\,nm$, a more important degradation of the subthreshold slope is observed. It is the consequence of both drain-induced barrier lowering and tunneling current which is properly accounted for with the Wigner function formalism.

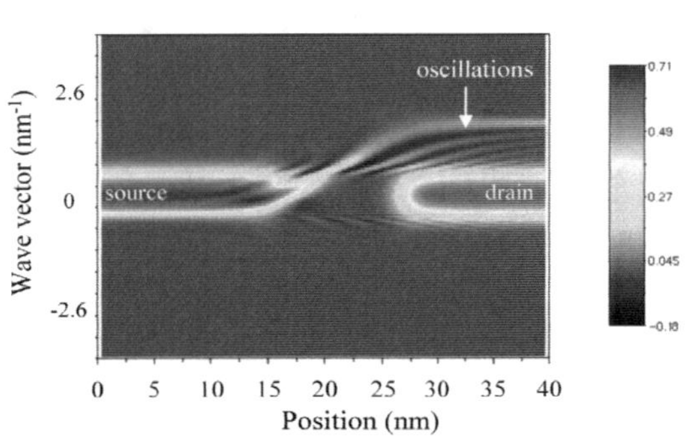


**Figure 4.** Cartography of the Wigner function related to the first energy subband for a gate bias $V_{gs} = 0.2\,V$ and a drain bias $V_{ds} = 0.4\,V$. The channel length is $L_c = 10\,nm$ and the square cross-section is $3\,nm \times 3\,nm$.

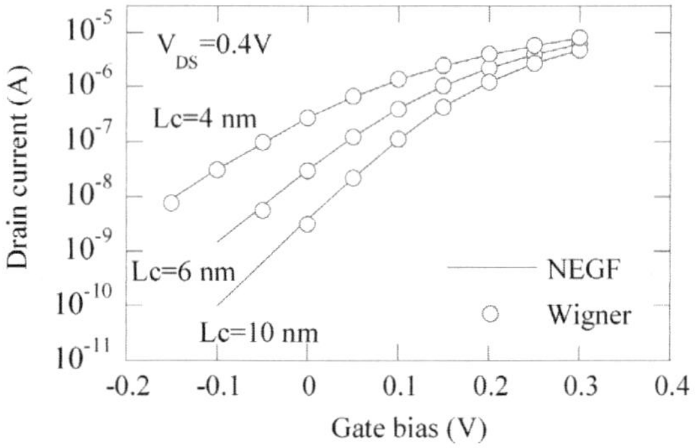


**Figure 5.** Current-Voltage characteristics of a gate-all-around silicon nanowire calculated using both the NEGF and Wigner function method. Channel lengths varying from $L_c = 10\,nm$ down to $L_c = 4\,nm$ are simulated.

## 4. Conclusion

In summary, we have developed an efficient numerical method to calculate electron transport in semiconductor devices. A direct solution of Wigner transport equation coupled with a Schrödinger/Poisson algorithm is demonstrated. Our simulator is used to calculate electrical characteristics in silicon nanowire with a channel length varying from $10\,nm$ down to $4\,nm$. The results clearly demonstrate the presence of expected quantum effects in the transport direction and their impact on electrical performances. Finally, this work shows that alternatively to the NEGF approach, the Wigner function-based formalism is well suited to investigate quantum transport in semiconductor nanostructures using a determinist resolution of transport equation.

## 5. Acknowledgments

Authors would like to acknowledge Prof. J.-L. Autran and Prof. D. Munteanu from university of Provence Aix-Marseille for fruitful discussions about NEGF method.

## References

[1] Dupre C and al 2008 *IEEE International Electron Devices Meeting 2008* pp 1–4
[2] Lherbier A, Persson M P, Niquet Y M, Triozon F and Roche S 2008 *Phys. Rev. B* **77** 085301–085305
[3] Poli S, Pala M, Poiroux T, Deleonibus S and Baccarani G 2008 *IEEE Transactions on Electron Devices,* **55** 2968–2976
[4] Autran J L and Munteanu D 2008 *Journal of Computational and Theoretical Nanoscience* **5** 1120–1127
[5] Jacoboni C and Bordone P 2004 *Rep. Prog. Phys.* **67** 1033–1071
[6] Querlioz D, Saint-Martin J, Huet K, Bournel A, Aubry-Fortuna V, Chassat C, Galdin-Retailleau S and Dollfus P 2007 *IEEE Transactions on Electron Devices,* **54** 2232–2242
[7] Frensley W 1990 *Rev. Mod. Phys.* **62** 745–791

# Marshmallowing of Nanopillar Arrays by Field Emission

**J Park, H Qin[$], H-S Kim[&], and R H Blick**

Electrical and Computer Engineering, University of Wisconsin-Madison, Madison, WI 53706, USA

E-mail: jonghoopark@wisc.edu

**Abstract.** We have fabricated mechanically flexible field electron emitters formed by highly-doped silicon nanopillars on a silicon membrane. Electron beam induced deposition of carbon-based contaminants is employed to probe the spatial activity of electron emission from the nanopillars. The experimental configuration provides a powerful tool to investigate the physics of the field electron emission (FEE). In contrast to the general assumption that field emission only occurs at the tips of nanoscale emitters, we found that the emission from the nanopillars' sidewalls is as strong as from their tips.

## 1. Introduction

In recent years, there has been increasing interest in field emission, well described by Fowler-Nordheim tunneling, mostly due to the progress in nanotechnology. The application of large-area emitters includes microwave sources, X-ray generation, and multiple electron-beam lithography [1]. There are also recent attempts to develop large-area emitters on mechanically flexible substrates [2]. In this current work, we fabricated nanopillars on a highly doped thin silicon membrane to look into the emission process from these pillars in greater detail. Electron beam induced deposition of carbon-based contaminants is employed to probe where the electron emission on the nanopillars surfaces actually happens [2, 3].

## 2. Device fabrication and measurement setup

The device fabrication is started with n-type silicon-on-insulator (SOI) wafers consisting of a 3 $\mu$m layer of silicon on a 1.1 $\mu$m insulating layer of silicon dioxide. The resistivity of the SOI is of the order of 12 $\Omega$cm. The SOI layer is thinned down by thermal oxidation, which creates a silicon dioxide layer on top. A silicon nitride is deposited on both sides of the wafer via low pressure chemical vapor deposition (LPCVD). As shown in figure 1(a), a small window on the top side aligned with a larger window on the backside defined by optical lithography, reactive ion etching and wet chemical etching. By using electron-beam lithography (EBL) and a reactive ion etching process, array of silicon

[$] Suzhou Institute of Nano-Tech and Nano-Bionics, Chinese Academy of Sciences, 398 Ruoshui Road, Dushu Lake Higher Education District, Industrial Park, Suzhou City, Jiangsu 215125, P. R. China

[&] Intel Corp., USA

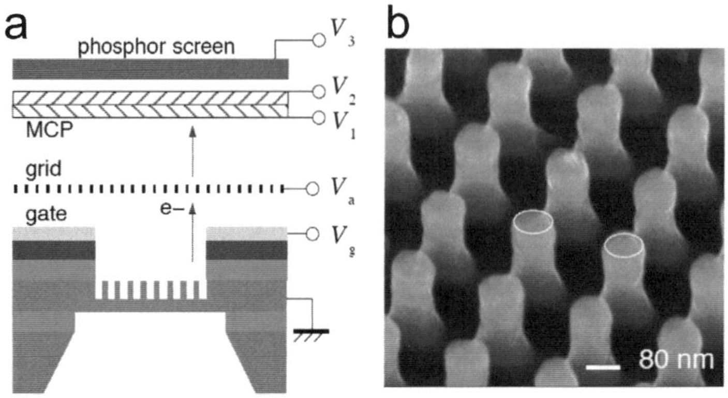


**Figure 1.** (a) Schematic and circuit of the device containing nanopillars on a thin silicon membrane. (b) A close up of individual nanopillars.

nanopillars were formed into the SOI layer. An anisotropic silicon etching process in potassium hydroxide (KOH) solution was used to etch the backside window to form thin SOI membranes. Finally, each device is dipped into a buffered hydrofluoric etchant obtaining clean nanopillars.

The resulting nanopillars-membrane (NPM) samples are of square shape with a side length of 35 $\mu$m. On each device, 16 such identical membranes were fabricated into four 2 x 2 arrays. On each membrane, there are about 17,600 silicon nanopillars defined in a 'frame'. A close-up of the nanopillars is shown in Fig. 1(b): each pillar has a diameter of 80 nm, a height of 300 nm, with a pitch distance of 200 nm to the next pillars.

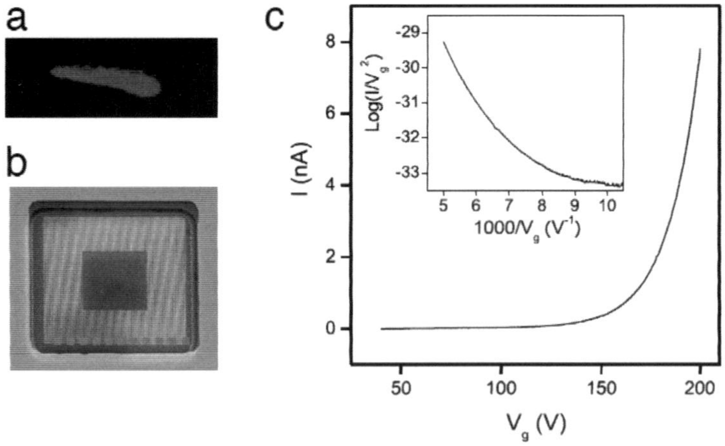


**Figure 2.** (a) Field emission pattern from the rim of the nanopillar array recorded on the fluorescent screen. (b) The emission sites are marked by the red dashed line on the edge of the nanopillar area. (c) Field emission current vs. anode voltage in linear representation. In the inset, Fowler-Nordheim plot revealing the deviation from ideal Fowler-Nordheim behaviour.

In order to obtain a sufficiently large electric field over the nanopillars, a local gate electrode is formed around the membrane by evaporating gold. Furthermore, a grid electrode is placed about 7 mm above the gate electrode to guide the electrons into a fluorescent screen which is located about 15 mm above the grid electrode.

## 3. Results

The amount and location of the carbon deposition on the nanopillars surfaces is a direct evidence of the emitted electrons. Applying a high gate voltage, the electric field at the nanopillars' tops becomes strong enough to facilitate field emission through the silicon-vacuum barrier. The extraction voltage of which is mainly determined by silicon's work function, type of doping, doping concentration and the geometry of the nanopillars. An emission pattern recorded from the fluorescent screen using the FEE setup (Fig. 1 (a)) is shown in Fig. 2 (a). This line shape of emission pattern stems from one of the edges of the nanopillar array, indicated by the red dashed line in the SEM graph of Fig. 2(c). In the inset of Fig. 2(c), the nonlinear curve represents a strong deviation from the ideal Fowler-Nordheim characteristic. This is partly due to the fact that multiple nanopillars with different electric field thresholds contribute to the overall emission current.

Carbon deposition on nanopillars induced by field emission is shown in Fig. 3. The nanopillars are covered by carbon contaminates. There is no obvious carbon contamination found on the membranes, indicating no electron emission occurs on the flat area. As the deposition continues, electron field emission persists and the pillars are more and more covered by a layer of carbon contaminants, indicating electron emission through the tips as well as through the sidewalls.

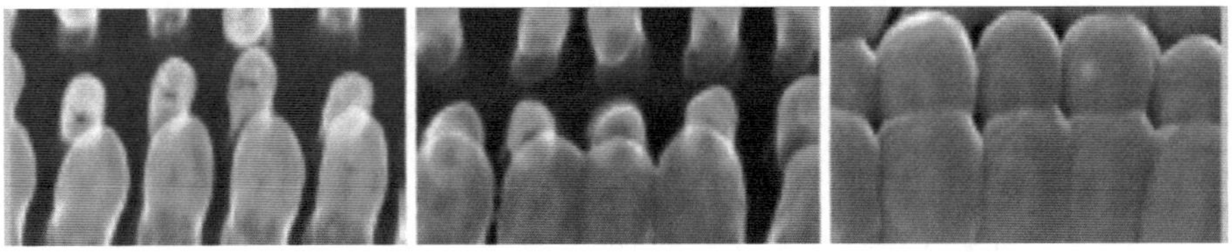

**Figure 3.** Towards the right-hand-side we show the marshmallowing effect. The pillars are more and more covered by a layer of carbon contaminants, indicating electron emission through the tips as well as through the sidewalls.

## 4. Conclusion

The marshmallowing reveals emission from all over the nanopillars' surfaces. This suggest that while carbon contaminants cause only a slight change in the electric field distribution, the emission from the nanopillars' sidewalls is as strong as from their tips. We can make the point that the common belief of strong field emission and hence EBD only from nanopillars' tips is an oversimplification.

## References

[1] *Field emission in Vacuum Microelectronics*, Fursey G N, Kluwer Academic-Plenum Publishers, New York (2005)
[2] Qin H, Kim H-S, and Blick R H 2008 *Nanotechnology* **19**, 095504
[3] Koenig D R, Scheible D V, and Blick R H 2004 *Appl. Phys. Lett.* **85**, 157

Journal of Physics: Conference Series **193** (2009) 012040 doi:10.1088/1742-6596/193/1/012040

# Figures of merit in high-frequency and high-power GaN HEMTs

**F. A. Marino, N. Faralli, D. K. Ferry, S. M. Goodnick and M. Saraniti**

Department of Electrical Engineering, *Journal of Physics: Conference Series*, IOP Publishing, Arizona State University, 12 Tempe, AZ, 85287 USA

E-mail: Fabio.Marino@asu.edu

**Abstract.**

The most important metrics for the high-frequency and high-power performance of microwave transistors are the cut-off frequency $f_T$, and the *Johnson figure of merit* $FoM_{Johnson}$. We have simulated a state-of-the-art, high-frequency and high-power $GaN$ HEMT using our full band Cellular Monte Carlo (CMC) simulator, in order to study the RF performance and compare different methods to obtain such metrics. The current gain as a function of the frequency, was so obtained both by the Fourier decomposition (FD) method and the analytical formula proposed by Akis. A cut-off frequency $f_T$ of 150 $GHz$ was found in both the transit time analysis given by the analytical approach, and the transient Fourier analysis, which matches well with the 153 $GHz$ value measured experimentally. Furthermore, through some physical considerations, we derived the relation between the $FoM_{Johnson}$ as a function of the breakdown voltage, $V_{BD}$, and the cut-off frequency, $f_T$. Using this relation and assuming a breakdown voltage of $80V$ as measured experimentally, a Johnson figure of merit of around $20 \times 10^{12} V/s$ was found for the HEMT device analyzed in this work.

## 1. Introduction

Several figures of merit are often used to evaluate the performance of microwave devices. The most important metrics for the high-frequency and high-power performance of microwave transistors are the cut-off frequency, $f_T$, and the *Johnson figure of merit*, $FoM_{Johnson}$ [1]. The cut-off frequency is related to the short circuit current gain, $h_{21}$, which is defined as the ratio of the small-signal output current to the small-signal input current of the device when the output terminals are shorted. The Johnson figure of merit takes into account the breakdown electric field, $E_{BD}$, and saturated electron drift velocity, $v_{sat}$, in defining a measure for the power handling capability at high frequencies.

We have simulated a state-of-the-art, high-frequency and high-power $GaN$ HEMT, reported by Palacios *et al.* [2], using our full band Cellular Monte Carlo (CMC) simulator, in order to study the RF performance and compare with the prediction of the analytical model for the cut-off frequency calculations proposed by Akis $et\ al.$ [3]. Furthermore, through some physical considerations, we derived the relation between the $FoM_{Johnson}$ in function of the breakdown voltage $V_{BD}$ and the cut-off frequency $f_T$. Using this equation, the Johnson figure of merit can be connected directly to microwave measurements in a meaningful manner and calculated easily by standard experimental measurements.

## 2. GaN HEMT with InGaN back barrier

The layout of the device described in this work is composed of an $AlGaN/GaN$ heterostructure with an $InGaN$ back barrier grown on a semi-insulating $SiC$ substrate by metal-organic chemical vapor

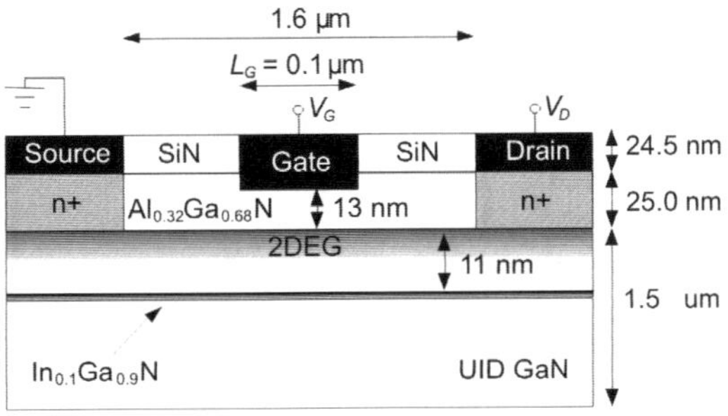


**Figure 1.** Schematic cross-section of the simulated InGaN back barrier HEMT

deposition. The heterostructure is unintentionally doped and consists, from bottom to top, of $1.5\mu m$ of $GaN$ followed by $1nm$ of $In_{0.1}Ga_{0.9}N$ capped by $11nm$ of $GaN$ and $25nm$ of $Al_{0.32}Ga_{0.64}N$ as shown in Figure 1. The $AlGaN$ layer is recessed in order to have a gate-to-channel distance of $13nm$. The source-gate and gate-drain separations are both $0.75\mu m$, while the gate length is set to $0.1\mu m$. Highly doped regions are created under the source-drain electrodes down to the $GaN$ channel to emulate metal spikes and to control contact resistances.

The presence of the $InGaN$ layer enhances the confinement of the electrons in the channel. In fact, the large polarization-induced electric field in the $InGaN$ layer raises the conduction band edge of the $GaN$ buffer with respect to the $GaN$ channel, creating a potential barrier against carrier diffusion in the bulk layer. More details on that device can be found in [2], where all the characteristics of the sample used to obtain the experimental data are reported.

## 3. RF analysis of AlGaN/GaN HEMT

The basic idea of the small-signal frequency, or RF, analysis consists of applying a small perturbation to one electrode of the device in steady-state, while keeping all other parameters constant. The small-signal parameters can then be derived from the Fourier analysis of the transient response. The response in the transient regime is a function of the device characteristics, such as the geometry, the doping profile, the transport properties of the device and the steady-state operating point (i.e. the DC components of $V_{DS}$ and $V_{GS}$).

In our study, the operating point for the RF analysis ($V_{GS} = 0V$, $V_{DS} = 6V$) has been chosen in order to compare simulation results with experiment [2]. The current gain as a function of the frequency is shown in Figure 2. The right plot, representing the simulation results, is obtained by the Fourier decomposition (FD) method, where a small step voltage is applied to the input port of the device in steady state, and the transient short circuit current response is Fourier analyzed to obtain the cut-off frequency.

## 4. Cut-off frequency

An alternative method to calculate the current gain cut-off frequency, $f_T$, is to relate this to the transient time at carriers across the effective gate length [3]:

$$f_T = \left[ 2\pi \int_{L_{eff}} \frac{1}{v_{ave}(x)} dx \right]^{-1}, \tag{1}$$

where $L_{eff}$ is an effective gate length and $v_{ave}(x)$ is the average simulated velocity along the channel.

The simulated electron drift velocity profile along the channel used in the calculation of the cut-off frequency according to Eq. (1) is illustrated in Figure 3.

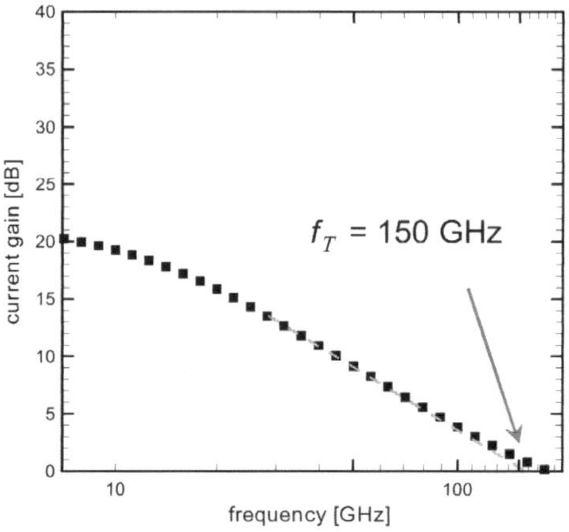


**Figure 2.** Current gain as a function of the frequency obtained by the CMC simulation.

A cut-off frequency $f_T$ of 150 $GHz$ was found in both the transit time analysis given by Eq.(1), and simulation results using transient Fourier analysis, which matches well with the 153 $GHz$ value measured experimentally [2].

## 5. Johnson figure of merit

In general, widebandgap materials like $GaN$ have desirable properties for high power applications due to their large band gap and high thermal conductivity. The Johnson figure of merit mentioned earlier, characterizes the high frequency performance of power devices, and is proportional to the saturation velocity and critical electric field for impact ionization initiated breakdown [1]

$$FoM_{Johnson} = \frac{v_{sat} E_{BD}}{2\pi}, \tag{2}$$

where $v_{sat}$ is the saturation velocity and $E_{BD}$ is the electric field at which impact ionization initiates breakdown. Unfortunately, this figure of merit is difficult to determine experimentally as both $v_{sat}$ and $E_{BD}$ are intrinsic properties of a device, although easily found from simulation. However, these quantities can be connected to microwave measurements in a meaningful manner. Generally, the cutoff frequency, $f_T$, is related to the effective saturation velocity through the well-known formula

$$f_T = \frac{v_{sat}}{2\pi L_G}, \tag{3}$$

where $L_G$ is the gate length (or effective gate length). We also find from detailed simulations, that the field under the gate varies almost linearly over the length of the effective gate. Breakdown usually occurs near the drain end of the gate, where the highest field occurs. Since there is also a potential drop across the space between the gate and the contacts, we can write the breakdown voltage in terms of the breakdown field as

$$V_{BD} = \alpha \frac{E_{BD} L_G}{2} \tag{4}$$

where $\alpha$ is an adjustable parameter that relates the voltage drop across the gate to the total voltage applied to the device. Hence, we can rewrite the figure of merit in terms of these experimentally determined parameters as

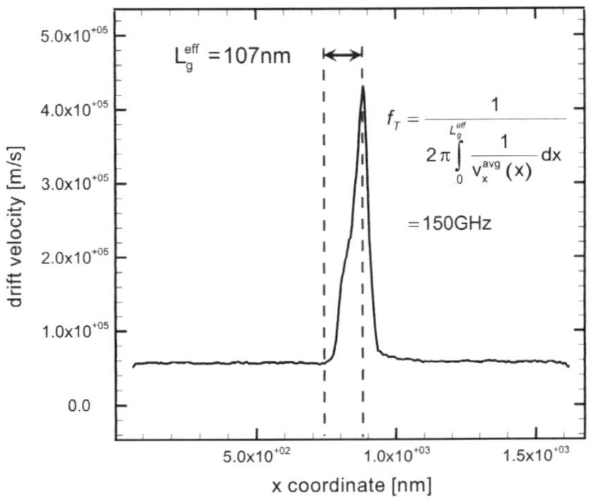


**Figure 3.** Electron drift velocity profile along the channel resulting by the CMC simulation

$$FoM_{Johnson} = \frac{1}{2\pi}(2\pi L_G f_T)\frac{2V_{BD}}{\alpha L_G} = \frac{2}{\alpha} f_T V_{BD}. \tag{5}$$

The $\alpha$ parameter can be extracted directly from the simulations using Eq. (4), and calculating the breakdown voltage as an integral of the field across the channel when the impact ionization first occurs.

Using a value previously published for $GaN$ for Eq. (2) [4], and values found in simulations of $GaN$ HEMTs, we find that $\alpha$ is appropriately 1.25, although this factor could be ignored without much loss of relevance of the figure of merit since its close to unity. Following Eq. (5) and assuming a breakdown voltage of 80V [2], a $FoM_{Johnson}$ around $20 \times 10^{12} V/s$ was found for the HEMT device analyzed in this work.

## 6. Conclusion

In this paper, the most important metrics to evaluate RF performance of $GaN$ HEMT devices were investigated, through comparison between experimental, simulation and analytical method. The Fourier decomposition method was used to evaluate the cut-off frequency in our full band CMC simulation code, which includes the full details of the band structure and the phonon spectra. Thermal simulations were performed with commercial software in order to determine the corrections needed to model thermal effects with our particle-based CMC simulator. Our studies indicate that both the transit time analysis given by Eq.(1), and simulation results using transient Fourier analysis, matches well with the experimental data. We also derived an analytical model to determine the Johnson figure of merit and extracted this quantity from standard experimental data.

## References

[1] A. Johnson, "Physical limitations on frequency and power parameters of transistors," *RCA Review*, vol. 26, pp. 163 – 177, 1965.

[2] T. Palacios, Y. Dora, A. Chakraborty, C. Sanabria, S. Keller, S. P. DenBaars, and U. K. Mishra, "Optimization of AlGaN/GaN HEMTs for high frequency operation," *Physica Status Solidi A*, vol. 203, no. 7, pp. 1845 – 1850, May 2006.

[3] R. Akis, J. Ayubi-Moak, N. Faralli, D. K. Ferry, S. M. Goodnick, and M. Saraniti, "The upper limit of the cutoff frequency in ultrashort gate-length InGaAs/InAlAs HEMTs: A new definition of effective gate length," *IEEE Electron Device Letters*, vol. 29, no. 4, pp. 306 – 308, April 2008.

[4] P. Das and D. Ferry, "Hot electron microwave conductivity of wide bandgap semiconductors," *Japanese Journal of Applied Physics*, vol. 19, pp. 851 – 855, 1976.

# Avalanche dynamics in silicon avalanche single- and few-photon sensitive photodiode

**J Blazej, I Prochazka**

Czech Technical University in Prague, Brehova 7, 115 19 Prague 1, Czech Republic

E-mail: blazej@fjfi.cvut.cz

**Abstract**. We are presenting the results of the study of the Single Photon Avalanche Diode (SPAD) avalanche pulse response rise-time and its dependence on several key parameters. We were investigating the unique properties of K14 type SPAD with its high delay uniformity of 200 μm active area, the character of avalanche, and the correlation between the avalanche build-up time and the photon number involved in the avalanche trigger. The detection chip was operated with bias higher then breakdown voltage, ie. in Geiger mode. The detection chip was operated in a passive quenching circuit with active gating. This set-up enabled us to monitor both the diode reverse current using an electrometer and a fast digitizing oscilloscope. The electrometer reading enabled to estimate the photon number per detection event, the avalanche build up was recorded on the oscilloscope and processed by custom designed waveform analysis package. The correlation of avalanche build up to the photon number, bias above break, photon absorption location, optical pulse length and photon energy was investigated in detail. The experimental results are presented.

## 1. Introduction

The single photon sensitive avalanche photodiodes acting as photon counting devices has been pioneered by Sergio Cova and co-workers [1]. This detector, called Single Photon Avalanche Diode (SPAD), is an avalanche photodiode structure prepared using a conventional planar technology on silicon. Single photon sensitivity is achieved by biasing the diode above the junction breakdown voltage. In this metastable stage the first absorbed photon is capable of triggering the avalanche multiplication of carriers; a fast rise-time current pulse is generated. The leading edge of the current pulse marks the event of the photon absorption with picosecond accuracy. The current increase is terminated by an external circuit connected to the diode, where the typical value of the gain achieved exceeds $1\times10^9$. Such an operation mode is called a Geiger mode in analogy with nuclear technique. The operation of the detector may be controlled – gated, by an external electrical signal.

In our experiments we have concentrated on the operation of the SPAD diodes in the regime of detecting single or several photons per pulse. The existing solid state photon counting detectors have been dedicated for high timing resolution and stability of single photon events. However, the high timing stability is maintained for individual single photons detection, only [2]. If more than one photon is absorbed within the detector time resolution, the detection delay will be significantly affected. This fact is restricting the application of the solid state photon counters to cases where single photons may be guaranteed in principle. The example of such an application is the time resolved spectroscopy, single molecules detection and similar. However, there is a demand for detection of optical pulses consisting of several up to several thousands of photons with picosecond resolution and stability. No conventional optical detector or photon counter can fulfill these requirements. The standard SPAD based photon counters work in a purely digital mode - uniform output signal is generated once a photon is detected. If the input signal consists of several photons, the first absorbed one triggers the avalanche. For multiple photon signals, the detection delay will be shorter in comparison to the single photon events. The detection delay dependence on the photon number is called the "detection time walk". To enable the detector operation in both the single and multi-photon signal regime with a minimal time walk, a time walk compensation technique has been developed [3] in nineties. The operation of this device was based on unique properties of the K14 SPAD detectors [4]. The procedure is based on the fact that the avalanche rise-time of the K14 SPAD chip is depending on the input optical signal strength. The circuit is using this information to compensate the time walk of the detector in this dynamical range.

## 2. Experiment

The SPAD chips were tested and operated in an active gated and passively quenched circuit. The scheme enables to bias the chip to the voltage on just below its breakdown voltage. The SPAD bias is increased for a short time (100 ns typically) for additional step of the negative gate pulse.

The picosecond laser diode Hamamatsu C4725 providing 42 ps wide pulses at 778 nm was used as a signal source. The laser output was attenuated by means of a stack of calibrated neutral density filters and focused on the detector active area. The precision X-Y-Z stage enabled us to investigate focusing and position sensitive effects of the detection process. The optical signal intensity was measured using the detection chip in a linear mode with unity gain and monitoring its reverse current by the electrometer Keithley 610C and operating the power stabilized laser source at a high repetition rate of 1 MHz.

The detector was operated in an active gated and passive quenched circuit; its pulse output was monitored on the digital oscilloscope LeCroy SDA 9000 with the 9 GHz analogue bandwidth and 40 Gs/s sampling frequency. The individual current rise-time waveforms were recorded with a resolution of 40 Gs/s and processed off-line using a custom software package. Independent measurement series were completed in the configurations of optical signal either focused to the active area center or illuminating the entire sensitive area.

## 3. Results and conclusion

At first, the "thick" SPAD C-30902S manufactured by Perkin Elmer (formerly RCA) was investigated. The breakdown voltage is 238 Volts. This relatively slow structure allows clearly see the difference in avalanche temporal profile. Examples of the response to a single photon are recorded in figure 1 together with the response to the short optical pulse consisting of 3000 and 10 thousands photons. The chip was pulse biased 5 Volts above its breakdown voltage. Different avalanche buildup shape – pulse fall times is apparent.

The last avalanche photodiode chip under test was the K14 structure made by our University. This is the "thin" SPAD structure on silicon, the active area diameter is with 200 µm. The K14 type SPADs are characterized by several special features in comparison to other SPADs. It is the exceptionally high active area uniformity in both the timing jitter and mainly of the detection delay over an entire sensitive area. In the series of experiments, the input signal strength was adjusted using calibrated neutral density filters to the values 1, 3, 10,... up to 10 000 photons per event. The avalanche current response pulses were recorded, stored and off-line processed using the custom designed software package. The times when the avalanche current exceeded the pre-selected level were evaluated. The results are summarized in figure 2. The difference between detection delay related to the trigger level of 10 mV are plotted for two different optical configurations. In the first configuration the detected photons were absorbed within a small area close to the sensitive area center. In the second configuration the detected photons were absorbed uniformly over the active area. The error in vertical axis are plotted, in horizontal axis the data are blurred by the photon counting statistics – Poisson distribution of attenuated laser light.

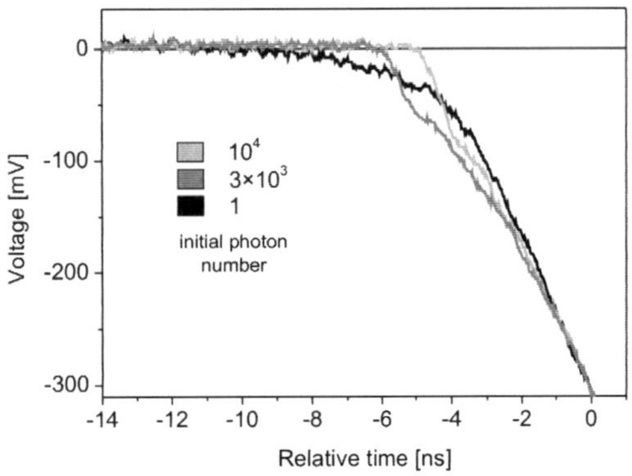


**Figure 1**. The oscillogram of the response of the C-30902s SPAD to single, 3000, and 10000 photon level.

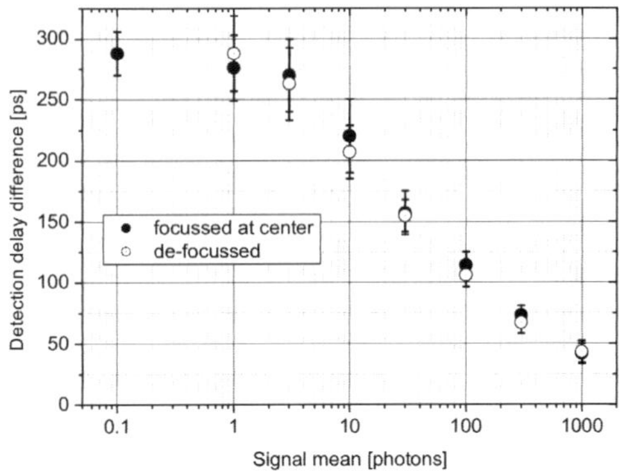


**Figure 2**. The detection delay for various photon numbers, K14 SPAD biased 5 V above breakdown voltage, optical pulses 48 ps long, trigger level 10 mV.

For higher photon counts, the SPAD avalanche buildup consists of two parts. In the initial part the current rise-time is much faster than exponential. The amplitude of this initial phase is roughly proportional to the initial photon number. It is expected, that this effect is caused by injecting of a number of carriers instead of just one for the case of single photon detection.

In both tested detection structures the avalanche current rise-time contains information about photon number detected in the dynamical range 1 photon to more than 3 000 photons per pulse. This is valid for optical pulses which are much shorter in comparison to the avalanche current rise-time. The avalanche pulse rise-time measurement permits to determine the photon number in involved in the detection process and to determine the detection time walk. However, for the "thick" SPAD detection structures the avalanche rise-time variations effects are combined with the additional effects of detection delay non-uniformity over an entire detection area. In contrast to it the "thin" SPAD structure K14 is suitable for photon number estimate and detection time walk compensation on the basis of avalanche current rise-time monitoring. The measured values of the rise-time difference is 600 ps for the photon number range of 1 to 10 000, it means 150 ps per one decade. This value makes possible to estimate the photon number involved in the detection on the basis of shot-by-shot output pulse rise-time measurements and the detection delay might be compensated, both in a dynamical range $1{:}10^4$.

**Acknowledgments**
This research has been supported by several provided by the Grant Agency of the Czech Republic, Ministry of Education of the Czech Republic and the international grant agencies, as well. The recent publication has been supported by MSM6840770015.

**References**
[1] Cova S. et al. 2004 *J. of Modern Optics* **51** 1267
[2] Prochazka I. et al 2008 Photon counting detector for space applications optimized for both gated and non-gated operation *Proc. SPIE* vol 7138 p 71380C
[3] Kirchner G. et al 2007 Time walk compensated SPAD: multiple photon versus single photon operation *Proc. SPIE* vol 3218 p 106
[4] Blazej J. 2004 *J. of Modern Optics* **51** 1491

# Dynamics of electron entanglement in semiconductor nanostructures

**Paolo Bordone**

Dipartimento di Fisica, Università di Modena e Reggio Emilia, and S3 Research Center, CNR-INFM, Via Campi 213/A, I-41125 Modena, Italy

E-mail: bordone@unimore.it

**Abstract.** Quantum entanglement, the most remarkable feature of quantum physics, is recognized as a resource for quantum information processing. The quest for quantum-computing devices has also produced great interest in entanglement formation in solid-state systems involving quantum effects. Indeed, the functionality of an increasing number of nanodevices is influenced by quantum correlations. In this work a theoretical approach developed in recent years to study the entanglement of fermions interacting via a Coulomb potential is presented, together with a number of applications to specific situations of physical interest in the field of charge transport in semiconductor nanostructures.

## 1. Introduction

Quantum entanglement undoubtedly represents one of the most peculiar aspects of quantum mechanics as it can be viewed as the furthest departure of the quantum world from the classical one [1, 2]. Quantum entanglement between two charge carriers can be created when they interact through the Coulomb potential. Indeed, after a scattering the two-particle system is in general described by a two-particle state that is not separable in two single-particle states. This entanglement building up is an intrinsically dynamical process and its analysis is not only useful to understand the nature of the scattering process itself, but it can also contribute significantly to design quantum information processing devices. In fact, on one hand, controlled entanglement has been recognized as the fundamental resource for quantum computation and communication [2], on the other hand entanglement with the environment (i.e., decoherence) represents the main threat to the proper functioning of a feasible quantum computer [1].

For these reasons, the study of the entanglement dynamics in scattering events in solid-state systems has became more and more relevant in recent years [3, 4, 5, 6] and different proposals to produce entangled states between charge carriers, have been presented [7, 8, 9].

In this context, the entanglement dynamics has been evaluated in 1D- and 2D-scattering events between two carriers explicitely considered as indistinguishable. To this aim, numerical approaches have been developed which allow one to evaluate the mechanisms governing entanglement creation and their connection with the characteristic physical parameters and initial condition of the systems. Estimations of the entanglement time-evolution show a deep relation between spin and space degrees of freedom of the states, even for interactions not involving spin-orbit terms [10, 11].

Entanglement and its effects have been investigated in semiconductors devices realized with semiconductor quantum dots (QD) by solving the multi-particle open-boundary Schrödinger equation. In particular numerical simulations show that the production of entanglement in QD structures may be controlled through external fields [12].

Dynamics of entanglement has also been analyzed for carrier-carrier collisions in low-dimensional semiconductor structures in presence of an external periodic potential. In particular the latter has been shown to provide a way to control electron decoherence due to carrier-carrier scattering, this model being of interest to study the building up of quantum correlations in electron transport assisted by surface acoustic waves (SAW).

## 2. Entanglement creation for identical particles

A quantitative evaluation of the entanglement dynamics for electron-electron collisions in 1D and 2D systems has been performed applying a method, suitable for indistinguishable particles, based on the Slater rank and the von Neumann entropy of the reduced density matrix [10]. The quantum entanglement of both spatial and spin degrees of freedom can be computed for various initial conditions of the system. Here, for the sake of brevity, we analyze only a 2D case. We consider an electron freely propagating in a 2D system interacting through a Coulomb potential with a second carrier confined in a harmonic potential. The latter can be viewed as a simple model of shallow impurity. Spin orbit effects are not taken into account. At the initial time one of the two particles (incoming electron) is represented by a minimum uncertainty wavepacket, while the bound electron is in the ground state of a 2D harmonic oscillator. Considering the spin degrees of freedom, we take four initial spin configurations, as given in the following quantum states: $|\Psi\rangle$ same spin, $|\Upsilon\rangle$ different spin, $|\Phi\rangle_-$ singlet, $|\Phi\rangle_+$ triplet;

$$\begin{aligned}
|\Psi\rangle &= \frac{1}{\sqrt{2}}\Big(|\psi\,\phi\rangle - |\phi\,\psi\rangle\Big)|\uparrow\uparrow\rangle, \\
|\Upsilon\rangle &= \frac{1}{\sqrt{2}}\Big(|\psi\,\phi\rangle|\uparrow\downarrow\rangle - |\phi\,\psi\rangle|\downarrow\uparrow\rangle\Big), \\
|\Phi\rangle_{\mp} &= \frac{1}{2}\Big(|\psi\,\phi\rangle \pm |\phi\,\psi\rangle\Big)\Big(|\uparrow\downarrow\rangle \mp |\downarrow\uparrow\rangle\Big),
\end{aligned}$$

where the ket $|\uparrow\rangle$ ($|\downarrow\rangle$) indicates a spin up (down) state.

In our approach we solve numerically the time-dependent Schröndinger equation for the 2D two-particle wavefunction in the center of the mass coordinate system. In order to quantify numerically the entanglement we have implemented the theoretical criterion, given by Schliemann, based on the Schmidt decomposition theorem for two-fermion state [13, 14]. Through a unitary transformation the antisymmetric wavefunction is expressed into a combination of Slater determinants with a minimum number of non vanishing terms with coefficients $z_i$. This number, the Slater rank, gives the measure of the entanglement that can still be equivalently expressed as the von Neumann entropy of the one particle reduced density operator $\rho$ as:

$$\varepsilon = -\mathrm{Tr}[\rho \ln \rho] = \ln 2 - 2\sum_{i=1}^{N} |z_i|^2 \ln 2|z_i|^2 \tag{1}$$

$2N$ is the dimension of the single particle Hilbert space and corresponds, in the present case, to twice the number of discretization points in the real space. For identical particles the von Neumann entropy reaches its minimum value $\varepsilon = \ln 2$ in case of non entangled states, whose Slater rank is equal to 1. This value indicates the uncertainty coming from indistinguishability of the particles, but the correlations due to exchange symmetry cannot represent a resource for quantum information [14].

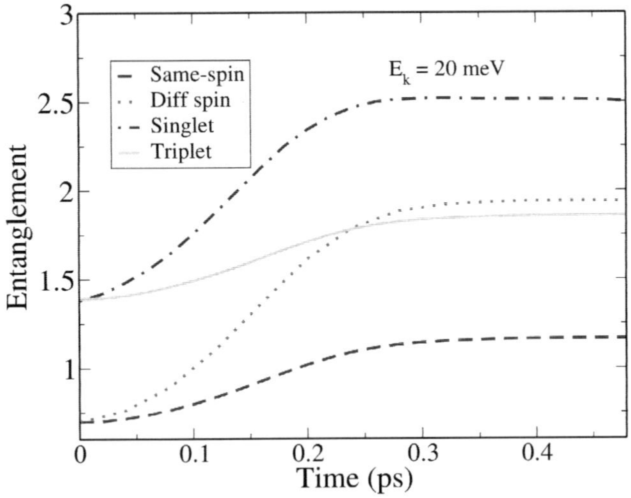


**Figure 1.** Entanglement as a function of time: two electrons with same spin $|\Psi\rangle$(dashed line), with different spin $|\Upsilon\rangle$(dotted line), singlet $|\Phi\rangle_-$ (dash-dotted line) and triplet $|\Phi\rangle_+$ (solid line).

Numerical result have been obtained adopting GaAs material parameters but can be considered representative of a more general behavior. At initial time, when the centers of the wavepackets are distant enough so that the Coulomb energy is negligible, the von Neumann entropy for the states $|\Psi\rangle$ and $|\Upsilon\rangle$ is $\ln 2$ as expected. In fact no correlation is present a part from the one due to the exchange symmetry and therefore $|\Psi\rangle$ and $|\Upsilon\rangle$ must be considered as non entangled. As the particles get closer their quantum correlation builds up and entanglement reaches a stationary value once particles get closer. In Fig.1 we report the time evolution of the entanglement for an initial kinetic energy of the incoming electron $E_k = 20$ meV, and for all the considered spin configurations. Note that at the initial time the von Neumann entropy of the singlet $|\Phi\rangle_-$ and the triplet state $|\Phi\rangle_+$ is $2\ln 2$. In fact such states have to be considered as initially entangled, being their Slater rank greater than 1: note that $|\Phi\rangle_-$ and $|\Phi\rangle_+$ cannot be factorized in terms of a single Slater determinant.

From our simulations we found a dependence of the entanglement dynamics upon the spin components of the state vector describing the system at initial time, even if we have neglected spin-orbit effects.

## 3. Effects of scattering resonance on carrier-carrier entanglement in quantum dots

Here we address numerically the problem of entanglement generation for a two-electron scattering in a 1D double-barrier resonant tunnelling device. Specifically, we consider an electron incoming from a source lead and described by a scattering state, while the other is localized in a bound state between the two small potential barriers, that separate the internal region from the two contacts. The system is depicted in the left panel of Fig. 2. The $N$ bound energies of the so-formed QD will be indicated as $E_n$ (with $n = 0, 1, \ldots, N$ in order of increasing energy), respectively.

Let us consider the envelope function of two spinless electrons in 1D $\psi(x_1, x_2)$, describing the state with a particle in a QD bound state $\chi_0$ and a second electron incoming from the left lead with energy $E_{IN}$, and scattered by the structure potential $V_s$, sketched in the left panel of Fig. 2, and by the Coulomb interaction with the bound particle. The two-particle scattering state is obtained by solving by means of QTBM [16] the time-independent open-boundary Schrödinger equation $\mathbf{H}\psi = E\psi$ in the square domain $x_1, x_2 \in [0, L]$ with $\mathbf{H}$ the two-particle Hamiltonian of the system. As a consequence of the scattering, quantum correlations appear between the energy levels $E_n$ of the electron bound in the potential well and the energies $T_n$ of the scattered electron. The calculation of the entanglement formation is performed, also in this case, by means

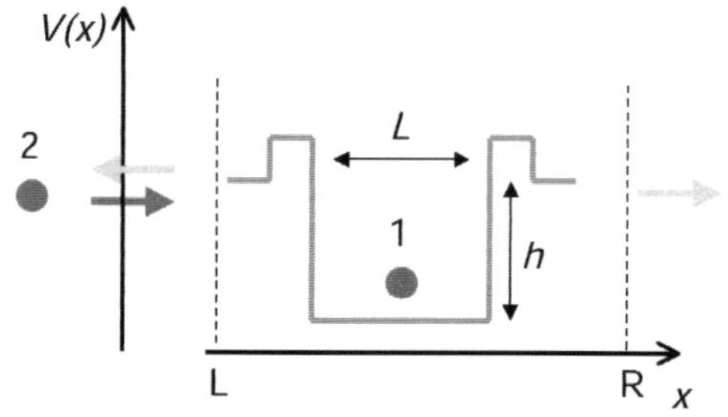


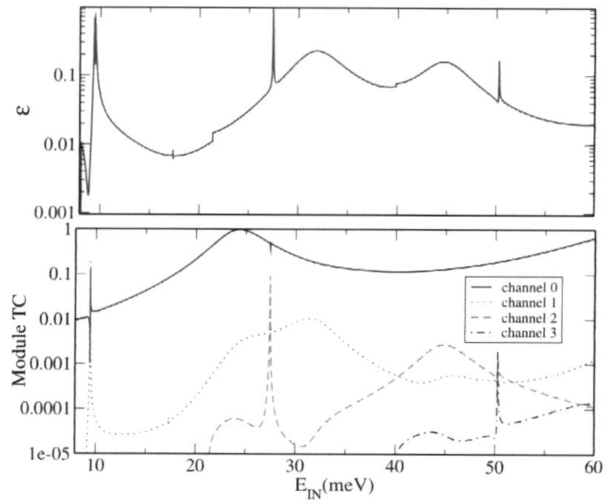


**Figure 2.** Left panel: Potential profile $V(x)$ of the simulated system: $V(x)$ is zero in both left and right contacts.The bound electron (1) is initialized in the QD ground state while the scattered electron (2) arrives from the left lead. Right top panel: the entanglement $\varepsilon$ as a function of the initial kinetic energy of the incoming electron $E_{IN}$. Right bottom panel: modulus of the TC of the four channels as a function of $E_{IN}$: channel 0 (solid line), 1 (dotted line), 2 (dashed line) and 3 (dash-dotted line).

of the von Neumann entropy [10] that can here be expressed as:

$$\varepsilon = -\sum_{n=0}^{M} |c_n|^2 \ln |c_n|^2. \tag{2}$$

where $c_n$ are the transmission probability amplitudes of two-particle state.

In the top right panel of Fig. 2 we report the entanglement as a function of the kinetic energy of the incoming particle $E_{IN}$. In our calculations we consider a potential well 150 meV deep and 40 nm wide. For $E_{IN}$ around 10 meV, with only two possible scattering channels, a Fano resonance [17] of the transmission coefficient (TC) is observed for channel 0. In this case entanglement shows a peculiar behavior. It presents a minimum between two close maxima (where von Neumann entropy is equal to $\ln 2$), indicating the maximum amount of quantum correlations between the transmitted and QD electron in a two-channel scattering. It is worth noting that the TC of channel 1 presents, in correspondence of the Fano resonance of channel 0, a Breit-Wigner resonance. On the other hand, the broad Breit-Wigner resonance for $E_{IN} \simeq 25$ meV of channel 0, does not influence the TCs of other channels (1 and 2), which does not show, at the specific energy, resonances of any kind and moreover the entanglement curve does not display maxima. Such a behavior is ascribed to the different mechanisms characterizing the two types of resonances. When the initial energy of the incoming electron increases new channels become effective in the scattering process thus influencing the entanglement curve. Our results suggest that the production and detection of entanglement in quantum dot structures may be controlled by the manipulation of Fano resonances through external fields [4].

## 4. Carrier-carrier entanglement in presence of a periodic potential

Let us now consider two electrons, interacting via the Coulomb potential, in a 1D channel and localized into two next minima of a periodic potential. To be more specific, we use a sinusoidal potential suitable for modeling SAW. The two particles are explicitly considered as indistinguishable and have the same spin. Two different physical conditions will be examined:

low-energy particles occupying the ground state of two next potential minima and high-energy particles taken in bound excited states. In our approach we solve numerically the time-dependent two-particle Schrödinger equation and then compute, at each time step, the bipartite entanglement which gives a measure of the correlation between the two electrons due to their mutual Coulomb interaction [10, 13]. We assume that the Hamiltonian describes a couple of electrons interacting with the traveling piezoelectric potential of the SAW, in the reference frame moving with the SAW itself [12].

We consider two electrons with the same spin so that the quantum state describing the system at the initial time $t_0 = 0$ is

$$\Phi_n(x_a, x_b) = \frac{1}{\sqrt{2}}\Big(\varphi_n(x_a)\varphi_n(x_b - \lambda) - \varphi_n(x_b)\varphi_n(x_a - \lambda)\Big) \tag{3}$$

where $n \geq 0$, with $\varphi_0(x)$ the ground state, and $\varphi_n(x)$ the $n$-th excited state, of the single-particle Hamiltonian in a closed domain with eigenvalues indicated by $E_n$. To be more specific, $\varphi_n(x)$ and $\varphi_n(x - \lambda)$ describe two electrons trapped in two next minima of a SAW with wavelength $\lambda$. To evaluate the system dynamics, we solve numerically the time-dependent Schrödinger equation for the two-particle spatial wavefunction. At each time step we compute the state of the system and from this we obtain the two-particle density matrix. The latter is used to compute the one-particle reduced density matrix by tracing over the degrees of freedom of one of the two electrons. Finally the entanglement is calculated in terms of the von Neumann entropy. We want to compare the behavior of the entanglement when the electrons are, at the initial

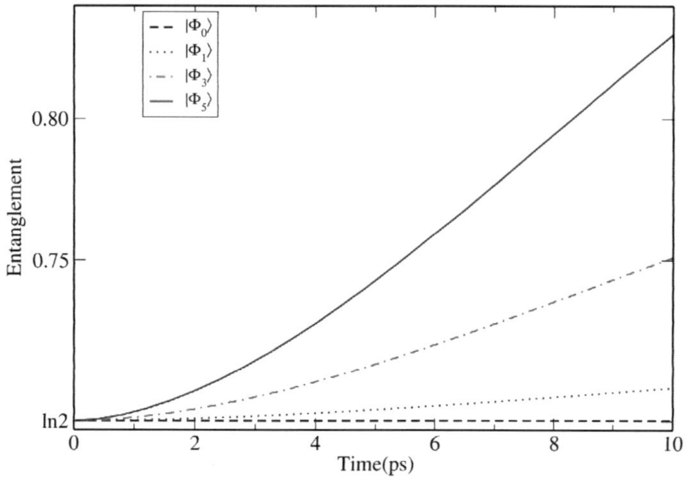


**Figure 3.** Entanglement dynamics of $\Phi_0$ (dashed line) with $E_0$=-47.17 meV, $\Phi_1$ (dotted line) with $E_1$=-41.54 meV, $\Phi_3$ (dash-dotted line) with $E_3$=-30.59 meV, and $\Phi_5$ (solid line)with $E_5$=-20.03 meV.

time, in the ground state of the single-particle Hamiltonian, or in one of its excited eigenstates. The results are reported in Fig. 3. When the ground state is considered, at initial time no correlation is present, apart from that due to the exchange symmetry, and the entanglement does not vary significantly with time. One can therefore conclude that during the time evolution of the two-particle wavefunction the Coulomb interaction does not become sufficiently strong to build up significative quantum correlations between the two electrons. On the other hand, when higher-energy states are involved, the quantum correlation is larger. Now, unlike the previous case, the entanglement is not negligible and reaches values significantly different from the initial one, ln2, at times of the order of ten ps. The reasons for this behavior can be related to the real-space two-particle wavefunction. Its spatial spreading is larger for high-energy states and the single-electron probability densities localized in the two next minima get closer. This leads to an increase of the Coulomb interaction between the particles that is responsible for the quantum correlation. The results suggest that for electrons captured in high-energy bound states of the SAW a description in terms of single-particle states is less accurate as time goes by. In particular, for times of the order of a few tens of ps, as required for a single quantum operation in SAW-based quantum logic gates[18], only the low-energy bound carriers can safely be assumed to keep coherence, the effect of their mutual quantum correlations being negligible.

## 5. Conclusions

The attempt to shed more light on the phenomenon of entanglement between electrons, with special attention to semiconductors systems, is the motivation of this work. To this purpose, numerical approaches have been developed allowing to investigate the processes ruling entanglement creation and their connection with the physical parameters and initial conditions of the systems. The entanglement dynamics has been evaluated in 2D-scattering events between two indistinguishable carriers, showing a deep relation between spin and space degrees of freedom of the states, even for interaction not involving spin-orbit terms. Entanglement generation for a two-electron scattering in a 1D double-barrier resonant tunneling device has been investigated. Results suggest that the production and detection of entanglement in such structures may be controlled by suitably tuning the resonances of the system through external fields. Finally carrier-carrier quantum correlations in a 1D semiconductor in presence of a periodic potential modeling SAW have been studied. In this context, the validity of the single-particle approximation to describe the acoustically driven charge transport in semiconductors quantum wires has been discussed.

## Acknowledgments

All the work presented in this paper has been realized in collaboration with F. Buscemi and A. Bertoni. C. Jacoboni is thanked for useful discussions. Support from CNR-INFM Progetto Supercalcolo 2008 CINECA is acknowledged.

## References

[1] Giulini D, Joos E, Kiefer C, Kupsch J, Stamatescu I -O, Zeh H D, *Decoherence and the Apparence of a Classical World in Quantum Theory* (Springer, Berlin, 1996).
[2] Peres A, *Quantum Theory: Concepts and Methods* (Kluwer Academic Publishers, Dordrecht, 1995).
[3] Tal A and Kurizki G 2005 *Phys. Rev. Lett.* **94** 160503.
[4] Buscemi F, Bordone P and Bertoni A 2007 *Phys. Rev. B* **76**, 195317.
[5] Gunlycke D, Jefferson J H, Rejec T, Ramsak A, Pettifor D G and Briggs G A D 2006 *J. Phys.: Condens. Matter* **18** S851-S866.
[6] Costa A T, Jr. and Bose S and Omar Y 2006 *Phys. Rev. Lett.* **96** 230501.
[7] Bertoni A, Bordone P, Brunetti R, Jacoboni C, and Reggiani S 2000 *Phys. Rev. Lett.* **84** 5912.
[8] Oliver W D, Yamaguchi F, Yamamoto Y 2002 *Phys. Rev. Lett.* **88** 037901(R).
[9] Ramsak A, Mravlje J, Zitko R, and Bonca J 2006 *Phys. Rev. B* **74** 241305(R).
[10] Buscemi F, Bordone P, and Bertoni A 2006 *Phys. Rev.A* **73** 052312.
[11] Buscemi F, Bordone P, and Bertoni A 2007 *Phys. Rev.A* **75** 032301.
[12] Buscemi F, Bordone P, and Bertoni A 2009 *J. Phys.:Condens. Matter.* **21** 305303.
[13] Schliemann J, Cirac J I, Kus M, Lewenstein M and Loss D 2001 *Phys. Rev. A.* **64**:022303.
[14] Ghirardi G and Marinatto L 2004 *Phys. Rev. A.* **70** 012109.
[15] Lama L and Leon J 2006 *Phys. Rev. A.* **73** 052322.
[16] Lent C S and Kirkner D J 1990 *J. Appl. Phys.* **67** 6353.
[17] Fano U 1961 *Phys. Rev.* **124** 1866.
[18] Kataoka M, Astley M R, Thorn A L, Oi D K L, Barnes C H W, Ford C J B, Anderson D, Jones G A C, Farrer I, Ritchie D A and Pepper M 2009 *Phys. Rev. Lett.* **102** 156801.

Journal of Physics: Conference Series **193** (2009) 012043 doi:10.1088/1742-6596/193/1/012043

# Lateral electron tunnelling spectroscopy in etched GaAs/AlGaAs-based nanostructures

**S F Fischer[1,2], J L Deborde[2], U Kunze[2] and D Reuter[3], A D Wieck[3]**
[2]Electronic Materials and Nanoelectronics, Ruhr-University of Bochum, Germany
[3]Applied Solid State Physics, Ruhr-University of Bochum, Germany

E-mail: saskia.fischer@rub.de

**Abstract.** Tunnelling spectroscopy between laterally positioned low-dimensional electron systems is investigated. We demonstrate that a reliable highly transparent tunnelling barrier can be fabricated by means of dynamic ploughing with the tip of an atomic force microscope and wet etching into the surface of a GaAs/AlGaAs heterostructure. This barrier allows to perform a tunnelling spectroscopy with dc bias voltages up to 30 mV. Band edge effects due to tunnel coupling were observed for two-dimensional electron gases (2DEGs) separated by such lateral tunnel barrier. Tunneling spectroscopy of a 2DEG and a laterally adjacent electron waveguide enable to monitor the one-dimensional (1D) density of states. Firstly, a lateral dual electron waveguide with 1D subband spacings of more than 5 meV was implemented showing significant features of 1D mode coupling in tunnelling spectroscopy.

## 1. Introduction

Using the principle of coherent quantum mechanical tunnelling, an electron waveguide coupler was initially conceived as an electrical analogue to an optical dual-channel directional coupler [1]. The device consists of two electron waveguides (EWGs) that over a certain length come in very close proximity to each other. However, the exploitation of a dual EWG as a logical device relies on the exact coupling characteristics. In particular, the implementation of a suitably thin but high potential barrier enabling the application of finite dc bias voltages of several milli-volts represents a major challenge. The present work is concerned with lateral tunnelling spectroscopy between 2D and 1D low-dimensional electron systems in etched GaAs/AlGaAs nanostructures. Here, we demonstrate the advantage of a new combination of two lithography techniques: The indirect patterning technique based on an atomic force microscope (AFM), and the electron-beam lithography (EBL) with high-resolution negative tone resist. AFM lithography proves to be a reliable technique to produce thin and deep grooves on a semiconductor surface, inducing tunnelling and insulating boundaries in a shallow buried 2DEG of a GaAs/AlGaAs heterostructure [2]. With EBL quantum wires with large 1D-subband spacings can be produced [3]. Three kinds of *lateral* electron tunnelling devices were successfully fabricated: A 2D-to-2D, a 2D-to-1D and a 1D-to-1D tunnelling device. The manifestation of coupling effects between the electron waveguides are of great interest for the implementation of a quantum directional coupler [1].

[1] To whom any correspondence should be addressed.

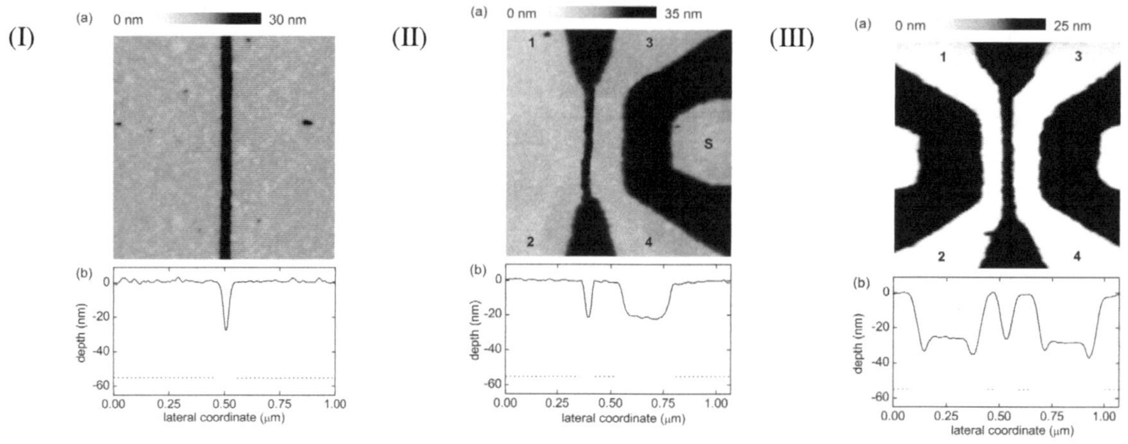


**Figure 1.** Top row: Atomic force micrographs of the surface of the nanostructures before deposition of Au top gates. The dark areas represent the etched regions. Bottom row: Cross sections. The dotted lines schematically represent the 2DEG and its depletion under the etched regions. (I) 2D-to-2D-, (II) 2D-to-1D- and (III) 1D-to-1D-tunnelling device. Leads to ohmic contacts are labelled 1 to 4.

## 2. Experimental Details

Two Si-doped GaAs/AlGaAs-heterostructures hosting 2DEGs situated 55 nm below the surface were used. The Hall electron densities were 4.4 and 3.7x10$^{11}$ cm$^{-2}$ and the electron mobilities were 10 and 9.4 x 10$^{5}$ cm$^{2}$/Vs, respectively, measured at 4.2 K in the dark. Tunnel barriers were prepared by indirect dynamic ploughing with an AFM [4]. For indirect patterning, a standard positive photoresist (Shipley 2510) strongly diluted with thinner (1:40 or 1:50) is deposited with a thickness of 5-7 nm. Using a drive amplitude of the AFM tip much larger than the imaging amplitude the resist can be plastically deformed. This pattern can be transferred into the substrate by subsequent wet-chemical etching [2] with a citric acid solution. Lines etched along the [1-10]-direction give U-shaped grooves in GaAs, whereas a [110] orientation gives V-shaped grooves which was used in this work. Electron wave guides were defined using electron-beam lithography (EBL) with a Leo Gemini 982 (at 2 kV and 0.8 μC/cm$^{2}$) and Elphy Plus Lithography software. We used the hexaacetate p-methylcalix[6]arene (MC6AOAc) derivative from the negative resist Calixarene which offers a very high lateral resolution [5]. 1 wt.% solution of Calixarene in monochlorbenzene was spun during 30 seconds at 3000 rpm on the sample, followed by a 30 minutes bake at 170°C resulting in a thickness of 25 ±2 nm. After EBL exposure, the resist is developed 30 s in a xylene bath, then stopped by isopropanol (10 s) and baked for 30 min at 110°C under nitrogen atmosphere. The remaining resist frames a high-quality mask for the subsequent wet-chemical etching with citric acid solution. The resist is finally removed in an oxygen plasma (5 min). Ni/AuGe/Ni/Au ohmic contacts were prepared by photolithography, electron-beam evaporation, subsequent lift-off processing and alloying by rapid thermal annealing. Finally, the Schottky Au top gate was fabricated by photolithography, thermal evaporation and lift-off processing.

## 3. Lateral tunnelling spectroscopy

The electrical measurements were performed at 4.2 K with a lock-in amplifier. The differential conductance $g_D = dI_D/dV_D$ describes the variation of the ac drain current $I_D$ with the ac drain voltage $V_{D,ac}$ (40-100 μV, 433 Hz) and was measured as a function of the dc top gate voltage $V_G$. A dc drain voltage $V_D$ (up to ±30 mV) as applied across the tunnel barrier for the tunnelling spectroscopy (tunnel spectrum) and along the EWGs for transport spectroscopy of the 1D subband spectra.

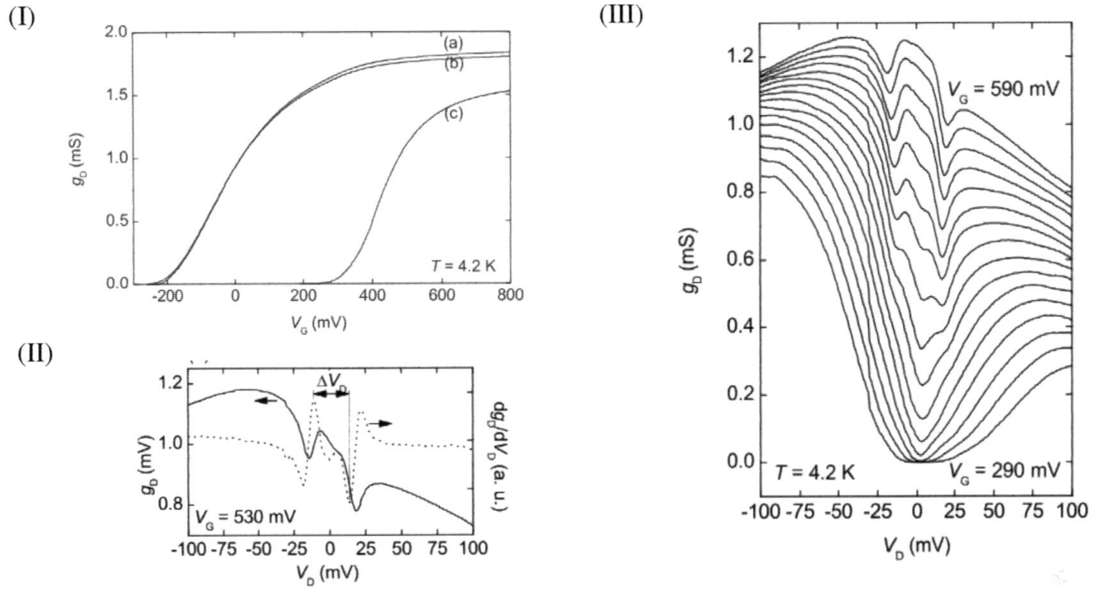


**Figure 2.** 2D-to-2D tunnelling device: Differential drain conductance $g_D$ at 4.2 K measured as a function of: (I) the topgate voltage $V_G$ , (a) and (b) of the 2DEGs each side of the potential barrier, (c) across the potential barrier, (II) of the the dc drain voltage $V_D$ across the barrier (dotted line: calculated transconductance $dg_D/dV_D$) and (III) (a) of the $V_D$ bias at gate voltages $V_G$ ranging from 290 to 590 mV with 20 mV interval.

### *3.1.1. 2D-2D tunneling*

The characteristic of the differential drain conductance $g_D$ across the barrier also denoted as tunnelling conductance was measured as a function of the bias drain voltage $V_D$ as shown in Fig. 2. $g_D$ generally increases with increasing top gate voltage $V_G$ and it drops with larger increasing dc drain voltages which is attributed to the heating of charge carriers in an electric field. For the drain bias interval $|V_D| \leq \pm 25$ mV, the behaviour of the tunnelling conductance strongly depends on the applied top gate voltage $V_G$. The basic behaviour of the curves is symmetric with respect to zero bias. However, due to the measuring assembly, the increase of the negative drain voltage $V_D$ acts like an increase of the effective gate voltage $V_{G,eff}$ along the transistor channel leading to an increase of $g_D$. When the barrier width $W$ is of about the Fermi wavelength $\lambda_F \sim W$ , the wavefunctions of both 2DEGs, which penetrate into the barrier, overlap each other and tunnelling can occur. For small gate voltages $V_G \leq 330$ mV and at zero drain bias voltage the tunnelling conductance is zero. An increase of source-drain bias voltage $|V_D|$ induces a reduction of the effective barrier height such as $\Delta\varphi = \frac{1}{2} e\Delta V_D$ and the transmission probability is increased. The tunnelling probability also increases as the top gate voltage $V_G$ gets larger. For large $V_G \geq 430$ mV, minima appear in $g_D$ nearly symmetric for positive and negative $V_D$. The minima in $g_D$ reflect Fermi-edge effects due to the tunnelling process across the barrier [6,7]. For small $V_D$ the tunnelling probability remains nearly constant, in turn the $I(V)$-curve should be linear and $g_D$ should show a quasi-plateau near $V_D$=0. In Fig. 2 this quasi-plateau is not horizontal because of the asymmetric contact geometry. Further, the drain bias gap between the minima is not constant with increasing gate voltage because of the increasing influence of the series resistance. With larger $V_D$, the barrier disappears and the 2DEG properties dominate the conductance characteristics. As the drain voltage interval between the two drops of the conductance is directly proportional to the Fermi energy of the 2DEG, $e\Delta V_D = 2E_F$, we can determine the density of the 2D electron sheet from $n_S = E_F m^*/(\pi\hbar^2)$.

Journal of Physics: Conference Series **193** (2009) 012043 doi:10.1088/1742-6596/193/1/012043

(I)

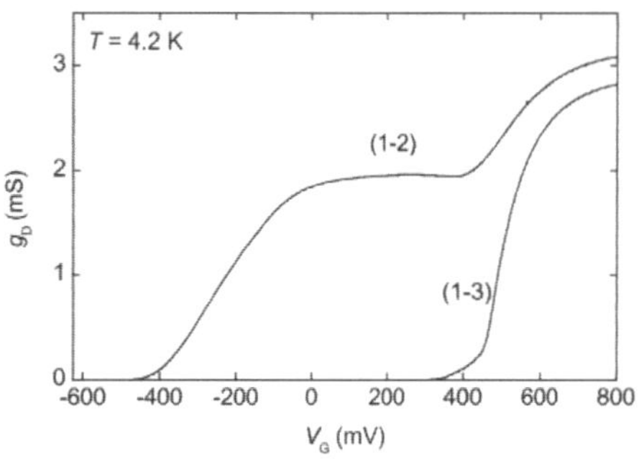


(III)

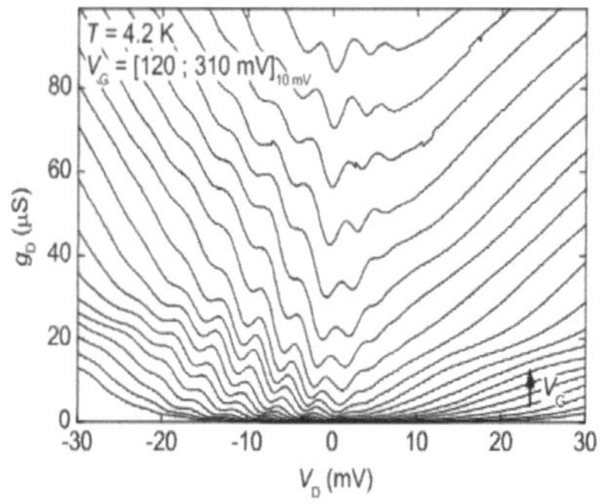


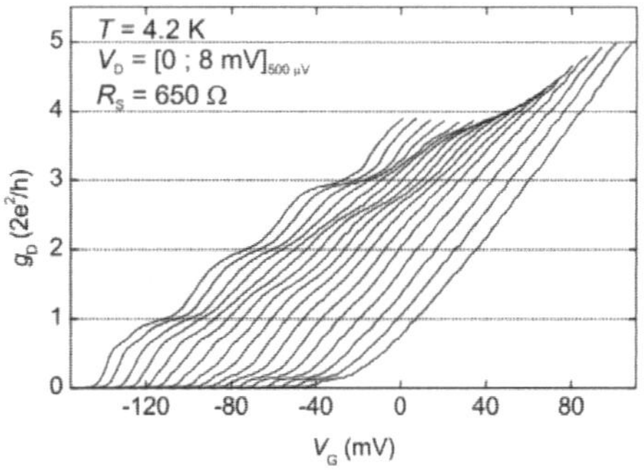


(II)

**Figure 3.** 2D-to-1D tunnelling device: differential conductance $g_D$ of (I) the 2DEG (contacts 1-2) and across the barrier (contact 1-3), (II) of the EWG for different $V_D$. The curves have been offset along $V_G$ for clarity and are corrected for $R_S$=650 Ω, and (III) measured across the tunnel barrier as a function of the applied $V_D$. $V_G$ was varied between 120 mV and 310 mV with a step of $\Delta V_G$ = 10 mV.

For a top gate voltage of 520 mV, the 2D electron density of $n_S$=3.38·10$^{11}$ cm$^{-2}$ compares well with 3.7·10$^{11}$ cm$^{-2}$ as extracted from Shubnikov-de Haas measurements on the heterostructure without a top-gate. Critical in the tunnel spectroscopy is the very high sensitivity of the tunnel barrier to the gate voltage and to the drain voltage. As $|V_D|$ increases the effective height of the barrier is reduced. The differential conductance rises because of the dramatic increase of the transmission probability leading to a V-shaped tunnelling conductance characteristic.

*3.1.2. 2D-1D tunnelling*

Fig. 3 shows measurements of the 2D-to-1D tunnelling device. The drain conductance of the 2D channel (contacts 1-2) exhibits a two-step characteristic arising from the barrier height controlled by the gate voltage. The first saturation at about 2 mS reflects the series resistance $R_S$≈500 Ω of the reservoirs and the alloyed contacts. At $V_G$ ≈ 450 mV, the barrier is completely removed, and the effective width of the 2D channel increases strongly. This reduces the saturation range to about 320 Ω. The conductance measured across the barrier (contacts 1-3) correspondingly shows a gradual increase from $V_G$≈ 300 mV to a saturation value determined by the series resistance $R_S$ ≈ 350 Ω of the common reservoirs. The electron wavefunction "leaks" out of the 1D waveguide channel through the tunnel barrier into the adjacent 2D electron bath and vice versa. Tunnelling spectroscopy was performed by measuring the two-terminal drain conductance $g_D$ (across contacts 1-3) as a function of the drain voltage $V_D$. A modulation voltage $V_D$ = 400 μV was imposed upon the 2DEG (contact 1), and the ac current $I_D$ is monitored by lock-in technique on contact 3. Figure 3 shows a series of tunnelling spectra for a top-gate-voltage interval 120mV ≤ $V_G$ ≤ 310 mV. A positive bias voltage refers to that applied at the 2DEG (contact 1). The gradual increase of the conductance with increasing gate voltage indicates the dependence of the barrier transmission with $V_G$ as observed in 2D-2D tunnelling. Each curve exhibits oscillations superimposed on a slowly varying background: With increasing $|V_D|$ the effective barrier height is diminished, increasing the tunneling transmission and thereby the total conductance. For $|V_D|$ > 25 mV, these oscillations are averaged out. Oscillations in the tunnelling conductance directly reflect the density of states of the 1D subband system in the waveguide and can be discussed by the semiclassical model of Eugster and del Alamo *et al.* [7]. Momentum conservation along the

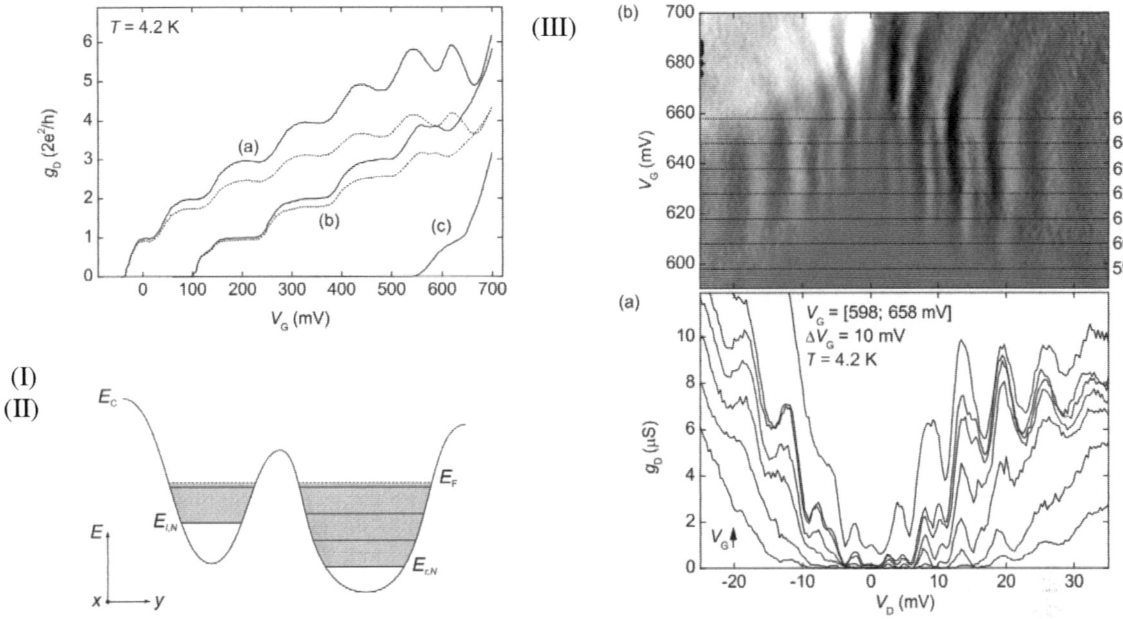


**Figure 4.** (I) Differential drain conductance $g_D$ as a function of the topgate voltage $V_G$ at T = 4.2 K measured (a) along the wide waveguide (contacts 3-4), (b) along the narrow waveguide (contacts 1-2) and (c) across the potential barrier (contacts 1-3). In (a) and (b), the solid curves represent the conductance corrected for the series resistance: $R_{S,W}$= 900 Ω (wide EWG), $R_{S,N}$ ,= 750 Ω (narrow EWG); the dotted curves are as measured. (II) Sketch of the potential landscape of a 1D-to-1D tunnelling device at a finite $V_G$ and $V_D$=0. (III) (a) $g_D$ at T = 4.2 K across the barrier (contacts 1-3 on Fig. 38) as a function of the applied dc voltage $V_D$ for 598 ≤ $V_G$ ≤ 658 mV, with 10 mV interval. (b) Grey-scale plot of the numerically derived tunneling transconductance d $g_D$ /d$V_D$ as a function of $V_D$ bias and $V_G$ . The dotted lines mark the corresponding measurements on (a).

waveguide (x-direction) requires that $k_{x,2D} = k_{x,1D}$ is fulfilled in the event of 2D-to-1D tunnelling and vice versa. However, in the model [7] a square-shaped barrier is regarded. In our experiment the effective barrier height depends on the applied drain voltage $V_D$ leading to the V-shaped background of the tunnelling conductance characteristic. A peak in the tunnelling conductance arises if the electrochemical potential of the 2DEG $E_F$ is in resonance with the edge of a 1D subband $E_N$ because the density of states is highest near the bottom of the 1D subband. As the 1D-DOS diminishes with increasing energy, the conductance decreases. From the oscillation period we estimate an average 1D subband separation of $\Delta E_{N,N+1}$=4.0±0.5 meV which agrees well with that obtained from transport spectroscopy (not shown here). However, because of the layout design of the device under investigation tunnelling spectroscopy of the lowest 1D subbands is not possible. First, the top-gate covers the whole device and hinders the influence of the in-plane gates. This does not give independent control of the tunnel barrier height and the electron density in the 1D waveguide. Second, the etch depth of the nanostructure, thereby the strength of the 1D confinement is limited by the subsequent AFM lithography step. This could be overcome by decreasing the width of the EWG.

### *3.1.3. 1D-1D tunnelling*

Figure 4 shows the differential drain conductance for each EWG exhibiting exhibit clear quantization with plateaux in units of $2e^2$/h. Both EWGs have the same etch depth, however different widths infer a different 1D confinement: The narrow wire has a larger subband spacing and a larger threshold voltage. From transport spectroscopy the 1D subband spacings are determined to $\Delta E_{1,2}$ =11.8 meV±0.5

meV (narrow EWG) and $\Delta E_{1,2}$=8.7 meV±0.5 meV (wide EWG). Fig. 3 pictures a series of measured tunnelling conductance spectra for several top-gate voltages. For different cooling cycles the results are concordant. The positive bias refers to the potential at the right wire, contact 3. Sweeping the drain voltage corresponds to displacing 1D energy subband ladder of one wire relative to the other. For topgate voltages $V_G$ of 595 mV and above, the tunnelling spectra exhibit oscillations similar to the ones observed in the 2D-to-1D tunnelling spectroscopy. Here, they map the 1D density of states by a resonantly enhanced tunnelling current for degenerate 1D subbands. Again, the asymmetry of the background with respect to $V_D$ = 0 is due to the measuring assembly. As previously described, $V_G$ influences the tunnelling transmission as detected in the conductance magnitude. Finally, two kinds of oscillations are observed: For the smaller gate voltages, the tunnelling oscillations go in pairs and, with increasing $V_G$, they merge to an oscillation with a larger magnitude and with and twice the period. This is shown in the grey-scale plot of the transconductance (Fig. 4) in which a full tunnelling oscillation is represented by the association of a dark stripe (ascending $g_D$) and a bright stripe (descending $g_D$). Two assumptions are considered: (i) only the top-most partially occupied subband contributes significantly [8], and (ii) when 1D subband edges come into resonance the electron wave functions couple leading to an splitting of the subband degeneracy into antisymmetric and symmetric states, has recently been demonstrated on vertically stacked tunnel-coupled EWGs [9]. A splitting of $\Delta E$=2.2±0.5 meV is measured coinciding well with the observations of anticrossings from transport spectroscopy (not shown here [8]). As $V_G$ is made more positive, the tunnelling barrier is reduced. The anticrossings disappear and the two tunnelling oscillations corresponding to the antisymmetric and symmetric coupling states merge and the resulting pattern directly translates the 1D density of states of the biased waveguide. The measured spacing $\Delta E_{N,N+1}$=5.5±0.5 meV agrees with the spacings between the subbands of higher energy of the waveguide as determined by transport spectroscopy.

## 4. Summary and Outlook

The AFM lithography with subsequent wet-chemical etching allows to fabricate a groove of desired depth-to-width aspect ratio in the surface of a semiconductor for lateral tunnelling devices. By combining it with EBL lateral tunnelling in electron waveguide devices of large subband spacings (~ 10 meV) can be achieved. Our results are promising for future investigations on lateral tunnelling also with respect to other material systems. Implementation via top-down fabrication is relatively easy compared to cleaved-edge-overgrowth [10] or flip-chip techniques [11]. Future developments might include distinct fingergates and sidegates for an independent control of the barrier height and EWG electron densities. Opportunities of lateral tunnelling in the *single-mode* regime exist for a quantum directional coupler, investigation of the 0.7-conductance anomaly, and spin-filtering applications.

We gratefully acknowledge the financial support by the BMBF nanoQUIT research programm.

## 5. References

[1] del Alamo J A, Eugster C G 1992 *Appl. Phys. Lett.* 56, p 78
[2] Klehn B, Kunze U 1999 *J Appl. Phys. Lett.* 85, p 3897; *ibid* 1998 Superlatt. Microst. 23, p. 441
[3] Apetrii G, Fischer S F, Kunze U, Reuter D, Wieck A D 2002 *Sem. Sci. Techn.* 17, p 735
[4] Knop M, Richter M, Maßmann R, Wieser U, Kunze U, Reuter D, Riedesel C, Wieck A D 2005 *Sem. Sci. Techn.* 20, p 814
[5] Fujita J, Ohnishi Y, Ochiai Y, Namura E, Matsui N 1996 *J Vac.. Sci. Techn.* 14, p 4272
[6] Eugster C G, del Alamo J A, Melloch M R, Rooks M J, 1993 *Phys. Rev. B.* 48, p 15057
[7] Eugster C G, del Alamo J A, Melloch M R, Rooks M J, 1994 *Appl. Phys. Lett.* 64, p 3157
[8] The full experimental details and transport spectroscopy will be given elsewhere.
[9] Fischer S F, Apetrii G, Kunze U, Schuh D, Abstreiter G, 2006 *Nature Physics* 2, p 91
[10] Auslaender OM, Yacoby A, de Piciotto R, Baldwin KW, Pfeiffer LN, West KW, 2002 Science 308, p88
[11] Bielejec E, Reno JL, Lyo SK, Lilly MP, 2008 Solid State Comm. 147, p 79

Journal of Physics: Conference Series **193** (2009) 012044 doi:10.1088/1742-6596/193/1/012044

# Dynamics of electron-electron entanglement in pulsed sinusoidal potentials

**F Buscemi[1,2], P Bordone[3,2] and A Bertoni[2]**

[1] ARCES, Alma Mater Studiorum, University of Bologna, Via Toffano 2/2, 40125 Bologna, Italy

[2] S3 Research Center, CNR-INFM, Via Campi 213/A, I-Modena 41125, Italy

[3] Dipartimento di Fisica, Università di Modena e Reggio Emilia, I-41125 Modena, Italy

E-mail: `fabrizio.buscemi@unimore.it`

**Abstract.** The surface acoustic wave (SAW) assisted charge transport represents a highly controllable mean to inject and drive electrons along quasi one-dimensional channels. By means of numerical simulations, we analyze the effect of the removal of SAW for a short time interval onto the entanglement stemming from the Coulomb scattering between two charged carriers bound by the SAW potential itself. We find that the particles get quickly entangled, thus confirming that the SAW driven transport in low-dimensional semiconductor structures permits to suppress the effects of electron-electron correlation due to the Coulomb interaction.

## 1. Introduction

The investigation of the entanglement formation in scattering events in semiconductor structures is not only useful for understanding the nature of the scattering process itself but it is also recognized as an important resource for the design of quantum information processing devices [1, 2, 3, 4, 5]. In the last years many theoretical efforts have been carried out [3, 4, 6, 7] and different proposals to produce entangled states between charged carriers have been presented [1, 2, 5].

The aingle-particle transport through surface acoustic waves (SAW), originally introduced in the context of metrological application for defining a new standard of electric current [8, 9, 10], constitutes also a viable means to implement quantum gate networks in semiconductor nanodevices [1, 11, 12, 13]. A number of devices exploiting SAW-electron interaction have been designed, realized and also proposed as basic building blocks for quantum computing applications, where the use of SAW has been shown to constitute a highly controllable mean to inject and drive electrons along quantum wires[12].

In this work, we analyze the time evolution of the entanglement during the scattering of two electrons interacting with an external periodic potential mimcking SAW and removed for a finite time interval. In particular we intend to examine the effect of the removal of such potential on the spreading of the wavefunction initially confined in a SAW minimum. Furthermore, we investigate how the turning off and on of the sinusoidal potential affects the building up of quantum correlations between the particles. This could provide us a clearer vision on the validity of the single-particle approximation, which leads to neglect the quantum correlations between the carriers in the study of SAW-assisted for the case of the SAW transport in low-dimensional semiconductor structures [14].

## 2. Physical model and results

In this section we investigate the entanglement between two electrons initially trapped into two next neighbor minima of a sinusoidal potential, when the latter is removed for a short time interval. In our approach the two particles are confined in a 1D-channel and the coupling originates from their mutual Coulomb interaction. The Hamiltonian of the system has the form

$$H(x_a, x_b) = H_{off-on}(x_a) + H_{off-on}(x_b) + \frac{e^2}{\epsilon\sqrt{(x_a - x_b)^2 + d^2}}, \tag{1}$$

where $\epsilon$ is the silicon dieletric constant and $d$ represent the thickness of the wire, here fixed at 1 nm. $H_{off-on}(x_a)$ is the single-particle Hamiltonian given by

$$H_{off-on}(x) = -\frac{\hbar^2}{2m}\frac{\partial^2}{\partial x^2} + \alpha\,[\theta(-t) + \theta(t - \Delta T)]\left(\sin(\frac{2\pi}{\lambda}x) + 1\right), \tag{2}$$

with $\alpha$ representing the amplitude of the oscillation and $\lambda$ the wavelength of the potential. Here we assume that, at the initial time $t = 0$, the external potential is abruptly switched off and then turned on again after a time interval$\Delta\,T$. As stated in the introduction, the investigation of the above model can be of interest in the contest of electron transport in quantum wires assisted by SAW. In fact, we can assume that the Hamiltonian given in Eq. (2) describes the effect of the removal of SAW potential for a time$\Delta\,T$ (as obtained by a brief switching off and on of the SAW transducers or the transit of the electron in a SAW-free region) onto the dynamics of a couple of electrons.

In our numerical calculation we take the SAW sinusoidal potential with amplitude $\alpha = 50$ meV and wavelength $\lambda = 110$ nm. In fact, we found that for longer wavelengths the switching off and on of the external potential does not lead to significant effects on the entanglement since the two particles remain at a distance at which the Coulomb interaction is rather weak. We consider two electrons with the same spin (up). The quantum state describing the system at the initial time is

$$|\Phi\rangle = \frac{1}{\sqrt{2}}\Big(|\chi\,\varphi\rangle - |\varphi\,\chi\rangle\Big)|\uparrow\uparrow\rangle. \tag{3}$$

The wavefunctions corresponding to the states $|\chi\rangle$ and $|\varphi\rangle$ are two eingestates of the SAW potential describing electrons trapped into two next minima. In Eq. (3) the ket $|\uparrow\rangle$ indicates spin up state.

To obtain the system evolution we solve the time-dependent Schröndiger equation for the two-particle wavefunction of Eq. (3). Once the real-space wavefunction is found at a given time step, we compute the two-particle density matrix $\rho = |\Psi\rangle\langle\Psi|$ and from the latter, we calculate the one-particle reduced density matrix $\rho_r$ by tracing over the degrees of freedom of one of the two electrons. $\rho_r$ is then used to evaluate the entanglement. In fact, it is well known that for a two-fermion system a good correlation measure is given by the von Neumann entropy of $\rho_r$ [16, 15]:

$$\varepsilon = -\mathrm{Tr}[\rho_r \ln \rho_r] = \sum_{i=1} |z_i|^2 \ln |z_i|^2, \tag{4}$$

where $|z_i|^2$ are the eigenvalues of the matrix $\rho_r$.

In Fig. 1 we report the time evolution of the entanglement for various$\Delta\,T$, ranging from 0.3 to 0.6 ps. We observe that $\varepsilon$ increases with time and exhibits a sort of double step structure. Finally, it reaches a stationary value that strictly depends upon$\Delta\,T$. In other words, the electrons get more correlated for larger$\Delta\,T$. To get a better insight into the process, we need to take into account the dynamics of the real-space two-particle wavefunction. Its initial peaks, describing the localized states of the two electrons, spread as the external potential is removed.

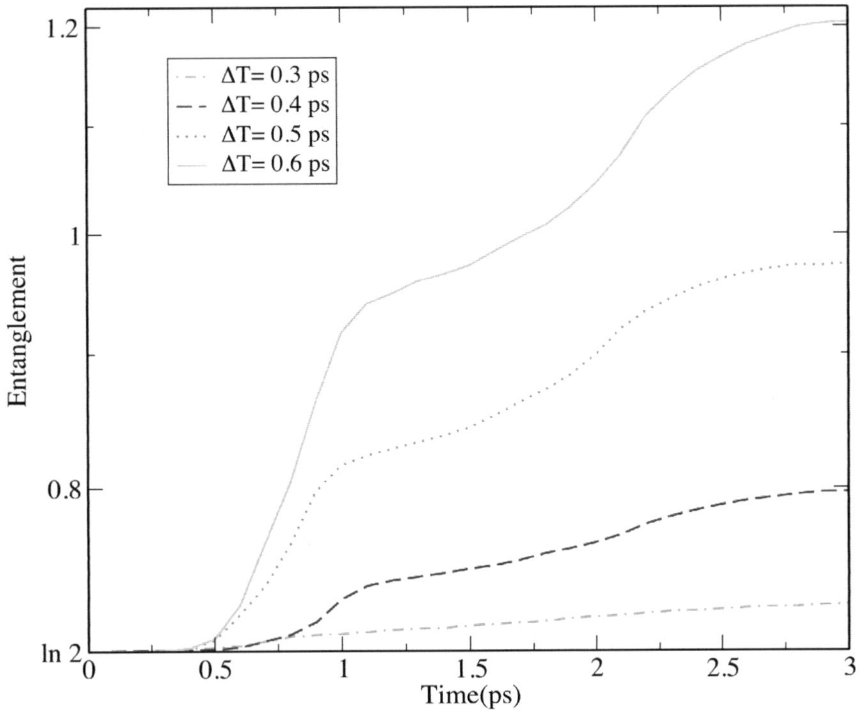


**Figure 1.** Entanglement as a function of time for four different values of the time interval $\Delta T$ without the sinusoidal potential. The amplitude of the oscillation is 50 meV and the wavelength 110 nm in order to mimic a SAW potential.

Therefore, for sufficiently long$\Delta$ $T$ each peak can overlap with the other (see Fig. 2). This gives rise to a strong building up of quantum correlations, due to Coulomb interaction. Such an effect is obviously stronger when the sine-like potential is turned off for a longer time. After the potential is turned on again, i.e. for $t > \Delta T$, the wave packet is not smooth anymore, but it shows rapid oscillations. This affects the spatial overlap described above, which is responsible of the entanglement formation, thus leading to the peculiar double step structure observed in Fig. 1. In the limit of long$\Delta$ $T$, the two peaks of two-particle spatial wavefunction spread and quantum correlation between the electrons increases with time until maximum entanglement condition is reached.

## 3. Conclusions

In this paper we addressed the effect of the SAW removal onto the entanglement created by the Coulomb interaction between two electrons localized into two next neighbor minima of a standing SAW potential of wavelength comparable or smaller than the ones of Refs. [8, 9, 10]. We found that the entanglement increases with time and reaches a stationary value that is "freezed" when the potential is restored. In particular, the particles get more correlated for longer switching-off time intervals, due to the larger overlap between the spreading real-space wave functions. The appearance of a significative entanglement implies that this system cannot be described in terms of single-particle states. On the contrary, it represents a suitable candidate for the controlled generation of bipartite entanglement.

Finally, the results presented in this paper can be also considered as an additional theoretical confirmation that the use of SAW to drive electrons along quantum wires prevents the spreading of the spatial wavefunction and allows to suppress the effects of electron-electron correlation due to the Coulomb interaction. On the other hand, if the travelling periodic potential is instantaneously turned off and then on, the electrons get quickly entangled.

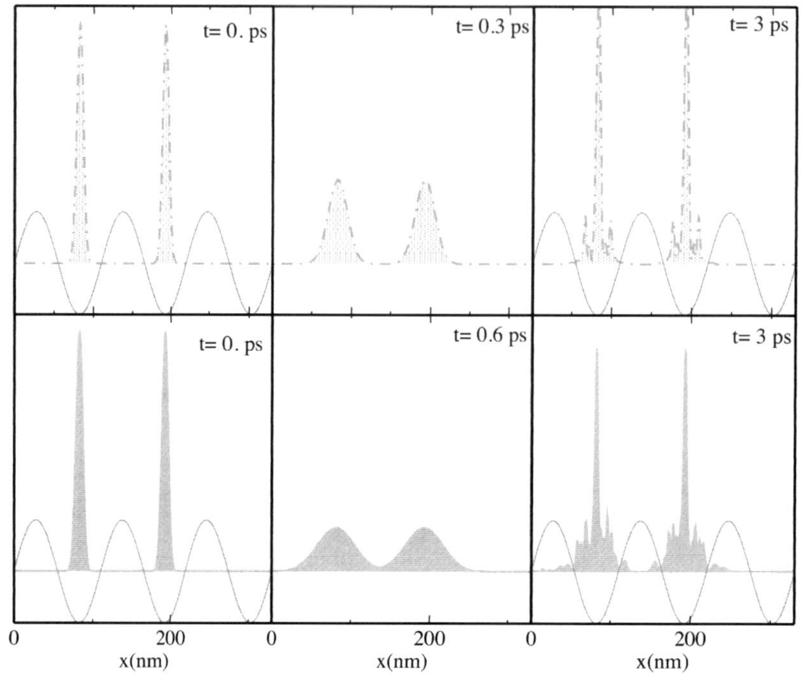


**Figure 2.** Single-particle probability density $\int |\langle x_a\, x_b|\chi\varphi\rangle|^2 dx_a$ evaluated at different time steps for two values of$\Delta$ $T$: 0.3 ps (dash-dotted line in the upper row of graphs) and 0.6 ps (solid-shaded curve in the lower row graphs). The thin solid line sketches the SAW sinusoidal potential, when present.

## Acknowledgments
The authors are pleased to thank Carlo Jacoboni for useful discussions. We acknowledge support from CNR-INFM Progetto Supercalcolo 2008 CINECA

## References
[1] Bertoni A, Bordone P, Brunetti R, Jacoboni C, and Reggiani S 2006 *Phys. Rev. Lett.* **4** 5912
[2] Oliver W D,Yamaguchi F and Yamamoto Y 2002 *Phys. Rev. Lett.* **88** 037901 (R)
[3] Buscemi F, Bordone P and Bertoni A 2007 *Phys. Rev. B* **76** 195317.
[4] Tal A and Kurizk G 2005 *Phys. Rev. Lett.* **94** 160503.
[5] Ramsak A, Mravlje J, Zitko R and Bonca J 2006, *Phys. Rev. B* **74** 241305(R)
[6] Gunlycke D, Jefferson J H, Rejec T, Ramsak A, Pettifor D G and Briggs G A D 2006 *J. Phys.: Condens. Matter* **18** S851-S866
[7] Costa A T, Bose Jr and S and Omar Y 2006 *Phys. Rev. Lett.* **96** 230501
[8] Shilton J M, Talyanskii V I, Pepper M, Ritchie D A, Frost J E F, Ford C J B, Smith C G and Jones G A C 1996 *J. Phys.: Condens. Matter* **8** L531
[9] Talyanskii V I, Shilton J M, Pepper M, Smith C G, Ford C J B, E.H. Linfield, D.A. Ritchie, and G A C Jones 1997 *Phys. Rev. B* **56** 15180
[10] Cunningham J, Talyanskii V I, Shilton J M, Pepper M, Simmons M Y and Ritchie D A 1999 *Phys. Rev. B* **60** 4850
[11] Barnes C H W, Shilton J M and Robinson A M 2000 *Phys. Rev. B* **62** 8410
[12] Rosini M, Bertoni A, Bordone P and Jacoboni C 2004 *J. Comp. Elec.* **3** 443
[13] Rodriquez R, OI D K L, Kataoka M, Barnes C H W, Ohshima T and Ekert A K 2005 *Phys. Rev. B* **72** 085329
[14] Buscemi F, Bordone P and Bertoni A 2009 *J. Phys.: Condens. Matter* **21** 305303.
[15] Buscemi F, Bordone P and Bertoni A 2006 *Phys. Rev. A* **73** 052312
[16] Schliemann J, Cirac J I, Kus M, Lewenstein M and Loss D 2001 *Phys. Rev. A* **64** 022303

# Non-equilibrium dephasing in ballistic interferometers

**Y. Yamauchi[1], M. Hashisaka[1], S. Nakamura[1], K. Chida[1], S. Kasai[1], T. Ono[1], R. Leturcq[2], K. Ensslin[3], D. C. Driscoll[4], A. C. Gossard[4], K. Kobayashi[1]**

[1]*Institute for Chemical Research, Kyoto University, Uji, Kyoto 611-0011, Japan*
[2]*Institute of Electronics, Microelectronics and Nanotechnology, CNRS - UMR 8520, Department ISEN, Avenue Poincaré 59652 Villeneuve d'Ascq, France*
[3]*Solid State Physics Laboratory, ETH Zürich, CH-8093 Zürich, Switzerland*
[4]*Materials Department, University of California, Santa Barbara, California 93106, USA*

Email: yamauchi@ssc1.kuicr.kyoto-u.ac.jp

**Abstract**. It was recently reported that the interference visibility in electronic Mach-Zehnder interferometers behaves unexpectedly in the non-equilibrium regime. Here we report an experiment on non-equilibrium dephasing in two different kinds of electron interferometers. Our results indicate that there exist some universal factors in the dephasing phenomenon in ballistic systems.

## 1. Introduction

Electron interference in ballistic interferometers has been one of the central topics in mesoscopic physics since 1980's. The Aharonov-Bohm (AB) effect is a typical example of the interference, and has been widely studied for many years. Recently, the AB effect was proposed to be used as a method to detect the electron entanglement in an electron Mach-Zehnder interferometer (EMZI) [1]. In the experiments to investigate coherence in an EMZI, however, there have been observed a few unexpected phenomena [2,3,4]. The main discovery was the "lobe structure", which is a rapid decrease of the interference visibility with a phase reversal of the interference signal (AB oscillation) when the drain-source bias for the EMZI is varied. This has invoked researchers to reconsider decoherence of the quantum interference in non-equilibrium ballistic systems. Several theoretical attempts propose that Coulomb interactions are responsible for the lobe structure to emerge and imply that this can be observed in other ballistic interferometers besides EMZI's [5,6]. However, experimental evidences to support this claim have been lacking.

Here, we report the measurement of the dephasing of the AB oscillation in non-equilibrium region in an AB ring (ABR). The drain-source bias dependence of the interference visibility shows a lobe-like structure whose energy scale is much larger than those observed at EMZI's. The numerical analysis for the observed structures shows that the energy scale is strongly dependent on the sample size. We also discuss the relevance of the Gaussian phase averaging under high magnetic field [3].

## 2. Measurement setup and Sample

The AB ring sample was fabricated on a GaAs/AlGaAs two-dimensional electron gas 34 nm under the surface by the local oxidation technique using an atomic force microscope (AFM) (see Fig. 1a) [7,8] .

Measurements were performed by using standard lock-in technique with 5 μV excitation signal at 37 Hz around zero magnetic field (up to 50 mT) and in the integer quantum hall (IQH) regime (with the Landau level filling ~10). The schematic picture of the electron channels in each region is shown in Fig. 1b. The sample was placed in a dilution refrigerator with a base electron temperature of 125 mK [9].

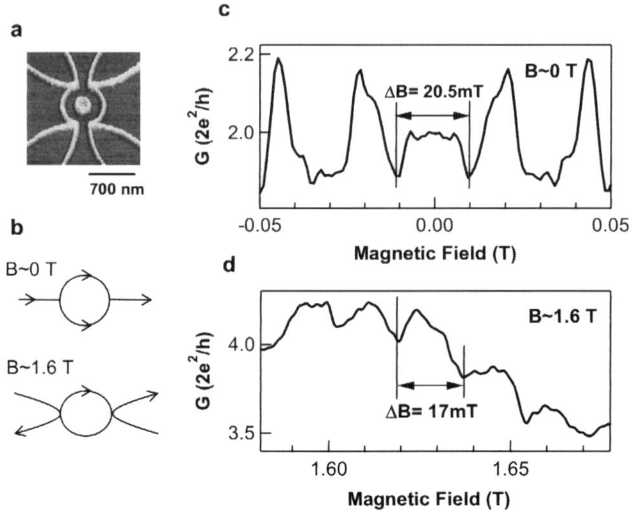


Fig. 1. (a) AFM image of the sample, white lines are oxidized area. The ring diameter of the outer edge is ~0.7μm. (b) Schematic images of the channel around zero magnetic field (upper) and in the IQHE regime (lower). (c) and (d) Obtained AB oscillation.

## 3. Results and discussion

Figures 1c and 1d show the conductance of the AB ring obtained at zero-bias voltage as a function of the magnetic field around zero magnetic field and in the IQH regime, respectively. Clear AB oscillations are seen in both regimes. The corresponding diameter of the AB ring obtained from the period of the AB oscillation is 0.51 μm at weak magnetic field, and 0.56 μm under higher magnetic field. The difference reflects the formation of the edge channel along the outer side of the ring in the IQH regime, which makes the path of the interferometer longer.

In order to obtain the interference visibility, we calculated the difference between the conductance at a peak ($G_{max}$) and the one at a dip ($G_{min}$) of the zero bias AB oscillations (See Fig. 2a). The visibility ($v$) is defined by $v = (G_{max} - G_{min})/(G_{max} + G_{min})$. Interestingly the bias dependence of the visibility shows a structure that looks similar to the well-known lobe structure observed in EMZI's (Fig. 2b). The oscillation phase is reversed around 300 μV with a decaying oscillation amplitude as the bias voltage increases. This observation is totally unexpected from the conventional single-particle picture, where the interference visibility is insensitive to the bias voltage [2]. We also succeeded in obtaining the lobe-like structure in the ABR in the IQHE regime, as well as in a Fabry-Pérot interferometer (FPI) (Fig. 3b. See Ref. [10] for more detail). Note that all our lobe structures have only one side lobe.

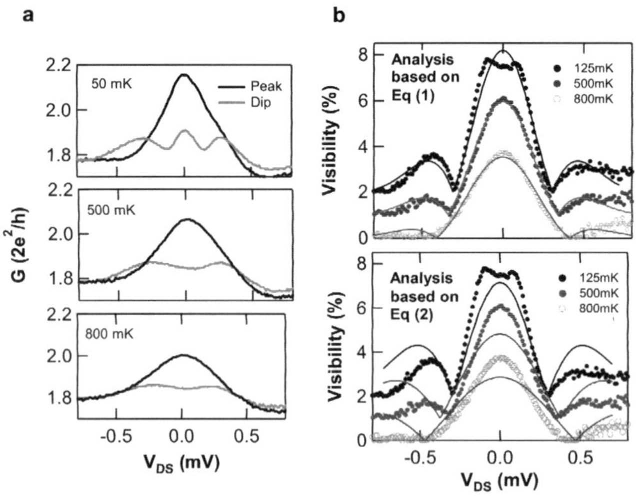


Fig. 2. (a) Differential conductance profile at the peak position of the AB oscillations at zero bias voltage (black line) and at the dip position (gray line). (b) Lobe-like structure at zero magnetic field. The visibility has an exponential dependence on the electron temperature.

Interestingly, the energy scale to characterize these lobe structures is much larger than those observed at EMZI. To evaluate the energy scale quantitatively, the obtained structures are fitted by the following function [4].

Journal of Physics: Conference Series **193** (2009) 012045 doi:10.1088/1742-6596/193/1/012045

$$v = v_0 \left|\cos\left(\frac{\pi eV}{\varepsilon_L}\right)\right| \exp\left(\frac{-(eV)^2}{2\varepsilon_0^{\ 2}}\right) \tag{1}$$

This empirical function is a product of an oscillation term and a relaxation term, and it is dominated by two different energy scales, $\varepsilon_0$ for the dephasing and $\varepsilon_L$ for the phase reversal. Equation (1) is in a good agreement with the experimentally obtained visibility (upper panel of Fig. 2b and Figs 3a, 3b). Although the relation between the two energy scales is still unclear, we found that they are strongly dependent on the sample size [10]. Typically, the energy scales have the relation, $\varepsilon_L, \varepsilon_0 \sim \alpha / L$, where $L$ is the arm length of the interferometer and the value $\alpha$ is around 200 μeV·μm.

It is interesting to compare our result with another type of fitting function proposed in Ref. [3],

$$v = v_0 \left|1 - \frac{(eV)^2}{\varepsilon_G^{\ 2}}\right| \exp\left(\frac{-(eV)^2}{2\varepsilon_G^{\ 2}}\right) \tag{2}$$

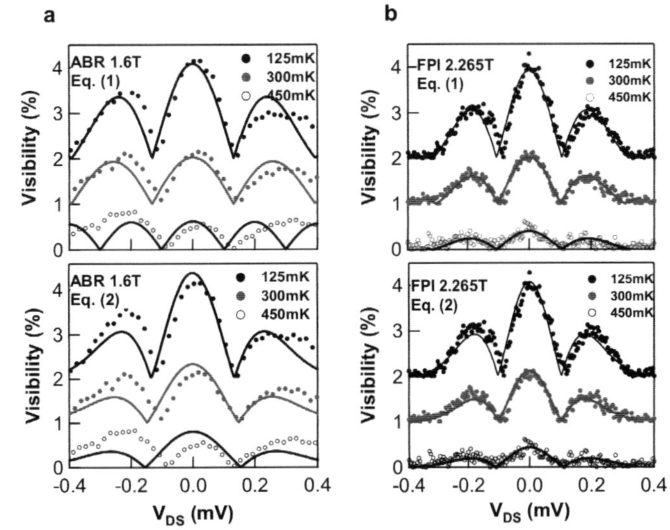


Fig. 3. (a) Fitting result of the lobe-like structures obtained in the AB ring in the IQHE regime (filling factor ~10). (b) Fitting result of lobe structures in the FPI in the IQHE regime (filling factor ~4). Note that in both cases the visibility at the lobe structures has an exponential temperature dependence, and is decaying more rapid compared to the lobe-like structures obtained at zero magnetic field.

which is obtained based on the assumption of "Gaussian phase averaging": the oscillation phase is distributed obeying the Gaussian distribution with the phase variance proportional to the bias voltage. This function is characterized by only one energy scale $\varepsilon_G$. In Ref [3], this function is applied to the lobe structures obtained in the IQHE regime, and it explains well the shape of the lobe structures. In the same way, we tried to apply the function to the lobe-like structures that we observed in the IQHE regime. These results are shown in the lower panels of Fig. 3. These lobe-like structures are explained well by Eq (2) as well as Eq (1). From the fitting, the value $\varepsilon_G$ is around 150 μeV in the AB ring and 100 μeV in the Fabry-Perot interferometer. Thus these results may be compatible with the Gaussian phase averaging as one of a universal and essential factor of the lobe structure under strong magnetic field.

On the other hand, at zero magnetic field, this fitting function does not yield a good agreement with the lobe-like structures (lower panel of Fig. 2b). There may be a mechanism for the dephasing at zero magnetic field which is different from that in the IQHE [10].

## 4. Conclusion

In conclusion, these results imply that the lobe structure can be universally observed in ballistic interferometers, and not only in EMZI. The energy scale of the lobe structure depends strongly on the arm length of the interferometer. This supports that the coulomb interaction between electrons in the interferometer arm are at the origin of the lobe structure. While the model based on Gaussian phase averaging might also be responsible for the lobe structure in the IQH regime, it fails in explaining quantitatively the lobe structure at zero magnetic field.

**Acknowledgement**

We appreciate fruitful comments from P. Roche, E. V. Sukhorukov, H. W. Lee and A. Helzel. This work is supported by KAKENHI, Yamada Science Foundation, and Matsuo Science Foundation.

## References

[1] Neder I, Ofek N, Chung Y, Heiblum M, Mahalu D, Umansky V 2007 *Nature* (London) **448** 333
[2] Neder I, Heiblum M, Levinson Y, Mahalu D, Umansky V 2006 *Phys. Rev. Lett.* **96** 016804
[3] Roulleau P, Portier F, Glattli D C, Roche P, Cavanna A, Faini G, Gennser U, Mailly D 2007 *Phys. Rev.* B **76** 161309
[4] Litvin L V, Helzel A, Tranitz H –P, Wegscheider W, Strunk C 2008 *Phys. Rev.* B **78** 075303
[5] Neder I and Ginossar E 2008 *Phys. Rev. Lett.* **100** 196806
[6] Youn S –C, Lee H –W and Sim H -S 2008 *Phys. Rev. Lett.* **100** 196807
[7] Held R, Vancura T, Heinzel T, Ensslin K, Holland M, Wegscheider W 1998 *Appl. Phys. Lett.* **73** 262
[8] Leturcq R, Bianchetti R, Götz G, Ihn T, Ensslin K, Driscoll D C, Gossard A C 2006 *Physica* E **35** 327
[9] Hashisaka M, Yoshiaki Y, Nakamura S, Kasai S, Ono T, Kobayashi K 2008 *Phys. Rev.* B **78** 241303(R)
[10] Yamauchi Y, Hashisaka M, Nakamura S, Chida K, Kasai S, Ono T, Leturcq R, Ennslin K, Driscoll D C, Gossard A C, Kobayashi K 2009 *Phys. Rev. B* **79** 161306(R)

Journal of Physics: Conference Series **193** (2009) 012046 doi:10.1088/1742-6596/193/1/012046

# Dynamical polarization effect of nuclear spin bath dragged by electron spin resonance in double quantum dot integrated with micro-magnet.

**T. Obata[1], M. Pioro-Ladrière[1,2], Y. Tokura[1,3], R. Brunner[1], Y.-S. Shin[1], T. Kubo[1], K. Yoshida[1], T.Taniyama[4], and S. Tarucha[1,5,6]**

[1]Quantum Spin Information Project, ICORP, Japan Science and Technology Agency, Atsugi-shi, Kanagawa, Japan

[2]Department de Physique, Universit'e de Sherbrooke, Sherbrooke, Quebec, J1K-2R1, Canada

[3]NTT Basic Research Laboratories, NTT Corporation, Atsugi-shi, Kanagawa, Japan

[4]Materials and processes for innovative next-generation devices project, PRESTO, Japan Science and Technology Agency; Materials and Structures Laboratory, Tokyo Institute of Technology 4259 Nagatsuta, Yokohama, Japan

[5]QPEC & Department of Applied Physics, University of Tokyo, Hongo, Bunkyo-ku, Tokyo, Japan

[6]Institute for Nano Quantum Information Electronics, the University of Tokyo, Komaba, Meguro-ku, Tokyo, Japan

E-mail: obata@tarucha.jst.go.jp

**Abstract.** We studied on Overhauser shift of electron dipole spin resonance (EDSR) peaks by using a double quantum dot integrated with a micro-magnet. Two EDSR peaks are well resolved, reflecting electron spin flip events at different resonance conditions between two dots, which depend on the in-plane field at the two dots produced by a micro magnet. One of the two peaks is significantly higher than the other and shows a larger Overhauser shift, indicating that an electron spin flip process local to the dot causes dynamical polarization of local nuclear spins to the same dot. After the nuclear spin polarization is saturated, we observed the decay of the Overhauser shift by repeatedly measuring the EDSR peak with a minimum microwave power. The decay time constant is much longer than by other groups. We discuss the possible reason for this difference.

## 1. Introduction

Quantum information processing by use of quantum dots in GaAs semiconductor is now on the most exciting stage towards the realization of spin QUBIT. In this context, dephasing due to nuclear spin fluctuations is an emerging problem to be sorted out, and intensive studies have been performed to understand and to suppress it [1, 2, 3, 4, 6]. Larmor precession of electron spins becomes inhomogeneous reflecting the statistically fluctuating nuclear field, and this is observed as dephasing. It is possible to partially eliminate the effect of the nuclear spin fluctuation using a spin echo technique [7] but it is more valuable to squeeze the fluctuations and to understand the role of the hyperfine interaction [4]. In order to examine the nuclear spin effect, we measured the Overhauser shift of electric dipole spin resonance (EDSR) for a lateral double quantum dot made in an AlGaAs/GaAs hetero structure and integrated with micro-magnet (Sec 2). The

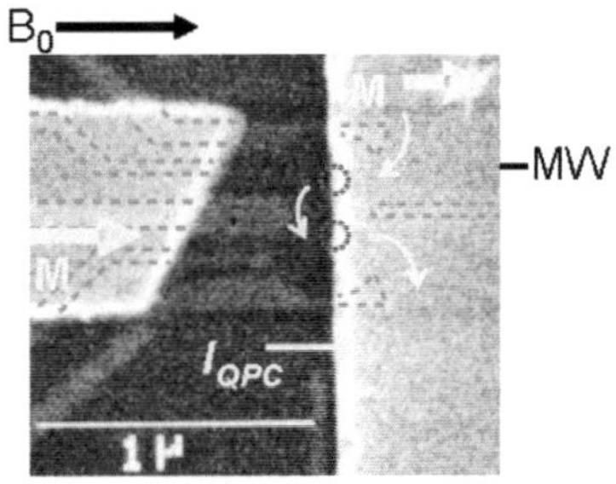


**Figure 1.** SEM photograph of the lateral dot. Two split Cobalt magnets are put on top insulated by EB resist. More detailed explanations, please refer [10]

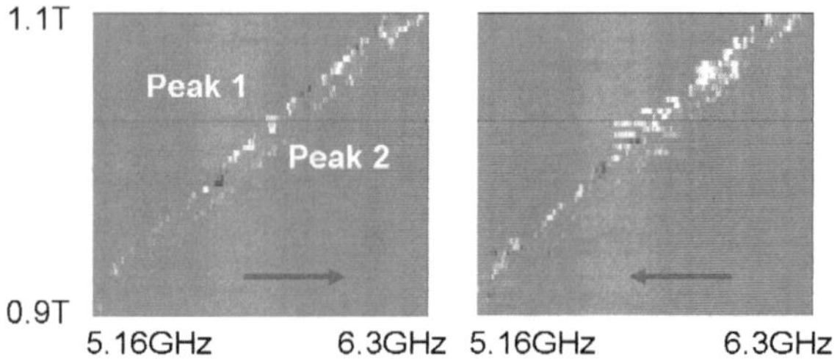


**Figure 2.** The frequency sweep direction dependence of EDSR signals. When we sweep up and down the MW frequency, we measure clear and broadened EDSR signals, respectively.

Overhauser shift is attributed to a dynamic nuclear spin polarization (DNP) effect [5]. We swept down microwave (MW) frequency and up magnetic field to measure the EDSR signals. This way of measurement is efficient to generate DNP [6]. Then we examined the decay of the Overhauser field as a function of wait time(Sec 3). The decay time constant is different. We compared the values with those by other groups (Sec 4).

## 2. Experiment of Dynamical Nuclear Spin polarization

We micro-fabricated a lateral double quantum dot (DQD) structure defined in a two-dimensional electron gas (2DEG) at a AlGaAs/GaAs interface by Schottky gates in Fig. 1(a). The 2DEG is located 100 nm below the surface. Two micro-magnets were placed on the surface with an insulating layer in between. We apply an in-plane magnetic field $B_0$ and then the magnets are magnetized in-plane as well. These magnets generate a static out-of-plane stray slanting field and a static in-plane stray field, both of which are slightly different between the two dots. EDSR is generated by applying a MW electric field to the right magnet to laterally accelerate an electron in each dot with MW frequency $\nu$ equivalent to the Zeeman energy. The Zeeman energy is given by the sum of an external magnetic field and a strong in-plane field, so that addressability to an electron in each dot is due to the in-plane stray field local to the dot [8, 9]. We measured the trans-conductance signal of current $I_{QPC}$ flowing through a nearby quantum point contact (QPC). Figure 2 shows the EDSR results for sweeping up and down MW frequency, repeatedly, at various $B_0$ fields from 0.9 to 1.1T. Two EDSR peaks are well resolved, and we label the peak at the lower $B_0$ as peak 1 and the other as peak 2, respectively. Both peaks fall on the straight lines with the same slope, from which we derived the Landè g-factor of 0.39 ± 0.1.

When comparing the two panels in Fig. 2, we find that the peak is much wider in the right, i.e. for the down sweep of MW frequency. This hysteretic behavior is assigned to DNP [13, 15, 6], which accompanies Overhauser shift. The local field $B_{local}$ indeed by the polarization of nuclear spins is given as $B_{local} = B_0 - A_n P$, where $P = N_\uparrow/(N_\uparrow + N_\downarrow) - 1/2$ is the polarization of nuclear spin bath and $A_n = 5T$ is the coupling contact of hyperfine interaction. The EDSR condition is finally given by $\mathbf{g}\mu_e B_{local} = h\nu$.

When sweeping down $\nu$ or sweeping up $B_0$ in Fig. 3(a), the polarization $P$ or DNP increases so as to sustain the EDSR condition. We measured the MW power dependence of DNP. By sweeping up $B_0$ the effect of DNP on EDSR becomes stronger with increasing MW power as shown in Fig. 3(a). Each $B_0$ field sweep from 0.98 to 1.08T took 10 minutes and it was repeated as increasing the MW power in a step of 0.2 dB. The MW frequency was set at 5.6 GHz.

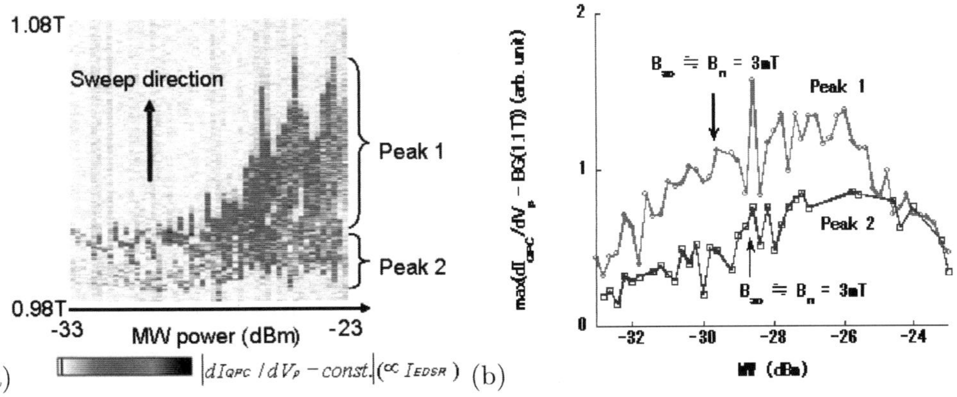


**Figure 3.** The power dependence of DNP. (a) The power dependence of Overhauser shift. (b) The power dependence of peak heights. As increasing the MW power, the peak height is getting higher saturates (b). Once the peak saturates, it starts to drag the peak (a). An interesting contrast between the Overhauser shift and the peak height is clearly measured. The base line is adjusted to 0 to avoid the effects by heating and photon assisted tunneling.

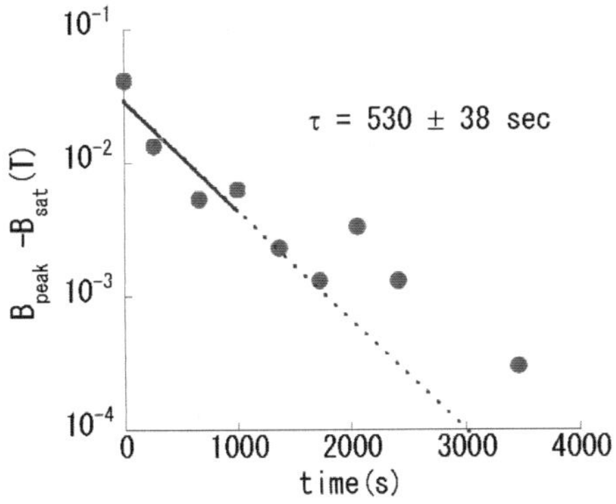


**Figure 4.** The time evolution of the reduced peak position from the saturation field $B_{sat}$ of 1.009 T. The peak position shows exponential decay [2, 6]. An exponential fitting function is adopted to the data points from 40 mT to 4 mT by taking account of an experimental error.

Two peaks were initially observed at 1.0, and 0.99 T, respectively, and suffered from DNP more strongly as the MW power was increased (note that peak 2 shows small amplitude of DNP). The DNP effect is significantly larger for peak 1 than for peak 2. We plot the maximum height of the two peaks in Fig. 3(b). Both peaks become gradually large, saturated, and finally small. In Fig. 3(b) the peak height starts to increase with increasing MW power exceeding - 30 dBm, and more or less simultaneously DNP starts to increase to apparently induces the Overhauser shift as seen in Fig. 3(a). From the critical MW power of -30 dBm we calculate the MW magnetic field of a few mT at the dot. This field is comparable to the amplitude of nuclear field fluctuation [8, 12]. This sounds reasonable because DNP can only be visible if the MW magnetic field overcomes the fluctuating Overhauser field. On the other hand, as the MW power is further increased exceeding -25 dBm, the EDSR peak height starts to decrease. This is probably because photon assisted tunneling between quantum dots and reservoirs or heating degrades the EDSR signal.

## 3. Decay of Overhauser shift and possible mechanism.

After observation of the largest Overhauser shift, we repeatedly measured EDSR in the down sweep of $B_0$ with the lowest MW power of -33 dBm. This power is weak enough to neglect the DNP effect. Each $B_0$ sweep took 6 minutes. Because the Overhauser field should decay during the $B_0$ sweep, we calibrated the elapse time before the EDSR peak was observed by taking into account the sweep rate and the peak position. The elapse time is counted from the time when the last peak of the polarization was measured in Fig. 3(a). We plotted the time evolution of the peak 1 position, $B_{peak}$ in Fig. 4 (The other peak has no resolution for the time evolution. ). The origin of the measurement is taken at the time when the last peak point on the top right corner in Fig. 3 (a). $B_{peak}$ decays with a time constant of 530 sec, which reflects nuclear spin dynamics or the relaxation time of nuclear spin polarization [11]. Decay time of Overhauser field previously reported is $\sim$ 10 seconds [2, 6], much shorter than the present value. In addition, typical Overhauser shifts in the reports [2, 6] are from 10 to 20 mT and a few times smaller than our results.

There are several reports about the decay time of nuclear spin polarization in GaAs systems, but the values are sometimes different. A short decay time of order of 10 seconds is reported in Ref. [2] and [6], while a decay time longer than a minute and the other is reported in Refs. [13] and [14]. There is also a report on observation of a fast decay followed by a slow decay. We consider that we only observed the slow decay, although the reason is not very clear.

The possible reason of the difference is due to the fact that nuclear spins are only polarized in one of the two dots, whereas nuclear spins in both dots are polarized in previous reports [2, 6]. For example in Ref. [2], nuclear spins are polarized through a flip-flop process between electron and nuclear spins when electrons transit the double dot thereby affecting nuclear spins over the two dots. In Ref [6] an Overhauser shift measured in ESR is explained by considering that both electron spins are sometimes flipped in two dots. Then the nuclear spin bath should be polarized differently between two dots, and it can accelerate the decay of nuclear spin polarization. In order to examine the decay mechanism in detail it is necessary to tune symmetry or asymmetry of nuclear spin polarization between two dots. For this purpose we have to modulate the magnetic field quickly over the range of EDSR for both dots. So we are developing a faster magnetic field handling system to polarize nuclear spins in both dots almost simultaneously.

## 4. Conclusion

We measured the Overhauser shift of EDSR in QD. The higher peak exhibit larger Overhauser shift than the other. The relaxation time of Overhauser field is so long as $\sim$ min but recently is reported much shorter. We compare our result with those previously reported. It is important to study carefully in detail to understand the DNP mechanism and the origin of the difference of relaxation time constant. Further studies are needed.

## Acknowledgments

The authors thank L.M. Vandersypen for fruitful discussions.

## References

[1] D. Klauser, W. A. Coish, and D. Loss, Phys. Rev. B **73**, 205302 (2006).
[2] D. J. Reilly *et al.*, Science **321**, 781 (2008).
[3] J. Danon and Y. V. Nazarov, Phys. Rev. Lett. **100**, 056603 (2008).
[4] M.S. Rudner and L.S. Levitov, Phys. Rev. Lett. **99** 036602 (2007).
[5] A. Abragam, Principles of Nuclear Magnetism, (Oxford University Press, 1961).
[6] I. T. Vink *et al.*, Nature phys on-line **16** August 2009, nphys1366.
[7] F.H.Koppens, K.J. Nowack, and L.M. Vandersypen ,Phys. Rev. Lett. **100** (2008) 236802.
[8] M. Pioro-Ladriere *et al.*, Nature Physics **4** 776 (2008).
[9] Y. Tokura et al., Phys. Rev. Lett. **96** 047202 (2006), M. Pioro-Ladriere et al., App. Phys. Lett. **90** 024105 (2007).
[10] M. Pioro-Ladriere et al., in preparation.
[11] N. Kumada *et al.* Phys. Rev. Lett **94**, 096802 (2005).
[12] E.I. Rashba, Phys. Rev. B **78**, 195302 (2008).
[13] K. Ono and S. Tarucha, Phys. Rev. Lett. **92**, 256803 (2004).
[14] S. Teraoka *et al.*, Physica E **21**, 928 (2004).
[15] J. Baugh *et al.*, Phys. Rev. Lett. **99**, 096804 (2007).

Journal of Physics: Conference Series **193** (2009) 012047 doi:10.1088/1742-6596/193/1/012047

# Perturbative vs Non-Perturbative impurity scattering in a narrow Si Nanowire GAA transistor: A NEGF study

**Antonio Martinez, Andrew R Brown, Natalia Seoane and Asen Asenov**

Dept. Electronics and Electrical Engineering, University of Glasgow, G12 8LT

E-mail: antonio@elec.gla.ac.uk

**Abstract**. In this paper we study the effect of impurity scattering on the performance of a Si gate-all-around nanowire transistors. The non-equilibrium Green function formalism is used in order to describe the carrier transport. Impurity scattering is introduced using two different formalisms, one that considers the impurity potential as a small perturbation by introducing self energies and the other in which the impurity potential is described exactly by included the impurity potential through the Poisson equation. The non-perturbative method does not use a fitting parameter but the perturbative one uses a phenomenological constant that can be calibrated to match the result of the non-perturbative method. We confirm Ohms-law-type behaviour by using the perturbative approach for larger channel lengths.

## 1. Introduction

The revolution in device miniaturisation has led to nanometre-dimension transistors. Short channel effects plague the bulk MOSFET architecture, which is at the core of the current Silicon VLSI technology. These effects degrade the transistor performance, making it impossible to continue the scaling of devices. SOI, and 3D architectures such as FinFETs and nanowires, offer a realistic alternative solution. These devices are not affected by short channel effects and possess a better electrostatic integrity compared to bulk MOSFET architectures. At nanometre dimensions, quantum mechanical effects play an important role; volume inversion, threshold voltage shift and increased leakage are some of the consequences at the device level of the underlying quantum rules. Those effects are linked to the wave structure of the electron, which appears when the dimensions of the system became of the order as the de Broglie wavelength. The interaction of the electron with the device becomes more like a wave interaction or scattering with the electrostatic potential.

Groups of dopants in enclosed domains can not be considered as continuous doping when the average number of them in the device is on the order of a hundred or less, and when their location has a direct impact on the $I_D$-$V_G$ characteristics. Work **[1]** on the description of random discrete dopants has been carried out using different levels of sophistication and approaches. Classical and semi-classical simulation techniques have been deployed from the micrometre scale to tens-of-nanometres. Recently, non-perturbative quantum mechanical simulations of nanowire transistors **[2, 3, 4]** based on the Non-Equilibrium Green's function (NEGF) formalism have been carried out. The discrete dopants are introduced as an electronic charge associated to a volume. The advantage of this technique is the straightforward incorporation of decoherence, which could be used to treat all types of perturbative scattering mechanisms such as phonon and impurity scattering. Recently **[5]** a perturbative approach using the self-consistent Born approximation has been used in order to estimate the effect of impurity scattering in a Si nanowire transistor. This model uses an empirical parameter. In this work we calibrated this parameter from our non-perturbative calculations. In section 2 we describe the different approaches used to treat scattering using the NEGF formalism. The simulated results for a thin Si nanowire transistor are presented in section 3. The last section reviews the conclusions of this work.

## 2. Impurity scattering in the NEGF formalism

The NEGF formalism was developed to treat many particle systems under external field at finite temperature. In the sixties, with the success of the perturbative quantum field theory based on the propagators or Green function, Schwinger [**6**] wrote a set of functional equations for the full Green function at zero temperature. These equations could be solved with the use of approximations and partial summation of terms. The extension of this formalism to finite temperature and non-equilibrium problems was pioneer by Schwinger, Kadanov, Baym and Keldysh [**7, 8, 9**]. The single particle Green function allows the calculation of the macroscopic physical magnitudes of interest such as the electron and current density without the hazard of calculating the many-particle wave function. In the finite temperature case the Green function becomes a 2x2 matrix [**7, 9**] where the four elements split into two types: two that have pure dynamical information (the *retarded* $G^R$ and the *advanced* $G^A$ Green function), and two that have statistical information (the *less-than* $G^<$ and the *greater-than* $G^>$ green function). All are called Green functions but the former two are the only Green functions from a mathematical point-of-view. The steady state versions of the equations for $G^R$ and $G^<$ are:

$$\left(E - T - U - \Sigma_C^R(E) - \Sigma_S^R(G^<, G^>)\right)G^R(E) = I \tag{1}$$

$$G^<(E) = G^R(E)\left(\Sigma_C^<(E) + \Sigma_S^<(G^<)\right)G^A(E) \tag{2}$$

where $E$, $T$ and $U$ are the total, kinetic and potential energies of the carrier and the retarded self energies for the contact and scattering are denoted by $\Sigma_C^R$, $\Sigma_S^R$ respectively (the structure of $\Sigma_S^R$ includes the Pauli exclusion principle [**8**]) $\Sigma^<$ stands for the *less-than* self energy. Eqs 1 and 2 are matrix equations in the spatial indices, which are implicit. The $I$ on the right hand side of Eq. 1 is the identity matrix, which indicates that the $G^R$ is truly a Green function in the mathematical sense. Eqs 1 and 2 are solved for the relevant energy interval, which usually consist in 600 to 1000 energy points. When scattering is considered the equations became a coupled set of non-linear matrix equations, which are very computational demanding. The non-linearity appears in the dependence on $G^<$ and $G^>$ of the self-energies. The electron and current densities are calculated by the diagonal and the first off-diagonal block of $G^<$. A recursive algorithm [**10**] is deployed in order to calculate just the diagonal and off-diagonal elements, dramatically reducing the computation time. Eqs. 1 and 2 are supplemented with boundary conditions. In the case of Eq. 1, open boundary conditions are imposed through the contact self-energies. Note that these equations are now matrix equations in the spatial indices (not shown), therefore matrix $G^R$ could be solved by inversion. This boundary condition allows injection of electrons into the system and makes the contacts less reflective. The boundary condition for $G^<$ or $G^>$ at the contact is called the equilibrium condition or KMS (Kubo-Martin-Schwinger) [**11**] which guarantees that, in the contacts, $G^<$ and $G^>$ reduce to the corresponding equilibrium values i.e $G^< = Imaginary(G^R)\, f_C(E)$, where $f_C(E)$ is the Fermi-Dirac distribution function at the corresponding contact. This means that the electron density in the reservoirs is the local density-of-states multiplied by the equilibrium probability (Fermi-Dirac) of finding an electron there. In the presence of decoherent processes such as phonon and impurity scattering, which are accounted for through self-energies, the energy distribution of an electron inside the simulated region is no longer a simple addition of the Fermi-Dirac distributions of the contacts, and we are in the presence of a genuinely non-equilibrium distribution like those considered in high-field Monte Carlo simulations. One advantage of considering the impurity scattering as self-energies is that it can account for the broadening in the density states, which is not possible in the non-perturbative case. In the perturbative approach we follow the approach of [**5**], in which the *less-than* self-energy is written as:

$$\Sigma^{<}_{imp}(E) = C\, G^{<}(E) \tag{3}$$

where $C$ is a constant which depends on the doping concentration in the source and drain and the strength of the interaction, a formal description of which can be found in [**5**]. In this work the $C$ parameter is varied to produce the non-perturbative results, which have no fitting parameter. The impurity scattering is assumed to be elastic (energy conserved). In the non-perturbative case the impurities are introduced as a charge homogeneously distributed in a discretisation volume ($0.2\times0.2\times0.2$ nm$^3$). It is important to mention that the static dielectric constant is approximately valid at a distance larger than two Angstroms away from the impurity location.

## 3. Simulation results

A very narrow gate-all-around (GAA) nanowire with $2.2\times2.2$ nm$^2$ cross-section and a channel length of 6nm has been used in the simulation. The source/drain doping is $10^{20}$ cm$^{-3}$ and the channel is undoped. First we have simulated the impurity scattering through a non-perturbative approach, i.e. we have introduced the impurity scattering through discrete dopants, which are randomly distributed through the source/drain (S/D) extension regions as is shown in Fig. 1.

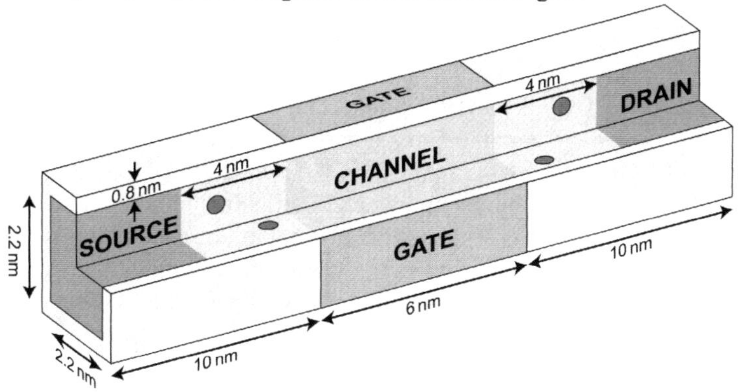


**Figure 1.** Schematic of the simulated nanowire.

Thirty nanowire transistors, which differ in the random locations of the discrete dopants, have been simulated at a low drain bias (1mV) in order to ensure the device is working in the linear response region. Fig. 2 shows the $I_D$-$V_G$ characteristics of the simulated transistors, and the corresponding average characteristic. At high gate bias the effective resistance of the source/drain extensions have been extracted by using the ballistic $I_D$-$V_G$ device characteristic. This resistance is $1.68\times10^4$ Ω.

The perturbative approach [**5**], which includes density-of-states broadening and decoherence has also been used to study the effect of impurity scattering. This model has been applied to include scattering through self-energies in 4nm regions of the source and drain. The $I_D$-$V_G$ characteristic of the transistor with incoherent scattering is shown in Fig. 3 as well as the curves for the atomistic average and ballistic transistor.

The $I_D$-$V_G$ of the transistor with incoherent scattering (short S/D) gives similar results to the atomistic average $I_D$-$V_G$. The perturbative approach is able to reproduce the atomistic average $I_D$-$V_G$ characteristic by varying the $C$ parameter at all the gate voltages. We have also calculated the $I_D$-$V_G$ characteristic for a device in which scattering is included in a 10 nm region of the source and drain (long S/D) using the calibrated $C$ parameter. The on-current for this device reduces to 1/3 of the ballistic current, compared to 2/3 for the shorter channel, as can be seen in Fig. 3. This should be expected if a linear dependence of the resistance as function of the wire length is assumed. For the

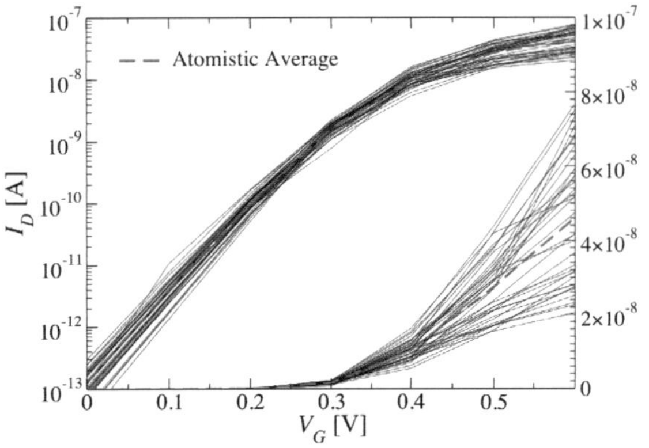


**Figure 2.** $I_D$-$V_G$ curves of the simulated device with random discrete dopants in the source/drain.

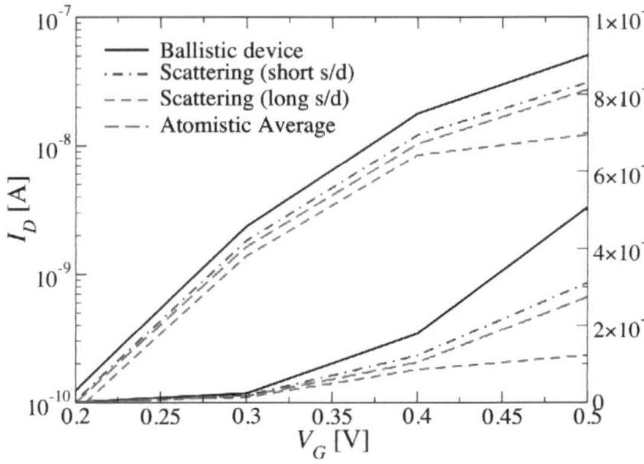


**Figure 3.** $I_D$-$V_G$ curves of perturbative and non-perturbative (atomistic) devices

large S/D device, the scattering reduces the on-current to 30 % of the current in the smooth device. Note that the effect is mainly in the on-state.

## 4. Conclusions

In this work we have employed the NEGF formalism to simulate the impurity scattering for a GAA Si nanowire transistor using two very different models. One model is non-perturbative (atomistic) and introduces the impurity scattering through discrete charges and the other is perturbative and introduces the impurity scattering through the imaginary part of the self-energy. From the atomistic simulation, by averaging all the $I_D$-$V_G$ curves we have extracted the series resistance associated with the impurity scattering. The perturbative method has been calibrated to the atomistic method and used to simulated impurity scattering in a large S/D region, producing Ohmic-type behaviour. The perturbative method can be used to include the series resistances in the heavily doped source/drain regions using a mode-space approach, which is less computationally expensive than the fully-3D real-space approach.

## References

[1] A. Asenov, S. Roy, A. R. Brown, G. Roy, C. Alexander, C. Riddet, C. Millar, B. Cheng, A. Martinez, N. Seoane, D. Reid, M. F. Bukhori, X. Wang and U. Kovac, *International Electron Devices Meeting (IEDM)* Technical Digest, p.421 (2008)

[2] A. Martinez , M. Bescond, J. R. Barker, A. Svizhenko, M. P. Anantram, C. Millar and A. Asenov, *IEEE Trans. Electron Dev.* Vol. 54, p.2213 (2007).

[3] A. Martinez, N. Seoane, A. R. Brown, J. R. Barker and A. Asenov , *IEEE Trans. Nanotechnol, in press*

[4] N. Seoane, A. Martinez, A. R. Brown, J. R. Barker and A Asenov, *IEEE Trans. Electron Dev.*, Vol.56, No.7, p.1388 (2009)

[5] J. H. Oh, D. Ahn, Y.S. Yu, and S.W. Hwang, *Phys. Rev. B.* 77, 035313 (2008)

[6] J. Schwinger, *Proc. Natl. Acad. Sci. USA,* Vol.37, pp.452–455 (1951), *Proc. Natl. Acad. Sci. USA,* Vol.37, pp.455–460 (1951)

[7] J. Schwinger, *J.Math.Phys.*, Vol. 2, No. 3, p.407 (1961)

[8] L. P. Kadanoff and G. Baym, *Quantum Statistical Mechanics: Green's Function Methods in Equilibrium and Non-Equilibrium Problems.* New York: W.A. Benjamin (1962)

[9] L. V. Keldysh, *Soviet Phys. JETP*, Vol. 20, No. 4, p.1018 (1965)

[10] A. Svizhenko, M. P. Anantram, T. R. Govindam, B. Biegel, and R. Venugopal, *J. Appl. Phys.*, Vol.91, p.2343, (2002)

[11] P. C. Martin and J. Schwinger, *Phys. Rev.* **115,** 1342–1373 (1959)

Journal of Physics: Conference Series **193** (2009) 012048 doi:10.1088/1742-6596/193/1/012048

# Multiple microwave resonance interaction in silicon single-electron transistors

**Morteza Erfani[1], David G Hasko[1], Wan Sik Cho[2], Jung-Bum Choi[2]**

[1]Cavendish Laboratory, University of Cambridge, J. J. Thomson Avenue, CB3 0HE Cambridge, United Kingdom

[2]Department of Physics and Institute for NanoScience and Technology, Chungbuk National University, Cheongju, 361-763, South Korea

E-mail: `me287@cam.ac.uk`

**Abstract.** Silicon single-electron transistors are promising nanodevices for implementation of quantum logic gates. Single frequency continuous wave microwave spectroscopy, carried out under cryogenic conditions on dc biased degenerately doped transistors have previously shown many high quality factor resonances in the drain current modulation, indicating the presence of long-lived excitations. However, such measurements are unable to indicate coupled behaviour between the systems undergoing resonance. Coupled behaviour is a necessary precursor for quantum gate operation. In this work, we investigate the response of devices to two frequency continuous wave microwave spectroscopy, which is a potential method for induced coupling.

## 1. Introduction

It is well-known that the source-drain current of a dc biased degenerately doped single electron transistor (SET) can be influenced by microwave irradiation [1]. Continuous wave (CW) microwave spectroscopy at liquid helium temperature shows a large number of resonances with quality factors (Q-values) up to $\sim 10^5$, which are thought to result from resonantly driven electrons tunnelling between sites where the electron is localised by the donor distribution. Equivalent experiments on conventional silicon MOSFETs do not show these large number of high Q-value resonances [2, 3]. The displacement of the tunnelling electron in the silicon SET modifies the electrostatic environment inside the SET island which changes the magnitude of the tunnel barriers to the source and drain leads. This change in the tunnel barrier transmission results in an increase or decrease of the SET current at resonance, despite the fixed bias condition. A typical high quality factor of $10^5$ at 1 GHz indicates an energy relaxation time of 10 $\mu$s, suggesting that this driven electron tunnelling mechanism is well isolated from the environment. The quantum mechanical nature of these resonances is indicated by their very large number ($\gg 10^4$ in the range of 1 - 10 GHz) and their very high Q-values (see figure 1).

## 2. Single frequency microwave spectroscopy

In the CW microwave spectroscopy, the resonance frequency of a peak gives the energy splitting for the electron occupying the donor sites involved in the tunneling event. Also the Q-value of each resonance indicates the relaxation time in the absence of inhomogeneous broadening and characterises the energy loss of the system. To investigate the coherent dynamics of a resonance,

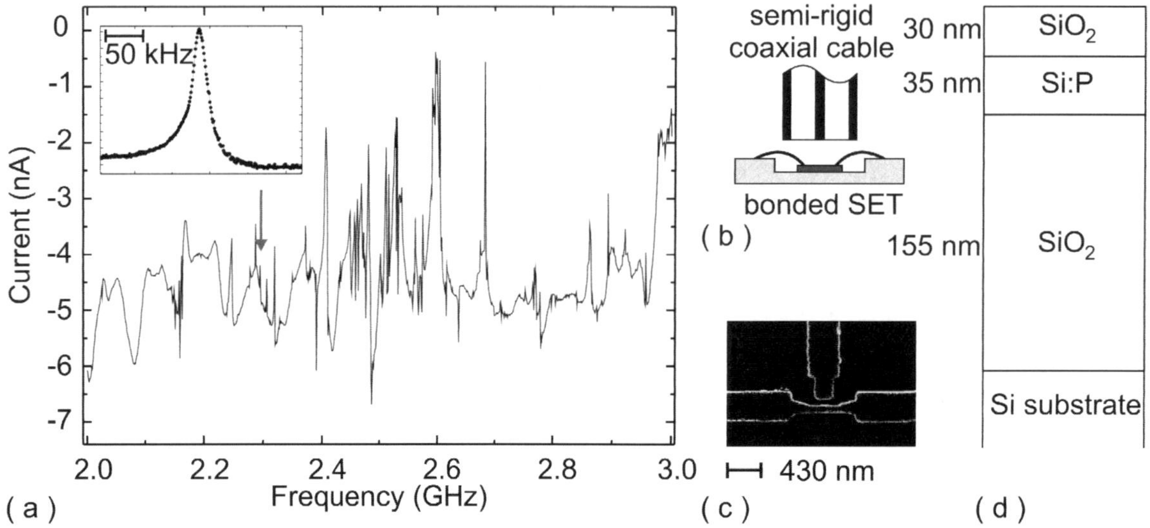


**Figure 1.** (a)Wideband monochromatic CW microwave spectroscopy showing the large number of resonances in the range from 2 - 3 GHz. The inset shows a resonance (indicated by the arrow) with the Q-value of $\sim 10^5$ in a view expanded by 1000 times.(b) Cross-sectional schematic view of the microwave coupling scheme. (c) SEM image of a typical device. (d) Schematic cross-section of the material structure.

pulsed excitation of the system is needed. Rabi oscillations have previously been observed in CW-pulsed spectroscopy, giving a timescale for the rate and duration of coherent processes [4]. A Rabi frequency ($\Omega$) of $\sim$ 1 MHz was observed for resonances in the frequency range ($\omega$) of a few GHz, so that the condition $\hbar\Omega \ll \hbar\omega$ ensures suppression of the fluctuation of electron occupation numbers for the localised sites. Each resonance can be treated as a separate qubit in a Hilbert space characterised by a spatial Rabi oscillation of an electron between the pair of localised states. To extract qubit information, necessary for a computational process, the change in SET current must be detected in response to one or more additional microwave pulses.

## 3. Two frequency microwave spectroscopy

In practice, quantum information processing requires controlled quantum coherent coupling between qubits to enable gate operation [5]. Coupling may be demonstrated by 2-D CW microwave spectroscopy. Previously, both direct and indirect methods for the coupling of microwave radiation to a nanodevice have been used in 1-D CW and pulsed-CW microwave spectroscopy. In this work, microwave signals were supplied successively by these two techniques to the same device. It was shown that resonances mostly appeared at the same frequency, but were modified in shape and amplitude. The indirect microwave coupling method uses a polished end of a semi-rigid coaxial cable, terminated ~0.5 mm away from the nano-device to provide impedance matching. The directly coupled microwave signal is capacitively connected to the source-drain leads of the SET. This is more effcient than the indirect method but leads to a greater electrical noise in the nano-device. The strength of coupling between the microwave field and the system undergoing spatial Rabi oscillations depends on their relative direction. The electric field directions in the nano-device due to the directly coupled and indirectly coupled microwave signals are mutually perpendicular. Therefore, the same CW spectroscopy measurement may look different in the two coupling methods.

We report on experimental measurements at a temperature of 4 K on a SET with a doping

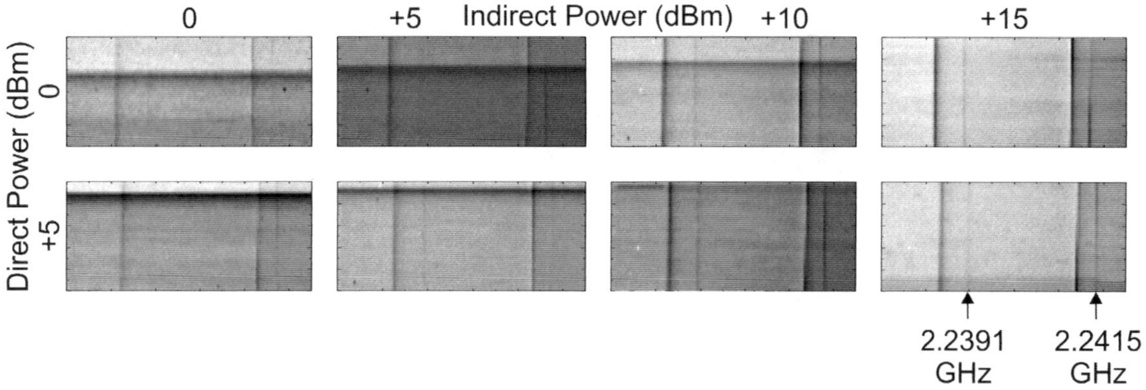


**Figure 2.** Two frequency spectroscopy for an indirectly coupled microwave signal (2.2376 - 2.2420 GHz horizontal axis) and a directly coupled microwave signal (2.2406 - 2.2411 GHz vertical axis). The position and amplitude of resonances depend on the power of the microwave signal. Amplitude of the resonances indicated by the two arrows vary dramatically with different power combinations.

concentration of $N_D = 3 \times 10^{19}$ cm$^{-3}$. In a frequency range of 4.4 MHz centred at 2.24 GHz, there are 4 high Q-value resonances. An indirectly coupled microwave signal was swept over this range. Another fixed frequency signal was directly coupled and stepped for each sweep of the indirectly coupled microwave signal over part of the same frequency range. The two sources were independent ie phase locking was not used. The same measurement was repeated for different power outputs on the direct and indirect coupling. The results for this measurement are shown in the grey scale map in figure 2. Displacement of electrons within the SET, say due to a change in the gate voltage, is known to cause small shifts in the resonant frequencies. This is seen as a slope on the trajectory of a resonant feature with respect to the swept x-axis (the indirectly coupled signal). The trajectories show different characteristic behaviours in different powers. The peak amplitude associated to each trajectory alters with power. This effect is so strong on two of the resonances such that they have too low amplitudes to be detected at certain power combinations. The centre frequencies of these resonances are at around 2.2391 and 2.2415 GHz which are indicated by two arrows in figure 2. Also, the slope of the trajectory varies with power. At certain power combinations some resonances show anticrossing behaviour, most noticeably at the centre of the resonance is shown in figure 3 in an expanded view. A sharp resonance is also detectable on the vertical axis. The centre frequency of this resonance shows a strong dependency on the combination of the two power outputs. It starts from 2.2409 GHz at 0 dBm for both signals and increases by increasing the indirect power up to +10 dBm. It then passes out of the frequency range at +15 dBm. This peak appears back on the frequency range by increasing the direct power to +5 dBm and again shifts upwards with increasing indirect power.

The dependency of a resonance frequency both on power and frequency of a second microwave signal, indicates an energy exchange mechanism inside the nano-device caused by the two frequencies. This can be a useful phenomenon for the purpose of conditional gate operation, as the resonance condition of the excitation depends on a second signal. In addition, the anticrossing behaviour observed under certain conditions suggests a coherent interaction between two systems undergoing simple harmonic motion. These two systems must be located within the transistor structure and can be manipulated rapidly in the time domain by the application of pulsed microwave signals at frequencies designed to select specific resonances. Such an approach to selective manipulations can be faster and more highly selective than one based on changing gate voltages.

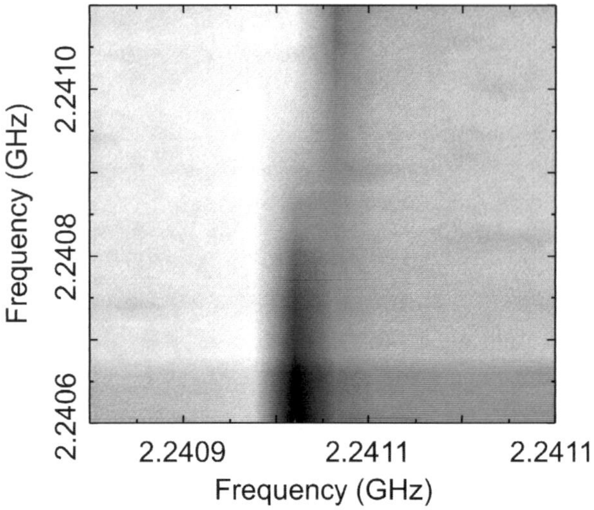


**Figure 3.** Two frequency spectroscopy for an indirectly coupled microwave signal (+15 dBm horizontal axis) and a directly coupled microwave signal (+5 dBm vertical axis) showing an anticrossing behaviour.

## 4. Conclusion and future work

In previous works, degenerately doped silicon SETs have been widely investigated as the basis for single electron memory and classical logic applications [6–10]. For these purposes, the quantization of the electronic charge and the ability to sense and manipulate charge at the single electron level, were the main technology requirements. However, the limited transconductance and power gain, as well as the lack of control of the offset charge were found to be significant limitations. By contrast, in a multi-qubit system based on the large number of microwave resonances in a single SET, the transconductance and power gain are unimportant. In a practical computer device, more than one microwave signal must be applied at the same time (or at least in close temporal proximity) and coupling between qubits must occur (preferably on demand).

## Acknowledgments

This work was supported by the Korea Science and Technology Foundation through the Global Partnership Research Program.

## References

[1] Creswell L A, Hasko D G and Williams D A 2009 *J. Appl. Phys.* **105** 104508

[2] Xiao M, Martin I, Yablonovitch E and Jiang H W 2004 *Nature* **430** 435

[3] Prati E, Fanciulli M, Calderoni A, Ferrari G and Sampietro M 2008 *J. Appl. Phys.* **103** 104502

[4] Creswell L A, Hasko D G and Williams D A 2007 *Microelectron. Eng.* **84** 1614

[5] Nielsen M A and Chuang I L 2000 *Quantum Computation and Quantum Information* (Cambridge: Cambridge University Press) p 676

[6] Gorman J, Hasko D G and Williams D A 2005 *Phys. Rev. Lett.* **95** 090502

[7] Hasko D G *et al.* 2008 *Appl. Phys. Lett.* **93** 192116

[8] Stone N J and Ahmed H 2000 *Appl. Phys. Lett.* **77** 744

[9] Irvine A C, Durrani Z A K and Ahmed H 2000 *J. Appl. Phys.* **87** 8594

[10] Stone N J and Ahmed H 1998 *Appl. Phys. Lett.* **73** 2134

# An user-friendly software tool for the solution of the time-dependent Schrödinger and Gross-Pitaevskii equations

**Thomas Serafini and Andrea Bertoni**
S3 National Research Center, INFM-CNR, 41125 Modena, Italy
E-mail: andrea.bertoni@unimore.it

**Abstract.** In this work we present TDStool, a general-purpose easy-to-use software tool for the solution of the time-dependent Schrödinger equation in 2D and 3D domains with arbitrary time-dependent potentials. The numerical algorithms adopted in the code, namely Fourier split-step and box-integration methods, are sketched and the main characteristics of the tool are illustrated. As an example, the dynamics of a single electron in systems of two and three coupled quantum dots is obtained. The code is released as an open-source project and has a build-in graphical interface for the visualization of the results.

## 1. Introduction

The astonishing development in semiconductor growth, characterization and processing technologies opened new horizons in the field of carriers control, up to the single electron level[1]. In fact, single-particle quantum interference is routinely achieved in advanced research laboratories and single electrons are not only confined, but also moved by varying electromagnetic fields[2]. The theoretical modeling of such systems is usually tackled with numerical instruments developed on a case-by-case basis. Although this allows for a great specificity, the development work is very expensive.

Here we present a novel software tool for the numerical solution of the time-dependent (TD) Schrödinger equation. We named it TDStool and released it as an open-source project (binaries are available for Windows and Linux systems), free for non-commercial use[3]. The main characteristics of TDStool are: 2D and 3D solution domain; arbitrary TD electric potential landscape; uniform magnetic field; nonuniform (x-y-z separable) discretization grid; clean graphical user interface for input setup and results visualization; comprehensive documentation. Two different numerical algorithms are available, namely Fourier split-step and box-integration method.

In addition to the linear Schrödinger case, the nonlinear system dynamics determined by TD Gross- Pitaevskii equation[4] can be simulated. In fact, many recent experimental results on the manipulation of quantum ultracold condensates exposed the quantum coherence of their dynamics[5] and confirmed the extra potential-like term $g|\psi(\mathbf{r})|^2$ in their equation of motion, proportional to the local particle density.

In the next section we sketch the numerical algorithms adopted in the code. Then, in the remaining two sections, we present two examples of applications, namely the single-particle dynamics of a carrier confined in a system of three coupled GaAs-based semiconductor quantum dots, and the quantum evolution of the probability density initially in one dot of a two-dot system, and subject to the TD piezoelectric potential of a surface acoustic wave (SAW).

## 2. Numerical methods

For the linear TD Schrödinger equation

$$i\frac{\partial}{\partial t}\psi(\mathbf{r},t) = [-\frac{1}{2}\nabla^2 + U(\mathbf{r},t)]\psi(\mathbf{r},t) \tag{1}$$

we allow a nonuniform but x-y-z separable grid and use the *box integration method*[6]. Here, for the sake of brevity, we neglect the magnetic field term and take $\hbar = 1, m = 1$. The equation is integrated on each box $C$ of the domain. The kinetic term represents the only non trivial part of the derivation, since it is the only one not diagonal in the real-space representation. By using the Gauss theorem, we can write $\int_C \nabla^2\psi\, ds = \oint_{\partial C} \nabla\psi\bar{\mathbf{n}}\, dl$. The latter term, containing the derivatives of $\psi$ on the box edges ($\mathbf{n}$ is the versor orthogonal to them), can be easily obtained through a finite-difference approximation. Finally, a Cranck-Nicholson scheme is used for the time derivative and the resulting sparse linear system is solved through a specific library routine[7].

In order to solve the TD Gross-Pitaevskii equation

$$\frac{\partial}{\partial t}\psi(\mathbf{r},t) = i[\frac{1}{2}\nabla^2 - U(\mathbf{r},t) - g|\psi(\mathbf{r},t)|^2]\psi(\mathbf{r},t) \tag{2}$$

(with a coupling constant $g$), or the Schrödinger equation on a uniform grid, we use a *split-step Fourier method*[8]. The Hamiltonian is split in a linear part $\mathcal{L} = \frac{i}{2}\nabla^2$ and a nonlinear part $\mathcal{N} = -i\,(U + g|\psi|^2)$, and the evolution operator $\mathcal{U}$ is decomposed by using a second-order approximation according

$$\mathcal{U}(t, t+dt) \approx e^{\frac{1}{2}\mathcal{N}(t+\frac{dt}{2})dt} e^{\mathcal{L}dt} e^{\frac{1}{2}\mathcal{N}(t+\frac{dt}{2})dt} + O(dt^3)\,. \tag{3}$$

It is easy to see that the wave function evolution obtained with the above approximation is equivalent to first perform a step of $\frac{dt}{2}$ with the nonlinear operator $\mathcal{N}$ at time $t + \frac{dt}{2}$, then a $dt$ step with $\mathcal{L}$, finally another $\frac{dt}{2}$ step with $\mathcal{N}(t + \frac{dt}{2})$. Now, the problem comes down to compute the separate evolutions generated by $\mathcal{N}$ and $\mathcal{L}$. It is clear that the operator $\mathcal{N}$ is diagonal in the real-space representation, thus its effect is local on each point of the wave function. On the other hand, the eigenvectors of the operator $\mathcal{L}$ are known analytically since they are the set of plane waves. The linear evolution of $\psi$ can be obtained in a straightforward way on the reciprocal $k$ space. In order to do so, after the first nonlinear step the wave function is Fourier transformed through an efficient library routine[9]. Then, the linear evolution is applied directly since its effect is local in $k$ space. Finally, $\psi$ is anti-Fourier transformed before the application of the second nonlinear step.

## 3. Carrier dynamics in a three-dot system

As a first example, we consider a system of three 2D quantum dots[10] with circular Gaussian confinement, as depicted in Fig. 1. We take the electron at the initial time as a Gaussian wave packet with zero kinetic energy, centered and fully contained within the upper dot. As the simulation evolves, the wave function is partially transferred into the other dots, due to their effective coupling. Four steps of the time evolution are shown in the first row of Fig. 4, where the gray shaded background represents the potential (darker means lower) and the foreground red plot represents $|\psi|^2$. We stress that the colormap figures are generated directly by TDStool and are reported here as simple screen-shots in order to better illustrate its graphical capabilities.

Journal of Physics: Conference Series **193** (2009) 012049 doi:10.1088/1742-6596/193/1/012049

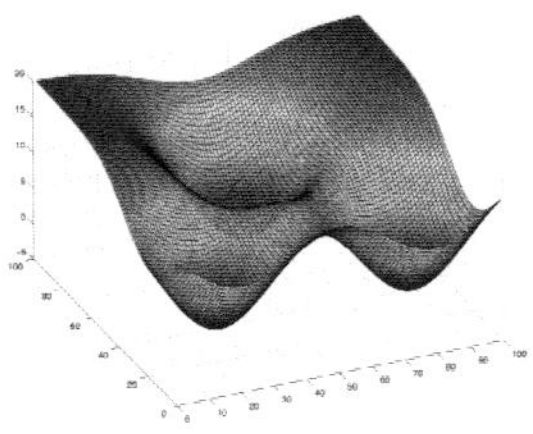

**Figure 1.** Confining potential of the three-dot system. The dots have a Gaussian profile with a maximum depth of $-20$ meV and $\sigma = 40$ nm. They are initially centered ion the vertexes of an equilateral triangle with 300 nm sides. The upper dot is displaced towards the left in the different simulations.

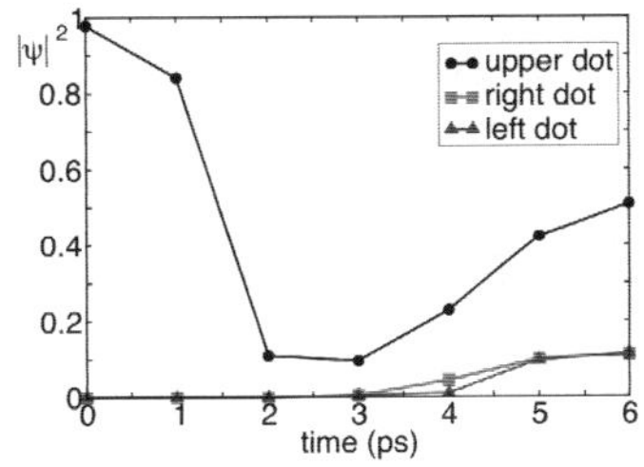


**Figure 2.** Time evolution of the dots occupancy for the case with no displacement of the upper dot. Note that $\psi$ is integrated over a region of the order of $\sigma$ centered in the potential minimum and the part of the wave function outside this region (e.g. in on the inter-dot barrier) is not included.

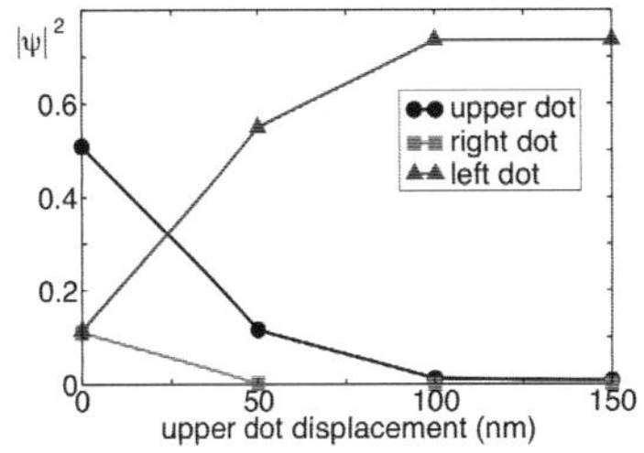


**Figure 3.** Dots occupancy at the final simulation time $t = 6$ps as a function of the displacement of the upper dot with respect to the central position of Fig. 1. As the displacement increases, the electron tends to transfer in the left dot, i.e. the one whose coupling with the upper one is increased.

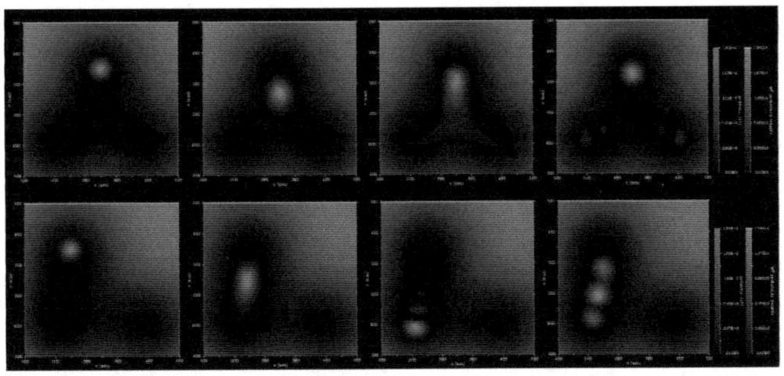

**Figure 4.** Screen-shots at four different time-steps, namely 0, 2ps, 4ps, 6ps, representing the time-evolution of $|\psi|^2$ (red, in the foreground) and the three-dot static potential (gray, in the background). The first and second rows represent the case with symmetric and non-symmetric (100 nm to the left) position of the upper dot, respectively.

We repeat the simulation with the upper dot moved towards the left in step of 50 nm. Figure 2 shows the time evolution of the dots occupancy when the dot is displaced by 50 nm and the second row of Fig. 4 shows the $|\psi|^2$ evolution when the displacement is 100 nm. As expected, at a given time, the more the upper dot is moved towards the left, the higher is the left dot occupancy with respect to the right one. This is illustrated in Fig. 3.

## 4. Electron in a double dot wit SAW

As an example of a simulation with a time-dependent potential we take a 2D double dot represented by the same confining potential described in the previous section. At the initial time, a Gaussian wave packet is centered in the left dot. A first simulation without the SAW potential confirms that the two dots are essentially decoupled and the electron dynamics is confined in

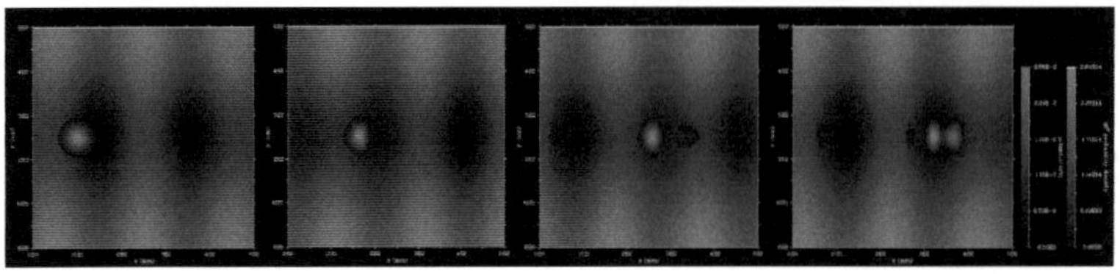

**Figure 5.** Screen-shots at four different time steps (0, 2, 4, 6 ps) representing the evolution of $|\psi|^2$ and of the time-dependent potential (see Fig. 4 caption). The latter potential is formed by the sinusoidal SAW moving towards the right ($\lambda = 200nm$, amplitude=10 meV and velocity=$3\times10^4$ m/s)) and the two Gaussian dots. In this case, after 6 ps the electron is transferred, almost entirely, in the right dot.

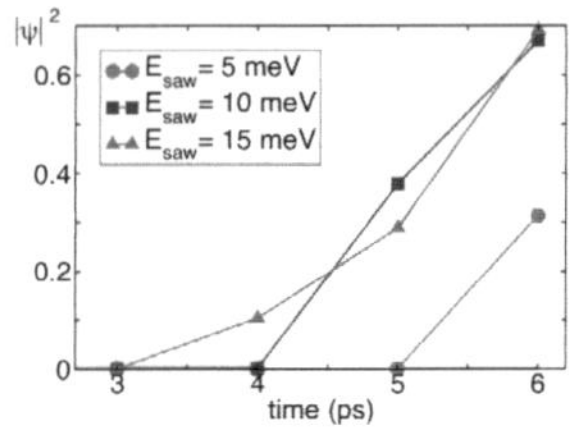


**Figure 6.** Evolution of right-dot occupancy in the two-dot system. As expected, the SAW-induced transfer of the electron is more efficient for higher SAW amplitudes.

the initial dot (the Gaussian wave packet is not an eigenstate of the single-dot potential). A sinusoidal potential that mimics the piezoelectric potential of a SAW[11] is introduced, with the parameters given in Fig. 5 caption. The latter figure shows $|\psi|^2$ at four different time steps, together with the SAW and dots potential in the background. In general, for increasing SAW amplitudes a larger amount of the wave function is moved into the right dot by the SAW, this showing how SAWs could effectively induce a coupling between the two dots. However, we do not find a monotonic behavior and at certain times the right dot occupancy is lower for higher SAW amplitudes (crossing at 5 ps in Fig. 6). This can be ascribed to higher-energy components that are excited in the second case and are able to escape the SAW confinement.

## References

We are pleased to thank Guido Goldoni and Massimo Rudan for most helpful discussions and suggestions.

[1] Ihn T 2004 *Electronic Quantum Transport in Mesoscopic Semiconductor Structures*, Springer, New York
[2] wan der Wiel W G, De Franceschi S, Elzerman J M, Fujisawa T, Tarucha S, Kouwenhoven L P 2003 *Rev. Mod. Phys.* **75**, 1
[3] http://www.s3.infm.it/tdstool
[4] Pitaevskii L P and Stringari S 2003 *Bose-Einstein Condensation*, Claredon Press, Oxford
[5] Cataliotti F S, et al. 2001 *Science* **293**, 843
[6] Eymard R, Gallouët T, Herbin R 2000 *The Finite Volume Method*, in *Handbook for Numerical Analysis*, Ciarlet P., Lions J L eds. (North Holland).
[7] Schenk O and Gärtner K, 2006 *Elec. Trans. Numer. Anal.*, **23** 158.
[8] Ablowitz M J, Prinari B, Trubatch A D 2004 *Discrete and Continuous Nonlinear Schrödinger Systems* Cambridge University Press.
[9] Frigo M and Johnson S G 2005 *Proceedings of the IEEE* **93**, 216.
[10] Gaudreau L, Studenikin S A, Sachrajda A S, Zawadzki P, Kam A, Lapointe J, Korkusinski M and Hawrylak P 2006 *Phys. Rev. Lett.* **97** 036807
[11] Bertoni A and Reggiani S 2004 *Semicond. Sci. Technol.* **19** 113
Gell J R, Ward M B, Young R J, Stevenson R M, Atkinson P, Anderson D, Jones G A C, Ritchie D A and Shields A J 2008 *Appl. Phys. Lett.* **93** 081115

# Nonequilibrium dynamics and coherent control of BCS superconductors driven by ultrashort THz pulses

**T Papenkort[1], T Kuhn[1] and V M Axt[2]**

[1] Institut für Festkörpertheorie, Westfälische Wilhelms-Universität Münster, Wilhelm-Klemm-Straße 10, 48149 Münster, Germany

[2] Institut für Theoretische Physik III, Universität Bayreuth, 95440 Bayreuth, Germany

E-mail: t.papenkort@uni-muenster.de

**Abstract.** The nonequilibrium dynamics of BCS superconductors after excitation with short THz pulses slightly above the superconducting gap is studied theoretically. Calculated pump-probe spectra show how the gap is influenced by this optical excitation. For ultrashort pump pulses a nonadiabatic regime emerges, in which the BCS order parameter exhibits oscillations. These oscillations do not show up in pump-probe spectra. However they may be made visible by using a coherent control type excitation with two phase-locked pump pulses.

## 1. Introduction

Superconductors have been studied since many years using optical techniques such as absorption or pump-probe spectroscopy [1; 2]. Absorption spectra of conventional superconductors under linear response conditions have successfully been calculated on the basis of the BCS theory [3; 4]. Here we apply this theory to the analysis of the nonequilibrium dynamics induced by the excitation with strong THz pulses. Our calculations are based on the mean-field BCS theory extended by a coupling to the electromagnetic field. By numerically solving the equations of motion both for the normal and the anomalous density matrix elements we thereby obtain information on the temporal evolution of microscopic variables like quasiparticle occupations and coherences and at the same time we can analyze how the microscopic dynamics is reflected in observable quantities like the absorption spectrum detected by a delayed probe pulse [5].

## 2. Model system

We use the standard BCS model in mean-field approximation with the Hamiltonian given by

$$H_{\mathrm{BCS}} = \sum\nolimits_{\mathbf{k},\sigma} \varepsilon_{\mathbf{k}} c^{\dagger}_{\mathbf{k},\sigma} c_{\mathbf{k},\sigma} - \sum\nolimits_{\mathbf{k}} \left( \Delta \, c^{\dagger}_{\mathbf{k}\uparrow} c^{\dagger}_{-\mathbf{k}\downarrow} + \Delta^* \, c_{-\mathbf{k}\downarrow} c_{\mathbf{k}\uparrow} \right), \tag{1}$$

where $c^{\dagger}_{\mathbf{k},\sigma}$ and $c_{\mathbf{k},\sigma}$ are electron creation and annihilation operators, $\hbar\mathbf{k}$ and $\sigma$ denote momentum and spin and $\varepsilon_{\mathbf{k}}$ is the dispersion relation for quasi-free electrons with effective mass $m$. The BCS order parameter $\Delta$ is defined by $\Delta = \sum_{\mathbf{k}} W_{\mathbf{k}} \langle c_{-\mathbf{k}\downarrow} c_{\mathbf{k}\uparrow} \rangle$. The BCS interaction only affects electrons close to the Fermi level and its strength $W_{\mathbf{k}}$ is assumed to be constant for these electrons. Under thermal equilibrium conditions $\Delta$ is real and the energy gap is given by

$2\Delta$, thus $\Delta$ can be deduced from the experiment. Under nonequilibrium conditions the order parameter becomes a dynamical variable and is in general complex.

For a fixed value of $\Delta$ the BCS Hamiltonian can be diagonalized by a Bogoliubov transformation [6] introducing new fermionic quasiparticle operators $a^\dagger_{\mathbf{k}\sigma}$ and $a_{\mathbf{k}\sigma}$ according to $a^\dagger_{\mathbf{k}\uparrow} = u^*_{\mathbf{k}} c^\dagger_{\mathbf{k}\uparrow} - v^*_{\mathbf{k}} c_{-\mathbf{k}\downarrow}$ and $a^\dagger_{\mathbf{k}\downarrow} = u^*_{\mathbf{k}} c^\dagger_{-\mathbf{k}\downarrow} + v^*_{\mathbf{k}} c_{\mathbf{k}\uparrow}$. The parameters $u_{\mathbf{k}}$ and $v_{\mathbf{k}}$ depend on $\Delta$ and the Hamiltonian (1) then simply reads

$$H_{\mathrm{BCS}} = \sum\nolimits_{\mathbf{k},\sigma} E_{\mathbf{k}}\, a^\dagger_{\mathbf{k},\sigma} a_{\mathbf{k},\sigma} + \mathrm{const} \qquad \text{with} \qquad E_{\mathbf{k}} = \sqrt{|\Delta|^2 + (\varepsilon_{\mathbf{k}} - E_{\mathrm{F}})^2}, \tag{2}$$

$E_{\mathrm{F}}$ being the Fermi energy. As the transformation depends on $\Delta$, which as a dynamical variable changes in time, this does not mean that the problem has been rewritten into an easily-solvable diagonal one. However it provides us with a way to better understand the microscopic state of the system [6]. In the electron basis even the ground state is quite complicated and excited states cannot easily be analyzed. In contrast, in the quasiparticle picture the ground state is just the vacuum state and an excitation of the superconductor can be understood as the creation of a pair of quasiparticles. The order parameter in this picture is given by

$$\Delta = \sum\nolimits_{\mathbf{k}} W_{\mathbf{k}} \Big[ u^*_{\mathbf{k}} v_{\mathbf{k}} \Big(1 - \langle a^\dagger_{\mathbf{k}\uparrow} a_{\mathbf{k}\uparrow}\rangle - \langle a^\dagger_{\mathbf{k}\downarrow} a_{\mathbf{k}\downarrow}\rangle\Big) + u^{*2}_{\mathbf{k}} \langle a_{\mathbf{k}\downarrow} a_{\mathbf{k}\uparrow}\rangle - v^2_{\mathbf{k}} \langle a^\dagger_{\mathbf{k}\uparrow} a^\dagger_{\mathbf{k}\downarrow}\rangle \Big]. \tag{3}$$

We immediately see that quasiparticle occupations (normal density matrices) lead to a reduction of the order parameter. In addition, however, the order parameter is also affected by quasiparticle coherences (anomalous density matrices).

To describe the pulse-induced dynamics the BCS Hamiltonian has to be complemented by the coupling to the electromagnetic field, which is done in the standard way for a vector potential in Coulomb gauge [5]. Due to the mean-field approximation we then obtain a closed set of equations of motion for the dynamical variables $\langle a^\dagger_{\mathbf{k}\sigma} a_{\mathbf{k}'\sigma}\rangle$ and $\langle a^\dagger_{\mathbf{k}\sigma} a^\dagger_{\mathbf{k}'\,-\sigma}\rangle$; the equations and further details can be found in Ref. [5].

## 3. Pulse-induced gap reduction

Based on this model simulations have been performed using material parameters for superconducting aluminum [7], but the effects do not depend qualitatively on the parameters used [5; 8]. Similar calculations have also been made for a model with d-wave symmetry of the order parameter typical for high-$T_C$ superconductors, and they also yield qualitatively similar results [9]. The THz pulses are of Gaussian shape and centered around 0.4 meV, which is slightly above the gap energy of 0.35 meV.

The pump pulse modifies the order parameter, as can be seen in Fig. 1(a). Here we have plotted $\lim_{t\to\infty} 2|\Delta(t)| =: 2\Delta_\infty$ versus the integrated intensity of pump pulses of different full width at half maximum (FWHM). Figure 1(b) shows absorption spectra (real part of the complex conductivity) for a weak probe pulse. The pump-probe delay has been taken to be sufficiently long to avoid a temporal overlap of the pulses. The spectra belong to different pump-pulse widths and intensities as indicated in the plot. The circled numbers link the excitation conditions to the corresponding points in part (a). In all spectra, we observe a well-defined energy gap terminated by a sharp peak. As can be seen by comparing with part (a), the gap visible in the absorption spectra agrees with $2\Delta_\infty$, which therefore can indeed be interpreted as the energy gap, although the superconductor is not in an equilibrium state. The sharp peak in the spectra is due to a van-Hove singularity in the the density of states at the band edge.

In the quasiparticle picture, a pump pulse creates quasiparticles which results in the reduction of the energy gap. For sufficiently low pulse intensities the number of excited quasiparticles is expected to increase linearly with the intensity. Indeed, the intensity dependence in Fig. 1(a)

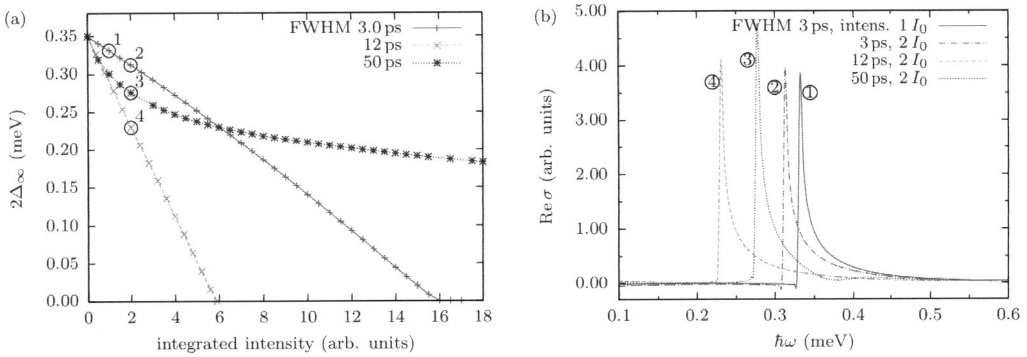


**Figure 1.** (a) Long-time value of $2|\Delta|$ after excitation with pump pulses of different intensity and pulse width. (b) Absorption spectrum of a weak probe pulse applied after the excitation by a pump pulse. The pump pulse parameters correspond to the marked points in (a).

starts linearly for all different pulse widths, but the slopes are different. This is because for the shortest pulse the spectral width is so large that a considerable amount of the intensity is below the gap and therefore cannot create quasiparticles. Since for a given value of the integrated intensity the peak intensity for this pulse is quite high compared to the longer pulses, here we observe a small quadratic term in the intensity dependence resulting from two-photon processes. The 50 ps pulse, on the other hand, creates quasiparticles with a rather sharp energy distribution. Thus for higher intensities the quasiparticle creation saturates because of Pauli blocking leading to a flattening of the curve. In the pump-probe spectrum corresponding to this pulse Pauli blocking is seen as a hole-burning: the absorption is slightly reduced just below the pump energy of 0.4 meV.

## 4. The nonadiabatic regime

While for longer pulses the order parameter drops monotonically from its equilibrium value to the final value $\Delta_\infty$, for short pump pulses we observe an oscillation in the time dependence of $|\Delta(t)|$ as shown in Fig. 2(a). The reason is that a long pulse adiabatically creates a quasiparticle occupation corresponding to the instantaneous value of the gap. For short pulses instead a nonadiabatic regime is reached where in addition quasiparticle coherences are excited [8; 10]. If, in the quasiparticle basis corresponding to the instantaneous value of $\Delta$, only an occupation exists, the state is stationary because the Hamiltonian is diagonal. But if quasiparticle coherences exist as well, the state is no longer stationary and the oscillation appears. For an unperturbed mean-field BCS system (i.e., without the coupling to the electromagnetic field) analytical solutions for the temporal evolution after preparation of a nonequilibrium initial state has been found [11]. Here the order parameter takes the form

$$|\Delta(t)| = \Delta_\infty + at^{-1/2}\cos(2\Delta_\infty t/\hbar + \phi), \quad (4)$$

where $a$ and $\phi$ depend on the initial conditions. This formula fits very well to our numerical results for the time after the pump pulse.

Surprisingly, however, it turns out that the oscillation of $|\Delta|$ does not show up in pump-probe spectra. The energy gap visible in the absorption spectrum of the probe pulse is independent of the pump-probe delay time and agrees with $2\Delta_\infty$. The reason is that the probe-induced current extends over many periods of the oscillation; the oscillation hence averages out. This leads to the question of how to detect this oscillation in measurable quantities. We have found that this is possible by application of two phase-locked pump pulses [8]: The reaction of the superconductor

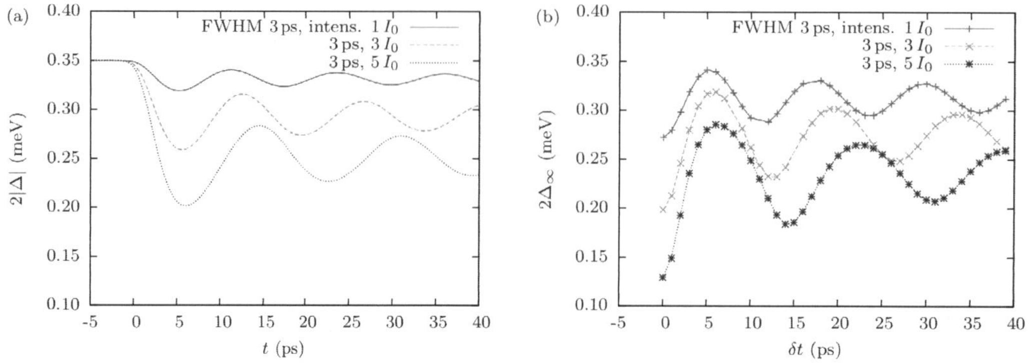


**Figure 2.** (a) Time dependence of $2|\Delta|$ after excitation with a 3 ps pump pulse with different intensities. (b) Long-time value $2\Delta_\infty$ after two pump pulses as a function of the time delay between the pulses. Pulse 1 has the intensity as in (a), pulse 2 always the intensity $I_0$.

to the second pump pulse depends on the instantaneous value of $\Delta$ at the time when the second pulse hits the system. The instantaneous value of $\Delta$ after the first pulse thus reflects itself in the final value $\Delta_\infty$ after the second pulse, which can be detected by a subsequent probe pulse. This is demonstrated in Fig. 2(b), which shows the the long-time value $\Delta_\infty$ after a second pulse versus the delay between the two pump pulses. Here the intensity of the first pump pulse is the same as for the corresponding curve in Fig. 2(a), whereas the second pump pulse in all cases has the same intensity $I_0$. The oscillation of $|\Delta(t)|$ now is clearly visible in $\Delta_\infty$ and thus can be measured by a subsequent probe pulse.

## 5. Conclusion

We have presented numerical simulations of a BCS superconductor driven by short pulses with frequencies slightly above the energy gap. The optical excitation lowers the energy gap, which can be observed in the absorption spectrum of a delayed probe pulse. The fast oscillation of the order parameter caused by short pump pulses, however, is not visible with this technique. Instead it may be revealed in the probe spectra by using a coherent control-type excitation with two phase-locked pump pulses.

## References

[1] Palmer L H and Tinkham M 1968 *Phys. Rev.* **165** 588–595

[2] Averitt R D, Rodriguez G, Lobad A I, Siders J L W, Trugman S A and Taylor A J 2001 *Phys. Rev. B* **63** 140502(R)

[3] Mattis D C and Bardeen J 1958 *Phys. Rev.* **111** 412–417

[4] Zimmermann W, Brandt E H, Bauer M, Seider E and Genzel L 1991 *Physica C* **183** 99–104

[5] Papenkort T, Axt V M and Kuhn T 2007 *Phys. Rev. B* **76** 224522

[6] Kuper C G 1968 *An Introduction to the Theory of Superconductivity* (Clarendon Press, Oxford)

[7] Poole Jr C P, Farach H A and Creswick R J 1995 *Superconductivity* (Academic, New York)

[8] Papenkort T, Kuhn T and Axt V M 2008 *Phys. Rev. B* **78** 132505

[9] Unterhinninghofen J, Manske D and Knorr A 2008 *Phys. Rev. B* **77** 180509

[10] Papenkort T, Kuhn T and Axt V M 2009 *Phys. Stat. Sol. B* **246** 325

[11] Yuzbashyan E A, Tsyplyatyev O and Altshuler B L 2006 *Phys. Rev. Lett.* **96** 97005

# Dynamics of photo-excited carriers in anatase $TiO_2$ thin film investigated by pump-probe method

**Taki Hashimoto, Keisuke Takahashi, Hideyuki Kunugita, Hiroshi Sakama, and Kazuhiro Ema**

Department of Engineering and Applied Sciences Sophia University 7-1, Kioi-cho, Chiyoda-ku, Tokyo 102-8554, Japan

taki-h@sophia.ac.jp

**Abstract**. We measured transient absorption spectra with pump-probe technique to investigate the dynamics of photo-excited carriers at 300 K and 12 K in anatase $TiO_2$ epitaxial film. Although the photoluminescence (PL) originated from self-trapped excitons (STEs) vanished at 300 K, we observed the change in probe signal. On the other hand, the intensity of the transient absorption signal at 12 K was much weaker than that at 300K. Therefore the carriers at 300 K become trapped at the surface, leading to a pump-probe signal, while at 12 K, the remained carriers inside the bulk arise from STE recombination. These results suggest that the pump-probe signal is derived from the trap state(s) near the surface.

## 1. Introduction

Anatase $TiO_2$ is well known to show high photocatalytic activity, and numerous studies have taken place [1] [2]. It is thought that photo-excited carriers create activated species by reacting with water and oxygen at the crystal surface, and such activated species decompose pollution (organic) substance. However, it is not known that why photo-excited carriers in anatase $TiO_2$ have long life time (a few hours) [3], and how both electrons and holes reach the surface. To understand the dynamics of photo-excited carriers, pump-probe [4] and PL measurement[5] have been done. However, the former were performed only at room temperature, and no study has compared pump-probe and PL results. In this study, we investigated the dynamics of photo-excited carriers by pump-probe technique at 12 and 300 K, and observed the changes in transient absorption and PL spectra before and after annealing in oxygen atmosphere.

## 2. Experiment

### 2.1. Sample

Anatase $TiO_2$ thin film was grown on $LaSrAlO_4$ (LSAO; a = 0.3755 nm) substrates by pulsed laser deposition (PLD) method using $TiO_2$ ceramic targets and a KrF excimer laser (248 nm wavelength) in the presence of oxygen. The growth temperature was 700°C. The film thickness was ~200 nm. The sample was cut into two pieces, and one of them was annealed 10 hours at 600°C in oxygen atmosphere to investigate the effect of oxygen vacancies and the change of crystalline.

### 2.2. Experimental setup

The schematic setup of the pump-probe system is shown Figure. 1. We used a regenerative amplified Ti:sapphire laser with a repetition rate of 100 kHz. The laser output (800 nm) was divided into two beams. The pump beam was the third harmonic (266 nm), and the probe beam was a white light continuum generated by focused on a sapphire crystal. The signal of probe beam was filtered by a spectrometer and detected by a Si:photodiode using a lock-in technique. The time resolution of our setup is about 1 ps. Measurements were performed at 12 K (cooled by cryostat) and 300 K (in air).

For PL measurements, samples were mounted on the cryostat and excited at 3.7eV using the second harmonic of the output from an optical parametric amplifier pumped by a regenerative amplifier. The time-resolved PL was measured using a streak camera.

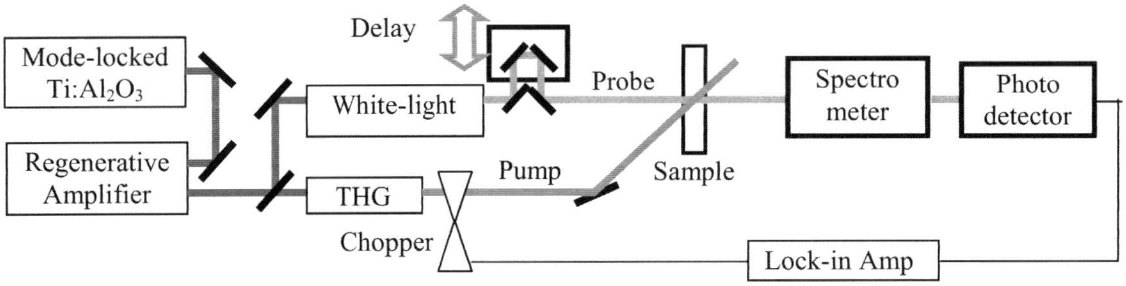


**Fig.1** Pump-probe system

## 3. Results and Discussion

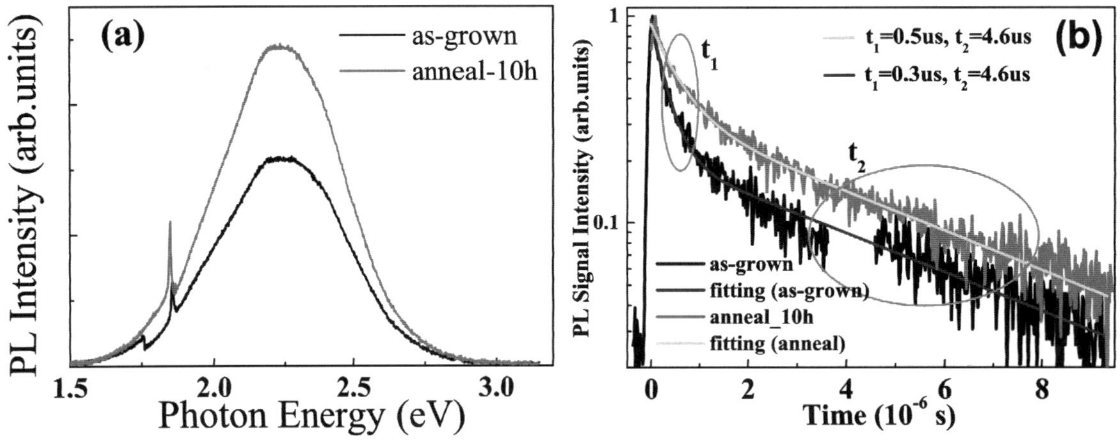


**Fig. 2** $TiO_2$ thin films at 12 K: (a) PL spectra; (b) Time-resolved PL decay curve, where the time constants $t_1$ and $t_2$ are for direct and indirect formation STE, respectively. The spike at 1.85eV of PL spectra is the second diffraction of the excitation light.

Figure. 2 shows the PL spectra (a) and time-resolved PL decay curves (b) at 12 K before and after annealing. The PL of anatase $TiO_2$ is originated from the recombination of self-trapped excitons (STEs) [6]. It is known that the relaxation processes to the STE state proceeds in two ways: the fast component (< 1 μs) of the PL decay curves is due to the direct formation of STEs by free electrons: the slow component (> 1 μs) of the curves is due to indirect formation of STEs by electrons hopping between oxygen vacancies [7]. If the number of oxygen vacancies decreases because of annealing in oxygen atmosphere, the decay time for indirect formation of STEs should increase. If the crystalline improves, STEs should form more effectively and PL intensity should increase. In our experiment only the fast component became slower and the total luminescence intensity increased after annealing. These results indicate that with annealing, the quantity of oxygen vacancies inside the sample remains constant while the crystalline improves. PL intensity decreases as temperature increases, and PL is not observed at 300 K in our system, consistent with the report by Tang *et al.* for a single crystal [5].

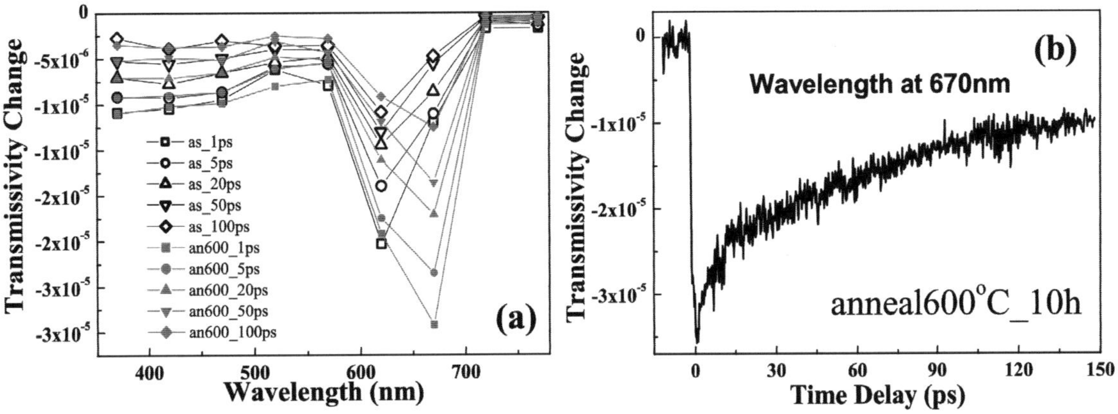


**Fig.3** $TiO_2$ thin films at 300 K: (a) Transient absorption spectra, where "as" is the as-grown sample and "an600" is the annealed sample; (b) Decay curve of annealed sample at 670 nm.

Figure.3 shows transient absorption spectra at 300 K and the decay profile at 670 nm. The presence of induced absorption indicates that photo-excited carriers exist at 300 K. Therefore, this absorption change is not derived from STEs. The peak positions of the as-grown and annealed sample are at 620 nm and 670 nm, respectively. Rothenberger *et al.* and Colombo *et al.* observed a similar broad transient absorption spectrum centered at 620 nm in $TiO_2$ particles, and they assigned the absorption to trapped electrons at the surface [8][9][10]. Thus, we can conclude that the signal in this study is strongly influenced by the surface condition. The signal at 670 nm is increased after annealing, leading to peak shift of the signal. To our knowledge, there is no previous report of such observation. This observation of peak shift and/or signal enhancement at 670 nm suggests that the oxygen vacancies at the surface are significantly affected by annealing in oxygen atmosphere as well as by improvement in crystalline.

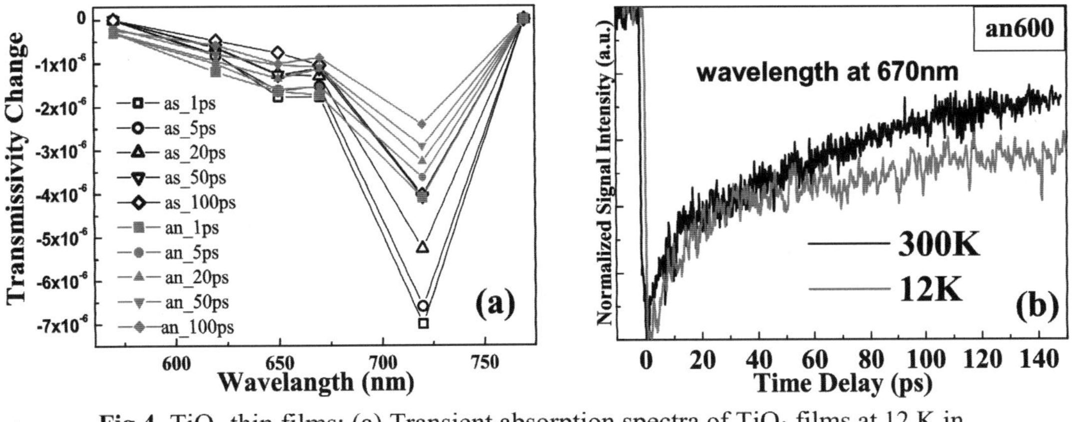


**Fig.4.** $TiO_2$ thin films: (a) Transient absorption spectra of $TiO_2$ films at 12 K in vacuum; (b) Normalized decay curves of annealed sample at 670 nm at 12 and 300 K.

Figure.4 shows transient absorption spectra at 12 K and the decay profile at 670 nm. The signal intensity is almost one order of magnitude weaker than at 300 K. In general, the diffusion rate of photo-excited carriers decreases as temperature decreases. Thus, at 12 K, many carriers are expected to remain in the bulk. Our observation of a long-lived PL signal, on the order of up to microseconds (Figure. 2b), supports this model.

Journal of Physics: Conference Series **193** (2009) 012051 doi:10.1088/1742-6596/193/1/012051

As shown in Figure.3a and 4a, the shift in peak position to lower energy is greater at 12K than at 300 K. To explain this result, we must consider differences in both temperature and atmosphere. Furube *et al.* also observed that for $TiO_2$ powders, the peak positions measured at room temperature in air and vacuum differ [11], but shifts of their study is opposite to our results.

We can reproduce the decay curves of the pump-probe signal (Figure. 4b) by fitting with double exponential decays; and the fast component shows the same decay time, ~15 ps, at both temperatures, but the slow component shows different decay times, ~600 and ~150 ps, at 12 and 300 K, respectively. This difference can also be attributed to differences in both measurement temperature and atmosphere (air/vacuum).

The peak position at 12 K is the same for both as-grown and annealed samples. We are planning further investigations to understand this result.

## 4. Conclusion

We measured transient absorption spectra at 12 K in vacuum and at 300 K in air by pump-probe method, for both as-grown and annealed samples. We confirmed that photo-excited carriers exist at 300 K. We believe the signal come under the influence of near the sample surface. The signal intensity is much weaker at 12 K than at 300 K, presumably because the diffusion rate is much lower at 12 K than at 300 K. At 12 K, many photo-excited carriers remain in the bulk. After hopping between internal oxygen vacancies for a long time (on the order of nanoseconds to microseconds), they finally undergo STE recombination. On the other hand, at 300 K, photo-excited carriers are more diffusive and become trapped at the surface state.

## References

[1] Chen D, Jiang Z, Geng J, Wang Q and Yang D 2007 *Ind. Eng. Chem. Res.* **46** 2743
[2] Liu Z, Sun D D, Guo P and Leckie J O 2007 *Chem. Eur. J.* **13** 1851
[3] Itoh C and Wada A 2005 *Phys. Stat. Sol. C* **2** 629
[4] Yoshihara T, Katoh R, Furube A, Tamaki Y, Murai M, Hara K, Murata S, Arakawa H and Tachiya M 2004 *J. Phys. Chem. B* **108** 3817
[5] Tang H, Berger H, Schmid P E, Lévy F and Burri G 1993 *Solid State Communications* **87** 847
[6] Tang H, Levy F, Berger H, Schmid P E 1995 *Phys. Rev. B* **52** 7771
[7] Harada N, Goto M, Iijima K, Ichikawa N, Sakama H, Kunugita H and Ema K *Jpn. Phys* **46** 4170
[8] Rothenberger G, Moser J, Gratzel M, Serpone N and Sharma D K 1985 *J. Am. Chem. Soc* **107** 8054
[9] Colombo D P and Bowman R M 1995 *J. Phys. Chem.* **99** 11752
[10] Cavaleri J J, Colombo D P Jr and Bowman R M, 1998 *J. Phys. Chem. B* **102** 1343
[11] Furube A, Asahi T, Masuhara H, Yamashita H and Anpo M 1999 *J.Phys. Chem. B* **103** 3120

Journal of Physics: Conference Series **193** (2009) 012052 doi:10.1088/1742-6596/193/1/012052

# First observation of $E_2$ coherent phonon modes in CdS using a noncollinear optical parametric amplifier

**Daisuke Suzuki, Hideyuki Kunugita, and Kazuhiro Ema**

Department of Engineering and Applied Sciences, Sophia University, 7-1, Kioi-cho, Chiyoda-ku, Tokyo, 102-8554, Japan

E-mail:daisuk-s@sophia.ac.jp

**Abstract**. We report the first observation of $E_2$ coherent phonon modes in CdS using a noncollinear optical parametric amplifier. We studied the intensity, temperature and excitation energy dependences of the high- and low-frequency $E_2$ modes. We obtain that generation mechanism under non- and near resonant regions is impulsive stimulated Raman scattering (ISRS) and $E_2$ modes are barely affected by Fröhlich interaction. As temperature increases, $E_2$ modes are affected by anharmonic effect. The observed phonon lifetime of the low- and high-$E_2$ modes in CdS is 243 ± 37 and 11.4 ± 0.4 ps, respectively, at 83 K. From fitting, we can estimate the lifetime of the high-$E_2$ mode at the low-temperature limit to be 49.3 ps.

## 1. Introduction

Coherent phonon means coherent lattice vibration. To excite coherent phonon, it is necessary that laser pulse duration is shorter than the oscillation period. Recent development in femtosecond laser technology has made it possible to excite coherent phonon.

Many studies of coherent longitudinal optical (LO) phonon in semiconductors have been reported [1, 2]. Among them, $E_2$ (non-polar) modes were studied only in ZnO and GaN [4, 5, 9]. To our knowledge, coherent phonon modes in CdS have not been repoted, although optical properties of CdS have been studied extensively [6, 7, 8]. $E_2$ modes are expected to play a role in metrology, because the low-$E_2$ mode has a very long lifetime and enables highly accurate determination of frequency. There is also a possibility that we can observe the enhancement of amplitude of coherent phonon for $E_2$ modes even in intense excitation, because these two modes interact weakly with photoexcited carriers and are expected to show no saturation, contrary to the case for Bi [3].

In this study, we observed the high- and low-$E_2$ modes for the first time in the wide-gap semiconductor CdS, and studied the intensity and temperature dependence of dephasing for these modes.

## 2. Experiment

We performed a degenerative pump-probe experiment using a home-built noncollinear optical parametric amplifier (NOPA) that produces 35-60 fs pulses. The central pulse energy can be tuned from 1.94 to 2.38 eV and the typical spectral width is 70 meV.

The sample was a $10 \times 10 \times 1$ mm$^3$ single wurtzite structural crystal grown along the [0001] axis. Figure 1. shows relationship between pump energies and absorption spectra of our CdS sample at 83 K (a) and 300 K (b). Pump-induced transmittance changes were measured by polarization-sensitive electro-optic sampling as a function of time delay between pump and probe. The electric field of the pump was set along the $[2\bar{1}\bar{1}0]$ axis and that of the probe was set 45 degrees with respect to the same axis. The high- and low-$E_2$ modes are expected to be observed under these conditions [4]. All experiments were performed at 83 K except for the temperature-dependence experiments.

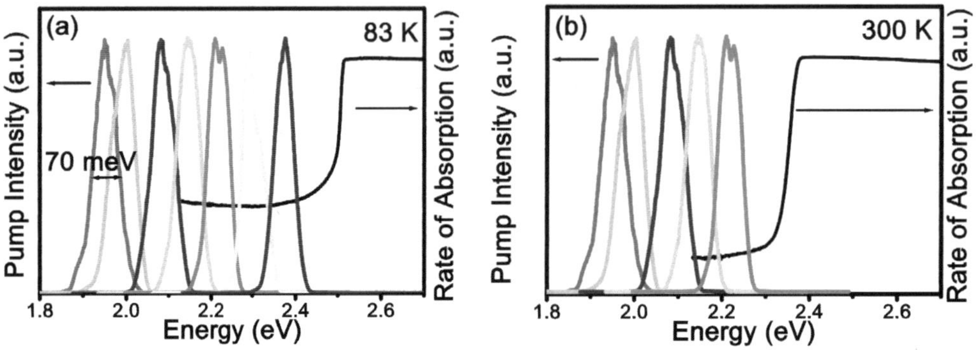


Figure 1. Relationship between pump energies and absorption spectra: (a) At 83 K and (b) At 300 K.

## 3. Result

### 3.1. Observation of $E_2$ modes

Figure 2. (a) shows typical transmittance change measured by the pump-probe method using the 2.21 eV pulse. We observe the two oscillating components on the slowly varying background, as shown in the inset.

Figure 2. (b) shows fast Fourier transform (FFT) spectra for each time span. The two peaks at 1.27 and 7.67 THz correspond to the high- and low-$E_2$ mode, respectively [8]. The lifetime of the high-$E_2$ mode is less than 20 ps, whereas that of the low-$E_2$ mode is relatively long.

The frequency, amplitude, and decay time of the two modes was determined by fitting using the following equation. The background was removed using a band-pass filter for low- or high-mode, respectively.

$$\Delta T/T = A_{h,l}\exp(-t/\tau_{h,l})\sin(2\pi f_{h,l}t + \alpha_{h,l}) \quad (1)$$

The estimated phonon lifetimes of the low- and high-$E_2$ mode in CdS are 243 ± 37 and 11.4 ± 0.4 ps, respectively. The former is longer than that in ZnO at 5 K (211 ps) [9].

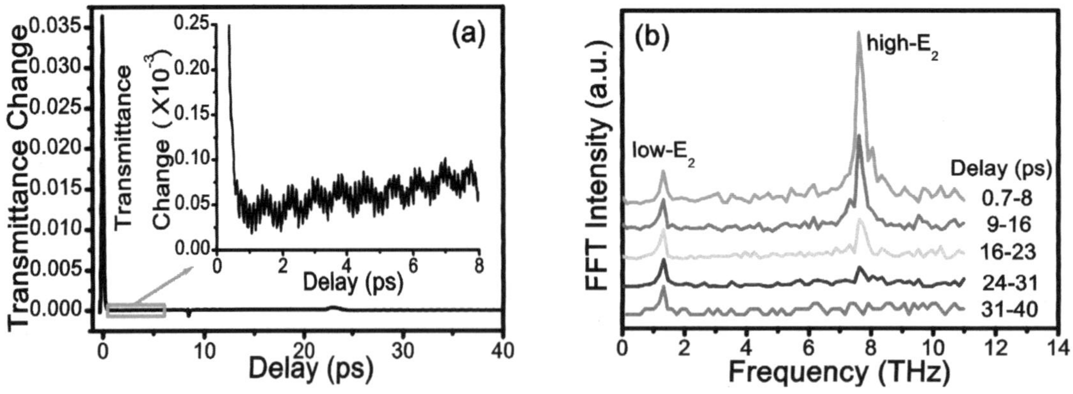


Figure 2. (a) Transmission change (raw data) as a function of the delay between pump and probe pulses. The inset shows appearance of phonon oscillations. (b) FFT spectra for each span.

### 3.2. Generation Mechanism and excitation intensity dependence of coherent $E_2$ modes

Figure 3. (a) shows the pump polarization dependence of the high-$E_2$ mode using 2.3 eV pulse. The observed phonon oscillation obeyed the Raman selection rule for $E_2$ mode [4]. A similar result was obtained for the low-$E_2$ mode using various excitation energies. We see that, in CdS, generation mechanism of coherent phonon below and near the band-gap is ISRS mechanism.

$E_2$ modes are supposed to be little influenced by the carrier directly because $E_2$ modes are not polar mode. Figure 3. (b) shows phonon dephasing rate for the high-$E_2$ mode as a function of pump intensity using 2.3 eV pulse. Unlike the case for the LO modes [5], the dephasing rate does not increase. A similar result was obtained for the low-$E_2$ mode and using various excitation energies.

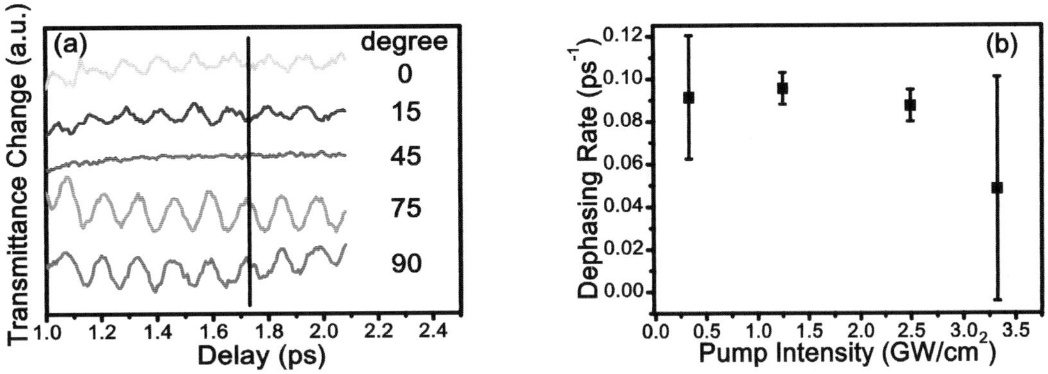


Figure 3. High-$E_2$ mode excited by 2.3 eV: (a) Transient transmittance change as a function of time delay at the angle of pump polarization φ defined with respect to $[2\bar{1}\bar{1}0]$. (b) Dephasing rate as a function of pump intensity

### 3.3. Temperature dependence of dephasing rate

Figure 4. shows the effect of temperature on dephasing rate for the low-$E_2$ mode using 2.07 eV pulse (a) and high-$E_2$ mode using 2.21 eV pulse (b). Because of phonon-phonon interaction, dephasing times become shorter from 11 ps to 3 ps for the high-$E_2$ mode and from 189 ps to 23 ps for the low-$E_2$ modes with increasing temperature. This decrease in phonon lifetime with increasing temperature is seen at various excitation energies. It is thought that phonon lifetime is limited as a result of phonon-phonon interaction and related to phonon dispersion [10]. Because there are many possible branches of decay for the high-$E_2$ mode but only a few for the low-$E_2$ mode, the high-$E_2$ mode is expected to be more affected by anharmonic effect, and thus to have a shorter lifetime, than the low-$E_2$ mode. However, we cannot find the reason why the low-$E_2$ mode shows superlinear dependence on temperature while the high-$E_2$ mode shows linear dependence.

The temperature dependence of the dephasing rate for the high-$E_2$ mode can be fitted using the following equation

$$\Gamma(T) = \Gamma_0 + \Gamma_{DN}(1 + n_1 + n_2) + \Gamma_{UP}(n_3 - n_4)$$
$$n_{1,2,3,4} = [\exp(hf_{1,2,3,4} / k_B T) - 1]^{-1} \tag{2}$$

Where $\Gamma_0$ is temperature-independence term related to defect and/or impurity scattering, and $\Gamma_{DN}$ and $\Gamma_{UP}$ are third-order anharmonic coefficients related to down-conversion process and up-conversion process, respectively [8]. From phonon dispersion curve for CdS [11], we take, for the high-$E_2$ mode, the frequency that is $f_1 = 3.19$ THz, $f_2 = 4.48$ THz for the down-conversion channel and $f_3 = 1.03$ THz, $f_4 = 8.7$ THz for the up-conversion channel. We cannot identify the channel to fit for the low-$E_2$ mode from our result, but the lifetime of that mode possibly approach nanosecond at the low temperature limit.

From this calculation, we can estimate that the lifetime of the high-$E_2$ mode in CdS at low-temperature is 49.3 ps, which is longer than that in ZnO [12].

Although the lifetimes of both modes are affected by anharmonic effect, we did not observe frequency softening [9].

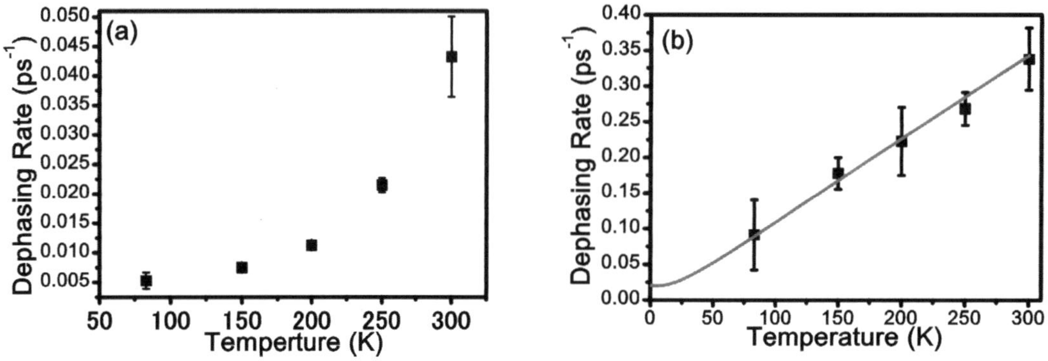


Figure 4. Dephasing rate for $E_2$ modes as a function of temperature: (a) Low-$E_2$ mode; (b) High-$E_2$ mode

## 4. Conclusion

We studied the contribution of intensity, temperature and excitation energy to the high- and low-$E_2$ modes in CdS. We determined that generation mechanism under non-resonant region is ISRS and $E_2$ modes are not much affected by Fröhlich interaction. At high temperatures, $E_2$ modes are affected by anharmonic effect. The measured lifetime of the low- and high-$E_2$ mode in CdS are $243 \pm 37$ and $11.4 \pm 0.4$ ps, respectively, at 83 K in our measurement. From fitting, we estimate the lifetime of the high-$E_2$ mode at low-temperature limit to be 49.3 ps.

## References

[1] Merlin R 1997 *Solid State Communications* **102** 207
[2] Cho G C, Dekorsy T, Bakker H J, Hövel R and Kurz H 1996 *Phys. Rev. Lett.* **77** 4062
[3] DeCamp M F, Reis D A, Bucksbaum P H and Merlin R 2001 *Phys. Rev. B* **64** 092301
[4] Lee I H, Yee K J, Lee G, Oh E and Kim D S 2003 *J. Appl. Phys* **93** 4939
[5] Yee K J, Lee K G, Oh E and Kim D S 2002 *Phys. Rev. Lett.* **88** 105501
[6] Wiesner P and Heim U 1975 *Phys. Rev. B* **11** 3071
[7] Itoh Y, Yamabe K, Adachi S and Hamaguchi C 1979 *J. Phys. Soc. Japan* **46** 542
[8] Bairamov R J and Ramadas A K 1976 *Phys. Rev. B* **13** 5518
[9] Aku-Leh C, Zhao J, Merlin R, Menéndez J, Cardona M 2005 *Phys. Rev. B* **71** 205211
[10] Serrano J, Romero A H, Manjón F J, Lauck R, Cardona M and Rubio A 2004 *Phys. Rev. B* **69** 094306
[11] Nusimovici M A, Balkanski M and Birman J L 1970 *Phys. Rev. B* **1** 595
[12] Cuscó R, Alarcón-Lladó E, Ibáñez J, Artús L, Jiménez J, Wang B and Callahan M J 2007 *Phys. Rev. B* **75** 165202

EDISON 16 IOP Publishing
Journal of Physics: Conference Series **193** (2009) 012053 doi:10.1088/1742-6596/193/1/012053

# Photo-excited carrier relaxation dynamics in InN films

**Masafumi Hashimoto[1], Kazuya Fukunaga[1], Kazuaki Kouyama[1,3], Jumpei Kamimura[1], Hideyuki Kunugita[1,2,3], Kazuhiro Ema[1,2,3], Akihiko Kikuchi[1,2,3] and Katsumi Kishino[1,2,3]**

[1]Department of Engineering and Applied Sciences Sophia University 7-1, Kioi-cho, Chiyoda-ku, Tokyo 102-8554, Japan
[2]Sophia Nanotechnology Research Center
[3]Core Research for Evolutional Science and Technology (CREST), Japan Science and Technology Agency

E-mail to first author : masafu-h@sophia.ac.jp

**Abstract**. Photo-excited carrier relaxation dynamics in InN films were investigated using femtosecond transient measurements. We measured two samples with different background carrier densities, and discussed the influence of the carrier densities on the relaxation dynamics. We also found that the slow decay component disappears under the optimized condition because of the effects of equilibrium band filling and bandgap renormalization.

## 1. Introduction

Among III-V semiconductors, due to its high electron mobility and high saturation drift velocity, InN is one of the most important materials for its potential applications in future devices, e.g. high-speed electronic and optical switching devices. Therefore, a study of its intrinsic physical properties is essential. Recently, InN films with high crystal quality and low electron concentration have been obtained using molecular beam epitaxy. As a result, the basic properties of InN, such as bandgap energy, are gradually becoming more understandable [1]. However, the study of photo-excited carrier relaxation dynamics in InN, which is the most important issue for realization of high-speed devices, is still not that much understood [2-5].

In this paper, we report the fast and slow carrier dynamics of high-quality wurtzite InN films with different background carrier densities. We performed degenerate pump-probe transmission measurements at room temperature. The dependence of the relaxation dynamics on the intrinsic carrier density and the excitation wavelength has been studied. We have observed that the amplitude ratio of the two decay components has the background carrier density dependence. We have also found that at longer excitation wavelengths the slow component becomes the induced absorption signals. These signals are caused by the bandgap renormalization (BGR) due to the photo-excited carriers [6]. When the excitation wavelength and the background carrier density are optimized, the slow component disappears in the pump-probe decay signal. This unique behavior exhibits a useful prospect for the high-speed device application.

## 2. Samples

We prepared two InN films (sample A and B) grown on sapphire (0001) substrates using RF-plasma-assisted molecular beam epitaxy with a 400-nm-thick GaN buffer layer. The height of the samples was about 1 μm. The background carrier densities of the samples A and B were 1.4 × $10^{18}$ $cm^{-3}$ and 8.5 × $10^{18}$ $cm^{-3}$, respectively, which were determined by the Hall measurement at room temperature.

## 3. Experiments

Degenerate pump-probe transient transmission measurements were performed using 180-fs pulses with 76 MHz repetition rate. Using an optical parametric oscillator, which generates a tunable infrared pulse, the excitation wavelength was tuned from 1100 nm (1.127 eV) to 1600 nm (0.775 eV), which is higher than the bandgap energy of InN ($E_g$ ~ 0.64 eV at room temperature). The pump intensity was set to be 1.0 × $10^{17}$ $cm^{-3}$, lower than the background carrier density. In order to increase the signal-to-noise ratio, an optical chopper and an acousto-optic modulator were utilized to modulate the pump and probe beam at 690 Hz and 190 kHz, respectively. The polarization of the pump and probe beam was parallel.

## 4. Results and Discussions

Figure 1 shows the normalized transient transmission changes measured at different excitation wavelengths.

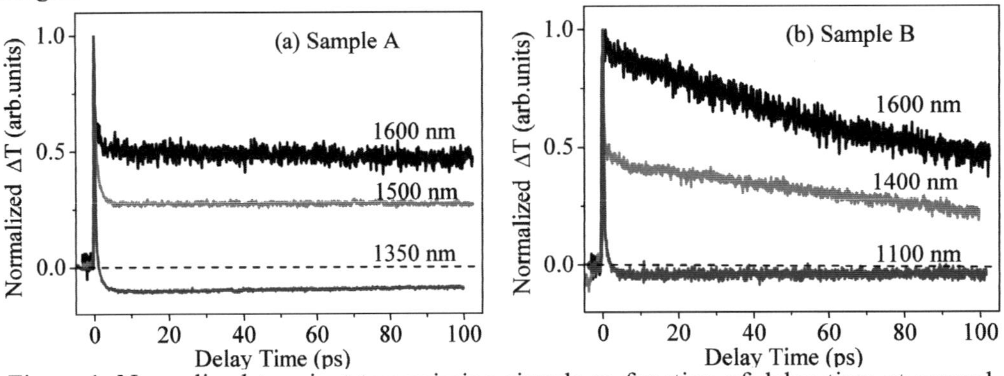


Figure 1. Normalized transient transmission signals as function of delay time at several excitation wavelengths with a photo-excited carrier density of 1.0 × $10^{17}$ $cm^{-3}$ for (a) sample A and (b) sample B.

The instantaneous increase in the transmission near zero time delay is due to the phase space filling induced by the pump pulses. After the instantaneous response there are the fast (~ ps) and the slow (~ ns) decay components. The experimental curves can be fitted to double exponential functions. The well-fitted slow decay time constants at 1600 nm (0.775 eV) are $\tau_{slow}$ = 1.1 ns for sample A and $\tau_{slow}$ = 145 ps for sample B. These slow time constants were found to be not that much dependent on the excitation wavelength. On the other hand, the fast decay constants seem to be slightly dependent on the excitation wavelength. However the fitted fast decay constants for all curves are less than 1 ps ranging from 300 fs to 900 fs, which are of the same order as the excitation pulse width. Therefore we are not concerned about this dependence, but rather concentrate on two other more distinct features in the curves; (1) the amplitude ratio between the fast and the slow components is apparently dependent on the excitation wavelength and (2) the slow decay components at shorter excitation wavelengths (1350 nm for sample A and 1100 nm for sample B) are negative.

First of all, we confirm that the fast decay component is originated from the carrier thermalization process as a result of electron-LO phonon ($E_{LO}$ = 74 meV [7]) scattering. It is to be noted that only the curve of sample B at 1600 nm (0.775 eV) has no fast component. The reason for this is believed to be because the excess energy of photo-excited carriers in this case is not large enough to emit the LO phonon. Figure 2 shows the square of the absorption coefficient against the photon energy.

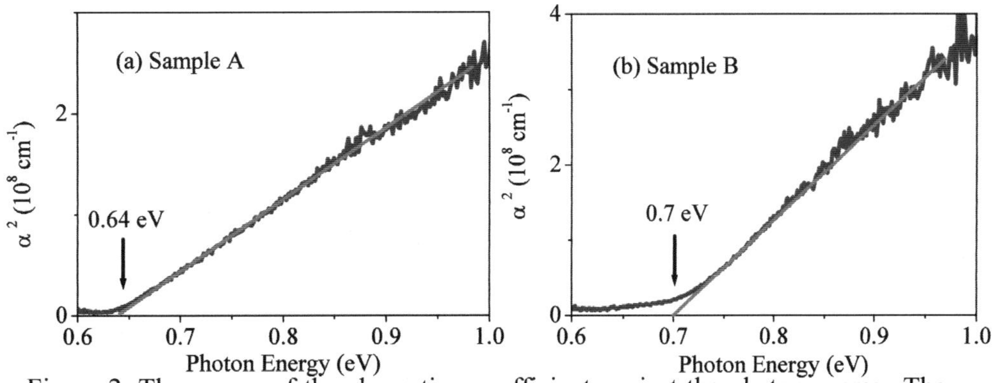


Figure 2. The square of the absorption coefficient against the photon energy. The extrapolated absorption edge is approximately 0.64 and 0.7 eV for sample A and B, respectively.

We can see that the absorption edge energy (almost equal to the Fermi energy) of sample A and B are $E_A \sim 0.64$ eV (~ 1940 nm) and $E_B \sim 0.7$ eV (~ 1770 nm), respectively. Since the sample B has a large background carrier density, the Fermi energy is located at a higher energy in the conduction band. In this case the excess energy (0.775 eV–$E_B$) is not much larger than $E_{LO}$, therefore the scattering ratio is too small to appear in the fast component. As the excitation energy increases for sample B, the fast component becomes dominant. For sample A, on the other hand, the Fermi energy is low enough to have the excess energy much higher than $E_{LO}$. Therefore all curves have the fast components. This provides strong evidence that the fast component is due to the LO phonon scattering process.

After the fast LO phonon scattering process subsides, the absorption saturation signals still remain because the photo-generated carriers increase the occupation number in the conduction band at the probe wavelength. These saturation signals survive until the photo-generated carriers recombine, which is the origin of the slow component. Because the increase in the electron occupation number rapidly decreases with the energy, the amplitude of the slow component is decreased as the probe wavelength is longer. This is the reason why the amplitude ratio of the slow and the fast components has the excitation wavelength dependence.

The carrier recombination includes radiative and nonradiative processes. Chen *et al.* proposed that the defect-related nonradiative recombination is the dominant process at room temperature [8]. Therefore the difference in the slow decay time between sample A and B is probably attributed to the difference in the number of the nonradiative carrier recombination centers. This indicates that the crystalline quality of sample A is greatly improved by decreasing the background carrier density.

We have also investigated the induced absorption signals. In the slow component, the negative signals which indicate the induced absorption are observed at shorter excitation wavelengths (see figure 1). The induced absorption signals are considered to be caused by the BGR, i.e. the many body effect of the photo-excited carriers. As a result of BGR, the density of state increases at the probe photon wavelength, and then the absorption increases. This effect is dominant only at shorter excitation wavelengths, because at longer excitation wavelengths the positive signals due to the absorption saturation discussed above is much more dominant than the BGR effect. Based on this, we can expect an existence of a balance point between the positive signals and the negative signals; i.e. the cancelling out of the band filling effect and the BGR effect. Indeed we have obtained the decay curves which have no slow decay component at 1400 nm for sample A and at 1150 nm for sample B as shown in figure 3. We have confirmed that none of the decay curves have a slow component when the excitation intensity is changed from $5.5 \times 10^{16}$ cm$^{-3}$ to $3.0 \times 10^{17}$ cm$^{-3}$.

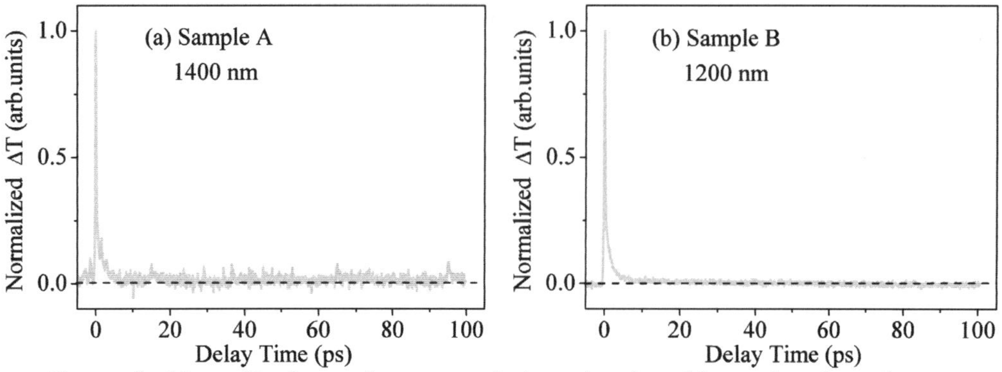


Figure 3. Normalized transient transmission signals without the slow decay components. These decay curves originate from the balance of the band filling effect and the BGR effect.

We believe that the observation of only the ultrafast decay component at room temperature will open the possibility of applications for high-speed electronics and switching devices.

**References**

[1] Walukiewicz W, Ager III J W, Yu K M, Liliental-Weber Z, Wu J, Li S X, Jones R E and Denlinger J D 2006 *J. Phys.*D **39** R83

[2] Wen Y -C, Chen C -Y, Shen C -H, Gwo S and Sun C -K 2006 *Appl. Phys. Lett.* **89** 232114

[3] Sun S -Z, Wen Y -C, Guol S -H, Lee H -M, Gwo S and Sun C -K 2008 *J. Appl. Phys* **103** 123513

[4] Jang D -J, Lin G -T, Wu C -L, Hsiao C -L, Tu L W and Lee M -E 2007 *Appl. Phys. Lett.* **91** 092108

[5] Tsai T -R, Chang C -F and Gwo S 2007 *Appl. Phys. Lett.* **90** 252111

[6] Sun C -K, Vallée F, Keller S, Bowers J E and DenBaars S P 1997 *Appl. Phys. Lett.* **70** 2004

[7] Davydov V Y, Emtsev V V, Goncharuk I N, Smirnov A N, Petrikov V D, Mamutin V V, Vekshin V A, Ivanov S V, Smirmov M B and Inushima T 2009 *Appl. Phys. Lett.* **75** 3297

[8] Chen F, Cartwright A N, Lu H and Schaff W J 2004 *Physica.* E **20** 308

Journal of Physics: Conference Series **193** (2009) 012054 doi:10.1088/1742-6596/193/1/012054

# Optical properties of $In_xAl_{1-x}N$ film and nanocolumns

**Masafumi Hashimoto[1], Jumpei Kamimura[1], Hideyuki Kunugita[1,2,3], Kazuhiro Ema[1,2,3], Akihiko Kikuchi[1,2,3], and Katsumi Kishino[1,2,3]**

[1]Department of Engineering and Applied Sciences Sophia University 7-1, Kioi-cho, Chiyoda-ku, Tokyo 102-8554, Japan
[2]Sophia Nanotechnology Research Center
[3]Core Research for Evolutional Science and Technology (CREST), Japan Science and Technology Agency

E-mail to first author : masafu-h@sophia.ac.jp

**Abstract.** We analyzed photoluminescence (PL) spectra of In-rich $In_xAl_{1-x}N$ nanocolumns (NCs; x = 0.80) and film (x = 0.82) using the free-electron recombination to bound (FERB) model. The FERB model fits well to the PL spectra of the In-rich $In_xAl_{1-x}N$ film and NCs, which implies that light-emitting properties of In-rich $In_xAl_{1-x}N$ are similar to those of InN. We obtained bandgap energies and electron carrier densities from the results of fitting the PL spectra.

## 1. Introduction

Because its bandgap energy is widely tunable (over the range 0.69–6.2 eV), indium aluminum nitride ($In_xAl_{1-x}N$) has great potential for application in light-emitting diodes, solar cells, and so on. However, its basic optical properties are not well understood, as it is difficult to grow because of its high immiscibility and the difference in thermal stability between AlN and InN. Recently, the basic properties of Al-rich $In_xAl_{1-x}N$ have been studied [1] because of its applicability as a distributed Bragg reflector or buffer layer during crystal growth. On the other hand, few reports exist on In-rich $In_xAl_{1-x}N$, although its light-emitting properties need to be clarified to realize light-emitting devices in the near-infrared region.

In this paper, we report the optical properties of In-rich $In_xAl_{1-x}N$, determined by optical absorption (OA) and photoluminescence (PL) spectroscopies of a high-quality $In_xAl_{1-x}N$ film (x = 0.82) and nanocolumns (NCs; x = 0.80) [2]. NCs are c-axis oriented column crystals. NCs of nitride semiconductors are known to be almost dislocation free and to show strong PL [3,4].

For analysis of luminescence properties of highly-doped semiconductors, the free-electron recombination to bound (FERB) model [5] is generally used. Even in nondoped InN, it has accumulated electrons on its surface and interface [6] and its Fermi level is located in the conduction band. Therefore, the FERB model is applicable to InN [7]. We used the FERB model to clarify the light-emitting properties of In-rich $In_xAl_{1-x}N$, and obtained several basic parameters by fitting the calculation to the observed PL spectra. We then discuss the adequacy of the FERB model and the differences between film and NCs.

## 2. Samples and Experimental Setup

### 2.1. Samples

All samples used in this study were grown by radio-frequency molecular beam epitaxy. For the film, the thickness is 370 nm, x = 0.82, the substrate is sapphire (0001), and the buffer layer is about 100 nm gallium nitride (GaN). The background carrier density was measured by Hall effect measurement (van der Pouw method). For the NCs, the height and width are approximately 1000 and 100 nm, respectively, x = 0.80, the substrate is Si (111), and the buffer layer is a few tens of nanometers of GaN. The background carrier density cannot be measured because of the unique style of the sample. Figure 1 shows a scanning electron microscope (SEM) image of the NCs sample.

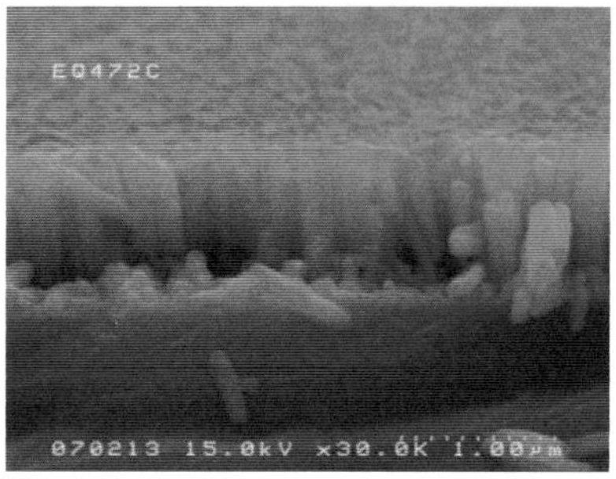


Figure 1. SEM image of NCs (bird's eye view).

### 2.2. Experimental setup

PL spectra were obtained using a mode-locked Ti:sapphire laser at a repetition rate of 76 MHz and excitation wavelength (energy) of 835 nm (~1.48 eV). OA measurements were performed using a halogen lamp. The spectra were detected using an InGaAs photodetector along with a monochromator.

## 3. Results

Figure 2a shows PL spectra of the film over the temperature range 10–300 K. Each spectrum has a shoulder on the high-energy side and a Gaussian tail on the low-energy side.

Figure 2b shows PL spectra of the NCs over the temperature range 10–300 K. Again, each spectrum has a Gaussian tail on the low-energy side, although the tail length differs from that for the film. The reason for this is discussed later.

Figure 3 shows OA spectra of the film. The PL peak energy and optical absorption edge differ by about 300 meV because of the Burstein–Moss shift [8,9]. This indicates that the film has a high carrier density, consistent with the Hall effect measurement, $n = 1.94 \times 10^{18}$ cm$^{-3}$.

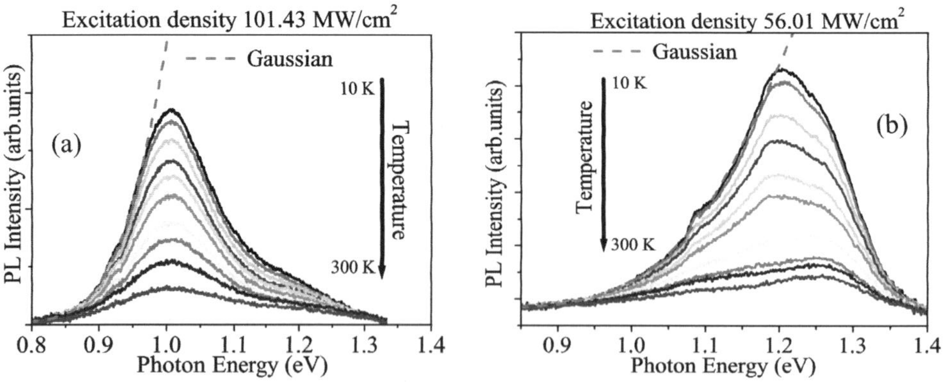


Figure 2. PL spectra: (a) Film; PL intensity decreases with increasing temperature. (b) NCs; the dashed line is the Gaussian function.

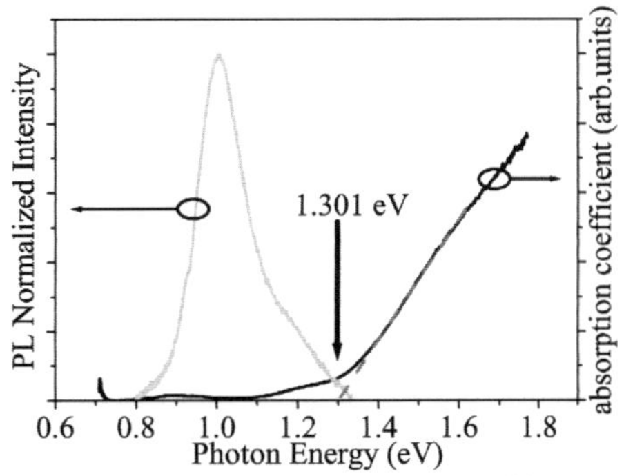


Figure 3. OA spectra of film. Burstein–Moss shift was observed.

## 4. Discussion

The FERB model is generally applicable to semiconductors if the following two conditions are satisfied: (1) The semiconductor has a high carrier density (>$10^{18}$ $cm^{-3}$ [5,10]). (2) The Thomas–Fermi screening length $R_s$ is smaller than the Bohr radius of electrons ($a_{Be}$) and larger than the Bohr radius of holes ($a_{Bh}$)—that is, $a_{Be} > R_s > a_{Bh}$. If these conditions are satisfied, the PL spectra have low-energy Gaussian tails because of localized impurity states.

In fact, the low-energy tails of our observed PL spectra show Gaussian tails, which indicates the applicability of the FERB model to our results. Because carrier density can be estimated by fitting the observed PL spectra to the FERB model, we can confirm that the estimated values satisfy conditions 1 and 2 as follows.

First, we calculate $a_{Be}$ and $a_{Bh}$, using electron and hole effective masses ($m_e^*$ and $m_h^*$) and electric permittivity ($\varepsilon$):

$$a_{Be} = a_0 \frac{\varepsilon}{m_e^*}, \quad a_{Bh} = a_0 \frac{\varepsilon}{m_h^*} \tag{4.1}$$

where $a_0$ is the Bohr radius, and $m_e^*$, $m_h^*$, and $\varepsilon$ were deduced by linear approximation between the values for InN and AlN. Thus, $a_{Be}$ and $a_{Bh}$ are determined only by indium concentration x. For the film, $a_{Be} = 8.1 \times 10^{-7}$ cm and $a_{Bh} = 3.2 \times 10^{-8}$ cm; for the NCs, $a_{Be} = 7.6 \times 10^{-7}$ cm and $a_{Bh} = 3.1 \times 10^{-8}$ cm.

We fitted PL spectra at 10 K for both the film and NCs to the FERB model, taking care to fit to the low-energy side of the PL spectra. Figure 4 shows the PL spectra and fitted results.

From the fitting results, we determined that the bandgap energy and carrier density for the film are 0.98 eV and $2.1 \times 10^{19}$ $cm^{-3}$ and those for the NCs are 1.29 eV and $1.5 \times 10^{20}$ $cm^{-3}$, respectively. Using these carrier densities, we calculated $R_s$ by the following equation:

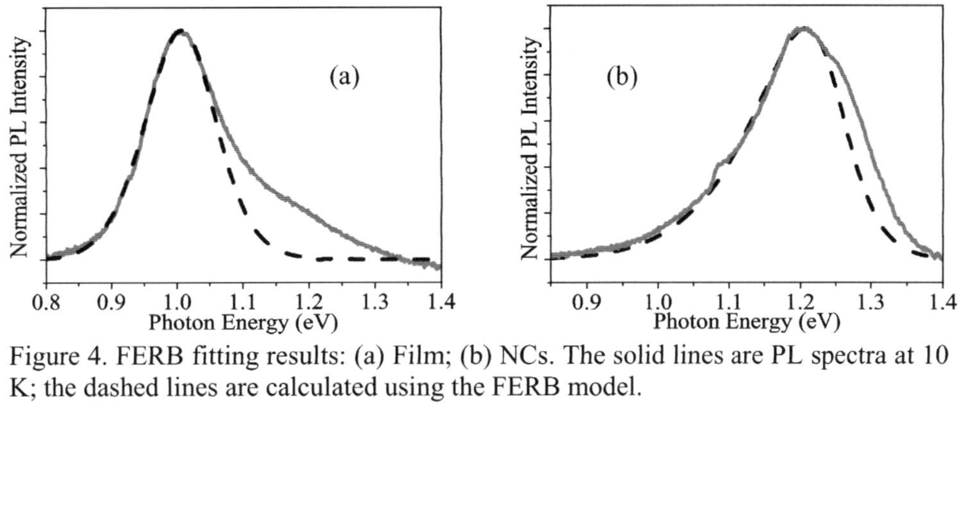


Figure 4. FERB fitting results: (a) Film; (b) NCs. The solid lines are PL spectra at 10 K; the dashed lines are calculated using the FERB model.

$$R_s \cong \frac{a_{Be}}{2}\left(na_{Be}^3\right)^{-1/6} \tag{4.2}$$

Thus, $R_s = 2.7 \times 10^{-7}$ cm and $1.9 \times 10^{-7}$ cm for the film and NCs, respectively, and we can confirm that the parameters, $R_s$, $a_{Be}$, and $a_{Bh}$, are consistent with condition 2, $a_{Be} > R_s > a_{Bh}$. Because the applicability of the FERB model to our In-rich $In_xAl_{1-x}N$ samples is verified, we believe that the estimated bandgaps are reliable.

The length of the low-energy tail for the NCs is longer than for the film. As the FERB model shows, increasing carrier density leads to decreasing $R_s$ and increasing impurity potential, resulting in increased tail length [5]. Another possible reason for the long tail is that each NC may have slightly different bandgap energies. If that were the case, however, the high-energy side of the PL spectra should also show an Gaussian form. Therefore, we consider that the inhomogeneous broadening of the bandgap energy in NCs can be neglected.

The estimated value of the carrier density for the film is about 10 times larger than that obtained by the Hall effect measurement. However, it is possible that the Hall effect measures carrier density only where current flows easily, whereas analysis of PL spectra is equivalent to extracting information inside a laser spot. Therefore, we believe that the difference is due to the principles of these measurement methods. We also believe that the high electron density for the NCs is due to increased surface area, because In-rich $In_xAl_{1-x}N$ accumulates electrons on its surface and interface [11].

## 5. Conclusion

We reported the optical properties of In-rich $In_xAl_{1-x}N$ (film: x = 0.82, NCs: x = 0.80) using PL and OA spectroscopies. We found the FERB model to be appropriate for analyzing PL spectra of In-rich $In_xAl_{1-x}N$. Analyzing PL spectra by this model, we found that light-emitting properties of In-rich $In_xAl_{1-x}N$ are similar to those of InN. Bandgap energy and electron carrier density were also obtained by fitting the PL spectra. Carrier density is higher for NCs than for the film. We believe that this indicates a relationship between surface/interface and carrier accumulation.

## References

[1] Carlin J -F and Ilegems M 2003 *Appl. Phys. Lett.* **83** 668
[2] Kamimura J, Kouno T, Ishizawa S, Kikuchi A and Kishino K 2007 *J. Cryst. Growth* **300** 160
[3] Kikuchi A, Yamano K, Tada M and Kishino K 2004 *Phys. Stat. Sol.* (b) **241** 2754
[4] Sekiguchi H, Nakazato T, Kikuchi A and Kishino K 2007 *J. Cryst. Growth* **300** 259
[5] Arnaudov B, Paskova T, Paskov P P, Magnusson B, Valcheva E, Monemar B, Lu H, Schaff W J, Amano H and Akasaki I 2004 *Superlattices and Microstructures* **36** 563
[6] Piper L F J, Veal T D, McConville C F, Lu H and Schaff W J 2006 *Appl. Phys. Lett.* **88** 252109
[7] Fukunaga K, Hashimoto M, Kunugita H, Kamimura J, Kikuchi A, Kishino K and Ema K 2009 *J. Lumin*, in press
[8] Burstein E 1954 *Phys. Rev.* **93** 632
[9] Moss T S 1954 *Proc. Phys. Soc.* B **67** 775
[10] Arnaudov B, Paskova T, Goldys E M, Evtimova S and Monemar B 2001 *Phys. Rev.* B **64** 045213
[11] King P D C, Veal T D, Adikimenakis A, Lu H, Bailey L R, Ilipoulos E, Georgakilas A, Schaff W J and McConville C F 2008 *Appl. Phys. Lett.* **92** 172105

# Anderson localization of light in a random configuration of nanocolumns

**Yuta Inose** [1,2], **Tomi Ohtsuki** [1,2,3], **Hideyuki Kunugita** [1,2,3], **Kazuhiro Ema** [1,2,3], **Masaru Sakai** [1,2,3], **Akihiko Kikuchi** [1,2,3] **and Katsumi Kishino** [1,2,3]

[1] Department of Engineering and Applied Sciences, Sophia University, 7-1 Kioi-cho, Chiyoda-ku, Tokyo 102-8554, Japan
[2] CREST, Japan Science and Technology Agency
[3] Sophia Nanotechnology Research Center, Sophia University

E-mail: y-inose@sophia.ac.jp

**Abstract.** We show pseudogap maps of Anderson localization of light adopting the parameters of self-organized nanocolumn samples, which consist of random arrays of parallel nanosized columnar semiconductor crystals. The maps indicate the parametric dependence of the localization effect. To obtain the maps, we simulated light propagation in open random media using the two-dimensional finite-difference time-domain method and analyzed the simulation results by Fourier transformation. We found that the shape of the pseudogaps is close to the one of bandgaps in photonic crystals. We conclude that strong localization of light occurs because of interference by average Bragg diffraction, not strong Mie resonant peaks.

## 1. Introduction

The possibility of Anderson localization of light, the electromagnetic wave localization similar to electron localization in disordered solids, in random dielectric systems has been discussed [1, 2]. However, theoretical and experimental studies for light propagation in random media are much more difficult than those in photonic crystals. As a result, the light localization effect is not well understood. Here we obtain pseudogap maps for self-organized nanocolumn samples [3] composed of nanosized columnar semiconductor crystals. A scanning electron microscope (SEM) image of nanocolumns is shown in figure 1.

The column diameter varies in the range from about 50 to 400 nm, and the height is around 1 μm. Nanocolumns have been fabricated using gallium nitride (GaN), indium nitride (InN), and a mixed crystal of the two, indium gallium nitride (InGaN). The band gap energies of GaN and InN are about 3.50 eV (near-ultraviolet emission) and 0.65 eV (near-infrared emission), respectively. Therefore, we can control the emission wavelength to be within near ultraviolet to near infrared by adjusting the composition of $In_xGa_{1-x}N$. Because nanocolumns show stronger photoluminescence than thin-films [4], they have attracted much attention as novel materials for visible light-emitting devices. We have conducted a numerical study on the localization of electromagnetic waves in nanocolumn ensembles.

Conventional nanocolumn samples are random media where the positioning of the columns is disordered, as shown in figure 1. To observe the localization effect in such systems, the radius of the nanocolumns ($r$), the filling fraction of nanocolumns ($\Phi$), and the angular frequency of electromagnetic

Journal of Physics: Conference Series **193** (2009) 012055 doi:10.1088/1742-6596/193/1/012055

waves ($\omega$) must be properly tuned. We have simulated light propagation in nanocolumn ensembles with various parameters and studied the parametric dependence of the localization behavior.

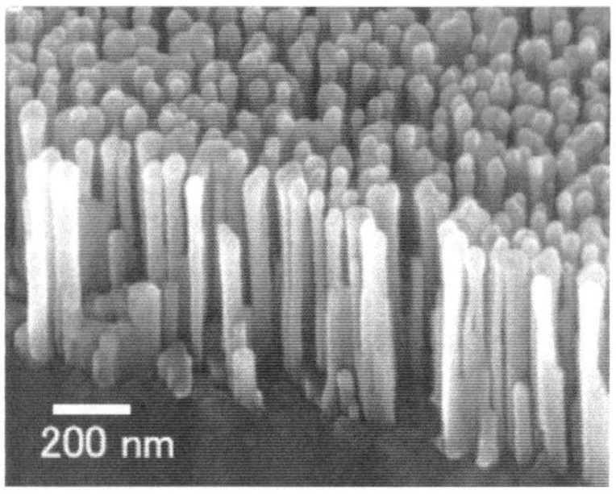


**Figure 1.** SEM image of GaN nanocolumns.

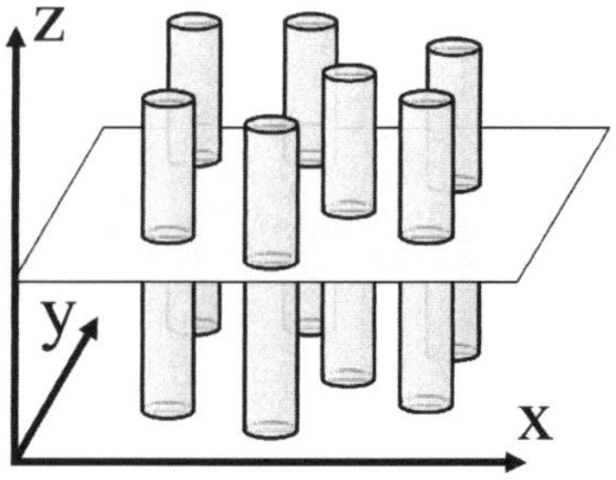


**Figure 2.** Schematic of the 2D FDTD model for nanocolumns.

## 2. Calculation

We have used the two-dimensional (2D) finite-difference time-domain (FDTD) method [5] to simulate the temporal propagation of electromagnetic waves in nanocolumn ensembles allocated within a finite area. The FDTD method is a numerical method in which Maxwell's equations are discretized using central difference approximations of the space and time partial derivatives. A schematic view of the 2D FDTD model is illustrated in figure 2. The electromagnetic waves propagate in the *x-y* plane. Maxwell's equations in the case of transverse-magnetic (TM) field and transverse-electric (TE) field can be described by

$$\varepsilon(\boldsymbol{r})\varepsilon_0\frac{\partial}{\partial t}E_z(\boldsymbol{r},t) = \frac{\partial}{\partial x}H_y(\boldsymbol{r},t) - \frac{\partial}{\partial y}H_x(\boldsymbol{r},t) \tag{1a}$$

$$\mu_0\frac{\partial}{\partial t}H_x(\boldsymbol{r},t) = -\frac{\partial}{\partial y}E_z(\boldsymbol{r},t) \tag{1b}$$

$$\mu_0\frac{\partial}{\partial t}H_y(\boldsymbol{r},t) = \frac{\partial}{\partial x}E_z(\boldsymbol{r},t), \tag{1c}$$

$$\mu_0\frac{\partial}{\partial t}H_z(\boldsymbol{r},t) = -\frac{\partial}{\partial x}E_y(\boldsymbol{r},t) + \frac{\partial}{\partial y}E_x(\boldsymbol{r},t) \tag{2a}$$

$$\varepsilon(\boldsymbol{r})\varepsilon_0\frac{\partial}{\partial t}E_x(\boldsymbol{r},t) = \frac{\partial}{\partial y}H_z(\boldsymbol{r},t) \tag{2b}$$

$$\varepsilon(\boldsymbol{r})\varepsilon_0\frac{\partial}{\partial t}E_y(\boldsymbol{r},t) = -\frac{\partial}{\partial x}H_z(\boldsymbol{r},t), \tag{2c}$$

respectively, where $\varepsilon_0$ and $\mu_0$ are the electric permittivity and magnetic permeability of vacuum, respectively, and $\varepsilon(\boldsymbol{r})$ is the relative electric permittivity. The refractive index $n(\boldsymbol{r})$ is related to $\varepsilon(\boldsymbol{r})$ by $\varepsilon(\boldsymbol{r}) = n^2(\boldsymbol{r})$. The space increment is $\Delta x = \Delta y = 5\,\text{nm}$, and the time increment is $\Delta t = 1 \times 10^{-17}\,\text{s} < \Delta x/\left(c\sqrt{2}\right)$, where $c$ is the velocity of light in vacuum. To model open systems, we used Berenger's perfectly matched layer [6] for the boundary conditions in the FDTD simulation.

The sample area is 4.5 µm square and consists of random arrays of parallel nanocolumns with a constant radius ($r = 50\,\text{nm}$) and a constant refractive index ($n = 2.35$, the value for GaN in the visible range). Nanocolumns do not touch each other and they are embedded in vacuum. Figure 3 shows an example of a simulated system with $\Phi = 0.4$.

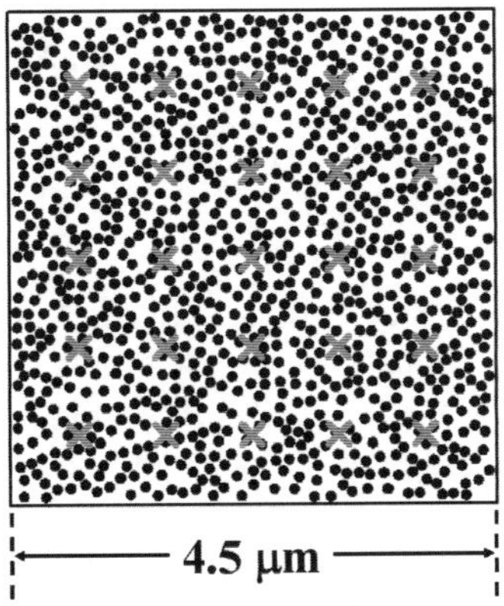


**Figure 3.** Example of a simulated system with $\Phi = 0.4$. The circles are nanocolumns, and the crosses are antennas. In the actual simulation, 400 antennas are evenly spaced in the system.

We have adopted several values of $\Phi$ ranging from 0.05 to 0.90, at intervals of 0.05, and calculated for the respective systems. In each case, we have irradiated a Gaussian white pulse ($\delta$-function pulse) onto the whole sample area at $t = 0$. After irradiation, light waves gradually escape through the boundaries because the simulation system is open. In the FDTD simulation, the temporal evolution of the electromagnetic field at each location is recorded via an array of 400 antennas evenly spaced in the system. From the recorded electromagnetic signals, we can obtain the localized light spectrum by averaging several power spectra that are Fourier transformed from each signal recorded in a time window $[T_1, T_2]$. Figure 4 shows an example of a localized light spectrum with TM mode, where $\Phi = 0.4$ and the time window is [2 ps, 4 ps]. The intensity was normalized by the intensity of incident light. We found five strong localization ranges designated as "frequency windows." The shape of this spectrum depends only on $T_1$, $T_2$, and $\Phi$. In the case of other random configurations with the same parameters, the change in the shape is very small. Furthermore, we can obtain the Q-factor spectrum, which characterizes the localization effect, by taking the localized light spectra of different time windows and deriving the intensity decay times as a function of the frequency.

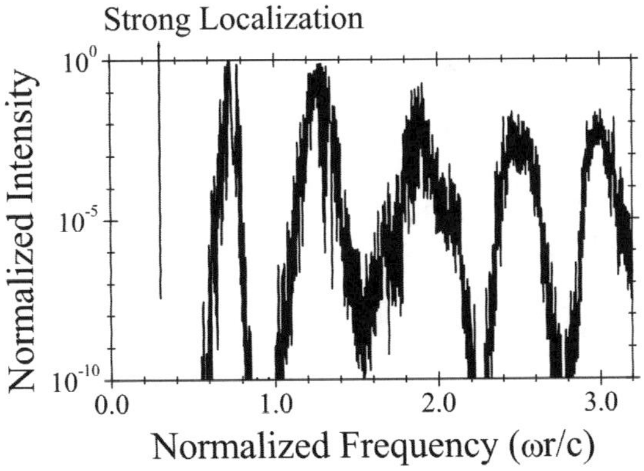


**Figure 4.** Example of a localized light spectrum with TM mode, where $\Phi = 0.4$ and the time window is [2 ps, 4 ps] (note the logarithmic vertical scale).

## 3. Results and Discussion

From the simulation and analysis above, we have succeeded in obtaining pseudogap maps using these Q-factor spectra. Figures 5 and 6 show the pseudogap maps for a random medium. The maps indicate the parametric dependence of the localization effect. This is the first detailed report on Anderson localization of light. We found that the shapes of the pseudogap maps are similar to one another.

We also calculated photonic band structures by adopting the same parameters using the plane wave expansion method [7] and compared them with the present calculation for random media. The gap map for a hexagonal photonic crystal with TM mode is shown in figure 5, where the areas enclosed by the dashed lines indicate the photonic band gaps. In the case of TE polarization, photonic band gaps are not formed. We conclude that the shape of the pseudogap map with TM mode is close to the gap map for the corresponding photonic crystal. It has been pointed out that the frequency windows do not correspond to the maxima of the scattering cross section in the Mie resonances of a single cylinder, but the detailed physical origin has not been clarified [8, 9]. In the case of photonic crystals, band gaps are formed if Bragg diffraction conditions are satisfied. Therefore, from figure 5 we conclude that light localization phenomenon is very effective under the condition of "average Bragg diffraction" and that the influence of Mie scattering is much smaller than the influence of average Bragg diffraction.

We have observed random lasing with TE polarization, $\Phi = 0.45$ and $\omega r/c = 1.13$ in a self-organized GaN nanocolumn sample at the emission wavelength [10]. This phenomenon occurs because the localization effect in the random medium plays the role of an optical resonator and nanocolumns also act as an efficient gain medium at the resonant wavelength. In fact, we find that the optimal parameter $\omega r/c$ for strong light localization with $\Phi = 0.45$ ranges from about 1.05 to 1.35 from figure 6. On the other hand, we have not observed random lasing in other samples which do not indicate high-Q at the emission wavelength on figure 6. In this way, we can estimate optimal conditions for obtaining high-Q samples from pseudogap maps.

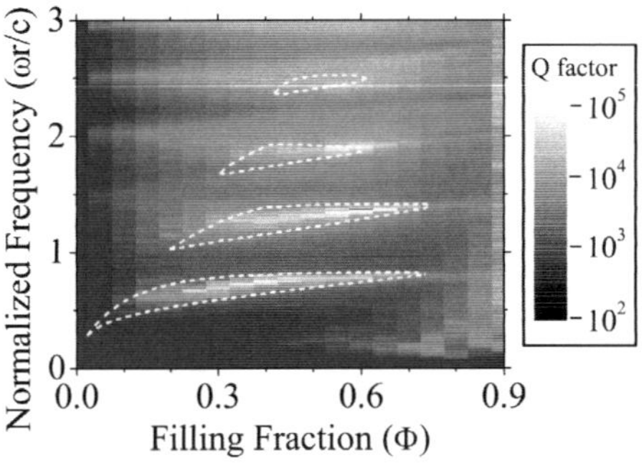


**Figure 5.** Pseudogap map for a random medium with TM polarization, and gap map for a hexagonal photonic crystal (photonic band gaps are inside dashed lines).

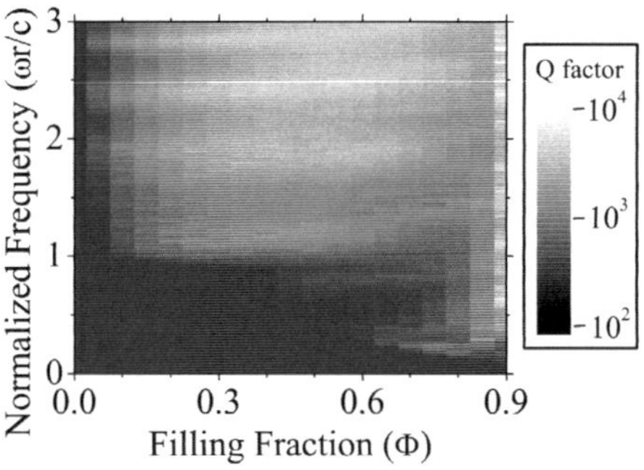


**Figure 6.** Pseudogap map for a random medium with TE polarization.

## References

[1] John S 1984 *Phys. Rev. Lett.* **53** 2169
John S 1987 *Phys. Rev. Lett.* **58** 2486
[2] Cao H, Zhao Y G, Ong H C, Ho S T, Dai J Y, Wu J Y, Chang R P H 1998 *Appl. Phys. Lett.* **73** 3656
Cao H, Zhao Y G, Ho S T, Seelig E W, Wang Q H and Chang R P H 1999 *Phys. Rev. Lett.* **82** 2278
[3] Yoshizawa M, Kikuchi A, Mori M, Fujita N and Kishino K 1997 *Jpn. J. Appl. Phys.* Part 2 **36** L459-62
[4] Kikuchi A, Yamano K, Tada M and Kishino K 2004 *phys. stat. sol.* (b) **241** pp 2754-8
[5] Taflove A and Hagness S C 1995 *Computational Electrodynamics: The Finite-Difference Time-Domain Method* (Norwood, MA: Artech House)
[6] Berenger J P 1995 *J. Comput. Phys.* **114** 185
[7] Pliha M and Maradudin A A 1991 *Phys. Rev.* B **44** 8565
[8] Sigalas M M, Soukoulis C M, Chan C T and Turner D 1996 *Phys. Rev.* B **53** 8340
[9] Sebbah P and Vanneste C 2002 *Phys. Rev.* B **66** 144202
[10] Sakai M, Kishino K, Kikuchi A, Sekiguchi H, Inose Y, Ema K and Ohtsuki T 2008 *Int. Workshop on Nitride semiconductors (Montreux, Switzerland)* We4-D4

Journal of Physics: Conference Series **193** (2009) 012056 doi:10.1088/1742-6596/193/1/012056

# Second harmonic generation from photonic structured GaN nanowalls

**Takahiro Soya**[1], **Yuta Inose**[1,3], **Hideyuki Kunugita**[1,2,3], **Kazuhiro Ema**[1,2,3], **Kouji Yamano**[1,3], **Akihiko Kikuchi**[1,2,3] **and Katsumi Kishino**[1,2,3]

[1] Department of Engineering and Applied Sciences, Sophia University 7-1, Kioi-cho, Chiyoda-ku, Tokyo 102-8554, Japan
[2] Sophia Nanotechnology Research Center
[3] Core Research for Evolutional Science and Technology (CREST), Japan Science and Technology Agency

E-mail : t-soya@sophia.ac.jp

**Abstract**. We observed large enhancement of reflected second harmonic generation (SHG) using the one-dimensional photonic effect in regularly arranged InGaN/GaN single-quantum-well nanowalls. Using the effect when both fundamental and SH resonate with the photonic mode, we obtained enhancement of about 40 times compared with conditions far from resonance.

## 1. Introduction

Second-order susceptibilities $\chi^{(2)}$ of GaN and InGaN are higher than those of conventional second harmonic (SH) crystals such as KDP and $LiNbO_3$ [1]. Nevertheless, the efficiency of SH generation (SHG) in a bulk GaN or InGaN/GaN quantum-well (QW) sample is very small because these are highly dispersive materials, preventing fulfillment of the phase-matching condition.

To increase the SH intensity, we use the photonic effect in an InGaN/GaN single-quantum-well (SQW) nanowall. A nanowall is a plate-like crystal with a thickness of several hundred nanometers and height of a few micrometers [2]. Regularly arrayed nanowalls are considered to be a one-dimensional photonic crystal (1-D PC). We have observed significantly enhanced SHG when both the fundamental and generated SH field are resonant with photonic eigenstates.

## 2. Principle of photonic enhancement in SHG

A schematic of the experimental configuration is shown in figure 1. A femtosecond pulse (fundamental frequency: $\omega$) was irradiated on the upper surface of the sample along the x axis, and an SH beam ($2\omega$) contained in the reflected fundamental was detected. The direction of the fundamental is specified by the incident angle $\theta$. The principle of SH enhancement is as follows. There are three situations. First, if the in-plane wave vector of fundamental light $[\mathbf{k}_{//}(\omega) = \mathbf{k}(\omega)\sin\theta]$ matches the wave vector of the 1-D PC mode in the nanowall, the fundamental EM field is strongly confined within the PC. Second, if the in-plane wave vector of the generated SH light $[\mathbf{k}_{//}(2\omega) = \mathbf{k}(2\omega)\sin\theta]$ matches that of the 1-D PC mode, the SH light can be efficiently coupled back in the surrounding vacuum. Third, the previous two resonance conditions synchronize,

Journal of Physics: Conference Series **193** (2009) 012056 doi:10.1088/1742-6596/193/1/012056

which causes significant enhancement. These three situations are called the resonant-nonresonant (R-NR), nonresonant-resonant (NR-R), and resonant-resonant (R-R) conditions, respectively [3,4].

Cowan *et al.* have theoretically shown about $10^6$-fold enhancement in a 2-D GaAs PC slab compared to an untextured sample [3]. Moreover, Torres *et al.* have experimentally obtained about $10^5$-fold enhancement using a 1-D GaN PC slab [5].

When the crystal c axis is the z direction, considering the crystallite symmetry of GaN and InGaN (class of crystal symmetry $6mm$), the induced polarization inside the nanowall is taken as

$$\begin{aligned}
\mathbf{P}_x(2\omega) &= 2\varepsilon_0\chi^{(2)}_{xzx}E_x(\omega)E_z(\omega)\\
\mathbf{P}_y(2\omega) &= 2\varepsilon_0\chi^{(2)}_{yzy}E_y(\omega)E_z(\omega)\\
\mathbf{P}_z(2\omega) &= \varepsilon_0\chi^{(2)}_{zxx}E_x(\omega)^2+\varepsilon_0\chi^{(2)}_{zyy}E_y(\omega)^2+\varepsilon_0\chi^{(2)}_{zzz}E_z(\omega)^2
\end{aligned} \qquad (1)$$

where $E_i(\omega)$ is the i-component of the fundamental electric field. Equation (1) shows that SH is generated with $p$ polarization when the fundamental was irradiated on the sample with either $p$ or $s$ polarization.

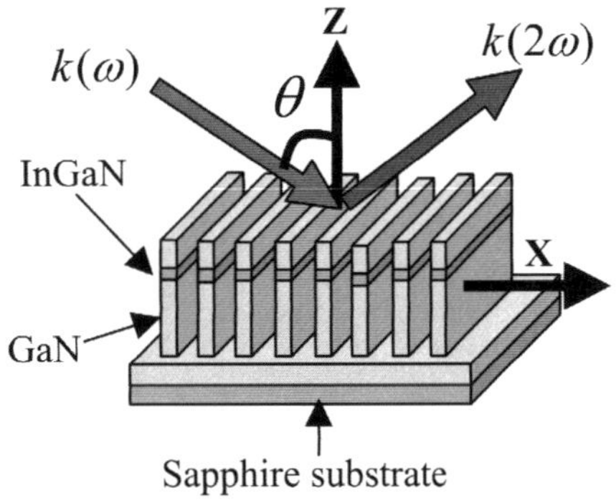


Figure 1. Schematic representation of the experimental configuration

(a) top view (b) bird's-eye view

Figure 2. SEM images of the nanowall sample

## 3. Experimental set-up

Figure 2 shows top view and bird's-eye view scanning electron microscopy (SEM) images of our nanowall sample. The nanowall's periodicity and width are 504 nm and 236 nm, respectively. The patterned area is 150 × 150 $\mu$m$^2$. Pulses (pulse width ≈ 100 fs) from a mode-locked Ti:sapphire laser were used for fundamental light and focused onto the sample with a spot size of ~180 $\mu$m. The polarization of the fundamental and SH was determined using a half-wavelength plate and a polarizer. The angle of incidence was fixed, and the fundamental wavelength dependency of the SH intensity was measured in the range from about 770 nm to 915 nm.

## 4. Experimental result

### *4.1. R-NR condition*

Figure 3 shows an example of the photonic band structure (PBS) ($\theta = 20°$) calculated by the transfer matrix method. Because the InGaN layer is thin (a few nanometer wide), the PBS was calculated assuming that the nanowall was composed only of GaN. The refractive index dispersion of GaN was used in the calculation. The vertical axis is the normalized frequency, and the horizontal axis is the in-plane wavenumber $k_x = k\sin\theta$. The squares show the PC mode for the *s*-polarized fundamental. The solid line shows the dispersion of light in vacuum in the case of $\theta = 20°$. The SH intensity is resonantly enhanced when the dispersion of incident light crosses the squares. This corresponds to the R-NR condition. In this case, we can expect an increase in SH intensity at 2.85 eV.

Figure 4 shows the observed SH intensity at $\theta = 20°$. The fundamental was *s*-polarized, and a *p*-polarized SH signal was detected. We observed a clear enhancement of SH intensity at 2.85 eV.

### *4.2. NR-R and R-R conditions*

Figure 5 shows the PBS in the case of $\theta = 23.4°$. The squares show the PC mode for the *s*-polarized fundamental. the triangles show the *p*-polarized mode plotted at half the frequency and in-plane wavenumber; therefore, they show the points where the generated SH beam resonates with the PC mode. The solid line shows the dispersion of light in vacuum in the case of $\theta = 23.4°$. The SH intensity is resonantly enhanced when the dispersion of incident light crosses the squares and/or the triangles. In this case, we can expect an increase at 2.92 eV (NR-R condition) and substantial enhancement at 2.78 eV (R-R condition).

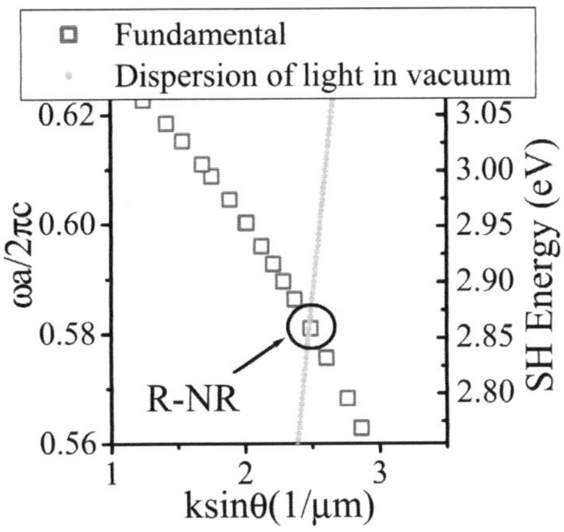


Figure 3. Calculated PBS in the case of $\theta = 20°$

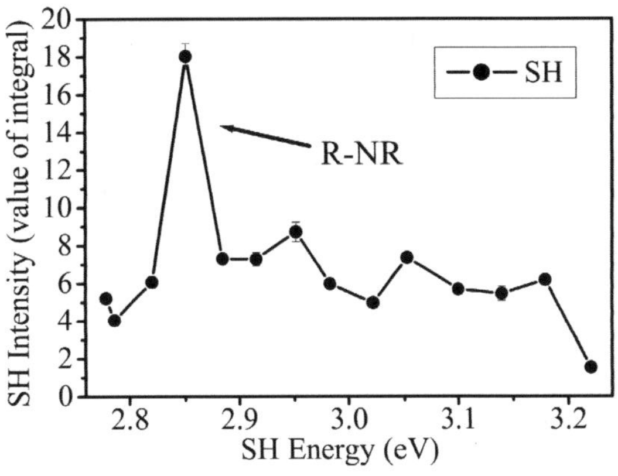


Figure 4. SH intensity at $\theta = 20°$

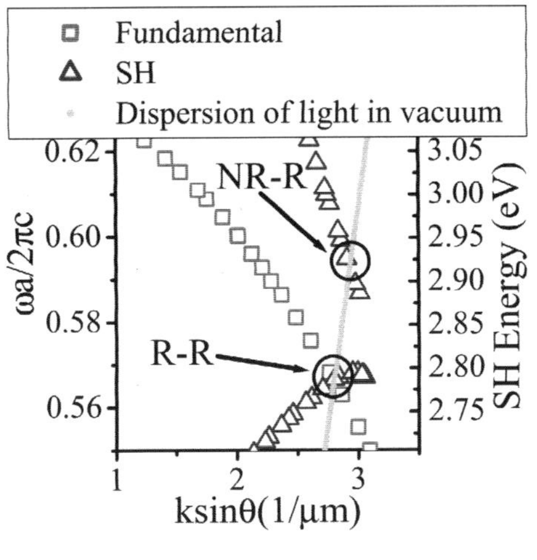


Figure 5. Calculated PBS in the case of $\theta = 23.4°$

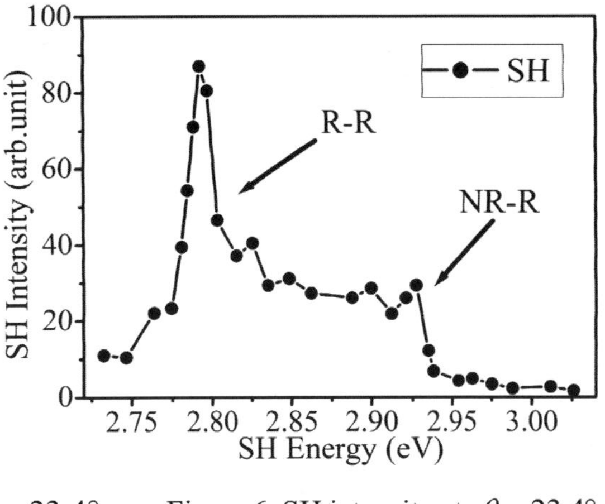


Figure 6. SH intensity at $\theta = 23.4°$

Figure 6 shows the observed SH intensity at $\theta = 23.4°$. We observed a small enhancement of SH intensity at 2.92 eV and a significant enhancement at 2.78 eV. At the R-R point, the degree of SH enhancement was about 40 times that of about 3 eV, far from the resonance.

Figure 7 shows the fundamental polarization dependency of SH intensity in the R-R condition. SH intensity decreased remarkably when the polarization of the fundamental was changed to *p* polarization, because the R-R condition is satisfied only for the fundamental light with *s* polarization.

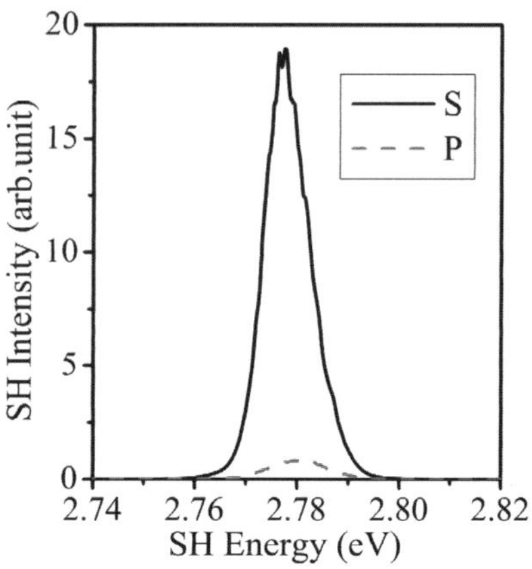


Figure 7. Fundamental polarization dependency of SH intensity in R-R condition

## 5. Discussion

At first glance, it seems that our enhancement factor of ~40 is much smaller than that observed by Torres *et al.* [4]. However, our enhancement factor is not a comparison between a film and a nanowall, but between the R-R point and a nonresonant point. There may be an enhancement even at the nonresonant point, because the wavelength region measured is not significantly away from the resonance point. In fact, Torres *et al.* also measured the change in SH intensity as a function of the fundamental wavelength [6], and their result is almost the same as ours.

It should be mentioned that the upper surface of our sample is not flat (figure 2), and thus, the generated SH is diffused because of scattering. Therefore, we could not detect all of the generated SH. In fact, the reflected fundamental light of the nanowall was more extensive than that of our film sample. Of course, imperfection of the period of the nanowall also influences the enhancement factor.

Several improvements could be made in our sample to obtain a stronger SH signal. If the upper surface of the nanowall is flat, a greater amount of the fundamental will be incident on the PC core, and the influence of scattering will be smaller in the radiation process of the SH beam. In addition, Cowan *et al.* showed that SH intensity is also influenced by the quality factor $Q$ of the resonant mode [3], which is decided by the sample's structure in the z direction. Progress in controlling the growth of nanowalls will lead to achieving a perfect period and the design of nanowalls with a high $Q$.

## 6. Conclusion

We have observed enhancement of reflected SH intensity using the 1-D PC effect in an InGaN/GaN SQW nanowall. By calculating the PBS, we found the R-R point at which both the fundamental and SH were resonant with the PC mode. In the R-R condition, we observed significantly enhanced SH signal. There is still great potential for further increasing the SH intensity by improving the sample structure.

## References

[1] Boyd R 1992 *Nonliner Optics* (Academic, Boston, MA) p 50
[2] Kikuchi A *et al.* 2007 MRS Fall Meeting, Q4.6, Boston,USA
[3] Cowan A R and Young J F 2002 *Phys. Rev.* B **65** 085106
[4] Torres J, Le Vassor d'Yerville M, Coquillat D, Centeno E, and Albert J P 2005 *Phy. Rev.* B **71** 195326
[5] Bouchoule S, Boubanga-Tombet S, Le Gratiet L, Le Vassor d'Yerville M, Torres J, Chen Y, and Coquillat D 2007 *J. Appl. Phys.* **101** 043103
[6] Torres J, Coquillat D *et al.* 2005 *Phy. Rev.* B **69** 085105

# Femtosecond electro-optic effect in (Cd,Mn)Te single crystals

**A S Cross,[1] D Kochanowska,[2] M Witkowska-Baran,[2] A Mycielski,[2] M Mikulics,[3] D Grützmacher[3] and Roman Sobolewski[1]**

[1]Department of Electrical and Computer Engineering and Laboratory for Laser Energetics, University of Rochester, Rochester, NY, 14627-0231, USA
[2]Institut of Physics, Polish Academy of Sciences, PL-02668 Warszawa, Poland
[3]Institute of Bio- and Nanosystems, Research Centre Jülich, D-52425 Jülich, Germany

roman.sobolewski@rochester.edu

**Abstract**. We present our studies on the femtosecond electro-optic (EO) effect in high quality $Cd_{0.88}Mn_{0.12}Te$ (CMT) single crystals, grown using a modified vertical Bridgman method. Our time-resolved experiments were performed using a coplanar transmission line setup, which incorporated a free-standing, low-temperature-grown GaAs photoconductive switch as a sub-picosecond electrical pulse generator and the tested CMT crystal as an active EO transducer. We determined the *r*-coefficient ($r_{41}$) of CMT at the high (THz) frequency range to be 29.4 pm/V, an order of magnitude greater than that measured for the same sample at low (MHz) frequencies. The observed suppression of the EO coefficient at low frequencies was due to effective electric field screening by free holes in the crystal volume. Comparison to the standard EO sampling scheme based on the $LiTaO_3$ transducer showed that CMT exhibited the same sub-picosecond temporal resolution with significantly greater voltage sensitivity. We also did not observe, the so-called dielectric loading typical for $LiTaO_3$-based samplers.

## 1. Introduction

The (Cd,Mn)Te system is among the most-studied, II-VI, diluted-magnetic semiconductors, due to its strong magnetic properties and the stable zinc-blend structure for Mn concentrations up to 0.70, resulting in the very wide energy band gap tuning range. Applications of the (Cd,Mn)Te compound vary from spintronics and Faraday isolators to x- and γ-ray detectors [1].

An electro-optic sampling (EOS) technique is a powerful tool allowing a minimally invasive means of characterizing performance of ultra-high-speed (THz bandwidth) electronic and optoelectronic devices, such as novel transistors, diodes, photodetectors, and transmission lines by exploiting the linear EO Pockels effect [2]. The EOS system enables one to precisely resolve sub-picosecond nonlinear transient effects and characterize ultrafast electrical pulses [3].

We present here EO experiments demonstrating that our $Cd_{0.88}Mn_{0.12}Te$ (CMT) single crystal is an excellent substitute for the commonly used $LiTaO_3$, and showing that the CMT EO-coefficient $r_{41}$ at THz frequencies (THz-f) is much greater than the low-frequency (low-f) values reported by Yariv [4] and Chen [5].

## 2. Configuration of EOS experiment and fabrication details

Our EOS system was realized using a 76-MHz repetition rate Ti:Sapphire laser providing 100-fs-wide pulses with an 800-nm wavelength. The laser beam was split by a 70/30 beam splitter into two beams:

Journal of Physics: Conference Series **193** (2009) 012057 doi:10.1088/1742-6596/193/1/012057

the excitation (pump) beam and sampling (probe) beam. The pump beam, modulated at 256 kHz by an acousto-optic (AO) modulator, illuminated our device under test (DUT), namely, a free-standing photoconductive switch incorporated into a coplanar strip (CS) transmission line [3]. The switches were produced from low-temperature grown (200-300 °C) GaAs (LT-GaAs) rapidly post-annealed at 600 °C. The 25x25 $\mu m^2$ LT-GaAs chips, removed from the native substrate, were placed into pre-etched 500-nm-deep wells in the MgO substrate [6]. Next, a Ti/Au CS line was deposited above the switch sides, securing it between 20-µm-wide signal and ground lines, separated by a 20-µm gap.

The sampling beam, on the other hand, traversed a computer controlled delay line, half-wave plate, polarizer, and was ultimately focused by a microscope objective upon our CMT EO transducer placed above the CS line near the DUT. The CMT crystals used in this study were fabricated using a modified vertical Bridgman method, in which a seed crystal of CdTe was placed at the bottom of a growth ampoule beneath a melt of Cd, Te and Mn. The ampoule was then slowly lowered from the furnace allowing the solution to crystallize into a zinc-blend structure. As-grown CMT is inherently *p*-type due to Cd vacancies and can have conductivity as high as 0.1 S/cm [7]. The crystals for EO transducers were cut along the (110) face to provide the greatest Pockels effect.

In the presence of an electrical field transient, the EO transducer acts as an optical intensity modulator for the transmitted sampling beam pulses with the output directed to a polarizing beam splitter and two photodetectors connected to a lock-in amplifier synchronized to the AO modulation frequency of the excitation beam. The orientation of the transducer was such that the EO-axis was aligned parallel to the electric field, and 45° with respect to the probe beam polarization.

### 3. Electro-optic measurements and analysis

Figure 1 presents an ultrafast EOS measurement taken using the $Cd_{0.88}Mn_{0.12}Te$ crystal as an EO transducer. We observe a very clean pulse with the full-width-at-half-maximum (FWHM) of ~1 ps and excellent signal-to-noise ratio. Since the free-standing LT-GaAs switch used in our experiments as the DUT is well-known for generating electrical transients as short as 360 fs [3], our CMT transducer faithfully recorded the photoswitch signal, which dispersed somewhat due to approximately 250-µm propagation distance between the switch position on the CS line and the sampling spot on the CMT crystal.

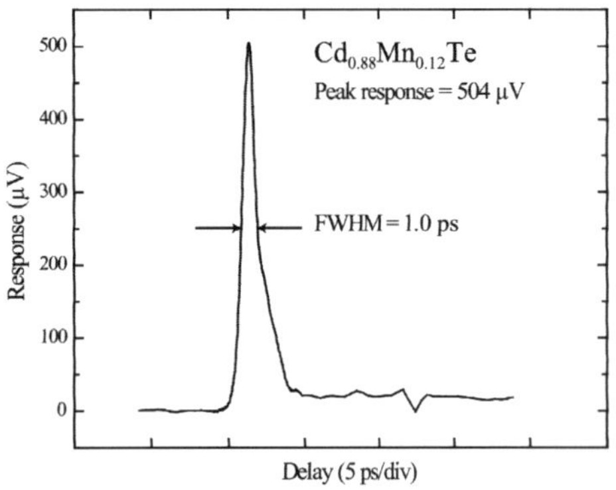


Fig. 1. Ultrafast EOS response with CMT transducer. The LT-GaAs switch operated under 4-mW pump illumination and 20-V bias.

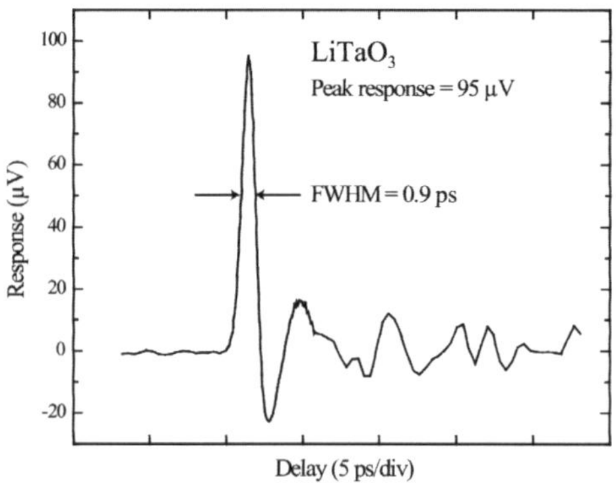


Fig. 2. Ultrafast EOS response with $LiTaO_3$ transducer. The LT-GaAs switch operated under 4-mW pump illumination and 20-V bias.

For comparison purposes, we repeated the same experiment with the standard $LiTaO_3$ EO transducer rather than the CMT crystal. Figure 2 shows the recorded electrical transient. As expected, the FWHM of the measured pulses are approximately the same. The post-pulse oscillations are characteristic for $LiTaO_3$ sensors and reflect the so-called dielectric loading due to the large dielectric mismatch between the $LiTaO_3$ crystal and the CS line fabricated on MgO. What is surprising is that under the same experimental conditions (the same signal generated by the DUT), the pulse recorded by CMT had roughly five times greater response amplitude over the $LiTaO_3$ result. Literature, however, indicates that CMT has a relatively small Pockels effect ($r_{41}$ ~4.5 pm/V) [4], resulting in a calculated EOS response smaller than that from a $LiTaO_3$ crystal, clearly conflicting with our measurements.

The EO measurement reflects the optical retardation $\Gamma$ of the polarized probe beam. The EOS system directly measures the change in probe light intensity ($\Delta V$, as measured by the photodetectors) at each position along the optical delay path, for which the differential transfer function is given by:

$$\frac{\Delta V}{V_0} = \frac{1}{2}\sin\left(\pi V / V_\pi\right) \approx \frac{\pi V}{2V_\pi}, \tag{1}$$

where $V$ is the absolute measurement of the signal pulse, $V_0$ is the DC component (31.5 mV/μW times the probe light power), and $V_\pi$ is defined as the voltage required retarding the polarized probe by π. The retardations for $LiTaO_3$ and CMT under the current configuration are, respectively, given by:

$$\Gamma_{LTO} = \frac{2\pi\left(n_e - n_o\right)L_1}{\lambda} - \frac{\pi\left(n_e^3 r_{33} - n_o^3 r_{13}\right)L_2}{\lambda d}V\text{, and} \tag{2}$$

$$\Gamma_{CMT} = \frac{\sqrt{2}\pi\left(n_o^3 r_{41}\right)L_2}{\lambda d}V, \tag{3}$$

where $n_e$ and $n_o$ are the extra-ordinary and ordinary refractive indices, $\lambda$ is the probe wavelength, $d$ is the gap between transmission lines, the crystal thickness is $L_1$, and the interaction length between crystal and electric field is $L_2$. For the non-isotropic $x$-cut $LiTaO_3$ there are two significant EO coefficients, $r_{33}$ and $r_{13}$, while for the isotropic CMT, the only nonzero EO-coefficients are $r_{41} = r_{52} = r_{63}$. It is important to note that the first term in (2) is the intrinsic (static) birefringence of $LiTaO_3$, and in our experiments was compensated by a second $LiTaO_3$ crystal of identical thickness with crystal orientation perpendicular to the first one.

A significant factor in calculating the absolute value of the EO response is the coupling between the transmission line and EO crystal. In the case of $LiTaO_3$, the calibration factor $V_C$ is typically obtained by measuring $\Delta V$ at low frequencies by applying a voltage sine wave of known amplitude directly to the transmission line, bypassing the photoconductive switch. Knowing $r_{33}$ and $r_{13}$ [4] and the $V_C$ for the $LiTaO_3$ crystal, we can extract $L_2$ = 1.18±0.26 μm from our experimental data using (1) and (2). The low value of $L_2$ is further corroborated by the estimated air gap of 24 μm between the CS line and the crystal. The air gap was calculated by the impedance of coplanar strip lines within multilayer substrates [8] using the effective dielectric ($\varepsilon_{eff} = c^2/v^2$), where $c$ is the speed of light, and $v$ is the velocity of the pulse. Assuming the same air gap, the CS characteristic impedance and $\varepsilon_{eff}$ were, subsequently, calculated for the CMT transducer, as well as its corresponding $V_C$ value.

Plugging the above parameters into (1) and (3) and knowing that for $Cd_{0.88}Mn_{0.12}Te$ $n_o$ = 2.94 for $\lambda$ = 800 nm [9], we calculated the low-f $r_{41}$ to be 2.8±0.8 pm/V. The obtained value is in close approximation to the CdTe $r_{41}$ value of 4.5 pm/V presented by Yariv [4] and the $Cd_{0.75}Mn_{0.25}Te$ value ($r_{41}$ = 3.5±0.2 pm/V) calculated in [5], but is much too small to account for the very large signal presented in figure 1. Therefore, we need to use the data obtained directly from our ultrafast EO experiment (figure 1) as the calculation starting point. In such situation, we find that the THz-f measurement of $r_{41}$ is 29.4±6.1 pm/V, an order of magnitude greater than the earlier calculated value at low-f (MHz). Under the optimal configuration (incident polarization parallel to electric field [4]) $V_\pi$ is as low as $556\pm106\cdot d/L_2$ V for $Cd_{0.88}Mn_{0.12}Te$, significantly smaller than that for $LiTaO_3$ ($V_\pi = 3490\cdot d/L_2$ V) and ZnTe ($V_\pi = 2911\cdot d/L_2$ V) [10], the latter being the most often used crystal for free-space THz EO sensing.

To explain the discrepancy between the THz and MHz measurements we propose that free carriers present within our *p*-type crystal effectively screen the low-f voltage signal. The free-carrier screening frequency can be estimated by dividing the CMT crystal's conductivity $\sigma$ by its permittivity. For as-grown CMT, $\sigma$ is on the order of 0.1 S/cm [6], providing a free-carrier screening at 125 GHz. Thus, THz-bandwidth transients are much faster than the carrier screening process and CMT is capable of fully rendering the electric field transient, based on the EO effect. These results were corroborated by our low-temperature EO measurements performed on $Cd_{0.91}Mn_{0.09}Te$. At 10 K, the low-f EO coupling was measured to be ten times greater than at 300 K. This indicates that the free carriers suppressing the Pockels effect at room temperature were actually quashed by deep level traps as the temperature lowered. Thermally stimulated discharge current measurements [11] are planned in order to determine the significance and distribution of traps in our CMT crystals.

## 4. Conclusions

We have shown CMT to be an excellent EO transducer to replace the commonly used $LiTaO_3$ EO crystal. It is also very promising for free-space THz sensing. The calibration method used here allowed us to obtain the absolute value of $r_{41}$ for CMT and we found that at THz-f it is much greater than that reported for CdTe or ZnTe, and an order of magnitude greater than the measured low-f $r_{41}$. Substantial carrier screening present in our *p*-type samples was indentified as the cause for the discrepancy between the THz and MHz EO measurements. CMT has a very small $V_\pi$, exhibiting a much greater sensitivity over $LiTaO_3$ in a standard EO sampling set-up. Furthermore, due to its low dielectric constant, a smaller percentage of the signal is lost to reflections along the transmission line and the response signal is much cleaner due to the absence of dielectric loading.

A. S. C. acknowledges support from the Frank Horton Graduate Fellowship Program at the University of Rochester Laboratory for Laser Energetics. Work in Rochester was supported in part by the U.S. Department of Energy Office of Inertial Confinement Fusion under Cooperative Agreement No. DE-FC 52-08NA28302, the University of Rochester, and the New York State Energy Research and Development Authority. Work in Warsaw was supported by the Polish Ministry of Science and Higher Education through grant 3 T08A 046 30.

## References

[1] Mycielski A, Kowalczyk L, Gałązka R R, Sobolewski R, Wang D, Burger A, Sowińska M, Groza M, Siffert P and Szadkowski A *et al.* 2006 *J. Alloys Compd.* **423** 163
[2] Valdmanis J, Mourou G and Gabel C 1982 *Appl. Phys. Lett.* **41** 211
[3] Zheng X, Xu Y, Sobolewski R, Adam R, Mikulics M, Siegel M and Kordoš P 2003 *Appl. Opt.* **42** 1726
[4] Yariv A 1997 *Optical Electronics in Modern Communications* (New York:Oxford Univ. Press)
[5] Chen C and Whitaker J 2008 *Appl. Phys. Lett.* **92** 101119
[6] Adam R, Mikulics M, Förster A, Schelten J, Siegel M, Kordoš P, Zheng X, Wu S and Sobolewski R 2002 *Appl. Phys. Lett.* **81** 3487
[7] Mycielski A, Burger A, Sowinska M, Groza M, Szadkowski A, Wojnar P, Witkowska B, Kaliszek W and Siffert P 2005 *Phys. Stat. Sol.* C **2** 1578
[8] Chen E and Chou S 1997 *IEEE Trans. Mic. Theory Tech.* **45** 939
[9] Cross A, Wang D, Guarino G, Wu S, Mycielski A and Sobolewski R 2007 *Jour. Phys. Conf. Ser.* **92** 012015
[10] Pradarutti B, Matthäus G, Riehemann S, Notni G, Nolte S and Tünnermann A 2008 *Opt. Comm* **281** 5031
[11] Wrobel J M, Gubański A, Płaczek-Popko E, Rezmer J and Becla P 2008 *J. Appl. Phys.* **103** 063720

Journal of Physics: Conference Series **193** (2009) 012058 doi:10.1088/1742-6596/193/1/012058

# Femtosecond optical spectroscopy studies of high-pressure-grown (Al,Ga)N single crystals

**J. Zhang,[1] A. Belousov,[2] J. Karpinski,[2] B. Batlogg[2] and Roman Sobolewski[1]**

[1]Department of Electrical and Computer Engineering and Laboratory for Laser Energetics, University of Rochester, Rochester, NY, 14627-0231, USA
[2]Solid-State Physics Laboratory, ETH, CH-8093 Zurich, Switzerland

Roman.Sobolewski@rochester.edu

**Abstract**. We present our time-resolved studies of the carrier dynamics in high quality $Al_{0.86}Ga_{0.14}N$ single crystals, grown using a solution technique in a high nitrogen gas pressure autoclave. Our optical measurements were performed using a femtosecond, pump-probe spectroscopy, by invoking the two-photon-absorption-type process and scanning the normalized transient differential transmissivity signal ($\Delta T/T$) of the probe beam. Our time-resolved $\Delta T/T$ photoresponse consisted of a Gaussian-shape correlation signal with the pulse width of ~300 fs, which in two-color experiments was followed by a bi-exponential hot-electron relaxation transient. By plotting the autocorrelation signal amplitude versus the total energy of the absorbed pump and probe photons, we were able to very accurately determine the optical bandap of $Al_{0.86}Ga_{0.14}N$, as equal to 5.81±0.01 eV.

## 1. Introduction

The III-V nitrides, such as GaN, AlN and most recently the (Al,Ga)N system, have been widely investigated due to their excellent optical and electrical properties and potential applications in novel devices, such as ultraviolet (UV) and blue laser sources, and acoustic wave generators [1-3]. (Al,Ga)N compounds are also very promising in high-frequency, high-temperature, and high-power applications, as they exhibit very large bandgap $E_g$, low microwave losses, and high thermal conductivity. Recently, nonlinear optical properties of III-V semiconductors in the transparent region of their absorption spectrum have attracted widespread attention due to their ultrafast response and relatively low dependence on the light frequency [4, 5]. The optical nonlinearity of semiconductors is described by the third-order optical susceptibility $\chi^{(3)}$, which is a complex quantity. The most often studied is the two-photon-absorption (TPA) process, which corresponds to the imaginary part of $\chi^{(3)}$ and is widely applied to pulsewidth measurements and optical spectroscopic studies. In this communication, we report our femtosecond optical study of the TPA process in high-pressure grown (Al, Ga)N crystals.

## 2. Growth of (Al, Ga)N single crystals

The $Al_xGa_{1-x}N$ crystals with the Al content $0.37 < x < 0.9$ have been synthesized from a solution in Ga melt in a high-pressure nitrogen gas system consisting of a 40-mm-diameter pressure chamber with three-zone internal furnace, compressor, and the electronic system for stabilizing and programming both pressure and temperature. A set of three thermocouples was used to monitor and control the temperature and temperature gradient in the furnace. First, nitrogen gas was pumped up to 3 kbar.

Next, the pressure was transmitted into the growth chamber and increased up to the required value by a pressure intensifier. The furnace inside the pressure chamber was outfitted with BN isolation and a W/Re heater, so the maximum temperature of about 1750-1800 °C could be obtained. At such high temperature, the diffusion of N into Ga melt is increased, facilitating the $Al_xGa_{1-x}N$ crystal growth [6]. We used polycrystalline (Al,Ga)N pellets as precursors and placed them in the Ga melt in the upper part of the graphite crucible inside the growth apparatus. The synthesis process of the actual $Al_{0.86}Ga_{0.14}N$ samples studied in this work was conducted under a thermal gradient of about 20 K/cm during 6-7 days. The crystals grew in the colder part of the graphite crucible. In order to prevent the growth of pure GaN crystals, the applied combinations of pressure and temperature during the synthesis were always out of the stable region for the GaN phase. After processing, the crystals were etched from the remaining untreated Ga/Al melt using hydrochloric acid and *aqua regia*. They were colorless platelets up to 0.5 mm size and exhibited the hexagonal form and wurtzite structure [6].

## 3. Femtosecond spectroscopy and two-photon-absorption process

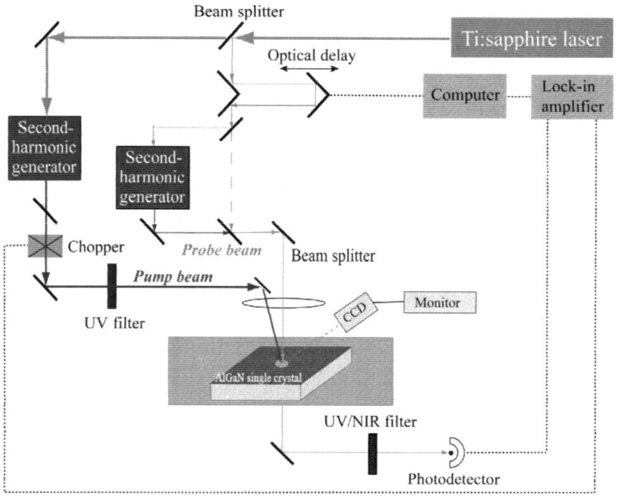


Fig. 1 Femtosecond optical spectroscopy pump-probe setup.

Figure 1 presents our femtosecond optical spectroscopy pump-probe setup, consisting of a commercial mode-locked Ti:sapphire laser (repetition rate 76 MHz) and two homemade second-harmonic generators. The UV pulses with the width of ~150 fs and photon energies ranging from 2.93 to 3.44 eV were used as the pump and focused on the $Al_{0.86}Ga_{0.14}N$ crystal surface with a spot diameter of ~100 μm and the incident angle of ~30°. The probe pulses were also ~150-fs wide and were either near-infrared (NIR) light, directly generated by the Ti:sapphire laser, or UV light after passing through the second-harmonic generator. The probe beam was aimed perpendicularly to the sample surface and its average power was on the same order as the pump. In both above configurations, pump and probe beams had always the photon energy much smaller than the optical bandgap of (Al,Ga)N. Both the one-color (UV) and two-color (UV-NIR) pump-probe spectroscopy experiments were performed in a transmission mode, by measuring the normalized differential transient transmissivity ($\Delta T/T$) signal of probe light as a function of the delay time between the pump and probe pulses.

Two-photon-absorption typically denotes the process of simultaneous absorption of two photons of identical or different energies in order to excite a system under test from one energy level (usually the ground state) to the second, higher one (excited state). According to the study of Sheik-Bahae *et al.* [7], the TPA coefficient $\beta$ for a direct-$E_g$ semiconductor at a photon energy $\hbar\omega$ can be written as:

$$\beta(\omega) = K\frac{\sqrt{E_p}}{n_o^2 E_g^3}F_2(\frac{\hbar\omega}{E_g}), \qquad \text{where} \quad F_2 = \frac{(2x-1)^{3/2}}{(2x)^5}, \tag{1}$$

where $K$ is a material independent constant, $E_p$ is the matrix element related to the inter-band momentum and is ~21 eV for most semiconductors, and $n_0$ is the refractive index.

In the regime of low attenuation of the incident pump and probe beams, the experimental $\Delta T/T$ transient measured in the Fig. 1 setup follows the TPA process and can be expressed as:

$$\Delta T/T = -\beta d P_{eff}, \tag{2}$$

where $d$ is sample thickness and $P_{eff}$ is the effective, absorbed pump beam power per pulse.

## 4. Experimental results and discussion

Figure 2 presents a typical time-resolved $\Delta T/T$ signal, measured in the one-color experiment by exciting our $Al_{0.86}Ga_{0.14}N$ crystal with 400-nm pump and probe light. The experimental $\Delta T/T$ waveform can be very accurately fitted as Gaussian-shape pulse with the full-width-at-half-maximum (FWHM) of 300 fs. Thus, a large transmission decrease observed around the zero delay represents the TPA autocorrelation process of simultaneous absorption of the pump and probe photons.

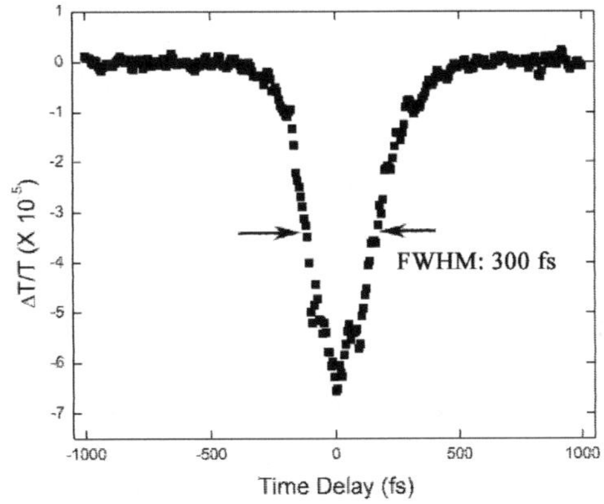


Fig. 2 TPA autocorrelation response of a $Al_{0.86}Ga_{0.14}N$ single crystal. The $\Delta T/T$ signal was measured using 400-nm-wavelength probe and pump pulses.

The two-color experimental arrangement, due to an ultra low absorption of the NIR probe light, allowed us to observe not only the sub-picosecond correlation signal at the zero delay time, but also a subsequent transient electron-phonon relaxation of optically excited carriers as is shown in figure 3(a). The inset in figure 3(a), which presents the $\Delta T/T$ signal on a much longer time scale, demonstrates that the carrier relaxation is a bi-exponential decay process (solid red line), consisting of an initial 12-ps electron-phonon time constant, followed by a slow, ~130-ps-long relaxation.

Figure 3(b) shows the initial correlation signal by numerically subtracting from the waveform plotted in figure 3(a) the carrier bi-exponential relaxation component. We note that, as in the case of one-color experiments, the transient has a Gaussian shape with FWHM of 310 fs. Since the TPA correlation, signal can only be observed when the simultaneously absorbed pump and probe photons have their total energy at least equal to the $Al_{0.86}Ga_{0.14}N$ bandgap, we must conclude that in the two-color case, the correlation effect actually involves three photons: one pump photon and two probe photons, as any other combination would give the total photon energy either way too large or too low.

The two-color measurements were repeatedly performed by varying the pump and probe wavelengths. This way we could observe the optical transition edge of our (Al,Ga)N samples in detail. We noted that at the 428-nm-pump wavelength (the corresponding probe had the wavelength of 856 nm) the correlation $\Delta T/T$ signal disappeared. By calculating the total energy of corresponding pump and probe photons, we could conclude that the $E_g$ of our $Al_{0.86}Ga_{0.14}N$ crystal was equal to 5.81±0.01 eV.

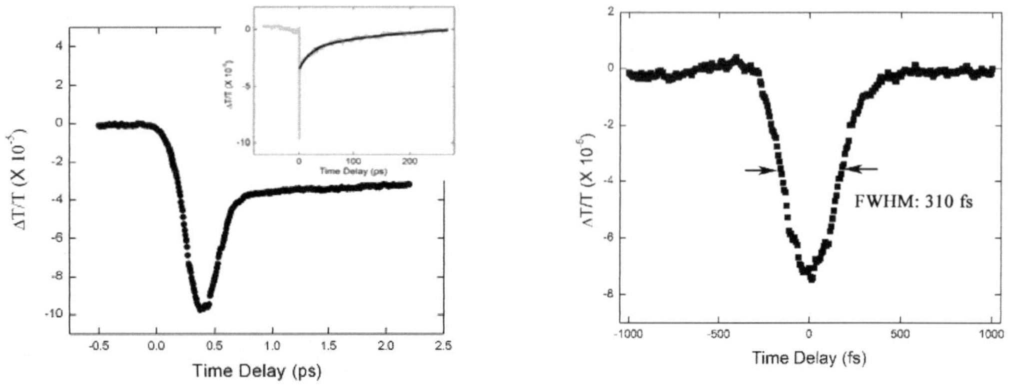


Fig. 3. (a) Time-resolved $\Delta T/T$ signal of a $Al_{0.86}Ga_{0.14}N$ crystal, measured using 380-nm-pump and 760-nm-probe beams. The inset shows the same waveform (grey dots), but in a much wider time window with the slow carrier relaxation fitted (solid line) as a bi-exponential decay. (b) Correlation transient extracted from the signal shown in (a) by subtracting the bi-exponential tail.

## 5. Conclusion

We have performed both one- and two-color time-resolved femtosecond pump-probe spectroscopy on $Al_{0.86}Ga_{0.14}N$ single crystals, synthesized using a high-pressure solution grown method. Since photon energies of our pump and probe beams were always much lower than the (Al,Ga)N $E_g$, the measured, ~300-fs-wide $\Delta T/T$ correlation signal was a result of a coherent (simultaneous) absorption of the pump and probe photons. In the two-color scheme, we also observed picosecond relaxation of photo-excited carriers that followed the initial ultrafast correlation signal. By varying the energies of the pump and probe photons coherently absorbed by our $Al_{0.86}Ga_{0.14}N$ crystal, we also determined its bandgap $E_g$ = 5.81±0.01 eV.

J. Z. acknowledges support from the Frank Horton Graduate Fellowship Program at the University of Rochester Laboratory for Laser Energetics.

## References

[1] Nakamura S, Senoh M, Nagahama S, Iwasa N, Yamada T, Matsushita T, Kiyoko H, Sugimoto Y, Kozaki T, Umemoto H, Sano M and Chocho K 1998 *Jpn. J. Appl. Phys.* **37**, 1020 - 22

[2] Pearton S J, Zolper J C, Shul R J and Ren F 1999 *J. Appl. Phys.* **86**, 1 - 78

[3] Wu S, Geiser P, Jun J, Karpinski J, Park J-R, and Sobolewski R 2006 *Appl. Phys. Lett.* **88**, 041917

[4] Sun C-K, Liang J-C, Wang J-C, Kao F-J, Keller S, Mack M P, Mishra U and DenBaars S P 2000 *Appl. Phys. Lett.* **76**, 439

[5] Rauscher C and Laenen R 1997 *J. Appl. Phys.* **81**, 2818

[6] Belousov A, Katrych S, Jun J, Karpinski J, Batlogg B, Günther D, Zhang J and Sobolewski R 2009 *J. Crys. Growth* **311**, 3971

[7] Sheik-Bahae M, Hutchings D C, Hagan D J, and Van Stryland E W 1991 *IEEE J. Quantum Electron.* **27**, 1296

# Light absorption related to hole transitions in quantum dots and impurity centers in quantum wells under external excitation

**D A Firsov[1,5], L E Vorobjev[1], V A Shalygin[1], A N Sofronov[1], V Yu Panevin[1], M A Vasil'eva[1], M Ya Vinnichenko[1], P Thumrongsilapa[1], S N Danilov[2], A E Zhukov[3], A I Yakimov[4], A V Dvurechenskii[4]**

[1]St. Petersburg State Polytechnic University, 195251 St. Petersburg, Russia

[2]University of Regensburg, 93040 Regensburg, Germany

[3]St. Petersburg Physics and Technology Centre for Research and Education of RAS, 195220 St. Petersburg, Russia

[4]Institute of Semiconductor Physics, Siberian Branch of Russian Academy of Science, 630090 Novosibirsk, Russia

E-mail: dmfir@rphf.spbstu.ru

**Abstract**. Optical absorption is studied in p-doped GaAs/AlGaAs quantum wells at lattice temperatures 4.2 - 300 K. Features in the absorption spectrum are associated to hole transitions from ground impurity level to resonant impurity states and subbands. Absorption modulation related to hole redistribution between impurity and subband states is studied. Spectra of intraband absorption are studied in Ge/Si quantum dots in equilibrium and under interband optical excitation. Filling the excited states with nonequilibrium holes leads to arising the additional peak in absorption spectrum related to hole transitions from the excited state.

## 1. Introduction

Optical transitions of charge carriers involving discrete energy states in nanostructures are very attractive in terms of the development of new light sources and detectors operating in mid-infrared and terahertz spectral ranges. Of particular interest are the resonant impurity states arising in quantum wells due to size quantization. Optical properties of resonant donor states in quantum wells have been studied rather extensively both theoretically [1] and experimentally [2]. The properties of resonant acceptor states are more poorly known and the first part of this paper is devoted to the influence of resonant acceptor states on intersubband light absorption. The acceptor state existence in quantum wells was experimentally confirmed with photoconductivity studies [3]. Optical phenomena related to the acceptors should be interesting because degeneracy of the valence band leads to lifting the polarization selection rules and both *x,y*- and *z*-polarized intersubband optical transitions become possible. Interest to the resonant acceptor states in quantum wells is also connected with the

[5] To whom any correspondence should be addressed.

development of terahertz laser on stressed p-Ge [4] which is based on intracenter hole transitions with participation of the resonant acceptor states.

Quantum dots (QD) possessing the discrete energy spectrum are suitable for infrared photodetector applications including devices with normal light incidence. Usually, intraband light absorption is studied in doped quantum dot structures and carrier transitions from the ground state are considered. However, in some cases optical carrier transitions starting from excited state play the significant role in the device operation. For example, such transitions can determine the optical losses in light emitting cascade QD structures. Nevertheless no experimental studies of these transitions have been carried out. Intraband light absorption in p-doped quantum dots including transitions from excited states will be considered in the second part. Optical transitions from excited states can be studied in the structures with different doping providing the different occupation of the ground and excited levels. In our studies we changed the QD occupation using interband optical excitation.

In the present paper we report the results of the optical absorption studies in p-doped GaAs/AlGaAs quantum wells and Ge/Si quantum dots. Absorption studies can give valuable information about energy spectrum of impurity states and QD levels. Additional information can be collected in strongly nonequilibrium conditions, therefore we measured both equilibrium absorption spectra and spectra under strong longitudinal electric field in quantum wells and interband optical excitation in quantum dots.

## 2. Intersubband light absorption in p-GaAs/AlGaAs quantum wells

The GaAs/$Al_{0.4}Ga_{0.6}As$ multi-quantum-well structure studied in the present work was grown by MBE. It contained 200 periods of 3.8 nm quantum wells separated with 7 nm barriers. Layer of 0.8 nm width in the center of each well was doped with Be to a concentration of $3.2 \cdot 10^{11} cm^{-2}$. The sample for optical investigations was prepared in multipass geometry with 12 light passes through QW layers. This sample configuration allowed to study the absorption of light polarized both in the plane of the structure (*s*-polarization) and at the angle of $45^0$ to the growth axis (*p*-polarization). Indium contacts were deposited on a sample surface and annealed. Equilibrium absorption spectra was measured with VERTEX 80v Fourier transform infrared spectrometer at lattice temperature $T = 4.2$ - 300 K. Spectra of the absorption modulation at the longitudinal electric field were obtained at $T = 94$ K using tunable $CO_2$-laser. Reference sample was used in order to get the value of absorbance.

Spectra obtained at $T = 4.2$ K are of particular interest because at this conditions all the holes occupy the ground impurity state and features in the spectrum can be explained with hole transitions from this state only. Equilibrium absorption spectra measured at $T = 4.2$ K are shown in figures 1 and 2. Both spectra can be presented as a sum of Lorenz contours.

We consider optical transitions from the localized impurity state, so the final state should have the

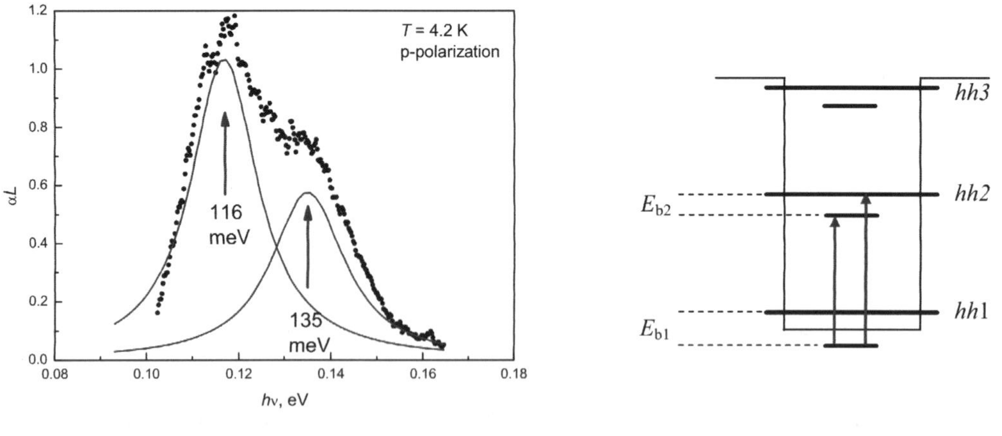


**Figure 1**. Equilibrium absorption spectrum for *p*-polarized light and result of its deconvolution; α is absorption coefficient, *L* is the optical path length. Diagram of hole transitions is shown on the right.

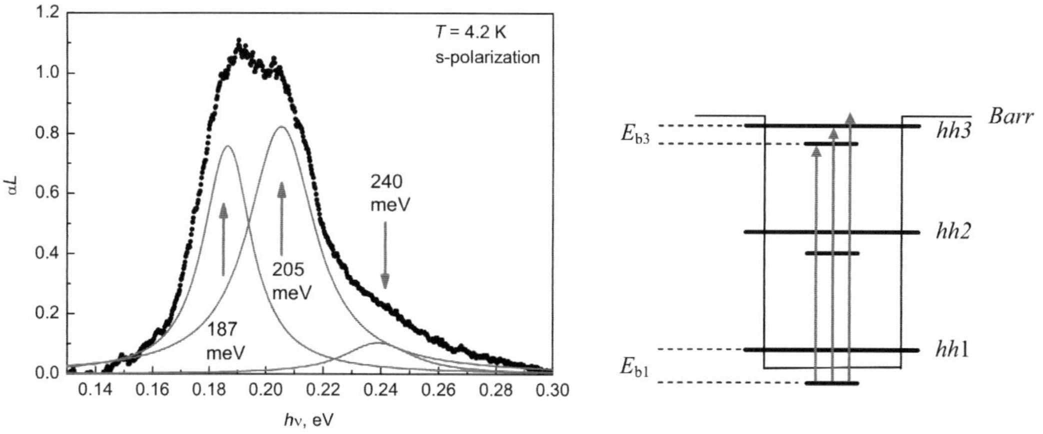


**Figure 2**. Equilibrium absorption spectrum for *s*-polarized light and result of its deconvolution. Diagram of hole transitions is shown on the right.

momentum near the center of Brillouin zone. Therefore we was able to estimate the energies of hole levels in QW using simple parabolic approximation without taking into account light and heavy hole mixing. We estimated energies for light (*lh*) and heavy (*hh*) hole subbands in meV: $E_{hh1} = 35$, $E_{lh1} = 88$, $E_{hh2} = 134$. It should be noted that quantum well is near the resonant one, namely there is one more level *hh*3 located in close proximity of the well top (barrier height $E_B = 231$ meV). The binding energy of Be ground state in 3.8 nm GaAs/AlGaAs QW $E_{b1} = 43$ meV [5]. We see that the position of peak at 135 meV in figure 1 corresponds well to the calculated energy $E_{hh2} - E_{hh1} + E_{b1} = 142$ meV. Peak at 116 meV can be related to the transitions to resonant acceptor state attached to *hh*2 level. Analogously, peaks at 205 and 187 meV in figure 2 can be associated to the transitions in hh3 subband and resonant impurity state attached to this subband, respectively. Weak peak at 240 meV corresponds well to the transitions in 3D continuum above the well because the calculations give $E_{Barr} - E_{hh1} + E_{b1} = 239$ meV. The obtained results allow to estimate the binding energies of resonant states attached to hh2 and hh3 subbands as approximately $E_{b2} = 135 - 116 = 19$ meV, $E_{b3} = 205 - 187 = 18$ meV. It should be noted that strong absorption of *s*-polarized light at intersubband carrier transitions was observed in contrast to the case of n-doped QW. Depopulation of the ground state and filling the *hh*1 state with the temperature reveal themselves in spectra giving the corresponding features.

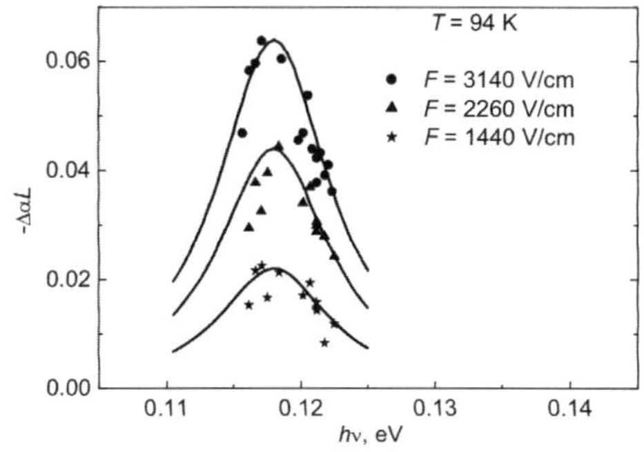


**Figure 3**. Spectra of the absorption modulation for p-polarized light at different electric fields *F*. Curves are plotted for eyes.

Spectra of absorption modulation at lateral electric field are shown in figure 3. These spectra present narrow peaks located in the region of the absorption of p-polarized light corresponding to the transitions from the ground localized impurity state to resonant one attached to *hh*2 subband. At $T = 94$ K impurity is partially ionized. Electric field leads to further impurity ionization decreasing the occupation of this state, so negative change of the absorption is observed.

Analyzing the field dependence of the absorption modulation and current-voltage characteristics one can determine the impurity ionization rate as a function of electric field.

Journal of Physics: Conference Series **193** (2009) 012059 doi:10.1088/1742-6596/193/1/012059

## 3. Optically induced intraband light absorption in Ge/Si quantum dots

Samples under investigations contained 10 layers of Ge quantum dots separated with 55 nm Si layers and were grown on p-doped Si substrate. Structures were δ-doped with boron to $8\cdot10^{11}$ $cm^{-2}$. Doped layers were placed 5 nm below QD layers. The growth procedure and QD characteristics are similar to ones described elsewhere [6]. The sample also was prepared in multipass geometry. Equilibrium absorption spectra were measured with VERTEX 80v Fourier transform infrared spectrometer. Spectra of the absorption modulation were obtained using interband optical excitation from YAG laser with frequency doubling (λ = 532 nm). Experimental results are shown in figure 4.

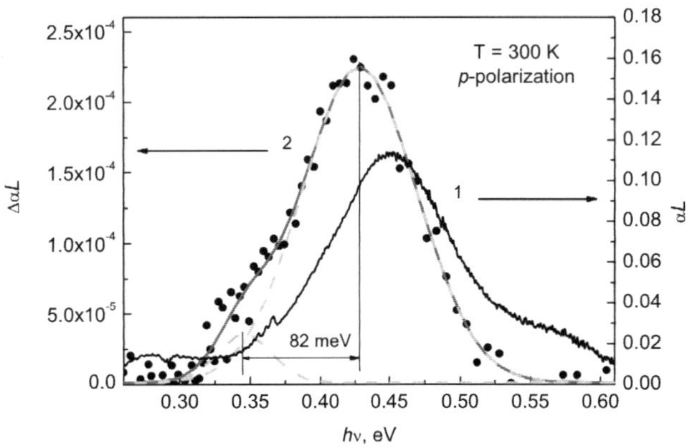


**Figure 4**. Equilibrium absorption spectra (1) and absorption modulation under interband optical excitation (2) for *p*-polarized light. Dashed lines present the result of modulation spectrum deconvolution. No absorption of *s*-polarized light was observed in this spectral range.

The observed peak in equilibrium absorption spectrum is related to hole transitions from the ground state to continuum. Spectrum of the absorption modulation contains two peaks. Position of the main peak corresponds to hole transitions from the ground state. So, at equilibrium the ground level is not fully occupied due to doping and nonequilibrium holes arising after interband excitation fill this level increasing the appropriate absorption. They also begin to fill the first excited state and optical transitions from this state to continuum become possible (see the small peak shifted by 82 meV, this value corresponds to the energy distance between the ground and first excited states [6]). In principle, these results can serve as a basis for the development of two-color QD photodetector.

## 4. Conclusion

In conclusion, we studied optical absorption in quantum wells related to intersubband and intracenter hole transitions and experimentally proved the presence of resonant acceptor states in GaAs/AlGaAs quantum wells. For the first time the intraband light absorption related to hole transitions from the first excited state was studied in Ge/Si quantum dots.

## Acknowledgements

This work was financially supported by Russian Foundation for Basic Researches, Russian Ministry of Education and Science, DAAD and DFG.

## References

[1] Blom A, Odnoblyudov M A, Yassievich I N and Chao K-A 2003 *Phys. Rev.* B **68** 165338
[2] Stehr D, Metzner C, Helm M, Roch T and Strasser G 2005 *Phys. Rev.Lett.* **95** 257401
[3] Aleshkin V Ya, Andreev B A, Gavrilenko V I, Erofeeva I V, Kozlov D V and Kuznetsov O A 2000 *Semiconductors* **34** 563
[4] Altukhov I V, Kagan M S, Korolev K A and Sinis V P 1994 *JETP Lett.* **59** 476
[5] Zheng W M, Halsall M P, Harmer P, Harrison P and Sterr M J 2002 *J. Appl. Phys.* **92** 6039
[6] Yakimov A I, Dvurechenskii A V, Nikiforov A I and Proskuryakov Yu Yu 2001 *J. Appl. Phys.* **89** 5676

# How fast electrons and photons mix: sub-cycle switching of intersubband cavity polaritons

**R Huber[1], A A Anappara[1,2], G Günter[1], A Sell[1], G Biasiol[3], L Sorba[2,3], A Tredicucci[2], S De Liberato[4], C Ciuti[4], and A Leitenstorfer[1]**

[1] Department of Physics, University of Konstanz, 78464 Konstanz, Germany
[2] Laboratorio NEST, CNR-INFM and Scuola Normale Superiore, 56127 Pisa, Italy
[3] Laboratorio Nazionale TASC CNR-INFM, 34012 Trieste, Italy
[4] Laboratoire Materieaux et Phenomenes Quantiques, CNRS and Universite Paris Diderot-Paris 7; Laboratoire Pierre Aigrain, ENS, 75005 Paris, France

E-mail: rupert.huber@uni-konstanz.de

**Abstract.** Intersubband cavity polaritons in a quantum well waveguide structure are optically generated within less than one cycle of light by a femtosecond near-infrared pulse. Mid-infrared probe transients trace the non-adiabatic switch-on of ultrastrong light-matter coupling and the conversion of bare photons into cavity polaritons directly in the time domain.

## 1. Introduction

Photonic microcavities have been exploited to tailor the interaction of light with material excitations. In the strong coupling regime, a photon may be absorbed and spontaneously reemitted by an elementary excitation many times before dissipation becomes effective. This process gives rise to new eigenstates of mixed light-matter character, so-called cavity polaritons. The characteristic energy anticrossing, known as vacuum-field Rabi splitting $2\hbar\Omega_R$, is a direct measure of the strength of light-matter interaction [1]. Recently intersubband resonances of semiconductor quantum wells (QW) hybridized with the mid-infrared photon mode of a planar waveguide have entered a new regime of ultrastrong interaction, where $\Omega_R$ amounts to a significant fraction of the bare eigenfrequencies $\omega_{12}$ themselves [2-5]. The resulting squeezed two-mode quantum vacuum is expected to give rise to a variety of novel quantum electrodynamical (QED) effects [6]. In particular, ultrafast switching of the Rabi frequency $\Omega_R$ has been predicted to release correlated photon pairs out of the vacuum, reminiscent of the intriguing, yet unobserved dynamic Casimir effect [6].

Up to now, however, non-adiabatic phenomena – classical or QED in nature – remained an academic curiosity since there has been no laboratory capable of controlling light-matter interaction on a sub-cycle time scale. Here, we report an all-optical pump – terahertz probe scheme for the first implementation of femtosecond control of ultrastrongly coupled cavity polaritons [7].

## 2. Multi-THz waveguide with semiconductor quantum wells

The sample contains 50 identical, undoped GaAs QWs separated by $Al_{0.33}Ga_{0.67}As$ barriers. The electronic wave functions are quantized along the growth direction forming subbands (Fig. 1). These are unpopulated in thermal equilibrium. Radiative transitions between subbands of quantum number $n = 1$ and $n = 2$ are activated if optical excitation promotes electrons from the valence into the conduction band. The intersubband resonance features a narrow absorption line centered about a photon energy of $\hbar\omega_{12}$ = 113 meV (wavelength $\lambda$ = 11 μm) and a strong dipole moment oriented along the growth direction. The multi-QW structure is designed as a planar step index waveguide for mid-infrared light [2]. Radiation is confined between a top-cladding ($Al_{0.33}Ga_{0.67}As$)-air interface ($n_{air} = 1$, $n_{QW} = 3.1$) on one side and a low refractive index AlAs layer ($n_{AlAs} = 2.9$) on the other. The effective thickness of the entire waveguide is chosen to be $\lambda/2$ at an internal angle of $\theta = 65°$. Photon modes with electric field components in growth direction (TM polarization) may resonantly couple to intersubband transitions provided the subbands are populated.

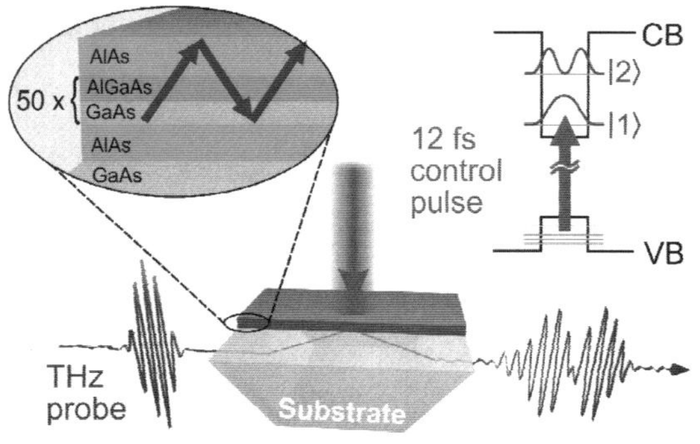


**Figure 1.** Schematic of the femtosecond optical switch: 50 undoped GaAs quantum wells of a thickness of 9 nm each, separated by $Al_{0.33}Ga_{0.67}As$ barriers are embedded into a planar waveguide based on total internal reflection. Intersubband transitions are activated by near-infrared 12-fs control pulses populating level $|1\rangle$. TM-polarized multi-THz transients guided through the prism-shaped substrate are reflected off the waveguide to probe the ultrafast buildup of intersubband cavity polaritons.

## 3. All-optical control of light-matter coupling: from zero to ultrastrong

We employ a low-noise Ti:sapphire laser system generating amplified 12-fs pulses (central photon energy: 1.55 eV) to photoinject electrons from the valence band into the lowest conduction subband of the QWs (Fig. 1), activating the intersubband oscillator. Since the inverse frequency of the latter transition amounts to 37 fs, switching occurs within less than half a cycle of light.

The subsequent ultrafast dynamics of the non-equilibrium cavity at room temperature is traced by multi-terahertz (THz) spectroscopy [8]: A second part of the laser output generates phase-locked THz pulses covering the spectral window from 80 meV to 150 meV by optical rectification in a 50-μm-thin GaSe emitter [9]. TM-polarized field transients are coupled through the prism-shaped substrate and internally reflected off the photoexcited area of the waveguide under incidence angles around $\theta = 65°$ (Fig. 1). The pulse front of the near-infrared pump is tilted to match the geometry of the THz phase surfaces. The oscillating electric field of the reflected THz transient is resolved in the time domain via phase-matched electro-optic sampling [9]. Fourier transformation provides amplitude and phase spectra in the mid infrared. The eigenmodes of the cavity are identified via their characteristic minima in the amplitude reflectivity [2].

In a first step, we demonstrate that light-matter coupling is continuously tunable via the control fluence $\Phi$. The spectra of Fig. 2 are recorded at a fixed delay $t_D$ = 20 ps between the near-infrared control and the multi-THz probe pulse. In equilibrium ($\Phi = 0$), a single reflectance minimum at $\hbar\omega_c$ = 113 meV (top curve) attests to the sole resonance of the unexcited cavity, the bare photon mode. For $\Phi > 0.05 \times \Phi_0$, $\Omega_R$ exceeds the widths (FWHM ~ 5 meV) of intersubband and cavity resonances and two cavity polariton branches are clearly discernible. At this point, the system crosses from the weak to the strong coupling regime. Further increase of the fluence enhances the separation of the minima to as much as 50 meV (see e.g. bottom curve of Fig. 3), corresponding to a fraction of 44% of the bare photon frequency. The apparent mode separation is not identical with the vacuum Rabi splitting at the anti-crossing point [2]. Only a quantitative simulation of the energy position of the polariton dips allows for extraction of $\Omega_R$. For a correct description of our data, the theory has to go beyond the rotating wave approximation and include antiresonant terms in the light-matter Hamiltonian. These contributions describe the simultaneous creation or annihilation of two excitations with opposite in-

Journal of Physics: Conference Series **193** (2009) 012060 doi:10.1088/1742-6596/193/1/012060

plane wave vector $k$ and give rise to a two-mode squeezed quantum vacuum [6]. By comparison with this theory we determine a maximum splitting $2\Omega_R = 0.18 \times \omega_{12}$ for our experiments. This value is comparable to the record achieved in doped structures [3,4] and clearly fulfills the criteria for ultrastrong coupling [6].

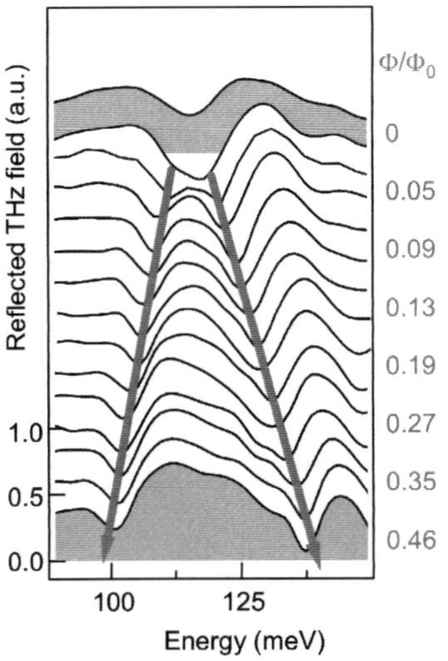


**Figure 2.** All-optical in-situ tuning of light-matter interaction. THz reflectance spectra measured at room temperature for various fluences $\Phi$ (vertically offset) of the control pulse ($t_D = 20$ ps). Minima indicate eigenmodes of the system. For $\Phi = 0$, only the bare photon mode is observed at $\hbar\omega_c$ = 113 meV; both branches of the intersubband cavity polaritons are discernible for $\Phi \geq 0.05\ \Phi_0$, where $\Phi_0 = 0.1$ mJ/cm$^2$.

## 4. Non-adiabatic switch-on of intersubband cavity polaritons

The central question is: How rapidly may ultrastrong coupling be activated? In order to address this issue, we repeat the experiments for various delay times $t_D$ between near-infrared pump and multi-THz probe pulses. The results are illustrated in Fig. 3a: For $t_D \leq -50$ fs the spectra are dominated by a single dip located at 113 meV which we may unequivocally assign to the photonic waveguide mode. Photoinjection of electrons into the lower subband induces dramatic changes of the spectra of order unity. The initial bare photon eigenstate is replaced, on a ten femtosecond scale, by two ultrastrongly coupled cavity polariton modes appearing simultaneously at energies of 94 meV and 143 meV, respectively (arrows in Fig. 3). Most remarkably, the new resonances do not develop by gradual bifurcation out of the bare cavity mode as in Fig. 2. In contrast, switching occurs discontinuously once subband $|1\rangle$ is populated.

Femtosecond activation of the light-matter Hamiltonian is predicted to give rise to yet unexplored effects on pre-existing photon states and the quantum vacuum itself [6]. This class of phenomena is best studied directly in the time domain (see Fig. 3b): When a few-cycle probe transient impinges on the unexcited modulator, part of its energy is reflected off the cavity surface. A second portion is evanescently coupled into the resonator, prepares a coherent photon state, and gets reemitted subsequently. This dynamics is encoded in a characteristic twin-pulse structure of the reflected transient (see curve (*ii*)). The most intriguing situation is realized if we turn light-matter coupling on while a coherent state of bare photons is still present inside the cavity: The control pulse (vertical arrow in curve (*iii*)) abruptly alters $\Omega_R$ during the free cavity decay. Remarkably, the emission of bare photons is interrupted on a time scale shorter than half an oscillation cycle of light – a compelling proof of non-adiabaticity. The subsequent field trace exhibits a characteristic two-mode beating (inset in Fig. 3b), the hallmark of coherent oscillations of both polariton branches. Thus we do not only control the eigenstates of the microcavity, but effectively convert a coherent photon population into ultrastrongly coupled cavity polaritons beating at the splitting frequency. We suggest the term "perturbed cavity decay" (PCD) to describe this novel sub-cycle phenomenon. A quantitative theory including anti-resonant terms of the light-matter Hamiltonian as well as the extremely non-equilibrium carrier dynamics is currently under way.

Journal of Physics: Conference Series **193** (2009) 012060 doi:10.1088/1742-6596/193/1/012060

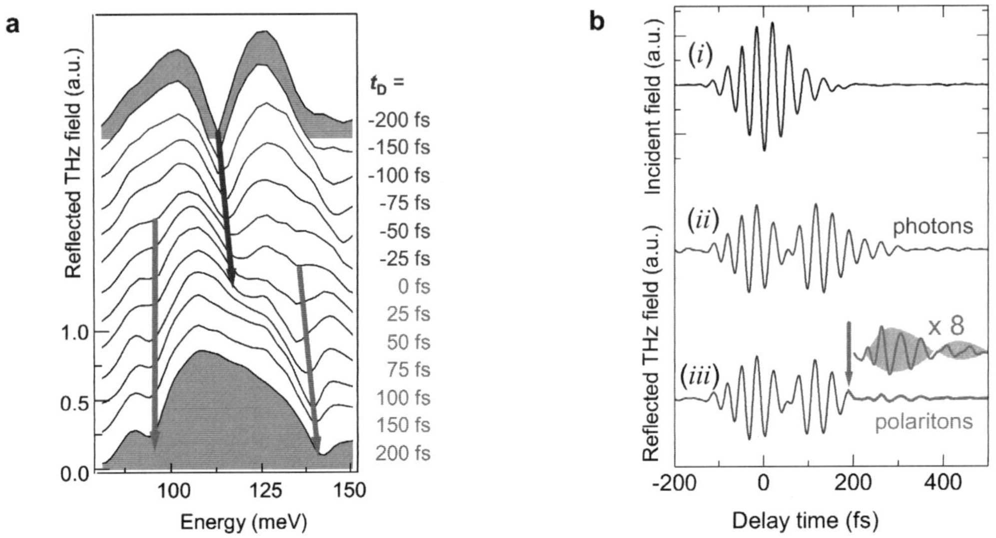


**Figure 3.** Two-dimensional optical pump – THz probe data. **a** Amplitude reflectivity spectra at $T_L$ = 300 K for various delay times $t_D$ (control fluence: $\Phi$ = 0.1 mJ/cm$^2$). The arrows serve as guides to the eye for the position of the cavity and polariton resonances, respectively. **b** Nonadiabatic conversion of bare photons into cavity polaritons, in the time domain: (*i*) incident THz transient, (*ii*) reflected transient without control pulse. The twin pulse structure is due to radiation directly reflected off the cavity surface (first burst) and bare photons reemitted from the cavity (second pulse). (*iii*) The near-infrared control pulse (vertical arrow) alters the cavity emission on a sub-cycle scale: The exponential bare cavity decay is replaced by the characteristic polariton beating (insert, magnified by factor 8).

## 5. Conclusions

We have demonstrated the first non-adiabatic switching scheme of cavity-polaritons, reaching the regime of ultrastrong light-matter coupling on a sub-cycle scale. The experiments provide a benchmark for latest theories in the ultrastrong coupling regime, point out a viable route towards novel QED phenomena such as the observation of Casimir-type vacuum radiation, and demonstrate a room-temperature optical switching device at the ultimate speed.


[1] Weisbuch C et al. 1992 *Phys. Rev. Lett.* **69** 3314

[2] Dini D et al. 2003 *Phys. Rev. Lett.* **90** 116401

[3] Anappara A A et al. 2009 *Phys. Rev.* B Rapid **79** 201303

[4] Todorov Y et al. 2009 *Phys. Rev. Lett.* **102** 186402

[5] Anappara A A et al. 2005 *Appl. Phys. Lett.* **87** 051105; 2006 *Appl. Phys. Lett.* **89** 171109

[6] Ciuti C, Bastard G, and Carusotto I 2005 *Phys. Rev.* B **72** 115303; De Liberato S et al. 2007 *Phys. Rev. Lett.* **98** 103602

[7] Günter G et al. 2009 *Nature* **458** 178

[8] Huber R et al. 2001 *Nature* **414** 286; 2005 *Phys. Rev. Lett.* **94** 0274011; Kübler C et al. 2007 *Phys. Rev. Lett.* **99** 116401

[9] Kübler C et al. 2005 *Semicond. Sci. Technol.* **20** 128 and references therein

Journal of Physics: Conference Series **193** (2009) 012061 doi:10.1088/1742-6596/193/1/012061

# Monolithic photonic crystal quantum-cascade laser

**A. Benz, Ch. Deutsch, G. Fasching and K. Unterrainer**
Photonics Institute and Center for Micro- and Nanostructures, Vienna University of Technology, Gusshausstrasse 29/387, A-1040 Vienna, Austria

**A. M. Andrews, P. Klang, W. Schrenk and G. Strasser**
Institute of Solid-State Electronics and Center for Micro- and Nanostructures, Vienna University of Technology, Floragasse 7/362, A-1040 Vienna, Austria

E-mail: `alexander.benz@tuwien.ac.at`

**Abstract.** We present the design and realization of active photonic crystal (PhC) terahertz quantum-cascade lasers. The devices consist of sub-wavelength isolated pillars which are embedded in a double-metal waveguide. The lasing is observed at flat-band regions not in the bandgap itself. A stable single-mode emission under all driving conditions is achieved.

## 1. Introduction

With the development of terahertz (THz) quantum-cascade lasers (QCLs), monolithic and coherent sources for the THz spectral region have become available [1]. The main applications are real-time imaging [2], heterodyne detection [3] or spectroscopy [4]. Due to the inhomogeneous gain broadening typically a multi-mode emission is observable [5]. For precise frequency control, the optical properties of the resonator have to be designed. The direct integration of photonic crystals (PhC) into the laser resonator allows to combine the full control over the dispersion relation with small modal volumes [6, 7].

Here, we will show the design and realization of two-dimensional PhC which are embedded in a high-confinement double-metal waveguide. The PhC consists of isolated, sub-wavelength pillars which are fabricated directly from the active region of the THz-QCLs. This allows their use as the gain region making any bulk gain region redundant [8, 9].

## 2. Active photonic crystal lasers

These lasers operate in the slow light regime next to the bandgaps, there are no states inside the bandgap as there are no defects incorporated. The calculated PhC bandstructure is illustrated in the inset of Fig. 1. This allows for a very precise frequency selection for the lasing mode as the slow-light regime works only for a narrow frequency range. Therefore it is easy to greatly reduce the number of emission modes [6, 7]. The optical gain in these structures is spatially separated, every single pillar is used to provide gain. It has to be pointed out that the pillars are of sub-wavelength dimension.

For the finite-difference time-domain (FDTD) simulations we use the *MIT Electromagnetic Equation Propagation*-package [10]. We arrange 37 infinitely high pillars in a two-dimensional (2D) array. The simulations predict two closely spaced modes, shown in Fig. 1. One corresponds

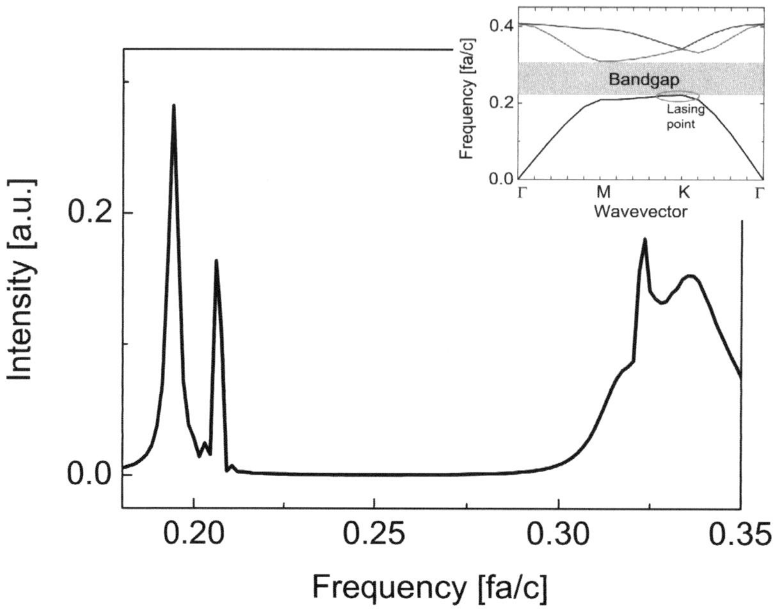


**Figure 1.** Calculated spectrum for the active PhC. The two modes correspond to the M- and K-point. The inset is showing a calculated bandstructure for the ideal PhC.

to the K-point, with a Q-factor of 1000, the other to the M-point, with a Q-factor of 60, in the bandstructure. For frequencies above 0.3 [fa/c] additional modes appear corresponding to higher bands of the PhC. In between, the full bandgap is clearly visible.

The active PhC lasers realized consist of 15 $\mu$m high, free-standing pillars which are embedded in a double-metal waveguide. All pillars are fabricated from the active region of a THz-QCL. The electric field distribution and the energy density for such a structure can be calculated very easily. The structure achieves a modal energy confinement of 95 %. The challenging task for the processing is the formation of the waveguide and the pumping of all the pillars as they are not interconnected. Details on the realization of these devices are given in Ref. [7]. The reason to attack this highly challenging processing is, apart from the beautiful physics, the theoretically predicted improvement in device performance. These 2D-PhC are predicted to give strong gain enhancement [11, 12, 13]. It is proportional to the overlap of the mode with the gain region multiplied by the time the mode spends inside the PhC. The group velocity and the modal overlap are both design parameters of the PhC, allowing to optimize both values at the same time.

The experimental data shows an excellent agreement with the theoretically predicted results. There is only one lasing mode visible corresponding to the K-point with the higher Q-factor. Measured spectra are showing the stable single-mode emission under all driving currents, an example for a device with a 26.6 $\mu$m period is shown in Fig. 2. The small discrepancy between the simulation and measurement is attributed to the uncertainty of the refractive index and the effects of the metallic waveguide. The predicted lasing mode is at 0.215 [fa/c] or 2.45 THz for this device.

These devices do not show any dependence of the emission frequency on the device size. This is not surprising as the lasing mode is defined by the photonic crystal itself and not by the position of any mirrors or facets. This concept shows already its high potential for the resonator tuning. It is easily possible to define the emission frequency of the device simply by changing the

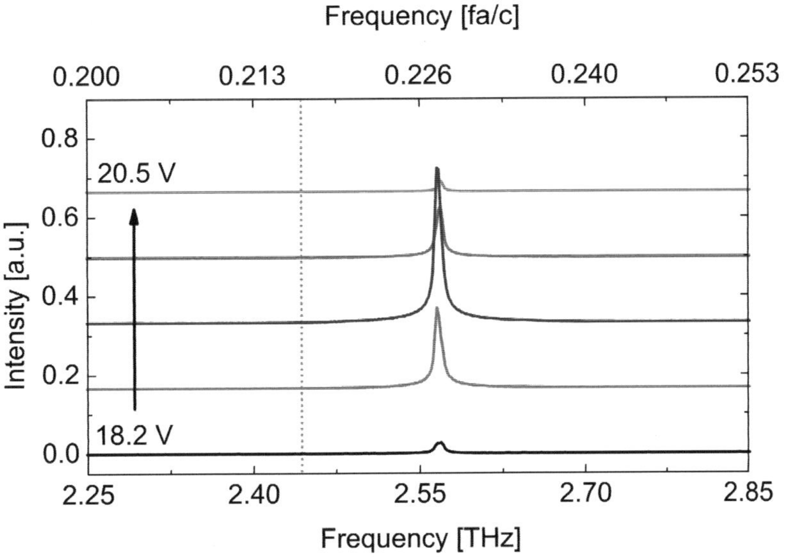


**Figure 2.** Measured spectrum for the active PhC with a period of 26.6 $\mu$m. The stable single-mode emission independently of the applied bias is clearly visible. The dotted, gray line represents the predicted lasing frequency.

period of the PhC without any need for a regrowth of the active region itself. The experimentally observed tuning range is on the order of 400 GHz or 15 % of the central wavelength [6, 7], which is much larger than the typical gain bandwidth in THz-QCLs of 150 GHz [5]. This is a strong evidence for the theoretically predicted gain enhancement.

## 3. Conclusion

In conclusion, we have designed and realized active PhC lasers based on a THz-QCL active region. The devices are lasing at the band edge of the PhC band structure. The group velocity is extremely low for these regions, resulting in strong optical feedback. Together with the excellent modal confinement in this type of resonators, we are able to achieve a tuning range of 400 GHz by varying the period of the PhC lithographically. These experimental results are a strong evidence for the predicted gain enhancement. In addition, the realized resonators have very small modal volumes. The typical diameter is the order of 240 $\mu$m with a filling factor of 33 %.

## Acknowledgments

This work was partly supported by the Austrian Scientific Fund FWF (SFB ADLIS, SFB IR-ON, DK CoQuS), the Austrian Nano initiative project (PLATON) and Austrian Society for Microelectronics (GMe).

## References

[1] Köhler R, Tredicucci A, Beere H E, Lienfield E H, Davis A G, Ritchie D A, Iotti R C and Rossi F 2002 *Nature (London)* **417** 156
[2] Lee A W M, Qin Q, Kumar S, Williams B S, Hu Q and Reno J L 2006 *Appl. Phys. Lett.* **89** 141125
[3] Gao J R, Hovenier J N, Yang Z Q, Baselmans J J A, Baryshev A, Williams B S, Kumar S, Hu Q and Reno J L 2005 *Appl. Phys. Lett.* **86** 244104
[4] Hübers H W, Pavlov S G, Richter H, Semenov A D, Mahler L, Tredicucci A, Beere H E and Ritchie D A 2006 *Appl. Phys. Lett.* **89** 061115

[5] Kröll J, Darmo J, Dhillon S S, Marcadet X, Calligaro M, Sirtori C and Unterrainer K 2007 *Nature (London)* **449** 698
[6] Zhang H, Dunbar L A, Scalari G, Houdre R and Faist J 2007 *Opt. Express* **15** 16818
[7] Benz A, Deutsch C, Fasching G, Unterrainer K, Andrews A M, Klang P, Schrenk W and Strasser G 2009 *Opt. Express* **17** 941
[8] Dunbar L A, Moreau V, Ferrini R, Houdre R, Sirigu L, Scalari G, Giovannini M, Hoyler N and Faist J 2005 *Opt. Express* **13** 8960
[9] Benz A, Fasching G, Deutsch C, Andrews A M, Unterrainer K, Klang P, Schrenk W and Strasser G 2007 *Opt. Express* **15** 12418
[10] Farjadpour A, Roundy D, Rodriguez A, Ibanescu M, Bermel P, Joannopoulos J D, Johnson S G and Burr G 2006 *Opt. Lett.* **31** 2972
[11] Sakoda K 1999 *Opt. Express* **4** 167
[12] Nojima S 1998 *Jpn. J. Appl. Phys.* **37** 6418
[13] Nojima S 2001 *J. Appl. Phys.* **90** 545

# Importance of Coulomb interactions in bound-to-continuum THz quantum cascade lasers

**C Jirauschek,[1,2] A Matyas,[1,2] and P Lugli[2]**

[1]*Emmy Noether Research Group "Modeling of Quantum Cascade Devices", Technische Universität München, D-80333 Munich, Germany*

[2]*Institute for Nanoelectronics, Technische Universität München, D-80333 Munich, Germany*

E-mail: jirauschek@tum.de

**Abstract**. We demonstrate the importance of Coulomb interactions in bound-to-continuum THz quantum cascade lasers, employing an ensemble Monte-Carlo analysis. In such structures, the electron-electron interactions between the closely spaced energy levels in the minibands tend to play a more important role than in resonant-phonon depopulation designs, where the energy levels are more energetically separated and LO phonon scattering prevails. Also a significant conduction band bending due to space charge effects is observed. Thus, especially for bound-to-continuum structures careful modelling of Coulomb interactions is crucial to obtain good agreement with experiment.

## 1. Introduction

THz quantum cascade lasers (QCLs) are commonly based either on the bound-to-continuum (BTC) or the resonant-phonon (RP) design. BTC structures feature closely spaced energy levels forming so-called minibands. In RP designs, the depletion transition is tuned to the longitudinal-optical (LO) energy to achieve efficient depletion of the lower laser level, and also the other energy levels tend to be more energetically separated. While the transport in such structures is (at least for elevated temperatures) governed by LO phonon scattering, Coulomb interactions play a pronounced role for the carrier transport in BTC designs and also lead to significant conduction band bending [1, 2, 3]. Thus, a careful modeling of Coulomb interactions is crucial especially for BTC devices. In the following, we focus on investigating the influence of Coulomb interactions in THz BTC QCLs, using an ensemble Monte-Carlo (EMC) transport simulation. Some widely-used approximations and assumptions which work fine for THz RP structures are shown to lead to erroneous results for THz BTC QCLs.

## 2. Method

Coulomb interactions in QCLs are due to the attraction or repulsion of charges in the structure, i.e., the interaction of electrons with other electrons and ionized impurities. In our simulation, this effect is considered in two ways [3, 4]: The quantized energy levels and wavefunctions are determined by solving the coupled Schrödinger-Poisson (SP) system to account for mean field effects. Here, the Coulomb interactions manifest themselves as space charge effects, leading to a conduction band bending. Furthermore, the electron-electron (e-e) and electron-impurity (e-i) interactions enter the

EMC transport simulation in the form of corresponding scattering events, which are evaluated based on Fermi's golden rule.

The EMC method allows for a self-consistent carrier transport and optical gain analysis [3]. As discussed above, the coupled Schrödinger-Poisson system is solved to obtain the quantized energy levels, based on which the carrier transport simulation is self-consistently performed. All essential scattering mechanisms (acoustic and LO phonons, e-e and e-i scattering, interface roughness) are included, and the Pauli exclusion principle is accounted for, as well as hot phonon effects [3, 4]. For e-e scattering, screening and spin effects are taken into account [5]. The evaluation of e-e scattering is much more computationally demanding than single-electron processes like electron-phonon interactions and thus has, up to now, not been included in comparable quantum mechanical simulations of QCLs [6, 7]. Since the quantized energy levels obtained by the SP solver depend on the space charges (i.e., the electron distributions obtained from EMC) and the EMC simulation is in turn based on these quantized energy levels, both simulations have to be performed iteratively until convergence of the carrier distribution as well as the conduction band bending is obtained, yielding so-called self-self-consistent solutions [1].

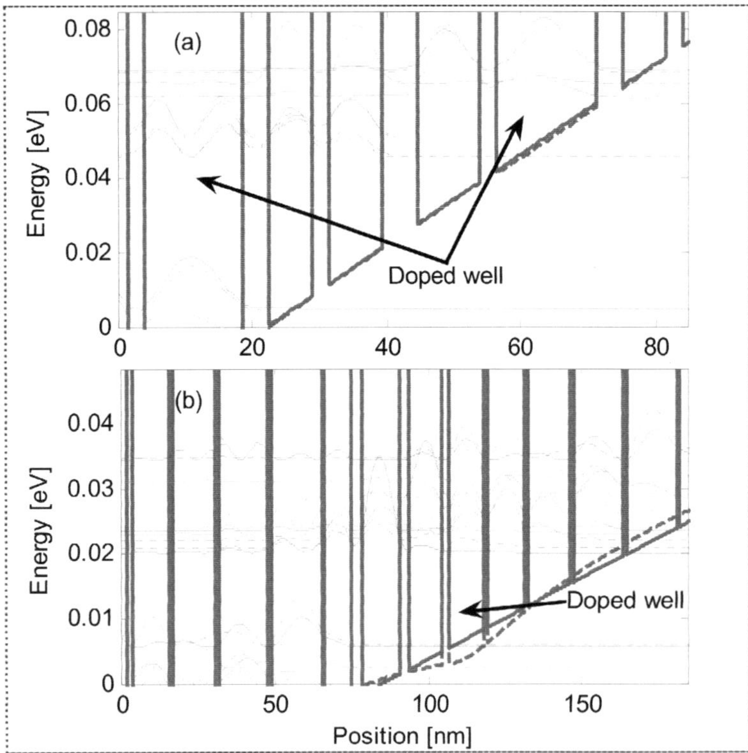


**Figure 1.** Conduction band profile and probability densities for space charge effects neglected (i.e., disabled Poisson solver) (solid lines) and a self-self-consistent simulation (dashed lines). (a) RP structure at 12.25 kV/cm and 30 K. (b) BTC structure at 2.5 kV/cm and 30 K.

## 3. Results

### *3.1. Influence of space charges*

In figure 1, the conduction band profile and the probability densities of the quantized energy levels are displayed for the two investigated QCLs, a 3.4 THz RP design (figure 1(a)) [8] and a 3.5 THz BTC design (figure 1(b)) [9]. We see that the conduction band bending due to space charge effects plays a secondary role in the investigated RP structure, while this effect is quite pronounced in the BTC design, although the doping level is comparable in both structures. Here, multiple iterations of SP and EMC simulation are necessary to obtain self-self-consistent solutions.

In the following, we investigate to what degree these space charge effects, leading to modified energy levels and wavefunctions in the SP solver, influence the outcome of the carrier transport analysis. Figure 2 shows the simulation results for the spectral gain profile of both the RP and BTC design. Here, a full EMC simulation is performed including all the scattering mechanisms, but space charge effects are considered with a varying degree of complexity. Shown are the results as obtained by completely neglecting space charge effects, i.e., disabling the Poisson solver, running the SP solver once in the beginning (assuming a thermal carrier distribution [2]) and then carrying out the EMC simulation, and iteratively performing the SP and EMC simulations until convergence is obtained. For the RP structure (figure 2(a)), the simulation results are relatively insensitive to space charge effects. In contrast, for the BTC structure (figure 2(b)), disabling the Poisson solver (dotted line) produces an almost vanishing gain, which is in clear contrast to the experimentally observed lasing of the structure [9]. This reflects the influence of space charge effects on the energy level alignment and the wave function shapes. Accounting for these effects (dashed and solid lines) produces realistic gain profiles centered around approximately 3.5 THz, in agreement with experiment. However, the exact shape of the gain profile is somewhat different for the non-iterative (dashed line) and iterative (solid line) approach, demonstrating the need of iterative SP and EMC simulations to obtain the self-self-consistent solution for such structures. The significant conduction band bending for the investigated BTC structure, shown in figure 1(b), is attributed to the fact that the positive charges due to the ionized dopants are concentrated in a single well per period, while the negative charges due to the electrons in the minibands are distributed across the whole period, leading to a pronounced net charge profile.

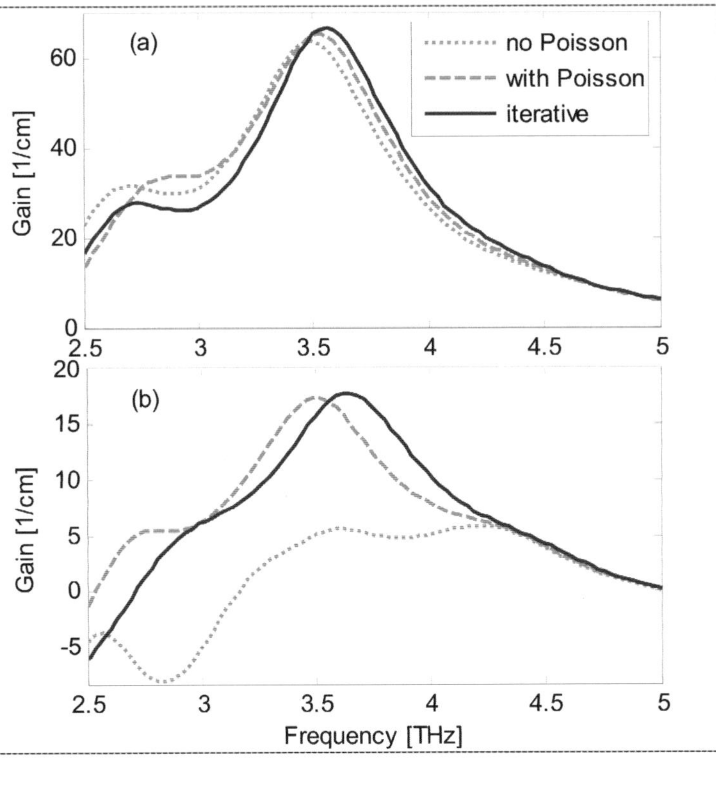


**Figure 2.** Simulated spectral gain versus frequency with disabled Poisson solver (dotted lines), running the Poisson solver once assuming a thermal carrier distribution (dashed lines), and a self-self-consistent solution obtained by iteratively running the SP and EMC simulations (solid lines). (a) RP structure at 12.25 kV/cm and 30 K. (b) BTC structure at 2.5 kV/cm and 30 K.

### *3.2. Influence of electron-electron scattering*

As we have seen in the previous section, space charge effects have a quite strong influence on the investigated BTC structure. In the following, we investigate if e-e scattering, i.e., the inclusion of Coulomb interactions beyond the mean field approximation, also plays an important role for the simulation of such devices. This question is especially interesting from a simulation point of view: Since the evaluation of e-e scattering is by far the most computationally demanding scattering process, it is usually neglected in quantum mechanical transport simulations of QCLs. This works well for RP structures, where good agreement with experiment is obtained [6,7]. Figure 3 shows the self-self-consistent EMC result for the BTC gain structure, now with e-e scattering taken into account as in figure 2 (solid line), and with e-e scattering neglected (dashed line). Omission of e-e scattering yields an unrealistic gain profile featuring two excessively narrowed spikes, demonstrating the significant role of this scattering process. Only the full simulation yields reasonable results. Specifically, the obtained gain bandwidth (full width at half-maximum) of 0.85 THz agrees very well with the experimental value of around 0.9 THz, deduced from electroluminescence measurements [9]. This shows that electron-electron scattering between the closely spaced energy levels in the minibands plays a pronounced role for the carrier transport in BTC designs.

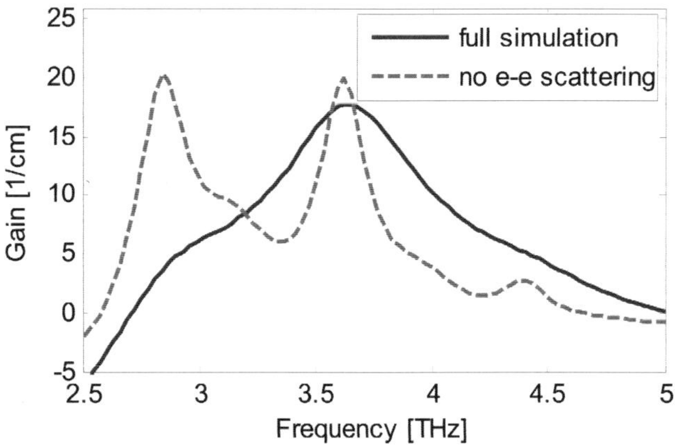


**Figure 3.** Simulated spectral gain versus frequency with (solid line) and without (dashed line) e-e scattering.

## 4. Conclusion

Employing an EMC analysis, we have investigated the importance of Coulomb interactions in THz quantum cascade lasers, and specifically THz BTC structures. Here, conduction band bending has a significant influence on the level alignment, calling for a self-self-consistent SP and EMC simulation. Also e-e scattering between the closely spaced energy levels in the minibands plays an important role and is the dominant gain broadening mechanism. The EMC analysis, where both effects can be adequately accounted for, is shown to yield meaningful results for the investigated BTC structure, which are in good agreement with experiment.

**Acknowledgments**

C.J. and A.M. acknowledge support from the Emmy Noether program of the German Research Foundation (DFG, JI115/1-1).

## References

[1] Jovanović V D et al. 2005 *Appl. Phys. Lett.* **86** 211117
[2] Jirauschek C 2009 *IEEE J. Quantum Electron.* **45** 1059
[3] Jirauschek C and Lugli P 2009 *J. Appl. Phys.* **105** 123102
[4] Jirauschek C, Scarpa G, Lugli P, Vitiello M S, and Scamarcio G 2007 *J. Appl. Phys.* **101** 086109
[5] Moško M, Mošková A and Cambel V 1995 *Phys. Rev. B* **51** 16860
[6] Kubis T, Yeh C, and Vogl P 2008 *phys. stat. sol. (c)* **5** 232
[7] Nelander R and Wacker A 2008 *Appl. Phys. Lett.* **92** 081102
[8] Williams B. S., Callebaut H., Kumar S., Hu Q., and Reno J. L. 2003 *Appl. Phys. Lett.* **82** 1015
[9] Scalari G, Ajili L, Faist J, Beere H, Linfield E, Ritchie D, and Davies G 2003 *Appl. Phys. Lett.* **82** 3165

Journal of Physics: Conference Series **193** (2009) 012063 doi:10.1088/1742-6596/193/1/012063

# How periodic are terahertz quantum cascade lasers?

**T Kubis and P Vogl**

Walter Schottky Institute, Technische Universität München, Am Coulombwall 3, 85748 Garching, Germany

E-mail: tillmann.kubis@wsi.tum.de

**Abstract**. We apply a novel non-equilibrium Green's function method for open quantum devices to analyze quantum cascade lasers. We find the carrier distribution in typical resonant phonon THz-QCLs to develop a periodicity that differs from the geometric periodicity of the QCL. We propose a design improvement that thermalizes electrons at threshold bias and thereby pins the electron density to the QCL periodicity.

## 1. Introduction

Quantum cascade lasers (QCLs) have been studied experimentally [1] and theoretically. The theoretical approaches used most commonly are the density matrix method [2, 3], the Boltzmann equation [4, 5], and the non-equilibrium Green's function method (NEGF) [6, 7]. All of these theoretical studies share the common approximation of invoking field periodic boundary conditions.

We have developed a self consistent non-equilibrium Green's function method (NEGF) for stationary charge transport in open THz-QCL devices that are attached to dissipative leads [8]. Importantly, our method differs from previous models by not imposing field-periodic boundary conditions. Instead, the electrons can enter and leave the device via travelling eigenstates of the semi-infinite leads. This allows us to capture non-periodic phenomena as well as investigate spatially extended, energetically high lying states.

The primary goal of this paper is to illustrate the formation of electron distributions in resonant phonon THz-QCLs that are commensurable or incommensurable with the QCL device periods. We show the impact of such distributions on the device current density and the optical gain. Finally, we propose a design modification that pins the electrons to the QCL periodicity.

## 2. Method

We use the non-equilibrium Green's function method (NEGF) to calculate stationary electronic transport and optical gain up to the threshold current in a single conduction band with an effective mass $m^*(z)$. We assume the QCLs to be homogeneous in the lateral $x,y$ directions and to be in contact with two equilibrium reservoirs at $z=0$ and $z=L$, respectively. Thereby, we consider charge transport as a scattering problem from source to drain with the QCL structure being the active device. The source and drain leads are assumed to consist of QCL periods by themselves, albeit with zero electric field and with the carriers obeying equilibrium Fermi statistics. In this way, we mimic adjacent QCL periods. The electrons enter the device in propagating eigenstates of the respective infinitely extended lead Hamiltonians [8]. We take into account inelastic acoustic and polar-optical phonon scattering, scattering by charged impurities, interface roughness, and by electron-electron interaction in the

Hartree approximation. The scattering self-energies are determined in the self-consistent Born approximation including their full nonlocal momentum and energy dependence. The electrostatic potential $\Phi$ is determined self-consistently with the Green's functions by solving the Poisson equation under the condition of global charge neutrality and unambiguous bias value. We calculate the optical absorption coefficient in linear optical response, taking into account the calculated laser states and non-equilibrium state occupations, but ignoring vertex corrections to the self energies. Details of the method are given in [8].

## 3. Results and Discussion

We consider electron transport in the resonant phonon THz-QCL structure of [9] which consists of 271 identical periods of GaAs and $Al_{0.15}Ga_{0.85}As$ layers of the widths (30) 92 (55) 80 (27) 66 (41) <u>155</u> Å. The values in parentheses indicate the $Al_{0.15}Ga_{0.85}As$ barriers and the underlined well is the only doped region with a doping density of $1.9\times10^{10}$ cm$^{-2}$. We calculate charge transport in terms of two models. In one model, we treat a single device period as active device, whereas the other model includes explicitly two device periods. In both models, the remaining periods are included in the leads as described above. In figure 1, we depict the calculated (black) and experimental (gray dots) I-V characteristics of this QCL. Both models agree with one another and reproduce the experimental data of the QCL up to the threshold bias. In the remainder of this section, we focus on the two-periods model.

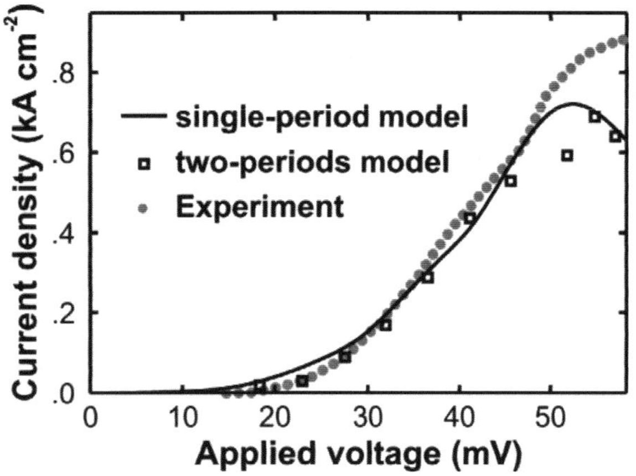


**Figure 1.** Calculated current density as a function of applied bias voltage per period of the QCL of [9] at 40K and a sheet doping density of $1.9\times10^{10}$ cm$^{-2}$ and comparison to experiment [9]. The results of the single-period model (line) and two-periods model (squares) agree well with experiment (dots).

We will show now that for low carrier concentration, the electronic states follow the QCL periodicity, in sharp contrast to the carrier density that deviates from periodicity. In figure 2 (a), we show the contour plot of the spectral function at vanishing in-plane momentum in two adjacent periods of the QCL for a bias of 52 mV/period. The maxima of this spectral function correspond to resonant electronic states. The states associated with the first and second QCL period are labeled by numbers and primed numbers, respectively. In both periods, the injector states (5 and 5') are seen to be aligned with the upper laser states (4 and 4'), whereas the lower laser states (3 and 3') are aligned with the collector well states (2 and 2'). We point out that the energy difference between the lower laser levels (3 and 3') and the lowest collector states (1 and 1') match the energy of an LO-phonon *in both periods*. Nevertheless, we find the carrier distribution to deviate from the geometric QCL periodicity. This can be seen most clearly by a contour graph of the local energy resolved current density $j(z,E)$ (as defined, e.g., in [6]). Figure 2 (b) shows this quantity for the same bias as in part (a). The function $j(z,E)$ shows spatially constant (i.e. horizontal) stripes in regions where the electrons propagate without dissipating energy. Disruptions of these horizontal stripes mark positions where LO-phonons get emitted. The figure shows that the number of emitted LO phonons is no longer equal for adjacent QCL periods. When the electrons have passed the first period and traversed a potential drop of 52 meV, they have emitted only one LO-phonon of energy 36 meV. The elastic and inelastic scattering mechanisms are not able to dissipate the remaining 16 meV within this QCL period. This is a

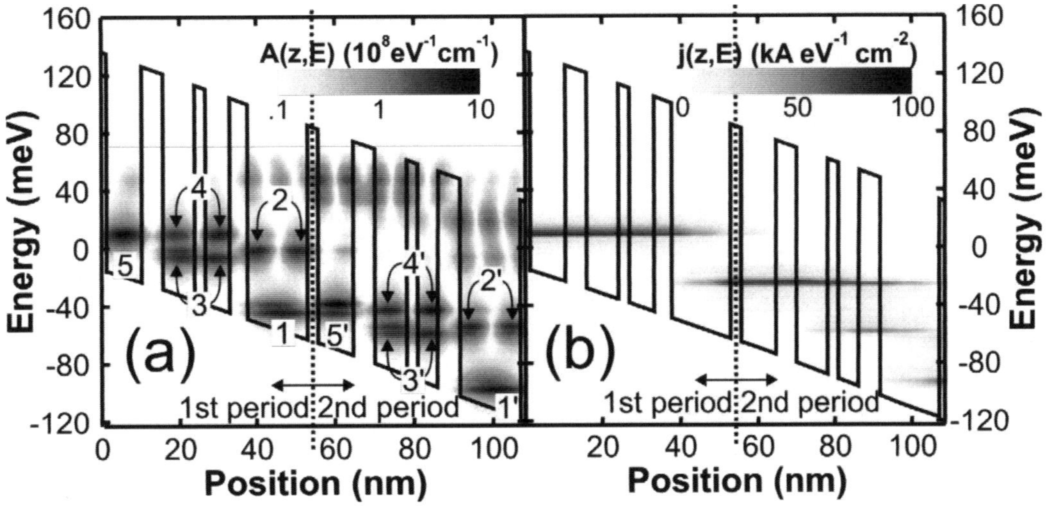


**Figure 2.** (a) Calculated conduction band profile (line) and contour plot of the energy and spatially resolved spectral function *A(z,E)* at vanishing in-plane momentum in two adjacent periods of the QCL at a bias voltage of 52 mV per period in the relevant energy range between -121 meV and 78 meV. The zero in energy marks the chemical potential of the source. The dotted line marks the boundary between first and second period. (b) Calculated spatially and energy resolved current density *j(z,E)* of the QCL in (a).

consequence of the good state alignment that supports efficient coherent multi-barrier tunneling [8]. Consequently, the electrons enter the second period with an in-plane kinetic energy of 16 meV. This can be seen in figure 2 (b); the energy of the leftmost current stripe coincides with states 5 and 4, whereas the following current stripe lies above the corresponding states 5' and 4'. Thus, these propagating electrons are now able to emit an LO-phonon, ending up in and occupying the lower laser level 3'. This occupancy leads to the build-up of the current stripe near $z=80$ nm in Figure 2 (b) that is absent in the first period. The electrons can now tunnel resonantly from 3' into state 2' and scatter into the lowest collector state 1' by the emission of an additional LO-phonon. Thus, the electrons have emitted a total of 3 LO phonons ($3\times36=108$ meV) across 2 QCL periods (voltage drop of $2\times52=104$ meV) and are finally fully thermalized. The remaining small energy discrepancy can be gained from absorbing or emitting acoustic phonons. This process is repeated in the subsequent QCL periods such that we obtain a commensurable charge distribution with period two. Since the detailed energy balance depends on the applied bias voltage, the carrier density and current distribution may even become incommensurable with the geometric periods.

A consequence of this incomplete carrier thermalization is a significant reduction in the occupation inversion and the optical gain in every other period. Concretely, we find a drop of approximately 65% in the second period at the voltage shown in figure 2. We would like to point out that the periodic nature of the spectral function in figure 2 (a) is intimately related to the small carrier density that does not distort the conduction band profile in spite of the different charge density in the second QCL period. This explains the good agreement between the two models (one-period and two-periods) we have obtained in figure 1.

In order to estimate whether the inelastic electron-electron (e-e) scattering can relax the electrons and restore the periodicity of the carrier distribution to a single QCL period, we have calculated the Green's functions including the e-e scattering in the static $GW_0$ approximation. However, we find this type of e-e interaction to be unable to thermalize the non-equilibrium subband distribution sufficiently strongly [8].

Journal of Physics: Conference Series **193** (2009) 012063 doi:10.1088/1742-6596/193/1/012063

## 4. QCL design for fully thermalized structure

We have seen the heated carrier distribution in the second period of the QCL in figure 2 to allow nonradiative transitions from the upper into the lower laser level. In order to pin the electron distribution to the QCL periodicity and thereby to prevent these nonradiative transitions, we propose a new QCL structure. We augment each QCL period from [9] by two additional layers, a 3 nm $Al_{.15}Ga_{.85}As$ barrier, followed by a 17.1 nm wide GaAs quantum well namely. This leads to a threshold bias of ~72 mV per period which results in a potential drop that is commensurable with the energy of two LO-phonons. The purpose of these additional layers is to allow the electrons to dissipate the 16 meV discussed above within the same QCL period rather than causing a nonperiodic density distribution. Figure 3 shows a contour plot of the energy and spatially resolved current density *j(z,E)*. Also shown in this figure are contour lines of the spectral function at vanishing in-plane momentum.

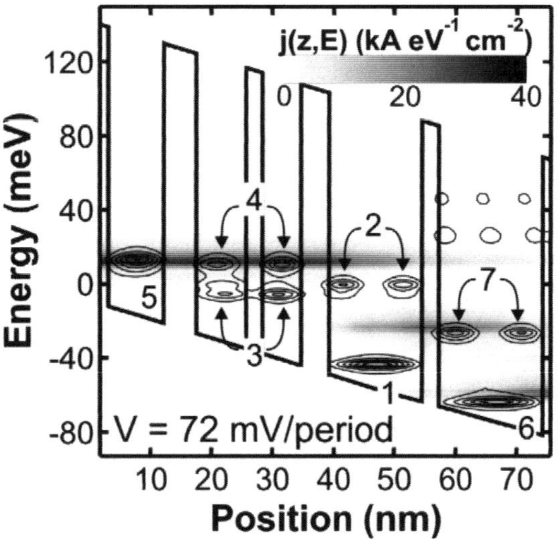


**Figure 3.** Calculated conduction band profile (thick line) and contour plot of the energy and spatially resolved current density *j(z,E)* in a single period of the proposed new QCL design described in the main text. The thin lines depict contour lines of the spectral function *A(z,E)* at vanishing in-plane momentum. The zero in energy marks the chemical potential of the source.

Most electrons that arrive in state 1 have a surplus of 16 meV of kinetic energy in the in-plane motion. Therefore, their total energy suffices to scatter into state 6 by the resonant emission of a second LO-phonon. This can be seen in figure 3 for positions $z=70$ nm. Here, the dominant maximum of *j(z,E)* coincides with the state 6. Thus, the main portion of the carrier distribution at the end of the period is reset to its original value. In this way, the electrons are almost completely thermalized and nonradiative transitions between the laser states of this and the following QCL period are suppressed.

**Acknowledgments**

The authors acknowledge financial support from the Deutsche Forschungsgemeinschaft (SFB 631), the Österreichische Fonds zur Förderung der Wissenschaft (SFB IRON) and the Nano Initiative Munich.

**References**

[1] Belkin M A, Fan J A, Hormoz S, Capasso F, Khanna S P, Lachab M, Davies A G and Linfield E H 2008 Opt. Express **16** 3242
[2] Iotti R C and Rossi F 2001 *Appl. Phys. Lett.* **78** 2902
[3] Willenberg H, Döhler G H and Faist J 2003 *Phys. Rev.* B **67** 085315
[4] Jirauschek C and Lugli P 2008 *phys. stat. sol.* (c) **5** 221
[5] Gao X, Botez D and Knezevic I 2007 *J. Appl. Phys.* **101** 063101
[6] Wacker A 2008 *phys. stat. sol.* (b) **5** 215
[7] Zheng X, Chen W, Stroscio M and Register L F 2006 *Phys. Rev.* B **73** 245304
[8] Kubis T, Yeh C, Vogl P, Benz A, Fasching G and Deutsch C 2009 *Phys. Rev.* B **79** 195323
[9] Andrews A M, Benz A, Deutsch C, Fasching G, Unterrainer K, Klang P, Schrenk W and Strasser G 2008 *Mater. Sci. Eng.* B **147** 152

Journal of Physics: Conference Series **193** (2009) 012064 doi:10.1088/1742-6596/193/1/012064

# Terahertz wide range tunable cyclotron resonance p-Ge laser

**O A Klimenko[1,2,3], Yu A Mityagin[1], S A Savinov[1], V N Murzin[1], N V Dyakonova[2], P Solignac[2] and W Knap[2]**

[1] P.N. Lebedev Physical Institute of RAS, 53 Leninskiy prospekt 119991 Moscow, Russia

[2] GES-UMR 5650 CNRS - Université Montpellier 2, 34090 Montpellier, France

E-mail: oleg.klimenko@mail.ru

**Abstract.** We present a Terahertz (THz) p-Ge laser operating in a cyclotron resonance (CR) mode. Thanks to the use of the optimum crystallographic orientation of p-Ge with respect to crossed $\boldsymbol{E}\perp\boldsymbol{B}$ fields, the laser line frequency is continuously tunable in the range 1.2-2.8 THz (40-90 $cm^{-1}$) with the linewidth being 6 GHz (0.2 $cm^{-1}$). We also show that due to such a wide CR radiation range, the laser can serve as a base for a THz spectrometer.

At present, THz region has a gap in frequencies of radiating sources. Powerful operating in continuous mode carcinotrons reach 1.5 THz [1], whereas QCLs cover frequencies from approximately 2.5 THz and higher [2]. Meanwhile, there are only discrete set of frequencies between them generated by bulky gaseous lasers [3, 4] and very few discrete QCLs' lines [2]. Thus, as far as this gap is desired for many fields such as security, astrophysics, medicine, and others, the fabrication of a portable relatively cheap source covering the whole gap becomes necessary.

In this work we present a THz wide range tunable p-Ge laser operating in a cyclotron resonance (CR) mode. The emission line is continuously tunable by magnetic field in the range 1.2-2.8 THz (40-90 $cm^{-1}$) with the linewidth about 6 GHz (0.2 $cm^{-1}$).

Spontaneous CR emission of light holes in p-Ge was first observed by Komiyama [5]. Stimulated emission was discovered by Ivanov and Vasil'ev [6] from a low impurity sample (acceptor concentration $N_A \leq 10^{13}$ $cm^{-3}$). Later Mityagin et al. [7] discovered CR emission from a sample of higher doping level ($N_A = 7 \cdot 10^{13}$ $cm^{-3}$) in a region of electric and magnetic fields different from that found by Ivanov's group.

Certain applied electric and magnetic fields allow holes to be accumulated in the light subband, which is split into Landau levels. Due to scattering processes and interaction with the heavy subband, population inversion and non equidistance between Landau levels occur that leads to CR emission. The picture is complicated by high anisotropy of heavy holes' effective mass, which strongly effects

[3] To whom any correspondence should be addressed. ⍰

Journal of Physics: Conference Series **193** (2009) 012064 doi:10.1088/1742-6596/193/1/012064

on hole scattering from the light to the heavy subband. Thus the choice of the acceptor concentration and the orientation of crystallographic axes with respect of the applied fields are of great importance and determine the CR emission domain in ***E***, ***B*** axes that defines the radiation frequency range. Detailed quantum-mechanical calculations and experimental studies carried out by Mityagin et al. [8] showed that to obtain the most suitable conditions for CR generation, one should apply $\boldsymbol{B}\uparrow\downarrow[110]$ and ensure the angle between ***E*** and $[1\bar{1}0]$ axis to be 50-90°. The orientation of the applied electric field, $\boldsymbol{E}_{ap}$, should take into account the Hall field, $\boldsymbol{E}_H$, which for $B$=2-4.5 T and chosen crystal dimensions $(3\times5(\parallel E_{ap})\times50(\parallel B)$ mm$^3$) takes on a value comparable with $\boldsymbol{E}_{ap}$ and turns the real electric field inside the crystal on angles 40-55°. Therefore, the angle between $\boldsymbol{E}_{ap}$ and the $[1\bar{1}0]$ axis, φ, should be 10-35°.

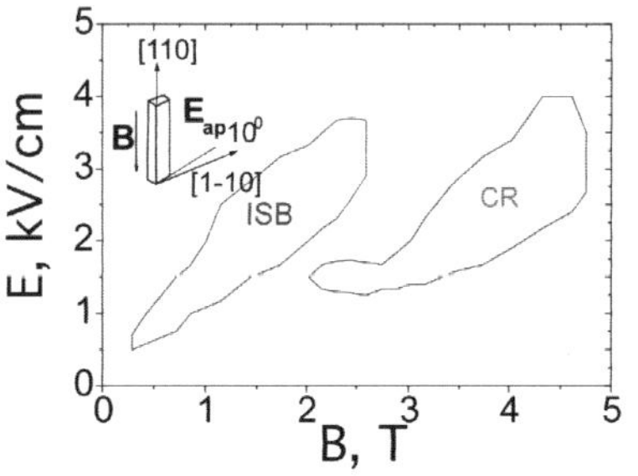


**Figure 1.** Generation areas of two types (ISB – the intersubband radiation, CR – the cyclotron resonance radiation) in E, B axes of the Ge:Ga crystal ($N_{Ga}=7\cdot10^{13}$, $3\times5\times50$ mm$^3$) at EB fields applied $\mathbf{B}\parallel[110]$ axis and **E** with the angle φ=10° to $[1\bar{1}0]$ axis.

The CR p-Ge laser uses as the gain medium a doped by Ga with the concentration $N_A=7\cdot10^{13}$ cm$^{-3}$ Ge crystal, which was fabricated as a rectangular parallelepiped with polished sides and dimensions of $3\times5\times50$ mm$^3$. Both $3\times50$ mm$^2$ facets were covered by 0.2 mm thick In layer to provide good electric contact. The magnetic field was applied along the 50 mm edge and was antiparallel to the [110] crystallographic axis $\boldsymbol{B}\uparrow\downarrow[110]$. The electric field was applied along 6 mm edge, which formed with the $[1\bar{1}0]$ axis an angle 10° as in the inset of figure 1.

In figure 1 the stimulated emission areas of the crystal in B, E domain are presented. The chosen crystallographic orientation provided CR emission for a very large magnetic field range 2-4.8 T that leads to radiation frequencies 1.2-2.8 THz. Beside the CR, at lower magnetic fields the crystal generates radiation of another type – intersubband (ISB). The ISB emission is produced by radiative transitions of holes from the light to the heavy subband. The observed ISB radiation had approximately 10 times higher integral power than the CR one, but of a larger linewidth (300-600 GHz, or 10-20 cm$^{-1}$). It means that to realize a tunable narrow-band source a very complicated selective resonator with gradual tuning of the frequency is required. Since the narrow CR line is easily tuned by the magnetic field without any supplementary devices, it is the main emission mechanism used in p-Ge laser presented in this work the while ISB regime serves for preliminary alignment of the optical system beyond the laser.

The general view of the CR p-Ge laser is presented on figure 2. The whole assembled system represents itself as a tube with a superconducting magnetic coil at the bottom that easily enters into a standard helium bottle (RH 65, for example). The system divides into external and internal parts. The external part consists of the internal tube bearing the coil, the external covering wires tube whereon there is a bonding the laser and the bottle clamping device, a joining both tubes head with a connector for the magnet power supply. The inner part consists of the crystal holder joined with a stainless steel polished waveguide, and head with connector for high voltage (HV) pulses and the waveguide output that can be aligned whether vertically, or horizontally. The crystal is fixed in the holder with two flat

**Figure 2.** The CR p-Ge laser in disassembled (left) and assembled (right) conditions.

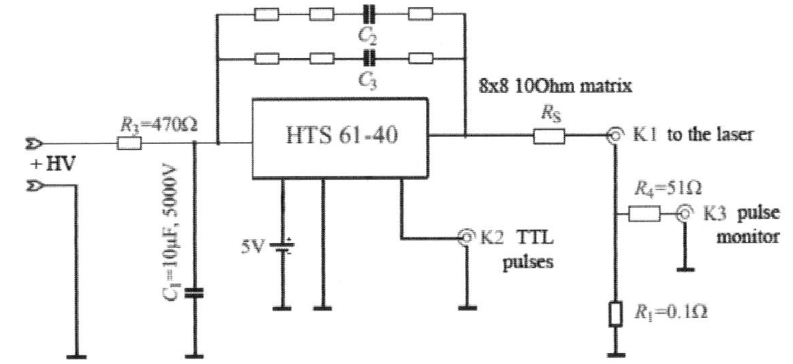


**Figure 3.** The principal scheme of the high voltage pulses generator.

brass mirrors forming a Fabry-Perot resonator. One of the mirrors has a hole emitting the radiation directly into the waveguide. Two parts of the laser are joined by the heads so that the holder enters into the magnetic coil and the crystal is located in the centre of the coil. The magnetic field uniformity along the full crystal forms 99%.

The electric field in the crystal is formed by a high voltage pulses generator (HVPG) producing 0.5-2 kV 50-200 A pulses passing through the crystal. The HVPG was designed especially for the laser and assembled in a well insulating box. Its principal scheme is presented on figure 3. A HV dc source charges through R3 the capacitor C1. An external TTL pulse opens at short time the switch Behlke HTS 61-40, and C1 discharges via the output resistance matrix $R_S$=10 Ohm and the p-Ge laser. As far as the p-Ge impedance is highly nonlinear, a part of HV pulse reflects and flies back to the switch. $R_S$ protects the switch from the fly-back pulses and provides some matching with the laser since during the pulse the p-Ge impedance takes on a value 10 Ohm, approximately. Capacitors C2 and C3 connected in series with resistances serve as high frequency filters and diminish noise on the rising edge. Outlet K3 allows to observe the generated pulse shape on an oscilloscope.

Figure 4 shows the laser spectra measured by a Fourier interferometer Bruker IFS 66. For every fixed magnetic field the electric field was adjusted to have maximum output power. Since characteristic lasing pulse duration and repetition rate are about 1 μs and 10 Hz, respectively, the spectrometer worked in a step scan mode. For every moving mirror position the detector signal was averaged over three points by a boxcar and then analyzed by the spectrometer. The emission spectrum has a single narrow line with the linewidth not wider than the spectrometer resolution limit 7.5 GHz

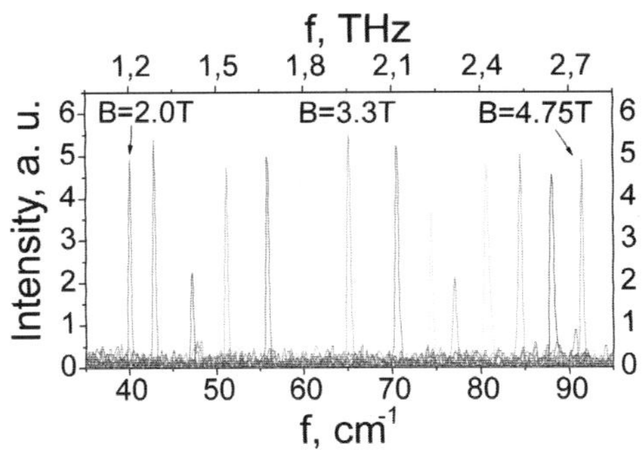


**Figure 4.** The CR p-Ge laser spectra measured by a Fourier spectrometer with resolution 7.5 GHz for magnetic fields B=2.0T; 2.15T; 2.35T; 2.55T; 2.8T; 3.0T; 3.3T; 3.6T; 3.8T; 3.95T; 4.15T; 4.35T; 4.55T; 4.75T.

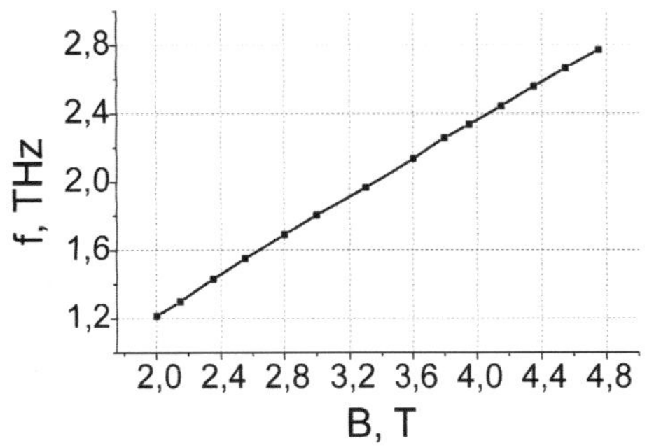


**Figure 5.** The CR p-Ge laser radiation frequency as a function of magnetic field.

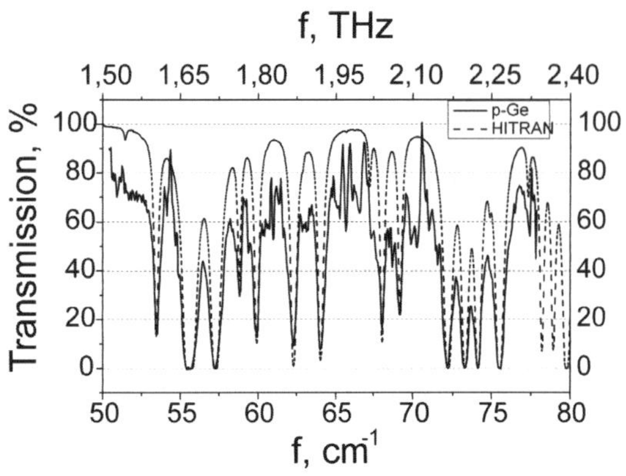


**Figure 6.** Atmospheric water vapor absorption spectrum. Solid – measured with the CR p-Ge laser based spectrometer, dashed – HITRAN data.

(0.25 $cm^{-1}$). The line position tuned with magnetic field as it shown on figure 5. The radiation was observed in the frequency range 1.2-2.8 THz (40-90 $cm^{-1}$).

An atmospheric water vapour absorption spectrum (see figure 6) was measured to demonstrate that the laser can serve as a base for a THz spectrometer. The spectrometer consisted of the CR p-Ge laser, a Si bolometer as the detector and a transparent cell with controlled pressure. For this type of measurements, electric field was fixed at 2.2 kV/cm, whereas magnetic field was slowly swept from 2.5 to 4 T. The spectrum was compared with HITRAN data for 296 K, 1 atm pressure and 20 cm optical path. The comparison enabled to estimate p-Ge laser linewidth as 6 GHz (0.2 $cm^{-1}$).

In conclusion, there was fabricated a CR p-Ge laser that is continuously tunable in a range 1.2-2.8 THz and can serve as a base for a THz spectrometer. The main parameters of the laser are presented in table 1.

**Table 1.** The main parameters of the CR p-Ge laser.

| | | | |
|---|---|---|---|
| Frequency range | 1.2 – 2.8 THz | Operation temperature | 4.2 K |
| Emission bandwidth | 6 GHz | Magnetic field strength | 2 – 5 T |
| Output power | 10 mW | Voltage amplitude | 0.5 – 2 kV |
| Operation mode | pulsed | Current amplitude | 50-200 A |
| Pulse duration | 0.3 – 3 μs | Active crystal dimensions | 3x5x50 $mm^3$ |
| Repetition rate | ≤15 Hz | Liquid helium consumption: | |
| Maximum filling factor | $10^{-5}$ | Operating mode | 1 liter/hour |
| | | Standby | 0.1 liter/hour |

**Acknowledgements.** We would like to thank prof. Gavrilenko V. I. (IPM RAS) for helpful discussions and C. L'Heronet for fabrication of all mechanical parts of the laser.

**References**

[1]. Gorshunov B, Volkov A, Spektor I, Prokhorov A, Mukhin A, Dressel M, Uchida S and Loid A 2005 *Int. J. IRMMW* **26** 1217
[2]. Tredicucci A, Köhler R, Mahler L, Beere H E, Linfield E H and Ritchie D A 2005 *Semicond. Sci. Technol.* **20** S222
[3]. Jacobson S 1989 *Infrared Phys.* **29** 853
[4]. Hodges D T 1978 *Infrared Phys.* **18** 375
[5]. Komiyama S 1982 *Phys. Rev. Lett.* **48** 271
[6]. Ivanov Yu L and Vasil'ev Yu B 1983 *Sov. Tech. Phys. Lett.* **9** 264
[7]. Mityagin Yu A, Murzin V N, Stoklitskii S A and Trofimov I E 1987 *JETP Lett.* **46** 144
[8]. Mityagin Yu A, Mursin V N and Stoklitsky S A 1991 *Optical and Quantum Electronics* **23** S287

Journal of Physics: Conference Series **193** (2009) 012065 doi:10.1088/1742-6596/193/1/012065

# A new aluminum-free material system for intersubband emitters and detectors

**Aaron Maxwell Andrews [1], Michele Nobile [1], Hermann Detz [1], Pavel Klang [1], Elvis Mujagić [1], Werner Schrenk [1], and Gottfried Strasser [1,2]**

[1] Center for Micro- and Nanostructures and Institute for Solid-State Electronics, Vienna University of Technology, Floragasse 7, 1040 Wien, Austria

[2] University at Buffalo, The State University of New York, 332 Bonner Hall, Buffalo, NY 14260, USA

aaron.maxwell.andrews@tuwien.ac.at

**Abstract**. We report on the InGaAs/GaAsSb material system lattice-matched to InP for intersubband devices. Due to the much lower electron effective mass in the GaAsSb barrier material, this system is very promising for the realization of high performance intersubband devices, like quantum-cascade lasers and quantum well infrared photodetectors. Type I intersubband absorption in InGaAs/GaAsSb multi-quantum wells samples has been studied experimentally and the conduction band offset at the InGaAs/GaAsSb hetero-interface has been determined to be 360 meV. Additionally, we realized the first Al-free QCL based on the same material system, emitting at a wavelength of 11.3 μm.

## 1. Introduction

Modern mid-infrared and far-infrared optoelectronic devices, such as quantum cascade lasers (QCLs) [1] and quantum well infrared photodetectors (QWIPs) [2], exploit intersubband (ISB) transitions in semiconductor quantum wells (QWs). High power and high temperature performance QCLs have been demonstrated in InP-based [3, 4] and GaAs-based materials system [5, 6]. For the shorter wavelengths (2.75-4 μm), strain compensated InGaAs/AlGaAs and InAs/AlSb based QCLs have been reported [7, 8], while GaAs/AlGaAs QCLs have been shown for the terahertz range (60-300 μm) [9]. An aluminum-free material system, where the effective mass of the electrons in the barrier acting layers is much lower, would be desirable. Higher optical gain and improved device performance are expected to be the consequence of this [10, 11]. Besides the development of II-VI materials in recent years [12], a very promising candidate to this aim is the InGaAs/GaAsSb material system lattice-matched to InP, which shows a type II alignment, with the InGaAs being the well for the electrons and the GaAsSb for the holes [13, 14]. We report on ISB absorption in InGaAs/GaAsSb multiple quantum wells, where the transitions between the quantized states within the conduction band are investigated. Using the measured CBO we designed and fabricated the first aluminum-free QCL.

## 2. Intersubband Absorption

The samples consisting of 30 periods of $In_{0.53}Ga_{0.47}As/GaAs_{0.51}Sb_{0.49}$ QWs were grown lattice-matched to InP substrates by solid-source molecular beam epitaxy (MBE) [15]. The thicknesses of the InGaAs wells range between 4.5 and 12 nm. In order to supply electrons in the wells, the wells were Si doped in the center to a sheet density of $5x10^{11}$ $cm^{-2}$. The thickness of the GaAsSb barriers is 50 nm.

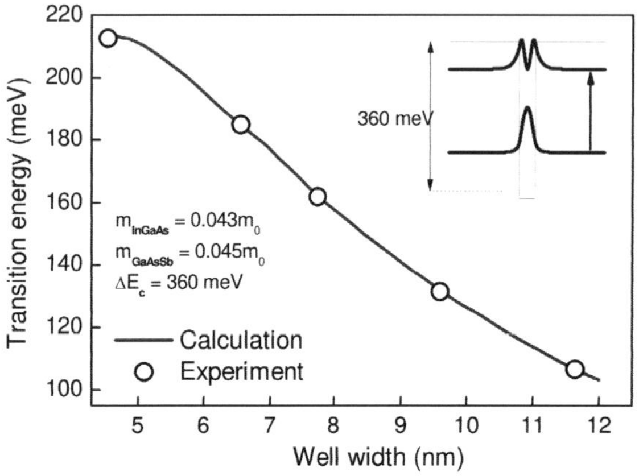


**Figure 1.** Well width dependence of the ISB transition energy for a InGaAs/GaAsSb QW. The circles indicate the experimental values of the ISB resonances, while the solid line is the calculation. Insert: Schematic of the conduction band with the moduli squared of the wavefunctions of the levels which are involved.

For the ISB absorption measurements, the samples have been cleaved into 3x10 $mm^2$ pieces and the facets have been 45° polished in order to allow multi-pass waveguide geometry [16]. The light from a broadband source is focused onto one of the polished facets and the light coming out from the opposite facet is refocused on a liquid-nitrogen cooled HgCdTe detector. The signal of the detector is then dispersed in a Fourier-transform infrared (FTIR) spectrometer. The incoming beam was prepared in the TM-polarization (the electric field of the optical wave is along the growth axis) and TE-polarization (the electric field of the optical wave is in the plane of the wells) and all measurements have been performed at room temperature.

The ISB absorption spectrum is obtained by taking the negative natural logarithm (-ln) of the transmission ratio TTM/TTE. Very clear ISB absorption peaks are observed in a broad spectral region, where the transition energy decreases from 214-107 meV (5.8-11.6 μm) by increasing the well width from 4.5-12 nm. In addition, the ISB absorption peaks show a full-width at half-maximum (FWHM) of ~ 17 meV, which is very close to the value obtained for state-of-the-art bound-to-bound transitions in InGaAs/InAlAs QWs [15, 16].

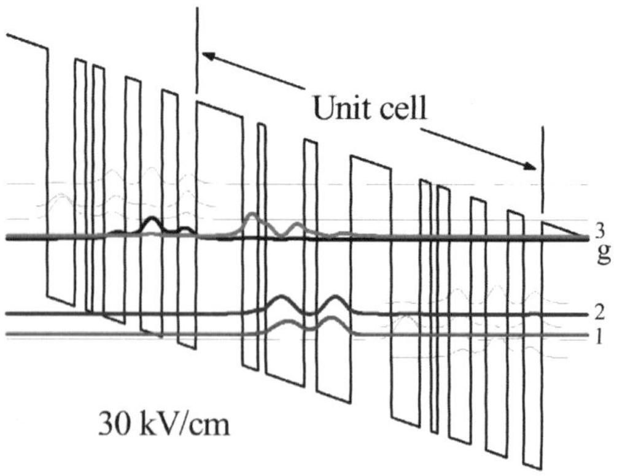


**Figure 2.** Conduction band energy diagram of a portion of the quantum cascade laser active region under an applied electric field of 30 kV/cm. The moduli squared of the relevant wave functions are shown. Indicated are also the ground state of the injector (labeled with g), the upper and lower laser levels (labeled with 3 and 2 respectively).

The ISB transition energies have been modeled using the envelope function approximation considering the nonparabolicity of the conduction band [16]. Since the sheet electron density is quite high in the QWs, we solved Schrödinger's and Poisson's equations self-consistently, including the exchange-correlation potential. The collective shift of the ISB resonance peak due to depolarization and excitonic shift has also been considered. Since the wells are decoupled because of the thick GaAsSb barriers (50 nm), the calculations have been performed assuming a single QW only. The values of the electron effective mass used in the calculations are $0.043m_0$ and $0.045m_0$ for the InGaAs wells and GaAsSb barriers respectively [17, 18]. By performing a least squares fitting between the theoretically calculated transition energies and the experimental ISB resonances, we obtain a CBO at the InGaAs/GaAsSb hetero-interface of 360 meV, in agreement with Hu et al. [13]. An excellent agreement is obtained between the theoretical calculation and the experimental results over the whole range of well thicknesses, as shown in Figure 1.

## 3. Quantum Cascade Laser

The QCL samples have been grown lattice-matched on an n-doped ($2x10^{17}$ $cm^{-3}$) InP substrate by MBE. The QCL active region has been designed following a 3-well scheme [5] and consists of 40 repetitions of an injector/active cell superlattice. The layer thicknesses in angstroms are as follows: **81**/27/**13**/67/**22**/59/**70**/50/**19**/12/**19**/<u>38</u>/**27**/<u>38</u>/**28**/32. The $GaAs_{0.51}Sb_{0.49}$ barrier layers are in bold, while the underlined layers are Si doped ($4x10^{17}$ $cm^{-3}$). As shown in Figure 2, the ground state of the injector (g) and the upper laser level (3) align at a field of 30 kV/cm. The calculated energy separation between the levels 3 and 2 is 114 meV, which corresponds to an emission wavelength of 10.9 μm. The active region is placed between upper and lower $In_{0.53}Ga_{0.47}As$ layers, 600 and 100 nm respectively, doped $1x10^{17}$ $cm^{-3}$. The growth is completed by a heavily doped contact layer.

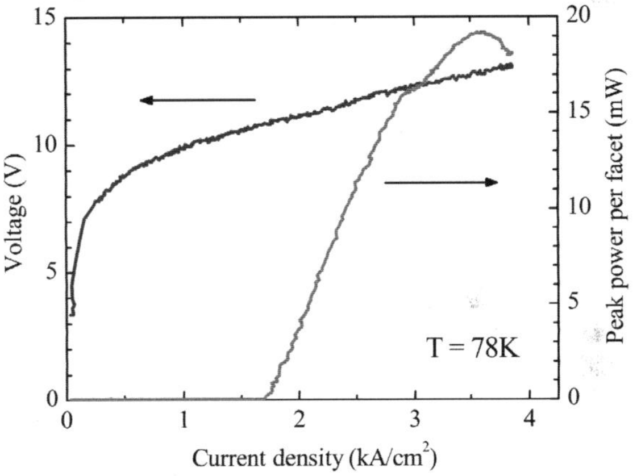


**Figure 3.** Light-current-voltage curve of a 60μm-wide, 2mm-long ridge InGaAs/GaAsSb quantum cascade laser at a temperature of 78K.

The samples have been processed into 60 μm-wide ridges by reactive ion etching. Afterwards, a 300 nm-thick SiN layer and a Ti/Au top ohmic contact were deposited in a way that an aperture is left on the top of the ridges, in order to form a low-loss air-waveguide. The back contact of the lasers is a standard Ge/Au/Ni/Au ohmic contact.

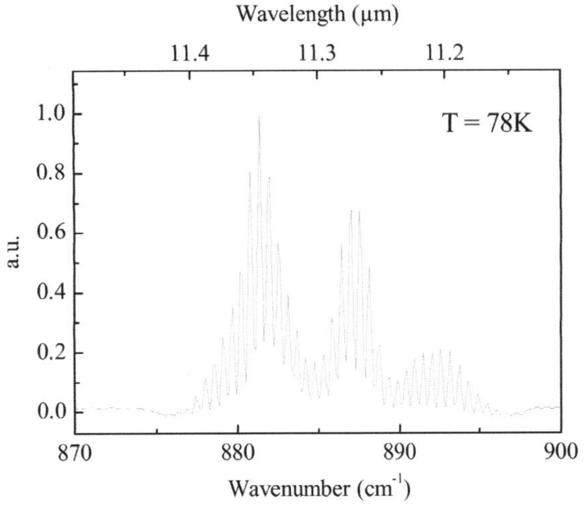


**Figure 4.** Spectrum of the light emitted by a 60 μm-wide ridge InGaAs/GaAsSb quantum cascade laser. The spectrum has been taken at a current density of 3.6 kA/cm², corresponding to the maximum optical output power in the light-current curve.

For the optical measurements, the lasers were cleaved into 2 mm-long bars and mounted on the copper cold finger of a cryostat equipped with ZnSe windows, kept at a temperature of 78K. The lasers are operated in pulsed mode, with a pulse width of 100 ns and a repetition frequency of 5 kHz. The light-current-voltage (LIV) curves and the spectra of the emitted light are recorded by a deuterated triglycine sulfate (DTGS) detector and a Fourier transform infrared (FTIR) spectrometer. In Figure 3, the LIV of a 60µm-wide ridge InGaAs/GaAsSb QCL is presented. The threshold current density and the maximum optical output power are 1.7 kA/cm$^2$ and 20 mW, respectively, at a temperature of 78 K. In Figure 4, the spectrum of the light emitted is shown.

## 4. Conclusions

We observed room temperature ISB absorption in InGaAs/GaAsSb MQWs lattice-matched to InP. The ISB absorption of five samples with different well widths covers a broad spectral region (5.8 - 11.6 µm). By using a value for the CBO at the InGaAs/GaAsSb hetero-interface of 360 meV, the experimental results are in excellent agreement with the theoretical calculations. Furthermore, we demonstrated the first aluminum-free QCL, realized in the InGaAs/GaAsSb material system lattice-matched to InP. A threshold current density and maximum optical output power of 1.7 kA/cm$^2$ and 20 mW respectively at 11.3 µm emission wavelength are reported at a temperature of 78K.

The authors acknowledge the support by the Austrian FWF ADLIS, the Austrian Nano-initiative project PLATON and the NYSTAR award.

## References

[1] J. Faist, F. Capasso, D. L. Sivco, C. Sirtori, A. L. Hutchinson and A. Y. Cho, *Science* **264**, 553 (1994)
[2] H. Schneider and H. C. Liu, *Quantum Well Infrared Photodetectors: Physics and Applications* (Springer, Berlin, 2007)
[3] A. Tahraoui, A. Matlis, S. Slivken, J. Diaz and M. Razeghi, *Appl. Phys. Lett.* **78**, 416 (2001)
[4] T. Aellen, M. Beck, N. Hoyler, M. Giovannini, E. Gini and J. Faist, *J. Appl. Phys.* **100**, 043101 (2006)
[5] H. Page, C. Becker, A. Robertson, G. Glastre, V. Ortiz, and C. Sirtori, *Appl. Phys. Lett.* **78**, 3529 (2001)
[6] C. Pflügl, W. Schrenk, S. Anders, G. Strasser, C. Becker, C. Sirtori, Y. Bonetti and A. Muller, *Appl. Phys. Lett.* **83**, 4698 (2003)
[7] J. Faist, F. Capasso, D. L. Sivco, A. L. Hutchinson, Sung-Nee G. Chu and A. Y. Cho, *Appl. Phys. Lett.* **72**, 680 (1998)
[8] J. Devenson, O. Cathabard, R. Teissier and A. N. Baranov, *Appl. Phys. Lett.* **91**, 251102 (2007)
[9] R. Köhler, A. Tredicucci, F. Beltram, H. E. Beere, E. H. Linfield, A. G. Davis, D. A. Ritchie, R. C. Iotti and F. Rossi, *Nature (London)* **417**, 156 (2002)
[10] C. Sirtori, F. Capasso, J. Faist, S. Scandolo, *Phys. Rev. B* **50**, 8663 (1994)
[11] J. Faist, F. Capasso, D.L. Sivco, A.L. Hutchinson, C. Sirtori, S.N.G. Chu, A.Y. Cho, *Appl. Phys. Lett.* **65**, 2901 (1994)
[12] B. S. Li, A. Shen, W. O. Charles, Q. Zhang and M. C. Tamargo, *Appl. Phys. Lett.* **92**, 261104 (2008)
[13] J. Hu, X. G. Xu, J. A. H. Stotz, S. P. Watkins, A. E. Curzon, M. L. W. Thewalt, N. Matine and C. R. Bolognesi, *Appl. Phys. Lett.* **73**, 2799 (1998)
[14] J. F. Klem, S. R. Kurtz and A. Datye, *J. Cryst. Growth* **111**, 628 (1991)
[15] M. Nobile, H. Detz, E. Mujagić, A.M. Andrews, P. Klang, W. Schrenk and G. Strasser, *Appl. Phys. Lett.* **95**, July 27$^{th}$ 2009
[16] H. C. Liu and F.Capasso, Intersubband Transitions in Quantum Wells: Physics and Device Applications I (Academic, San Diego, 2000), Chap. 1.
[17] I. Vurgaftman, J. R. Meyer and L. R. Ram-Mohan, *J. Appl. Phys.* **89**, 5815 (2001)
[18] P. Devlin, H. M. Heravi and J. C. Woolley, *Can. J. Phys.* **59**, 939 (1981)

# THz Photoresponse and Magnetotransport of detectors made of HgCdTe/HgTe quantum well structures

**F.Gouider[1], M.Bugár[1], J. Könemann[2], Yu.B.Vasilyev[3], C.Brüne[4], H.Buhmann[4] and G.Nachtwei[1].**

[1]Institute of Applied Physics, Technical University Braunschweig, D-38106, Germany
[2]Physikalisch-Technische Bundesanstalt, D-38116 Braunschweig, Germany
[3]A.F.Ioffe Physical Technical Institute, RU-194021 St.Petersburg, Russia
[4]Julius-Maximilians-University Würzburg, D-97074, Germany

Email: f.gouider@tu-bs.de

**Abstract.** In addition to space- and time-resolved measurements of the electrical generation of excited electrons in quantum Hall (QH) devices the dissipation in these devices induced by Terahertz (THz) laser impulses is investigated. For these tasks a THz laser system ( *p*-Ge-Laser [1]) is applied. This laser uses transitions between Landau levels of light holes and emits laser impulses with an adjustable duration from 300 ns to some μs, tunably in the wavelength range 120μm < λ < 180μm. The THz laser radiation is used in order to stimulate charge carriers optically over the Landau-gap. The response of the sample to the laser impulses is in 2D-samples in the QH regime measured in order to receive data of the relaxation of the charge carriers. In this presentation we present photoconduction measurements of the HgTe/HgCdTe-(MCT) - quantum well devices in Hall bar, Corbino as well as combined Corbino- Hall bar geometry in the QH- regime. The materiel system HgTe/HgCdTe is characterized by a small effective mass (compared to GaAs [2], in our case $0.023m_0$) and accordingly smaller magnetic fields for the appropriate cyclotron energy.

## 1. Introduction

Although the THz range of the electromagnetic spectrum bears so far severe difficulties for applications in the material research, medical technology, astro-and atmosphere physics, safety engineering and data communication, this range is little investigated concerning the basical physics. It exists a so-called "Terahertz gap" for frequencies between some hundred GHz (millimeter waves) and THz (submillimeter waves) unite. The upper critical frequency for electronically produced waves is under a THz (e.g. $f$ =712GHz for resonant tunnel diode [4]) with, the upper border for the wavelength with optical generation with quantum cascade lasers (QCL) against it with for instance 160μm [5] (corresponds to a lower critical frequency of somewhat less than 2THz). The production of QCL is very complex (epitaxy of many layers), in order to make a economical and broad application possible. For the wavelength coverage directly in THz gap are so far only as laboratory solutions present monochromatic generators (e.g. optically pumped molecular gas lasers [6] and electrically pumped germanium lasers [7] like also used for our measurements). Apart from the generation of the THz radiation also the long-distance transmission of the signals and detection are problematic. $CO_2$ molecules absorb strongly radiation with $\lambda \approx 15\mu m$, and rotation excitations of water absorb clearly at wavelengths of $\lambda > 20\mu m$. Thus a transmission in the THz range realizable at justifiable expenditure is limited for the indoor radio nets to distances of about 10m [3]. That is sufficient for a number of conceivable applications. For the receipt of THz or far infrared radiation there are meanwhile various detectors available (e.g. bolometers, photoconductors, Schottky diodes and pyroelectric detectors), also with reasonably high sensitivity. The latter, however, goes along in bolometer detectors usually with rather long response times (low working frequencies) and requires a complex cooling [8]. In semiconductor detectors, the photoconduction effects (e.g. endowed germanium detectors such as Ge: Ga and narrow Gap semiconductors such as InSb or HgCdTe) or charge carrier avalanches (e.g. silicon Hot Electron bolometers (HEB)) use, can respond and response times in the nanosecond range have been observed [8]. However these detectors are only a little tunable (semiconductor detectors) or posses a very wide band concerning the spectral selectivity (bolometers).

A possibility of operating QH detectors at clearly smaller magnetic fields offers the material system HgTe/HgCdTe. At 2DEG in HgTe quantum wells the QHE could be observed at magnetic fields as low as 1T. The effective mass of the electrons in this HgTe quantum well is smaller than for GaAs quantum wells (GaAs: $m^*/m_0$ = 0,067). Thus, comparable Landau gaps can be observed at magnetic fields being approximately 1/3 of those needed for GaAs-based samples.

## 2. Experimental

The sample and the *p*-Ge laser with the superconducting magnet for the laser are placed on one holder, which is inserted into the cryostat with a Helium bath. The sample area is surrounded by another superconducting magnet, which consists of two coils. The outer one is wound from NbTi filaments surrounded by stabilization matrix of copper; the inner coil is made from $Nb_3Sn$. The whole cryostat with the magnet is controlled by commercial equipment (Oxford Instruments). There is a control of the temperature and the liquid He level (controlled by a He needle valve) in the inner cryostat part and, finally, the liquid He level in the outer part (necessary for cooling of the sample magnet). For the generation of THz waves we use a *p*-Ge laser system. The principle of the *p*-Ge laser function is based on the light hole population inversion between the Landau levels, which are created after placing of a *p*-doped ($p \approx 10^{14}$cm$^{-3}$) Germanium crystal to the magnetic field. In addition, an electric field applied perpendicular to the magnetic field causes a shift of the center of the cyclotron orbits to negative $k_{||}$-values to the state (0;-*E*/*B*). When the holes reach the optical phonon energy $\hbar\omega_{op}$ (by increasing the magnetic field), holes start to be markedly scattered by these phonons. Therefore, carriers in the outer Landau level interact strongly with optical phonons resulting in a small velocity after the phonon emission. The population of the lower Landau level increases and a population inversion between Landau levels (which is not influenced by carriers scattered by optical phonons) occurs. The emitted photons are amplifyed by the Cu resonator. The output from laser is focused to the sample area.The gain of the system is ~ 0.05cm$^{-1}$. The energy range of the laser is 7-11meV (corresponding to wavelengths $\lambda$=110-180μm, frequencies $f$ =1.8-2.7 THz or wave numbers $\nu$=60-90cm$^{-1}$). The electrical pumping pulses with voltages around 1.5kV have pulse times ranging from 0.3 to 50 μs. The peak power of emitted laser pulses is about 1W per pulse with a repetition rate of about 1Hz.

In this report we present photoconduction measurements of the HgTe/HgCdTe- (MCT) - quantum well structures in Hall bar, Corbino as well as combined Corbino- Hall bar geometry in the QH- regime. The wafers were prepared by Molecular Beam Epitaxy (MBE) technique.

Three specific properties of HgTe quantum well (QW) structures have been found in systems with direct gap. The first is the nonparabolicity of the band structure, which causes an increase of the effective electron mass with increasing electron density [9]. Second, the size of the forbidden band depends on the quantum well width. When the HgTe QW is thinner than 6.3nm, the structure behaves as a standard semiconductor. However, for QWs wider than 6.3nm the band structure is inverted and the system becomes a semimetal. And finally, in a semimetallic case, holes exist simultaneously with electrons and form a 2D electron-hole system without any spatial separation of the 2D electons and 2D holes.

For the device fabrication, photolithography was used. Here, a light sensitive chemical substance (photoresist) was selectively exposed to light through a patterned glass plate (photomask). The fabrication started by cutting small rectangular pieces from the wafers, followed by a 2-solvent cleaning procedure. After cleaning, the photoresist was spun on the wafer pieces. Then, ultraviolet light was radiated onto the wafers through the mesa-photomask. After developing the photoresist on the samples in a developer solution, the desired mesa patterns were replicated on the samples. By placing the samples in a chemical solution, the mesa patterns were etched. After etching, a similar procedure was applied with the photomask for the ohmic contacts. Then, the contact material was evaporated on the samples. After lift-off and annealing steps the device fabrication was completed.

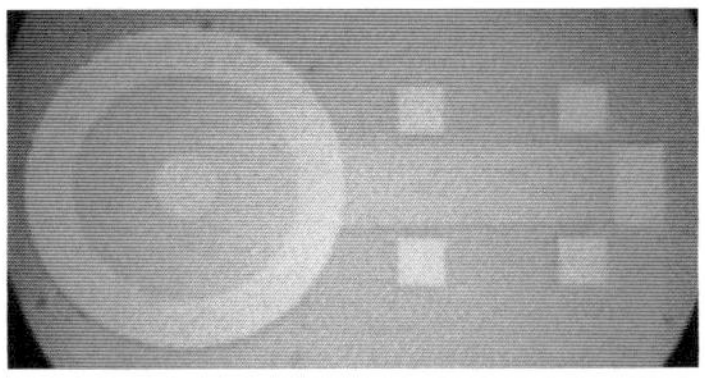

**Figure 1.** Combined Hall bar-Corbino device.

## 3. Results and discussion

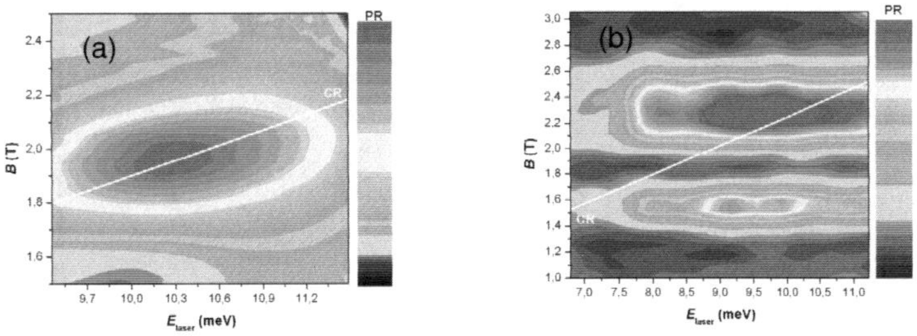


**Figure 2.** Photoresponse as a function of the magnetic field and the laser energy. (a) Corbino and (b) combined Hall bar- Corbino geometry.

In samples with semiconducting QW properties we cannot resolve photoresponse (PR) signals. In contrast, the QWs with semimetallic properties show a clearly observable PR. In Fig.2 we show the PR of the sample with 10nm and 12nm quantum well width as a function of the magnetic field and the laser energy in a coloured plot. For a Corbino device, the PR follows the line which shows the magnetic field of the cyclotron resonance (CR) as a function of the laser energy. From this we conclude a dominant contribution of the CR to the PR (Fig.2 (a), $d_{QW}$= 10nm). A similar plot is shown for a combined Hall bar-Corbino device (Fig.2 (b), $d_{QW}$= 12nm). For the filling factor $\nu = 4$, there is a crossing of the calculated dependence of the magnetic fields for the CR as a function of the laser energy and the measured maximum of the PR. Thus, for $\nu = 4$ we cannot clearly separate the bolometric from the CR contribution to the PR. In contrast, at $\nu = 6$ there is no such crossing as for $\nu = 4$. Therefore, here the PR is dominantly bolometric.

**References**

[1] Y.B. Vasilyev, Y.L. Ivanov, Sov. Phys. Techn. Phys. **31**, 359 (1986).
[2] G. Nachtwei, F. Gouider, C. Stellmach, G. Vasile, Y.B. VAsilyev,G. Hein,and R. R. Gerhardts, Phys. Rew. B **78**, 174305 (2008).
[3] J.M. Guldbakke: Diplomarbeit TU Braunschweig, 2004
[4] E.R. Brown, J.R. Söderström, C.D. Parker, L.J. Mahoney, K.M. Molvar, T.C.McGill: Appl.Phys.Lett. **58**, 2291 (1991)
[5] G. Scalari, S. Blaser, M. Graf, L. Ajili, J. Faist, H. Beere, E. Linfield, D. Ritchie, G. Davies: 27th International Conference on the Physics of Semiconductors (ICPS-27), Flagstaff/ AZ (USA), AIP Conference Proceedings **772**, 1573 (2005)
[6] G. Abstreiter, P. Kneschaurek, J.P. Kotthaus, and F.J. Koch: Phys. Rev. Lett. **32**, 104 (1974)
[7] Yu.L. Ivanov, Yu.B. Vasilyev: Sov. Tech. Phys. Letters **9**, 264 (1983)
[8] H. Kuzmany: "Festkörperspektroskopie", Springer-Verlag 1989, S. 166 ff. and P.L. Richards: J. Appl. Phys. **76**, 1 (1994)
[9] Z.D.Kvon, S.N.Danilov, N.N.Mikhailov, S.A.Dvoretsky, W.Prettl, S.D.Ganichev; Physica E **40**, 1885 (2008)

# Analytical model for the TeraHertz current noise in nanometric Schottky-barrier diodes and heterostructure barrier varactors

**F. Z. Mahi, A. Helmaoui**

Physics of Semiconductor Devices Laboratory (LPDS), University of Bechar, Algeria

E-mail: fati_zo_mahi2002@yahoo.fr

**L. Varani**

Institute of Electronics of the South (IES - CNRS UMR 5214), University of Montpellier, France

**P. Shiktorov, E. Starikov and V. Gruzhinskis**

Semiconductor Physics Institute, Vilnius, Lithuania

**Abstract.** In this paper we propose an analytical model for the calculation of the spectral density of current fluctuations in Heterostructure-barrier varactors . The structures of the calculated spectra are analyzed in terms of physical processes useful to optimize the device parameters for the extraction of the high-order harmonics.

## 1. Introduction

The varactor devices, like the Schottky barrier diode (SBD) and the heterostructure barrier varactor (HBV) diode, are nonlinear elements able to generate high-frequency signals by frequency multiplication [1]. More recently, the HBV diodes have emerged as interesting competitors of SBD as frequency multipliers due to their symmetric capacity-voltage (C-V) and antisymmetric current-voltage (I-V) characteristics [2,3], which lead to produce only odd harmonics compared to the SBD [1]. Under high-frequency large-signal operation, a critical parameter assessing their electrical performances is the intrinsic current noise, which constitutes also an important limit for the extraction of the high-order harmonics at the basis of terahertz generation. In this contribution we present an extension of the analytical model proposed in Ref. [4] which gives a description of the resonances appearing in the high frequency region.

## 2. Analytical Model

We apply the analytical model described in Ref. [4] to calculate the intrinsic high-frequency noise spectrum in a typical $n^+n-$ barrier $-nn^+$ HBV structure shown in Fig. 1. Here $l_i$ is the length of the i-regions with $N_2$ and $N_4$ the dopings of the first and second $n$ regions, respectively, $N_1$ and $N_3$ the dopings of the first and second $n^+$ regions, respectively, and $l_d$ the length of the depletion region (see, e.g. [3]).

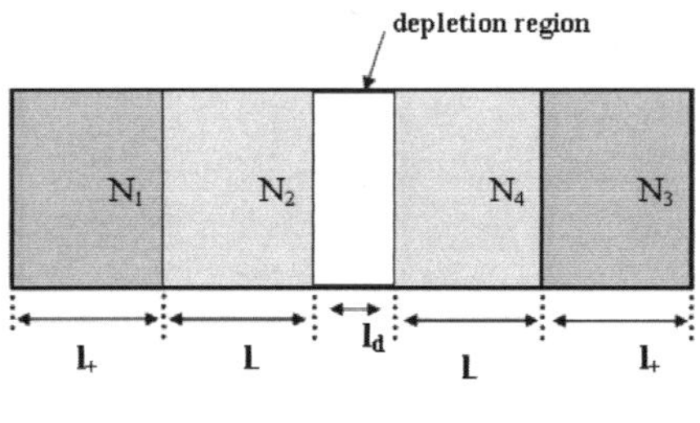


**Figure 1.** Schematic view of the $n^+n$-barrier-$nn^+$ HBV with: $l_+$ the length of the $n^+$ regions, $l_-$ the length of $n$ regions and $l_d$ the length of the depletion region.

In accordance with Ref. [4] the Langevin-force induced initial perturbations of the collective shifts of free carriers with respect to the donor levels in the i-regions are equal to $\delta x_i$, so that the $\delta E_i$ fluctuation self-consistent with these shifts can be written as:

$$\delta E_i = \begin{cases} \Delta E - (\frac{e}{\epsilon\epsilon_0})N_1\delta x_1 \\ \Delta E - (\frac{e}{\epsilon\epsilon_0})N_2\delta x_2 \\ \Delta E \\ \Delta E - (\frac{e}{\epsilon\epsilon_0})N_3\delta x_3 \\ \Delta E - (\frac{e}{\epsilon\epsilon_0})N_4\delta x_4 \end{cases} \text{for regions} \begin{cases} 1^{st}\ n^+ \text{ region} \\ 1^{st}\ n \text{ region} \\ \text{depletion region} \\ 2^{nd}\ n^+ \text{ region} \\ 2^{nd}\ n \text{ region} \end{cases} \tag{1}$$

Here the homogeneous part of the global perturbation$\Delta\ E = \frac{e}{\epsilon\epsilon_0 L}\sum_{i=1}^{4} N_i l_i \delta x_i$ is determined under constant applied voltage operation and the variation of the voltage drop equals to zero $\delta U = 0$, where $i = 1,2,3,4$ is the number of regions. According to the symmetry of HBV structure $l_1 = l_3 = l_+$, $l_2 = l_4 = l_-$ and $L = 2l_+ + 2l_- + l_d$ is the total length of the HBV [2,3]. To describe evolution of these shifts we take the advantage of the Langevin approach formulated by the following system of equations:

$$\frac{d^2}{dt^2}\delta x_i + \nu_i \frac{d}{dt}\delta x_i = \frac{e}{m}\delta E_i(x) + f_i \tag{2}$$

where $f_i$ is the Langevin force in the i-region. For the numerical results we assume that $N_1 = N_3 = N_+$, $N_2 = N_4 = N_-$ and the plasma frequency is defined as $\omega_i^2 = (e^2/\epsilon\epsilon_0 m)N_i$. The previous system of equations (1) and (2) can be rewritten in matrix form as:

$$\begin{pmatrix} a_{11} & a_{21} & a_{31} & a_{41} \\ a_{12} & a_{22} & a_{32} & a_{42} \\ a_{13} & a_{23} & a_{33} & a_{43} \\ a_{14} & a_{24} & a_{34} & a_{44} \end{pmatrix} \begin{bmatrix} \delta x_1(\omega) \\ \delta x_2(\omega) \\ \delta x_3(\omega) \\ \delta x_4(\omega) \end{bmatrix} = \begin{bmatrix} f_1(\omega) \\ f_2(\omega) \\ f_3(\omega) \\ f_4(\omega) \end{bmatrix} \tag{3}$$

The expressions of the matrix elements are: $a_{11} = a_{33} = -\omega^2 + i\omega\nu_+ + \omega_+^2(1 - r_+)$, $a_{12} = a_{14} = a_{24} = a_{32} = a_{34} = a_{42} = -r_-\omega_-^2$, $a_{13} = a_{21} = a_{23} = a_{31} = a_{41} = a_{43} = -r_+\omega_+^2$, $a_{22} = a_{44} = -\omega^2 + i\omega\nu_- + \omega_-^2(1 - r_-)$, $r_+ = \frac{l_+}{L}$ and $r_- = \frac{l_-}{L}$ are the relative lengths of the $n^+$, $n$ regions. In accordance with the conservation law of the total current, the current fluctuation $\delta J$ in the external circuit is determined by the rate of change of the surface density $\delta\sigma_M = \epsilon\epsilon_0\Delta E$ [4]:

$$\delta J = \frac{eA}{L}\sum_{i=1}^{4} N_i l_i \frac{d}{dt}\delta x_i \tag{4}$$

For uncorrelated (in time and space) Langevin forces their spectral density normalized to the total number of free carriers in the $i$-th region takes the form:

$$S_{ff}^{i} = 4kT\frac{1}{Am}\frac{\nu_i}{N_i l_i} \tag{5}$$

By using the equations (4) and (5) and the $b_{ij}$ elements of the inverse matrix of equation (3) the spectral density can be represented as:

$$S_{ii}(\omega) = \left(\omega \frac{eA}{L}\right)^2 \sum_{j=1}^{4} \left| \sum_{i=1}^{4} N_i l_i b_{i,j} \right|^2 S_{ff}^{j} \tag{6}$$

Let us stress that by keeping in equation (3) only the elements $a_{11}, a_{12}, a_{21}, a_{22}$ we obtain the analytical model for the SBD noise developed in [4].

## 3. Results and discussion

The intrinsic current noise obtained for a GaAs SBD is illustrated in Figure 2, with a carrier momentum relaxation rate $\nu = 1.364 \times 10^{12}$ s$^{-1}$ similar to the value obtained by Monte Carlo simulation [5]. The results of figure 2 obtained by this analytical model are compared by the results of SBD device in article [5].

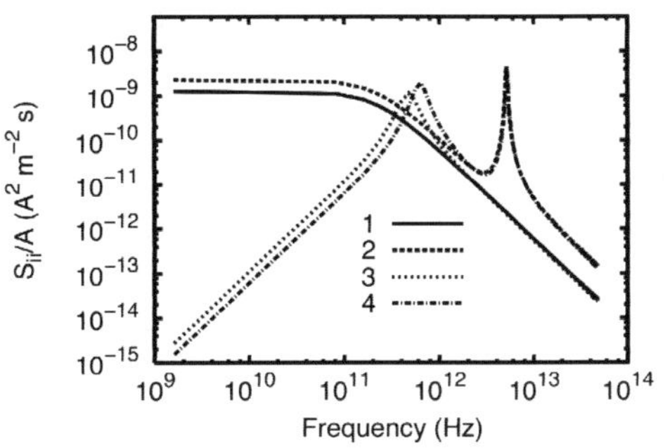


**Figure 2.** Spectral density of current fluctuations per unit surface in a GaAs $n^+n$-metal SBD with: $l_+ = l_- =0.1$ $\mu$m and $N_-=5 \times 10^{16}$ cm$^{-3}$. Curve 1: homogeneous material ($N_+ = N_-$) and $l_d = 0$; curve 2: diode with $l_d$=0 and $N_+ = 10N_-$; curve 3: $N_+ = N_-$ and depletion region $l_d = 0.01$ $\mu$m and curve 4: diode with $l_d = 0.01$ $\mu$m and $N_+ = 10N_-$.

For the case of a complete SBD (curve 4 in Fig.2), the spectral density exhibits an initial growth proportional to $\omega^2$ followed by two resonant peaks in the low and high frequency regions corresponding respectively to the carriers reflected by the SBD barrier and to the hybrid plasma resonance frequency.

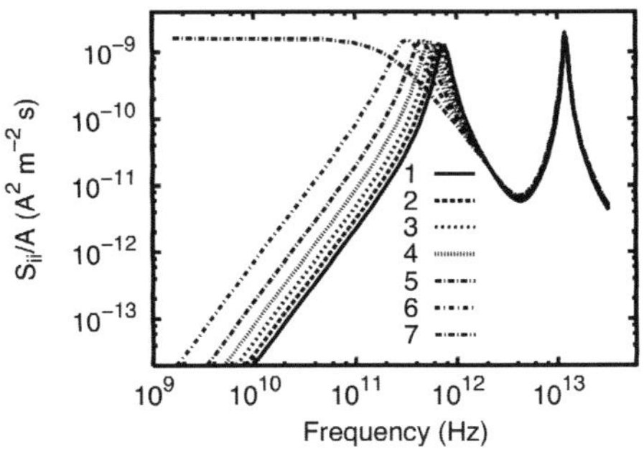


**Figure 3.** Spectral density of current fluctuations in a GaAs $n^+n$-metal SBD with: $l_+ = 0.05\mu$m, $l_- = 0.15\mu$m, $N_- = 5 \times 10^{16}$cm$^{-3}$. Curve 1: applied voltage is $U = 0.7$ V; Curve 2: $U = 0.75$ V; Curve 3: $U = 0.8$ V; Curve 4: $U = 0.85$ V; Curve 5: $U = 0.9$ V; Curve 6: $U = 0.95$ V and Curve 7: $U = 1$ V.

In Fig. 3 the absence and presence of the depletion region is controlled by the applied voltage $U$. We observe a red-shift of the low-frequency resonance peak by increasing the applied voltage. The HBV spectral density is similar to the spectrum of the SBD with a difference in the position of the resonance peaks. For a complete HBV structure (curve 4 of Fig.4) the two resonance peaks appear near 1 and 20 THz compared to the SBD structure (curve 4 of Fig.2) where the peaks appear at 0.3 and 3 THz. Therefore the resonance frequency position increases significantly.

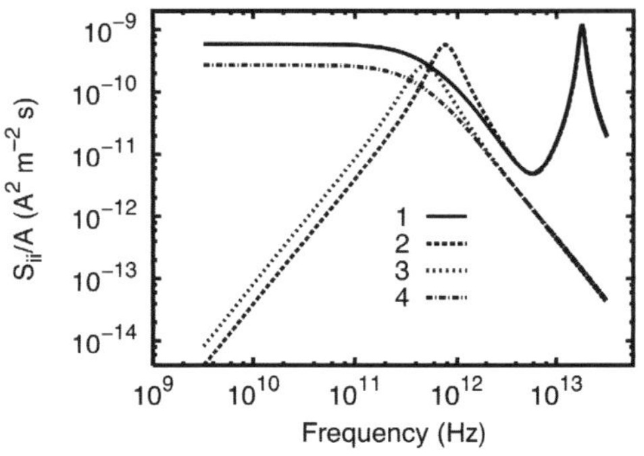


**Figure 4.** Spectra density of current fluctuations in a $n^+n$-barrier-$nn^+$ HBV with: $l_1 = l_3 = 0.5$ $\mu$m, $l_2 = l_4 = 0.3$ $\mu$m, $N_2 = N_4 = 1 \times 10^{17}$ cm$^{-3}$. Curve 1: $N_1 = N_3 = 50N_2$and $l_d = 0$; curve 2: $l_d$ =23nm and $N_1 = N_3 = 50N_2$; curve 3: $N_1 = N_3 = N_2$ with $l_d = 23$ nm and curve 4: $l_d = 0$ and $N_1 = N_3 = N_2$

Figure 5 illustrates the spectral density of $i = 6$ regions compared with $i = 4$ regions in figure 4 (i.e. we apply the analytical model for $i = 6$). The presence of a $n^+, n^-, n^{--}$ regions leads to the appearance of an additional peak associated with the appearence of extra $n^-n^{--}$ homojunction.

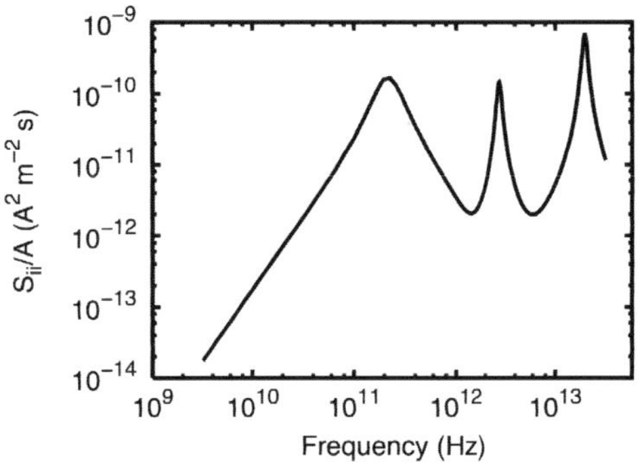


**Figure 5.** Spectral density of current fluctuations per unit surface in a HBV structure with $i = 6$ regions $n^+n^-n^{--}$-barrier-$n^{--}n^-n^+$ with: $l_1 = l_4 = 0.3$ $\mu$m, $l_2 = l_5 = 0.5$ $\mu$m, $l_3 = l_6 = 0.3$ $\mu$m, $N_3 = N_6 = 1 \times 10^{16}$ cm$^{-3}$, $N_2 = N_5 = 1 \times 10^{17}$ cm$^{-3}$, $N_1 = N_4 = 5 \times 10^{18}$ cm$^{-3}$ and $l_d = 23$ nm.

## 4. Conclusion

We have proposed an analytical model of the intrinsic noise in HBV structures which describes the high-frequency part of the noise spectrum. The low-frequency shot noise is not considered in the model. The model proposed for the HBV can calculate the intrinsic noise of $i$ regions of any structure and the results can be useful for optimizing the device parameters for the extraction of the high-order harmonics.

The financial support of the Lithuanian State Science and Studies Fundation contract n. P-01/2007 is acknowledged.

## References

[1] J. Vukusic, T. Bryllert, T. Arezoo Emadi, M. Sadeghi and J. Stake 2007 *IEEE Electron Dev. Lett* **28** 340–342
[2] M. Saglam, B.Schumann, V. Mullerwiebus, A. Megej, U. Auer, M. Rodriguez-Girones, R. Judaschke, E. J. Tegude and H.L. Hartnagel 2002 *Electron Lett* **38** 657–8
[3] X. Mélique, A. Maestrini, J. M. Goutoule, P. Goy, P. Mounaix, M. Trier, O. Vanbésien, G. Beaudin and D. Lippens 1999 *European Microwave conference-Munich* $\mathbf{29}^{th}$ 113–115.
[4] P.Shiktorov, E. Starikov, V. Gruzinskis, L. Reggiani, L. Varani and J. C Vaissière 2005 *IEEE Electron Dev. Lett* **26** 2–4
[5] P.Shiktorov, E. Starikov, V. Gruzinskis, L. Reggiani, L. Varani and J. C Vaissière 2004 *IEEE Electron Dev. Lett* **25** 1–3.

# Spectral Narrowing of Terahertz Emission from Super-grating Dual-Gate Plasmon-Resonant High-Electron Mobility Transistors

**A. El Moutaouakil, T. Watanabe, C. Haibo, T. Komori, T. Nishimura, T. Suemitsu and T. Otsuji**
Research Institute of Electrical Communication (RIEC), Tohoku University, 2-1-1 Katahira, Aoba-Ku, 980-8577, Sendai – Japan

Email: amine@riec.tohoku.ac.jp

**Abstract**. Coherent terahertz emission from an original InGaP/InGaAs/GaAs plasmon-resonant high-electron-mobility transistor, having a super-grating dual-gate structure, is investigated. We report total emission from multi-mode of plasmons: thermally excited incoherent modes and instability-driven coherent modes, using the Fourier Transform Infrared (FTIR) spectroscopic measurement. We discriminate the coherent emission of ultrafast instability-driven plasmon self-oscillation from incoherent hot plasmons using the Electro-Optic Sampling (EOS) method. The resultant emission shows a clear optimal plasmon-resonant mode with sharp peak at 3.2 THz for an optimal bias condition of $V_{ds}$ = 2V. It indicates the spectral narrowing effect on the super-grating-gate structure.

## 1. Introduction

The Terahertz (THz) technology has attracted increasingly interest in nowadays industrial applications. The oscillation of the density of electrons in a two dimensional system can become unstable against the generation of electromagnetic radiation at specific bias conditions, which can be used for tunable detection/generation, and photomixing functions in the terahertz domain [1–6].

Recently, Otsuji et al. have proposed an original HEMT (High Electron Mobility Transistor) based on InGaP/InGaAs/GaAs with doubly interdigitated grating gates, succeeding in broadband generation of terahertz radiations at room temperature due to incoherent/coherent excitation of the multimode plasmons [5-7]. The drain bias causes dispersion on the sheet electron density along the channel, resulting in spectral broadening [5]. The super-grating gate structure, in which the dispersion of the carrier density is compensated for the grating-finger size (cavity size), can unify the emission frequency under an optimal bias condition [8]. In the present paper, we report on experimental verification of the effect of the super-grating-gate structure and demonstrate the coherent terahertz emission from our original device.

## 2. Device structures and operation principle

### 2.1. Device structure

The device [7] was fabricated with InGaP/InGaAs/GaAs material systems as shown in Figure 1. The 2D plasmon layer is formed within a quantum well in the InGaAs channel layer. The metallic grating gate was formed with 65-nm thick Ti/Au/Ti by a standard lift-off process. The double interdigitated grating geometry was designed with gate 1 (G1): $L_{G1}$ = 70 nm, and gate 2: $L_{G2}$ = 1850 nm with 100 nm in space (See figure 1).

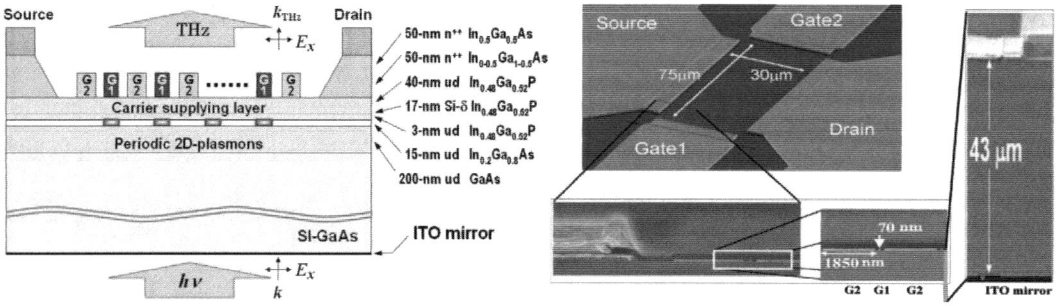


Fig.1: Left: device cross-section for typical GaAs-based heterostructure material systems, right: SEM image of the new grating gate device, 15 fingers for G1 and 16 fingers for G2.

### 2.2. Operation principle

Suppose that the plasmons cavities under G1 are highly charged ($10^{12}$ cm$^{-2}$) while the regions under G2 are lightly charged ($\leq 10^{11}$ cm$^{-2}$). If a specific drain-to-source bias is applied to promote a uniform channel current, each cavity boundary may excite the plasmons, promoting current-driven instability of THz self oscillations. The hot electrons also may thermally excite the plasmons. Such coherent (former) and incoherent (latter) plasmons can be transformed to radiative THz electromagnetic radiation via broadband antenna functions of the grating gate structure. The drain bias causes the distribution of sheet carrier density along the channel, causing dispersion of plasma resonant frequencies. This spectral broadening effect can be compensated by introducing a super-grating gate structure; the grating finger size is varied along with the channel so as to unify the plasmon resonant frequencies over all the plasmon cavities under the grating fingers (see Figure 2.) [8].

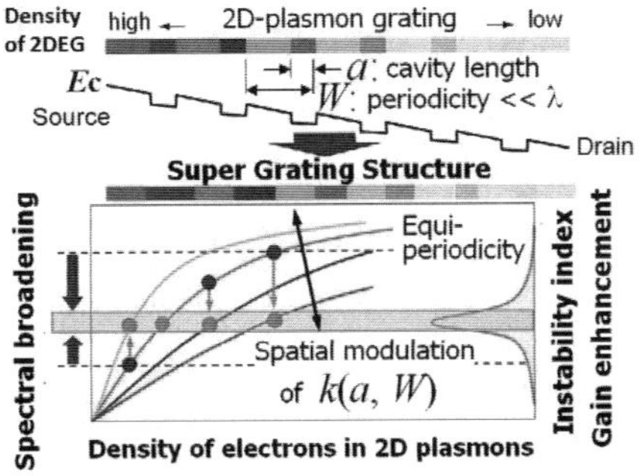


Fig.2: Super-grating gate structure.

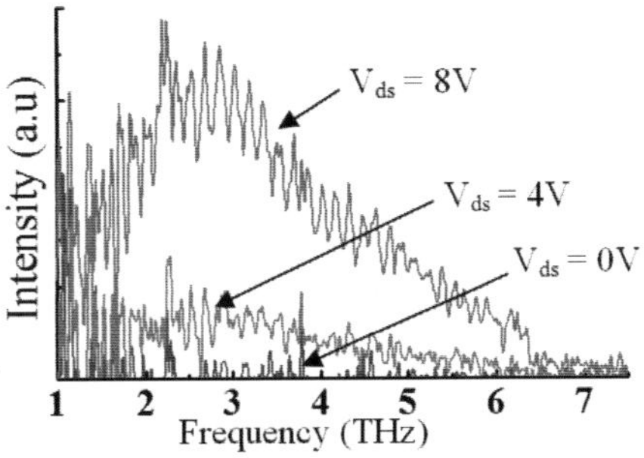


Fig. 3: Measured FTIR spectra.

## 3. Experimental results and discussion

First, the measurements using the Fourier-transformed far-infrared spectroscopic (FTIR) system were carried out. The experimental procedure is fully detailed in literature [7]. The FTIR measured spectra for the InGaP/InGaAs/GaAs HEMT is showing relatively broad emission spectra ranging from 1 to 6.5 THz (Figure 3) independently of the drain bias $V_{ds}$. Previous discussion concluded that the emission includes coherent and incoherent plasmon modes as mentioned in 2 [7]. The latter may conceal the narrowing effect. In order to discriminate the coherent mode of self oscillation from the incoherent mode, we introduced the time-resolved measurement: Electro-Optic Sampling (EOS) [7-9].

At zero bias (Figure 4.(a)), a coherent photoconductive emission line around 2 THz is observed (width around 1 THz). For 2-V bias (Figure 4.(b)), the resonant frequency increases up to 3.2 THz, and the line narrowing down to 0.4 THz is observed. This result shows that for the actual dimensions of our double grating structure, 2-V bias condition gives the optimal situation to compensate for the dispersion of plasmon resonant frequency.

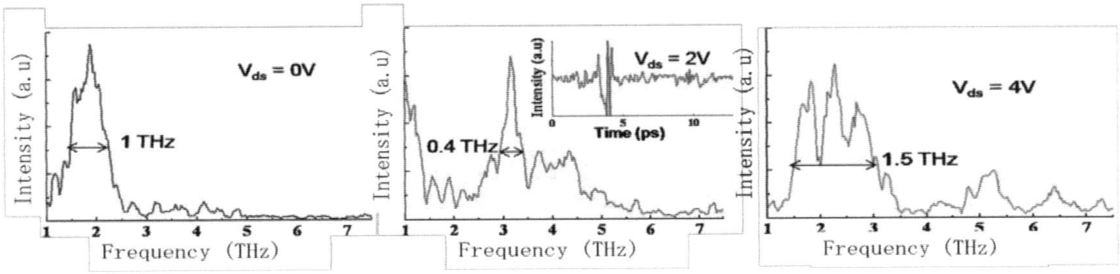


(a) (b) (c)

Fig. 4: Measured EOS spectra: (a) Emission at $V_{ds}$ = 0V. (b) Emission at $V_{ds}$ = 2V. Inset (b) EOS signal for $V_{ds}$ = 2V. (c) Emission at $V_{ds}$ = 4V.

When $V_{ds}$ increases up to 4-V (Figure 4.(c)), out of the detuning range, we drive the device to much more deeper slope range of the sheet carrier density distributions along with the channel axis. Since the plasma frequency is in proportion to the square root of the sheet carrier density while be in inverse proportion to the cavity size (grating finger size), when the drain bias is detuned to disperse the sheet carrier density from an optimal distribution, the super-grating structure cannot compensate out the distribution anymore [9]. This may cause the spectral broadening. Besides, when the drain bias increases more, the carrier density in the channel, in particular near the drain side, will decrease, which may decrease the plasmon resonant frequencies. As a consequence, it is qualitatively interpreted in the result, shown in Figure 2(c), that the excess drain bias may cause the emission spectral broadening to the lower frequency region than the original spectral peak frequency position. There exists a detuning range around $V_{ds}$ = 2.0 V to preserve the spectral narrowing effect in a specific bandwidth.

## 4. Conclusion

Coherent terahertz emission from an original InGaP/InGaAs/GaAs plasmon-resonant high-electron-mobility transistor, having a super-grating dual-gate structure, was investigated. The spectral narrowing effect on the super-grating-gate structure has been successfully demonstrated using both FTIR and EOS measurement systems.

## Acknowledgements

This work is financially supported in part by the Grant-in-Aid for Basic Research (S), JSPS, Japan. The device process for this work was carried out in part at the Laboratory for Nanoelectronics and Spintronics, RIEC at Tohoku University.

## References

[1] M. Dyakonov and M. Shur, *Phys. Rev. Lett.* 71, 2465 (1993).

[2] V. Ryzhii, I. Khmyrova, A. Satou, P. O. Vaccaro, T. Aida, and M. Shur, *J. Appl. Phys.* 92, 5756 (2002).

[3] W. Knap, J. Łusakowski, T. Parenty, S. Bollaert, A. Cappy, V.V. Popov and M.S. Shur, *Appl. Phys. Lett.* 84,2331 (2004)

[4] W. Knap, Y. Deng, S. Rumyantsev, J.-Q. Lu, M. S. Shur, C. A. Saylor and L. C. Brunel, *Appl. Phys. Lett.* 80, 3434 (2002).

[5] T. Otsuji, Y. M. Meziani, T. Nishimura, T. Suemitsu, W. Knap, E. Sano, T. Asano and V.V. Popov, *J. Phys.: Condensed Matter* 20, 384206 (2008).

[6] Y. M. Meziani, H. Handa, W. Knap, T. Otsuji, E. Sano, V. V. Popov, G. M. Tsymbalov, D. Coquillat, and F. Teppe, *Appl. Phys. Lett.* 92, 201108 (2008).

[7] T. Otsuji, Y. M. Meziani, M. Hanabe, T. Ishibashi, T. Uno and E. Sano, *Appl. Phys. Lett.* 89, 263502 (2006).

[8] T. Nishimura, N. Magome, H. C. Kang and T, Otsuji, *IEICE Trans. Electron.* E92-C, 696 (2009).

[9] Y. M. Meziani, T. Otsuji, M. Hanabe, T. Ishibashi, T. Uno and E. Sano, *Appl. Phys. Lett.* 90, 061105 (2007).

Journal of Physics: Conference Series **193** (2009) 012069 doi:10.1088/1742-6596/193/1/012069

# Monte Carlo simulation of THz radiation detection in GaAs $n^+nn^+$ diodes with inhomogeneously doped $n$-region

**V. Gružinskis, E. Starikov, P. Shiktorov and S. Ašmontas**

Semiconductor Physics Institute, Gostauto 11, LT-011108 Vilnius, Lithuania

E-mail: jane@pav.pfi.lt

**Abstract.** A proper profile of nonuniform doping of GaAs $n^+nn^+$ structure to reach the maximum detection sensitivity is proposed. It is shown that the main role in formation of longitudinal transport asymmetry and THz radiation detection is played by optical phonon emission process. It is found that the detection current independs of frequency in spectral range 0.02 - 1 THz, reaches maximum near 2.2 THz (10 and 80 K) or 3 THz (300 K) and then becomes to decrease due to the inertia of optical phonon emission process.

## 1. Introduction

To detect the microwave power semiconductor diodes with asymmetric current-voltage relation (CVR) are widely used. Usually, such an asymmetry is obtained by forming various barriers (*pin*, Schottky, heterojunction, etc.) in the active region. Unfortunately, in going to terahertz (THz) region the sensitivity of barrier-structures rapidly decreases due to time-delay related with large values of the barrier capacitance and its resistance. In contrast, as it was experimentally demonstrated recently [1], these restrictions can be overcame in asymmetricaly narrowed $nn^+$ GaAs structure. Due to the resistance gradient along the axis of narrowing active region there appears nonuniform electric field with a peak placed in the narrowest place. The height and width of the peak depend on the flowing current direction, thus leading to asymmetric CVR.

The novel idea proposed in this report is to form the longitudinal gradient of resistance necessary for the CVR asymmetry by using nonuniform doping of the active region and keeping the constant cross-section area along the structure. In the framework of such an appoach, a proper choise of $n$-region doping profile will allow one to obtain asymmetric CVR and microwave-THz power detection by using more simple geometry than previous one. Therefore, the aim of this work is to investigate an electron transport and THz radiation detection in GaAs $n^+nn^+$ diodes with inhomogeneously doped $n$-region by using Monte Carlo particle method.

## 2. Simulation method and detector model

The simulations are performed using Monte Carlo particle (MCP) technique [2]. Electron motion in GaAs is simulated with account of scatterings by acoustic, optical and intervalley phonons and scatterings by ionized impurities as well. Three valley (Γ, L and X) GaAs model is used [3].

The investigated detector consists of two 0.3 $\mu$m $n^+$-layers with 0.5 $\mu$m $n$-layer between them. The $n^+$-layers are uniformly doped by donors up to $N^+ = 10^{18}$ cm$^{-3}$. The $n$-layer consists of

20 $\delta x$ = 0.025 $\mu$m thick minilayers. The doping of each minilayer is given by formula $N = N^{+}\exp\text{-}[\delta x(21\text{-n})/\text{b}]^2$, where n = 1,2,...20 is minilayer number (the number is increasing with increased coordinate $x$) and b = 0.19024 $\mu$m. So formed doping profile is shown in Fig. 1. The number of simulated particles depending on case is from 60,000 to 150,000 and time step in all cases is of $10^{-15}$ s.

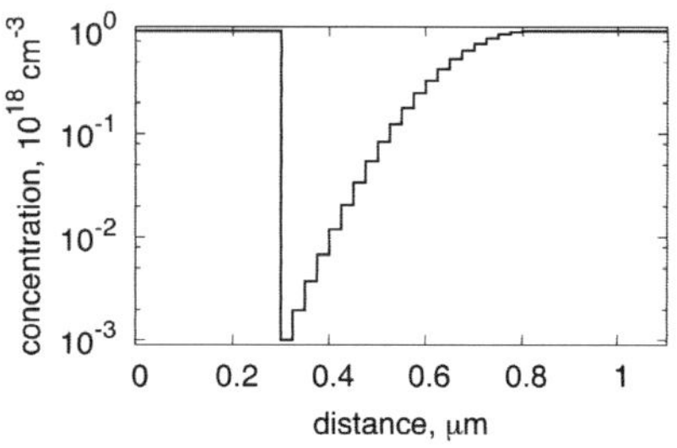


**Figure 1.** Doping profile of GaAs $n^+nn^+$ structure.

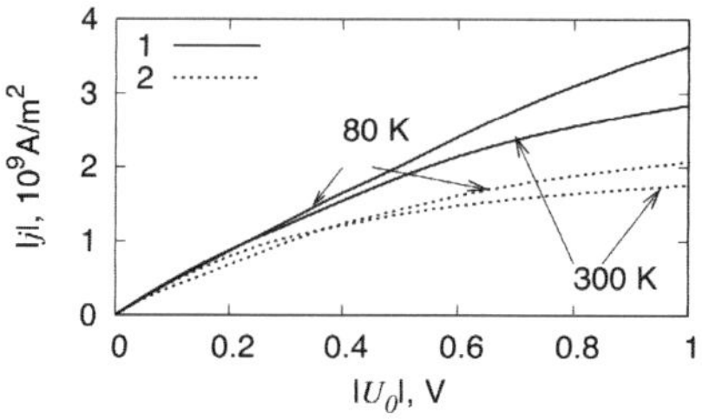


**Figure 2.** Static CVR of GaAs structure for positive (1) and negative (2) current direction.

## 3. Static characteristics

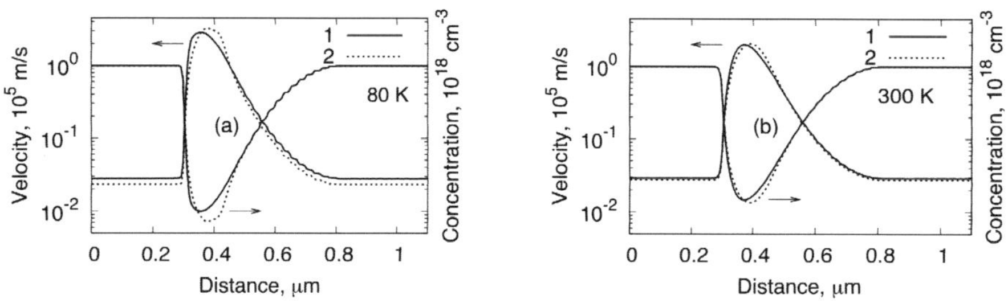


**Figure 3.** Electron velocity and concentration profiles in GaAs structure at 80 K (a) and 300 K (b), when $|U_0| = 0.1$ V. 1 - positive current direction, 2 - negative current direction.

The current-voltage relation (CVR) is calculated by simulation of time dependent current at given constant bias. The current value is calculated by averaging in time the stationary part of current trajectory. So calculated CVR for 80 and 300 K are presented in Fig. 2. To show the CVR asymmetry the absolute current values $|j|$ are plotted versus absolute voltage $|U_0|$. The solid lines 1 in Figs. 2-4 correspond to the positive current direction i.e. current flows in direction of increased coordinate $x$. Also it should be stressed that due to the negative electron charge the electrons flow in the direction opposite to the current. For CVR asymmetry analysis the profiles of various physical quantities inside the detector at $|U_0| = 0.1$ V and 80 and 300 K are presented below. Fig. 3(a,b) shows the electron velocity absolute value and concentration profiles. One can see that higher velocity in high concentration regions gives higher current in positive direction. The CVR assymetry can be explained by analysis of electron dynamics in the detector. In this case the electrons are scattered by impurities and acoustics and optical phonons. The electron dynamics is mostly affected by optical phonon scattering because this scattering changes electron velocity and energy at the same time. Electrons entering the $n$-region from the left immediately enter the high field region and are near ballistically accelerated to higher velocity and energy, than electrons entering from the right side where electric field gradient is much smaller due to lower concentration gradient and electrons are accelerated slowly with high scattering by impurities and acoustic phonons. Electrons moving from left to right have to experience more scattering events by optical phonon emission than ones moving in opposite direction. This is illustrated in Fig. 4(a,b) where local optical phonon emission $\nu_L$ profiles are shown.

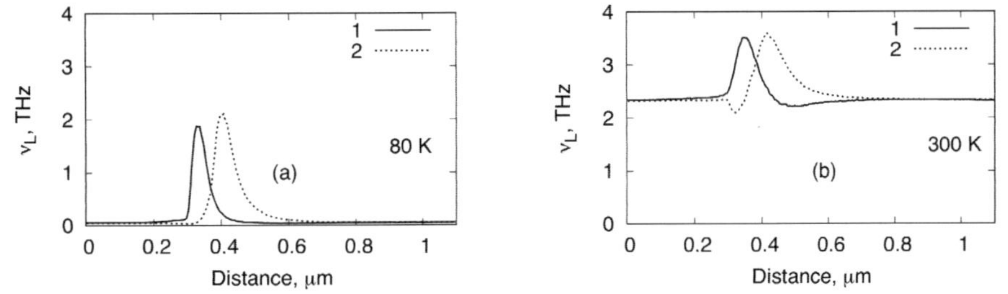


**Figure 4.** Local optical phonon emission $\nu_L$ profiles in GaAs structure at 80 K (a) and 300 K (b), when $|U_0| = 0.1$ V. 1 - positive current direction, 2 - negative current direction.

It is easy to see that electrons moving from left to right experience more optical phonon emission events (Fig. 4(a), compare lines 1 and 2), wich results in higher resistivity than in the electron opposite movement case. At 300 K the asymmetry is much lower due to the increased scattering by acoustic phonons and optical phonon absorbtion. Thus, we have shown that major influence to the CVR asymmetry in the $n^+nn^+$ GaAs detector with the doping gradient in $n$-region gives the optical phonon emission. Below we will investigate the characteristics of the detector in high frequency electric field.

## 4. THz radiation detection

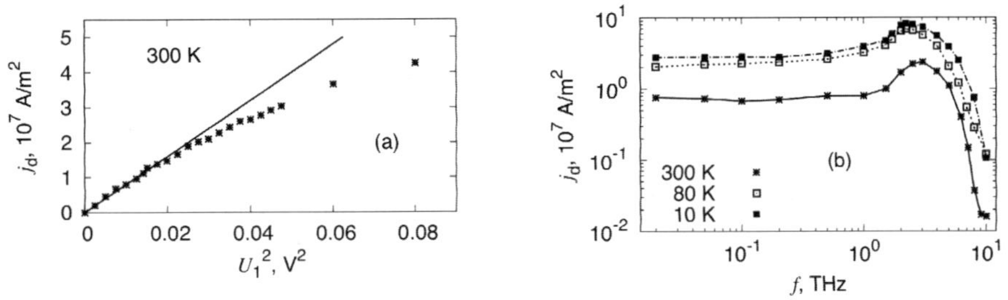


**Figure 5.** Detector current $j_d$ dependence on ${U_1}^2$ at 300 K and $f = 1$ THz (a) and $j_d$ spectra at $U_1 = 0.1$ V and 10, 80 and 300 K (b). The straight line in (a) shows the linear part of $j_d$.

The detector current $j_d$ response to the radiation is simulated applying to the structure an alternative bias $U(t) = U_1\cos(2\pi ft)$, where $U_1$ is voltage induced in structure by radiation and $f$ is the radiation frequency. As in the case of static characteristics the long current response trajectories are simulated. The $j_d$ value is obtained by averaging in time the stationary part of trajectory. The wide band detectors are used in radiation power measurements. Therefore the detected signal should be the linear function of radiation power, which is proportional to the ${U_1}^2$. To find the $U_1$ range where $j_d$ is linear function of power the $j_d$ dependence on ${U_1}^2$ is calculated at $f = 1$ THz and 300 K (Fig. 5(a)). It is found that linear dependence of $j_d$ on radiation power takes place when $U_1$ is in the range from 0 to 0.15 V (see Fig. 5(a)). This range weekly depends on temperature and is approximately the same for 10 and 80 K. Choosing $U_1$ = 0.1 V the $j_d$ dependence on frequency $f$ (from 0.02 to 10 Thz) is simulated at 10, 80 and 300 K (Fig. 5(b)). At lower temperatures (10 and 80 K) $j_d$ is higher due to the reduced scattering rates by acoustics phonons and optical phonon absorbtion as compared with that at 300 K (Fig. 5(b)). In the frequency range 0.02 - 1 THz the $j_d$ spectra are flat for all temperatures. At frequencies over 1 THz the $j_d$ is increased with frequency increase and after reaching the maximum becomes to decrease. At low temperatures (10 and 80 K) the $j_d$ maximum is at 2.2 THz, while at 300 K the maximum is at 3.0 THz (Fig. 5(b)). Such a behavior of $j_d$ spectrum can be explained using optical phonon emission rate $\nu$ dependencies on electron energy (Fig. 6(a)). At low temperatures (10 and 80 K) the $\nu$ dependencies on energy are practically the same while at 300 K the optical phonon emission rate is significantly higher: at energy 0.1 eV and low temperature $\nu \approx 4.8$ THz, while at 300 K $\nu \approx 6.5$ THz. It should be noted that the ratio of these frequencies is practically the same as the frequency ratio of low temperature and 300 K $j_d$ spectra maxima (Fig. 5 (b)). The $j_d$ decrease at high frequencies can be explained

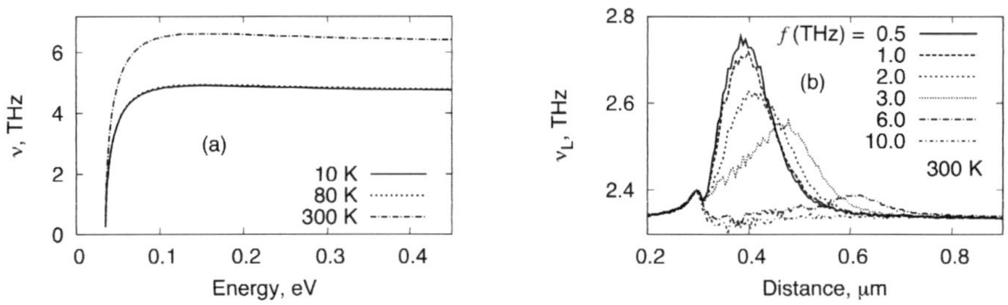


**Figure 6.** Optical phonon emision rate $\nu$ dependence on energy in GaAs at various temperatures (a) and local optical phonon emission rate $\nu_L$ profiles in GaAs structure at 300 K, $U_1 = 0.1$ V and various radiation frequencies $f$ (b).

by optical phonon emission inertia: electron gains the energy from the field over optical phonon energy during one half of period and looses it during another half of period with no optical phonon emission. So the radiation frequency increase should result in decrease of local optical emission rate $\nu_L$ inside the detector. It is demonstrated in Fig. 6(b) where the $\nu_L$ profiles in detector are simulated at 0.5, 1.0, 2.0, 3.0, 6.0 and 10 THz frequencies. One can see from Fig. 6(b) that local optical phonon emision rate in $n$-region of detector is decreased at increased radiation frequency. Also it is seen from this figure that $\nu_L$ maximum is aproaching the higher electron concentrations (compare Fig. 1 and Fig. 6(b)) at frequency increased. It means that more electrons experience the maximum optical emission rate. So it is the reason for $j_d$ maxima formation in $j_d$ spectra (see Fig. 5(b)).

Finally, the upper analysis shows that optical phonon emission plays the main role in wide band THz radiation detection by GaAs $n^+nn^+$ diodes with inhomogeneously doped $n$-region.

*This work is supported by the Lithuanian State Science and Studies Fundation contract No P-01/2007.*

## 5. References

[1] Sužiedėlis A, Ašmontas S, Požėla J, Kundrotas J, Širmulis E, Gradauskas J, Kozič A, Kazlauskaitė V and Arbinderis T 2008 *Acta Phys. Pol.* A **107** 803
[2] Mitin V, Gruzinskis V, Starikov E and Shiktorov P 1994 *J. Appl. Phys.* **75** 935
[3] Brennan K and Hess K 1984 *Solid-State Electron.* **27** 347

# 3D-integration of a log spiral antenna onto a dual grating-gate plasmon-resonant terahertz emitter for high-directivity radiation

**H-C Kang[1], T Nishimura[1], T Komori[1], T Mori[2], N Watanabe[2], T Asano[2], and T Otsuji[1]**
[1]Research Institute of Electrical Communication (RIEC), Tohoku University, 2-1-1, Katahira, Aoba-Ku, Sendai 980-8577, Japan
[2]Graduate School of Information Science and Electrical Engineering, Kyushu University, 744, Motooka, Nishi-ku, Fukuoka-city 819-0395, Japan
E-mail: kang@riec.tohoku.ac.jp

**Abstract.** We report a high-directivity plasmon-resonant terahertz (THz) emitter (PRE) incorporating a 3D-integrated antenna complex. The emitter structure is based on a high electron mobility transistor having unique doubly interdigitated grating gates as a broadband THz antenna that can convert non-radiative plasmons to radiative electromagnetic waves. Due to the sub-wavelength aperture of practical grating-antenna dimension, however, present structure of PRE exhibits undesirable diffraction effect, resulting in poor directivity. We developed a new device structure featuring a 3D-integration of tightly-coupled multiple-antenna complex to improve the directivity. The directivity of a new device was dramatically improved by a factor of 5.7 over the frequencies from 1.8 to 4.0 THz.

## 1. Introduction

Two-dimensional (2D) plasmons in high-electron mobility transistors (HEMT's) have attracted attention due to their nature of instability promoting terahertz (THz) electromagnetic (EM) radiation, which is expected to realize integrated THz sources [1-4]. We have recently proposed a plasmon-resonant THz emitter (PRE) [3, 4], which is based on a InGaP/InGaAs/GaAs HEMT incorporating original doubly interdigitated grating gates (DGG's) acting as a broadband THz antenna (Fig. 1(a)). Due to the sub-wavelength aperture (~100 $\mu m^2$) of the DGG's region which is limited by the gate-to-channel breakdown voltage, however, the PRE exhibits undesirable EM diffraction, resulting in a poor directivity. On the other hand, self-complimentary antennae like log-spiral antennae (LSA's) [5] are well populated for broadband THz radiation, which has been utilized in a stand-alone manner [6] or in combination with THz discrete optics [7]. In this work, to improve the radiation directivity of the PRE, we propose a device structure featuring 3-D integration of a tightly-coupled DGG's-LSA complex.

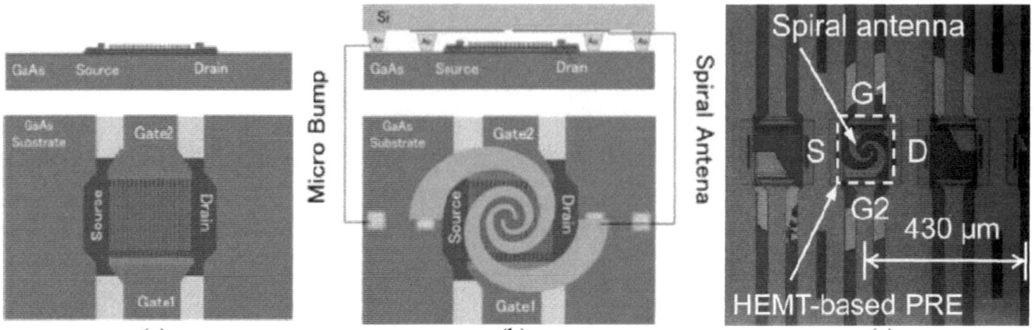


(a) (b) (c)

**Fig. 1.** HEMT-based PRE having the DGG's. (a) conventional structure (grating aperture of $\sim 70\mu m \times \sim 50\mu m$, (b) proposed structure with a LSA tightly stacked onto the PRE, and (c) an infrared microscopic image viewed from the bottom.

## 2. 3D-Integraion of Complex Antenna Systems

A fine patterned planar LSA is dedicated on a separate Si substrate, which is tightly stacked onto the PRE chip via micro-bump interconnection [8] (Fig. 1(b)). The distance between the DGG's and LSA plane is only 5 μm, which is far shorter by orders than the wavelength of EM waves emitted from the PRE. The micro-bump interconnection, thus, efficiently feeds the EM radiation induced at the active regions of the PRE into the LSA throughout the DGG's. Fig. 1(c) shows an infrared microscopic image viewed from the bottom of the LSA-integrated PRE.

A log-spiral antenna is geometrically defined by the equations: $r_1 = r_i e^{a\phi}$ and $r_2 = r_i e^{a(\phi-\pi/2)}$ that make up one spiral arm, where $r_1$ outer radius, $r_2$ inner radius, $r_i$ the initial outer radius, $a$ the wrap tightness, and $\phi$ spiral turns. Based on the semi-empirical formulae [5], those geometric parameters are designed to be $r_i$ = 14.14 μm, $a$ = 0.395 rad$^{-1}$, $\phi$ =2.75π rad, $r_1$ = 430 μm, and $r_2$ = 231 μm so as to cover the original radiation bandwidth of PRE ranging from 0.5 to 5 THz. The antenna structure is fabricated lithographically with a Ti (30 nm)/Al (250 nm) metallization directly on a Si substrate.

Theoretical calculation suggests that the radiation intensity of the LSA itself to the normal direction ($\theta = 0°$) is enhanced by a factor of 2.5 from an isotropic radiation. Taking account of a numerical aperture of the LSA (ideally 0.5), the radiation intensity of the simplified model of the PRE with a free-standing LSA is expected to be enhanced by a factor of $\lesssim 1.25$ with respect to the PRE without LSA. When the directivity is evaluated as the half-power beam width (HPBW) [9], corresponding directivity of the simplified model is also calculated to be improved by a factor of ~2.0 with respect to the PRE without LSA.

## 3. Experimental Results and Discussions

We characterized the THz-EM radiation property of the fabricated samples (LSA-integrated PRE's and PRE's without LSA) by using a Fourier-transform far-infrared spectrometer (FTIR) [4]. The

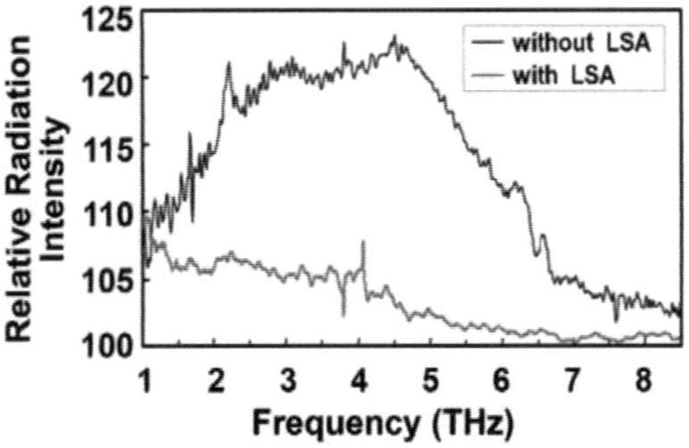


**Fig. 2.** Measured spectra at $\theta = 0°$ as a relative intensity where the background thermal radiation is scaled as 100.

**Table 1** Directivity performance at frequencies from 1.8 to 4.0 THz.

| Performances / Devices | HPBW (deg.) | Directivity (dBi) | *Relative intensity at 0° |
|---|---|---|---|
| LSA-integrated PRE | ~14 | ~23 | ~106 |
| PRE without LSA | ~80 | ~8 | 112 ~ 121 |

(*Background thermal radiation is scaled as 100.)

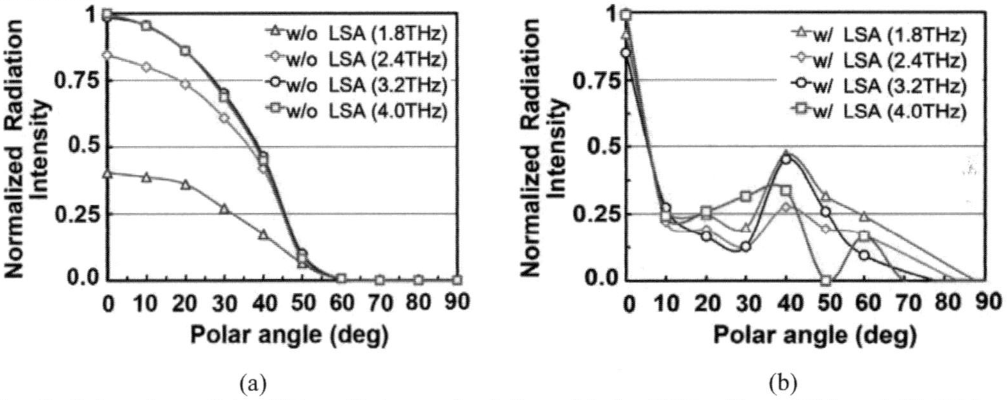


**Fig. 3.** Polar plots of the THz radiation emitted from (a) the PRE without LSA and (b) LSA-integrated PRE at 1.8, 2.4, 3.2, and 4.0 THz. The vertical axis for each is normalized to the maximum intensity of the sample.

effectively measurable lower cutoff frequency of the FTIR is $\lesssim$1 THz. The radiation spectra were measured at polar angles $\theta$ from 0 (normal to the surface) to 90°. Figure 2 shows a typical result at $\theta = 0°$ as the *relative* intensity where the background thermal (black-body) radiation is scaled as "100" so that the baseline at 100 means no excess radiation. The original PRE without LSA shows broadband radiation from 1 to 6 THz, which is up-shifting by ~1 THz from the nominal bandwidth of the LSA. The spectrum of the LSA-integrated PRE almost traces the designed bandwidth of the LSA although the radiation intensity is unexpectedly attenuated. Such a radiation loss is thought to be caused by the impedance mismatch and excess losses along with the DGG's-LSA propagation originated from undesirable high contact resistance among the micro-bump interconnection. Improvement on the interconnection process to recover the original radiation gain is the future subject.

Figure 3 shows the polar-angle dependence of the radiation intensity at typical four different frequencies: 1.8, 2.4, 3.2 and 4.0 THz. The vertical scale is normalized to the maximum radiation intensity for each type of the sample among those frequencies. As seen in Fig. 3(a), the PRE without LSA exhibits a poor directivity ~8 dBi (Table 1) with a broad distribution from 0 to ~40° at all the frequencies; corresponding HPBW is ~ 80° (Table 1). In contrast, as seen in Fig. 3(b), the radiation pattern of the LSA-integrated PRE exhibits an extraordinary sharp main-lobe in the normal direction ($\theta = 0°$) and relatively broad side-lobe(s) around $\theta \approx 40°$ at all the frequencies from 1.8 to 4.0 THz. The side-lobe is suppressed below 50% of the main peak. As a result, the directivity is greatly improved to ~23 dBi; corresponding HPBW is $\sim 14^{\circ}$ (Table 1). The improvement factor in HPBW is 5.7, which is about three times as high as the estimated value for the simplified model mentioned in section 2.

One can consider a left-handed metamaterial effect on parasitics of the tightly-coupled DGG's-LSA interconnection to explain the causes of such extraordinary directivity. The physical structure of connecting portion can be modelled as a serial capacitor and a shunt inductive/resistive element along with the antenna pattern. Due to the propagation angle the capacitance value may change depending on the distance while the shunt inductance will be monotonically decreased with frequency. This might cause a specific frequency response with electromagnetic bandgap(s) in specific range(s) of the polar angles, which may modulate the original frequency response and radiation pattern. Further investigations and accurate modelling/calculation should be followed, which will be the future subject.

## 4. Summary

We proposed a high-directivity THz PRE featuring a LSA-integrated 3D antenna complex structure. Although the impedance mismatch caused by undesirable interconnection losses deteriorates the radiation intensity almost by one third of the PRE without LSA, extraordinary sharp radiation directivity was successfully obtained over the broadband frequencies from 1.8 to 4.0 THz. The tightly-coupled DGG's-LSA complex metamaterial structure is considered to work for manipulating extraordinary propagation of EM waves.

## Acknowledgement

This work was financially supported in part by SCOPE program from MIC, Japan and by the Grant in Aid for Basic Research (S) from the JSPS, Japan.

## References

[1] D. C. Tsui, E. Gornik, and R. A. Logan, *Solid State Commun.* **35**, 875 (1980).
[2] M. Dyakonov and M. Shur, *Phys. Rev. Lett.* **71,** 2465-2468 (1993).
[3] T. Otsuji, M. Hanabe, T. Nishimura and E. Sano, *Opt. Express* **14,** 4815-4825 (2006).
[4] T. Otsuji, Y. M. Meziani, T. Nishimura, T. Suemitsu, W. Knap, E. Sano, T. Asano, V. V. Popov, *J. Phys.: Condens. Matter* **20**, 384206-1-11 (2008).
[5] J. D. Dyson, *IRE Trans. Ant. Propagat.* **AP-7,** 181-187 (1959).
[6] S. Verghese, K. A. McIntosh, E. R. Brown, *IEEE Trans. MTT*, **45,** 8, 1301-1309 (1997).
[7] A. D. Semenov, H.-W. Hubers, J. Shubert, G. N. Gol'tsman, A. I. Elantiev, B. M. Vornov, E. M. Gershenzon, *J. Appl. Phys.* **88,** 11, 6758-6767 (2000).
[8] N. Watanabe, T. Kojima and T. Asano, *IEDM Tech. Dig.* 687-690 (2005).
[9] C. A. Balanis, *Antenna Theory: Analysis and Design 3rd ed.*, John Wiley&Sons, Ins. (2005).

Journal of Physics: Conference Series **193** (2009) 012071 doi:10.1088/1742-6596/193/1/012071

# Enhancement of terahertz radiation by CW infrared laser excitation in a doubly interdigitated grating gates transistors

**Y M Meziani[1], T. Nishimura[2], H Tsuda[2], T Suemitsu[2], W. Knap[3], V V Popov[4] and T Otsuji[2]**

[1]Dpto. de Fisica Aplicada, Universidad de Salamanca, Pza de la Merced s/n, 37008 Salamanca, Spain.
[2]Research Institute of Electrical Communication, Tohoku University, Sendai 980-8577, Japan.
[3]Groupe d'Etude des Semiconducteurs, Univ. Montpellier 2 and CNRS, UMR5650 Montpellier 34095, France.
[4]Kotelnikov Institute of Radio Engineering and Electronics (Saratov Branch), Saratov 410019, Russia.

E-mail: meziani@usal.es

**Abstract**. We report on a broadband terahertz emission from a doubly interdigitated grating gates high electron mobility transistor. The observed emission was explained as due to the excitation of multi mode of plasmons: thermally excited incoherent modes and instability-driven coherent modes. The experiment was performed using Fourier spectrometer system coupled with high sensitive 4K Silicon bolometer under the vacuum. To enhance the efficiency, the device was subjected, from the backside, to a CW 1.5 μm laser beam. Dependence of the emission on the gate bias was observed and interpreted as due to the self-oscillation of the plasma waves.

## 1. Introduction

The oscillation of electrons density in a two dimensional system can become unstable against the generation of electromagnetic radiation within terahertz range [1-2]. This phenomenon is known as the instability of the plasma waves and can be used for tunable detection/generation, and photomixing of terahertz radiation. The oscillation frequency of the plasma waves depends on the sheet electron density ($n_s$) and the length of the transistor gate ($L$). Terahertz (THz) range can be reached for sufficiently high $n_s$ and submicron size $L$. There are two types of excitation of the plasma waves: (i) incoherent type related to the thermal excitation of the plasmons by hot electrons and (ii) coherent excitation involving either the plasma wave instability mechanism such as Dyakonov-Shur [1] or the electron transit-time effect [3]. The first mechanism was studied earlier and emission of THz radiation was observed from selectively doped GaAs/GaAlAs heterostructure as well as Si metal-oxide semiconductor [4]. Room temperature terahertz emission interpreted in terms of coherent excitation was observed from GaInAs and GaN/AlGaN transistors [5]. Emission of THz radiation has been measured recently by using doubly interdigitated grating gate high electron mobility transistors (HEMT's) [4] and related to incoherent excitation.

Here, we report on broadband terahertz emission from a doubly interdigitated grating gates HEMT. The observed emission was explained as due to the excitation of multi mode of plasmons: thermally excited incoherent modes and instability-driven coherent modes. The experiment was performed using Fourier spectrometer system coupled with high sensitive 4K Silicon bolometer under the vacuum. To enhance the efficiency, the device was subjected, from the backside, to a CW 1.5 μm laser beam. Dependence of the emission on the gate bias was observed and interpreted as due to the self-oscillation mechanism of the plasma waves.

## 2. Device structure

The device structure is based on a InGaP/InGaAs/GaAs material system and incorporates (i) dual interdigitated grating gates (G1 & G2) that periodically localize the 2D plasmon in sub-micron regions with 100 nm interval and (ii) a vertical cavity structure in between the 2D plasmon and a THz mirror at the backside. The 2D plasmon layer is formed with a quantum well in the InGaAs channel layer. The grating gate was formed with 65-nm thick Ti/Au/Ti by a standard lift-off process (Fig. 1). Table 1 describes the geometrical parameters of the samples. The GaAs substrate was thinned down to a thickness of 43 μm, and 100 nm thick ITO (indium titanium oxide) transparent metal was sputtered on the optically polished back surface to form the vertical cavity, which make possible the irradiation from the backside. The reflectivity of the ITO mirror in the THz range (0.6 to 6 THz) is about 90% and the transmittance is 83% for 1.5 μm. More details can be found in [7, 8].

**Table 1.** Description of the geometrical parameters of the samples

| | $L_{G1}$ (*nm*) | $L_{G2}$ (*nm*) | # fingers ($L_{G1}/L_{G2}$) | d (*nm*) | L/W (*μm*) |
|---|---|---|---|---|---|
| **Sample1** | 70 | 350 | 60/61 | 80 | 30/75 |
| **Sample2** | 100 | 1850 | 15/16 | 80 | 30/75 |

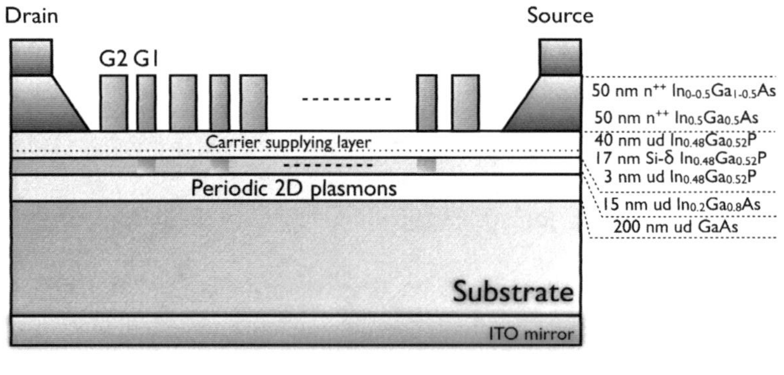


**Figure 1.** Device cross-section for typical GaAs-based heterostructure material system.

## 3. Results and discussions

### 3.1. *Emission of terahertz rays.*

The samples were placed in the source position of the vacuum cavity of the Fourier-transformed far-infrared spectrometer. The radiation intensity was measured using a high sensitive Silicon bolometer. The background spectrum was measured first (signal measured without any current flowing through the sample). It contains the information on the 300 K blackbody emission modified by the spectral functions of all the elements inside the spectrometer. Then we measured the spectrum with current flowing thought the sample. The signal shown here is the ratio between both signals. Figure 2 shows the observed signal from both samples. A broadband signal with maximum at 2.5 THz for sample 1 and 3 THz for sample 2 is observed. For both samples the signal dies off abruptly around 6.5 THz, which is believed to be related to the Reststrahlen band of optical phonon modes of the GaAs-based materials [4,9]. The detected signal was interpreted as due to the thermal excitation of the plasmons by means of injection of hot electron in the channel (incoherent excitation) [4]. However, dependence of the emission on drain-to-source voltage ($V_{DS}$) shows that some features of coherent excitation could be possible since a threshold property is observed (Fig.3) for Vg1/Vg2=-1V/0V. Its known that the coherent excitation by means of Plasmon instability could lead to threshold behaviour [1,3]. However, incoherent excitation gives a super-linear behaviour (Vg1/Vg2=0/0V and 0V/-0.5V).

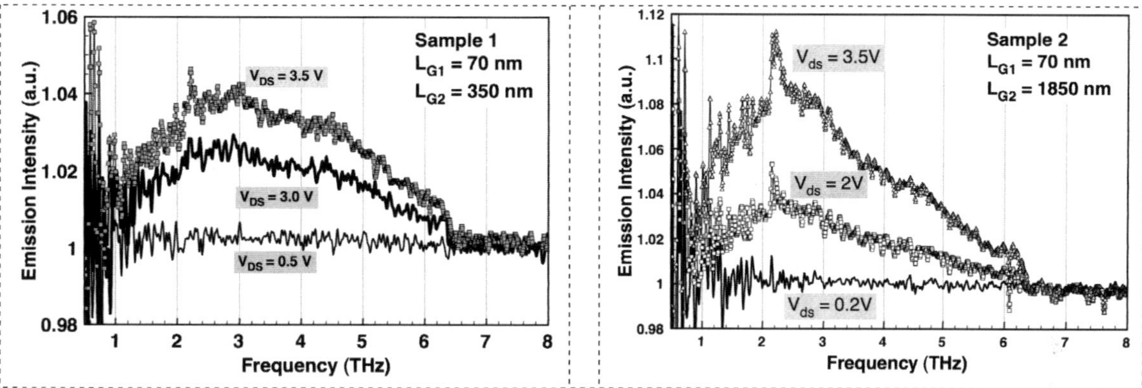


**Figure 2.** FTIR measured signal from the doubly interdigitated grating gates HEMT.

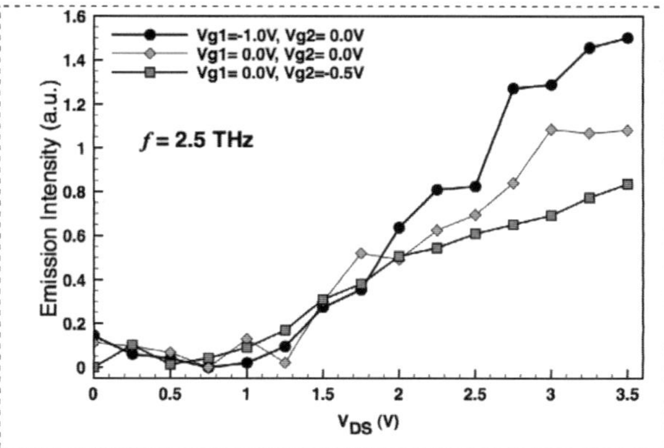


**Figure 3.** Emission intensity vs. drain-to-source bias at 2.5THz for sample 2.

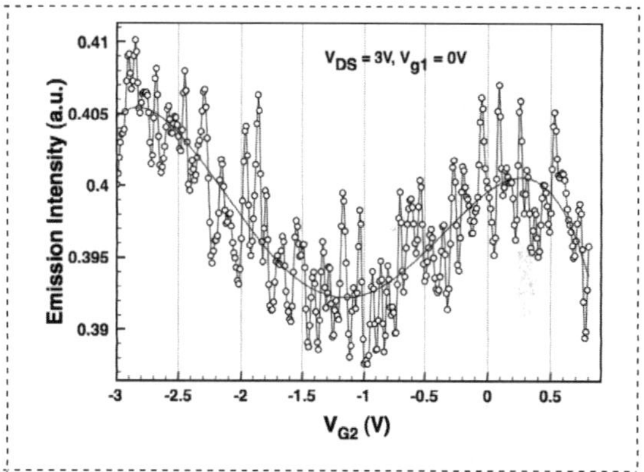


**Figure 4.** Emission intensity vs. $V_{g2}$ for sample 1 for fixed $V_{DS}$ = 3V and $V_{g1}$ = 0V

### *3.2. Optical enhancement of terahertz radiations*

The devices were subjected to 1500 nm CW laser from the back surface at room temperature. The beam was linearly polarized along the channel axis and the power was around 3 mW. A high sensitive 4K Silicon bolometer placed in the front of the sample was used to detect the THz radiation emitted from the device. A window of detection from 0.5 to 3 THz is possible by using a 100 $cm^{-1}$ far-infrared cut-on type filter. The photon energy of the irradiated laser (0.8 eV) is much lower than all the band gap energies of this material system. However, the electrons are weakly photo-excited at the InGaAs/GaAs heterointerface via two-photon processes due to the existence of deep-level trap centers. From I-V characteristics with and without illuminations, the concentration of photoelectrons was estimated to be around $3x10^{11}$ $cm^{-2}$. Figure 4 shows the detected signal as a function of gate 2 bias and for $V_{DS}$ = 3 V and Vg1 = 0V. A maximum of signal is observed for Vg2 around -2.8 V (close to threshold bias).

When the device is illuminated, the photoelectrons are generated under the depleted G2 gate due to the applied bias (close to threshold) and to more unoccupied states. If a specific drain-to-source bias is applied to promote a uniform slope along the source-to-drain direction on the energy band, photoelectrons under the depleted gate (G2) are unidirectionally injected to one side of the adjacent plasmon cavity under the other gate (G1). This may excite the plasmons under an asymmetric cavity boundary [2, 10]. The periodically localized 2D plasmon can be coupled with those in neighbour regions and make in-phase resonant oscillation so that the 2D-plasmon grating can convert the non-radiative plasma waves to radiative electromagnetic waves. It is noted that the laser irradiation may excite the plasmon not only in the regions under G1 but also in the regions under G2 if the cavity size and carrier density of the regions under G2 also satisfies the resonant conditions. The discrepancy between the gate biases of G1 and G2 can enhance the oscillations of the plasma waves via increasing the drift velocity of the photoelectrons injected in the plasmon cavities (self-oscillation mechanism) [11]. The second peak at Vg2 around 0.25 V can also be interpreted as a self oscillation of the plasma waves: even the gate biases for G1 and G2 are similar, the sheet electron density in the cavities under both regions are different due to the difference of the threshold voltage. The self-oscillation of the plasma waves can also occur at this condition.

## 4. Conclusion

We report on a broadband emission from a doubly interdigitated grating gate high electron mobility transistor at room temperature. The emission was explained as due to coherent/incoherent excitation of the plasma waves. Enhancement of the terahertz radiation by optical stimulation has been also shown and related to self-oscillation mechanism.

**Acknowledgements:** This work was financially supported in part by the SCOPE Programme from the MIC, Japan, and by the Grant in Aid for Basic Research (S) from the JSPS, Japan. Y.M.M. acknowledges the support from RamonY Cajal program in Spain and Spanish MICINN through grant Number/Reference TEC2008-02281. V.V.P. acknowledges a support from the Russian Foundation for Basic Research (Grants 08-02-92497 and 09-02-00395) and from the Russian Academy of Sciences Programme "Basic Research in Nanotechnology and Nanomaterials". W.K from Montpellier University acknowledge the region of Languedoc-Roussillon Terahertz Platform as well as GDR-E Teraherz Sources and Detectors project

## References

[1] Dyakonov M and Shur M 1993 *Phys. Rev. Lett.* **71** 2465.
[2] Dyakonov M and Shur M 1996 *IEEE Trans. Electron Dev.* **43** 380.
[3] Ryzhii V, Satou Aand Shur M 2006 *IEICE Trans. Electron.* **E89-C** 1012. Ryzhii V, Satou Aand Shur M 2005 *Phys. Status Solidi A* **202** R113.
[4] Meziani Y M, Handa H, Knap W, Otsuji T, Sano E, Popov V V, Tsymbalov G M, Coquillat D and Teppe F 2008 *Appl. Phys. Lett.* **92** 201108.
[5] Knap W, Lusakowski J, Parenty T, Bollaert S, Cappy A, Popov V V, and Shur M S 2004 *Appl. Phys. Lett.* **84** 2331.
[6] Dyakonova N, Teppe F, Lusakowski J, Knap W, Levinshtein M, Dmitriev A P, Shur M S, Bollaert S, and Cappy A 2005 *J. Appl. Phys.* **97** 114313.
[7] Otsuji T, Meziani Y M, Hanabe M, Ishibashi T, Uno T and Sano E 2006 *Appl. Phys. Lett.* **89** 263502.
[8] Meziani Y M, Otsuji T, Hanabe M, Ishibashi T, Uno T and Sano E 2007 *Appl. Phys. Lett.* **90** 061105.
[9] Perara A G U, Matsik S G, Liu H C, Gao M, Buchanan M, Schaff W J and Yeo W 2000 *Appl. Phys. Lett.* **77** 741
[10] Hanabe M, Otsuji T, Ishibashi T, Uno T and Ryzhii V 2005 *Jpn. J. Appl. Phys.* **44**, 3842
[11] Meziani Y M, Hanabe M, Otsuji T, and Sano E 2007 *Jpn. J. Appl. Phys.* **46**, 2409

# Tunable room temperature Terahertz sources based on two dimensional plasma instability in GaN HEMTs

**A El Fatimy[1,*] , T Suemitsu[1], T Otsuji[1], N Dyakonova[2], W Knap[1,2], Y M Meziani[3], S. Vandenbrouk[4], K. Madjour[4], D.Théron[4], and Ch. Gaquiere[4], P.Prystawko[5], C. Skierbiszewski[5]**

[1] Research Institute of Electrical Communication (RIEC), Tohoku University, 2-1-1 Katahira, Aoba-Ku, 980-8577, Sendai – Japan

[2] Groupe d'Etude des Semiconducteurs, UMR CNRS 5650, Université Montpellier 2, 34095 Montpellier, France

[3] Dpto. de Fisica Aplicada, Universidad de Salamanca, Plaza de la Merced s/n, 37008 Salamanca, Spain

[4] Institut d'Electronique et de Microélectronique du Nord, UMR CNRS 8520, 59655 Villeneuve d'Acsq, France

[5] Institutes of High Pressure Physics, PAS, Unipress, Sokolowska 29/37, 01-142 Warsaw, Poland

Email: a.elfatimy@riec.tohoku.ac.jp

**Abstract**. In this work, we report on room temperature terahertz radiation from sub-micron size GaN/AlGaN based high electron mobility transistors (HEMTs). They could successfully replace the standard Fourier Transform spectrometer source and were investigated with a standard Si-bolometer as a detector. The relatively broad (~1THz) emission line was observed. The maxima were found to be tunable by the gate voltage between 0.75 and 2.1 THz. The observed emission was interpreted as due to the current driven plasma waves instability in the two-dimensional electron gas. The emitted power from a single device reached 150 nW, showing possible application of these transistors as compact sources for terahertz spectroscopy and imaging.

[*] Also at: Cardiff School of Physics & Astronomy, Cardiff University, Queens Buildings, The Parade, Cardiff, United Kingdom
Email: ElFatimyA@cardiff.ac.uk

Journal of Physics: Conference Series **193** (2009) 012072 doi:10.1088/1742-6596/193/1/012072

## 1. Introduction

The channel of a High Electron Mobility Transistor (HEMT) can act as a resonator for the plasma waves propagating in 2D electron gas. The plasma frequency increases with decreasing the channel length and can reach the terahertz (THz) range for nanometer size transistors. As it was predicted by Dyakonov and Shur (DS theory) [1,2], when a current flows through a HEMT, the steady state can become unstable against generation of plasma waves leading to the emission of an electromagnetic radiation at THz frequencies. Three features that are intrinsic for this type of THz emission: threshold-like appearance, appropriate (THz) frequency range and tunability by the gate voltage.

Results of several experiments [3-7] have been tentatively interpreted in terms of DS theory [1]. Although the emission obtained from the nanotransistors really has been in the predicted frequency range and has appeared in the threshold manner, the tunability by the gate voltage has never been observed. In this work, we present the room temperature emission of terahertz radiations from GaN/AlGaN based nanometric high electron mobility transistors. We show that in our structures the emission frequency can be tuned by the gate voltage in agreement with theoretically dependence.

## 2. Experimental setup and results

The measurements were carried out at room temperature. The emission signal was excited by square source-drain voltage pulses with amplitude $V_{ds}$, pulse frequency 30 Hz and duty cycle 0.5. The registration of the emission was performed by Si bolometer whose frequency range of 0.2–4.5 THz was determined by incorporated filters. The total integrated radiation intensity has been measured by placing the samples directly in front of the bolometer. The spectral analysis of the device radiation was performed using vacuum Fourier Transform spectrometer (FTIR). The standard spectrometer source was replaced by our GaN/AlGaN THz source. THz emission was studied for the gate voltage range (-4Volt to 0Volt). Details about of GaN/AlGaN samples can be found in Ref. [8].

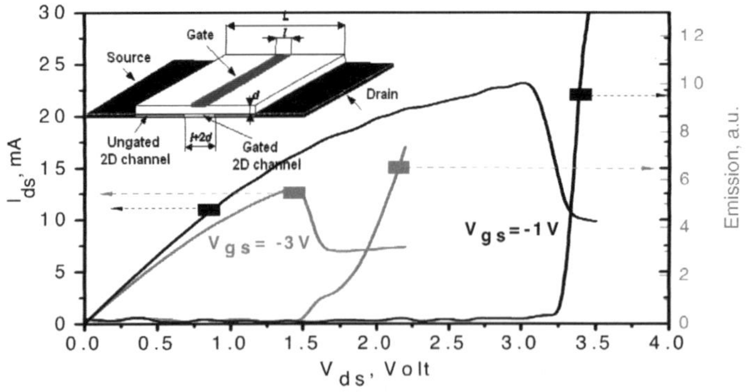


**Fig.1:** Output characteristics and the emission intensity for Sample 1 (150nm) at gate biases $V_{gs}$=-3V and for Sample 2 (250nm) at $V_{gs}$ -1V.

The signal in the fig.1 is shown for two values of the gate voltage. The emission spectrum shows a threshold–like behaviour. The THz emission intensity increases drastically. At the same time a decrease of the drain current is observed. With a more negative gate bias, the total drain current drops and the corresponding onset of the emission are shifted towards lower values of $V_{ds}$. This result has been already observed and related to the plasma-wave instability at room temperature for transistors based on different materials [7]. Using the calibration data of the bolometer we roughly estimated the emission power, which was typically around 150nW. The strongest emission was observed for the sample 2 (250 nm gate length). Our result shows a possible application of the GaNAlGaN transistors as compact sources for terahertz spectroscopy and imaging [9].

Both $V_{gs}$ and $V_{ds}$ biases were fixed (within the range of the emission) and the THz radiation spectra was taken using FFT spectrometer. In Fig. 2-a, we show typical emission spectra. A maximum of the signal is changing from 0.75 THz to 2.2 THz for $V_{gs}$=-3.5 V to $V_{gs}$= 0 V. It can be clearly seen that the signal is tuned by the gate bias. The figure 2-b shows the dependence of the resonant frequency as a function of the gate bias for 150-nm and 250-nm gate length GaN/AlGaN HEMT's. The dotted points are extracted from the experiments and the line is simulation result. One can see that the position of the maximum changes from 2.2 THz at $V_{gs}$= 0V to 0.75 THz at $V_{gs}$=-3.5V.

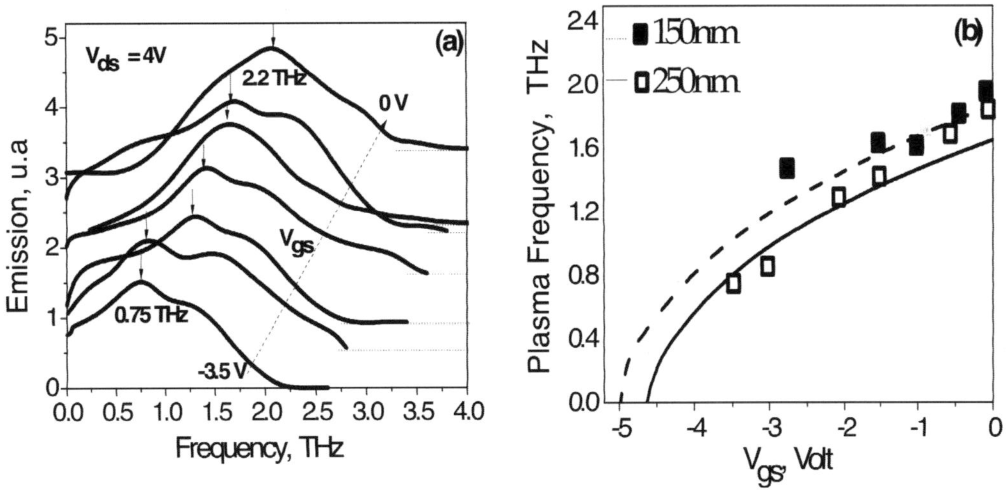


**Fig.2: (a):** The Emission spectrum for 150nm GaN and 250nm GaN at $V_{ds}$= 4V. To highlight the tuning of the spectrum by the gate bias, the arrows show the maxima of spectrum. For clarity, curves are vertically shifted; **(b):** Maxima of resonant gate voltages corresponding to different frequencies of the radiation source for 150 and 250nm gate length.

## 3. Discussion

The results demonstrate THz emission that appears when the drain current exceeds a certain threshold value. At the same time a drop of the current is observed. This may suggests that the generation of THz radiation is related to negative differential resistance mechanism (NDR). In fact a strong heating

Journal of Physics: Conference Series **193** (2009) 012072 doi:10.1088/1742-6596/193/1/012072

of electrons may lead to a real space transfer and/or to population of higher valleys. This could result in a negative differential resistance, which could in principle persist up to THz frequencies. Some possible mechanisms were recently discussed, Mateos et al [10].

The solid line on Fig.2-b shows the results of calculations according to simple equations of Ref. [1] with the effective gate length taken as geometrical one plus two spacer thickness [11]. One can see that the simple model gives a good quantitative description of experimental data. This result supports the plasma wave instability (rather than NDR) origin of the THz emission.

In conclusion: Efficient (150nW) room temperature THz emission from GaN/AlGaN HEMt was reported. Observed emission was frequency-tunable by the gate voltage and changed also with gate length – in agreement with theoretical predictions. These observations strongly support the interpretation of the observed emission as due to plasma wave instability and resonant Plasmon excitation.

**Acknowledgement**

This work was financially supported in part by the SCOPE Program from the MIC, Japan, and by the Grant in Aid for Basic Scientific Research (S) from the JSPS, Japan. The authors acknowledge also French-Japanese research program "Sakura.". Y.M.M. acknowledges the support from RamonY Cajal program in Spain and Spanish MICINN through grant Number/Reference TEC2008-02281. The authors from Montpellier University acknowledge the CNRS guided GDR/GDR-E projects "*Semiconductor sources and detectors of THz frequencies*" and the support from Region Languedoc-Roussillon through the "Terahertz Platform" project.

**References**

[1] M. Dyakonov and M. Shur, Phys. Rev. Lett. **71**, 2465 (1993)

[2] M. Dyakonov and M. Shur, IEEE **43**, 380(1996)

[3] N. Dyakonova, F. Teppe et al, J. Appl. Phys., V **97**, N 11, p.4313 , (2005)

[4] J. Lusakowski, W. Knap et al J. Appl. Phys., V **97**, N 6, p.64307 ,( 2005)

[5] W. Knap, J. Lusakowski et al Appl.Phys. Lett., V. 84, N 13, pp.2331-2333,( 2004)

[6] A.El Fatimy, W. Knap et al, phys. stat. sol. (c), Vol **5**, 244– 248 (2008)

[7] N. Dyakonova, A. El Fatimy et al, Appl. Phys. Lett. **88**, 141906 (2006)

[8] A. El Fatimy, Thesis, Hall, 00182726, (2007)

[9] A. El Fatimy et al, Optics Communications, Volume **282**, Issue 15, Pages 3055-3058
1 August (2009)

[10] J.Mateos, S. Pérez, et al, AIP Conf. Proc. **800**, 423 (2005)

[11] W. Knap, Y. Deng, et al, Appl.Phys. Lett, V. **81**, N 24, pp.4637-4638, (2002)

# Fano profile in the intersubband terahertz response of photoexcited GaAs/AlGaAs quantum wells

**M. Wagner[1], D. Golde[2], D. Stehr[1], H. Schneider[1], M. Helm[1], A. M. Andrews[3], T. Roch[3], G. Strasser[3], M. Kira[2], and S. W. Koch[2]**
[1]Institute of Ion Beam Physics and Materials Research, Forschungszentrum Dresden-Rossendorf, P.O. Box 510119, 01314 Dresden, Germany
[2]Department of Physics and Material Sciences Center, Philipps-University, 35032 Marburg, Germany
[3]Micro- & Nanostructure Center, TU Wien, Floragasse 7, 1040 Vienna, Austria
E-mail: m.wagner@fzd.de

**Abstract.** In our work we probe the conduction intersubband transition of an undoped GaAs/$Al_{0.34}Ga_{0.66}As$ multiple quantum well via broadband terahertz pulses after resonant photoexcitation at the 1s heavy-hole exciton. The pump-induced change in the transmitted terahertz field shows a strong beating. In the frequency domain this results in an asymmetric Fano-like line shape for the intersubband resonance and an additional broad low-frequency peak. However, the total THz absorption shows only the single symmetric peak of the intersubband transition. In our microscopic theory these signatures unambiguously originate from the phase sensitive superposition of ponderomotive and terahertz intersubband currents.

## 1. Introduction

Terahertz (THz) spectroscopy on semiconductor heterostructures has revealed new insight in low-energy excitations and their dynamics. A microscopic analysis of these experiments has to take into account the true THz transitions, but also the so-called ponderomotive current. This current results from the classical, electric-field induced charge acceleration according to $\hbar\dot{\mathbf{k}} = -e\,\mathbf{E}_{\mathrm{THz}}$ [1] that causes a wiggling motion of the carriers. In recent experiments the influence of the ponderomotive current could only be seen rather indirectly [2, 3].
Here, we study the intersubband transition of multi quantum wells via THz spectroscopy and directly observe signatures of the ponderomotive current [4]. In the differential transmission spectra we measure a characteristic Fano-like line shape originating from the superposition of ponderomotive and true THz intersubband currents. Usually, asymmetric Fano line shapes result from the interference between a dicrete level and a continuum of states [5, 6].

## 2. Experimental

Spectrally broadband THz pulses are generated in a 55 $\mu$m thin $z$-cut GaSe crystal (see Fig. 1(a)) by phase-matched difference frequency mixing [7] within ultrashort 12 fs optical pulses from a 78 MHz Ti:sapphire oscillator (Femtolasers: Femtosource Scientific sPro). The THz beam is focused on the multi quantum well (MQW) sample. The THz field has a strong component normal to the MQW plane to couple to the intersubband transition. Phase-matched electro-optic sampling [8] allows for field-resolved detection in a second 30 $\mu$m thin $z$-cut GaSe crystal.

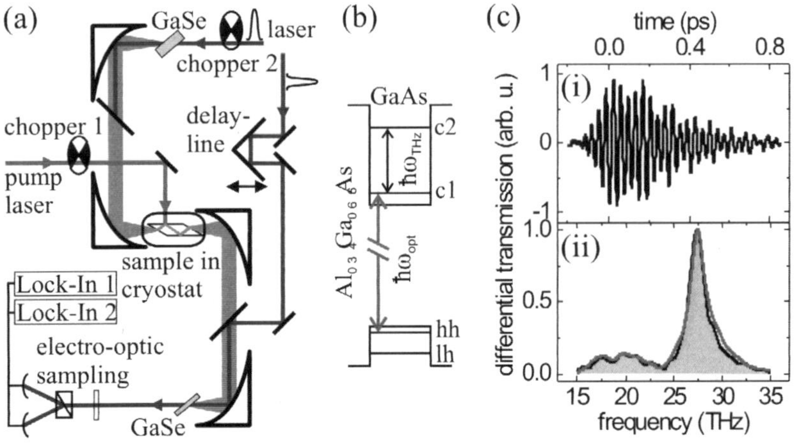


**Figure 1.** (a) Optical interband pump intersubband THz probe experiment. (b) THz pulses probe the c1-to-c2 transition after photoexcitation of the hh-c1 transition in the MQW. (c)(i) Time-resolved differential transmission after ps excitation. (c)(ii) Measured (shaded area) and computed (red line) spectral amplitude.

The MQW sample consists of 60 8.2-nm-thick undoped GaAs quantum wells, separated by 19.6-nm-thick barriers of $Al_{0.34}Ga_{0.66}As$. It was prepared as a 38-degree wedged waveguide and an $Al_{0.34}Ga_{0.66}As$ spacer layer of 300 nm thickness was grown on top for improved overlap of THz beam and MQWs. It was kept at a temperature of 6 K in a liquid He flow cryostat.

For photoexcitation (see Fig. 1(b)) an optical interband pump pulse either from a spectrally narrow 2.5 ps or a broad-band 100 fs 78 MHz laser (both Spectra Physics: Tsunami) excites the MQWs resonantly at the 1$s$ heavy-hole exciton around $\hbar\omega_{opt} = 1.56$ eV. Only the first conduction subband is populated. The c1-to-c2 intersubband transition is then probed by THz pulses 25 ps after excitation. The photoexcited carrier density is around $2 \times 10^{10}$ cm$^{-2}$ per quantum well.

The transmitted THz transients with and without excitation are simultaneously measured. To this end both pump and THz probe beams are chopped at the same time at different frequencies. The first Lock-In amplifier locks on the modulation of the optical pump beam. Twice its signal yields the differential transmission $\Delta E(t)$, i.e. the pump-induced change in the transmitted THz field. The second Lock-In amplifier detects the mean value between transmission with and without excitation. The reference $E_{\text{ref}}(t)$ without excitation follows by subtracting $\frac{1}{2}\Delta E(t)$.

## 3. Results

After ps photoexcitation the time-resolved differential transmission $\Delta E(t)$ shows a strong beating (Fig. 1(c)(i)), superimposed on the dephasing of a reradiated THz-induced intersubband polarization [9]. The beating is connected to a broad low-frequency peak around 20 THz in the spectral amplitude $|\Delta E(\omega)|$ (Fig. 1(c)(ii)) while the intersubband resonance at 27.3 THz displays a Fano-like asymmetry with an undershoot at the low-frequency side.

To explain our experimental findings, the THz transmission of a quantum well positioned at $z = 0$ is computed ($z$ is the growth direction). The THz response follows from the wave equation $\left(\nabla^2 - (\frac{n_b}{c})^2 \frac{\partial^2}{\partial t^2}\right) E(\mathbf{r}, t) = \mu_0 \delta(z) \frac{\partial}{\partial t} \left(J_{\text{A}}(t) + J_{\text{THz}}(t)\right)$ ($n_b$ is the background refractive index). Here a delta function is used since the quantum well width is much smaller than the THz wavelength. On the right hand side of the equation the induced current density contains $J_{\text{THz}}$, the current due to the intersubband transition, and $J_{\text{A}}$, the THz field induced ponderomotive motion of the excited carriers. The solution yields the differential transmission $\Delta E \propto J_{\text{A}} + J_{\text{THz}}$, which can be interpreted as a field reemitted by the current density that is directly proportional to the induced current [10].

$J_{\text{A}}$ is defined via $J_{\text{A}} = -\sum_\lambda (e^2 n_\lambda / m_\lambda) A_{\text{THz}}$ with the effective mass $m_\lambda$ and the carrier density $n_\lambda$ in band $\lambda$ ($\lambda$ includes both bulk-band index and subband index). The vector potential $A_{\text{THz}}$ of the THz probe pulse is given by $E_{\text{THz}} = -\frac{\partial}{\partial t} A_{\text{THz}}$. Thus, $J_{\text{A}}$ directly follows $A_{\text{THz}}$ with an

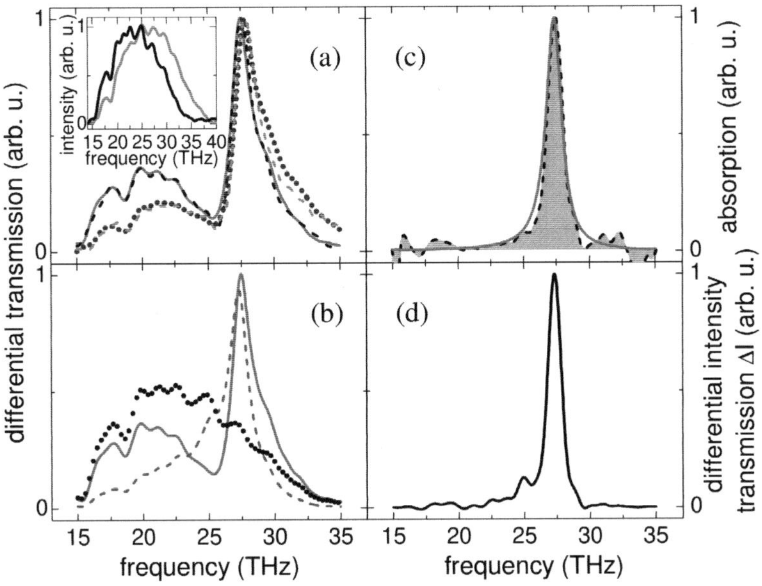


**Figure 2.** Experiment-theory comparison for fs excitation. (a) Measured (black and grey dashed line, resp.) and computed (red solid and blue dotted line, resp.) differential transmission for two different THz probe spectra (inset). (b) Decomposition of the computed current density (red solid line) into ponderomotive contribution $|J_{\mathrm{A}}|$ (black dotted line) and intersubband current $|J_{\mathrm{THz}}|$ (green dashed line). (c) Measured (shaded area) and computed (red line) THz absorption. (d) Simulated differential intensity change $\Delta I$.

opposite phase. Since $J_{\mathrm{A}}$ leads to a real-valued linear susceptibility $\chi_{\mathrm{A}}(\omega) = -\omega_p^2/\omega^2$ (with the plasma frequency $\omega_p$), it does not contribute to the absorption $\alpha(\omega) \propto \omega\,\mathrm{Im}[\chi(\omega)]$.
The THz current is given by $J_{\mathrm{THz}} = \frac{1}{S}\sum_{\lambda,l,l',\mathbf{k}} j^{\lambda}_{l',l}\, p^{\lambda}_{l,l',\mathbf{k}}$ with the quantization area $S$, the matrix elements $j^{\lambda}_{l,l'}$ and the microscopic intersubband polarization $p^{\lambda}_{l,l',\mathbf{k}}$ between subbands $l$ and $l'$ of bulk band $\lambda$. $j^{\lambda}_{l,l'} = -\frac{i\hbar e}{m_\lambda}\int \mathrm{d}z\, \xi^{*}_{\lambda,1}(\mathrm{z})\, \frac{\partial}{\partial \mathrm{z}}\xi_{\lambda,1'}(\mathrm{z})$ is the intersubband matrix element with the confinement wave function of the carriers $\xi_{\lambda,l}(z)$. $p^{\lambda}_{l,l',\mathbf{k}}$ is computed microscopically with an equation-of-motion approach. Regarding only the c1-to-c2 transition, the sum is restricted to $\lambda = c$ and $l, l' \in \{1, 2\}$. In analogy to the semiconductor Bloch equations [11], one finds for $p^{c}_{1,2,\mathbf{k}}$: $i\hbar\frac{\partial}{\partial t}p^{c}_{1,2,\mathbf{k}} \approx \left(\tilde{\varepsilon}^{c}_{2,\mathbf{k}} - \tilde{\varepsilon}^{c}_{1,\mathbf{k}}\right)p^{c}_{1,2,\mathbf{k}} - i\gamma\, p^{c}_{1,2,\mathbf{k}} + \left(f^{c1}_{\mathbf{k}} - f^{c2}_{\mathbf{k}}\right)\left[j^{c}_{2,1}A_{\mathrm{THz}} - \sum_{\mathbf{q}\neq\mathbf{k}} V_{\mathbf{k}-\mathbf{q}}\, p^{c}_{1,2,\mathbf{q}}\right]$, with the renormalized single-particle energies $\tilde{\varepsilon}^{c}_{l,\mathbf{k}}$, the optically excited carrier distributions $f^{c,l}_{\mathbf{k}}$ in the conduction bands and the Coulomb matrix element $V_{\mathbf{q}}$. Optically excited interband coherences have already decayed and do not contribute. All scattering effects are modeled by a phenomenological dephasing constant $\gamma$ to match the measured 440 fs decay.
We compute the single-particle energies via $\mathbf{k}\cdot\mathbf{p}$-perturbation theory. For the THz-response, we include the first two conduction subbands and the first heavy-hole and light-hole subband. The experimental THz reference transient serves as input for the time-dependent fields.
For ps excitation (Fig. 1(c)(ii)) the theoretical result (red line) as the absolute value of the computed current density, i. e. the differential transmission $|\Delta E(\omega)|$, is in excellent agreement with the measurement (shaded area). In the case of broadband fs excitation the low-frequency peak and the Fano-like asymmetry of the intersubband resonance are enhanced (Fig. 2(a), measured: black dashed line; computed: red solid line). These features are also affected by the THz probe spectrum. Shifting its center away from the previous 23.5 THz (Fig. 2(a) inset, black line) to 26.5 THz (inset, grey line) decreases the broad low-frequency peak while the asymmetry of the resonance increases (Fig. 2(a), measured: grey dashed line; computed: blue dotted line). In Fig. 2(b) the total current density for the low-frequency THz probe spectrum (red solid line) is broken up in its individual contributions, i. e. the ponderomotive current density $J_{\mathrm{A}}$ (black dotted line) and the intersubband current $J_{\mathrm{THz}}$ (green dashed line). $J_{\mathrm{A}}$ follows mainly the THz probe spectrum, while $J_{\mathrm{THz}}$ contains the intersubband resonance weighted with the THz probe

spectrum. The observed Fano-like line shape in the differential transmission results from a phase sensitive superposition of $J_{\mathrm{A}}$ and $J_{\mathrm{THz}}$: $|J_{\mathrm{total}}| = |J_{\mathrm{A}} + J_{\mathrm{THz}}|$. Below the intersubband resonance both contributions partially compensate each other whereas they interfere constructively above. Thus, the resonance is narrowed at the low-frequency side and broadened at the high-frequency side, leading to an asymmetric Fano-like line shape. By shifting the THz probe spectrum mainly $J_{\mathrm{A}}$ is affected and shifted accordingly, explaining the observed behavior in Fig. 2(a). However, there is no Fano signature in the THz absorption $\alpha(\omega) = 2\mathrm{Im}[-\mathrm{i}\Delta\mathrm{E}(\omega)/\mathrm{E}_{\mathrm{ref}}(\omega)]$ (Fig. 2(c), shown for the low-frequency THz spectrum).

In the case of fs excitation we saw an enhanced ponderomotive contribution. The fs pulses are spectrally broader than the linewidth of the 1s exciton resonance and thus they are partially (46 %) transmitted through the quantum well film and absorbed in the GaAs substrate. Since the ponderomotive current $J_{\mathrm{A}}$ is proportional to the carrier density, $J_{\mathrm{A}}$ is increased, but not $J_{\mathrm{THz}}$. Consequently the carrier density in the wells plus substrate is 1.85 times larger than in the wells alone, which can be modeled by increasing $J_{\mathrm{A}}$ relative to $J_{\mathrm{THz}}$ by 1.85.

It would be interesting to check whether the ponderomotive contribution could be seen in standard Fourier transform infrared (FTIR) spectroscopy while modulating the carrier density, e.g. by photoexcitation. Since FTIR spectroscopy detects only intensities one would record a differential intensity change $\Delta I(\omega)$ of the transmission without and with excitation, respectively, of the form $\Delta I(\omega) = I_{\text{without exc.}}(\omega) - I_{\text{with exc.}}(\omega) \propto |E_{\mathrm{ref}}(\omega)|^2 - |E_{\mathrm{ref}}(\omega) + \Delta E(\omega)|^2$. In the case of fs photoexcitation Fig. 2(d) presents the simulation of $\Delta I(\omega)$ based on our field-resolved data for $E_{\mathrm{ref}}(t)$ and $\Delta E(t)$. It turns out that the ponderomotive contribution has vanished almost completely compared to a field-resolved measurement.

## 4. Summary

In summary, we have observed strong Fano signatures in the differential transmission spectra of broadband THz pulses probing an undoped GaAs/AlGaAs multiple quantum well after resonant photoexcitation. In our microscopic theory, the Fano-like line shape of the intersubband resonance is unambiguously attributed to a phase sensitive interference of the sharp intersubband resonance with the effective continuum provided by the light-matter interaction through the ponderomotive contribution. Finally, the applied field-resolved method is better suited than intensity-based measurements for revealing the ponderomotive influence to the THz response.

## Acknowledgments

The authors from Forschungszentrum Dresden-Rossendorf are grateful to Dr. Stephan Winnerl for critical discussions. The Marburg work is supported by the Quantum Optics in Semiconductors DFG Research Group. The Vienna group is supported by the Austrian FWF.

## References

[1] Krieger J B and Iafrate G J 1986 *Phys. Rev.* B **33** 5494
[2] Danielson J R, Lee Y-S, Prineas J P, Steiner J T, Kira M and Koch S W 2007 *Phys. Rev. Lett.* **99** 237401
[3] Leinß S, Kampfrath T, Volkmann K v, Wolf M, Steiner J T, Kira M, Koch S W, Leitenstorfer A and Huber R 2008 *Phys. Rev. Lett.* **101** 246401
[4] Golde D, Wagner M, Stehr D, Schneider H, Helm M, Andrews A M, Roch T, Strasser G, Kira M and Koch S W 2009 *Phys. Rev. Lett.* **102** 127403
[5] Fano U 1961 *Phys. Rev.* **124** 1866
[6] Faist J, Sirtori C, Capasso F, Chu S-N G, Pfeiffer L N and West K W 1996 *Opt. Lett.* **21** 985
[7] Kaindl R A, Smith D C, Joschko M, Hasselbeck M P, Woerner M and Elsaesser T 1998 *Opt. Lett.* **23** 861
[8] Kübler C, Huber R, Tübel S and Leitenstorfer A 2004 *Appl. Phys. Lett.* **85** 3360
[9] Heyman J N, Kersting R and Unterrainer K 1998 *Appl. Phys. Lett.* **72** 644
[10] Kira M and Koch S W 2006 *Prog. Quantum Electron.* **30** 155
[11] Haug H and Koch S W 2004 *Quantum Theory of the Optical and Electronic Properties of Semiconductors* 5th ed. (Singapore: World Scientific)

Journal of Physics: Conference Series **193** (2009) 012074 doi:10.1088/1742-6596/193/1/012074

# Terahertz detection in a double-grating-gate heterotransistor

**D Coquillat[1], S Nadar[1], F Teppe[1], N Dyakonova[1], S Boubanga-Tombet[1], W Knap[1], T Nishimura[2], Y M Meziani[3], T Otsuji[2], V V Popov[2,4,5] and G M Tsymbalov[4]**

[1]GES-UMR5650, Université Montpellier 2 and CNRS, 34095 Montpellier, France
[2]Research Institute of Electrical Communication, Tohoku University, 2-1-1 Katahira, Aoba-Ku, Sendai 980-8577, Japan
[3]Dpto. de Fisica Aplicada, Universidad de Salamanca, Pza de la Merced s/n, 37008 Salamanca, Spain
[4]Kotelnikov Institute of Radio Engineering and Electronics (Saratov Branch), 410019 Saratov, Russia

E-mail: popov@soire.renet.ru

**Abstract**. We observed a photovoltaic non-resonant terahertz photoresponse in a InGaAs/GaAs heterostructure with a large area double-grating-gate at room temperature. Semi-quantitative estimation of the characteristic detection length combined with self-consistent calculations of the electric fields exited in the structure by incoming terahertz radiation allowed us to interpret this detection as coming from the depleted regions of the channel.

## 1. Introduction

It is known that hydrodynamic nonlinearities in two-dimensional (2D) electron channels can be used for detection of terahertz (THz) radiation [1]. Resonant THz detection is related to plasmonic nonlinearities in 2D electron channel whereas a broadband non-resonant THz detection is produced by nonlinearities of the oscillating electron currents in depleted regions of the channel where plasmon oscillations are overdamped. High external detectivity can be obtained only if high-quality plasmon resonances are excited in the resonant THz detector and strong THz electric fields are induced in depleted regions of the channel in the non-resonant THz detector. The metal-grating-gate coupler is a conventional tool for coupling plasmons in 2D electron channel to THz radiation [2]. In this paper we demonstrate non-resonant THz detection in a double-grating-gate InGaAs/GaAs heterotransistor at room temperature and interpret this detection as coming from the depleted regions of the channel.

## 2. Experimental

The device structure is based on a high-electron mobility transistor (HEMT) and incorporates doubly interdigitated grating gates (G1 and G2) [3]. The 2D plasmon layer is formed with a quantum well at the heterointerface between a 15-nm-thick undoped InGaAs channel layer and a 60-nm-thick, Si-$\delta$ doped InGaP carrier-supplying layer. The grating gate is formed with 65-nm-thick Ti/Au/Ti by a standard lift-off process. The metal fingers of each grating gate G1 and G2 ($L_{G1}$ and $L_{G2}$, respectively)

[5] Corresponding author

are of the same length but the finger lengths $L_{G1}$ and $L_{G2}$ are different for different grating gates G1 and G2. Four different structures with a double-grating-gate were fabricated. Grating gate G1 was designed to have fingers of length $L_{G1}$=100 nm in each structure, while grating gate G2 was designed to have fingers of different lengths from $L_{G2}$=300 nm to $L_{G2}$=1800 nm for different structures. The spacing between the grating-gate fingers is 100 nm for all four structures. The gate-to-channel separation is $d$ = 65.5 nm and the threshold (depletion) voltages at gates G1 and G2 are – 3.5 V and – 3.0 V, respectively. The whole length of the structure between the source and drain contacts covered by the double-grating gate was about 80 $\mu$m with the channel width of 30 $\mu$m.

**Table 1.** Measured photoresponse for different samples.

| Sample # | 1 | 2 | 3 | 4 | 5 | 6 | 7 | 8 |
|---|---|---|---|---|---|---|---|---|
| $L_{G\text{-biased}}$(nm) $/L_{G\text{-unbiased}}$(nm) | 100 /1800 | 100 /1300 | 100 /800 | 100 /300 | 300 /100 | 800 /100 | 1300 /100 | 1800 /100 |
| Photoresponse per period (μV) | 0.5 | 0.4 | 0.6 | 1.15 | 1.8 | 2.1 | 2.3 | 2.85 |

The double-grating-gate structures were irradiated with terahertz beam at frequency 240 GHz. We measured a photovoltaic response between the source and drain contacts at room temperature. The greatest non-resonant photoresponse was observed when one of the grating gates, either G1 or G2, was biased to the threshold voltage. The photoresponse per period of the structure is shown in table 1 for eight different samples. The photoresponse grows with increasing the length of a depleted portion of the channel per the structure period (under a biased grating-gate finger) for samples ## 5-8 and also grows with decreasing the length of undepleted portion of the channel per the structure period (under unbiased grating-gate finger) for samples ## 1-4. Note that an absolute value of the photoresponse is relatively small because fabrication arrangements were not undertaken to bring designed asymmetry into the structure unit cell which had to be normally required for ensuring stronger photoresponse.

## 3. Discussion

The theory of non-resonant terahertz detection due to hydrodynamic nonlinearity of the 2D electron fluid has been proposed in [1] and then developed in more advanced form in [4] taking into account a finite electron temperature. However, this theory assumes asymmetric excitation of the channel, where a predefined THz electric field is applied to the input of the channel, and then describes how this field propagates along the channel producing detection photoresponse. Such theory can not be used directly for qualitative analysis of the system under study where the oscillating currents are excited in every point of the channel simultaneously by normally incident THz wave. Therefore, in this paper we use a semi-quantitative approach for describing the THz photoresponse in the double-grating-gate structure.

According to [1] the hydrodynamic nonlinearity in 2D electron channel comes from the product $n(x)v(x)$ in the hydrodynamic equation of continuity, where $n(x)$ and $v(x)$ are the oscillating electron density and velocity depending on the coordinate along the channel ($x$-coordinate). The oscillating electron density $n(x)$ in every $x$-point of the channel is proportional to the normal component of the electric field at that point, while the oscillating electron velocity $v(x)$ in every $x$-point is proportional to the in-plane component of the THz electric field at that point. Amplitudes of the normal and in-plane components of the electric field in the channel are proportional to each other because they are interrelated by the linear Maxwell equation (or Poisson equation in the electrostatic case). However, those components of the electric field are shifted in phase along the $x$-coordinate by $\pi/2$. Then assuming a harmonic dependence of $n(x)$ and $v(x)$ on the $x$-coordinate we can estimate this nonlinearity as $n(x)v(x) \sim |E|^2\sin(2\pi x/\Delta)\cos(2\pi x/\Delta)$, where $|E|$ is the amplitude of the normal (or in-plane) component of the electric field in the channel and $\Delta$ is the spatial period of its variation along the $x$-coordinate. It is known [1,4] that a non-resonant detection response comes predominantly from depleted parts of the channel. We can describe the net effect of detection coming from an active (depleted) channel portion of length $w$ by the characteristic detection length

$$L_D = \frac{|E|^2}{|E_0|^2}\int_0^w \sin\left(\frac{2\pi}{\Delta}x\right)\cos\left(\frac{2\pi}{\Delta}x\right)dx = \frac{\Delta}{2\pi}\frac{|E|^2}{|E_0|^2}\sin^2\left(\frac{2\pi}{\Delta}w\right),$$

where $E_0$ is the electric field of incoming THz wave. The value of $L_D$ exhibits maximum for $w = \Delta/4$, which means that the active part of the channel is effectively short-circuited at its one end whereas it is open-circuited at the other end.

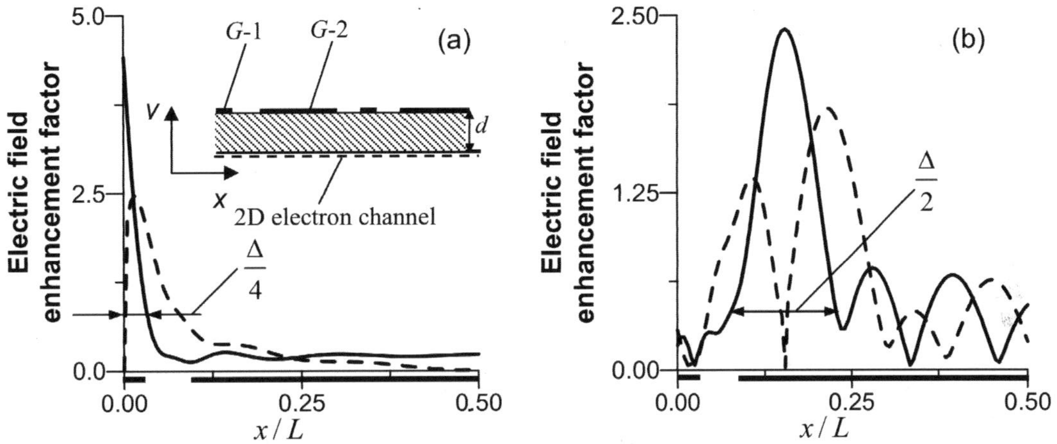


**Figure 1.** In-plane (solid curves) and normal (dashed curves) THz electric field distributions in the 2D electron channel for (a) sample #1 and (b) sample #8 over a half of the structure period $L/2$ at frequency 240 GHz. Inset in panel (a) shows a schematic of the double-grating-gate structure. Location of the grating-gate fingers are also indicated by thick black bars under the abscissa axes. The origin is located under the centre of the G1 finger.

We calculated the normal and in-plane electric field distributions over the structure period in a first-principle electromagnetic approach [5] for all eight samples specified in table 1. The electric field distributions for samples ## 1 and 8 are shown in figure 1. The regions of strong electric field with the electric-field-enhancement factor, $|E|/|E_0|$, exceeding unity are seen in figure 1 in depleted parts of the channel. These regions with strong electric field can be thought as having the length of $\Delta/2$. It means that each region of strong electric field can be viewed as two oppositely-connected channels, which are short-circuited at the center of this region but open-circuited at the ends of this region. Then, for samples ## 5-8 we have $w \approx \Delta/4$ and, hence,

$$L_D \sim \frac{|E|^2}{|E_0|^2}\frac{\Delta}{2}$$

in this case. Whereas for samples ## 1-4 we have $w < \Delta/4$ and, hence,

$$L_D \sim \frac{|E|^2}{|E_0|^2}\frac{2w}{\Delta}w$$

for these samples. In the last formula the ratio $2w/\lambda$ is the form-factor describing a fraction of $\Delta/2$ occupied by the active (depleted) portion of the channel. Electrostatic modeling of the equilibrium electron density distribution show that in samples ## 1-4 with the grating-gate G1 biased to the threshold voltage only very short regions under the centers of 100-nm-wide fingers are depleted. Because of that we estimated the effective length of the active (depleted) region of the channel as $w \approx 20$ nm for samples ## 1-4. Actually, this value of $w \approx 20$ nm serves as a fitting parameter to match the detection length dependencies for samples ## 1-4 and those for samples ## 5-8. Calculated detection lengths for all eight different samples shown in figure 2 demonstrate good agreement with the measured photoresponse for all samples.

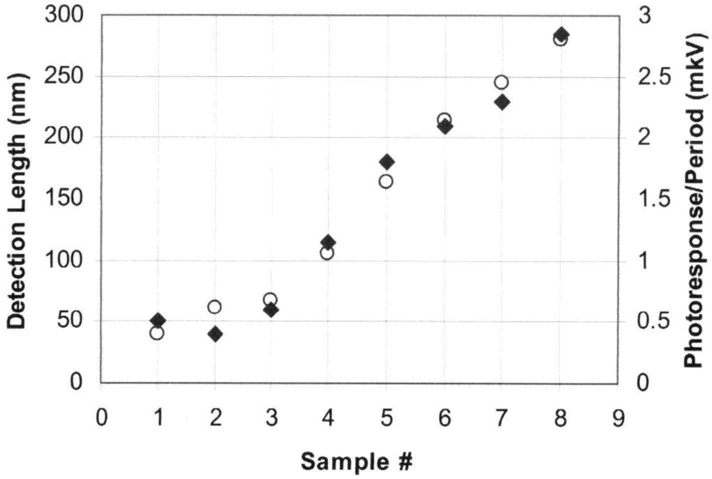


**Figure 2.** Detection length (open circles) and measured photoresponse (solid diamonds).

## 4. Conclusions

We have measured a THz photovoltaic response in a double-grating-gate InGaAs/GaAs heterotransistor at room temperature, which was interpreted as coming from depleted regions of the channel due to hydrodynamic nonlinearity. Large THz-electric-field enhancement factor achieved in the double-grating-gate structure was demonstrated. We believe that the photoresponse could be considerably enhanced by fabricating a double-grating-gate heterotransistor structure with asymmetric unit cell.

## Acknowledgments

This work has been supported by the GDR-E project "Semiconductor Sources and Detectors for Terahertz Frequencies" and PHC SAKURA "Research and Development of Terahertz Plasma-wave Transistors". The work at the Université Montpellier2 and CNRS was supported by the Region of Languedoc-Roussillon "Terahertz Platform". The work at the Kotelnikov Institute was supported by the Russian Foundation for Basic Research (Grant Nos. 08-02-92497 and 09-02-00395) and from the Russian Academy of Sciences Program "Fundamentals of Nanotechnologies and Nanomaterials". The work at the Tohoku University was supported by the SCOPE Program from the MIC, Japan, and by the Grant in Aid for Basic Research (S) from the JSPS, Japan. YM thanks the Ramon y Cajal Programme, Spain, for support.

## References

[1] Dyakonov M and Shur M 1996 *IEEE Trans. Electron Devices* **43** 380

[2] Popov V V, Shur M S, Tsymbalov G M and Fateev D V 2007 *Int. J. High Speed Electronics and Systems* **17** 557

[3] Otsuji T, Meziani Y M, Hanabe M, Ishibashi T, Uno T and E Sano 2006 *Appl. Phys. Lett.* **89** 263502

[4] Knap W, Kachorovskii V, Deng Y, Rumyantsev S, Lü J - Q , Gaska R, Shur M S, Simin G, Hu X, Asif Khan M, Saylor C. A. and Brunel L C 2002 *J. Appl. Phys.* **91** 9346

[5] Popov V V, Tsymbalov G M and Shur M S 2008 *Journal of Physics: Condensed Matter* **20** 384208

Journal of Physics: Conference Series **193** (2009) 012075 doi:10.1088/1742-6596/193/1/012075

# Room-temperature terahertz spectroscopy of optically excited plasma waves in HEMTs

**P. Nouvel, H. Marinchio, J. Torres, C. Palermo, L. Chusseau and L. Varani**

Institut d'Électronique du Sud, UMR 5214 CNRS, Université Montpellier Sud de France, 34095 Montpellier, France

**P. Shiktorov, E. Starikov, V. Gružinskis**

Semiconductor Physics Institute, Gostauto 11, LT-01108 Vilnius, Lithuania

E-mail: jeremi.torres@ies.univ-montp2.fr

**Abstract.** We report on systematic measurements of resonant plasma waves oscillations in several gate-length InGaAs HEMTs and compare them with numerical results from a specially developed model. A great concern of experiments has been to ensure that HEMTs were not subject to any spurious electronic oscillation that may interfere with the desired plasma-wave spectroscopy excited via a terahertz optical beating. The influence of geometrical HEMTs parameters such as gate-length or cap-layer length as well as biasing conditions is then explored extensively owing to many different devices. Plasma resonances up to the THz are observed.

## 1. Introduction

In a Field Effect Transistor (FET) channel of given gate-length $L_g$, excitation of plasma waves [1] is of particular interest because of the feasibility to control their oscillation frequency by varying the applied gate-voltage. For the submicron gate lengths, it has been shown [2] that the nonlinear properties of such waves can be exploited for selective tuneable terahertz (THz) detection and bridge the lack of cheap electronic detectors operating at room temperature. The THz detection was experimentally demonstrated only recently using high electron mobility transistors (HEMTs) at cryogenic temperature [3] and then it was also achieved at room temperature [4, 5]. Besides the direct THz illumination, an alternative approach involving the direct carrier excitation in the channel by means of an optical beam can activate plasma waves in transistor-like structures [6]. Recently, InGaAs on InP HEMTs were considered, thereby allowing for a direct carrier density modulation in the channel that is much more efficient in plasma wave activation [7]. As compared to direct THz detection, this kind of experiments have allowed an exact spectroscopic analysis of the two-dimensional electron gas (2DEG) undergoing plasma wave oscillations. Although the optical excitation of plasma waves in 2DEG of HEMTs has shown its efficiency and greatest advantages for spectroscopic applications, unsolved questions still exist on both the experimental and the theoretical sides that this paper is intended to answer.

## 2. Photoexcitation experiments

The key point of the set-up is the use of a prober to plot both static and dynamic characteristics of transistors (details on experimental configuration can be found in ref. [8]). The bias is applied to contacts using high frequency (HF) probes connected to bias-tees. The overall microwave bandpass of this system cover the 50 MHz - 40 GHz frequency range. In the experiments, HEMTs are connected in common source configuration and behave like HF single-stage transistor amplifiers.

The prescribed bias conditions required for plasma oscillations (i.e. $Z = 0$ at the source and $Z = \infty$ at the drain) make the resulting amplifier potentially unstable in the microwave domain because the input and output HF loads are uncontrolled and probably not appropriate [9]. We cured this problem simply by loading the HF outputs of the bias-tees by a 50 $\Omega$ impedance while keeping the dc output fed by voltage and current sources for the gate and the drain, respectively. The photoconductivity response, due to the generation of carriers by the difference-frequency term of the optical beating, is obtained by monitoring the modulation of the dc drain-to-source potential via a lock-in amplifier. The given photoresponse represents the actual measured photovoltage because the amplification of the lock-in amplifier and the gain of the low-noise pre-amplifier placed before the lock-in detection system are subtracted from the original measured signal.

Experiments were performed on HEMTs from InP technology schematically shown in Fig 1, with several gate-length values $L_g = 200$, 400, 800 and 1500 nm. Layers consist of an InP substrate, a 200 nm-thick $In_{0.52}Al_{0.48}As$ buffer, a 15nm $In_{0.7}Ga_{0.3}As$ channel, a 5nm-thick undoped $In_{0.52}Al_{0.48}As$ spacer, a silicon planar $\delta$-doping layer of $6 \times 10^{12}$ cm$^{-2}$, a 12 nm-thick $In_{0.52}Al_{0.48}As$ barrier layer, a 10nm-silicon-doped $In_{0.53}Ga_{0.47}As$ cap layer and a contact. The HEMT channel between source and drain contacts comprises three contacted regions. One is the gate itself with a $L_g$ length and two are covered by the cap layers with a $L_c$ length each. Finally two window regions of length $L_w$ each complete the source-drain length $L_{sd}$ (Fig. 1) given by $L_{sd} = L_g + 2L_c + 2L_w$.

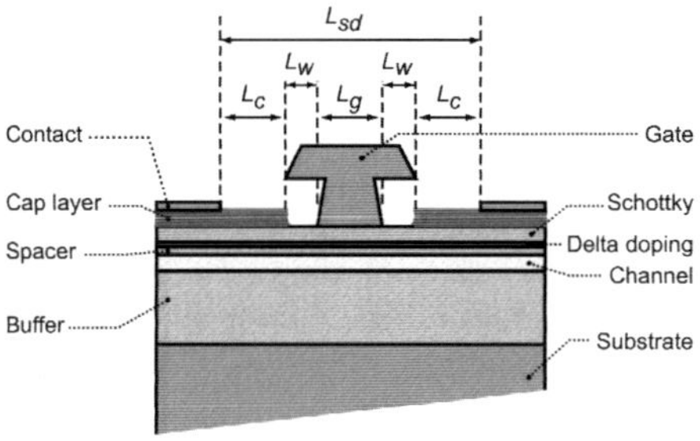


**Figure 1.** Scheme of the studied InGaAs HEMTs presenting the different layers lengths.

## 3. Cap-layers effect

Figure 2 displays the measured photoresponse versus optical beating frequency for three gate-length transistors ($L_g = 1500$ nm (a); $L_g = 400$ nm (b) and $L_g = 200$ nm (c)) at a fixed value of the swing voltage $V_0 = V_g - V_{th} \approx 170$ mV for three biasing points and at room temperature.

On each figure, several peaks are clearly observed. Similarly to previous findings [7], peaks corresponding to the fundamental and higher odd modes of plasma waves oscillations are clearly observed. The reason because only odd modes appear in the spectra come from the boundary condition (i.e. $Z = 0$ at the source and $Z = \infty$ at the drain), that form a $\lambda/4$ cavity for the plasma waves. For each curves, superimposed to the plasma wave peaks, a large and broadband frequency photoresponse appears. This continuous photoresponse comes from driving the transistors in their current saturation region and its value depends on the I-V characteristics of each transistor. Its value is around 1.4 mV for Fig. 2 (a), 19 mV for Fig. 2 (b) and 29 mV for Fig. 2 (c). The non-resonant detection, due to a mismatching between the plasma wave oscillation frequency in a cavity of given length and the optical beating frequency, is also included in the background continuum.

On one hand, the measured fundamental frequencies are 90 GHz for $L_g = 1500$ nm, 160 GHz for $L_g = 400$ nm and 200 GHz for $L_g = 200$ nm. On the other hand, the values predicted by the analytical formula [2] which only considers the gated channel, $f_0 = \sqrt{(}eV_0/m^*)/4L_g$ are approximatively 140 GHz, 540 GHz and 1000 GHz. We remark that by reducing the gate (while the total length $L_{sd}$ is kept constant), a strong disagreement between experimental and analytical results is observed. An important influence of the regions surrounding the gate is suggested. As a matter of fact, in a real transistor, it is important to investigate the effects of the cap layers on the plasma waves. In our devices, as described, the channel region under the gate cannot be considered as separated from the other parts of the transistor. The importance of the cap layer regions was theoretically predicted in [10] where it has been shown that increasing the length of these regions dramatically decreases the resonant frequencies. According to Ref. [10], the 2D electron gas in the sections of the channel strictly under the cap layers and under the gate exhibit similar collective dynamic behaviors.

Using the model detailed in [11], we have calculated, in Fig. 2 (bottom) the average photoresponses for the same transistors and swing voltages of Fig. 2 (top). A good agreement is found between numerical and experimental results as concerning the frequency of the plasma peaks as well as the amplitude of the photoresponse.

This confirms the fundamental role played by the cap layers in determining the dynamics of plasma waves in HEMT-like structures.

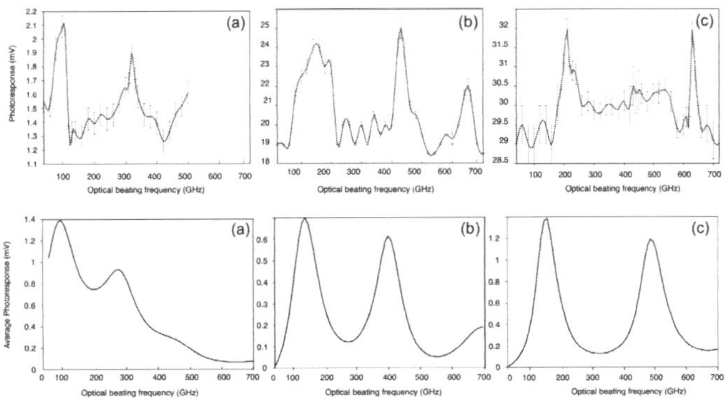


**Figure 2.** Room temperature measured (*top*) and calculated (*bottom*) photoresponses versus the optical beating frequency for three gate-length transistors at fixed swing voltage $V_0 \approx 170$ mV and three biasing point. *Experiments* : from (a) to (c) : $L_g = 1500$ nm : $V_d = 330$ mV ($I_d = 150$ $\mu$A); $L_g = 400$ nm : $V_d = 550$ mV ($I_d = 1.3$ mA); $Lg = 200$ nm : $V_d = 310$ mV ($I_d = 865$ $\mu$A). *Model* : $V_d = 350$ mV for different gate lengths: (a) $L_g = 1500$ nm, (b) $L_g = 400$ nm, (c) $L_g = 200$ nm. Error-bars are experimental data joined by eye guidelines.

## 4. Drain voltage effect

Figure 3 displays the measured photoresponses versus the optical beating frequency for two gate-length transistors $L_g = 200$ nm and $L_g = 400$ nm respectively, for a fixed value of the swing voltage $V_0 = 170$ mV and for several values of the applied $V_d$. It is remarkable that in the short gate transistor a photoresponse peak up to a frequency of about 1 THz is detected by our experimental set-up.

We have observed that, in general, the frequency of the plasma modes decreases with increasing $V_d$. For instance, this effect is particularly evident in Fig. 3, where $f_0$ varies from 250 to 190 GHz when $V_d$ varies from 190 to 500 mV. This frequency red-shift results from the increase of the electron drift velocity of the plasma wave as early discussed in ref. [11]. This effect predominates over the decrease of the effective gate length when the transistor is driven far into saturation regime which is expected to provide an opposite variation of the frequency [10]. Moreover the amplitude of the peaks practically increases exponentially with $V_d$. Such a behaviour can be due to: (i) the increase of the amplitude of the plasma oscillations by approaching the instability conditions because of the increase of the drift velocity, and/or (ii) the enhancement of these oscillations rectification process due to an increase of the devices non-linearities. Indeed, when the transistor is driven far-from-equilibrium and especially in saturation regime, important and highly non-uniform drift velocities are achieved. The non-linear convective term $v\,\partial v/\partial x$ [eq. (4) of ref. [8] ], neglected in the analytical calculation of the average response of optically or electronically excited plasma waves, becomes very important and dominates the rectification process. This explains why the DC response is exalted while no significant effect on the harmonic oscillations is noticed (see ref. [8] for more details) when the transistor is driven in its saturation regime.

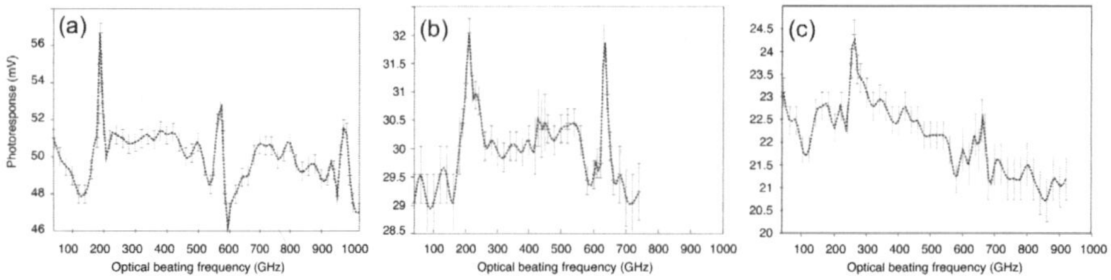


**Figure 3.** Photoresponses as functions of the optical beating frequency and versus the applied drain-source current for a 200 nm gate-length transistor with $V_0$ = 170 mV. (a) $V_d$ = 500 mV ($I_d$ = 970 $\mu$A), (b) $V_d$ = 310 mV ($I_d$ = 865 $\mu$A) and (c) $V_d$ = 190 mV ($I_d$ = 780 $\mu$A). Error-bars are experimental data joined by eye guidelines.

## 5. Conclusions

We have presented room temperature systematic experimental and theoretical investigations of plasma waves excited by an external THz optical beating in InGaAs HEMTs with different gate lengths and bias conditions.

A comparative analysis of experimental and numerical results of a home-made pseudo-2D HD model enabled us to demonstrate that HEMT contacts cannot be interpreted as point-like, and their planar geometry must be taken into account. The response spectrum of plasma waves at the beating frequency is thus determined by a resonant cavity which includes the contribution of the cap-layers.

We have also observed that the frequency of the plasma modes systematically decreases with increasing drain voltage (current). The theoretical model clarifies that the dominant effect on this frequency shift when our devices are biased deeply into saturation regime, is associated with the increase of the drift velocity in the channel.

[1] A. Chaplik, "Possible crystallization of charg-carriers in low-density inversion layers," in *Zh. Eksp. Teor. Fiz. (1972)*, vol. 62, no. 2. Sov. Phys. JETP 35, 395, 1972, p. 746.

[2] M. Dyakonov and M. Shur, "Shallow water analogy for a ballistic field effect transistor: New mechanism of plasma wave generation by dc current," *Phys. Rev. Lett.*, vol. 71, no. 15, pp. 2465–2468, Oct 1993.

[3] W. Knap, Y. Deng, S. Rumyantsev, and M. S. Shur, "Resonant detection of subterahertz and terahertz radiation by plasma waves in submicron field-effect transistors," *Appl. Phys. Lett.*, vol. 81, p. 4637, 2002.

[4] F. Teppe, D. Veksler, V. Y. Kachorovski, A. P. Dmitriev, X. Xie, X.-C. Zhang, S. Rumyantsev, W. Knap, and M. S. Shur, "Plasma wave resonant detection of femtosecond pulsed terahertz radiation by a nanometer field-effect transistor," *Appl. Phys. Lett.*, vol. 87, no. 2, p. 022102, 2005.

[5] A. M. Hashim, S. Kasai, and H. Hideki, "Observation of first and third harmonic responses in two-dimensional algaas/gaas hemt devices due to plasma wave interaction," *Superlattices and Microstructures*, vol. 44, pp. 754–760, 2008.

[6] T. Otsuji, M. Hanabe, and O. Ogawara, "Terahertz plasma wave resonance of two-dimensional electrons in ingap/ingaas/gaas high-electron-mobility transistors," *Appl. Phys. Lett.*, vol. 85, no. 11, pp. 2119–2121, 2004.

[7] J. Torres, P. Nouvel, A. Akwoue-Ondo, L. Chusseau, F. Teppe, A. Shchepetov, and S. Bollaert, "Tunable plasma wave resonant detection of optical beating in high electron mobility transistor," *Appl. Phys. Lett.*, vol. 89, no. 20, p. 201101, 2006.

[8] P. Nouvel, H. Marinchio, J. Torres, C. Palermo, L. Chusseau, D. Gasquet, L. Varani, P. Shiktorov, E. Starikov, and V. Gružinskis, "Terahertz spectroscopy of plasma waves in HEMTs," *J. Appl. Phys.*, vol. 106, p. 013717, 2009.

[9] D. M. Pozar, *Microwave Engineering*, 2nd ed. New York: John Wiley & Sons, 1998.

[10] V. Ryzhii, A. Satou, W. Knap, and M. S. Shur, "Plasma oscillations in high-electron-mobility transistors with recessed gate," *J. Appl. Phys.*, vol. 99, no. 8, p. 084507, 2006.

[11] H. Marinchio, G. Sabatini, C. Palermo, J. Pousset, J. Torres, L. Chusseau, L. Varani, P. Shiktorov, E. Starikov, and V. Gruzinskis, "Hydrodynamic modeling of optically excited terahertz plasma oscillations in nanometric field effect transistors," *Appl. Phys. Lett.*, vol. 94, no. 19, p. 192109, 2009.

Journal of Physics: Conference Series **193** (2009) 012076 doi:10.1088/1742-6596/193/1/012076

# Electronic, optical and thermal excitation of plasma waves in HEMTs: a theoretical study

**Hugues Marinchio, Giulio Sabatini, Christophe Palermo, Jérémi Torres, Laurent Chusseau and Luca Varani**

Institut d'Électronique du Sud (CNRS UMR 5214), Université Montpellier 2, 34095 Montpellier, France

**Pavel Shiktorov, Evguenij Starikov, Viktor Gružinskis**

Semiconductor Physics Institute, Gostauto 11, LT-01108 Vilnius, Lithuania

E-mail: hugues.marinchio@ies.univ-montp2.fr

**Abstract.** We investigate the influence of collective plasma modes in a field-effect transistor channel under different excitations and biasing conditions. First, we study the case of a device externally-excited by a harmonic optical beating or an electronic excitation and current-driven at the drain. The harmonic and continuous responses of the transistor are calculated using a pseudo-two-dimensional hydrodynamic approach. They show sharp resonances related to the first odd plasma modes, whose frequencies and amplitudes can be modified by playing on the drain bias. Then, through the generalized impedance-field method we calculate also the spectral density of drain voltage fluctuations in the absence of external excitations. Also these noise spectra exhibit peaks corresponding to the odd plasma modes.

## 1. Introduction

Two-dimensional plasma waves in quasi-ballistic nanometric transistors are considered a very promising physical phenomenon which can be exploited for the conception of terahertz (THz) devices operating at room temperature. Recent experiments have shown the possibility of exciting plasma modes in HEMTs by means of a THz radiation, useful for detection applications [1], or by means of a beating photoexcitation which could provide a possible THz emitter [2]. In addition, it has been shown that the high-frequency noise spectra of HEMTs contain excess noise related to thermal excitation of plasma modes [3].

We propose to investigate theoretically these three situations by using a numerical pseudo-two-dimensional hydrodynamic (HD) approach. This model allows us to simulate nanometric field-effect transistors from ohmic to saturation conditions by taking into account nonequilibrium conditions, presence of access regions, etc. [4]. The two cases of external excitation (electronic and optical) of plasma modes are studied by calculating the deterministic frequency responses of the transistor working in different operation modes. On the other hand, the thermal excitation is treated in the framework of the impedance-field method based on noise sources calculated by Monte-Carlo simulations [3].

Thus, this article constitutes a synthesis on the influence of plasma resonances on the high-frequency transistors operation with or without an external excitation, and a very useful tool to achieve a detailed comprehension of the plasma mechanism.

## 2. Numerical model

We couple the one-dimensional HD equations relative to the electron density $n(x,t)$, the mean velocity $v(x,t)$ and the mean energy $\epsilon(x,t)$, with an approximation of the Poisson equation described in [3, 4]:

$$\begin{cases} \dfrac{\partial n}{\partial t} = \dfrac{\partial (nv)}{\partial x} + G + \delta G \\ \dfrac{\partial v}{\partial t} = -v\dfrac{\partial v}{\partial x} - \dfrac{eE}{m^*} - \dfrac{1}{n}\dfrac{\partial \left(\overline{\delta v^2}\, n\right)}{\partial x} - v\nu_v + \tilde{f} \\ \dfrac{\partial \epsilon}{\partial t} = -v\dfrac{\partial \epsilon}{\partial x} - eEv - \dfrac{1}{n}\dfrac{\partial \left(\overline{\delta v \delta \epsilon n}\right)}{\partial x} - (\epsilon - \epsilon_0)\nu_\epsilon \\ \varepsilon_c \dfrac{\partial^2 V}{\partial x^2} + \varepsilon_s \dfrac{V_g + \delta V_g - V}{d\delta} = e\left(n - N'_D\right) \end{cases} \tag{1}$$

Here $e$ is the elementary charge and $E$ the $x$ electric field component. $\epsilon_0$ is the equilibrium mean energy. $\varepsilon_s$ and $\varepsilon_c$ are the dielectric constants of, respectively, the Schottky layer and the channel. $d$ is the gate-to-channel distance and $\delta$ the channel thickness. $N'_D$ is the effective donor concentration in the channel. The velocity and energy relaxation rates $\nu_v$, $\nu_\epsilon$, the electron effective mass $m^*$, the velocity variance $\overline{\delta v^2}$ and the energy-velocity covariance $\overline{\delta v \delta \epsilon}$ depend on the local mean energy and they were calculated by a Monte-Carlo simulation of InGaAs bulk material at room temperature.

Three source terms can be included in the equations: (i) a beating optical excitation through the term $G + \delta G = G_0[1 + \cos(2\pi f t)]$ [4] ; (ii) an electronic excitation on the gate which may describe the coupling between a THz radiation and an antenna connected to the gate [1], $\delta V_g = \Delta V_g \cos(2\pi f t)$; (iii) a local source of thermal fluctuations in the channel $\tilde{f}$ , described by the spectral density $S_{ff} = 4k_b T \nu_v / m^*$ obtained by Monte Carlo simulations [3].

## 3. External excitation of plasma modes

It has been experimentally shown that the plasma resonances can be efficiently excited by external stimulations [1, 2]. In this section, we study the resonant response of a transistor channel to an optical beating and an electronic oscillating voltage on the gate.

We simulate a $In_{0.53}Ga_{0.47}As$/InAlAs HEMT, with $d = 22$ nm and $\delta = 10$ nm, composed of a $L_g = 100$ nm long gated part ($N'_D = 1 \times 10^{18}$ cm$^{-3}$) surrounded by two 50-nm-long ungated sections ($N'_D = 5 \times 10^{18}$ cm$^{-3}$). The threshold voltage $V_{th}$ calculated by the Poisson equation in (1) gives a value $V_{th} = -\frac{eN'_D d\delta}{\varepsilon_s} = -300$ mV. By taking $V_g = 0$ mV, the swing voltage is $V_0 = V_g - V_{th} = 300$ mV.

In order to be in the optimal static conditions for plasma wave excitation, the source voltage and the drain current are fixed. Under this constant current operation, the response of the transistor to an external oscillating excitation can be described by the drain voltage harmonic and continuous variations, $\delta V(f) \cos(2\pi f t)$ and $\overline{\Delta V(f)}$, respectively [4].

### *3.1. Optical excitation*

In this case,$\Delta$ $V_g = 0$, $\tilde{f} = 0$ and $G_0$ is taken equal to $1 \times 10^{27}$ cm$^{-3}$/s which corresponds to typical conditions of photomixing experiments [5]. The harmonic and average photoresponses are reported on Fig. 1. For both quantities, two resonances corresponding to the first plasma modes are observed at THz frequencies. At increasing $V_{DS}$, the resonance frequency of the peaks decreases significantly. This can be explained by the global growth of the mean velocity in the channel leading to an increase of the plasma transit time along the channel. Moreover, we observe that the amplitude of the resonant photoresponse can be maximized by playing on the drain voltage. As discussed in [4], this effect is mainly due to the optimization of the wave increment of the forced oscillator constituted by the photoexcited channel.

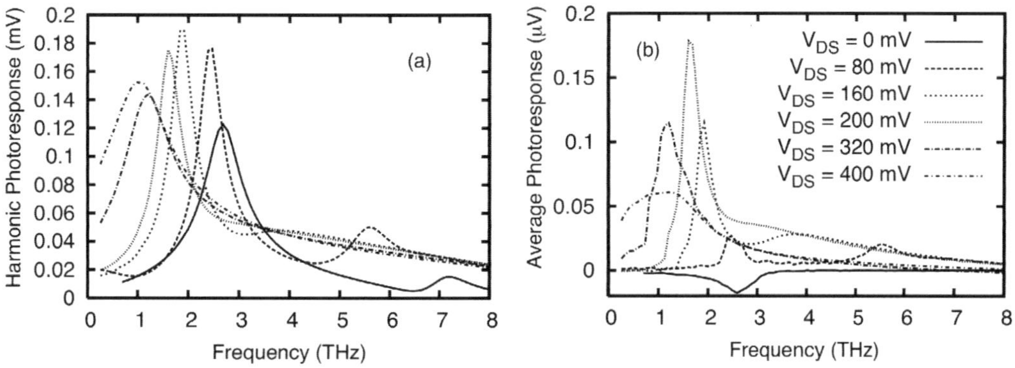


**Figure 1.** Calculated harmonic (a) and average (b) photoresponses under optical excitation as functions of the beating frequency and for the reported $V_{DS}$.

### *3.2. Electronic excitation*

In this subsection, $G_0 = 0$, $\tilde{f} = 0$ while$\Delta$ $V_g = 1$ mV. Figure 2 shows the frequency-dependent harmonic and average responses of the transistor to the harmonic oscillating gate voltage for different $V_{DS}$. The behavior is very similar to that observed in the case of optically excited channels: the same resonance frequencies are observed at the different drain voltages and the variation of resonance qualities with the drain bias is identical. Even if the carrier modulation in the channel is induced by two very different physical mechanisms, the excited plasma modes, whose properties only depends on static parameters (gate length, gate and drain biases), are similar.

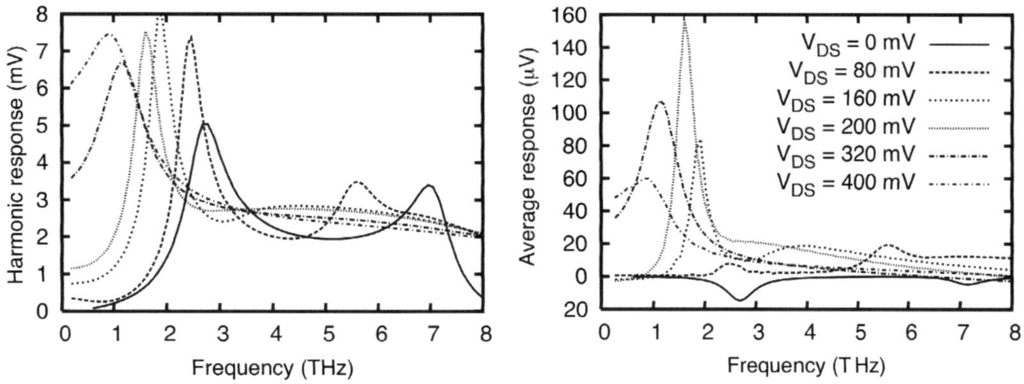


**Figure 2.** Calculated harmonic (a) and continuous (b) responses under electronic excitation as functions of the frequency of electronic excitation and for the reported $V_{DS}$.

The main difference between the two kinds of excitation stands into the fact that high order modes are better excited by an electronic excitation, and this especially at low voltages. Indeed, an optical illumination controls the carrier generation velocity while an oscillating gate-to-channel voltage is almost proportional to the electron concentration [see Eq.(1) for low $\partial^2 V/\partial x^2$]. Thus, the oscillating optical excitation is less efficient at high-frequency than the electronic stimulation.

## 4. Thermal excitation of plasma modes

In this section, we consider a similar HEMT without any external excitation. We solve the first two hydrodynamic equations taking $\nu_v = 3 \times 10^{12}$ s$^{-2}$, $m^* = 0.04\ m_0$, $G_0 = 0$ and$\Delta\ V_g = 0$.

The drain current is fixed at zero value and we calculate the spectral density of voltage oscillations at the drain in the framework of the generalized impedance-field method [3]. The result is reported on Fig. 3.

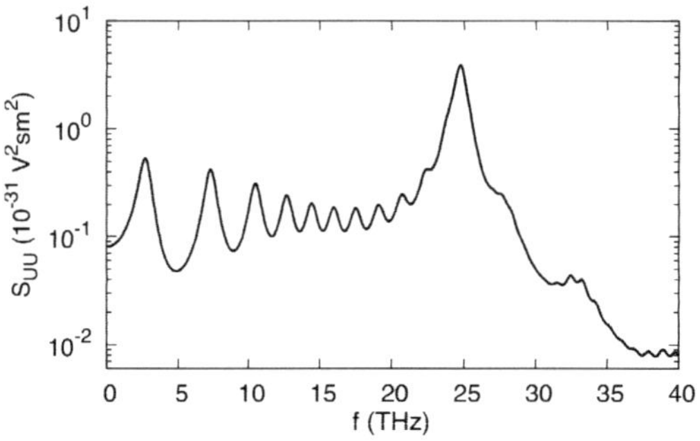


**Figure 3.** Spectral density of voltage fluctuations at the drain terminal as a function of frequency.

Since the noise sources are considered white and not correlated in space, they can excite the different odd plasma modes whatever their wavelength or frequencies. As a consequence, sharp peaks related to the odd plasma modes, already seen in Fig. 1 and 2, appear in the high-frequency noise spectra. The larger peak around 25 THz can be attributed to the 3D plasma resonance of the access region.

## 5. Conclusion

We studied the most common cases where plasma wave resonances manifest themselves in field effect transistor channels. This theoretical investigation is carried out by means of an original hydrodynamic pseudo-2D numerical model. We demonstrate that the first plasma resonances can be efficiently excited by an optical or electronic excitation. In this latter case, the high-order harmonics are significantly more stimulated. In both cases, the first resonance can be optimized and tuned by playing on the static bias. Finally, even in the absence of an external excitation, these plasma resonances are evidenced in the noise spectrum as an excess contribution.

## References

[1] Teppe F, Knap W, Veksler D, Shur M S, Dmitriev A P, Kachorovskii V Y and Rumyantsev S 2005 *Appl. Phys. Lett.* **87** 052107

[2] Otsuji T, Hanabe M and Ogawara O 2004 *Appl. Phys. Lett.* **85** 2119–2121

[3] Shiktorov P, Starikov E, Gružinskis V, Varani L, Sabatini G, Marinchio H and Reggiani L 2009 *J. of Stat. Mech.* **01** 01047

[4] Marinchio H, Sabatini G, Pousset J, Palermo C, Torres J, Chusseau L, Varani L, Starikov E and Shiktorov P 2009 *Appl. Phys. Lett.* **19** 192109

[5] Torres J, Marinchio H, Nouvel P, Sabatini G, Palermo C, Varani L, Chusseau L, Shiktorov P, Starikov E and Gruzhinskis V 2008 *IEEE J. Sel. Top. Quantum Electron.* **14** 491–497

Journal of Physics: Conference Series **193** (2009) 012077 doi:10.1088/1742-6596/193/1/012077

# Terahertz GaAs/AlGaAs- and InGaAs-based bow-tie diodes: spectral features and applications for imaging

**I Kašalynas[1], D Seliuta[1], V Tamošiūnas[1], J Macutkevič[1], S Balakauskas[1], G Valušis[1] and K Köhler[2]**

[1] Semiconductor Physics Institute, A. Goštauto 11, LT-01108 Vilnius, Lithuania

[2] Fraunhofer-Institut für Angewandte Festkörperphysik, Tullastr. 72, DE-79108 Freiburg, Germany

E-mail: irmantak@ktl.mii.lt

**Abstract**. Spectral features of broadband bow-tie diodes fabricated of InGaAs layers and modulation doped GaAs/AlGaAs structures are discussed. It is shown that sensitivity of both GaAs/AlGaAs- and InGaAs-based sensors at room temperature weakly depends on frequency in the range 0.01-1.0 THz ; the sensitivity is found to be 0.3 V/W and 10 V/W, respectively. The InGaAs bow tie diodes coupled with silicon lenses are demonstrated in terahertz imaging application, their feasibility for real-time THz imaging is considered as well.

## 1. Introduction

Compact sensors operating at room temperature (RT) and suitable for material inspection, security and medical aims still remains an issue in applications of terahertz (THz) frequencies. A particular attention gains broadband and fast sensors having potential in spectroscopy and real-time imaging. To provide operation free from temperature limitations, principle behind device performance cannot be related to any intra-subband transitions. One of the possible ways is to employ plasma waves that can be excited in semiconductor field effect transistors with nanometric gate [1]. It is shown recently that spectrally broad photoresponse signal excited by the external radiation of 0.6 THz in commercial Fujitsu FHX05X GaAs-based high electron mobility transistors with a gate length of 250 nm can successfully be used to record images at RT [2]. In order to achieve higher sensitivity and improve noise properties, and thus, imaging quality, silicon-based field effect transistors [3] seems to be the most suitable choice, in particular, as sensing elements in THz cameras [4].

In this communication, we consider another possible detection approach – asymmetric bow-tie diode application – in realization of compact RT THz imaging. We discuss spectral features of the bow-tie diode based on InGaAs layers and modulation doped GaAs/AlGaAs structures and show THz images recorded at room temperature. The diode uses concept of bow-tie antenna with structurally broken symmetry [5]: one of the two semiconductor leaves is metalized in order to couple incident radiation into the second one, where electrons are heated non-uniformly by the radiation. Consequently, the voltage signal across the leaves is induced without application of a bias voltage. A set of structures of different design has been grown by molecular beam epitaxy technique; for processing of diodes with apex ranging from 12 μm down to 800 nm we have used conventional optical and electron-beam lithography plus dry etching.

## 2. Samples and measurement techniques

The GaAs/AlGaAs structures consist of a 20-nm-thick GaAs cap layer, a Si-doped ($1\times10^{18}$ $cm^{-3}$) 80 nm $Al_{0.25}Ga_{0.75}As$ layer, an undoped $Al_{0.25}Ga_{0.75}As$ spacer of 45 nm, and 1 μm of undoped GaAs. Twenty surface-smoothing periods of undoped 3 nm-AlGaAs/1.5 nm-GaAs/0.5 μm-GaAs layers on top of the substrate are not depicted. The sheet electron density and the mobility at room temperature are $5.5\times10^{11}$ $cm^{-2}$ and 4700 $cm^2$/V·s, respectively. More detailed description can be found in Ref. 5.

The core constituent of InGaAs-based device is $In_{0.54}Ga_{0.46}As$ layer of 534 nm thickness grown on InP (001) substrate with one InAs monolayer in between. The electron concentration and mobility at room temperature are about $2\times10^{15}$ $cm^{-3}$ and 13 300 $cm^2$/V·s, respectively. More detailed description can be found in Ref. 6.

A pulsed microwave electric field at 10 GHz and 35 GHz with pulse duration of 2-20 μs and a repetition rate of 40 Hz was delivered by magnetron generators. The THz radiation source (from 0.584 THz up to 2.52 THz) was an optically pumped molecular laser operating in a continuous wave regime.

## 3. Broadband sensing at room temperature

Frequency dependence of the voltage sensitivity of both structures is shown in Fig. 1. As one can see the sensitivity at RT of both GaAs/AlGaAs- and InGaAs-based sensors weakly depends on frequency from 10 GHz up to 1 THz. The experimental points are well fitted by phenomenological approach with small deviations which we assign to slightly varying radiation coupling conditions with frequency. It is seen that the sensitivity of the InGaAs-based diodes is more than order of magnitude higher (close to 10 V/W) in comparison to the one based on GaAs/AlGaAs (close to 0.3 V/W). Therefore the first type diodes are much better suited for THz imaging aims. Within the range of 1–2.52 THz, the sensitivity of both type diodes drops down faster that the phenomenological approach predicts. It can be explained by a weaker coupling of the incident radiation as it is illustrated by modeling using finite-difference time-domain method [5,6].

One can note that with the decrease of the apex size down to 2 μm, for instance, in GaAs/AlGaAs structures, allows increase the sensitivity up to 2.5 V/W, which is nearly one order of magnitude higher compared to that of diodes with 12 μm apex size (0.32 V/W). It is worth noting, however, that further decrease of the apex size down to 1 μm or 800 nm gives no increase in the sensitivity. We attribute this effect with a weaker coupling of the incident radiation. In InGaAs-based devices the apex size decrease also increases the sensitivity, but the frequency dependence becomes distorted by bow-tie antenna-shape and geometry related resonances [7].

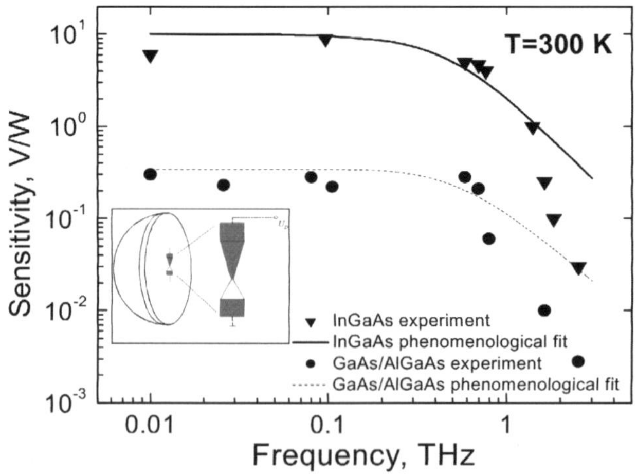


**Figure 1.** Sensitivity as a function of frequency of InGaAs and GaAs/AlGaAs-based bow tie diodes with apex size 12 μm. Points denote experimental data, lines – the phenomenological fit. Inset shows the shape of the diode attached to the silicon lens. Grey areas – parts covered by metal, white area – GaAs/AlGaAs or InGaAs layer.

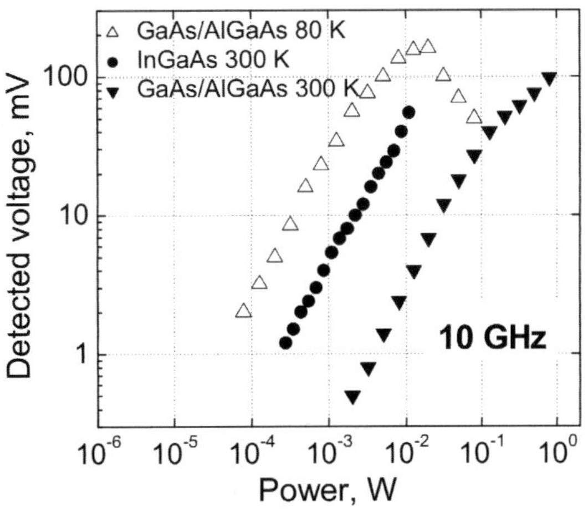


**Figure 2.** Voltage-power characteristics of GaAs/AlGaAs and InGaAs bow-tie diodes with 12 μm apex placed in 10 GHz frequency radiation at room and liquid nitrogen temperatures.

The dynamic range is illustrated in Fig. 2, where voltage-power characteristics of GaAs/AlGaAs and InGaAs bow-tie diodes with 12 μm apex placed in 10 GHz frequency radiation at RT and liquid nitrogen temperatures are given. As it is seen the dynamic range exceeds more than two orders of magnitude. The sensitivity of InGaAs-based diodes is around 6 V/W. One can note that at low temperatures GaAs/AlGaAs exhibits sensitivity of 40 V/W. The nonmonotonical behaviour observed above 10 mW can be attributed to the real-space transfer effects [8].

## 4. Application of the bow-tie diodes for terahertz imaging

To demonstrate THz imaging we have chosen InGaAs bow tie diodes coupled with silicon lenses. The object under test was a plastic capsule with pharmaceutical powder. The transmission spectra of the plastic capsule alone and filled with pharmaceutical powder are recorded using coherent time domain spectroscopy [9]. The results are shown in Fig. 3a. As it is seen, the medicine powder absorption is clearly distinguished in the frequency range of 0.25-1.5 THz. Thus, plenty of optically pumped molecular THz laser lines are suitable for THz-ray imaging of chosen pharmaceutical powder.

The THz recorded image in a transmission mode of the plastic capsule with pharmaceutical powder (its photo is presented as inset) enclosed in the paper envelope is presented in Fig. 3b. The image of the object mounted onto computer-driven X-Y stage was obtained pixel-by-pixel scanning it in a focal plane of the THz beam spot with radius of 0.95 mm (more parameters of the images are listed in the Fig. 3 caption). As one can see, the bow-tie diode clearly resolves concealed object in the THz light, it also reveals the powder level in the capsule. Shadowed pattern due to reflection between the envelope sheets is also clearly visible. Note that the intensity scale is logarithmic in order to resolve structure details of differently absorbing parts. The signal-to-noise ratio is more than 3000.

For real-time THz imaging it is of particular importance to know response time and noise equivalent power (NEP) of the device. The response time less than 7 ns, responsivity – 0.1 mA/W, and NEP – 5.8 nW/√Hz have been recently shown [10]. One deserves mentioning that for real-time imaging (for instance, at 50 frames/s capture rate) the InGaAs-based sensor demonstrates the dynamic range higher than 50 dB if a maximum not-saturating incident power of about 10 mW is applied.

As for comparison of bow-tie diodes with field effect nanotransistors one can note that the bow-tie diodes production is technologically less demanding, the diodes are not sensitive to the static electric charge and are more robust to overloads. More detailed comparison of parameters between bow-tie diodes and field effect transistors is given in Ref. 10.

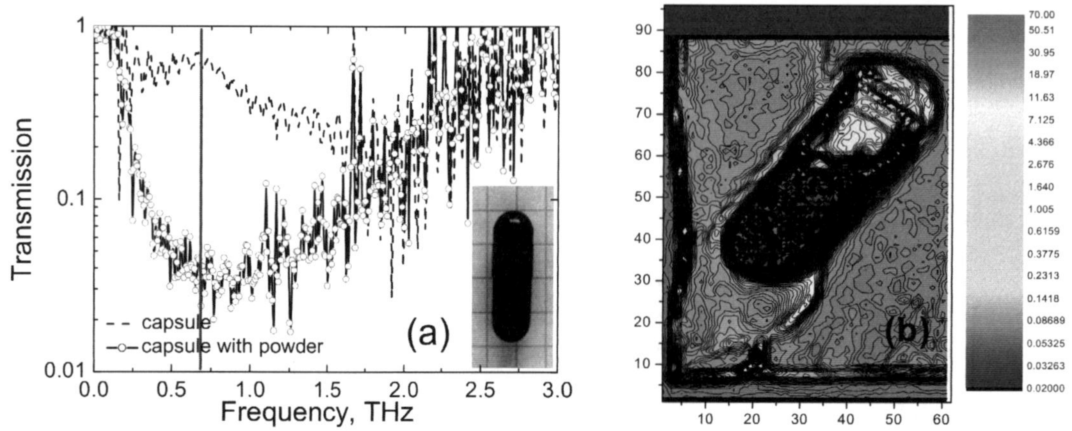


**Figure 3 (colour online) (a)** Terahertz transmission spectra of plastic capsule (photo of the capsule on 5 mm squared background is given in the inset) half-filled with pharmaceutical powder and its image **(b)** at 0.693 THz (in the spectral plot it is denoted by the vertical blue line). The capsule is enclosed in the paper envelope. Horizontal and vertical scales are in pixels, intensity is normalized to the maximal value and plotted in logarithmic scale. Imaging parameters: 61×85 pixels, each of 325 μm size, lock-in integration time - 50 ms. Note that the signal-to-noise ratio is about 3500.

**Acknowledgments**
We highly appreciate Ramūnas Adomavičius for the kind help in experiments of coherent terahertz time-domain spectroscopy. The research at the Semiconductor Physics Institute was conducted under the topic "Terahertz optoelectronics: devices and applications" (No. 179J), and, in part, supported by the Lithuanian State Science and Science Foundation under contract C-07004/C-17/2009.

**References**

[1] Knap W, Deng Y, Rumyantsev S, Lü J-Q, Shur M S, Saylor C A, and Brunel L C 2002 *Appl. Phys. Lett.* **80** 3433
[2] Lisauskas A, von Spiegel W, Boubanga-Tombet S, El Fatimy A, Coquillat D, Teppe F, Dyakonova N, Knap W and Roskos H G *Electron. Lett.* 2008 **44** 408
[3] Tauk R, Teppe F, Boubanga S, Coquillat D, Knap W, Meziani Y M, Gallon C, Boeuf F, Skotnicki T, Fenouillet-Beranger C, Maude D K, Rumyantsev S, and Shur M S 2006 *Appl. Phys. Lett.* **89** 253511
[4] Lisauskas A, Pfeiffer U, Öjefors E, Haring Bolivar P, Glaab D, and Roskos H G 2009 *J. Appl. Phys.* **105** 114511
[5] Seliuta D, Širmulis E, Tamošiūnas V, Balakauskas S, Ašmontas S, Sužiedėlis A, Gradauskas J, Valušis G, Lisauskas A, Roskos H G, and Köhler K 2004 *Electron. Lett.* **40** 631
[6] Seliuta D, Kašalynas I, Tamošiūnas V, Balakauskas S, Martūnas Z, Ašmontas S, Valušis G, Lisauskas A, Roskos H G , and Köhler K 2006 *Electron. Lett.* **42** 825
[7] Tamošiūnas V and Minkevičius L 2009 *unpublished*
[8] Juozapavičius A, Ardaravičius L, Sužiedėlis A, Kozič A, Gradauskas J, Kundrotas J, Seliuta D, Širmulis E, Ašmontas S, Valušis G, Roskos H G and Köhler K 2004 *Sem. Sci. Technol.*, **19** S436
[9] Grischkowsky D, Keiding S, van Exter M and Fattinger Ch 1990 *J. Opt. Soc. Am. B* **7** 2006
[10] Kašalynas I, Seliuta D, Simniškis R, Tamošiūnas V, Köhler K and Valušis G 2009 *Electron. Lett.* **45** 833

Journal of Physics: Conference Series **193** (2009) 012078 doi:10.1088/1742-6596/193/1/012078

# Nonlinear carrier transport within gigahertz-terahertz frequencies in spatially non-uniform InSb

**I Kašalynas, L Subačius, D Seliuta and G Valušis**

Semiconductor Physics Institute, A. Goštauto 11, LT-01108 Vilnius, Lithuania

E-mail: irmantak@ktl.mii.lt

**Abstract**. Influence of a dc electric field, microwave and terahertz (up to 1.63 THz) excitation on the carrier transport in moderately compensated *n*-InSb<Cr> containing graded *n*/*n*+ junction is investigated. Hot electron thermo-electromotive force and field dependences of the dc-conductivity were measured within 10-110 K. Peculiarities of electron gas heating and low-temperature electron transport can be attributed to the potential fluctuation effects due to random spatial distribution of impurity density in high-resistivity InSb<Cr> crystal.

## 1. Introduction

Non-equilibrium carrier transport phenomena caused by external illumination in *n*-type indium antimonide, can serve in many applications, for instance, in hot-electron bolometers, widely employed in a terahertz sensing systems. As a rule, a key-component of such detector is uniform InSb, which exhibits high carrier mobility at low temperatures allowing thus effective electron gas heating above the lattice temperature.

In a present communication, we concentrated our attention into compensated InSb containing graded *n*/*n*+ junction along the crystal investigating an influence of a gigahertz (GHz) and terahertz (THz) excitation on the carrier nonlinear transport. The compensated InSb due to the electron gas cooling effect in strong electric fields [1,2] exhibits low-level noise properties [3] and thus can be an attractive option in optimizing sensitivity of high-frequency devices. The structures with electron density of $10^{12}$–$10^{13}$ cm$^{-3}$ were studied measuring dc-current-voltage characteristics, thermo-electromotive force of hot carriers at different temperatures and its exposition characteristic at GHz- and THz-frequency excitation. The experiment were carried out within the temperature range of 10-110 K using magnetron as a source of GHz frequency radiation and an optically-pumped molecular laser as THz emitter.

## 2. The sample and experimental set-up

### 2.1. Compensated *n*-type indium antimonide

The sample was cut from single crystals of *n*-type indium antimonide doped with Cr and Zn. The crystal carrier mobility is of about $3.5\times10^5$ cm$^2$/Vs at 78 K. A dumb-bell shape of the sample was made by polishing the crystal. After all sample preparation processes the length *l*, width *w* and height *h* of active sample part were of about 3.3 mm, 0.4 mm, and 0.4 mm, respectively. Ohmic contacts were prepared using In + 2%Te alloy. The sample was then soldered to a copper frame and placed in a closed cycle cryostat having an optical port to admit the THz radiation. The sample temperature was controlled by a thermocouple at the point of sample attachment to the frame.

The distribution of carrier density along the sample was monitored using four contacts attached stationary to the sample at different temperatures (up to 4 K). More detailed carrier distribution was probed using a mobile probe at liquid nitrogen temperature.

### 2.2. An experimental set-up

An optically-pumped molecular laser was used as a THz radiation source. The laser was tuned to emit at 0.762 THz or 1.63 THz frequency. The spherical mirror was used to focus THz radiation to the sample placed behind the high pressure polyethylene window. The spot size diameter of the laser beam was of about 2 mm. The laser beam has linear polarization and the orientation of the electric field vector was along the length axis of the sample. For experiments at 9.4 GHz frequency the same sample was placed in the rectangular waveguide perpendicularly to its broad wall.

The sample dc-conductance is measured using a digitally controlled source-meter applying voltage below 1 V and dc-pulse technique generating up to 200 V, i.e. providing the electric fields higher than the one needed to start impact ionization process in InSb. The signal of the thermo-electromotive force (thermo-*emf)* was recorded by means of 350 MHz bandwidth oscilloscope and the lock-in amplifier.

## 3. Experimental Results and Discussions

The conductance of the sample measured in wide temperatures range is shown in Fig. 1. The conductance change reflects different processes related with change in carrier density and the mobility. The area of interest is in the range 50-10 K, where the sample dc-conductivity governed by mobility decreases with temperature due to grown electrons scattering on ionized impurities. Also, it is worth noting that the impact of potential fluctuation due to random inhomogeneities in the impurity density increases [4].

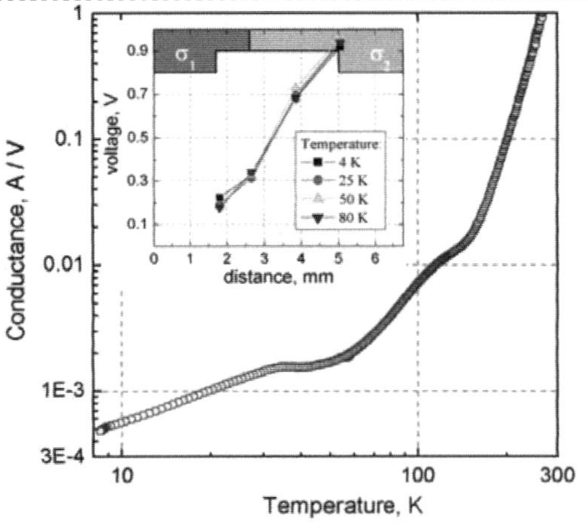


**Figure 1.** The sample conductance dependence on temperature at dc-current of about 10 μA. Inset: voltage distribution along the sample measured at fixed points at four temperatures under dc-voltage of about 1 V. A two-grey-tone outline drawing illustrates the contour of the sample with different $\sigma_1$ and $\sigma_2$ conductivity regions.

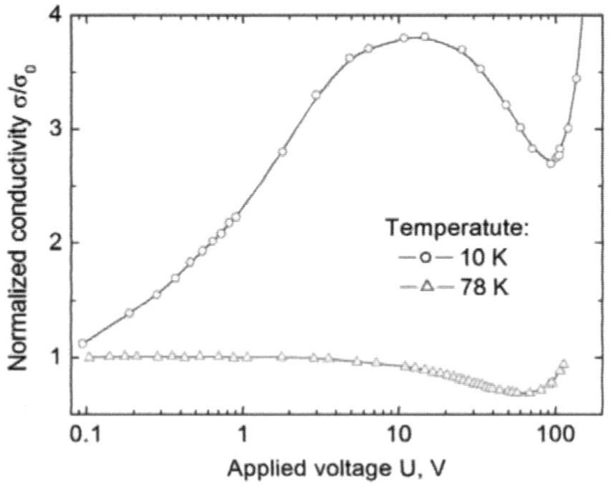


**Figure 2.** Variation of the sample conductivity with applied dc-voltage at 10 K and 78 K temperatures. The dc-conductivity is normalized to its value obtained at low bias voltage.

The amplitude of the potential barrier arising due to *n*/*n*+ junction presence is estimated at 78 K measuring the voltage distribution along the sample. It was found that the dc-conductivity of differently doped sample parts schematically shown in the inset of Fig. 1 is: $\sigma_1 = 1.7$ S/cm and $\sigma_2 = 0.83$ S/cm. The height of potential barrier is determined using the formula:

$$V_k = \frac{kT}{e}\ln\left(\frac{\sigma_1}{\sigma_2}\right), \qquad (1)$$

*k* is Boltzmann constant, *T* is temperature. The value of the potential barrier of about 4.7 meV is found. The dependence of the potential barrier on temperature is tested up to liquid helium temperature. The data is shown in the inset of Fig. 1. It is seen that the voltage distribution along the sample remains unchanged varying the temperature in 4-80 K. Therefore, it allows to assume that the ratio between $\sigma_1$ and $\sigma_2$ values remains the same within all above indicated temperatures.

The electric field dependence of sample dc-conductivity studied at different temperatures and electric field regions are shown in Fig. 2 (strong fields). At 10 K temperature, the conductivity increases because of increase in the mobility [4]. Later, within the voltage range of 10-80 V, the conductivity decreases due to electrons interaction with optical phonon and, finally, conductivity starts to increase once again due to impact ionization process. One can note that at temperature of 78 K and voltage below of 100 V, the conductivity modulation with electric field is weaker due to strengthened electron-electron interaction.

The sample conductance under condition of low dc-fields was studied at different temperatures. The result is shown in Fig. 3. It is seen that the conductance change with the electric field becomes evident if the sample temperature is reduced to 30 K. Further lowering of the temperature leads to more pronounced nonlinear current-voltage characteristic showing up at lower dc-voltage values.

The results of thermo-*emf* signal on temperature are shown in Fig. 4. The signal sign corresponds to the *emf* induced by hot electron effects. Within the temperature range of 20-50 K the signal change is almost linear in a log-log plot scale and its value is changed nearly by 3 orders of magnitude. It is worth indicating that 2 mW power of the THz laser generates thermo-*emf* signal value of about 0.6 mV at 10 K. Although the sample temperature was varied from 10 K up to 120 K, however, the *emf* signal induced with THz frequency radiation is observed only at temperatures below 50 K.

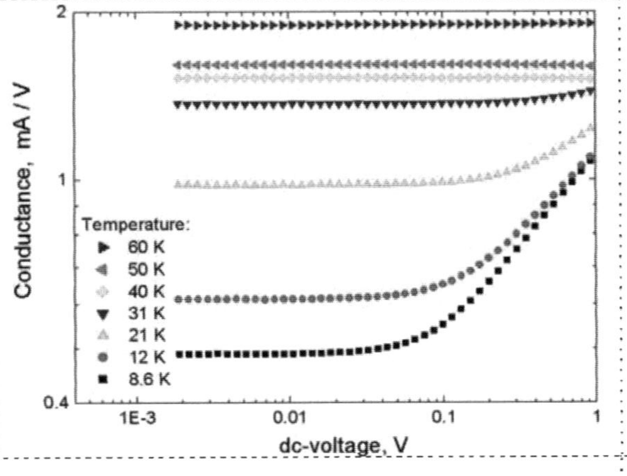


**Figure 3.** Variation of the sample conductance with applied dc-voltage at different temperatures.

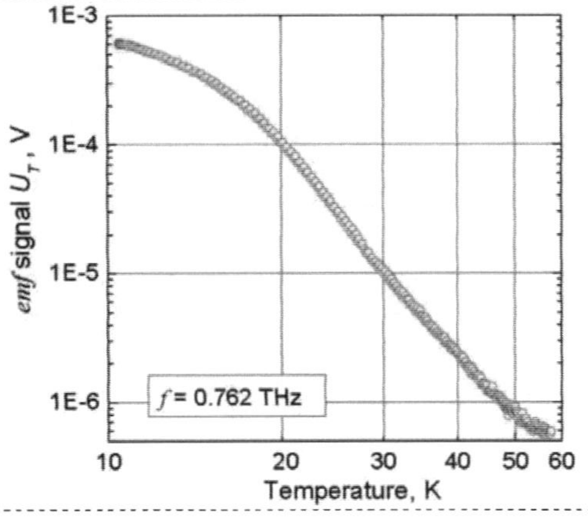


**Figure 4.** The thermo-*emf* signal dependence on sample temperature at 0.762 THz excitation frequency.

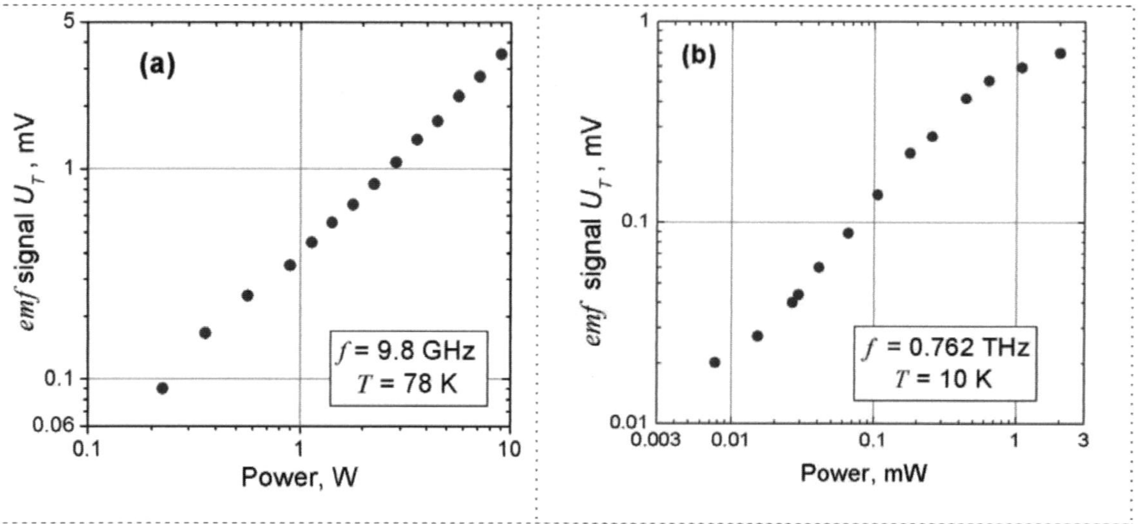


**Figure 5.** The thermo-*emf* signal dependence on radiation power at (a) 9.4 GHz and (b) 0.762 THz frequency, and respectively at 78 K and 10 K temperature. The results are plotted in log-log scale.

The measured thermo-*emf* signal exposition characteristics in GHz and THz range are shown in Fig. 5. The results show that signal vs. power of high frequency radiation has linear dependence in log-log scale. The functional behaviour can be understood in terms of warm electron physics, e.g., if electron energy relaxation time $\tau_\varepsilon$ is large ($\omega\tau_\varepsilon >> 1$) and an ac-electric field amplitude $E_m$ is sufficiently small (hot electron distribution function is close to the distribution function in a steady state) then the thermo-*emf* signal $U_T$ is [1]:

$$\frac{U_T}{V_k} = \frac{1}{3}\frac{e\bar{\mu}}{kT}E_m^2\tau_\varepsilon, \qquad (2)$$

here $\bar{\mu}$ is differential mobility averaged over the ac-electric field period, other notations are as in previous formula.

## 4. Conclusions

Hot electron thermo-electromotive force induced by GHz and THz radiation and the field dependences of the dc-conductivity have been studied experimentally within wide temperature range. Conditions of formation of the thermo-electromotive force signal in THz range are determined; peculiarities of the low-temperature transport and electron heating effects are explained to be associated with significant influence of the random potential fluctuations on the ionized impurity scattering mechanisms in high-resistivity InSb.

## References

[1] Ašmontas S and Subačius L 1982 *Semiconductors* **16** 40
[2] Ašmontas S, Subačius L, Požela J, and Valušis G 1988 *Solid St. Electronics* **31**, 701
[3] Ašmontas S, Liberis J, Subačius L, and Valušis G 1992 *Semicond. Sci. Technol.* **7** B331
[4] Meyer J R and Bartoli F J 1985 *Phys. Rev. B* **32** 1133

# Towards Wannier-Stark THz superlattice laser

**A A Andronov, E P Dodin, D I Zinchenko and Yu N Nozdrin**

Institute for Physics of Microstructures RAS, GSP-105, Nizhny Novgorod, Russia.

E-mail: andron@ipm.sci-nnov.ru

**Abstract**. Basing on measurements, computer simulations and calculations of transport properties in GaAs-GaAlAs superlattices (SL) with narrow (weak) barriers we propose and elaborate scheme of the SL Wannier-Stark (W.-S.) THz lasers. The laser transitions (tunable by applied voltage) are between levels of the first and the second W.-S. ladders corresponding to wells separated two-three periods apart and being nearby the resonant condition. The resonance provides high value of the transition matrix element that gives value of THz amplification coefficient up to 200-300 $cm^{-1}$ for the doping $5\text{–}10 \cdot 10^{15}$ $cm^{-3}$. This value is substantially higher than the one for THz quantum cascade lasers and allows to use in the lasers simple n+-SL-n+ waveguide. Some observations on transport in the SLs are given.

## 1. Introduction

At the beginning of 70s three proposals to use semiconducting SL to generate high frequency electromagnetic radiation appeared [1–3] – figure 1. These are: Esaki-Tsu oscillator (1) and the two of Kasarinov & Suris (2 and 3) laser proposals. The Esaki-Tsu oscillator is population inversion free system (cf. though [4]) with herewith low enough negative conductivity existing up to the Bloch frequency [5]. The Kasarinov & Suris systems are simple laser system with intrawell population inversion provided by electron injection from nearby well (2) or by **universal** interwell interlevel mechanism due to quasi equilibrium electron distribution in each well (3). This latter system may be considered as a system with optical transition between W.-S. ladders (shown in the case 2 in figure 1) and provides also high tunability by applied voltage. Unfortunately all these simple source schemes turn out to be unusable because of inherent static negative conductivity (NDC) leading to formation of inhomogeneous electric field domains.

To overcome domain formation and other issues the quantum cascade laser (QCL) scheme was introduced [6] leading to creation of mid IR [6] and THz [7] QCL. Since then outstanding advance in QCL occurred which is surveyed at this Conference by C. Sirtory. In all of the QCL schemes the injection inversion mechanism with almost fixed (weakly tunable) transition frequency was employed while the universal mechanism stayed intact.

In attempts to overcome domain formation in Esaki-Tsu scheme the weak barrier SLs with strong interminiband tunneling were introduced [8]. For these SLs there is a possibility to have dynamic NDC (though once again with small value like in the Esaki-Tsu case) nearby the Bloch frequency at the rising portion of SL current-voltage curve [8]. And in experimental study of transport in such SLs [9-11] we came across a possibility (a lasing scheme) to use such SL in the modified version of Kasarinov & Suris (2) scheme [12]. And this paper devoted to a discussion of this scheme and related transport measurement results.

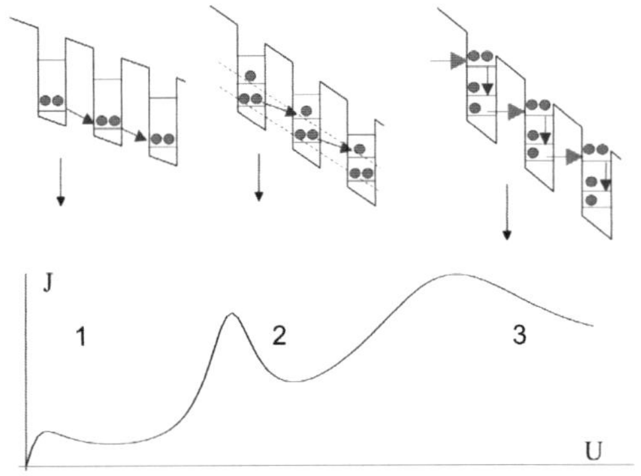


**Figure 1.** Earlier proposals for negative conductivity and lasing in SLs.

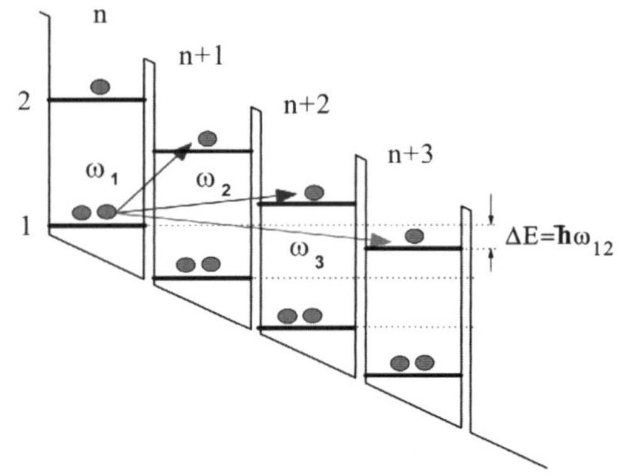


**Figure 2.** Scheme of energy levels and optical transitions in the SLs studied; circles represent relative population of levels.

## 2. The Wannier-Stark laser scheme

The essence of the scheme can be understood from figure 2 where different transitions between levels of the first and the second W.-S. ladders are shown. Experiments ([10-13] and discussed below) show the existence of long-range tunneling (several SL periods) in studied weak barrier SL near the resonance values of electric field which correspond to aligning of W.-S. levels originating from different minibands. In such resonant situation matrix elements of transitions between aligned levels become large due to long distance between wells where these levels are located. Simple consideration shows [12] that at the resonance matrix element of the transition is equal to a half of the distance between the resonant levels. On the other hand, there is natural inversion between levels of first and levels of second W.-S. ladders, when the level of the first ladder lies higher than (less populated) the level of the second ladder located in another well (figure 2, arrow labeled $\omega_3$).

At the same time due to the weak barriers in the SL current is a rising function of applied voltage and domain formation is prevented. Figure 3 and figure 4 present calculation results for the transition matrix element and amplification coefficient for situation close to that shown in figure 2 for SL parameters corresponding to one of the experimentally studied SL 816 but with doping level $10^{16}$ cm$^{-3}$ and scattering rate $\nu=5\cdot10^{12}$ sec$^{-1}$ (see also [12]). We see a very high amplification coefficient that permits to use in the appropriate THz lasers simple $n^+$–SL–$n^+$ waveguides. On the other hand with metal-SL-metal waveguide very low doped (or just undoped) SLs could be used in the lasers.

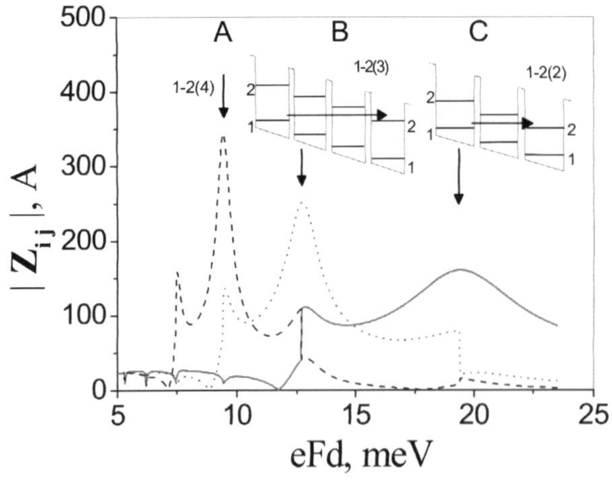


**Figure 3.** Calculated inter Wannier-Stark ladder transition matrix element $Z_{ij}$ versus voltage drop per period. Curves labeled (A), (B) and (C) correspond to transitions between 1$^{st}$ and 2$^{nd}$ levels separated by 4 (1->2(4)), by 3(1->2(3)) and 2 (1->2(2)) periods.

## 3. Transport and emission measurement results

We have studied several weak barrier GaAs/GaAlAs SL grown by MOCVD with number of periods up to 1000 (cf. [9 - 12]). Here we discuss results for the three samples: two undoped and one doped to about $10^{16}$ cm$^{-3}$. In the table 1 some properties of the SLs are given: here $d$, $w$ and $b$ are SL period, well and barrier widths, respectively, and $N_d$ is the number of SL periods. In figure 5 I-V curves measured on circular mesas for the undoped SLs are presented. The quality of these samples was high enough, as XRD and SIMS data indicate. As a result a lot of phenomena related to superlattice transport were observed (see [9 - 10]), and the main features in context of the present paper are "shoulders" at the rising portion of I-V curves. They are due (as it was established before [9, 11]) to alignment of levels of the first and the second W.-S. ladders situated several periods apart. These features establish existence of high wavefunction spread nearby the alignment voltage. No THz stimulated emission was observed on these SLs. We think that the absence of stimulated emission in this case of low doped SL is due to insufficient amplification coefficient to overcome losses in $n^+$-SL-$n^+$ waveguide. On the other hand, if one uses metal-metal waveguide for such low doped structures, amplification could overcome waveguide losses, but we haven't fabricated such structures yet.

**Table 1.** Properties of SL samples.

| SL | $d = w+b$, Å | $N_d$ | Doping, cm$^{-3}$ |
|---|---|---|---|
| 426 | 185+10 | 100 | undoped |
| 816 | 160+20 | 1000 | undoped |
| 1410 | 135+19 | 1000 | $1.5{\cdot}10^{16}$ |

We fabricated doped SL 1410 hoping to achieve amplification coefficient higher than the waveguide losses, as calculations in figure 4 predicted. However, the quality of the grown doped SL samples was not so good: low field mobility at 4 K is about 1500 cm$^2$/V·sec (up to 10000 cm$^2$/V·sec in the undoped SL 426) supposedly due to defects in barriers (as SIMS data indicate); increase in doping itself cannot explain such mobility reduction. In the I-V curve of the SL at 4K only "hint" to a possibility of the low field Esaki-Tsu type NDC was established while in the undoped SLs a lot of phenomena related to the NDC were observed [9,10]. No THz stimulated emission was observed in this case either.

## 4. Conclusion

We have discussed and elaborate the new SL intraband THz laser scheme which provides high laser emission tunability by applied voltage. In this scheme, the upper lasing level is the ground state of a quantum well in SL, while the lower level is the excited state of the well located 2-3 SL periods downstream (down the applied voltage). The levels are close to resonance condition that provides high matrix element for the transition and gives high amplification coefficient which permits to use simple $n^+$-SL-$n^+$ waveguides for lasing in doped SLs. It is important (as the transport experiments on the SLs show) that the lasing condition in the SLs correspond to rising portion of I –V curve what excludes inhomogeneous electric field formation under lasing.

Specifically the scheme was elaborated for THz range and for GaAs/GaAlAs SLs. At the same time it should work at other A3B5 material systems, at shorter wavelength and also at Si-GeSi [111] SL. Due to weak (short) barriers with low fraction of Ge the W.-S. laser scheme looks easier to realize compared to the Si-Ge/Si QCL schemes discussed nowadays.

Experiments conducted on both doped and undoped SL samples have shown the existence of features connected to the elaborated laser scheme, but we haven't seen the stimulated emission from these SLs yet.

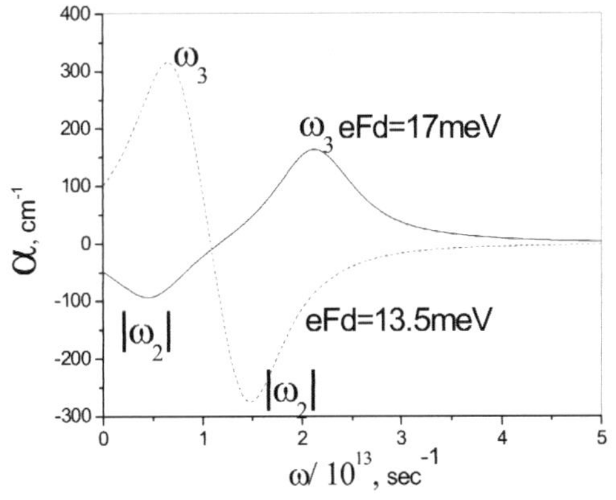


**Figure 4.** Frequency dependence of amplification coefficient for 2 different values of voltage drop on one SL period.

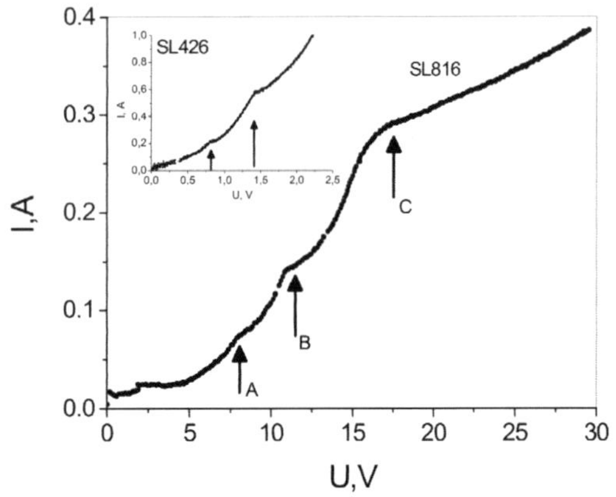


**Figure 5.** I-V curves for 426 and 816 SLs. Shoulders due to W.-S. level alignment: 1-2(4) – A, 1-2(3) – B and 1-2(2) – C are indicated by arrows.

**Acknowledgments**
The work is supported by grants of Russian Academy of Sciences Presidium and by RFBR (N 09-02-00380-a) and by Scientific School grant N НШ-2786.2008.2. The authors are indebted to: A. A. Marmaluk, A. A. Padalitza from Sigm Plus, Moscow, for SL growth; Yu. N. Drozdov for SL X-ray characterization; to M. N. Drozdov for SL SIMS measurements; to A. Yu. Klimov, V. V. Rogov and I. Yu. Shuleshova for SL sample processing.

**References**
[1] Esaki L and Tsu R 1970 *IBM J.Res.Dev.* **14** 61
[2] Kazarinov R F and Suris R A 1971 *Sov.Phys. Semicond.* **5** 707
[3] Kazarinov R F and Suris R A 1972 *Sov.Phys. Semicond.* **6** 120
[4] Willenberg H, Dohler G and Faist J 2003 *Phys.Rev.* B **67** 085315
[5] Ktitorov S A, Simin G S and Sindalovskii V Ya 1972 *Sov.Phys. Solid State* **13** 1872
[6] Faist J, Capasso F, Sivco D, Sirtori C, Hutchinson A and Cho A 1994 *Science* **264** 553
[7] Kohler R, Tredicucci A, Beltram F *et al* 2002 *Nature* **417** 156
[8] Andronov A A, Nefedov I M and Sosnin A V 2003 *Semiconductors* **37** 360
[9] Andronov A A, Dodin E P, Nozdrin Yu N and Zinchenko D I 2008 *Phys. stat. sol.* (c) **5** 190
[10] Andronov A A, Dodin E P, Nozdrin Yu N and Zinchenko D I 2009 *Semiconductors* **43**, 228
[11] Andronov A A, Dodin E P, Nozdrin Yu N and Zinchenko D I 2009 *Semiconductors* **43**, 236
[12] Andronov A A, Dodin E P, Nozdrin Yu N, Zinchenko D I, Marmaluk A A and Padalitza A A *ZhETF* (submitted)

# Detection mechanisms of terahertz radiation in a short gate $In_{0.3}Ga_{0.7}As/In_{0.53}Al_{0.47}As$ field effect transistor

**M L Orlov and L K Orlov**

Institute for Physics of Microstructures of the Russian Academy of Sciences (IPM RAS) 603950 Nizhny Novgorod, Russia

E-mail: orlovm@ipm.sci-nnov.ru

**Abstract**. The detection properties of some short-channel field-effect transistors (FETs) have been analyzed using the steady state output characteristics of these devices. The calculated dependences of voltage–power sensitivity on applied voltage are compared with the corresponding curves obtained from high-frequency measurements. It is shown that the nonmonotonic dependence of the FET photosensitivity on gate voltage that is observed in the frequency range of 400–750 GHz is not related to resonant excitation of 2D plasmons in the gated plasma but is due to the change in the distribution of stationary fields in the structure and, as a result, to the change in the efficiency of nonresonant nonlinearity procedures in the transistor's electron sub-system with an increase in the gate–channel voltage. This conclusion is confirmed by the analysis of the frequency dependences of photoresponse in the range under consideration, which do not exhibit resonant behavior at the frequencies corresponding to the peaks that are measured at a fixed frequency and for different gate voltages.

## 1. Introduction

Along with the development of the conventional technique for amplifying and transforming electromagnetic signals, some new concepts in this field are actively debated. One of them is based on application of resonant phenomena related to excitation of 2D plasmons in the gated electronic plasma of short channel transistors for detecting and generating terahertz electromagnetic signals [1]. In this context, many current studies are devoted to the search for peculiarities in the emission and detection characteristics of features of short-channel transistors, which would indicate resonant excitation of 2D plasmons in the transistor's transport channel.

At a low rectified signal, the detection caused by plasmon excitation (if this effect is present in the system under consideration) occurs against the background of other various nonresonant nonlinearity mechanisms such as concentration and carrier drift velocity nonlinearities. Hence, the expected resonant features in the transistor element characteristics can be reliably revealed only by simultaneous experimental and theoretical study of the volt-watt sensibility (VWS) dependences of a nonlinear element on both the external voltage and frequency of incident radiation. The most widespread practice is to study the parametric (most often current and field) dependences of the photoresponse of a short-channel high-electron-mobility transistor (HEMT) to terahertz radiation of a specified frequency.

However, in the absence of a pronounced effect of plasmon generation, explanation of the observed phenomena only within the Dyakonov–Shur instability theory [1,2,3] seems to be not quite correct. To describe reliably the nature of the experimentally observed nonmonotonic dependences of FET photosensitivity, along with the possible interaction of radiation with excited plasmons, one has to take into account other types of nonlinearities, whose competition with an increase in the external fields applied to a structure can also affect the shape of the curves studied [4]. The most important nonresonant mechanism, along with the nonlinearity of the carrier drift velocity [5], is the concentration nonlinearity [4, 5], which are most pronounced in the FET channel at high transmitted currents, corresponding, in particular, at channel cutoff. When the current flows, the gradient of the potential difference between the gate and channel induced by a source-to-drain bias may cause a coordinate dependence of the 2D transport channel-width, the position of energy levels in the electron spectrum and the equilibrium electron concentration and mobility. Therefore, to get a clearer idea about the mechanisms responsible for the experimental dependences, it is important to consider, along with the widely discussed plasma model [3, 4], the behavior of nonlinear FET response to a high-frequency perturbation on the basis of more conventional (in particular, quasi-static) models, which is facilitated by the choice of the frequency range under study, implying application of a frequency-tunable submillimeter generator based on a backward wave oscillator (BWO) and satisfying the condition $\omega\tau_{\mathrm{p}} << 1$, where $\tau_{\mathrm{p}}$ is the electron's elastic scattering time. The scheme of FET connection in a measuring circuit significantly differs from that for a conventional diode detector. This circumstance may cause peculiarities in the FET VWS that are not specific of diode structures and are related, in particular, to the change in the contribution of different nonlinearity processes to detection with an increase in the longitudinal and transverse field components in the transistor structure. For a short-channel transistor, this fact is extremely important because the longitudinal field component in the channel becomes comparable in magnitude with the transverse component, hindering the use of gradual-channel approximation for estimates.

## 2. Analysis of the transistor's nonlinear response on the basis of output characteristics

To reveal the general regularities and understand better the observed dependences, we analyze the VWS of some short-channel transistors on the basis of their output characteristics. The calculations are performed using the output characteristics of a short-channel $In_{0.7}Ga_{0.3}As/In_{0.47}Al_{0.53}As$ FET, which was previously considered in [3, 4]. For the most promising $In_{0.7}Ga_{0.3}As/In_{0.47}Al_{0.53}As$ HEMTs, the conductivity $\sigma$ is approximately 3.5 × $10^4$ $\mathrm{Sm}^{-1}$. As a result, at $n_s \geq 5\times10^{11}\mathrm{cm}^{-2}$, we have $\mu_{\mathrm{e}} = \sigma / en \leq 0.65\mathrm{m}^2\mathrm{V}^{-1}\mathrm{s}^{-1}$, which corresponds to the characteristic room-temperature values of electron mobility in the system under study. Accordingly, for $\tau_{\mathrm{p}} = m_{\mathrm{e}}\mu_{\mathrm{e}} / e$ at $m_{\mathrm{e}} \approx 0.03m_0$, we obtain $\tau_{\mathrm{p}} \leq 10^{-13}\mathrm{s}$. These relaxation times $\tau_{\mathrm{p}}$ at frequencies $f \approx$ 0.65 THz yield $\omega\tau_{\mathrm{p}} \leq 0.4$ at $T = 293\mathrm{K}$. The values $\omega\tau_{\mathrm{p}} << 1$ measured at room temperature justify the quasi-static approximation, which is used below to analyze the observed regularities in the effect of transistor detection of submillimeter ($f <$ 1THz) radiation.

To estimate the experimental VWS values of the $In_{0.7}Ga_{0.3}As/In_{0.53}Al_{0.47}As$ transistor, we will use the output characteristics of the transistors which were discussed previously in [3, 4]. In figure 1a, the experimental dependences $\Delta U(I_{\mathrm{D}})$, found in the high-frequency experiment (see, for example, [4]), are compared with the theoretical curves derived from analysis of the output characteristics (figure 1a) of transistor. Comparison of the theoretical and experimental dependences in figure 1a indicates along with the prepared above estimations on the possibility using a quasi-static approximation. The increase in VWS with an increase in channel current, experimentally observed at frequencies up to 1 THz, is most likely related to the nonresonant mechanisms responsible for the nonlinear shape of output characteristics of the transistor; this circumstance was noted, in particular, in [4, 5], where more specified models were analyzed.

In view of the search for peculiarities in the characteristics of a high-frequency transistor, related to resonant excitation of 2D plasmons in the electron subsystem, the subject of the most animated discussion is the dependences of the high-frequency photoresponse of a nonlinear element on the gate voltage $U_G$ [2, 3, 5], specifying the electron concentration and, correspondingly, the resonant plasmon frequency ω0 in the transistor channel.

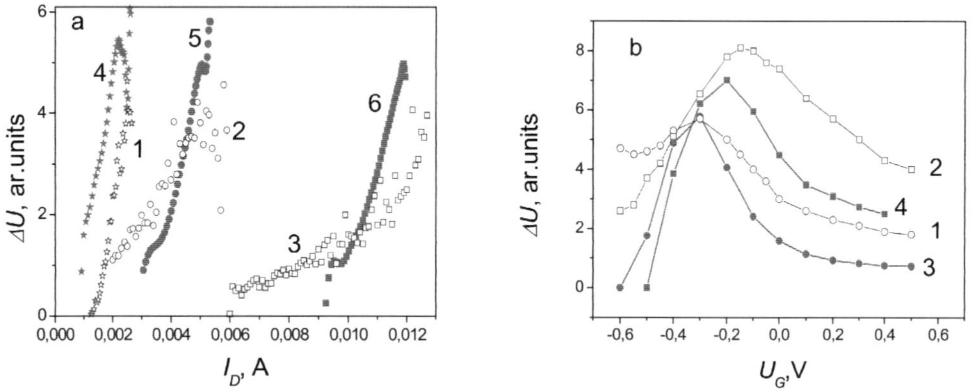


**Figure 1.** Nonlinear response of an InGaAs/InAlAs transistor: (a) the experimental [4] (*1–3*) and calculated (*4–6*) dependences of VWS on the current $I_D$ at $U_G$ = (*1, 4*) –0.2, (*2, 5*) –0.1, and (*3, 6*) 0 V and (b) the experimental [4] (*1, 2*) and calculated (*3, 4*) dependences of VWS on $U_G$; $U_D$ = (*1, 3*) 0.2 and (*2, 4*) 0.4 V.

However, a variation in the gate voltage not only changes the carrier concentration in the electron channel but leads also to a change in the field distribution along and across the transistor structure. Correspondingly, along with the concentration along the channel length, the contributions of different nonresonant nonlinearities to the total nonlinearity of the system change as well. In view of the aforesaid, the dependence $\Delta U(U_G)$ may be nonmonotonic far from the resonant frequency $\omega_0$. To estimate the possibility of such a situation, we calculated the dependences of nonlinear high-frequency transistor response on voltage $U_G$ in the quasi-static approximation, using the transistor's output characteristics. The dependences derived from the output characteristics $\Delta U(U_G)$ (figure 1b) are compared with the corresponding experimental dependences at high frequencies (400–700 GHz) [3, 4]. It follows from figure 1b that the dependence of the response of the system to a high-frequency perturbation, found in the quasi-static approximation based on the analysis of the transistor's transfer characteristics, on the gate voltage, has a maximum.

## 3. Frequency characteristics of an InGaAs/InAlAs transistor detector

Reliable information about the occurrence of the resonant plasma process in a system can be obtained only from analysis of the frequency dependences of the rectified signal at specified contact voltages. In particular, the plasma resonance suggested in [3, 5] could be confirmed by coincidence of the positions of resonant peaks in the dependences $\Delta U(U_G)$ and $\Delta U(\omega)$, all other factors being equal. A more convenient technique was used in [3], where radiation sources were high-frequency tunable BWO-based generators. The study of the rectified signal dependence with the radiation frequency is made possible with that experimental configuration.

Figure 2 shows the dependences of the photoresponse of an $In_{0.7}Ga_{0.3}As/In_{0.52}Al_{0.48}As$ transistor on the radiation frequency for drain voltages $U_D$ = 0.1, 0.2, and 0.3 V at a fixed gate voltage $U_G = 0V$. The solid line is the approximation of experimental points by the dependence of a second-order polynomial. Hereinafter, the power $W$ detected by the transistor is normalized to the power $W_0$ of the generator based on the BWO-74, measured at the same radiation frequencies using an M3-22A thermistor power meter. The corresponding emission spectrum of the generator near some fixed frequencies is shown in the inset in figure 2. The envelope of the frequency dependence of the generator signal is characterized by a wide band with close values of maximum emitted power at selected frequencies in the range of 580–650 GHz.

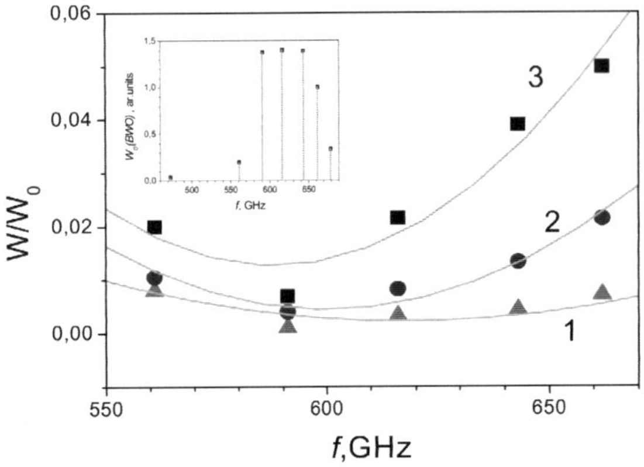


**Figure 2.** Dependences of the photoresponse of an $In_{0.7}Ga_{0.3}As/In_{0.52}Al_{0.48}As$ transistor on the generator frequency $f$ at the drain voltages $U_D$ = (*1*) 0.1 V, (*2*) 0.2 V, and (*3*) 0.3 V and gate voltage $U_G = 0V$. The curves of photosensitivity $W$ are normalized to the BWO power $W_0$, whose amplitude–frequency characteristic is shown in the inset.

The frequency dependences of a detected signal at different gate voltages are shown in figure 2. It follows from them that the transistor photosensitivity decreases with an increase in frequency in the range of 450–600 GHz, such a decrease is typical of nonlinear elements operating in the quasi_static mode (Schottky diodes, detectors based on heating photoconductiveity). Above 600 GHz (figure 2), the photosensitivity of the transistor detector increases with an increase in the radiation frequency. The dependence of the Schottky diode's VWS is known to decrease with an increase in frequency as $1/\omega^2$, and the photosensitivity of detectors based on heating photoconductivity decreases even more rapidly.

Obviously, if plasma model corresponded to the experiment, there would be a resonance at $\omega = \omega_0$ in the frequency dependence of the photocurrent. However, there are no resonant features in the observed dependences (figure 2) in the range from 550 to 700 GHz. The absence of resonant features in the curves in the noted range confirms the conclusion about dominant contribution of nonresonant nonlinearities to the formation of features observed in this frequency range in the field dependences of the photoresponse [3, 4, 6].

The above VWS analysis shows that features observed in the dependence of the high frequency response on $U_G$ [3,4] may be related not only to the plasma resonance, as was suggested in the overhelming majority of studies on this subject [3,5]. Comparing the rectifying characteristics received from the output characteristics and from high-frequency measurements one can see that observed nonmonotonic dependence of photosensitivity on gate voltage can be also related to change in field distribution and, as a result, to the change in the efficiency of nonresonant nonlinearity expression in the transistor's electron sub-system with an increase in the gate–channel voltage.

**References**

[1] Dyakonov M and Shur M 1996 *IEEE Trans. Electron. Dev.* **43** 380
[2] Knap W, Lusakowski J, Parenty T, Bollaert S, Cappy A, Popov V V and Shur M S 2004 *Appl. Phys. Lett.* **84** 2331
[3] Teppe F, Orlov M, El Fatimy A, Tiberj A, Knap W, Torres J, Gavrilenko V, Shchepetov A, Roelens Y and Bollaert S 2006 *Appl. Phys. Lett.* **89** 222109
[4] Orlov M L 2008 *Semiconductors* **42** №3 339-45
[5] Veksler D, Teppe F, Dmitriev A P, Kachorovskii V Yu, Knap W and Shur M S 2006 *Phys. Rev.* B **73** 125328
[6] Orlov M L, Panin A N and Orlov L K 2009 *Semicoductors* **43** 787

EDISON 16 IOP Publishing
Journal of Physics: Conference Series **193** (2009) 012081 doi:10.1088/1742-6596/193/1/012081

# One-dimensional approximation of Poisson equation for the description of multi-gate conducting channels of FETs and HEMTs

**P. Shiktorov, E. Starikov, V. Gružinskis**
Semiconductor Physics Institute, Gostauto 11, LT-01108 Vilnius, Lithuania

**H. Marinchio and L. Varani**
Institut d'Electronique du Sud (UMR 5214 CNRS), Université Montpellier 2, 34095 Montpellier, France

E-mail: pavel@pav.pfi.lt

**Abstract.** One-dimensional approach for simulation of multi-gate FETs and HEMTS is proposed and validated by comparison with full two-dimensional simulations.

## 1. Introduction

At the present time, a one-dimensional (1D) approximation for the two-dimensional Poisson equation is extremely useful for the theoretical description of carrier transport and noise in FET/HEMT channels as [1,2]:

$$\varepsilon_c \frac{\partial^2 \varphi}{\partial x^2} + \varepsilon_s \frac{U_g - \varphi(x)}{d'(x)\delta} = e[n(x) - N'_D(x)] \tag{1}$$

This expression takes into account the influence of the gate potential $U_g$ (second term in l.h.s. of Eq. 1) on the distribution of potential $\varphi(x)$ and electron concentration $n(x)$ along the channel. Here, $d'(x)$ is the local gate-to-channel distance, $\delta$ is the channel width, and $N'_D(x)$ is the effective channel donors density. One of the main advantages of such an approximation is that it recovers, as limiting relations, two representations of the Poisson equation widely used in the theoretical analysis of both $n^+nn^+$ structures and HEMT/FET channels: (i) the limit of an infinitesimally narrow channel ($\delta \to 0$) corresponding to the *gradual channel approximation*, and (ii) the opposite *ungated* case ($d'(x) \to \infty$) corresponding to the trivial case of the 1D Poisson equation. The spatial dependences of $d'(x)$ and $N'_D(x)$ in Eq. (1) allow us to describe all reasonable situations such as gated, ungated, $T$-gated channel regions as well as a difference in effective compensated donor concentration in different regions. In essence, Eq. (1) with $U_g$ = const describes a three-terminal device which includes source, drain and gate. Here we propose an extended interpretation of this equation to the multi-gate case when the gate potential in various channel regions is different, that is $U_g(x)$ in Eq. (1) depends on $x$.

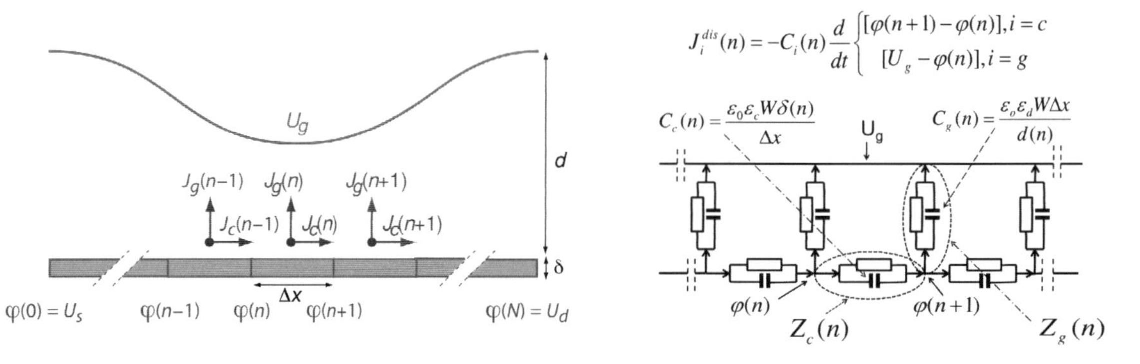


**Figure 1.** Discrete representation of the channel (left) and equivalent scheme of such a representation (right).

## 2. Analytical model

Here, by following ref. [1], we shall briefly discuss main assumptions used to derive Eq. (1). It is supposed that the channel of FET/HEMT structures can be represented as a set of discrete regions (cells) of the length$\Delta\ x$ connected sequentially one with another (see Fig. 1). In each cell the total current $J$ flowing along the channel can undergo the branching between two directions, namely, the longitudinal direction where $J_c$ flows along the channel and the transverse one where $J_g$ goes to the gate. In accordance with Kirchhoff law the incoming and outcoming longitudinal currents in the $n$-th cell, $J_c(n-1)$ and $J_c(n)$, respectively, are related with the leakage current to gate, $J_g(n)$, by the zero-sum rule:

$$J_c(n-1) - J_c(n) = J_g(n) \tag{2}$$

The connection in series of channel cells implies that the longitudinal component of the total current $J_c(n-1)$ which is outcoming current for the $(n-1)$-th cell is simultaneously the incoming current for the next $n$-th cell. It is evident that such an interpretation of the channel does not take into account the displacement currents which can appear between any two remote cells due to variation in time of the longitudinal component of the electric field in the dielectric surrounding the channel. The equivalent scheme of such a representation is shown in Fig. 1 (right), where the impedances $Z_c(n)$ and $Z_g(n)$ which characterize the $n$-th cell determine a local branching of the current between the channel and the gate. By separating the total currents $J_c(n)$ and $J_g(n)$ into two, namely, drift and displacement components such as: $J_i(n) = J_i^{dis}(n) + J_i^{cond}(n)$ where $i = c, g$, and going to the limit$\Delta\ x \to 0$, in accordance with ref. [1] we shall obtain two quasi-one-dimensional equations which allows us to describe separatively the electrodynamics and the carrier transport along the channel. The electrodynamic part is described by Eq. (1) which is the one-dimensional analogue of two-dimensional Poisson equation and it accounts for the gate influence, while the transport is described by the charge conservation law in the channel:

$$\frac{\partial}{\partial t}n(x) + \frac{\partial}{\partial x}j_c^{cond}(x) + \frac{j_g^{cond}(x)}{\delta(x)} = 0 \tag{3}$$

where $j_c^{cond}(x)$ and $j_g^{cond}(x)$ are densities of the drift component of currents which, respectively, flow along the channel and go into the gate.

Since the local impedances $Z_c(n)$ and $Z_g(n)$ can depend on a cell number $n$ , this means that such parameters of the model as the effective channel width $\delta$, the gate-to-channel distance $d$, the donor concentration in the channel $N_d$ can be considered as functions of $x$. As shown in [1-5], this allows us in the framework of Eq. (1) to consider a wide set of tasks which include both gated and ungated channel regions, T-gate, the longitudinal modulation of the channel doping. In essence, all these tasks are reduced to an analysis of three-terminal devices where one of the terminals is the gate which at any point has the same potential $U_g$.

However, both the discrete representation of the channel and the limit at$\Delta x \to 0$ will have no formal changes if one assumes that the gate potential depends on the cell number $n$ that is in Eq. (1) $U_g$ is function of $x$. This allows us to interpret in the framework of Eq. (1) the influence of an inhomogeneous potential at the gate on the carrier transport in the channel as an influence of a sequence of separated gates with different potentials. To avoid unphysical situations in points of gates where the potential has discontinuity, the gate regions with different potentials must be separated by ungated regions where $d(x) \to \infty$. Below we shall present a comparison of results obtained by Monte Carlo simulation performed in the framework of proposed here multi-gate model of FET/HEMT channel and by using full 2D Poisson equation [6].

## 3. Numerical results

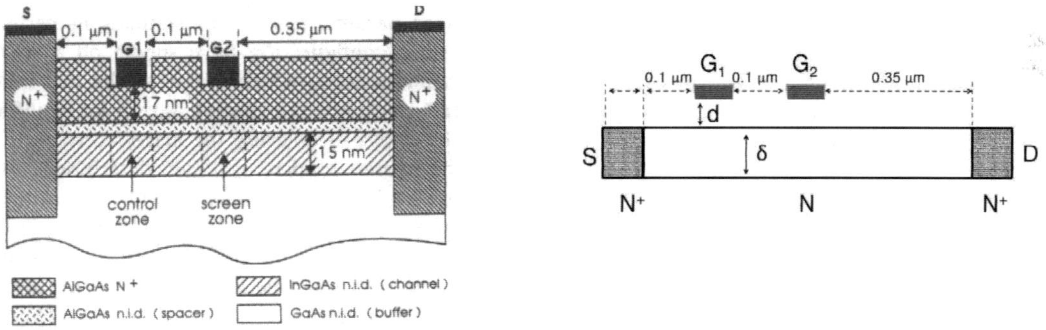


**Figure 2.** Schematic representation of 2D HEMT of ref. [6] (left) and our simplified 1D model of the same HEMT (right).

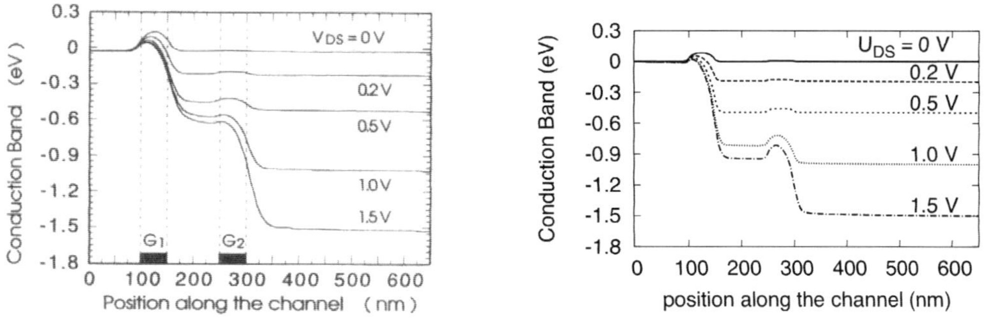


**Figure 3.** Conduction band profile in DG HEMT calculated by MC simulation for 2D (left) and 1D (right) HEMT representations.

The schematic representation of InGaAs HEMT in two-dimensional (2D) case taken from ref. [6] and one-dimensional (1D) approximation used in this work is shown in Fig. 2, left and right, respectively. The gate-to-channel distance $d$ = 22 nm, the channel width $\delta$ = 15 nm, the gate lengths $G_1 = G_2 = 50$ nm, $N^+ = 2 \times 10^{18}\ cm^{-3}$ (see [6] for more details).

Journal of Physics: Conference Series **193** (2009) 012081 doi:10.1088/1742-6596/193/1/012081

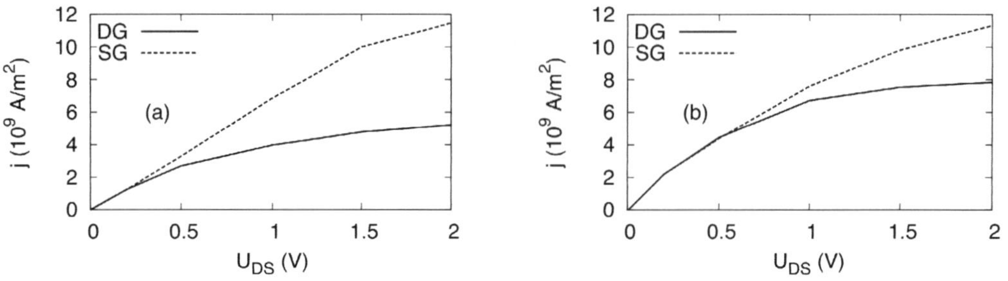


**Figure 4.** Current-voltage relation for double- and single-gate HEMT calculated by MC approach for 2D [6] (a) and 1D (b) representations.

As an example, Figs. 3 and 4 show results of Monte Carlo simulation of InGaAS HEMT near pinchoff operation at source-to-gate voltages $U_{G1} = -0.2$ V and $U_{G2} = 0.6$ V. The conduction band profiles along the channel are presented in Fig. 3 at increasing values of the applied drain-to-source voltage $U_{DS}$. At $U_{DS} < U_{G2}$ the applied voltage drops mainly at the first gate, while at $U_{DS} > U_{G2}$ the extra voltage drop takes place mainly on the second gate. The drain currents of double- and single gate (DG and SG) InGaAs HEMTs are compared in Fig. 4 for 2D and 1D cases (left and right, respectively). Here, the improvement of the saturation behavior in DG case at $U_{DS} > U_{G2}$ due to action of the second gate is evident. Thus, at the level of electrodynamics (description of potential, see Fig. 3) we find a good quantitative and qualitative agreement of both the approaches. For drift characteristics (see Fig. 4) we get a good qualitative coincidences of results but the quantitative agreement is not so good (probably, due to sufficiently complicate 2D structure (see Fig. 2, left) which cannot be taken into account by simplified 1D model).

## 4. Conclusions

In conclusion, let us emphasize that one-dimensional approximation given by Eq. (1) with different gate-to-source potential values in different regions of the channel separated by ungated areas provides a good qualitative and reasonable quantitative description of the potential and drift current in the channel of 2D multi-gate FET/HEMT structures. Let us note, that parameters of Eq. (1) such as the channel width $\delta(x)$, the gate-to-channel distance $d(x)$, the ionized donor concentration in the channel $N_D(x)$ must be considered as some "effective parameters" which take into account an influence of the transvers surrounding of the channel on formation of self-consistent distribution of free carriers in the channel. Depending on the task they either can have a simple geometrical meaning or can represent some effective feeting parameters.

*This work is supported by the Lithuanian State Science and Studies Fundation contract No P-01/2007.*

## 5. References

[1] Shiktorov P *et al.* 2009 *J. Stat. Mech.* 01047
[2] Marinchio H *et al.* 2008 *Phys. Stat. Sol.* c **5** 251
[3] Torres J et al 2008 *IEEE J. Select. Top. Quant. Electron.* **14** pp. 491-497
[4] Marinchio H et al 2009 *Appl. Phys. Lett.* **94** 192109
[5] Starikov E et al 2009 *AIP conf. proc.* **1129** 267-272
[6] Dollfus P and Hesto P 1993 *Solid-State Electron.* **36** 711

# Terahertz tunable detection in self-switching diodes based on high mobility semiconductors: InGaAs, InAs and InSb

**I. Iñiguez-de-la-Torre[1], H. Rodilla[1], J. Mateos[1], D. Pardo[1], A. M. Song[2] and T. González[1]**

[1]Departamento de Física Aplicada, Universidad de Salamanca,
Plaza de la Merced s/n, 37008 Salamanca, Spain
[2]School of Electrical and Electronic Engineering, University of Manchester,
Manchester M60 1QD, United Kingdom

E-mail: indy@usal.es

**Abstract**. In this work we report on the use of high mobility materials in the channel of self-switching diodes as potential candidates for terahertz operation. By means of Monte Carlo simulations we envisage the feasibility of tuneable-by-geometry detection in the terahertz range. The low effective mass of InAs and InSb in relation to InGaAs enhances ballistic transport inside the diode, thus improving the amplitude and quality factor of the resonance found in the detection spectra of self-switching diodes. The frequency of the resonant peak is also increased with the use of these narrow band gap semiconductors. The analysis of the noise spectra provides useful information about the origin of the resonance. By decreasing temperature below 300 K, a clear improvement in detection sensitivity is also achieved.

## 1. Introduction and purpose

The terahertz domain is reaching more and more relevance in security, medical and life science applications. Detection systems with high level of compactness, low noise, low cost and without the requisite of cryogenic cooling are essential parts for the development of new tools in this wide field of the technology roadmap. In this contribution we explore a novel device, so-called self-switching diode (SSD) [1] [geometry illustrated in figure 1(a)], which is able to work at frequencies reaching the terahertz domain. The SSD is fabricated with a very simple technological process. It only involves a single step of high resolution lithography, the etching of insulating trenches or recess lines with an asymmetric layout onto a semiconductor epilayer. The working principle is absolutely different from that of a normal diode. It is not based on any doping junction or barrier structure. Electron conduction through the channel between the two grooves is allowed only when the bias between the right and left terminal is positive; the induced positive voltages at the dielectric trenches attract electrons to the channel and the current can flow. Otherwise, when a negative voltage is applied to the right terminal, the induced electric field depletes the channel and the current flow is not possible [1]. With this diode-like *I-V* characteristic, RF rectification and detection can be achieved. SSDs based on silicon are also perfectly viable; nevertheless the use of high mobility semiconductors like InGaAs implies larger electron velocities and therefore higher operation frequencies. In fact, experiments have demonstrated microwave detection up to 110 GHz at room temperature [2] and up to 2.5 THz at 10 K (dropping in the response above 70 K) [3] for InGaAs SSDs. Monte Carlo (MC) simulations [4] are able to explain the behavior of the device and foresee working frequencies of above 1 THz. One of the most significant advantages of this device is its planar geometry. The integration with antennas to improve detection sensitivity is very simple, the parasitic capacitances between contacts with respect to conventional vertical devices of the same size is reduced and, finally, the feasibility to integrate many SSDs in parallel without the need of interconnects overcomes the high-impedance problems typical of nanodevices.

The aim of this contribution is to show, by means of MC simulations, that by using narrow band gap materials with lower effective mass than InGaAs (like InAs and InSb) ballistic transport is enhanced, so that it is possible to increase the cutoff frequency of SSDs and also the amplitude and quality factor of the resonant peak found in their detection spectra. The analysis of the current noise spectra calculated in SSDs made of InGaAs, InAs and InSb allows confirming the physical origin of such a peak, already explained in previous works [5-6]. The influence of decreasing temperature on the resonance features is also investigated as another possible improvement in device performance.

## 2. Tunable-by-geometry resonant RF detection. InGaAs-based SSDs.

By virtue of the non-linear behavior of the SSD [see *I-V* curve in the inset of figure 1(b)], several applications like rectification, detection, mixing, etc. are possible. In this case we study the DC response to AC excitations to examine the cutoff frequency of the diode. For this purpose we perform a simulation [4] in which the current is recorded while a sinusoidal input voltage signal, $V(t)=V_0 \,\mathrm{sen}(2\pi f t)$, of increasing frequency $f$ is applied between the contacts, as sketched in figure 1(a).

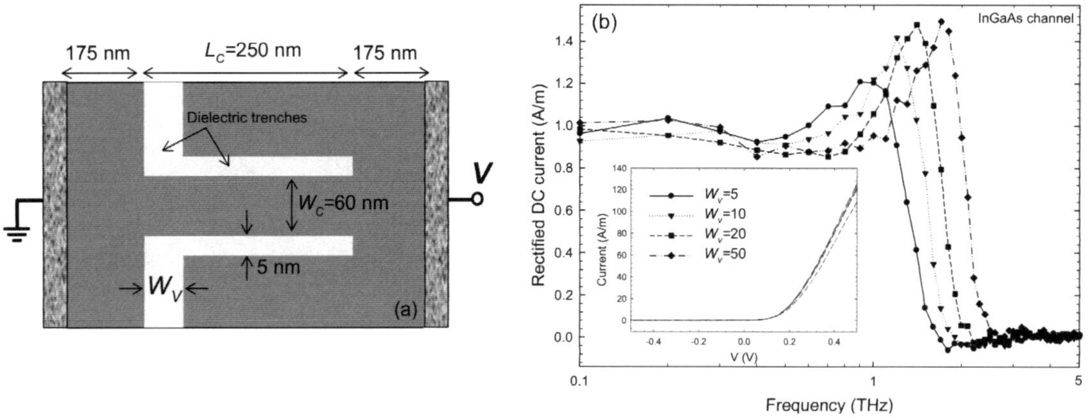


**Figure 1.** (a) Two-dimensional topology of the SSD. (b) DC response vs. frequency of a harmonic signal of amplitude 0.15 V applied to the right terminal of an InGaAs diode with the topology shown in (a) for different widths of the vertical trench $W_V$. The inset shows the corresponding *I-V* curves.

The average current as a function of $f$, sweeping the range 0.1-5 THz is shown in figure 1(b) for InGaAs SSDs in which different widths of the vertical trench $W_V$ have been considered. A pronounced peak not expected from the static *I-V* curve is observed after a flat region, just before the decay in the response. The frequency of the peak $f_P$ is shifted to higher values for wider vertical trenches $W_V$. The presence of such a resonance could be useful to design and fabricate passive detectors of THz signals with a certain degree of frequency selectivity tuned by the geometry of the diodes.

## 3. Enhanced resonance. InAs- and InSb-based SSDs.

In the last years, two very promising narrow band gap semiconductors, InAs and InSb, are being used in the fabrication of different devices. Using the same heterolayer of Sb-based HEMTs [7] (AlSb/InAs and AlInSb/InSb) an improvement in the operation frequencies of SSDs is expected as a consequence of an enhanced channel mobility. Performing the same type of simulations with our MC tool recently adapted for these new materials [7] we analyze diodes with identical geometry to those of Sec. 2. Figures 2(a) and 2(b) show the rectified DC current for InAs- and InSb-based SSDs, respectively. The more ballistic character of transport in these low-effective-mass materials enhances the resonance and shifts $f_P$ to higher frequencies. As observed, the resonance in the rectified DC current exhibits a remarkable quality factor, much higher than in InGaAs, with an amplitude more than a hundred times the low frequency value in the case of InSb SSDs.

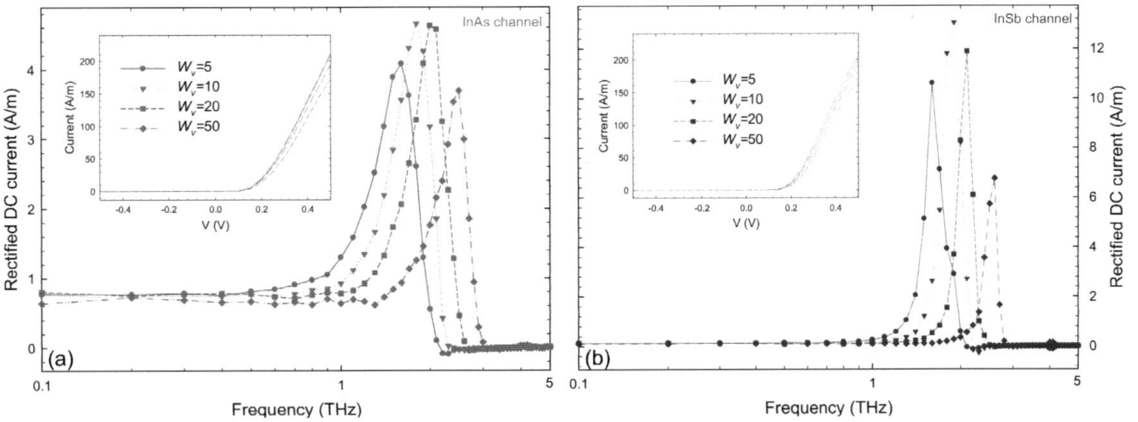


**Figure 2.** DC response vs. frequency of a harmonic signal of amplitude 0.15 V applied to the right terminal of a diode with the topology shown in figure 1(a) for different widths of the vertical trench $W_V$ and (a) InAs channel and (b) InSb channel. The insets show the corresponding *I-V* curves.

### 3.1. Origin of the resonance and noise spectra.

The physical origin of the resonance was studied in previous works [5-6]. It is attributed to a collective charge fluctuation phenomenon (associated to returning carries [5] or localized surface plasma (LSP) effects [6]) in the space-charge region around the active channel of the device. In addition, as explained in Ref. 5 there is a clear link between the peak in the rectified DC current and a similar one present in the noise spectral density. To corroborate this, in figure 3(a) we plot the noise spectra at equilibrium for SSDs based on the three III-V high mobility semiconductors for the case $W_V$=20 nm. Two peaks are observed. 3D plasma oscillations are at the origin of that appearing at the highest frequencies (above 3 THz and only determined by the electronic properties of the material). The other peak, at lower frequency (in the range 1.5-2.5 THz), is originated by the same phenomenon as in the rectified DC current [5]. In figure 3(b), the frequency of both noise peaks together with $f_P$ of

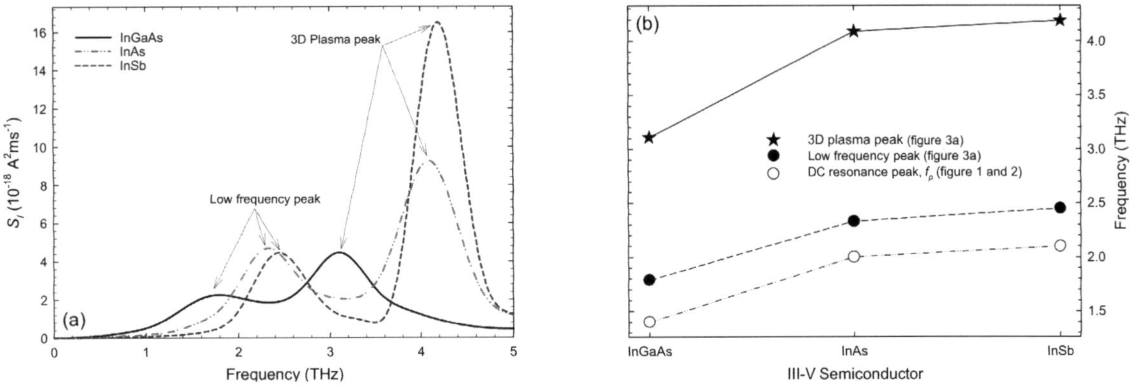


**Figure 3.** (a) Current noise spectra at equilibrium for InGaAs, InAs and InSb diodes ($W_V$=20 nm). (b) Comparison of the frequency of the peaks in the noise spectra with that of figures 1 and 2.

figures 1 and 2 are plotted for each material. The three frequencies exhibit a similar relative variation between the different materials, mainly affected by their effective masses and dielectric constants (0.014 and 17.65, respectively, for InSb, 0.023 and 15.15 for InAs and 0.042 and 13.88 for

$In_{0.47}Ga_{0.53}As$) thus revealing a common origin linked with plasma effects. Remarkably, the peak in the rectified DC current has the same tendency, certifying our previous conjecture that a noise mechanism is coupled via the particular geometry of the SSD to the DC to AC response and enhances the performance of the device.

## 4. Temperature effect

All the previous simulations were performed at 300 K (RT), temperature specially appropriate for potential commercial applications of SSDs ranging from telecommunications to health. Nevertheless, exploring the possible advantages of cooling the samples (space applications can be also foreseen as THz radiation plays a key role in radio astronomy) and problems related to heating is also interesting. A new set of simulations of InGaAs diodes at lower and higher temperatures have been carried out, with the results plotted in figure 4. A remarkable increase in the peak amplitude takes place when decreasing the temperature as a result of the enhancement in the ballistic nature of electron transport (the channel length of the simulated SSD is 250 nm while the mean free path in InGaAs at RT is about 200 nm and 1-2 μm for 77 K). For $T$=400 K and 500 K the resonant peak disappears.

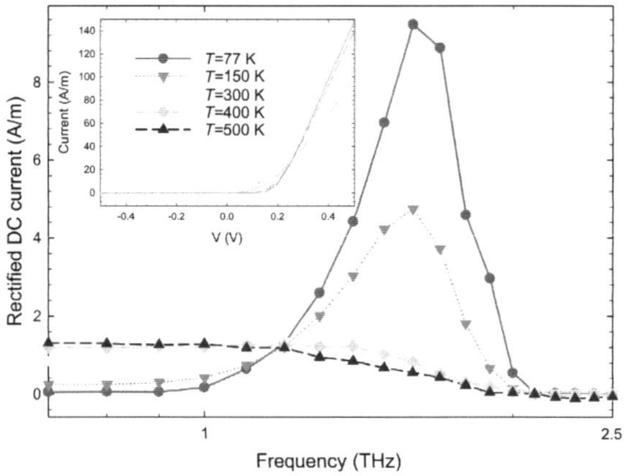


**Figure 4.** Influence of temperature on the response of an InGaAs SSD. $W_V$=20 nm. Inset: *I-V* curves.

## 5. Conclusions

In virtue of the particular topology and the rectifying behavior of SSDs (based on surface and electrostatic effects) a tunable-by-geometry resonance in the THz domain can be exploited for detection applications. The use of InAs channels (with much higher electron mobility than InGaAs) or even faster InSb channels provides a clear improvement of performance in terms of amplitude, quality factor and cutoff frequency. When the operation temperature is decreased to 77 K an outstanding increase of the resonant peak is achieved.

This work has been partially supported by the Dirección General de Investigación (MEC) and FEDER (project TEC2007-61259/MIC and Acción Integrada HF2007-0014) and by the Consejería de Educación, Junta de Castilla y León (projects SA019A08 and GR270).

### References

[1] Song A M *et al.* 2003 *Appl. Phys. Lett.* **83** 1881
[2] Balocco C *et al.* 2005 *Nano Letters* **5** 1423
[3] Balocco C *et al.* 2008 *J. Phys.: Condens. Matter* **20** 384203
[4] Mateos J *et al.* 2005 *Appl. Phys. Lett.* **86** 212103
[5] Iñiguez-de-la-Torre I *et al.* 2009 *Appl. Phys. Lett.* **94** 093512
[6] Xu K Y *et al.* 2008 *J. Appl. Phys.* **103** 113708
[7] Rodilla H *et al.* 2009 *J. Appl. Phys.* **105** 113705

# Plasma waves Terahertz detection by field effect transistor in Quantinzing magnetic fields

**S Boubanga-Tombet[1], F Teppe[1], W Knap[1,2], K Karpierz[2], J Lusakowski[2], M Grynberg[2], M I Dyakonov[1], J Lyonnet[1], J M Peiris[1]**

[1] Universite Montpellier II, CNRS,Place E. Bataillon, 34095 Montpellier, France
[2] Institute of Experimental Physics, University of Warsaw, Hoza 69, 00-681 Warsaw, Poland

E-mail: albon@ges.univ-montp2.fr

**Abstract.** Detection of THz radiation by a Field Effect Transistor InGaAs/InAlAs transistor was investigated at 4.2 K as a function of the magnetic field and gate voltage. We observed oscillations of the photovoltaic signal analogous to Shubnikov-de Haas oscillations, as well as their strong enhancement at the cyclotron resonance conditions. The results are successfully quantitatively described within the frame of a recent theory, taking into account a new source of nonlinearity related to Shubnikov-de Haas effect. We show that the detection is due to gated plasmon.

## 1. Introduction

The exploitation of nonlinear properties of the 2D plasma in Field Effect Transistors (FETs) channel is a promising approach proposed by M. Dyakonov and M. Shur [1], [2] as a new route to create compact solid state high frequency devices. These FETs can be used as terahertz (THz) resonant [3] and selective detectors at cryogenic temperatures, as well as broadband detectors at room temperature [4]. These transistors were found to have competitive Noise Equivalent Power in comparison with other kinds of commercially available detectors [5]. Detection of THz radiation by Si-CMOS based transistors was also demonstrated and already reported in many application systems such as THz cameras [6]. The detection signal results from the rectification of THz currents induced by the incident radiation in the transistor channel. The rectification takes place due to a nonlinear response of the gated 2DEG to the radiation. The main source of nonlinearity, explored until now, were changes in conductivity due to radiation-induced modulation of the carrier density in the channel. Nevertheless, in high magnetic fields the oscillatory effects like Shubnikov-de Haas (SdH) oscillations and cyclotron resonance (CR) can take place and have influence on the rectification related photovoltaic response.

The purpose of this work is to clarify the physical mechanism of Shubnikov de Haas (SdH) oscillations and Cyclotron Resonance (CR) related detection by FETs. We investigate the THz response in the transistor of a simple architecture, with the gate covering the major part of the channel. Fig 1 shows the transfers characteristics of the FET at different values of the drain voltage (0.1 V and 0.2 V) obtained at 4.2 K. In this simplified structure (see the inset of Fig 1), the effect of the gate bias on the observed signal can be clearly established and the results can be compared with the recent theory [7].

## 2. Results and discussion

Experiments were carried out on a pseudomorphic InGaAs/InAlAs HEMT . The measurements were performed with an optically pumped molecular CH3OH laser. The radiation of frequency 2.5 THz. The photoconductivity signal was measured using the standard lock-in technique. Two types of experiments were performed. First, the detection signal was measured as a function of the gate voltage Vg at fixed magnetic field B. Second, the magnetic field was swept at a fixed gate voltage. The results are shown in Figs. 2 - 3. In both kinds of experiments the photoresponse was found to be periodic in $B^{-1}$, i.e, it showed Shubnikov - de Haas oscillations. Its periodicity versus Vg and 1/B clearly indicates that the oscillations are related to the crossing between the Fermi level and the Landau levels (analogous to the SdH effect). Additionally, a strong enhancement of the signal is observed in magnetic fields close to cyclotron resonance magnetic field $B_c \approx 4.5$ T.

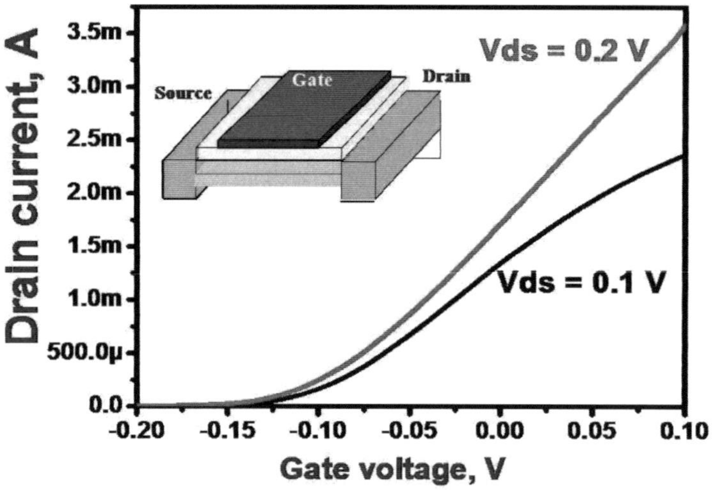


**Figure 1.** Transfert characteristics at different values of the drain voltage (0.1 V and 0.2 V) obtained at 4.4 K. The inset shows a schematic of the transistor geometry.

Fig. 2 shows experimental results of detection signal as a function of the gate voltage at different values of B/$B_c$ (0.44, 0.88, 1, 1.55). One can notice that the amplitude of oscillations grows as the magnetic field increases. This can be understood in terms of an increase of a nonlinearity related to SdH oscillations as described in Ref [7] and [8]. In the post-cyclotron resonance region (B > $B_c$), the oscillations of the signal are damped. This can be qualitatively explained by considering the plasma wave dispersion law in the magnetic field: k $=(\omega^2$-$\omega_c^2)^{1/2}$/s, where s=$(eU_0/m^*)^{1/2}$ is the plasma wave velocity at B=0, where $U_0$ = $V_g$ - $V_{th}$ is the gate voltage swing, $V_{th}$ is the thersold voltage, $m^*$ is the effective mass and e the electron charge. For $\omega_c > \omega$ the plasma wave vector becomes imaginary, plasma waves cannot propagate, and one can expect a reduction of the signal, as evidenced in Fig. 2.

Not only the oscillation amplitude of the signal, but also its background grows with B. One can clearly see in Fig.2 that this background has it's maximum close to the CR condition. This enhancement can also be explained within the model in Ref [7]

Fig. 3 shows the measured magnetic field dependence for fixed gate voltage (fixed electron density). Fig. 3a corresponds to a relatively high value of electron density in the channel, for which one can see strong oscillatory behavior of the photovoltaic signal. Fig. 3b corresponds

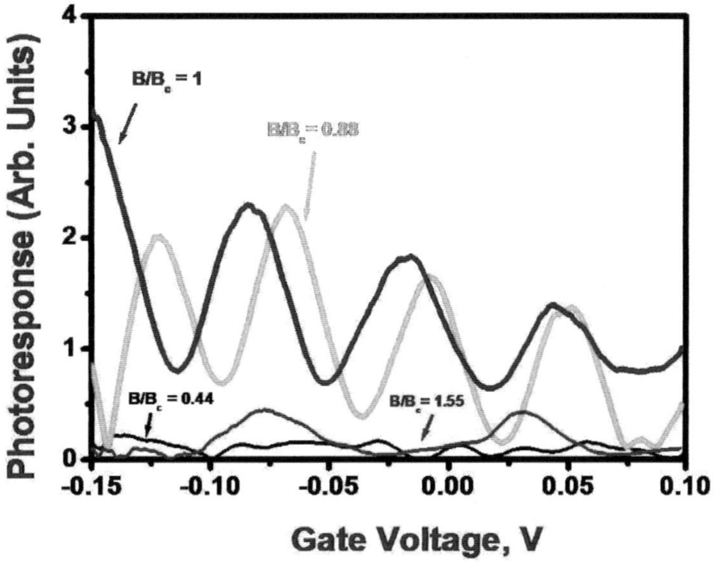


**Figure 2.** Experimental results of photoresponse at 2.5 THz as a function of the gate voltage for different values of the magnetic field at T = 4.2 K. Values of B/Bc are indicated.

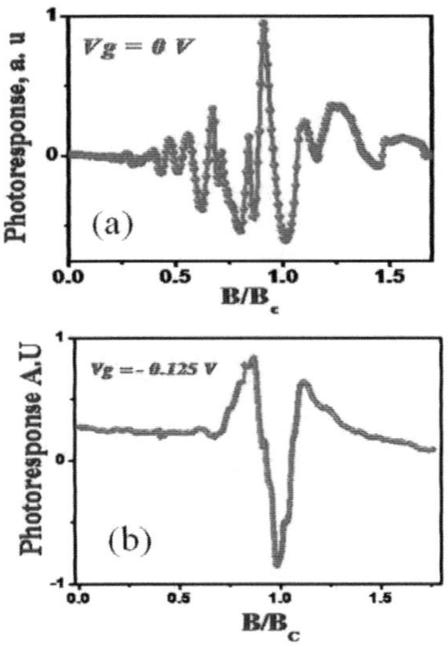


**Figure 3.** Figure a - Experimental photoresponse as a function of the magnetic field for higher values of the electron density (Vg = 0 V ) at 4.2 K. b - Same as in a for 3a lower carrier density (Vg = -0.125 V)

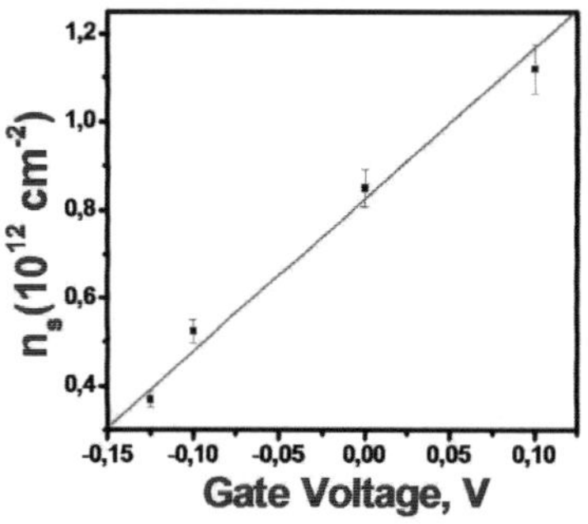


**Figure 4.** Electron density extracted from experimental curves at different values of gate voltage

Journal of Physics: Conference Series **193** (2009) 012083 doi:10.1088/1742-6596/193/1/012083

to a lower value of electron density in the channel with a very week oscillations signature. For B>$B_c$ the oscillation amplitude decreases (see Fig. 3a). This behavior is quite consistence with the results shown in Fig. 2, and is also described by the theory [7]. The strong dependence of the photoresponse on the gate voltage (see Fig 3a and Fig 3b) indicates that the rectification takes place in the gated part of the transistor channel.

The Fourier transform of the oscillations (of the photovoltage resulting from THz radiation) allowed us to extract electron density that participate in detection process. Indeed the carrier density is related to the period of SdH oscillations ($\Delta(1/B)$ ) by, $n = (e/\pi\hbar)^*\Delta(1/B)$ . Fig 4 shows values of electron density extracted from experimental curves at different values of gate voltage. One can clearly see that this density shift with the gate voltage. This result confirms that the detection signal is induced in the gated part of the channel.

## 3. Conclusion

In conclusion, we have performed magnetic field studies of plasma wave related detection in InGaAs/InAlAs field effect transistor. We observe oscillatory behavior of the photovoltaic signal, similar to Shubnikov-de Haas oscillations, as well as a strong enhancement of the signal at the cyclotron resonance conditions. The experimental data are quantitatively reproduced by a theoretical model [7]. Our results clearly show that the detection mechanism is related to the THz rectification by nonlinearities of the plasma in the transistor channel.

## 4. Acknowledgments

The transistor used in this work was produced by the IEMN, Lille. We acknowledge the help of A. Cappy, S. Bollaert and A. Shchepetov. This work was supported by CNRS, the GDR E project "Semiconductor sources and detectors of terahertz frequencies". We acknowledge the region of Languedoc-Roussillon through the "Terahertz Platform" project, the EU Grant No. MTKD-CT-2005-029671 and PL 162/THz/2006/02 program.

[1] Dyakonov M, and Shur M S 1993 *Phys. Rev. Lett.* **71** 2465

[2] Dyakonov M, and Shur M S 1996 *IEEE Trans. on Elec. Dev.* **43** 380

[3] Boubanga-Tombet S, Teppe F, Coquillat D, Nadar S, Dyakonova N, Videlier H, Knap W, Shchepetov A, Gards C, Roelens Y, Bollaert S, Seliuta D, Vadoklis R, and Valuis G 2008 *Appl. Phys. Lett.* **92** 212101

[4] Knap W, Kachorovskii V, Deng Y, Rumyantsev S, L J Q, Gaska R, Shur M S, Simin G, Hu X, Asif Khan M, Saylor C A, and Brunel L C 2002 *J. Appl. Phys* **91** 9346

[5] Tauk R, Teppe F, Boubanga S, Coquillat D, Knap W, Meziani Y M , Gallon C, Boeuf F, Skotnicki T, Fenouillet-Beranger C, Maude D K , Rumyantsev S, and Shur M S 2006 *Appl. Phys. Lett.* **89** 253511

[6] Lisauskas A, von Spiegel W, Boubanga-Tombet S, El Fatimy A, Coquillat D, Teppe F, Dyakonova N, Knap W, and Roskos H G, 2008 *Electon. Lett.* **44** 408

[7] Lifshits M and Dyakonov M I 2009 *http://arxiv.org/abs/0901.2712*

[8] Boubanga-Tombet S, Sakowicz M, Coquillat D, Teppe F, Knap W, Dyakonov M I, Karpierz K, Lusakowski J , Grynberg M, *Appl. Phys. Lett.*, in press

Journal of Physics: Conference Series **193** (2009) 012084 doi:10.1088/1742-6596/193/1/012084

# Terahertz study of ultrafast carrier dynamics in InGaN/GaN multiple quantum wells

**H P Porte, D Turchinovich, D G Cooke and P Uhd Jepsen**

DTU Fotonik - Department of Photonics Engineering Technical University of Denmark
Ørsteds Plads 343, DK 2800 Kongens Lyngby – Denmark

E-mail: hpor@fotonik.dtu.dk

**Abstract.** Ultrafast carrier dynamics in InGaN/GaN multiple quantum wells is measured by time-resolved terahertz spectroscopy. The built-in piezoelectric field is initially screened by photoexcited, polarized carriers, and is gradullay restored as the carriers recombine. We observe a nonexponential decay of the carrier density. Time-integrated photoluminescence spectra have shown a complete screening of the built-in piezoelectric field at high excitation fluences. We also observe that the terahertz conductivity spectra differs from simple Drude conductivity, describing the response of free carriers, and are well fitted by the Drude-Smith model.

## 1. Introduction

The understanding of ultrafast dynamics in group-III nitride based quantum wells is important for the development of wide bandgap semiconductor devices, such as LEDs and lasers in the blue-green spectral region [1]. In such systems the large, strain-induced built-in piezoelectric field has significant influence on the optical and electronic properties of AlGaN/GaN and InGaN/GaN quantum wells, described by the quantum confined Stark effect (QCSE) [2]. The QCSE will manifest itself by a strong dependence of the transition energy, absorption coefficient and recombination rate on the carrier density in the quantum well [3]. Whereas in a previous terahertz investigated, the dynamics of the absorption coefficient has been studied [4], we will concentrate on the carrier density decay dynamics by using time-resolved terahertz spectroscopy.

## 2. Experiments and Results

Our sample consists of a 10-layer $In_{0.2}Ga_{0.8}N$/GaN multiple quantum well (MQW) with a well thickness of 1.8 nm and a barrier width of 7 nm. The quantum well structure is sandwiched between a 180 nm thick capping layer and a 2 $\mu$m buffer layer. The sample is grown on a sapphire substrate by metalorganic vapor-phase epitaxy. The lattice mismatch between In and Ga causes strain in the material, which results in large built-in piezoelectric fields of 3.1 MV/cm in the quantum wells and 0.5 MV/cm (with opposite polarity) in the barriers [4, 5].

The optical properties of a biased system are well described by the QCSE [2]. The tilt of the bandgap (Fig. 1(a)) results in a smaller transition energy and the reduction of the spatial overlap between the electron and hole wavefunctions leads to a smaller transition probability. During excitation with an ultrashort laser pulse, a dipole moment is created due to the shifted position of the electron and hole wavefunctions. This dipole moment has a polarity opposite to that of the built-in piezoelectric field and thus it will partly screen this field. With strong excitation fluences

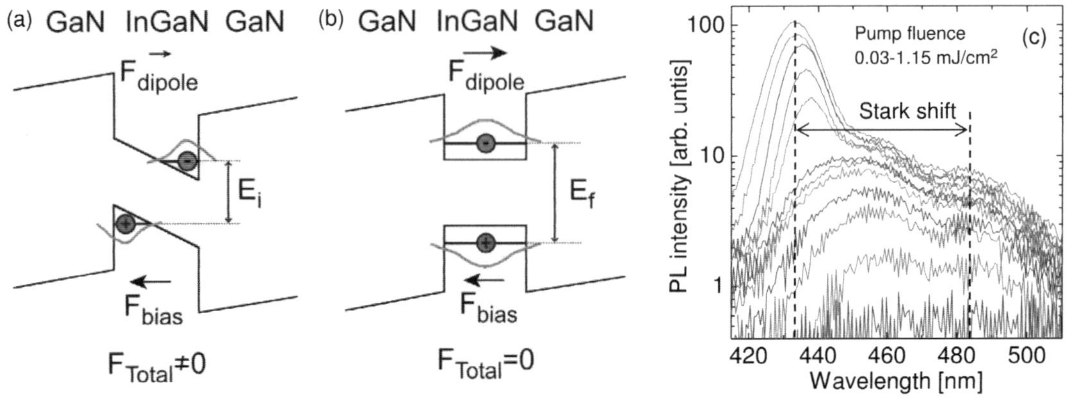


**Figure 1.** (a) Schematic of the InGaN/GaN QW bandstructure, at weak excitation fluence. (b) Schematic of the QW bandstructure, showing the screening of the piezoelectric field, at strong excitation fluence. $E_i$: initial transition energy, $E_f$: final transition energy after total screening of the piezoelectric field, $F_{dipole}$: dipole field induced by the electron-hole pairs, $F_{bias}$: build-in piezoelectric field, $F_{total}$: net electric field of the combined dipole and piezoelectric field. (c) Time-integrated PL spectra with excitation fluences in the range 0.03-1.15 mJ/cm$^2$.

complete screening of the built-in piezoelectric field can be achieved (Fig. 1(b)). The screening results in flattening of the bandstructure and hence in an increase of the optical transition energy and transition probability. Since every photoexcited carrier contributes to the screening, this process is highly dynamical. As the carriers recombine the reverse process will take place. Recombination of electron-hole pairs decreases the dipole field and thus causes the restoration of the built-in piezoelectric field. This may imply that during the recombination process the recombination rate will slow down due to increased spatial separation of the wavefunctions.

We use time-resolved THz spectroscopy to measure the decay of the photoexcited carriers. The setup is feeded by a 3.5 mJ/pulse, 45 fs duration, Ti:sapphire amplifier with central wavelength around 800 nm. The THz pulses, used as a probe for the carrier density, are generated by optical rectification in a 1 mm thick ZnTe crystal. The temporal shape of the THz pulses are detected after normal incidence transmission through the InGaN/GaN MQWs by free-space electrooptic sampling in a 2 mm thick ZnTe crystal. Part of the Ti:sapphire amplifier output is used to generate 400 nm pump pulses by frequency doubling in a beta-barium borate (BBO) crystal. The 400 nm wavelength assures that with single photon absorption, electron-hole pairs can only be generated in the InGaN quantum wells and not in the GaN barriers. We observed no significant influence of two-photon absorption in GaN on our experimental results.

To show the screening of the built-in piezoelectric field, time-integrated photoluminescence (PL) measurements were performed with excitation fluences in the range of 0.03-1.15 mJ/cm$^2$ (Fig. 1(c)). A strong Stark shift is observed between low and high excitation fluences. The saturation of the blue shift for the highest excitation fluences indicates that complete screening of the built-in piezoelectric field takes place. The large shoulder at longer wavelength for the highest excitation fluences is a result of using a time-integrated technique. As carriers recombine the piezoelectric field is restored, which results in a decrease of the transition energy and thus in a red-shift of the PL spectra.

The relative differential transmission of the THz probe between an excited and unexcited sample can be related to the THz conductivity within the measured 0.5-2.5 THz spectral bandwidth. An easier method to monitor changes in THz conductivity than measuring the full THz pulse is to measure only the peak of the THz pulse, for different pump-probe delay

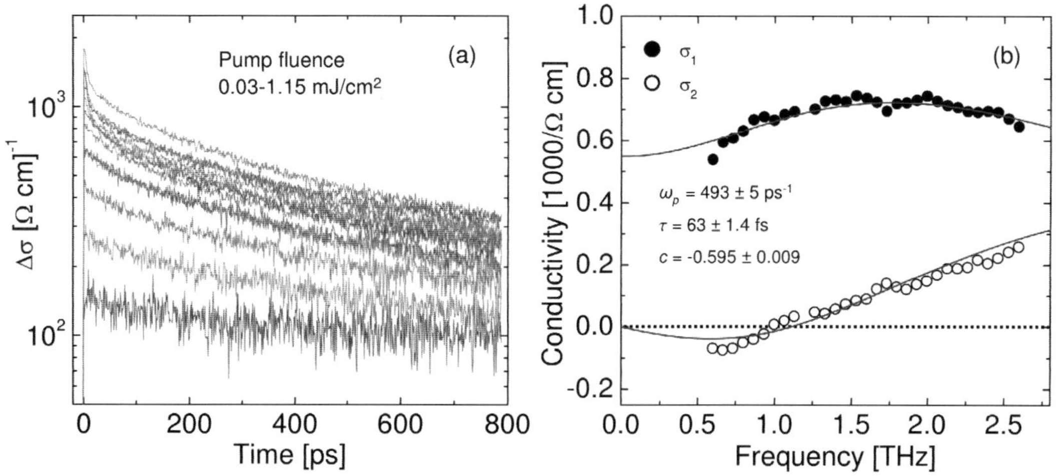


**Figure 2.** (a) Photoinduced conductivity dynamics in a 1.8 nm thick InGaN/GaN MQW with pump fluences in the range 0.03-1.15 mJ/cm$^2$. (b) Real and imaginary parts of the terahertz conductivity, 5 ps after optical excitation with a fluence of 0.14 mJ/cm$^2$. The blue are fits of the data to the Drude-Smith model.

times [7, 8, 9]. Fig. 2(a) shows the transient photoconductivity in InGaN/GaN MQWs for the same pump fluences as in Fig. 1(c). A nonexponential decay of the photoconductivity is observed, more enhanced for large pump fluences. At early pump-probe delay times and high pump fluences, the complete screening of the built-in piezoelectric field results in a large overlap of the electron and hole wavefunction and hence a large recombination rate. At later pump-probe delay times, the carrier density decreases due to electron-hole recombination, resulting in the restoration of the built-in piezoelectric field, which will decrease the recombination rate, causing a nonexponential carrier density decay. The same nonexponential decay dynamics are observed by time-resolved PL experiments in AlGaN/GaN quantum wells, where the QCSE is also present [6]. The dependence of the decay rate on the carrier density can also be observed by the faster initial decay at higher fluences (Fig. 2(a)).

Relating the induced photoconductivity dynamics directly to the carrier density dynamics implies a constant carrier mobility. Information about the carrier mobility can be obtained by time-resolving the THz pulses for various times after optical excitation and compare them with a reference pulse, transmitted through an unexcited sample. The Fourier transforms of the THz pulses can be related to the photoinduced complex THz conductivity, $\tilde{\sigma}(\omega)$ [8, 9, 10, 11]. The real, $\sigma_1$, and imaginary, $\sigma_2$, parts of the complex conductivity, 5 ps after optical excitation are shown in Fig. 2(b). The simple Drude-model, describing a conductivity response caused by free carriers, observed in bulk n-GaN [10] and n-GaN films [11], fails to explain the observed features here. Deviations from the Drude model in n-GaN films have also been reported [12]. The smaller conductivity for low frequencies and the zero-crossing of the imaginary part of the conductivity at non-zero frequency indicate a non-Drude conductivity response. The observed deviations from Drude conductivity can be described by a classical generalization of the Drude model, known as the Drude-Smith model [13]. THz conductivity spectra of various nanostructures have been successfully described by the Drude-Smith model [8, 9]. This model reads:

$$\sigma(\omega) = \frac{\epsilon_0 \omega_p^2 \tau}{1 - i\omega\tau}[1 + \sum_j \frac{c_j}{(1 - i\omega\tau)^j}]. \tag{1}$$

The first term is the Drude term, where $\epsilon_0$ is the permittivity of free space, $\omega_p$ is the plasma frequency and $\tau$ is the scattering time. The second term describes the persistence of the carriers initial velocity after a certain scattering event, where $c_j$ is the persistence of velocity coefficient for scattering event $j$. The blue curves in Fig. 2(b) are fits to the conductivity spectrum with the Drude-Smith model, where only the first scattering event is taken into account ($j$=1). The real and imaginary part of the conductivity are fitted to the same set of parameters simultaneously. From these fits we obtain, $\omega_p/2\pi = 493 \pm 5$ THz, $\tau = 63 \pm 1.4$ fs and $c = -0.595 \pm 0.009$. The conductivity spectra at later times after optical excitation have shown similar values for $\tau$ and $c$ (not shown here). The change in plasma frequency, $\omega_p$, for later times is caused by a change in carrier density. The time and carrier density independent scattering time, $\tau$, and the persistence of velocity coefficient, $c$, implies that the mobility is also independent of the carrier density. This means that the observed nonexponential decay of the photoconductivity (Fig. 2(a)), is completely caused by a nonexponential carrier density decay.

## 3. Conclusion

A nonexponential carrier density decay is observed in InGaN/GaN MQWs as a result of the QCSE, caused by the restoration of the built-in piezoelectric field as carriers recombine. Time-integrated PL spectra have shown the initial screening of the built-in piezoelectric field. The THz pulses are time-resolved to obtain THz conductivity spectra at various times after optical excitation. Fits to the Drude-Smith model with a non-zero value of $c$ show that the conductivity response differs from that of free carriers.

## Acknowledgement

We acknowledge Danish Advanced Technology Foundation (HTF) for financial support.

## References

[1] Nakamura S, Senoh M, Nagahama S, Iwasa N, Yamada T, Matsushita T, Kiyoku H and Sugimoto Y 1996, *Jpn. J. Appl. Phys.* **35** L74
[2] Miller D A B, Chemla D S, Damen T C, Gossard A C, Wiegmann W, Wood T H and Burrus C A 1984 *Phys. Rev. Lett.* **53** 2173
[3] Turchinovich D, Monozon B S and Jepsen P U 2006 *J. Appl. Phys.* **99** 013510
[4] Turchinovich D, Jepsen P U, Monozon B S, Koch M, Lahmann S, Rossow U and Hangleiter A 2006 *Phys. Rev.* B **68** 241307
[5] Im J S, Kollmer H, Off J, Scholz F and Hangleiter A 1999 *Mater. Sci. Eng.* B **59** 315
[6] Pinos A, Marcinkevičius S, Liu K, Shur M S, Kuokštis E, Tamulaitis G, Gaska R, Yang J and Sun W 2008 *Appl. Phys. Lett.* **92** 061907
[7] Lui K P H and Hegmann F A 2001 *Appl. Phys. Lett.* **78** 3478
[8] Beard M C, Turner G M, Murphy J E, Micic O I, Hanna M C and Nozik A J and Schmuttenmaer C A 2003 *Nano Lett.* **3** 1695
[9] Cooke D G, MacDonald A N, Hryciw A, Wang J, Li Q, Meldrum A and Hegmann F A 2006 *Phys. Rev.* B **73** 193311
[10] Zhang W, Azad A K and Grischkowsky D 2003 *Appl. Phys. Lett.* **82** 2841
[11] Nagashima T, Takata K, Nashima S, Harima H and Hangyo M 2005 *Jpn. J. Appl. Phys.* **44** 926
[12] Tsai T-R, Chen S-J, Chang C-F, Hsu S-H, Lin T-Y and Chi C-C 2006 *Opt. Express* **14** 4898
[13] Smith N V 2001 *Phys. Rev.* B **64** 155106

# Capture and release of carriers in InGaAs/GaAs quantum dots

**D Turchinovich[1], H P Porte[1], N Daghestani[2], K G Wilcox[2], E U Rafailov[2] and P Uhd Jepsen[1]**

[1] DTU Fotonik - Department of Photonics Engineering Technical University of Denmark Ørsteds Plads 343, DK 2800 Kongens Lyngby – Denmark

[2] School of Engineering, Physics and Mathematics, University of Dundee, Dundee DD1 4HN, UK

E-mail: dmtu@fotonik.dtu.dk

**Abstract.** We observe the ultrafast capture and release of charge carriers in InGaAs/GaAs quantum dots (QDs) at room-temperature with time-resolved terahertz spectroscopy. For excitation into the barrier states, a decay of the photoinduced conductivity, due to capture of carriers into the nonconducting QD states is observed. The increase of the decay time constant with increasing pump fluence is attributed to filling of the QD states. In the case of resonantly excitation into the QD ground state a maximum conductivity is reached 35 ps after photoexcitation, which is assigned to the release of carriers from the QDs into the wetting layer and barrier states.

## 1. Introduction

Ultrafast carrier dynamics in semiconductor quantum dots (QDs) has been widely studied for better understanding of QD-based optical devices, such as QD-lasers [1], QD-infrared photodetectors [2] and QD semiconductor saturable absorber mirrors (SESAM) [3]. Several ultrafast optical techniques, like time-resolved photoluminescence [4], optical pump-probe spectroscopy [5, 6] and time-resolved THz spectroscopy [6, 7, 8, 9] have been used. Previous THz-studies on QD systems have concentrated on the carrier capture into the QDs. In this work we have studied both the carrier capture into solid state QDs as well as the release of carriers from the QDs into the wetting layers (WLs) and barrier states (BSs) with an optical pump - terahertz probe technique.

## 2. Experiments and Results

The capture and release of carriers is measured in a QD SESAM structure. The sample is grown by molecular beam epitaxy. A schematic of the sample structure is shown in Fig. 1(a). It consists of 15 layers of $In_{0.5}Ga_{0.5}As$/GaAs QDs, separated by GaAs barrier layers with a total thickness of 611 nm and a 33-layer $Al_{0.9}Ga_{0.1}As$ Bragg reflector. The Bragg reflector has a 100-nm-wide stopband centered around 1060 nm (Fig. 1(b)). Besides the small signal reflectivity of the QD SESAM, the room-temperature photoemission spectra of the QD ground state (GS) is shown in Fig. 1(b). The emission spectrum has a central wavelength around 1017 nm, which is within the stopband of the Bragg reflector. The Bragg reflector is important in our experiment in the

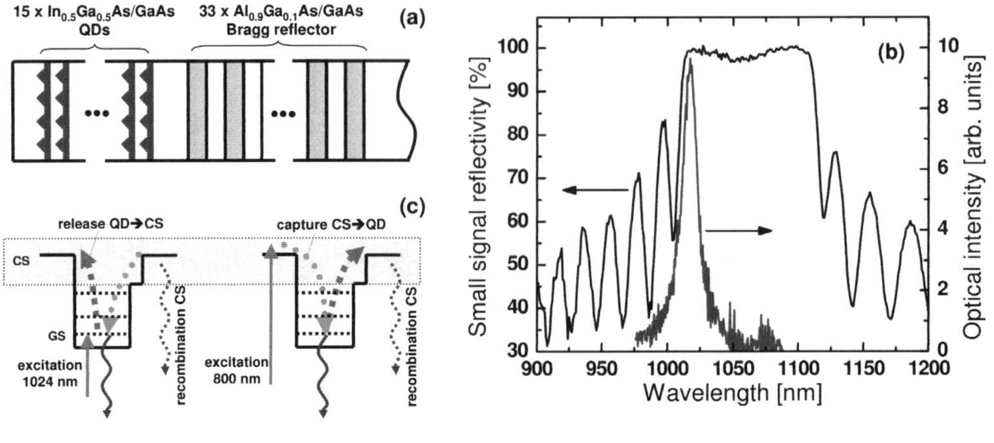


**Figure 1.** (a) Schematic of the studied InGaAs/GaAs quantum dot semiconductor saturable absorber mirror. (b) Small signal reflectivity (black curve) and room temperature optical emission spectra (blue curve) from the QD ground state. (c) Carrier dynamics processes in photoexcited quantum dots. CS: conducting state and GS: QD ground state.

case of resonantly excitation into the QD GS. The optical pump beam is reflected at the Bragg grating, preventing carrier excitation in the thick GaAs substrate by two-photon absorption, which could easily obscure the signal from the release of carriers from the QDs.

To measure ultrafast carrier dynamics in InGaAs/GaAs QDs on a picosecond timescale, time-resolved THz spectroscopy is used. For the experiment, the output of a regenerative Ti:sapphire amplifier with a central wavelength of 800 nm is used. Part of the beam is used as a pump, to excite carriers either directly into the GaAs BS or with 1024 nm central wavelength pulses, generated via frequency conversion in an optical parametric amplifier, into the QD GS. The 1024-nm pulses have a full-width-at-half-maximum (FWHM) of 45 nm, ensuring resonant excitation into the QD GS. The other part of the 800 nm laser beam is used to generate and detect the THz probe pulses. The THz probe pulses, generated by optical rectification in a nonlinear crystal, are transmitted through the sample, thereby probing the changes in conductivity, and then detected by free-space electrooptic sampling. The changes in conductivity of the photoexcited sample is measured by monitoring the relative differential transmission of the peak of the THz pulse, $-\Delta T/T_0$, for different pump-probe delay times [10]. The relative differential transmission can then be related to the change in conductivity.

The relevant ultrafast processes in QDs after optical excitation in the QD GS and the BS are shown in Fig. 1(c). Only carriers in the BS and WL, here referred to as the conducting state (CS) (marked in blue in Fig. 1(c)), will contribute to the conductivity. For simplicity only the conduction band with electron capture into the QDs, the release of electrons into the CS and carrier recombination in both QDs and CS are shown. Also important for the ultrafast carrier dynamics in QDs are intradot carrier relaxation, carrier-carrier and carrier-phonon scattering. All these process are temperature and carrier density dependent [5, 11].

To measure the carrier capture dynamics into the QDs, carriers were excited into the BS with 800 nm excitation. The time dependencies of the frequency integrated change in sheet conductivity,$\Delta$ $\sigma_{sheet}$, for pump fluences in the range 0.04-1.18 $\mu$J/cm$^2$ are shown in Fig. 2(a). A near instantaneous rise of the photoinduced conductivity is observed, followed by a fast decay and a long-lived contribution. The fast decay is attributed to carrier capture into the nonconducting QD states. To support this interpretation, the change in photoinduced conductivity of bulk GaAs is measured (black curve in Fig. 2(a)), which shows a much slower initial decay. The

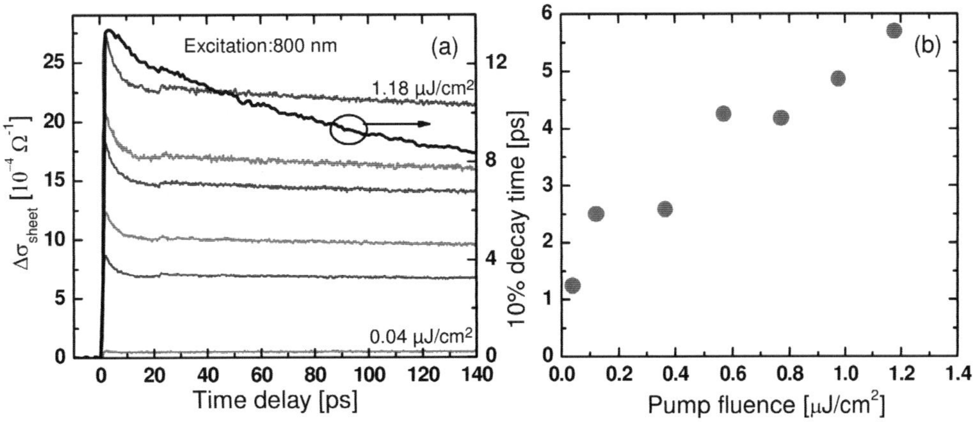


**Figure 2.** (a) Color lines: Dynamics of photoinduced sheet conductivity,Δ $\sigma_{sheet}$, of a InGaAs/GaAs QD SESAM at 800 nm optical excitation into the barrier layers, with pump fluences of 0.04-1.18 $\mu$J/cm$^2$. Black line: Dynamics ofΔ $\sigma_{sheet}$ in bulk GaAs, with 800 nm excitation with a pump fluence of 0.9 $\mu$J/cm$^2$. (b) 10% decay time of the traces in (a), as a function of pump fluence.

long-lived component is most likely caused by carriers excited in the Bragg reflector, which is accessible for 800 nm light. The Bragg reflector is a type-II $Al_{0.9}Ga_{0.1}As$ semiconductor, causing separation in space of electrons and holes, viaΓ -X transfer [12]. The little bump at around 22 ps (Fig. 2(a)) is an experimental artifact due to double reflection in a 2-mm thick beam attenuator.

The initial rise in conductivity depends linearly on excitation fluence (not shown here), as expected for one-photon absorption. From the fast decay of the conductivity, the trapping time into the QDs can be estimated. In Fig. 2(b), the time required for the conductivity to drop by 10% from its maximum, as function of pump fluence is plotted. A near-linear growth of the 10% decay time from 1.2 to 5.7 ps is observed, with increasing pump fluence. This growth in trapping time with increasing pump fluence is in agreement with observation made in Refs. [7, 8, 9, 10] and is attributed to filling of the QD trap states at higher excitation fluences, which decreases the trap efficiency. However, these observation are in contrast to the observed decrease in trapping time in Refs. [5, 6]. The different technique used in Ref. [5], where only the population of the QD GS is measured may explain the different observation. The suggested trapping mechanism electron-electron scattering may also provide carriers back into CS from the higher-lying QD states, thereby slowing down the decay dynamics of the conductivity. It is not clear at this point what causes the difference between Ref. [6] and our observations.

To observe the release of carriers from the QDs into the CS, carriers are resonantly excited into the QD GS at a wavelength of 1024 nm, with pump fluences in the range 2.1-22.8 $\mu$J/cm$^2$ (Fig. 3(a)). A non-instantaneous rise of the conductivity, which peaks after approximately 35 ps after photoexcitation, followed by a slow decay is observed. We interpret the slow rise time of the photoconductivity as release of carriers into the CS. After 35 ps the recombination in the BSs and WLs becomes the dominant process, causing the decay of the conductivity. The peak conductivity has a sub-linear dependence on pump fluence (Fig. 3(b)), suggesting that the rise in conductivity is caused by a saturated one-photon process, expected for resonantly excitation into the QD GS. From the fits to the saturable absorption [13] function, $\propto F_p \exp[-F_p/F_s]$, where $F_p$ is the pump fluence and $F_s$ is the saturation fluence, a saturation fluence of $42 \pm 6$ $\mu$J/cm$^2$ is obtained. This is in reasonable agreement with a value of $F_s = 25$ $\mu$J/cm$^2$, observed in a similar QD-based structure [3].

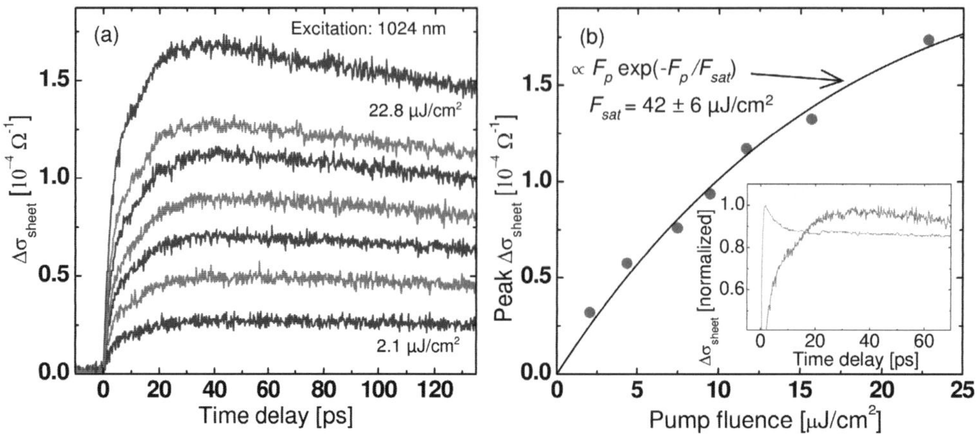


**Figure 3.** (a) Dynamics of photoinduced sheet conductivity,$\Delta$ $\sigma_{sheet}$, at resonant optical excitation of the QD GS at 1024 nm, with pump fluence of 2.1-22.8 $\mu$J/cm$^2$. (b) Maximum of $\Delta\sigma_{sheet}$ traces in (a), as a function of pump fluence. Solid line: Fit to the saturable absorption function with saturation fluence $F_s = 42 \pm 6$ $\mu$J/cm$^2$. Inset: Normalized$\Delta$ $\sigma_{sheet}$ traces for excitation at 800 nm into the barrier state with a fluence of 1.18 $\mu$J/cm$^2$ (red line) and at 1024 nm into the QD GS with a fluence of 22.8 $\mu$J/cm$^2$.

## 3. Conclusion

We have demonstrated the difference in photoconductivity dynamics of an InGaAs/GaAs SESAM, depending on whether carriers were optically excited resonantly into the QD GS or into the BS (inset of Fig. 3(b)). The dynamics can be explained by the dominating process, which is either carrier capture into the QDs or release of carriers into the CS.

## Acknowledgement

We acknowledge Danish Advanced Technology Foundation (HTF) and EU FP7 Programme (FAST-DOT), Grant No. 224338 for financial support; and to K. Iwaszczuk, D. G. Cooke and J. M. Hvam for valuable assistance and discussions.

## References

[1] Rafailov E U, Cataluna M A and Sibbett W 2007 *Nat. Photonics* **1** 395
[2] Liu H C, Gao M, McCaffrey J, Wasilewski Z R and Fafard S 2001 *Appl. Phys. Lett.* **78** 79
[3] Lagatsky A A, Bain F M, Brown C T A, Sibbett W, Livshits D A, Erbert G and Rafailov E U 2007 *Appl. Phys. Lett.* **91** 231111
[4] Morris D, Perret N and Fafard S 1999 *Appl. Phys. Lett.* **75** 3593
[5] Müller T, Schrey F F, Strasser G and Unterrainer K 2003 *Appl. Phys. Lett.* **83** 3572
[6] Yarotski D A, Averitt R D, Negre N, Crooker S A, Taylor A J, Donati G P, Stintz A, Lester L F and Malloy K J 2002 *J. Opt. Soc. Am.* B **19** 1480
[7] Turchinovich D, Pierz K and Jepsen P U 2003 *phys. stat. sol.* (c) **0** 1556
[8] Cooke D G, Hegmann F A, Mazur Y I, Ma W Q, Wang X, Wang Z M, Salamo G J, Xiao M, Mishima T D and Johnson M B 2004 *Appl. Phys. Lett.* **85** 3839
[9] Prasankumar R P, Scopatz A, Hilton D J, Taylor A J, Averitt R D, Zide J M and Gossard A C 2005 *Appl. Phys. Lett.* **86** 201107
[10] Lui K P H and Hegmann F A 2001 *Appl. Phys. Lett.* **78** 3478
[11] Nielsen T R, Gartner P and Jahnke F 2004 *Phys. Rev.* B **69** 235314
[12] Feldmann J, Nunnekamp J, Peter G, Göbel E, Kuhl J, Ploog K, Dawson P and Foxon C T 1990 *Phys. Rev.* B **42** 5809
[13] Haiml M, Grange R and Keller U 2004 *Appl. Phys.* B: Lasers Opt. **79** 331

# Advanced THz laser performance of shallow donors in axially stressed silicon crystal

**V N Shastin[1], R Kh Zhukavin[1], K A Kovalevsky[1], V V Tsyplenkov[1], S G Pavlov[2], H-W Hübers[2]**
[1]Institute for Physics of Microstructures, Russian Academy of Sciences (IPM RAS), Nizhny Novgorod, Russian Federation
[2]Institute of Planetary Research, Germany Aerospace Center (DLR), D-12489 Berlin, Germany

E-mail: shastin@ipm.sci-nnov.ru

**Abstract.** Terahertz stimulated emission (4–6 THz) from optically excited group-V donors (phosphor P, antimony Sb, arsenic As, bismuth Bi) in silicon crystal under compressive force has been studied. Measurements evidence that compressive force of 1–1.5 kbar for P, Sb and of 2–3 kbar for As, Bi applied along {100} crystallographic orientations results in the enhancement of the laser gain and enlarges the emission efficiency. At that laser threshold intensity can be decreased by two orders of magnitude. For As and Bi donors it is accompanied by the laser line switching while the $2p_0$ state turns out to be the upper laser state instead of the $2p_\pm$ one. The effect of the uniaxial stress on donor lasing originates from conduction band valley shift which split donor states changing their eigen-values and eigen-functions. According to the calculations of phonon-assisted relaxation rates appropriate donor modification increases the lifetime and makes pump efficiency of the upper laser states better.

## 1. Introduction.

At present the only known Si based sources of THz stimulated emission utilize intracenter optical transitions of group-V donor centers of substitution (phosphor P, antimony Sb, arsenic As, bismuth Bi) in silicon single crystal at low temperatures ($T \leq 20K$) [1]. Donor states are degenerated according to six-valley arrangement of silicon conduction band [2]. For s-type states corresponding six-fold degeneracy is partly lifted by short-range central cell potential localized at the site of a donor atom. Such effect known as a chemical shift depends on the nature of a donor and turns out to be significant for the $1s(A_1)$ ground state of the 1s multiplet. As a result different donors differ in phonon-assisted intracenter relaxation and it manifests itself in population of different impurity states and their laser performance. In fact under $CO_2$ laser excitation P and Sb donors laze on the transition ended by the 1s ($T_2$) states which for the As and Bi centers are completed by the transition to the 1s (E) subset ($T_2$, E, etc. denote the symmetry of the eigen functions). At the same time the odd parity $2p_0$ states occur to be the upper laser states in P and Sb while the $2p_\pm$ states carry out the same function in As and Bi centers [1]. Stated above concerns donors embedded in the unstrained silicon crystal. However the uniaxial crystal deformation results in substantial reconfiguration of donor states. Of course it becomes apparent in THz emission of donors. We focus on the experimental study of THz laser performance of shallow donors in silicon excited by $CO_2$ laser radiation (9.6-10.6μm) under the axial compressive stress applied along [100] crystallographic orientation to the host crystal. The theoretical discussion is shortened and will be extended somewhere.

## 2. Experiment.

Doped silicon crystals with donor concentration $N_d \approx 3\times10^{15}cm^{-3}$ and low compensation ($N_a/N_d<10^{-2}$) due to residual boron acceptors were grown using flout zone technique. For the THz experiment samples were shaped as regular rectangular parallelepipeds (~2–3×5–6×6–7mm$^3$) and lapped to provide internal reflection mode high Q cavity for THz frequency range.

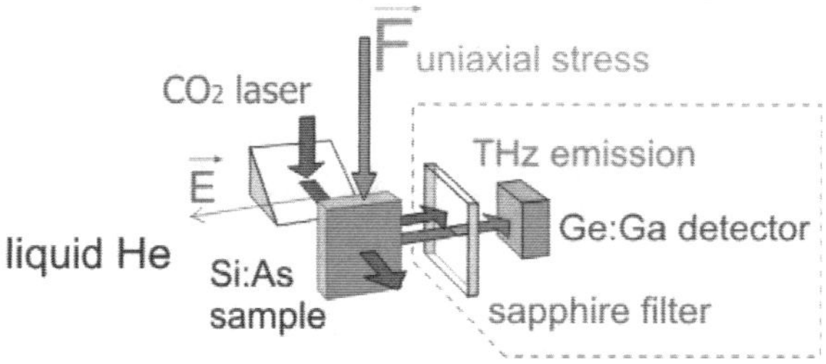


**Figure 1.** Experimental scheme.

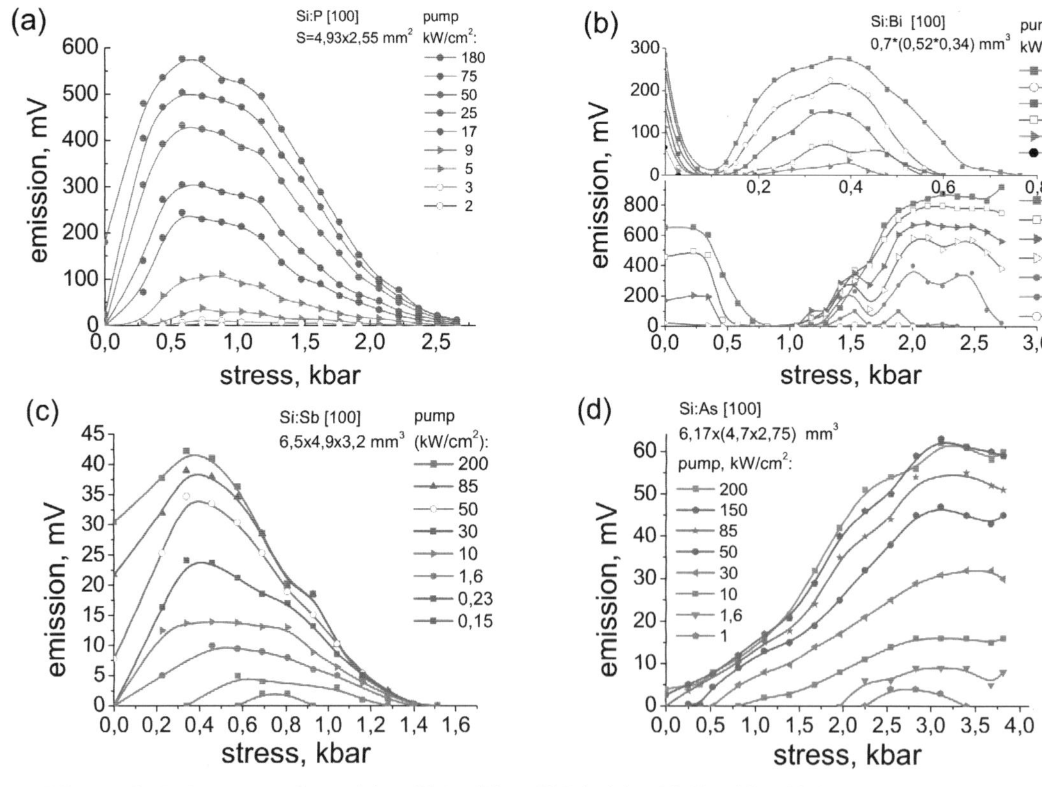


**Figure 2**. THz output from (a) – Si:P; (b) – Si:Bi; (c) –Si:Sb; (d) –Si:As versus compressive stress applied along [100] axis under different $CO_2$ laser (10.6μm) pump intensities. Pump intensity 200kW/cm$^2$ corresponds to photon flux density of $10^{25}$quant/cm$^2$s.

The experiment configuration is shown on Figure 1.THz output was provided due to the out-of-parallelism of reflecting facets and directional radiation pattern was broad. THz signal was detected by liquid helium cooled Ge:Ga photoconductor placed within of ~2cm in front of an investigated sample and was protected against pump radiation by one millimeter thick sapphire filter. Character dependencies of THz stimulated emission intensity from silicon doped by different group-V donors on compressive stress and carbon dioxide TEA laser pump intensity are presented on Figures 2 and 3.

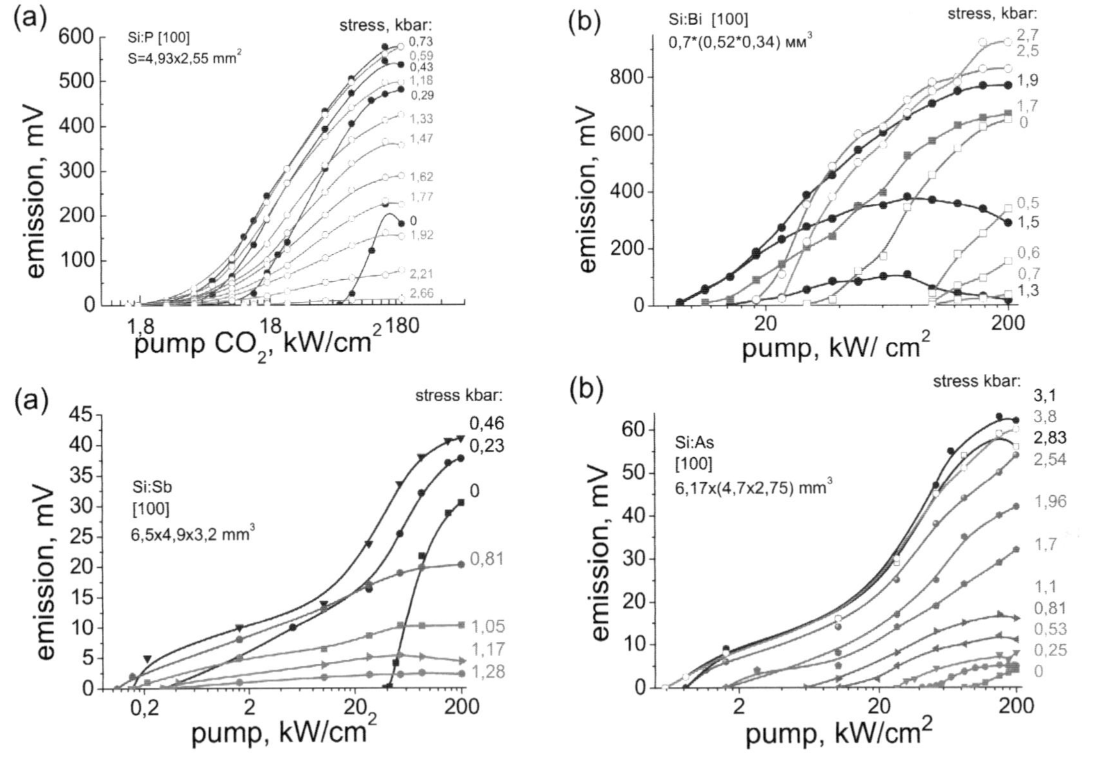


**Figure 3.** THz output on pump intensity (10.6μm) under different values of [100] compressive stress for (a) – Si:P; (b) – Si:Bi (c) –Si:Sb; (d) –Si:As samples.

## 3. Discussion.

Experiment exhibits strong dependence of THz laser performance from donors on the uniaxial compression of silicon crystal. That reveals itself both in THz output and laser threshold intensity. One can see remarkable decrease of laser threshold intensity for silicon samples under the stress applied. Note that for Si:As it is accompanied by a change-over of the upper laser states from the $2p_{\pm}(T_1\&T_2)$ with emission frequency $\nu$=200&212 cm$^{-1}$ to the $2p_0$ ($B_2$) with $\nu$=172cm$^{-1}$ if stress is applied along [100] and exceeds $F$=0.3 kbar [3]. For bismuth donors the same laser transition shift has been revealed if $F$>1.5 kbar [4]. One can see that noticeable increasing of the laser gain is observed for pump flux densities $10^{22}$–$10^{24}$quant/cm$^2$c (Si:P, Si:Sb) and $10^{23}$–$10^{25}$quant/cm$^2$s (Si:As, Si:Bi).According to the theoretical treatment stress induced rise of THz gain is explained by several reasons. First of all, conduction band valley shift turns off *f*-phonon relaxation of the $2p_0$ laser states and their lifetime (40ps in P, 50ps in Sb) become ~ 3–4 times longer [5]. Secondly, the lifting of state degeneracy can increase pump efficiency of the laser states by factor of 2–3. The letter depends on the energy of initially excited electrons and their relaxation routs via conduction band continuum and donor states (Figure 4). At that *TA-f* phonon (47.6 meV) emission plays a key function. Moreover the net gain includes negatively charged $D^-$ donor absorption [6]. Note that the photo-ionization of $D^-$ centers leads to a specific downward tendency of the THz output under pump flux density from $10^{25}$ down to $10^{24}$quant/cm$^2$s (Figure3). For pump intensities $10^{23}$–$10^{24}$quant/cm$^2$s $D^-$ center concentration ($N_a/N_d$<<1) is estimated to be of ~$10^{-2}N_d$ while their absorption cross-section is of~$3\times10^{-15}$cm$^2$ (see [7]). Thus $D^-$ absorption is high enough to prevent lasing for the pump less than $10^{24}$quant/cm$^2$s. The axial stress reduces $D^-$ concentration to a tenth part (compare [8]). Reasonable agreement between measurement results and calculations can be obtained if one takes into account all pointed factors. Important that stress induced enhancement of the laser gain raise hopes of CW laser operation from group-V donor centers in silicon.

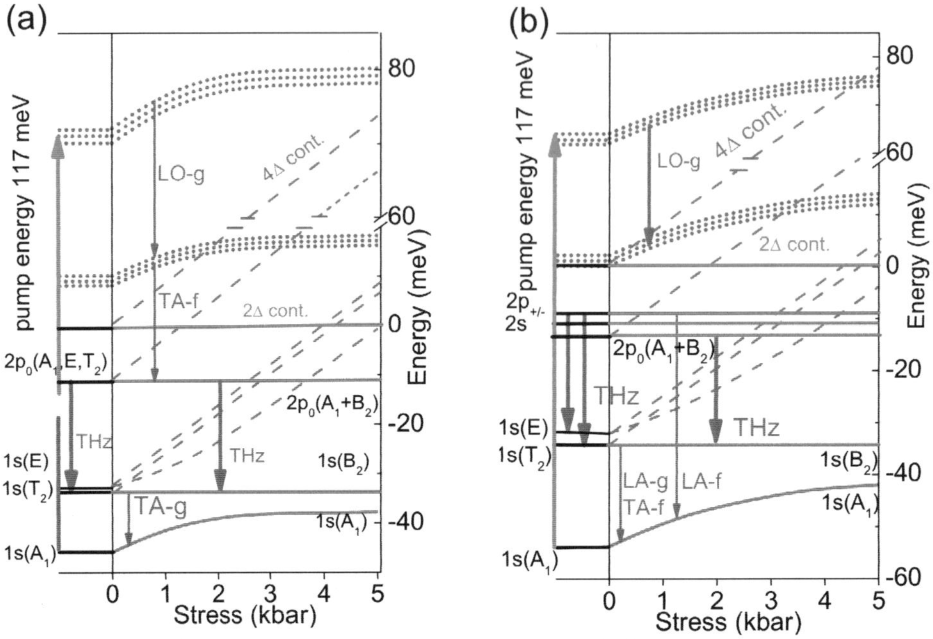


**Figure 4.** Layout of principal electron states, pump and THz laser transitions for (a) – phosphor and (b) – arsenic donors versus [100] stress. Phonon-assisted intervalley relaxation due to transverse acoustic *TA-f* and *TA-g*, longitudinal acoustic *LA-f* and optical *LO-g* phonon emission (umklapp process) are depicted by arrows.

## 4. Acknowledgments.

This work was supported by the Deutsche Forschungsgemeinschaft (DFG) and the Russian Foundation for Basic Research (RFBR-DFG 08-02-91951, RFBR 08-02-00333).

## 5. References.

[1]. Hübers H-W, Pavlov S G and. Shastin V N 2005 *Semicond. Sci. Technol.* **20** S211

[2]. Ramdas A K and Rodriguez S 1981 *Rep. Prog. Phys.* **44** 1297

[3]. Pavlov S G, Böttger U, and Hübers H-W, Zhukavin R Kh, Kovalevsky K A, Tsyplenkov V V, and Shastin V N, Abrosimov N V and Riemann H 2007 *Appl. Phys. Lett.* **90** 141109

[4]. Zhukavin R Kh, Pavlov S G, Hübers H-W, Kovalevsky K A, Tsyplenkov V V and Shastin V N 2006 *Proc.Int.Conf. on Joint Infrared and millimiter waves and Teraherz Electronic* (Shanghai, China) p.393

[5]. Tsyplenkov V V, Demidov E V, Kovalevsky K A, Shastin V N 2008 *Semiconductors* **42** 1016-22

[6]. Zhukavin R Kh, Pavlov S G, Kovalevsky K A, Hübers H-W, Riemann H, Shastin V N 2005 *J. Appl. Phys.* **97** 113708

[7]. Gershenzon E M, Melnikov A P, Rabinovich R I 1985 *in A.L. Efros and M. Pollak (eds.) Electron-Electron Interactions in Disordered systems, Elsevier Science Publishers, Amsterdam* 483

[8]. Olivera L E, Falicon L M 1986 *Phys.Rev.B*, **33** 6990-97

Journal of Physics: Conference Series **193** (2009) 012087 doi:10.1088/1742-6596/193/1/012087

# Relaxation of THz impurity photoconductivity in GaAs/InGaAsP and Ge/GeSi quantum well heterostructures

**S.V. Morozov, K.V. Marem'yanin, D.I. Kuritsin, S.M. Sergeev, A.V. Antonov, L.V. Gavrilenko, E.E.Orlova, I.V. Erofeeva, M.S. Joludev, V.I.Gavrilenko**

Institute for Physics of Microstructures of Russian Academy of Sciences, 603950 N.Novgorod, Russia
E-mail: more@ipm.sci-nnov.ru

**Abstract.** In *n*-GaAs/InGaAsP and *p*-Ge/GeSi quantum well heterostructures excited by both the picosecond broad band and the nanosecond narrow band pulses of THz radiation the relaxation times of the impurity photoconductivity were measured. In one of the sample Ge/GeSi this more narrow quantum well we observed the first time increases maximum relaxation time from 9 to 14 ns at the narrow band THZ radiation frequency decrease from 60 $cm^{-1}$ down to 25 $cm^{-1}$ that can be explained by a dispersion of shallow acceptor binding energies. The frequency of 60 $cm^{-1}$ nearly corresponds to the binding energy of acceptors situated nearby Ge QW center while that of 25 $cm^{-1}$ corresponds to ionization energy for the hole (in Ge QW) bound with a remote acceptor ion in the center of GeSi barrier. We explain the observed larger relaxation time for the acceptor with lower binding energy by an enhanced probability for the hole captured by an excited impurity state to be thermally activated again into the valence band.

## 1. Introduction

Nowadays there is a growing interest to the nonequilibrium phenomena and possibilities to produce the intraband population inversion in semiconductor nanostructures doped with shallow impurities. In order to design THz laser operating at the optical transitions involving shallow impurity states direct information on lifetimes of carriers both excited into conduction (valence) band and captured by excited impurity states is required. In this paper we study the relaxation of impurity photoconductivity at T = 4.2K in n-GaAs/$In_{0.1}Ga_{0.9}As_{0.8}P_{0.2}$ and p-Ge/$Ge_{0.9}Si_{0.1}$ quantum well (QW) heterostructures excited by broadband THz radiation (0.5 – 3 THz) generated by optical rectification of femtosecond laser pulses and by narrow band (0.2 $cm^{-1}$) tunable THz pulses of nanosecond duration.

## 2. Experimental

GaAs/$In_{0.1}Ga_{0.9}As_{0.8}P_{0.2}$ heterostructures under study were pseudomorphically grown by MOCVD technique on GaAs(001) semi-insulating substrates and contained 30 GaAs QWs separated by 400 A wide $In_{0.1}Ga_{0.9}As_{0.8}P_{0.2}$ barriers. QW widths $d_{QW}$ were 200 A in the structure #4147 and 90 A in #4236 one. Silicon delta-layers were introduced in the center of each QW. The donor concentration measured by Hall effect was $3.3*10^{10}$ $cm^{-2}$ per QW. Strained Ge/$Ge_{1-x}Si_x$ heterostructures with residual acceptors were CVD grown on Ge(111) substrates [1]. The structure #306 (x = 0.12) contained 162 Ge QWs 200 A wide and #308 one (x = 0.09) contained 80 Ge QWs 355 A wide. The residual acceptor concentration was about $10^9$ $cm^{-2}$ per QW. The photoconductivity spectra were measured using BOMEM DA3.36 Fourier-transform spectrometer at $T$ = 4.2 K. Two strip ohmic contacts 4 mm apart were deposited onto a sample surface 5×5 mm in size. The samples were biased by a d.c. voltage. Impurity photoconductivity spectra of GaAs/$In_{0.1}Ga_{0.9}As_{0.8}P_{0.2}$ samples contain a broad long-wave band in between 30 and 100 $cm^{-1}$ corresponding to transitions from the ground donor state into excited states and to the continuum (Fig.1,2). In the photoconductivity spectra of Ge/GeSi samples #306 and #308 broad band in between 20 and 40 $cm^{-1}$ and around 60 $cm^{-1}$ respectively were observed (Fig.3,4). The binding energy of shallow acceptor is known to depend on the impurity position in QW [1,2].

These spectra due to the built-in strain in the Ge/GeSi heterostructures differ significantly from the impurity photoconductivity spectrum of bulk *p*-Ge where the maximum is located nearby 100 $cm^{-1}$.

Investigations of impurity photoconductivity relaxation kinetics were carried out at the excitation by THz pulses with two techniques. In the first case picosecond pulses of broadband (0.5 to 3 THz). THz radiation were generated by optical rectification in ZnTe crystal of 100 fs long laser pulses ($\lambda$= 800 nm). In the second case the narrow-band THz difference frequency radiation was generated in GaP nonlinear crystal by mixing of two near IR laser beams [3]. The photoconductivity signal was registered with averaging by a digital oscilloscope with temporal resolution of 1 ns.

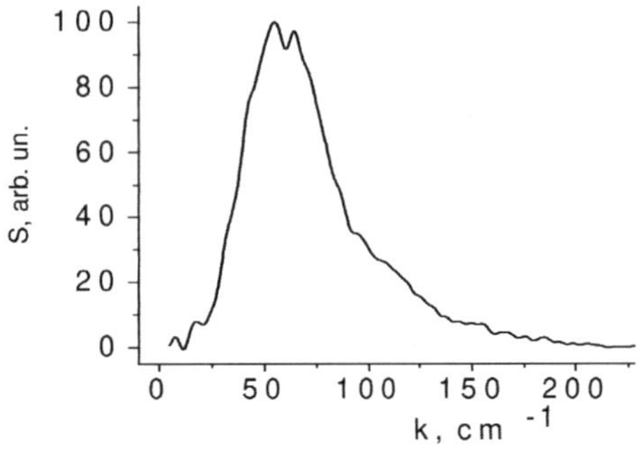


**Figure 1.** Photoconductivity spectrum of GaAs/$In_{0.1}Ga_{0.9}As_{0.8}P_{0.2}$ structure #4147 GaAs quantum wells 200 A wide. Bias voltage *U*=0.9 V.

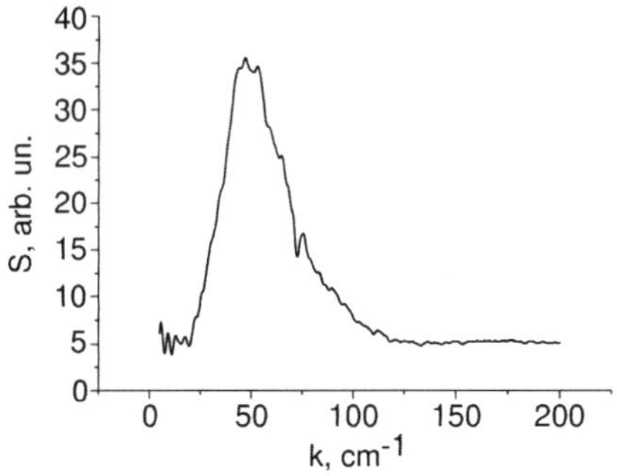


**Figure 2.** Photoconductivity spectrum of GaAs/$In_{0.1}Ga_{0.9}As_{0.8}P_{0.2}$ structure #4234 with GaAs quantum wells 90 A wide. Bias voltage *U*=0.9 V.

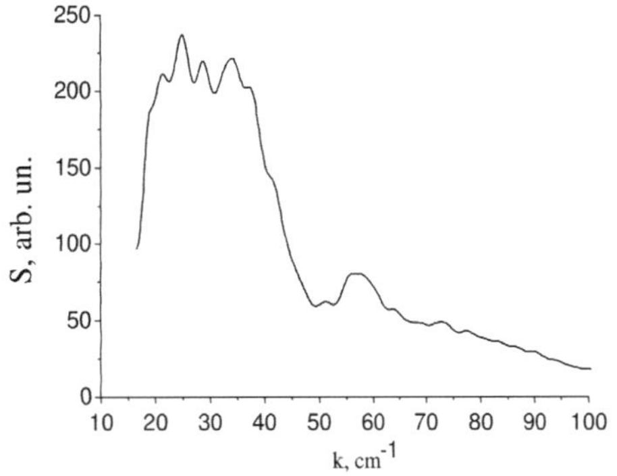


**Figure 3.** Photoconductivity spectrum of strained Ge/$Ge_{0.88}Si_{0.12}$ structure # 306 with Ge quantum wells 200 A wide. Bias voltage *U*=1.2 V.

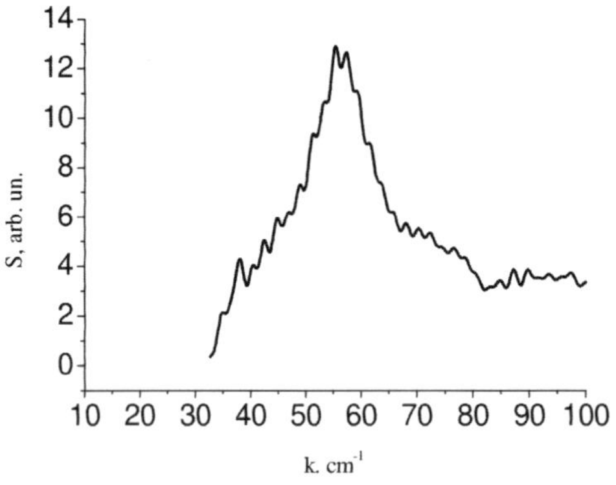


**Figure 4.** Photoconductivity spectrum of strained Ge/$Ge_{0.91}Si_{0.09}$ structure # 308 with Ge quantum wells 355 A wide. Bias voltage *U*=1.2 V.

## 3. Results and discussions

Typical oscillograms of the photoresponse on broad band THz radiation in two GaAs/InGaAsP samples are given in Fig.5,6. In both cases a significant decrease of the photoresponse relaxation time $\tau$ with the bias voltage *U* rise from 2 up to 10 V was observed: from 5 ns down to 2 ns in the sample #4147 and from 3 ns down to 2 ns for #4236 one. In Ge/GeSi samples the $\tau(U)$ dependence proved to be more complicated – Fig.7,8. For both samples #306 and #308 the increase of the bias voltage at first results in the rise of the decay time (from 4.2 to 9.4 ns and from 4.9 to 11.3 ns, correspondently). We relate this effect within the mechanism of the cascade carrier capture by an ionized impurity due to emission of acoustical phonons with the carrier heating by the applied electric field as well as with the destruction in the electric field the excited impurity states [4] (cf.[5]). The cascade acoustic phonon emission model of the carrier capture by ionized impurity [4] does not take into account the effect of impact ionization of shallow impurities. As it was shown in Ref.[5] the impurity breakdown results in a hot carrier "cooling" that leads to the relaxation time drop with the further increase of the electric field just we have observed in *p*-Ge/GeSi samples. We have not observed the relaxation time increase with the electric field prior to the impurity breakdown in *n*-GaAs/InGaAsP samples that can be attributed to their higher doping. Therefore in the latter case the Auger recombination of free carriers can be important nearby the breakdown electric field thus preventing from the above increase of the relaxation time with the carrier heating.

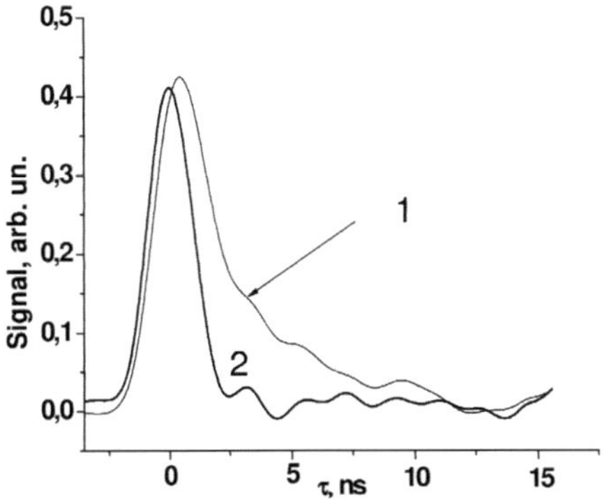


**Figure 5.** Oscillograms of the photoresponse on broad band THz radiation in GaAs/InGaAsP sample #4147 ($d_{QW}$ = 200 A) at $U$ = 2 V (curve 1) and at $U$ = 10 V (curve 2).

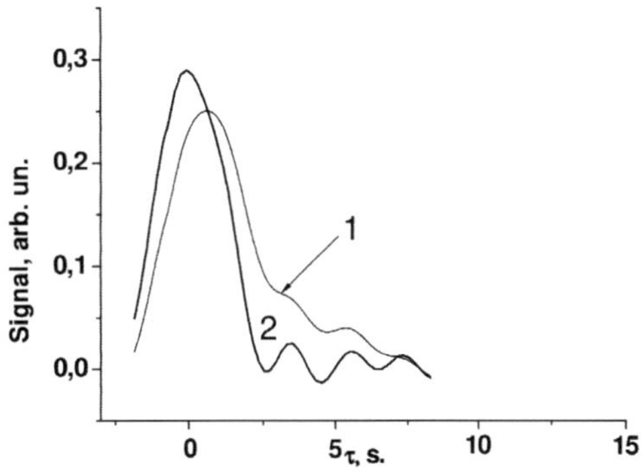


**Figure.6.** Oscillograms of the photoresponse on broad band THz radiation in GaAs/InGaAsP sample #4236 ($d_{QW}$ = 90 A) at $U$ = 2 V (curve 1) and $U$ = 10 V (curve 2).

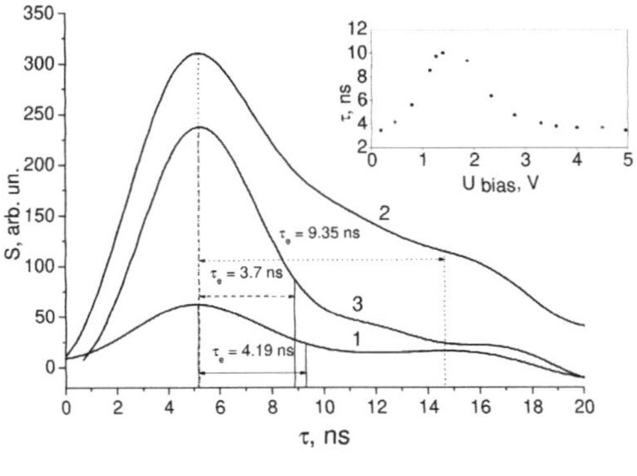


**Figure 7.** Oscillograms of the photoresponse on broad band THz radiation in Ge/GeSi sample #306 ($d_{QW}$ = 200 A) at $U$ = 0.78 V (curve 1), $U$ = 1.86 V (curve 2) and $U$ = 4.5 V (curve 3). In the insert: photoresponse relaxation time versus bias voltage.

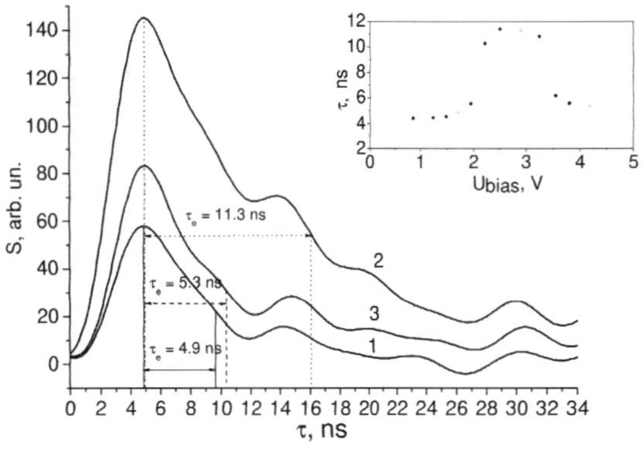


**Figure 8.** Oscillograms of the photoresponse on broad band THz radiation in Ge/GeSi sample #308 ($d_{QW}$ = 355 A) at $U$ = 1.7 V (curve 1), $U$ = 2.9 V (curve 2) and $U$ = 4.2 V (curve 3). In the insert: photoresponse relaxation time versus bias voltage.

A detailed study of the relaxation time on the radiation frequency was undertaken in *p*-Ge/GeSi sample #306 with QWs 200A wide using tunable narrow-band THz difference frequency radiation pulses about 5 ns in duration [3]. One can see in Fig.9 that the observed $\tau(U)$ dependences in the case of monochromatic excitation correspond fairly well to that obtained at the excitation by the broad band THz radiation (Fig.7), the relaxation time increase being observed up to the impurity breakdown voltage (see insert in Fig.9). As easy to see in Fig.9 the maximum (versus bias voltage) relaxation time increases from 9 to 14 ns at the radiation frequency $\lambda_{exi}$ decrease from 60 cm$^{-1}$ down to 25 cm$^{-1}$ that can be explained by a dispersion of shallow acceptor binding energies [1]. The frequency of 60 cm$^{-1}$ nearly corresponds to the binding energy of acceptors situated nearby Ge QW center while that of 25 cm$^{-1}$ corresponds to ionization energy for the hole (in Ge QW) bound with a remote acceptor ion in the center of GeSi barrier. We explain the observed larger relaxation time for the acceptor with lower binding energy by an enhanced probability for the hole captured by an excited impurity state to be thermally activated again into the valence band (rather than to relax to the ground state).

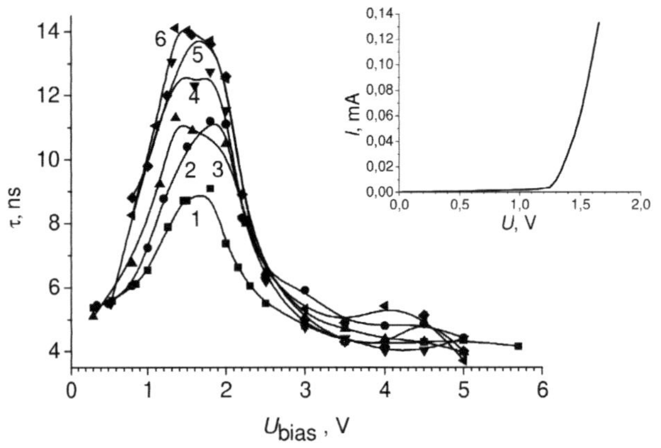


**Figure 9.** Dependences photoresponse relaxation time «$\tau$» at tunable narrow-band THz emission for Ge/GeSi sample #306 versus bias voltage $\tau(V)$ -$\lambda_{exi}(cm^{-1})$ : *1 (■) - 61; 2 (●) - 56; 3 (▲) - 42; 4 (▼) - 34; 5 (◆) - 30; 6 (◀) - 25 cm*$^{-1}$. In insert volt-ampere characteristic.

**Acknowledgements**
This work was financially supported by RFBR (grant 07-02- 01307), Russian Federal Agency for Science and Innovations (contract 02.518.11.7133) and by Russian Academy of Sciences.

References

[1] V.Ya.Aleshkin, V.I.Gavrilenko, I.V.Erofeeva, A.L.Korotkov, D.V.Kozlov, O.A.Kuznetsov, M.D.Moldavskaya. phys. stat. sol. (b), 1998, v.210, p.649.
[2] A. A. Reeder, J.-M. Mercy, B.D. McCombe. Effect of conferment on shallow donors and acceptors in GaAs/AlGaAs quantum wells. IEEE J. Quant. Electron. - 1998. – Vol.24, № 8. P. 1690-1697.
[3] V.Ya.Aleshkin, A.A.Antonov, S.V.Gaponov, A.A.Dubinov, Z.F.Krasil'nik, K.E.Kudryavtsev, A.G.Spivakov and A.N.Yablonskii. JETP Letters, vol.88, p.787 (2009).
[4] V.N.Abakumov, L.N.Kresshchuk and I.N.Yassievich. Sov. Phys. Semicond. **12**, 152 (1978).
[5] W.Quade, G.Hüpper, E.Schöll, K.Kuhn. Phys. Rev. B **49**, 13408 (1994).

Journal of Physics: Conference Series **193** (2009) 012088 doi:10.1088/1742-6596/193/1/012088

# THz Probe Studies of MBE Grown Epitaxial GaAs

**D. Köseoğlu, H.H. Güllü, and H. Altan**

Department of Physics, Middle East Technical University, 06531, Ankara, Turkey

Tel. (90 312) 210-3277, Fax (90 312) 210-5099,

email: devrim@newton.physics.metu.edu.tr

**Abstract**. We have built a THz time-domain spectrometer driven by a sub-15fs pulse duration mode-locked Ti:$Al_2O_3$ laser. Using THz time-domain spectroscopy with photoconductive antenna for THz generation and electro-optic sampling for detection as well as photoexcited THz spectroscopy, we measured the carrier concentrations and mobilities of epitaxially grown undoped GaAs samples to be used in photoconductive antenna production. The samples were grown at 600 °C to 1 μm effective layer thickness on top of a 650 μm SI-GaAs wafer. The resistivities, mobilities and the carrier concentrations were measured and calculated by the van der Pauw method under the magnetic field. These Hall effect measurements and the THz probe studies were compared with each other. The measurements and calculations obtained electronically are compared optically using the Drude Model for the conductivity and mobility.

## 1. Introduction

One of the major applications of THz time-domain spectroscopy (THz-TDS) is in material characterization. THz spectroscopy has been used to determine ultrafast carrier dynamics of doped semiconductors such as GaAs and silicon wafers.[1 2, 3, 4] An important focus is on the measurement of the dielectric constant of thin films[6]. Epitaxially grown GaAs films using MBE methods have attracted much attraction due to their projected use in the infrared detection systems.

In this study, we have demonstrated the efficiency of THz-TDS to measure the electron and hole mobilities and the carrier number densities of MBE grown GaAs. From the photoexcitation measurements, we found that the photoexcitation increases the conductivity of the GaAs samples in addition to changes in the refractive index. This paper is organized as follows: In section 2, we describe the experimental setups and procedure as well as the data collected by the Hall effect and THz spectroscopy measurements. In the section 3, we present the analysis of data, and the critical parameters of the epitaxial grown GaAs. In the section 4, the results are discussed.

## 2. Experiment

The epitaxial nominally undoped GaAs samples were grown at 600 °C to a 1 μm effective layer thickness on top of a 650 μm thick SI-GaAs wafer. The resistivity measurements were done by the van der Pauw technique along with the standard Hall Effect measurements[5, 6] to determine mobilities and the carrier concentrations. To obtain accurate data, very small (less than 1 $mm^2$) AuGe Ohmic contacts were devised at the corners of the square shaped (1.4 cm x 1.4 cm) GaAs sample. The resistivity, conductivity, Hall coefficient, charge carrier concentration, and mobility of the p-type GaAs samples were calculated to be 10.4 Ω.cm, 0.096 $\Omega.cm^{-1}$, 5800 $cm^3.C^{-1}$, $1.1x10^{15}$ $cm^{-3}$, and 550 $cm^2.V^{-1}.s^{-1}$ respectively.

The THz spectroscopy techniques, on the other hand, provide noncontact measurements with subpicosecond temporal resolution as an alternative method in the field of solid state electronics where it is difficult to use traditional probes. This makes THz spectroscopy an ideal tool for measuring the conductivities and obtaining data on charge carrier dynamics without use of electrical contacts.

THz-TDS is driven by a mode-locked Ti:$Al_2O_3$ laser with a center wavelength of $\lambda$ = 800 nm and a pulse duration of $\tau$ = 15 fs at a repetition rate of 75 MHz. 180 mW of the average beam power is split into two optical lines, where 36 mW is fed through to a LT-GaAs based photoconductive antenna (PCA) with a dipole gap of 6 μm is used for generation. The antenna bias was AC modulated at 2.5 kHz with $V_{p-p}$ = 10 V. For the detection of the THz beam, the electro-optic method is performed by use of a 2 mm thick <110> oriented ZnTe crystal, a quarter wave-plate, Wollaston prism (WP) and a balanced photodetector. The amplitude and phase of the signal was detected with the aid of a digital dual channel lock-in amplifier (Model SR830 DSP). The photoexcited measurements were made with the aid of 750 mW continuous laser at 808 nm pump wavelength. The sample was placed at 45° to the THz and probe beam. The THz beam was focused on the sample by using 10 cm focal length TPX lenses. The system shown is Figure 1.

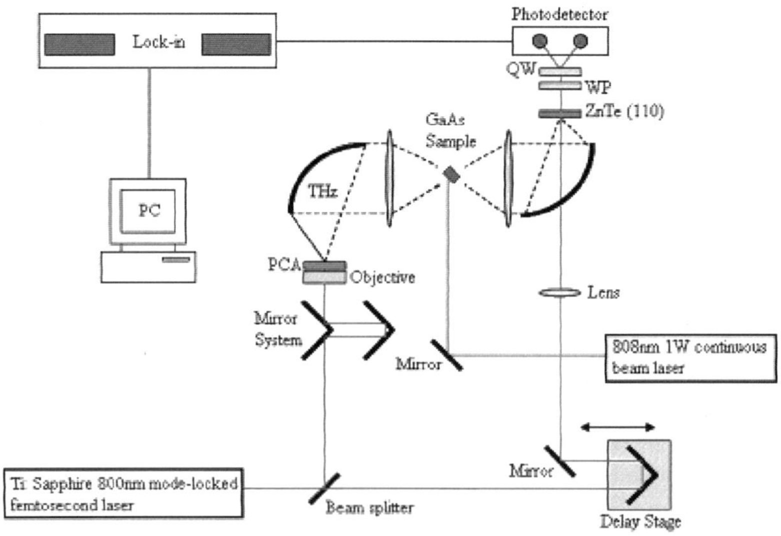


Figure 1: THz spectroscopy system for the probe measurements

The typical THz waveforms measured on the GaAs sample as well as photoexcited data is shown in Figure 2.

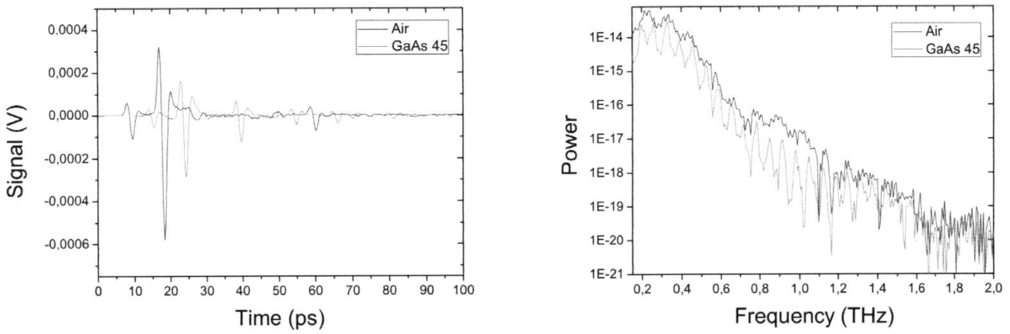


Figure 2a: THz-TDS scan and FFT of the direct spectroscopy of GaAs compared with air.

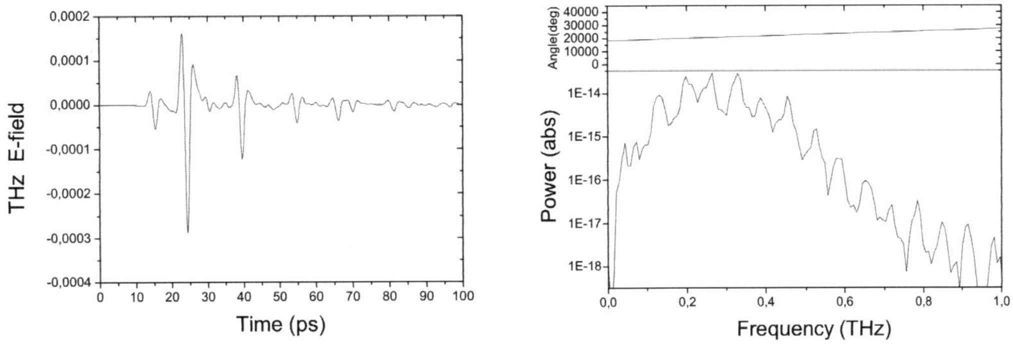


Figure 2b: THz-TDS of Time-Domain scan and FFT of the photoexcited spectroscopy

## 3. Analysis

Using the experimental data, the real ($n_r$) and complex ($n_i$) components of the refractive index can be calculated by

$$n_r = \frac{1}{kl}\left(\varphi(\omega,l) - \varphi(\omega)\right) \text{ and } n_i = \frac{1}{kl}\left(\ln\left(\frac{E(\omega,l)}{E(\omega)}\right)\right)$$

where $k$ is the wave number, $l$ is the thickness of the sample. The results of these calculations are shown in Figure 3.

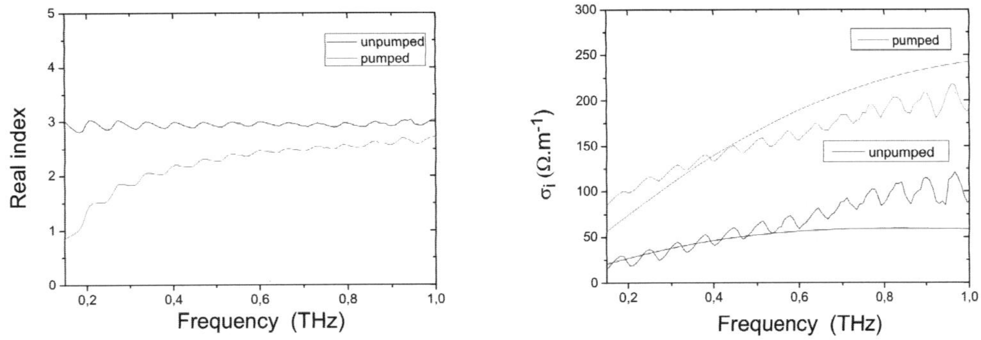


Figure 3a (left): The real part of the refractive index obtained by direct and photoexcited THz spectroscopy
Figure 3b (right): The conductivities obtained by direct and photoexcited THz spectroscopy with their Drude fits

The absorption coefficient of the sample can be obtained as $\alpha = \frac{1}{l}\left(\ln\left(\frac{I}{I_0}\right)\right)$ where $I$ is the intensity after transmission and $I_0$ is the intensity without sample. The complex absorption spectrum is obtained by the Fourier analysis and the imaginary index can be extracted by $\alpha = \frac{2n_i\omega}{c}$.

Any of the above parameters can be obtained experimentally, however, the other carrier dynamics, such as conductivity and mobility can be extracted from an appropriate conduction model. In order to calculate the conductivity and the mobility, the Drude model for electrical conductivity is appropriate to use in which the charge carriers can move freely, although their motion is subject to damping with time constant $\tau$. From the experimental values, the imaginary and real components of conductivity are calculated by $\sigma_i = \varepsilon_0\omega\left(\varepsilon_{dc} - n_r^2 + n_i^2\right)$ and $\sigma_r = 2n_r n_i \omega\varepsilon_0$. The extracted values for the mobility and carrier density are obtained by a fit to the experimental data shown in Figure 3 where the imaginary conductivity is determined by

$$\sigma_i = \frac{\varepsilon_0\omega\omega_p^2}{\omega^2 + \langle\tau\rangle^{-2}}.$$

In this equation, $1/\tau = v_c$ is the collision frequency $\omega_p = \left(\frac{Ne^2}{m\varepsilon_0}\right)^{1/2}$ is the plasma frequency, $N$ is the number density, $e$ is the charge of the free carriers, $m$ is the mass of the free carriers. From this model, the Drude parameters for direct THz spectroscopy are calculated to be N = $1.0\times10^{16}$ $cm^{-3}$, $v_C = 5.26\times10^{12}$ $s^{-1}$ where m = 0.45 $m_0$ [7] and for the photoexcited THz spectroscopy N = $1.0\times10^{16}$ $cm^{-3}$, $v_C = 8.33\times10^{13}$ $s^{-1}$ where m= 0.067 $m_0$.

Therefore, using $\mu = \frac{e}{m v_c}$, the calculated the mobility values of the GaAs sample from the direct and photo excitation measurements are 740 $cm^2\ V^{-1}\ s^{-1}$ and 3150 $cm^2\ V^{-1}\ s^{-1}$ respectively.

## 4. Conclusion:

We have demonstrated that linear and photoexcited THz-TDS measurements give an estimate of the carrier densities and both the hole and electron mobility values with good precision. Here we assumed single layer transmission in our calculations, inclusion of multilayer transmissions can improve the accuracy of the results. While Hall effect and van der Pauw measurements are more accurate, THz-TDS coupled with photoexcitation allows for measurement of electron mobilities of p-type samples.

**Acknowledgments:** We would like to thank Prof. Gengiz Beşikçi and his group for giving us the GaAs samples, and Prof. Mehmet Parlak and his group for aiding in the van der Pauw measurements. This work is supported by TÜBİTAK grant #107T742.

**References:**

[1] D. Grischkowsky, et al, J. of the Opt.Soc.Am.B: Optical Physics, 7 (10), pp2006-20015, (1990)
[2] M. van Exter, D. Grischkowsky, IEEE Trans. Microwave Theory and Tech, 38 (11), pp.1684-1691, (1990).
[3] M. van Exter, D. Grischkowsky, Applied Physics Letters, 56 (17), pp.1694-1696, (1990)
[4] M. van Exter, D. Grischkowsky, Physical Review B, 41 (7), pp.12140-12149, (1990)
[5] L.J. van der Pauw, Philips Res. Repts,.13, 1-9, (1958)
[6] ASTM International, F 76 – 86, (2002)
[7] Tobias Bauer, et al, Semicond. Sci. Techol. 18, 28-32, (2003)

# Miniband quantum transport in semiconductor nanodevices under broadband illumination

**Fabrizio Castellano**[1,2], **Rita C. Iotti**[2], **Fausto Rossi**[1]
[1]ETH Zurich, Wolfgand-Pauli-Strasse 16, Zurich, Switzerland
[2]Politecnico di Torino, Corso Duca degli Abruzzi 24, Torino, Italy
E-mail: castella@phys.ethz.ch

**Abstract.** A general theoretical framework for the modelling of miniband transport in semiconductor nanostructures under broadband illumination is presented. Our approach is based on the multi-subband three-dimensional Boltzmann transport equation and allows to investigate both the steady-state and the time-dependent dynamics. The model has been succesfully applied to the problem of the design and optimization of Terahertz quantum well infrared photodetectors and is now applied to a more generic superlattice system. We consider the carrier dynamics in multi-miniband superlattices under black-body illumination showing the existence of regimes in which the current in some minibands is inverted, eventually leading to absolute negative resistance.

Electron dynamics in photoexcited nanostructures has been a matter of intense study for a long time both for the fundamental physical insights that this kind of systems can give and for potential technological applications.

Among others, semiconductor superlattices has received considerable attention as they provide a relatively simple way to build systems in which the band structure of electrons can be precisely determined. This ability, along with the use of optically-active materials, allows to study interesting phenomena regarding the interactions between electrons and light. In particular, superlattices subject to coherent electromagnetic fields, like those typically provided by lasers, have been shown to give rise to a number of interesting phenomena including light-induced absolute negative resistance [1, 2, 3].

On the other hand superlattices subject to an incoherent and broadband photoexcitation rather than a laser source are also interesting as they can be used as a model for state-of-the-art quantum well infrared photodetectors (QWIPs) which basically consist in a sequence of quantum-wells and barriers (comprising many tens of periods) which are then exposed to background blackbody radiation and infrared light emitted by objects. The incoming photons will then promote bound electrons into the continuum which in turn will give rise to a current once the system is properly biased.

In the past we have focused our attention on the development of a general theoretical framework for the simulation of carrier transport photoexcited weakly-coupled superlattices and we have applied it to the simulation and design of a new architecture for THz QWIPs with improved temperature performance [4, 5]. The model is based on the multi-subband three-dimensional Boltzmann transport equation describing the dynamics of the electron population

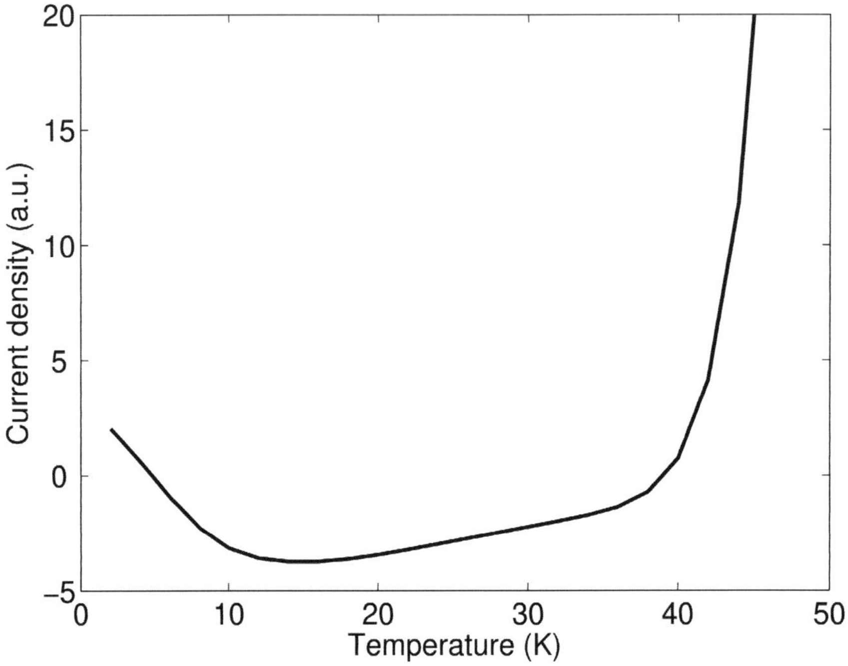


**Figure 1.** Total current as a function of temperature when the device is subject to a positive bias and irradiated with a 300K blackbody. The device shows absolute negative resistance between 5 and 40 K.

in a superlattice

$$\frac{\partial f_\nu(\mathbf{k})}{\partial t} = \frac{e}{\hbar}\mathbf{F} \cdot \nabla f_b(\mathbf{k}) + \sum_{\nu'=1}^{N} \int \left[ P_{\nu\nu'}(\mathbf{k},\mathbf{k}') f_{\nu'}(\mathbf{k}') - P_{\nu'\nu}(\mathbf{k}',\mathbf{k}) f_\nu(\mathbf{k}) \right] \mathrm{d}\mathbf{k}' \tag{1}$$

where $f_\nu(\mathbf{k})$ is the single-particle distribution function of electrons in a state with wavevector $\mathbf{k}$ in subband $\nu$, $P_{\nu'\nu}(\mathbf{k}',\mathbf{k})\mathrm{d}\mathbf{k}'$ is the probability per unit time that a scattering event bringing an electron from a state in band $b$ and wavevector $\mathbf{k}$ to a state in band $\nu'$ and wavevector $\mathbf{k}'$ occurs, and $\mathbf{F}$ is the external electric field providing the electron drift.

The knowledge of the distribution function then allows us to compute the mean values of all physical observables including the current density across the device.

The various scattering mechanisms affecting the electron dynamics are included into the global probabilities $P_{\nu'\nu}(\mathbf{k}',\mathbf{k})$ and may be separated into the following contributions

$$P_{\nu'\nu}(\mathbf{k}',\mathbf{k}) = P^{opt}_{\nu'\nu}(\mathbf{k}',\mathbf{k}) + P^{th}_{\nu'\nu}(\mathbf{k}',\mathbf{k}) \ , \tag{2}$$

where $P^{opt}_{\nu'\nu}(\mathbf{k}',\mathbf{k})$ is the electron-photon interaction part and $P^{th}_{\nu'\nu}(\mathbf{k}',\mathbf{k})$ accounts for all thermalization processes.

The aim of the model is to focus on electron-photon interaction thus the latter will be treated in a fully microscopic scheme by use of the Fermi's golden rule. Conversely, all other interactions will be described in terms of a phenomenological electronic mean-lifetime, $\tau$, guaranteeing the proper thermalization of the electron population in the absence of external electromagnetic fields. Such a mean lifetime therefore enters the model as a fitting parameter, representing the global strength of all non-radiative thermalization mechanisms. A detailed description of the

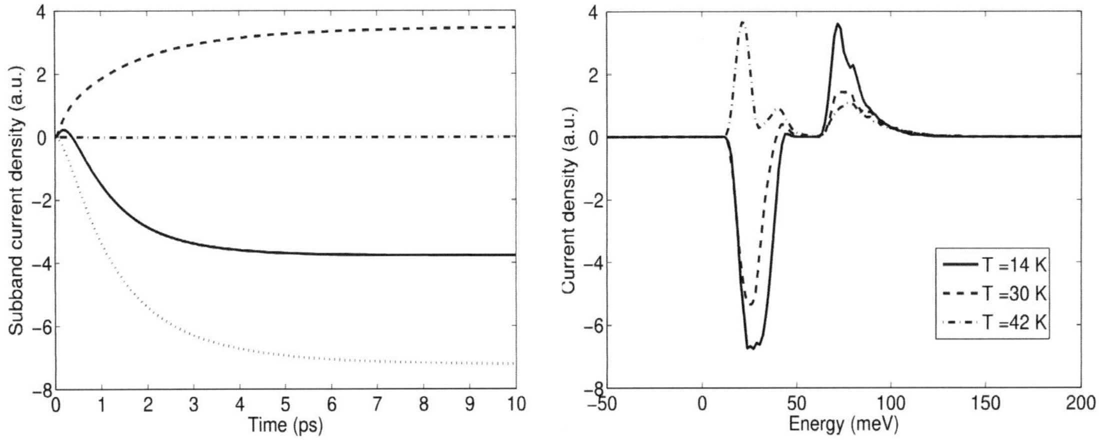


**Figure 2.** (Left) Time evolution of the total current (continuous line) at 14K, positive bias and 300K blackbody illumination. Dotted, dashed and dash-dot lines show the current contributions from the first, second and third excited minibands respectively.
(Right) Current contribution as a function of electron energy for different temperatures.

functional form of $P^{opt}$ and $P^{th}$ as well as the methods employed for the computation of the band structure can be found in [6].

The model is quite general and is capable of both stationary and time dependent simulations; in this paper it has been applied to a prototypical superlattice structure with 4nm wells and 12nm barriers with a barrier height of 120 meV. The system exhibits a single miniband below the barrier energy (which we shall call the ground miniband) plus a virtually infinite number of excited minibands at higher energies. Figure 1 shows the total current flowing through the device as a function of temperature, under a 300K blackbody radiation field and a positive applied bias. Since the applied voltage is low enough that the device response is linear, only normalized current values are shown. It can be clearly seen that between 5 and 40K the current output is negative and thus the device exhibits an absolute negative resistance. When temperature is increased over 40K the current starts to grow exponentially due to thermal excitation of carriers from the ground to upper minibands.

This peculiar behaviour can be analyzed in more detail by looking at the time evolution of the total current and to its energetic distribution. The left picture in Fig. 2 shows the current evolution at 14K (continuous line) as well as the contributions coming from the first 3 minibands (dotted, dashed and dash-dot lines). It can be seen that although the stationary value of the total current is negative, a fast sub-picosecond transient is present where the current starts as positive and is then inverted. By looking at the main miniband contributions, coming from the first two excited minibands, it can be seen that this inversion is due to electrons in the first excited miniband carring a negative current which is bigger, in absolute value, to the current coming from electrons in the second miniband.

On the right of Fig. 2 the same result is shown from a different perspective. The picture shows the relative current contributions from electrons at different energies, for three different device temperatures. It can be seen that current comes from two well separated energy regions corresponding to the first and second excited minibands. Again it can be noted that the first miniband carries a negative current for temperatures below 40K, then the current turns positive as temperature increases. On the other hand, the second miniband is carrying a positive current for every temperature.

This peculiar miniband-specific bahaviour can be explained by analytically solving the Boltzmann equation for an N-miniband system interacting with two bosonic thermal baths at different temperatures, namely phonons and temperature $T_0$ and blackbody photons at temperature $T_1$. For space reasons we shall omit the complete calculation and report only the final result. By considering the system evolution at short times it can be shown that a necessary condition for current inversion in a miniband is

$$\frac{\epsilon_\nu}{\epsilon_{\nu-1}} < \frac{T_1}{T_0} - 1 \tag{3}$$

where $\epsilon_\nu$ is the width of the $\nu$-th miniband.

This condition shows that there exists a critical device temperature $T_0$ below which the current in a miniband can be negative. For the model structure reported in this paper, the condition is met by the first excited miniband for temperatures below 40K and is never fulfilled by other minibands. Nevertheless since the first excited miniband gives the biggest contribution to the total current, a negative resistance is obtainedd in any case.

It is important to stress that this effect derives from the joint action of the two termal baths on the electron population. In fact, in case of absence of one of the two interactions the system would be driven to thermal equilibrium with a temperature equal to $T_0$ or $T_1$ depending on which mechanism is active. The presence of two themal baths at different temperatures drives the system to a stationary but non-equilibrium state characterized by a distribution function $f_\nu(\mathbf{k})$ exhibiting an inverted convexity with respect to the growth-direction wavector $k_z$, compared to the equilibriumstate. This convexity inversion is what finally determines the current inversion bacause of the presence of the $\frac{\partial}{\partial k_z}$ term in the drift operator in the Boltzmann equation, which changes sign if the convexity of the distribution function is reversed.

In conclusion, with this study we have shown that absolute negative resistance can be induced in superlattices not only by coherent optical excitation but also by means of a broadband incoherent stimulus like blackbody radiation. Anyway it is necessary to point out that in the simulations presented here, the current coming from the ground miniband has been neglected. This means that from an experimental point of view the observation of this phenomenon could be prevented by the presence of tunneling currents due to ground-state electrons directly crossing the barriers. Anyway, a careful structure design and, possibly, a different choice for the incident bradband field could lead to a measurable effect.

## References:

[1] B. J. Keay, S. Zeuner, S. J. Allen, K. D. Maranowski, A. C. Gossard, U. Battacharya, M. J. W. Rodwell, Phys. Rev. Lett. **75** 4102 (1995)
[2] Y. A. Romanov, J. Y. Romanova, L. G. Murokh, N. J. M. Horing, J. Appl. Phys. **89** 3836 (2001)
[3] Y. A. Romanov, J. Y. Romanova, L. G. Murokh, J. Appl. Phys. **99** 013707 (2006)
[4] F. Castellano, R.C. Iotti, and F. Rossi, Appl. Phys. Lett. **88**, 182111 (2006).
[5] F. Castellano, R.C. Iotti, and F. Rossi, J. Appl. Phys. **104**, 123104 (2008).
[6] F. Castellano, R.C. Iotti, and F. Rossi, Appl. Phys. Lett. **92**, 091108 (2008).

Journal of Physics: Conference Series **193** (2009) 012090 doi:10.1088/1742-6596/193/1/012090

# On the geometrical tunabililty of THz Gunn-like oscillations in InGaAs/InAlAs slot diodes

**S. Pérez, J. Mateos, D. Pardo and T. González**

Dpto. Física Aplicada, Universidad de Salamanca
Plaza de la Merced s/n, 37008 Salamanca – Spain

susana@usal.es

**Abstract**. By using an ensemble Monte Carlo simulation we study the influence of the bias, the recess-to-drain distance, recess length and δ-doping on the frequency of the ultra-fast Gunn-like oscillations $f_G$ found in InGaAs/InAlAs slot diodes. A minimum carrier concentration is needed under the recess for the process to start up. Thus, a δ-doping larger than $5\times10^{12}$ cm$^{-2}$ is necessary, but too high values of this parameter reduce the oscillation frequency. As expected, by shortening the devices (small recess and recess-to-drain lengths) $f_G$ increases. As concerns the DC voltage, the optimum conditions correspond to biases exceeding as less as possible the threshond voltage for the onset of intervally transfer.

## 1. Introduction

The THz frequency range lies between microwaves and infrared light in the electromagnetic spectrum; thus, the technology for producing T-ray sources is at the limits of electronics from one side and optical systems from the other. Indeed, no powerful radiation sources in this range have been available until the past few years [1]. Even nowadays it does not exist yet a compact source working at room temperature able to generate CW THz fields with a relatively high output power.

In a recent work [2] we showed the possibility to generate ultrafast (in the THz range) Gunn-like oscillations in InGaAs/InAlAs slot diodes (ungated recessed heterostructures, see figure 1). Monte Carlo simulations predict that when the bias surpasses a threshold voltage, associated to the Γ-L intervalley energy, the current exhibits an oscillatory behaviour at extremely high frequency (much higher than that obtained by the 'classical' Gunn effect).

In this work, we present a detailed MC study of the dependence of the oscillation frequency on applied bias, recess-to-drain length, recess length and δ-doping. For this sake, we perform calculations of the current noise spectra, which can give precise indications on the onset of Gunn oscillations. At the same time the instantaneous values of electron velocity, electric potential profile, valley occupations, etc. are monitored in order to optimize the parameters allowing to tune the frequency of the ultrafast Gunn-like oscillations $f_G$.

## 2. Device topology and Monte Carlo simulations

The layer structure of the simulated slot diodes is shown in figure 1. Note that it is the same used for the realization of lattice-matched InGaAs HEMTs [3]. The device of reference has a source-to-recess distance ($L_S$) of 200 nm, a recess length of ($L_R$) 200 nm and a recess-to-drain distance ($L_D$) of 550 nm.

Calculations are performed by using an ensemble MC simulator at room temperature, which is self-consistently coupled with a two dimensional Poisson solver. The material parameters based on a three valley system are reported in [3]. Ohmic boundary conditions are considered at the source and drain contacts, which are vertically placed adjacent to different materials. Accordingly, nonuniform potential and concentration profiles are considered along these contacts: those that would be obtained if real top electrodes were simulated [3]. The effect of degeneracy is accounted for by locally using a simple rejection technique, where electron heating and nonequilibrium screening effects are introduced by means of the local electron temperature. No other quantum effects are considered in the simulation in order to have reasonable CPU times. The validity of this approach, especially under high-field conditions, and that of the whole MC model was confirmed in previous works [3]. In order to detect the presence of ultrafast Gunn oscillations, special attention is devoted to the calculation of the current noise spectra due to their extreme sensitivity to microscopic features of carrier dynamics and the possibility to easily perform a frequency analysis of the electrical fluctuations.

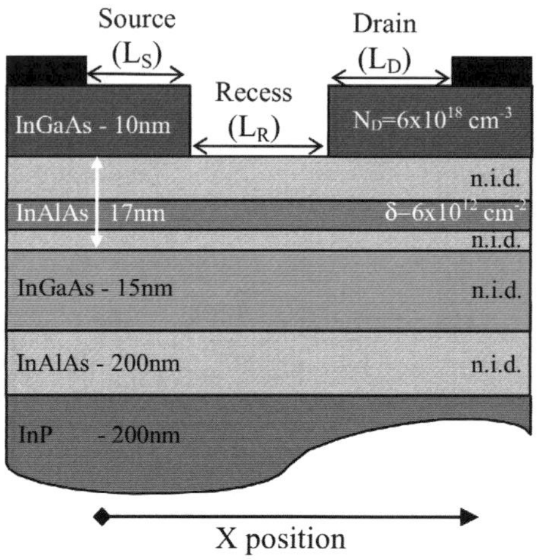


**Figure 1.** Geometry of the simulated InAlAs/InGaAs slot diode

## 3. Results

As explained in a recent work [2], a necessary condition for the onset of an oscillatory behaviour within the slot diode is the presence of two types of carriers in the channel: fast electrons in the Γ-valley and slow electrons in the upper valleys. Therefore, the Gunn-like phenomenon can only start when the bias applied to the slot diode surpasses a threshold voltage given by the Γ-L intervalley energy (around 0.6 V). When such voltage drop is reached within a small region at the drain side of the recess, the L valley concentration increases at this point and a zone depleted of Γ electrons appears nearby [see figure 2(a)]. The width of this depleted zone increases rapidly, following the motion of the fast Γ electrons, while only L electrons are injected into the drain region, thus leading to a decrease of the current flowing through the device. This evolution of the depletion region makes the resistance of the recess-drain region increase. When it becomes comparable to that of the region under the recess, it absorbs a part of the voltage drop, not allowing the electrons to gain the intervalley energy while moving under the recess. When this occurs, Γ-valley electrons start to be injected into the drain region,

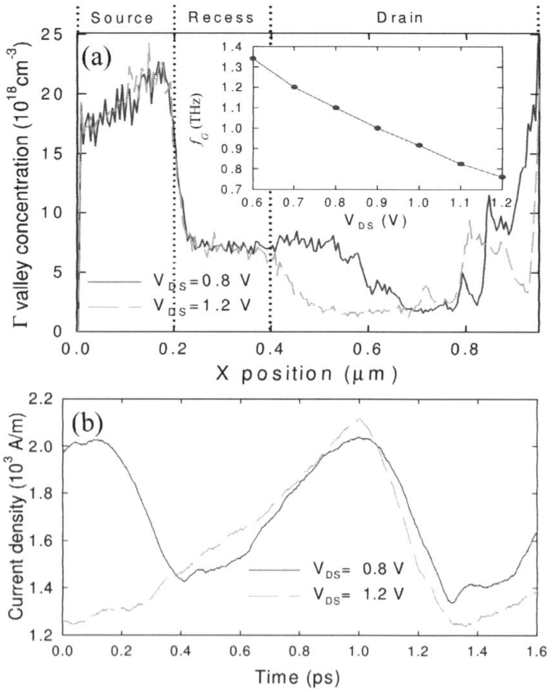


**Figure 2.** (a) Instantaneous profile of Γ valley electron concentration inside the device for different $V_{DS}$; the inset shows $f_G$ as a function of $V_{DS}$. (b) Time dependent current for the same $V_{DS}$.

filling the previously created depletion and providing a net increase of total concentration and current. The period of this ultra-fast Gunn oscillation is mainly determined by the transit time of Γ-valley electrons through the recess-drain region, and depends on the applied voltage, length of the recess, recess-drain distance and δ-doping. These dependencies will be studied in the following sections.

### 3.1. Influence of the bias voltage

The inset of figure 2 shows how $f_G$ decreases when increasing the bias from 1.34 THz for $V_{DS}$=0.6 V to 0.76 THz for $V_{DS}$=1.2 V. This is mainly due to the fact that for lower voltages L-valley electrons are not injected just at the drain side of the recess, but somewhat inside the drain region, thus decreasing the traveling time of the depletion/filling region (figure 2). However, the current modulation achieved under these conditions (amplitude of the oscillations) is lower than for higher voltages due to the less efficient L-valley electron injection.

### 3.2. Influence of the recess-to-drain distance, $L_D$

In figure 3(a) $f_G$ is plotted vs. $L_D$ when the bias is fixed to 1 V. For high values of $L_D$ ($L_D$>550 nm) $f_G$ perfectly scales as $1/L_D$, while for shorter devices the increase of $f_G$ is not as high as expected. This deviation from the simple transit time criterion for $L_D$<550 nm is a consequence of the increase of L-valley population at the drain side of the recess (due to a sharper voltage drop at that point), which decreases the velocity of the depletion/filling mechanism (with respect to that exhibited by Γ electrons) at the origin of the observed Gunn-like oscillations, figure 3(b).

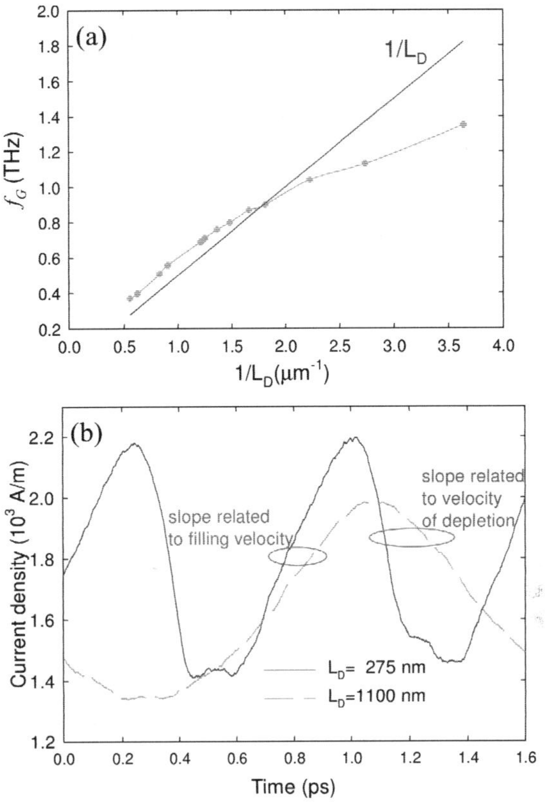


**Figure 3.** (a) $f_G$ as a function of the inverse of the recess-drain distance ($L_D$) and (b) time-varying current for $L_D$=275 and 1100 nm. $V_{DS}$=1.0 V.

### 3.3. Influence of the recess length, $L_R$

Ultrafast quasiballistic Γ electrons, at the origin of the ultra fast dynamics of the observed oscillations, appear in the channel of slot diodes thanks to (i) the presence of the recess that focuses the electric field and (ii) the effect of degeneracy, which significantly reduces the rate of scattering mechanisms and much increases the electron mean free path.

The inset of figure 4 presents the dependence of $f_G$ with the length of the recess $L_R$ for $V_{DS}$=1 V, and $L_D$=550 nm. $f_G$ decreases for longer $L_R$ due

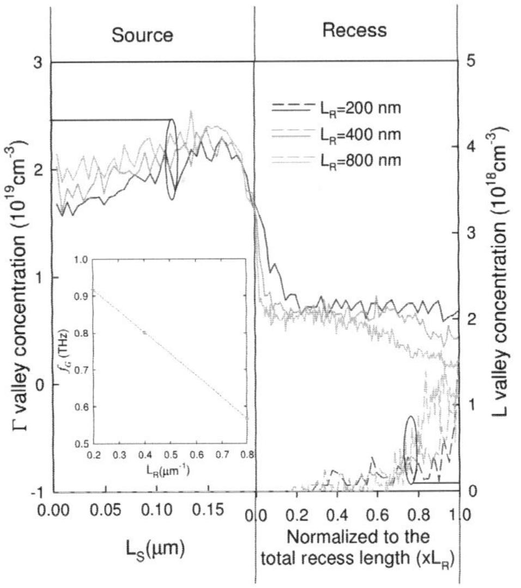


**Figure 4.** Instantaneous profile of Γ and L valley electron concentration inside the device for $L_R$=200, 400, 800 nm. $V_{DS}$=1V. The inset shows $f_G$ as a function of $L_R$.

to the attenuation of the "electron-launching effect" of fast Γ electrons into the drain region. The injection velocity decreases for longer $L_R$ due to a less efficient focusing of the electric field by the recess. The reduction of $f_G$ comes along with a degradation of the amplitude of the oscillations due to the smaller presence of Γ-valley electrons under the recess, as shown in figure 4. This happens because part of the electrons can jump to the L valley before reaching the drain side of the recess, thus degrading the modulation of the current associated to the injection of Γ-valley electrons.

### 3.4. Influence of the δ-doping

Another necessary condition for Gunn-like oscillations to appear is the presence of a large enough amount of carriers in the channel below the recess. Therefore, a minimum value of the the δ-doping ($\delta$=6x10$^{12}$ cm$^{-2}$ in our device) is required for observing current oscillations, figure 5(a). However, if the δ-doping is further increased, the oscillation frequency is degraded because the electron concentration in the channel becomes more homogeneous and the recess is not able to correctly concentrate the electric field, figure 5(b).

## 4. Conclusion

We have performed a thorough study of the influence of different topological parameters on the THz Gunn-like oscillations found in heterojunction slot diodes. High oscillation frequencies can be achieved by correctly choosing values of the δ-doping level and applied voltage, ensuring an enough carrier concentration under the recess and allowing for the onset of the oscillations, respectively. Additionally, both $L_D$ and $L_R$ must be as short as possible, the first for decreasing the length of transit of the depletion/filling region, and the second for enhancing the electric field concentration under the recess, thus improving the velocity of the Γ-valley electrons that determine the speed of the mechanism.

**Acknowledgements**

This work has been partially supported by The Ministerio de Educación y Ciencia (TEC2007-61259/MIC and HF 2007-0014) and by the Junta de Castilla y León through the projects SA019A08 and GR270.

## 5. References

[1] Köhler R, Tredicucci A, Beltram F, Beere H E, Linfield E H, Davies A G, Ritchie D A, Iotti R C, and Rossi F 2002 *Nature* **417** 156
[2] Pérez S, González T, Pardo D and Mateos J 2008 *J. Appl. Phys.* **103** 094516
[3] Mateos J, González T, Pardo D, Hoel V and Cappy A 2000 *IEEE Trans. Electron. Devices* **47** 1950

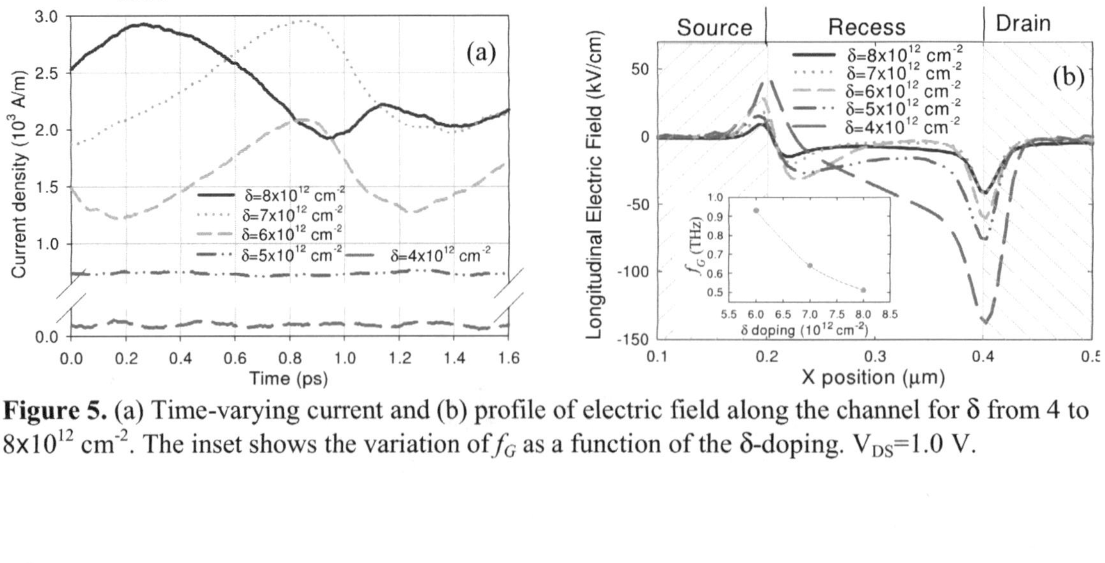


**Figure 5.** (a) Time-varying current and (b) profile of electric field along the channel for δ from 4 to 8x10$^{12}$ cm$^{-2}$. The inset shows the variation of $f_G$ as a function of the δ-doping. $V_{DS}$=1.0 V.

Journal of Physics: Conference Series **193** (2009) 012091 doi:10.1088/1742-6596/193/1/012091

# Plasmonic noise in Si and InGaAs semiconductor nanolayers

**J. Pousset[1], J.-F. Millithaler[1], L. Reggiani[1], G. Sabatini[2], C. Palermo[2], L. Varani[2], J. Mateos[3], T. González[3], S. Pérez[3], A. Bournel[4] and P. Dollfus[4]**

[1] Dipartimento di Ingegneria dell'Innovazione and CNISM, Università del Salento, Via Arnesano s/n, 73100 Lecce, Italy

[2] Institut d'Électronique du Sud, UMR CNRS 5214, Université Montpellier II, place Bataillon, 34095 Montpellier Cedex 5, France

[3] Departamento de Física Aplicada, Universidad de Salamanca, Pza. Merced s/n 37008 Salamanca, Spain

[4] Institut d'Électronique Fondamentale, UMR CNRS 8622, Université Paris-Sud, 91 405 Orsay Cedex, France

E-mail: `jeremy.pousset@unile.it`

**Abstract.** By Monte Carlo simulations we investigate the spectrum of voltage fluctuations of Si layers of variable thickness $W$ in the range $2 \div 100$ nm and variable length $L$ in the range $10 \div 1000$ nm embedded in an external dielectric medium. Calculations are performed at T = 300 K for different doping levels and in the presence of an external bias of increasing strength. For $W \geq 100$ nm and carrier densities of $5 \times 10^{17}$ and $5 \times 10^{18}$ cm$^{-3}$ the peaks agree with the value of the three dimensional (3D) plasma frequency. For $W \leq 10$ nm the results exhibit a plasma frequency that depends on $L$, thus implying that the oscillation mode is dispersive. The corresponding frequency covers a wide range of values $0.2 \div 10$ THz and is in agreement with the values predicted by existing analytical models. At sufficiently high bias the 2D plasma peak is washed out and we observe the onset of a peak in the subTHz region which, by analogy with the results obtained in n-InGaAs layers, is associated with transit time instabilities induced by current saturation conditions.

## 1. Introduction

Generation and detection of electromagnetic radiation in the TeraHertz (THz) domain is a subject in fast development because of its potential applications in different branches of advanced technologies, such as broad-band communications and high-resolution spectroscopy [1, 2]. As a consequence, the realization of affordable solid-state devices operating in the THz domain at room-temperature and with compact, powerful, and tunable characteristics is a mandatory issue. To this purpose, one of the most promising strategies lies in the plasmonic approach which exploits the plasma frequency of free carriers as a possible mechanism for detection/generation [3]. In this framework, through an analytical approach, the case of a two-dimensional (2D) electron layer constituted by the ungated channel of a nanometric transistor was considered in Ref. [4]. Moreover nanometric Si transistors have shown their ability to detect THz radiation using plasma waves [5]. The aim of this work is to investigate the same problem from a microscopic point of view thus testing the limits of applicability of the analytical approach and exploiting possible implementations. Previous theoretical studies have shown the presence and the main properties of electron plasma in InGaAs layers [6]. Although electron mobility in Si is about a factor of ten lower than in InGaAs, Si technology is more mature and its industry has a leading role. Therefore, here we mainly focus on electron plasma in Si layers. To this purpose, we consider an $n$-type Si layer, embedded in a symmetric dielectric, in the presence of an external bias of arbitrary strength and analyse the frequency spectrum of voltage fluctuations obtained from a Monte Carlo simulator coupled with a 2D Poisson solver

## 2. Models and results

The analytical model considers a 2D sheet of electrons, under collisionless conditions, in the $\vec{r} \equiv xy$-plane embedded in a dielectric along the $z$ direction and leads to a dispersive 2D plasma frequency [4]

$$f_{2D} = \frac{1}{2\pi}\sqrt{\frac{e^2 n_{2D}|\mathbf{k}|}{2\ m_0\ m\ \varepsilon_0\ \varepsilon_{diel}}} \tag{1}$$

where $|\mathbf{k}|$ is the modulus of the wavevector in the $k_x$,$k_y$ plane, $m_0$ and $m$ the free and effective electron masses, respectively, $n_{2D}$ the 2D carrier density, $e$ the electron charge, $\varepsilon_{diel}$ the relative dielectric constant of the outside dielectric, and $\varepsilon_0$ the vacuum permittivity. We notice that the 2D plasma frequency is dispersive (i.e. $k$ dependent) and depends on the external dielectric.

Plasma waves also exist in bulk (3D) and their dispersionless frequency is

$$f_{3D} = \frac{1}{2\pi}\sqrt{\frac{e^2 n_{3D}}{m_0\ m\ \varepsilon_0\ \varepsilon_{mat}}} \tag{2}$$

with $n_{3D}$ the 3D carrier density and $\varepsilon_{mat}$ the relative dielectric constant of the bulk material.

The numerical investigation of this problem is carried out by using a microscopic Monte Carlo approach coupled with a 2D Poisson solver. By evaluating the fluctuations of the voltage around its steady-state value in the center of the device, we calculate the spectral density of this quantity from the corresponding correlation function and look for its spectrum. We consider doped layers of Si and InGaAs at 300 K in which the spectral density exhibits a peak in the THz range. This peak is analyzed as function of geometrical parameters, doping concentration and applied voltage. To this purpose we take lengths $L$ in the range $10 \div 1000$ nm, thicknesses $W$ in the range $1 \div 100$ nm and carrier densities of $5 \times 10^{17}$ and $5 \times 10^{18}$ cm$^{-3}$. The layers are surrounded by a 10 $\mu$m-thick dielectric ($SiO_2$). For the case of Si, scattering with impurities is neglected.

### *2.1. Thermal equilibrium*

Figure 1 reports for the case of Si the frequency of the plasma peaks as a function of $L$ for different widths and a carrier density of $5 \times 10^{18}$ cm$^{-3}$. Symbols refer to calculations and lines to the theoretical 3D and 2D values for widths of 2 and 10 nm, respectively. For a width $W = 100$ nm, the frequency of the observed peak corresponds to that of the 3D plasma frequency. Because of its ellipsoidal manyvalley band structure, by neglecting nonparabolic effects the electron effective mass is given by: $m = \left[3m_l^{-1}m_t^2/(m_t + 2m_l)\right]^{2/5}$, with $m_t$ the transverse component and $m_l$ the longitudinal one. For the cases of the two thinnest layers, the frequency of the peak is found to decrease as the square root of the length. Furthermore, the frequency increases as the square root of the width. Both dependences are actually predicted by Eq. (1) when taking $n^{2D} = n_{3D} \cdot W$ and $k = \pi/2L$.

Figure 2 reports the frequency of the peaks as a function of the layer length for two carrier densities of $5 \times 10^{17}$ and $5 \times 10^{18}$ cm$^{-3}$, respectively, and a width $W = 10$ nm . The theoretical values of the respective 2D frequencies are also reported. Here the frequency of the peak is

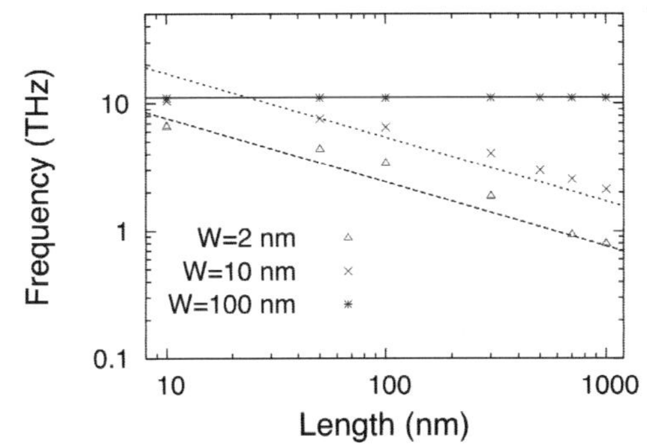


**Figure 1.** Frequency of the peak present in the spectral density of voltage fluctuations as a function of the layer length, whose carrier density is $n_{3D} = 5 \times 10^{18}$ cm$^{-3}$. Lines refer to theoretical 3D plasma frequency (solid line) and 2D plasma frequencies for $W = 2$ nm (dashed line) and 10 nm (dotted line), respectively.

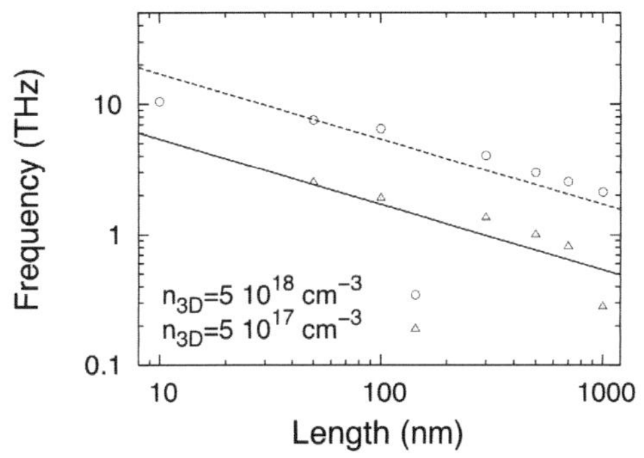


**Figure 2.** Frequency of the peak present in the spectral density of voltage fluctuations as a function of the layer length for $W = 10$ nm and for the reported carrier densities. Lines refer to the 2D plasma frequencies for $n_{3D} = 5 \times 10^{17}$ (solid line) and $5 \times 10^{18}$ cm$^{-3}$ (dashed line), respectively.

found to increase as the square root of carrier density in agreement with Eq. (1). Therefore, the presence of 2D and 3D plasma waves in Si layers has been evidenced through microscopic simulations and the dependences from geometric parameters and carrier density have been pointed out.

By simulating Si layers with $W$ in the range $1 \div 150$ nm, $L = 100$ nm and a carrier density of $5 \times 10^{18}$ cm$^{-3}$, we have investigated the cross-over between the 2D and 3D plasma peaks, which is found to be well described by the simple expression $f = f_{3D} f_{2D}/(f_{3D} + f_{2D}) = f_{3D}[\sqrt{W}/(\sqrt{W_0} + \sqrt{W})]$ with $W_0 = 4\varepsilon_{diel} L/(\pi \varepsilon_{mat}) = 44$ nm, the width of the cross-over.

### *2.2. Ohmic and saturation regime*

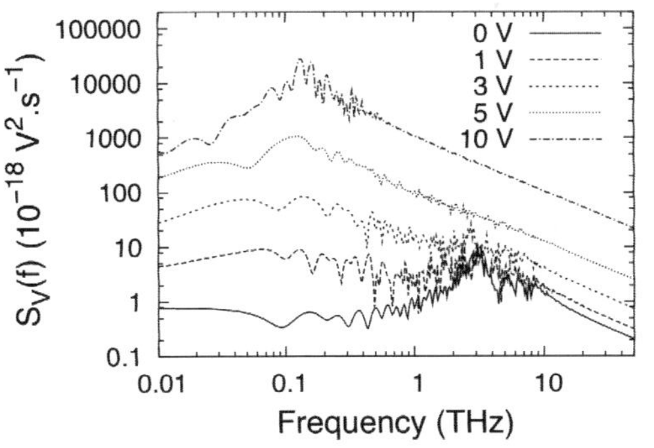


**Figure 3.** Spectral densities of voltage fluctuations in a Si layer with $n_{3D} = 5 \times 10^{18}$ cm$^{-3}$, $W = 10$ nm and $L = 1$ $\mu$m for different applied voltages.

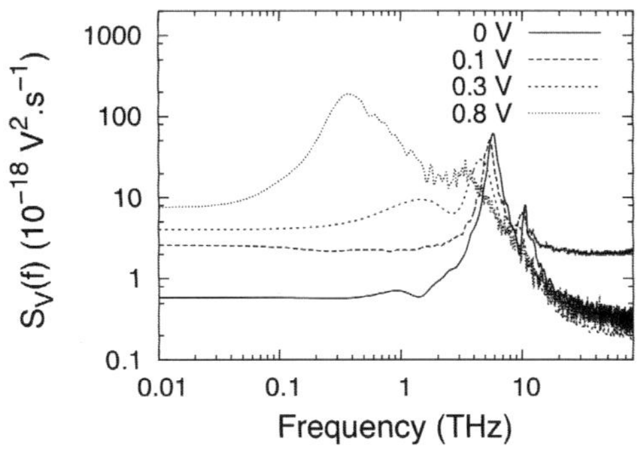


**Figure 4.** Spectral densities of voltage fluctuations in an InGaAs layer with $n_{3D} = 10^{18}$ cm$^{-3}$, $W = 1$ nm and $L = 500$ nm for different applied voltages.

We consider now the plasma peak as a function of an applied voltage and compare the case of n-Si with that of similar n-InGaAs structure. To study the transition from Ohmic to saturation regime, Fig. 3 reports the spectral density of voltage fluctuations $S_V$ extracted in the middle of Si layers for several applied voltages. At low bias, the peak related to the 2D plasma waves is recovered. At increasing voltages, the position and amplitude of this peak remains practically constant whereas the low frequecy plateau of the spetrum, related to diffusion noise, increases until the plasma peak is completely suppressed. At the highest voltage the onset of another peak is observed in the sub-THz region for a frequency which is found to scale as the inverse of $L$. The existence of this peak seems to be related to the presence of voltage oscillations induced by transit time effects coming from current saturation conditions. Figure 4 reports the results for the analogous case of an n-$In_{0.53}Ga_{0.47}As$ layer. Again, at increasing voltages the low frequency plateau of the spectrum starts increasing while the amplitude of the plasma peak remains practically constant. Then, because of the onset of negative differential mobility conditions, the spectrum exhibits the onset of a peak which is precursor of the establishment of current oscillations due to periodic travelling of the associated Gunn domains.

## 3. Conclusions

Through Monte-Carlo calculations of the spectral density of voltage fluctuations we investigated THz oscillations in Si nano-layers embedded in an external dielectric as a function of geometry, carrier density, and strength of applied voltage. The microscopic investigation of the characteristic frequency peaks of the voltage fluctuations evidences a rich scenario, only partially predicted by the exisisting 2D analytical model [4]. Under thermal equilibrium and Ohmic conditions, the results of simulations show that for thick layers, i.e. $W$ above 100 nm, 3D plasma oscillations appear independently of channel length, thus confirming an analogous study carried out in InGaAs [6]. By contrast, for thin layers, i.e. $W$ below 100 nm, we observed the transition from 3D to 2D plasma modes where the 2D plasma frequency decreases with increasing channel length in agreement with the predictions of the analytical model [4]. Under saturation conditions, the results of simulations show that the low frequency plateau of the spectrum, related to diffusion noise, totally covers the plasma peak and another peak, probably related to transit time effects of some instability induced by current saturation, appears at sub-THz frequencies. The analogous case of InGaAs layers evidences the same behaviour, with Gunn oscillations, associated with negative differential mobility, washing out completely plasma waves. The work was supported by: CNRS - GDR and GDR-E projects "Semiconductor sources and detectors of THz frequencies", "Ministère des affaires étrangères et europénnes" through a "Lavoisier" grant (J. Pousset), Region Languedoc-Roussillon - project "Plate-forme Technologique THz", the Dirección General de Investigación (MEC, Spain), FEDER through the project TEC2007-61259/MIC, Accion Integrada HF2007-0014 and by the Consejería de Educación of the Junta de Castilla y León (Spain) through the project SA019A08 and GR270.

## References

[1] Ryzhii V 2008 *J. Phys.: Condens. Matter* **20** 380301
[2] Woolard D L, Brown E R, Pepper M and Kemp M 2005 *Proc. IEEE* **93** 1722
[3] Fatimy A E, Tauk R, Boubanga S, Teppe F, Dyakonova N, Knap W, Lyonnet J, Meziani Y M, Otsuji T, Poisson M A, Morvan E, Bollaert S, Shchepetov A, Roelens Y, Gaquiere C, Theron D and Cappy A 2008 *phys. stat. sol.* 244–248
[4] Dyakonov M and Shur M 2005 *Appl. Phys. Lett.* **87**
[5] Tauk R, Teppe F, Boubanga S, Coquillat D, Knap W, Meziani Y M, Gallon C, Boeuf F, Skotnicki T, Fenouillet-Beranger C, Maude D K, Rumyantsev S and Shur M S 2006 *Appl. Phys. Lett.* **89** 253511
[6] Millithaler J F, Reggiani L, Pousset J, Varani L, Palermo C, Knap W, Mateos J, Gonzalez T, Perez S and Pardo D 2008 *Appl. Phys. Lett.* **92** 042113

Journal of Physics: Conference Series **193** (2009) 012092 doi:10.1088/1742-6596/193/1/012092

# Current transport modeling and experimental study of THz room temperature ballistic deflection transistors

**Vikas Kaushal[+], Martin Margala[+], Qiaoyan Yu[*], Paul Ampadu[*] Gregg Guarino[*], Roman Sobolewski[*]**

[+*] Department of Electrical and Computer Engineering

[+]University of Massachusetts Lowell, MA, USA 01854

[*]University of Rochester, NY, USA 14627

vikas_kaushal@student.uml.edu

**Abstract.** In this paper, two different theoretical models, Comsol Multiphysics™ (a Finite Element Analysis tool), and a field solver Atlas/Blaze from Silvaco, are compared qualitatively to study the effect of the deflector position, its size and electric field on the charge transport and its distribution along the channel, resulting in current outputs and leakages in ballistic deflection transistors (BDT). Silvaco simulations and experimental results were then used to study the lateral charge transport as a result of variation in electric field distribution, which controls the charge current along the channel in BDT. The electric field dependence of gain is also studied with experimental and theoretical results.

## 1. Introduction

The future electronics depends significantly on the scalable solid state transistors with much higher speed than the III-IV FETs and lower energy dissipation than current CMOS transistors. The extensive scaling leads to increased leakages through gates, which motivates researchers to explore novel devices and structures. In the same context, ballistic devices have been proposed by many groups [1,2]. These devices have emerged as a prominent and practical device for high speed and high current handling applications. The current transport in these ballistic devices shows a unique behavior as all the charge is moving in a single 2DEG plane. The lateral charge density depends on the electric field distribution along the 2DEG plane, which leads to the variation in currents at various device terminals. Since, the device size is being scaled down aggressively and the scattering is only exhibit at the boundaries, the charge distribution also depends on its geometry and dimensions significantly.

In this paper, section one gives the overview of BDT, which is followed by the second section, which explains the theoretical simulations of current transport in BDT with two different modeling tools, Comsol Multiphysics™ , an FEA software package and commercially available tool Atlas/Blaze from Silvaco. Third section explains the qualitative agreement between simulations and D.C. experimental results to describe the electric field distribution dependence of transconductance of BDT.

## 2. BDT operation and Structure

BDT, which is a 1-THz transistor, is proposed by our group in [3]. BDT is a low noise and low loss device which is based upon an electron steering and a ballistic deflection effect. The material system

Journal of Physics: Conference Series **193** (2009) 012092 doi:10.1088/1742-6596/193/1/012092

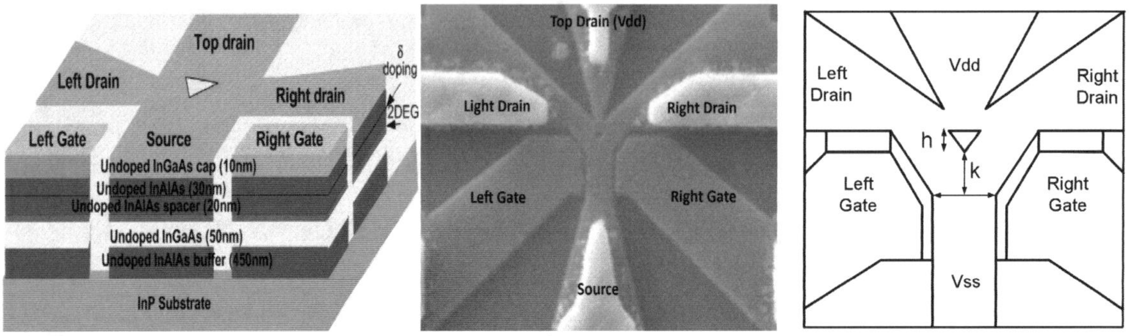


(a) (b) (c)

**Figure 1:** Cross section(a), SEM(b) and sketch ('h' and 'k' are deflector size and position respectively) of BDT.

chosen, provides a large mean free path and high mobility to support ballistic charge transport at room temperature. This transistor is unique in that no doping junction or barrier structure is employed. The theoretical value of $f_t$ obtained at room temperature is 1.02 THz for a specific geometry and dimension which can vary with dimension. Typical cross-section, SEM image and schematic of BDT are shown in Figure 1 (a), (b) and (c) respectively.

## 3. Modeling and Theoretical Results

Two different theoretical models are studied and compared qualitatively for the effect of deflector and electric field ($E$) on the charge transport along the planar channel, resulting in outputs and leakages. Two different selected simulations are Comsol Multiphysics™, a Finite Element Analysis (FEA) software package and commercially available tool Atlas/Blaze from Silvaco. For Comsol simulations, BDT was simulated using FEA, based on a simple conductive media model. The conductivity of the 2DEG was approximated based on a carrier density of $10^{12}$ cm$^{-2}$, and the InGaAs bulk electron mobility of 12,000 cm$^2$/(V.s), giving a value of 0.00192 S. The effect of applied gate voltage was modeled as depletion of the channel on the side of the negative electrode. The depletion was assumed to advance laterally across the channel, as a linear function of a gate voltage, until complete pinch-off occurs at the max gate voltage.

The function of the triangular deflector in this FEA model was particularly of interest. Although it does not act as a deflector in the ballistic electron sense, it redirects the flow of current. Its shape and position influence the transport of output current between the top, left, and right electrodes. Top drain leakage ($I_{TD}$) has been a major source of leakage in the BDT [4]. Figure 2(b) shows $I_{TD}$ variation with applied differential gate voltage at different deflector positions ($k$). Deflector size ($h$) selected for this specific case is 75nm. "$k$" increases upwards as shown in Figure 1. This study shows that when deflector is placed at $k=230nm$, leakage rises and when deflector touches the top drain port, leakage

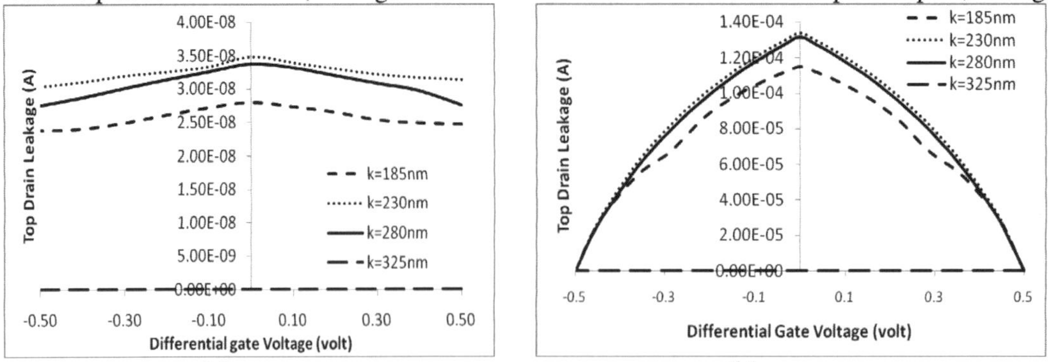


(a) (b)

**Figure 2:** Leakage variation with differential gate voltage at different deflector positions. Comsol simulation (b) is qualitatively in agreement with Silvaco simulation (a).

**Table 1** : Output and Leakage current variation with $h$ for deflector position $(k)=215nm$.

| $h(nm)$ | $I_{LD}(A)$ | $I_{RD}(A)$ | $I_{TD}(A)$ |
|---|---|---|---|
| 50 | 1.78E-04 | 1.78E-04 | 1.06E-04 |
| 75 | 1.44E-04 | 1.44E-04 | 1.15E-04 |
| 100 | 1.58E-04 | 1.58E-04 | 1.21E-04 |

**Table 2** : Output and Leakage current variation with $k$ for deflector size ($h$)=75nm.

| $k(nm)$ | $I_{LD}(A)$ | $I_{RD}(A)$ | $I_{TD}$ (A) |
|---|---|---|---|
| 185 | 1.44E-04 | 1.44E-04 | 1.15E-04 |
| 230 | 1.80E-04 | 1.80E-04 | 1.34E-04 |
| 280 | 1.95E-04 | 1.93E-04 | 1.33E-04 |
| 325 | 2.44E-04 | 2.44E-04 | 2.55E-12 |

becomes negligible. Ideally, no current should pass through top drain but because of the fact that the electric field distribution is uneven and electron pull by top drain due to high electric field is much higher in the channel, electron current towards top drain counts for the leakage. It is observed that placing deflector at approximately 25% lower than the exact centre of the device reduces leakage significantly, which indicates the high deflector-electron interactions essentially creating a barrier for electron to move towards top drain [5].

Current distribution is also noticed with '$h$' and it can be seen from Table 1 that for $h$=50$nm$, output obtained is maximum and leakage also reduced significantly. Table 2 shows a current distribution variation at different terminals with '$k$'. It is noticed that output rises with '$k$', and at the same time leakage also rises; leakage minimizes when deflector touches the top port. Fabrication with configuration of k=325 is avoided because of the removal of the whole top channel in that case.

This effect of '$k$' is also studied with Silvaco simulations. The structure used for the simulations is the stacked lattice matched layers of InGaAs and InAlAs on InP substrate (Figure 1a). The Si-planar doping layer with a doping density of $2 \times 10^{12} cm^{-2}$ is used. Hall measurements of this structure yielded an electron sheet density of $n_s = 9.832 \times 10^{11} cm^{-2}$ and electron mobility of $\mu_H = 1.206 \times 10^4 cm^2/Vs$ at 300$K$. Physical models chosen are Boltzmann, thus a statistical approach. Mobility models chosen are *conmob* and *fldmob* where *conmob* uses simple power law temperature dependence and *fldmob* is used to model velocity saturation effects. Shockley-Read-Hall model is used because it uses fixed minority carrier lifetimes for numerical solutions. Fig. 2(b) shows the effects of applied gate voltage on $I_{TD}$ at applied gate voltage. This study also showed that '$k$' is a significant in BDT and Silvaco simulations are in qualitative agreement with Comsol modeling. Silvaco simulations confirm that for $k=230nm$, leakage becomes maximum and becomes zero when deflector touches the top drain port (at $k=325nm$).

## 4. Theoretical and Experimental Results

Silvaco simulations are also used to study the current transport and distribution as a result of variation in electric field ($E$), which controls the flow of charge along the planar 2DEG channel. For this purpose, 2D Poisson equation is solved analytically for the entire high-field region. Electrostatic potentials between electrodes are differentiated to get $E$. Figure 3(a) shows the theoretical variation of the $E$ along the channel with drain voltage for a left gate bias of 0.3$V$. $E$ increases along the channel, and reaches a maximum at a point between the gate and the drain as most of the applied drain potential appears across this narrow region of small conductance. The $E$ is higher for higher values of $V_{DS}$.

Figure 3(b), which is a D.C. experimental result, confirms it, which shows that the transconductance increases with $V_{DS}$. It is also seen that the peak transconductance shifts to higher $V_{GS}$ with increased $V_{DS}$. This is easily understood since the voltage drop between the gate and the drains is much greater in the high-drain bias case, resulting in a high $E$. Due to variation in the transport of the charges at different terminals, $I_{LD}$ and $I_{RD}$ appear as outputs and $I_{TD}$ as leakage. BDTs with different dimensions are fabricated. Qualitatively, the results obtained are in agreement with the Silvaco simulations which confirm that $E$ near drains is higher which leads to high conductivity inside the channel at these areas. Electron velocity ($V_E$) also plays a major role for the device performance. The $V_E$ is significantly greater in the channel under high $V_{DS}$ due to greater confinement of electrons within the 2D system. When the applied $V_{DS}$ is low, the transistor is in linear operation regime, and the electron velocity is uniform. Electric field increases sharply due to velocity saturation. At this point, the average electron energy increases dramatically, leading to significant velocity overshoot.

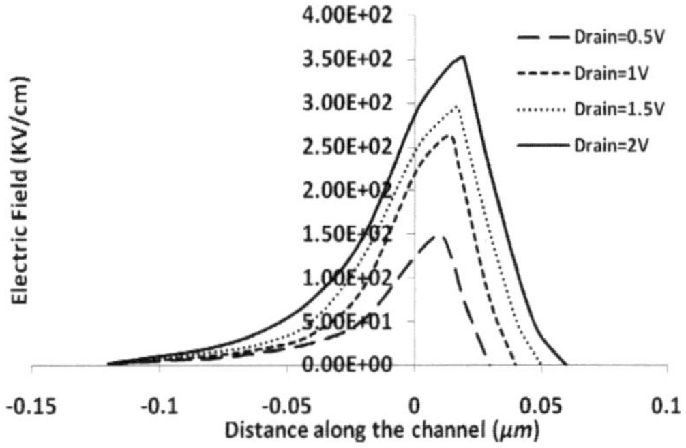


(a)

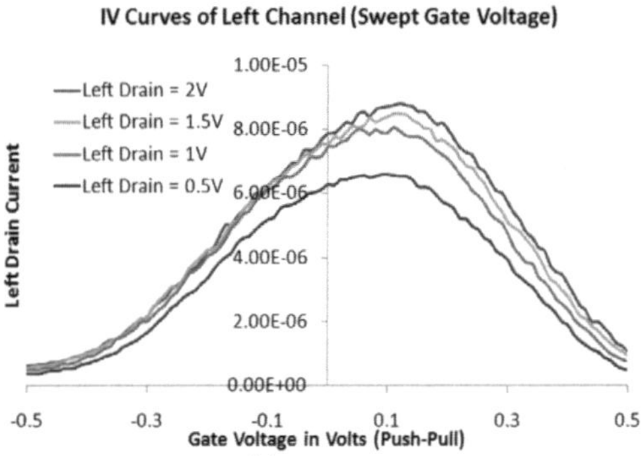


(b)

**Figure 3:** (a) Variation of the electric field along the channel for different drain voltages.Each drain is kept same for single event. (b) IV characteristic of the left drain port as a function of differential gate voltage (in reference to the left gate), an increase in drain voltage increases the transconductance, near linearly for voltage beyond 1V.

**Table 3:** Comparison of theoretical and experimental results for output and leakage with 75nm deflector height.

| *h=75(nm)* | $I_{LD}(A)$ | $I_{RD}(A)$ | $I_{TD}(A)$ |
|---|---|---|---|
| Simulation | 1.9308E-08 | 1.9474E-08 | 2.4674E-08 |
| Experiments | 2.565E-06 | 2.4965E-06 | 9.2374E-07 |

Table 3 gives the current transport and distribution comparison between simulated results and experimental data. It is observed that the experimental results give good device performance but the leakage was high. These results can be used to define the electric field contour inside the device and thus the control on electric field distribution can be achieved for optimum device performance.

## 5. Conclusions

In this paper, two different modeling tools are used to study the current transport and its distribution inside the channel for various deflector positions and sizes. Geometry is optimized for reduced leakage and better device outputs. It is confirmed that placing deflector at k=185*nm,* which is 25% below the exact centre of the device, reduces leakage significantly. It is also observed that for the deflector size of 50*nm*, the outputs are at maximum and the leakage reduces significantly. Fabricated devices were tested for the electric field distribution comparison with Silvaco simulations. For each drain voltage of 2V, maximum electric field of the order of 352.675*KV/cm* is obtained, which indicates that with higher drain voltage, transconductance will increase. Output and leakage currents are compared in order to achieve the geometry optimization for future fabrications and experiments.

## References

[1] A. M. Song, "Room-temperature ballistic nanodevices", Encyclopedia of Nanoscience and Nanotechnology, vol. 9, pp. 371-389, 2004.
[2] L. Bednarz, et al, "Theoretical and experimental characterization of Y-branch nanojunction rectifier up to 94 GHz", 2005 Microwave conference, vol. 1, pp. 4-8, Oct. 2005.
[3] Q. Diduck, M. Margala, M. Feldman, "A Terahertz transistor based on geometrical deflection of ballistic current", IEEE MTT-S International, 2006, pp.345-347.
[4] V. Kaushal, G. Guarino, Q. Yu, W. R. Donaldson, P. Ampadu, R. Sobolewski, M. Margala, "A Study of Effects of Deflector Position Variation on Leakage Currents in Ballistic Deflection Transistors", IEEE NMDC 2009.
[5] D. Huo, Q. Yu, P. Ampadu, "A Ballistic Nanoelectronic device Simulator", 2007 IEEE International Symposium on Nanoscale Architecture (NANOARCH 2007), pp. 38-45, 2007.

# Transport and excess noise in polyacenes under trap filling transition

**C Pennetta[1], L. Reggiani[1], A. Carbone[2]**
[1] Dipartimento di Ingegneria dell'Innovazione, Università del Salento
73100 Lecce, Italy
[2] Dipartimento di Fisica, Politecnico di Torino, 10129 Torino, Italy
E-mail: cecilia.pennetta@unisalento.it

**Abstract.** Recent measurements in organic semiconductors (polyacenes) have evidenced a strong superlinear increase of the current-voltage characteristics associated with a sharp peak of the relative spectral density of current noise. Both features occur at the trap filling transition (TFT), i.e. at voltages corresponding to the transition between the Ohmic regime and the space charge limited current (SCLC) regime. Here we discuss the interpretation of this behavior in terms of a contribution from trapping and detrapping (TD) processes of the injected carriers. We found an excellent agreement between the predictions of our model and the experimental results in tetracene, which allows us to conclude that TD processes are responsible for the significant enhancement of the relative current noise observed in polyacenes in the TFT region.

## 1. Introduction

Organic devices, based on polymeric materials or molecular semiconductors, successully compete with traditional electronic devices at least in terms of cost, flexibility and weight. The performance of organic devices is crucially controlled by the charge carries injected at the molecule-metal interfaces. In turns, injected carriers are drastically affected by the presence of trapping centers related to defect states. The effect of thermal and electrical stresses on charge carrier trapping and detrapping processes was widely investigated in the literature [1, 2, 3, 4, 5, 6, 7, 8, 9, 10, 11]. A smaller number of studies was devoted to noise in organic devices [12, 13].

In particular, recent transport and noise measurements in polyacenes [12] have evidenced a strong superlinear increase of the current-voltage, $I - V$, characteristics associated with a sharp peak of the relative spectral density of current noise. This noise peak occurs at voltages corresponding to the crossover between Ohmic and space charge limited currents (SCLC) regimes [12], at the so called trap filling transition (TFT). The interpretation of these experiments was addressed in terms of a continuous percolation between a Ohmic phase and an insulating (SCLC) phase [12]. According to this model, the noise peak was attributed to the clustering of insulating regions among the conducting material, bringing a narrowing of the current paths inside the film and thus a strong increase of current noise.

Here we discuss an alternative interpretation of the observed behavior in terms of TD processes of the injected carriers by deep traps. The phenomenological model considered here makes use of the information on the ionized traps fraction, extracted from the experimental I-V characteristics. To validate the model we consider the measurements performed on tetracene thin films [12]. Moreover, we will also report the results of numerical simulations based on a new percolation model which further support these conclusions.

## 2. Model and Results

We start by considering the $I - V$ characteristic measured [12] at 300 K for a tetracene thin film as reported in Fig. 1 (log-log plot). The film, of thickness $L = 0.65$ $\mu$m and cross-sectional area $A = 0.1$ cm$^2$, was sandwiched between Au and Al electrodes [12]. Charge transport is assumed to be mainly driven by holes injected from the Au contact. At increasing voltage, the characteristics show three regimes: Ohmic (linear $I - V$), TFT (strong superlinear $I - V$), SCLC (quadratic $I - V$). Now we assume that the current can be written as the sum of two asymptotic contributions, Ohmic and SCLC, respectively denoted as $I_\Omega$ and $I_{SCLC}$, where the last contribution is modulated by the fraction of filled (ionized) traps $u(V)$. Thus we take:

$$I(V) = I_\Omega(V) + u(V)I_{SCLC}(V) \quad (1)$$

where: $I_\Omega(V) = (n_0 e\mu A/L)V$, $e$ is the elementary charge, $n_0$ and $\mu$ respectively are the average density and the mobility of the free thermal carriers, $I_{SCLC}(V) = (9A\epsilon_0\epsilon_r\mu\Theta/8L^3)V^2$, with $\epsilon_0$ and $\epsilon_r = 3.5$ the vacuum permittivity and the relative dielectric constant of the material. The trapping parameter:$\Theta \equiv n/n_t = (n_v/gn_t)\exp[-E_t/k_bT]$ is the ratio between the average densities of free, $n$, and trapped, $n_t$, carriers ($n_v$, $g$, $E_t$, $k_B$ and $T$ respectively are the valence band density of states, the trap degeneracy, the trap energy level, the Boltzmann constant and the temperature).

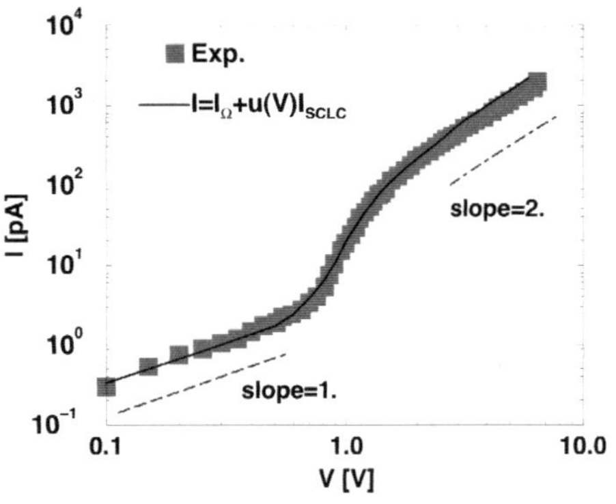


**Figure 1.** Black (red in color view) squares: Current-voltage characteristic of a tetracene thin film measured at 300K [12]. The two lines show the slopes of the I-V at the lowest (dashed) and highest (dot-dashed) voltage regions. The solid curve fits the data with the expression given in the figure (see text for details).

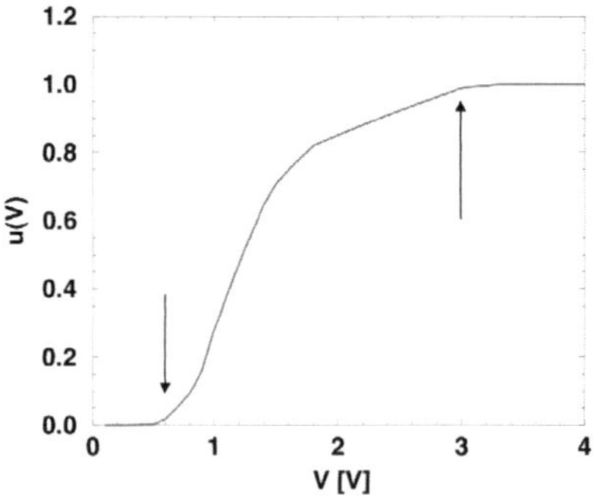


**Figure 2.** Fraction of filled traps vs voltage. The function $u(V)$ is obtained by inverting Eq.(1) and taking for $I$ the experimental values, while $I_\Omega$ and $I_{SCLC}$ are evaluated by considering respectively the low and high voltage limit. The two arrows point out the transition regions of the ohmic regime ($u \approx 0.$) and of the SCLC regime ($u \approx 1.$).

The function $u = u(V)$ is obtained by best fitting the experimental curve. The fit is shown in Fig. 1 while Fig. 2 reports the function $u(V)$. From the fit we find: $L/(Aen_0) = 3 \times 10^{11}$ $\Omega$ and $9A\epsilon_0\epsilon_r\mu\Theta/(8L^3) = 6 \times 10^{-11}$ A/V$^2$. Combining these results, we get for the effective mobility: $\mu_{eff} = \Theta\mu = 4.7 \times 10^{-10}$ cm$^2$/Vs, and $n_0\mu = 1.4 \times 10^6$ (mVs)$^{-1}$. In absence of an independent determination of $\mu$ or $\Theta$, we can only discuss the possible range of values for the relevant parameters of interest. In particular, by assuming $\mu = 1$ cm$^2$/Vs, a typical value for a single crystal, we find:$\Theta \approx 10^{-10}$ and $n_0 \approx 10^4$ cm$^{-3}$. By using the theoretical expression for $\Theta$ reported above and by taking $g = 2$, $n_v = 10^{21}$ cm$^{-3}$, $n_t = 10^{13} \div 10^{18}$ cm$^{-3}$ we find $E_t = 0.7 \div 1.0$ eV. We note that these values compare well with the values reported in the literature.

Figure 3 displays on a log-log plot the relative current noise spectral density, $S(f,V) \equiv S_I(f,V)/I^2$, measured at a frequency $f = 20$ Hz, in constant-voltage conditions, as a function of the applied voltage, for the same tetracene sample used to get the $I - V$ in Fig. 1 [12]. Consistently with the interpretation of the $I - V$ characteristics, we identify three regimes in the behavior of $S(f,V)$: Ohmic, TFT and SCLC. Accordingly we write:

$$S(f,V) = S_\Omega(f) + S_{TFT}(f,V) + S_{SCLC}(f,V) \tag{2}$$

where: $S_\Omega(f) = (1/ALn_0)(\alpha_\Omega/f)$, with $\alpha_\Omega$ the Ohmic Hooge parameter [14] (as well known, the Ohmic component is independent of voltage). On the other hand, according to Ref. [14], the SCLC component can be written as: $S_{SCLC}(f,V) = (4eL/5A\epsilon_0\epsilon_r\Theta)(\alpha_{SCLC}/f)(1/V)$, with $\alpha_{SCLC}$ the SCLC Hooge parameter. Finally, for the TFT regime we use the standard expression for the relative trapping-detrapping noise with a lorentzian spectrum:

$$S_{TFT}(f,V) = \frac{1}{N_t}\frac{4\tau}{[1+(2\pi f\tau)^2]}u(V)[1-u(V)] \equiv Bu(1-u) \tag{3}$$

where $N_t$ is total number of traps inside the volume of the device and $\tau$ the carrier lifetime. The parameter $B$ is chosen to reproduce the total excess noise measured at the given frequency in the TFT regime. This is obtained for $B = S_{max}/0.25 = 2.8 \times 10^{-10}$ s, where $S_{max}$ is the maximum experimental value of the noise. Furthermore, by defining as $V_{th}$ the threshold voltage for the onset of the SCLC regime and by taking into account that $u \approx 0$ in the Ohmic regime, we can rewrite $S(f,V)$ at a given frequency and for $V < V_{th}$ as:

$$S(f,V) = \Big[1 - u(V)\Big]\Big[S_\Omega + Bu(V)\Big] \tag{4}$$

The values obtained from the above expression are reported in Fig. 3 as a solid line with grey (orange in color view) diamonds. Moreover, Fig. 3 also shows as a solid grey (green) line the power-law fit with slope $-1$ to the experimental data in the high voltage region. Thus, Fig. 3 points out an excellent agreement between the predictions of our model and the experimental results in tetracene, which allow us to state that TD processes are responsible for the significant enhancement of the relative current noise observed in polyacenes in the TFT region.

This statement is further supported by the results, reported in Fig. 4, of numerical simulations based on a new percolation model. The model considers a two-dimensional square-lattice, binary resistors network [15] (size $N \times N$), where each elementary resistor $r_{i,j}$ is subjected to random transitions from two states of resistances $r_1$ and $r_2$, with probabilities: $W_{1\to2} = \exp[-(E_2 - qv_{i,j})/k_BT]$ and $W_{2\to1} = \exp[-E_1/k_BT]$, where $v_{i,j}$ is the voltage applied to $r_{i,j}$, $E_1$, $E_2$ are two activation energies and $q$ is an effective charge. It should be noted that the relative current noise of the network depends only on its size $N$ and on the three ratios: $f_r = r_1/r_2$, $a = (E_2 - E_1)/K_BT$ and $b = q/K_BT$. The results reported in Fig. 4 are obtained by assuming constant-voltage boundary conditions and by taking the following values for the parameters: $N = 75$, $f_r = 290$, $a = 2.5$ and $b = 210V^{-1}$.

In conclusion, we have developed a phenomenological model which provides a quantitative interpretation of the current-voltage characteristics and of the relative current noise in tetracene as function of the applied voltage. The excellent agreement with the experiments allow us to state that the sharp peak of noise in the TFT region basically arises from the fluctuating

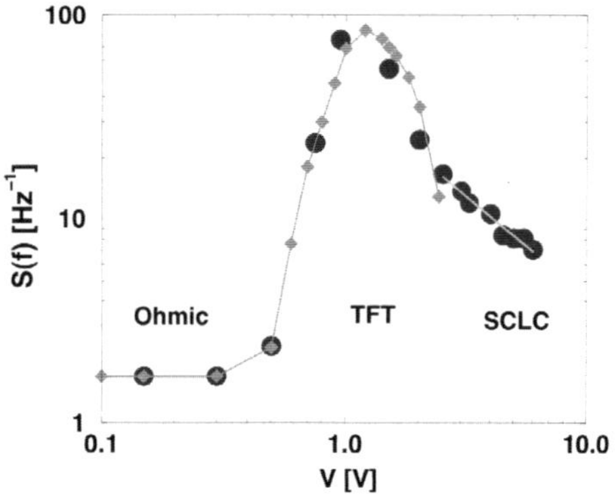


**Figure 3.** Black (blue in color view) circles: relative spectral density of current noise $S(f)$ measured vs voltage at 20 Hz and 300 K for the same tetracene film [12]. Line with grey (orange) diamonds: curve obtained from the phenomenological theory exposed in the text. Solid grey (green) line: SCLC noise component at high voltage decreasing as $1/V$.

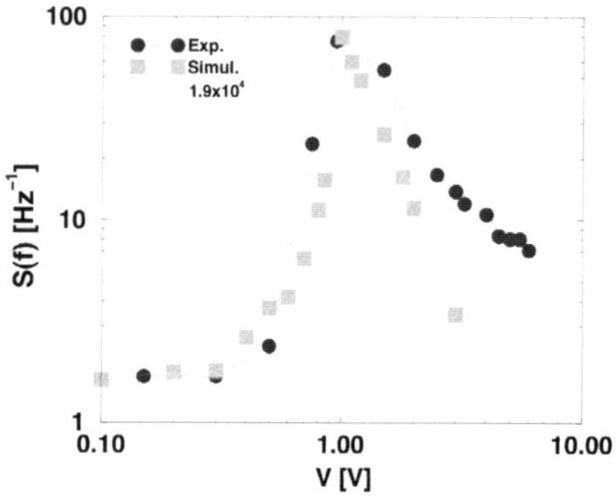


**Figure 4.** Black (blue in color view) circles: experimental values of $S(f)$ (the same of Fig. 3). Grey (green) squares: relative current noise $S_{sim}$ obtained from numerical simulations. $S_{sim}$ is calculated by the percolative model described in the text. The simulated values are normalized to $S_{norm} = 5.26 \times 10^{-5}$ to agree with the measured value of $S_{max}$.

occupancy of the traps due to TD processes. Furthermore, we have reported results of numerical simulations based on a new percolation model which furtherly support this conclusions. Finally, we remark that the measured current noise spectrum was found to be $1/f$-like rather than Lorentzian. However, the broadening of the trap level energy could account for the $1/f$-like power spectrum [14, 15].

### *2.1. Acknowledgments*
The support of D. Kotowski, B. Kutrezba-Kotowska and M. Tizzoni at the early stage of the work is gratefully acknowledged.

## 3. References
[1] Lang D V et al. 2004 *Phys. Rev. Lett.* **93**, 076601
[2] Miyadera T et al. 2008 *Appl. Phys. Lett.* **93**, 033304
[3] Koch N et al. 2005 *Appl. Surf. Sc.* **244**, 593
[4] Yang Y S et al. 2002 *Appl. Phys. Lett.* **80**, 1595
[5] Knipp D et al. 2003 *Appl. Phys. Lett.* **82**, 3907
[6] Dinelli F et al. 2004 *Phys. Rev. Lett.* **92**, 116802
[7] Schwalb C H et al. 2008 *Phys. Rev. Lett.* **101**, 146801
[8] De Boer R W I et al. 2004 *J. Appl. Phys.* **95**, 1196
[9] Kang J H D et al. 2005 *Appl. Phys. Lett.* **86**, 152115
[10] Chandra W et al. 2007 *Appl. Phys. Lett.* **90**, 153505
[11] Giulianini M et al. 2009 *Appl. Phys. Lett.* **94**, 083302 and Torres et al. 2004 *Appl. Phys. Lett.* **85**, 314
[12] Carbone A et al. 2005 *Phys. Rev. Lett.* **95**, 236601 and Carbone A 2006 *Eur. Phys. J.* B **50**, 77
[13] Jurchescu O D et al. 2008 *Appl. Phys. Lett.* **92**, 132103
[14] Kleinpenning T G M 1978 *Physica* B,C **94**, 141 and Hooge F N et al. 1981 *Rep. Prog. Phys.* **44**, 479
[15] Pennetta C et al. 2009 *J. Stat. Mech.* **p02053**

# Preliminary results of bench implementation for the study of terahertz amplification in gallium nitride quantum wells

**T. Laurent[1], P. Nouvel[1], J. Torres[1], L. Chusseau[1], C. Palermo[1], L. Varani[1], Y. Cordier[2], J.-P. Faurie[3], B. Beaumont[3], E. Starikov[4], P. Shiktorov[4] and V. Gružinskis[4]**

[1] Institut d'Électronique du Sud, UMR 5214 - CNRS / Université Montpellier 2, France
[2] Centre de Recherche sur l'Hétéro-Épitaxie et ses Applications, UPR 10 -CNRS, Sophia Antipolis, France
[3] Lumilog®, Vallauris, France
[4] Semiconductor Physics Institute, Vilnius, Lithuania

E-mail: thibault.laurent@ies.univ-montp2.fr

**Abstract.** We present continuous-wave terahertz spectroscopic results obtained with a photomixing emitter and a bolometric detection. The tested elements are candidates for a further cryogenic THz experiment which will consist to demonstrate the effective amplification by optical phonon transit time resonance in gallium nitride quantum wells.

## 1. Introduction

In the actual goal of realization of sources and detectors in the terahertz (THz) frequency domain, we carry out a new gallium nitride (GaN) quantum welled (QW) device using the optical phonon transit time resonance (OPTTR) phenomenon[1]. However, the experimental demonstration of effective wave generation needs the development of a spectroscopic bench working in the THz frequency range. Even if time-domain THz spectroscopy is currently rather common [2, 3], continuous-wave spectroscopy by photomixing remains marginal [4]. The reason of this lack is related to the efficiency of available sources that is significantly reduced in continuous-wave operation with respect to pulsed operation. Frequency-domain spectroscopy has some advantages in regards of time-domain spectroscopy among which the frequency resolution or the phase extraction using homodyne detection [5]. Alternatively it is possible to use a bolometric detection able to measure lower THz intensity for the verification of the effective amplification of our new QW GaN devices.

## 2. Terahertz amplification with optical phonon transit time resonance in GaN

The OPTTR is a well-known effect that takes its origins in sufficiently pure lattices at low temperatures when the mechanism of absorption of optical phonons with energy $\hbar\omega_0$ by free carriers is practically absent (this condition is achieved when $T \ll \hbar\omega_0/k_B$, where $k_B$ is the Boltzmann constant). The momentum space is by consequence divided into two parts, the so-called passive and active regions corresponding respectively to energies smaller or bigger than

the optical phonon energy.
As illustrated in figure 1 (where the two cases of 2D and 3D transport are represented), the passive region is characterized by a very low optical-phonon emission rate whereas the active region by a great interaction rate. These two regions are separated by a threshold corresponding to the optical phonon energy value. The cyclic movement of electrons due to the combined action of electric field ballistic acceleration and energy loss by emission of optical phonons produces a negative differential mobility in the THz frequency domain [6]. As a consequence, the frequency of this cyclic movement lineary depends on the applied electric field. The limitations in term of frequencies can be summarized by two conditions on the free flight time $\tau_E$ of each electron (time for electrons to reach the energy of the optical phonon) : $\tau_+ \ll \tau_E \ll \tau_-$, where $\tau_+$ and $\tau_-$ are respectively the scattering times in the active and passive regions.
An experimental verification of such a generation was obtained for the first time in InP samples by the group of Prof. L. E. Vorob'ev in St. Petersburg [7]. Nevertheless, recent publications [6] evidence that nitrides are promising materials for power generation in the THz frequency range because of the significant increase of the generation frequency band which extends from 200 GHz up to 7 THz at 77 K. However, from one hand, at the present time bulk nitride samples of sufficiently good quality are practically unavailable. Experimentally available samples are mostly heterostructures with 2D electron transport such as quantum-wells (QW) and heterolayers (HL). From another hand, it is worthwhile to emphasize that 2D transport has two evident advantages with respect to 3D case: (i) a possibility to increase the maximum generation frequency as shown in Fig. 2, that is the consequence of the sharper threshold of the optical phonon emission rate (see Fig. 2), and (ii) a possibility to obtain a greater amplification coefficient by the use of higher carrier densities and, at the same time, avoiding impurity scattering which tends to destroy the ballistic acceleration of electrons in the passive region.

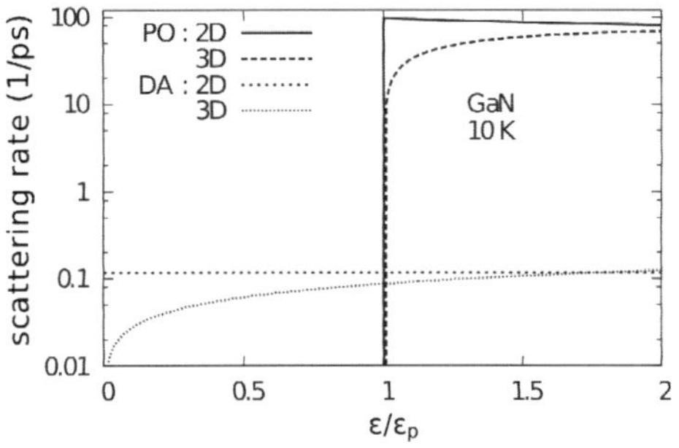


**Figure 1.** Polar optical (PO) and deformation acoustic (DA) phonon scattering rates as functions of the electron energy normalized to the optical phonon energy in GaN at 10 K for both bulk and QW GaN.

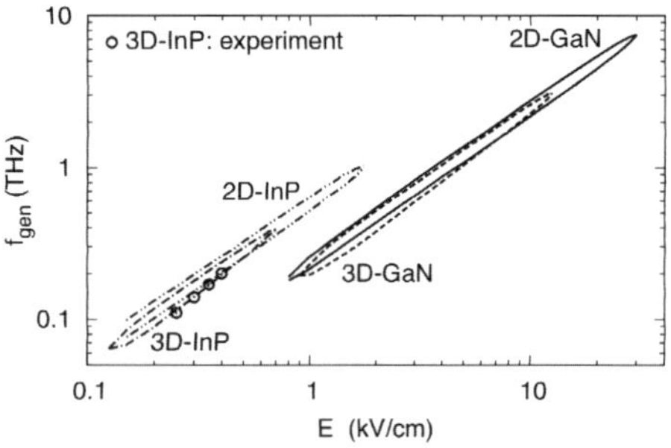


**Figure 2.** Generation frequency bands as functions of applied external electric field in 2D and 3D InP and GaN at 77 K. Symbols refer to experimental results of ref. [7].

In the following sections we describe the preliminary development of an original THz spectrometer devoted to the measurement of THz emission by GaN QW.

## 3. Experimental setup

For the measurement of the GaN-transmission in the THz region, we have built a continuous wave (CW) THz-spectrometer represented schematically in figure 3. We used a low-temperature-grown GaAs photoconductive antenna illuminated by the beating of two tunable CW-DFB lasers emitting around 780 nm. The lasers wavelength emission is fixed by controlling the current and temperature (using Peltier stages). As the nominal power required by our photomixing-device is around 60 mW [8], a tapered-semiconductor-amplifier was associated with the lasers in order to achieve this value. The THz generated field covers the range from 0.7 THz up to 1.7 THz. The developed experimental set-up uses two parabolic mirrors that collimate the THz field on the device under test (DUT). The detection is ensured by a 4 K cooled silicon bolometer associated with a lock-in amplifier synchronized with the electrical modulation of the photomixing device.

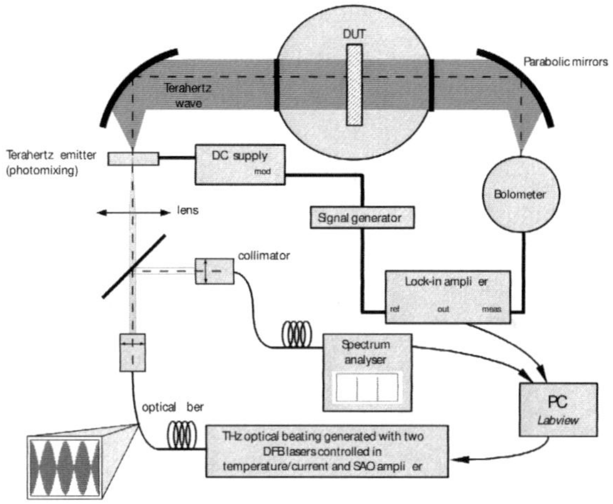


**Figure 3.** Experimental setup of THz spectroscopic bench (from 0.7 to 1.7 THz)

The procedure to obtain a spectrum in the frequency range of 0.7-1.7 THz with a 2 GHz step is the following. (i) A calibration spectrum is obtained without DUT then (ii) the measurement spectrum is obtained by placing the DUT in front of the THz beam. The resulting transmission spectrum is obtained by dividing the measurement spectrum by the calibration one.

## 4. Results

As shown on previous sections, the OPTTR phenomenon is effective when the GaN device is cooled down to low temperature. For this purpose, we measured several transmission spectra of candidate materials for the entrance windows of the cryostat. One example is given in figure 4 where the transmission spectrum was done through a 5 mm thick PTFE-slice (Polytetrafluoroethylene, also known as Teflon®). Several water-vapor-absorption peaks are clearly observed at approximatively 0.98, 1.2, 1.22, 1.33 and 1.6 THz. To validate our bench, we have compared these results with a spectrum obtained by time-domain-spectroscopy at the RIKEN institute [9]. The overall shape of this spectrum as well as the absorption peaks reproduce well the spectrum obtained by Riken. Nevertheless, we obtain a transmission rate which is around ten times lower that these of RIKEN: this discrepancy can be explained by some loss between calibration and measurement spectra.

Finally, the performances of our THz frequency domain spectrometer are summarized in the table 1.

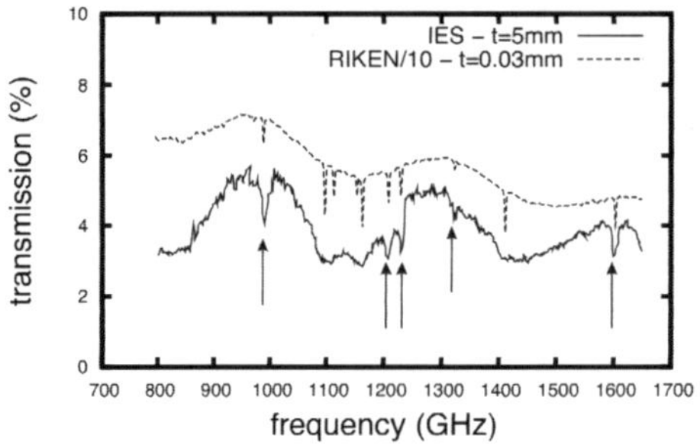


**Figure 4.** Transmission measurement of a 5 mm thick PTFE sample obtained with our frequency-domain spectrometer (continuous line) and compared with a 30 $\mu$m thick PTFE sample obtain with a time-domain spectrometer at the RIKEN's tera-photonics group, Japan (dashed line)[9].

| **Parameter** | **Value** | **Unit** |
|---|---|---|
| Spectral range | $0.7 - 1.7$ | THz |
| Frequency accuracy | $< 2$ | GHz |
| Power stability all over the band | $> 75$ | % |
| Average power | $\approx 10$ | nW |
| Signal to noise ratio | $> 20$ | dB |

**Table 1.** Typical performances of our THz spectrometric experimental bench

## 5. Conclusion

We realized a functional continuous-wave THz spectrometer in the $0.7 - 1.7$ THz range. We implemented a calibration procedure and realized transmission spectra of several materials that are candidates for the windows of the 4 K-cooled cryostat that will be used for the demonstration of the THz radiations generation and amplification by bidimensional GaN structures through the OPTTR phenomenon.

## Acknowledgments

We acknowledge the French National Research Agency (ANR) for funding the research under contract AITHER n°ANR-07-BLAN-0321.

## References

[1] Starikov E, Shiktorov P, Gruzinskis V, Reggiani L and Varani L 2007 *JOURNAL OF NANOELECTRONICS AND OPTOELECTRONICS* **2** 11–35 ISSN 1555-130X
[2] Laamiri Y, Garet F and Coutaz J L 2009 *Appl. Phys. Lett.* **94** ISSN 0003-6951
[3] Kadlec C, Kadlec F, Kuzel P, Blary K and Mounaix P 2008 *Optics Letters* **33** 2275–2277 ISSN 0146-9592
[4] Matton S, Rohart F, Bocquet R, Mouret G, Bigourd D, Cuisset A and Hindle F 2006 *Journal of molecular spectroscopy* **239** 182–189 ISSN 0022-2852
[5] Mouret G, Matton S, Bocquet R, Bigourd D, Hindle F, Cuisset A, Lampin J and Lippens D 2006 *Appl. Phys. Lett.* **88** ISSN 0003-6951
[6] Shiktorov P, Starikov E, Gružinskis V, Varani L and Reggiani L 2008 *Acta Physica Polonica A* **113** 795–802
[7] Vorob'ev L, Danilov S, Tulupenko V and Firsov D 2001 *JETP Letters* **73** 219–222 ISSN 0021-3640
[8] Geizutis A, Krotkus A, Bertulis K, Molis G, Adomavicus R, Urbanowicz A, Balakauskas S and Valaika S 2008 *Optical materials* **30** 786–788 ISSN 0925-3467 international Workshop on Advanced Spectroscopy and Optical Materials, Gdansk, Poland, Jun 11-14, 2006
[9] Ito H may 2009 Tera-photonics laboratory riken sendai, thz database web : http://www.riken.jp/thzdatabase/

Journal of Physics: Conference Series **193** (2009) 012095 doi:10.1088/1742-6596/193/1/012095

# Silicon MOSFETs As Room Temperature Terahertz Detectors

**H. Videlier[1], S. Nadar[1], N. Dyakonova[1], M. Sakowicz[1], T. Trinh Van Dam[1], F. Teppe[1], D. Coquillat[1], W. Knap[1], S. Denorme[2], T. Skotnicki[2], J. M. Peiris[3] and J. Lyonnet[3]**

[1] GES UMR 5650 Univ. Montpellier 2 and CNRS 34095 Montpellier, France.
[2] ST Microelectronics, BP 16, 38921 Crolles, France.
[3] Centrale de Technologie en Micro et Nanoélectronique, Univ. Montpellier 2, 34095 Montpellier, France.

E-mail: h.videlier@gmail.com

**Abstract.** Because of their competitive Noise Equivalent Power (NEP) at room temperature and their fast modulation frequency, Field Effect Transistors (FETs) are one of the most promising devices for room temperature multipixel THz detection and imaging. We report here on an experimental study of the sub-THz detection as a function of the gate length in Silicon FETs and discuss the physical basis of their optimisation.

**Backgrounds**

In 1990, Dyakonov and Shur proposed that two dimensional transistor channel could be used as a cavity for plasma waves excited by current flow or incident radiation [1, 2]. The frequency of emitted or detected electromagnetic radiation in such a way for transistors of nanometer size is in the THz frequency range. The detection of radiation is due to the nonlinear properties of the transistor channel: the ac current induced by the incoming radiation is rectified and a photoresponse in the form of a dc voltage between source and drain appears.

Here we deal with sub-THz detection into Silicon FETs. In this particular material, because of low electron mobility and relatively low incident frequency, plasma waves in the channel are overdamped [3]. As it was shown theoretically, this nonresonant detection is characterized mainly by a single parameter: a critical length, which describes the exponential decay of the plasma wave amplitude with the distance [4].

**Experimental details**

We studied fully depleted nMOS transistors fabricated on biaxially strained SOI (here tensile strain). The thickness of the buried oxide (BOX) was 20 nm, silicon layer thickness $T_{Si}$ was 8 nm. Gate stack consisted of 2.5 nm $HfZrO_2$ layer deposited by atomic layer deposition method and covered by a 5 nm TiN metal gate and 80 nm of polysilicon.

Studied devices had the channel width $W_g$ = 10 μm, the channel length, $L_g$, was in the range from 50 nm to 10 μm. The mobility for transistors with $W_g >> L_g$ was determined by the magneto resistance method [5,6]. The channel depletion threshold voltage is near 0.2 V, and channel resistance is decreasing linearly with its length.

Journal of Physics: Conference Series **193** (2009) 012095 doi:10.1088/1742-6596/193/1/012095

The photoresponse measurements were performed using a Back Wave Oscillator (BWO) at 230 GHz and output power of few mW. The FETs were mounted on an X-Y two-dimensional stage controlled by a computer and placed at a distance of 10 mm from the BWO output horn and the electric field of the incoming radiation was parallel to the source-drain direction. The spatial intensity information was preliminary collected at the same distance by a pyroelectric detector. The radiation intensity was mechanically chopped (300-1000 Hz range), and the open-circuit source drain voltage was measured using a lock-in technique. All measurements were done at room temperature.

**Results and discussion.**
The electron mobility was about 300 $cm^2.(V.s)^{-1}$ for gate length higher than 0.1 μm and then decreased to 180 $cm^2.(V.s)^{-1}$ for 55 nm gate length. This type of dependence of electron mobility on the gate length, $\mu(L_g)$, had been described and discussed previously in reference [5].

Figures 1 below shows the photoresponse of a whole set of FETs, with different gate lengths (varying from 50 nm, up to 10 μm), versus the gate swing voltage $U_0$ where $U_0 = U_g - U_{th}$ with $U_g$ is the gate voltage and $U_{th}$ the threshold voltage. One can see that the photoresponse is a decreasing function of the gate swing voltage for $U_0 \geq 0$. This behavior is characteristic for the non resonant detection and it has been observed for different materials (see, for example, [3]).

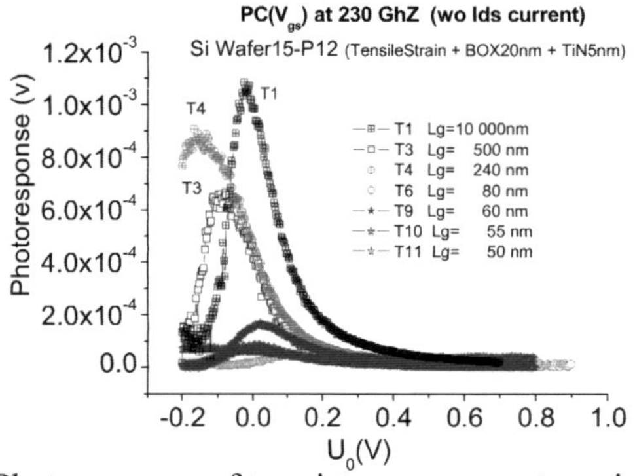


**Figure 1.** Photoresponse of transistors versus gate swing voltage.

One can also see that the gate length significantly influences the photoresponse amplitude: for the long devices the amplitude is much higher than for the short ones. The photoresponse versus gate length is presented in Figure 2.

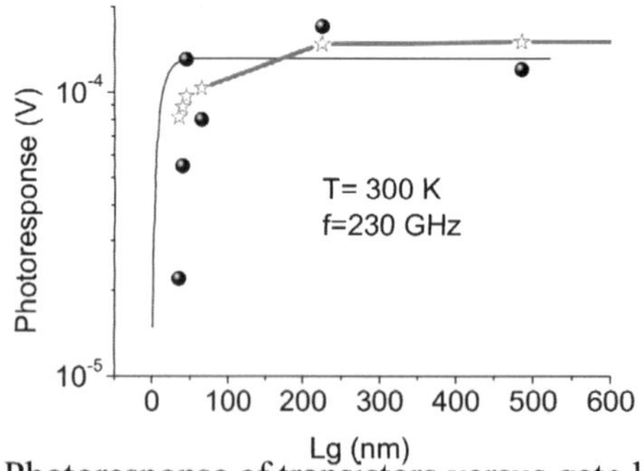


**Figure 2.** Photoresponse of transistors versus gate length.

The black points are experimental values of photoresponse, taken in the range 0.1-0.25 V of the gate voltage swing, where signal has a $1/U_0$ shape described by the broadband-detection theory [3]. At smaller values of gate voltage swing the resistance of the transistors strongly increases and results can be influenced by impedance mismatch with measurement equipment [7]. As it is seen, the photoresponse strongly increase with the gate voltage increasing, then at $L_g$=250 nm it becomes saturated.

The detection regime depends on the parameter $\omega\tau$ and on the gate length $L_g$ [4]. As the mobility in our samples ($\mu$< 350 $cm^2V^{-1}s^{-1}$) and the radiation frequency ($f \approx$ 230 GHz) are rather low, the parameter $\omega\tau$ is much lower than unity, and the plasma waves in the channel are overdamped along the channel.

In this case, photoinduced voltage is given by [4]:

$$U(x) = \frac{U_a^2}{4U_0}\left[1 - \exp(-2x/l_c)\right] \tag{1}$$

Where x is the distance from the source, $U_a$ the amplitude of the ac modulation due to the incident radiation on the FET, $U_0$ is the gate voltage swing and $l_c$ is the characteristic length for the decay of the ac voltage (and current) away from the source, also given by $l_c = s(2\tau/\omega)^{1/2}$ ,with the plasma waves velocity $s = (eU_0/m)^{1/2}$. Using the set of parameters $U_0 = 0.25$ V, $\mu = 0.03$ $cm^2.V.s^{-1}$, and $f = 230$ GHz, we obtain the $l_c = 100$ nm.

In the case of short gate transistors, $L_g < L_c$, the ac current due to the incident radiation goes through the gate-to-channel capacitance practically uniformly on the whole length of the gate, and just a part of the photoresponse dc voltage is built up. On the contrary, in the case of long gate transistors with $L_g >> L_c$, the ac current will leak to the gate before it achieves the drain, and the total photoresponse is built up. This can be illustrated by plotting the photoresponse signal versus gate length of transistors (see Figure 2).

The blue solid curve is a fit of experimental results, using equation 1, with a fixed critical length of $l_c = 100$ nm. The red curve with stars, is a fit of experimental results using equation 1 also, but taking into account that a) the effective gate length is 15 nm shorter than nominal ones (due to the implant pocket zones), and b) that the mobility also depends on the gate length ([6]). One can see that the theoretical estimation is in a relatively good agreement with experimental results.

**Conclusion**

In this paper we investigated THz detection by Silicon FETs. We found that the dependence of the photoresponse on the gate length is well described by the theory. We established, that for this type of detection (broadband at low frequency regime) the channel length should be at least several times the critical length which is defined by the incident frequency and carrier mobility. This supplementary information will allow to optimize the design of Terahertz detectors based on Si-MOSFETs technology. Because of their competitive NEP and their fast modulation frequency, these improved detectors could be used in arrays for real time imaging applications at room temperature (THz camera) [7].

**Acknowledgement**
This work is supported by ANR program TeraGaN, the French ministry of finance and industry, by CNRS guiding GDR and GDR-E projects "Semiconductor sources and detectors of THz frequencies", by the Region of Languedoc-Roussillon through the "Terahertz Platform" project, by STMicroelectronics through the "Nano2008" program, by the European Commission by "PullNano" program, by the CEA-LETI through the THz imaging on Si-MOSFETs project, and by the joint French-Lithuanian research program "Gilibert/EGIDE".

## References

[1] Dyakonov M and Shur M (1993) Shallow water analogy for a ballistic field effect transistor: New mechanism of plasma wave generation by dc current *Phys. Rev. Lett.* 71, pp.2465-2468
[2] Dyakonov M and Shur M (1996) *IEEE*, Vol.43, No.3
[3] Knap W et al (2002) Nonresonant detection of terahertz radiation in field effect transistors *J. Appl. Phys*, V 91, N 11, pp.9346-9353
[4] Knap W, Dyakonov M et al (2009) Field Effect Transistors for Terahertz Detection: Physics and First Imaging Applications, Review submitted to *International Journal of Infrared and Millimeter Waves*. arXiv:0907.2523 (July 2009)
[5] Lusakowski J, Knap W, Meziani Y, Cesso J.-P, El Fatimy A, Tauk R., Dyakonova N, Ghibaudo G, Boeuf F, SkotnickiT (2006) Electron mobility in quasi-ballistic Si MOSFETs, Solid-State Electronics, 50, 632–636.
[6] Tauk R, Lusakowski J et al (2007)Low electron mobility of field-effect transistor determined by modulated magnetoresistance, *J. Appl. Phys*. 102, 10371
[7] Öjefors E, Pfeiffer U R, Lisauskas A, and Roskos H G (2008) 0.65 THz Focal-Plane Array in a Quarter-Micron CMOS Process Technology, *IEEE Journal of solid-states circuits*, vol.44, no. 7

Journal of Physics: Conference Series **193** (2009) 012096 doi:10.1088/1742-6596/193/1/012096

# Hydrodynamic study of terahertz three-dimensional plasma resonances in InGaAs diodes

**Pierre Ziadé, Ziad Kallassy**

Laboratoire de Physique Appliquée, Université Libanaise, Faculté des Sciences 2, Fanar, Liban

**Hugues Marinchio, Giulio Sabatini, Christophe Palermo, and Luca Varani**

Institut d'Électronique du Sud (CNRS UMR 5214), Université Montpellier 2, 34095 Montpellier, France

E-mail: pierre.ziade@ies.univ-montp2.fr

**Abstract.** We investigate the presence of plasma resonances in InGaAs diodes under different optical excitation conditions. In particular, we study the case of diodes submitted to an optical photoexcitation presenting a beating in the terahertz frequency domain. The responses of the diodes are calculated using a hydrodynamic approach coupled to a one-dimensional Poisson solver. The results show clearly the presence of three-dimensional plasma resonances in the terahertz frequency domain. We also show that the resonances frequency and amplitude can be tuned by modifying the diode geometry and doping profile.

## 1. Introduction

Recent experiments have shown the possibility to use field effect transistors (III-V and Si FETs) as devices able to emit and/or detect terahertz radiations [1, 2, 3]. In all cases, the physical mechanism at the basis of the terahertz working operation is a two-dimensional electron gas plasma wave excited in the transistor channel. On the other hand, the possibility to use three-dimensional plasma waves has received much less attention in the literature even if a terahertz radiations detector based on the excitation of 3D plasma in bulk GaAs was recently proposed [4].

By means of hydrodynamic equations coupled to a one-dimensional Poisson solver, we have performed a systematic study of InGaAs $n^+ - n - n^+$ diodes submitted to an optical photoexcitation presenting a beating in the terahertz frequency domain, according to the experimental technique detailed in [3].

The numerical analysis has been performed by calculating the instantaneous response of carrier velocity, concentration, electric field and current density. In particular, we have studied the effect of the diode geometry and doping profile on the resonance frequency and its amplitude.

## 2. Numerical model

We couple the one-dimensional hydrodynamic equations relative to the electron density $n(x,t)$, the mean velocity $v(x,t)$ and the mean energy $\epsilon(x,t)$, with the one-dimensional Poisson equation:

$$\left\{\begin{aligned} \frac{\partial n_{3D}}{\partial t} &= \frac{\partial (n_{3D} v)}{\partial x} + G \\ \frac{\partial v}{\partial t} &= -v\frac{\partial v}{\partial x} - \frac{eE}{m(\epsilon)} - \frac{1}{n_{3D}}\frac{\partial\left(\overline{\delta v^2}(\epsilon)\, n_{3D}\right)}{\partial x} - v\nu_v(\epsilon) \\ \frac{\partial \epsilon}{\partial t} &= -v\frac{\partial \epsilon}{\partial x} - eEv - \frac{1}{n_{3D}}\frac{\partial\left(\overline{\delta v \delta \epsilon}(\epsilon)\, n_{3D}\right)}{\partial x} - (\epsilon - \epsilon_{equ})\nu_\epsilon(\epsilon) \\ \varepsilon\frac{\partial E}{\partial x} &= -e\,(n_{3D} - N_D) \end{aligned}\right. \tag{1}$$

where $e$ is the elementary charge, $E$ the electric field, $N_D$ the density of donors, $\varepsilon$ the dielectric constant of InGaAs, and $\epsilon_{equ}$ the equilibrium mean energy. The velocity and energy relaxation rates $\nu_v$, $\nu_\epsilon$, the electron effective mass $m$, the velocity variance $\overline{\delta v^2}$ and the energy-velocity covariance $\overline{\delta v \delta \epsilon}$ depend on the local mean energy and are calculated by a Monte Carlo simulation of InGaAs bulk material at room temperature. A beating optical excitation can be included in the equations through the term $G = G_0[1 + \cos(2\pi f t)]$ [3].

## 3. Study of the plasma resonances in a $n^+ - n - n^+$ diode

### *3.1. Frequency responses in n and $n^+$ regions of the diode*

The calculation is made for $n^+ - n - n^+$ diodes submitted to a constant voltage of 0.5 V, and to an optical beating of amplitude $G_0 = 10^{26}$ cm$^{-3}$s$^{-1}$. We consider a $n^+ - n - n^+$ diode with $n = 10^{16}$ cm$^{-3}$ and $n^+ = 10^{17}$ cm$^{-3}$ and where the lengths of the different regions are identical, equal to 500 nm. We calculate, through a Fourier transform, the frequency response of the local electric field taken at the center of both $n$ and $n^+$ regions, in order to characterize the resonances. The modulus of the electric field response is represented in Figure 1 which shows that the responses in both $n$ and $n^+$ regions of the diode have a resonance at a frequency within the terahertz domain, that is at 3 and 3.6 THz respectively.

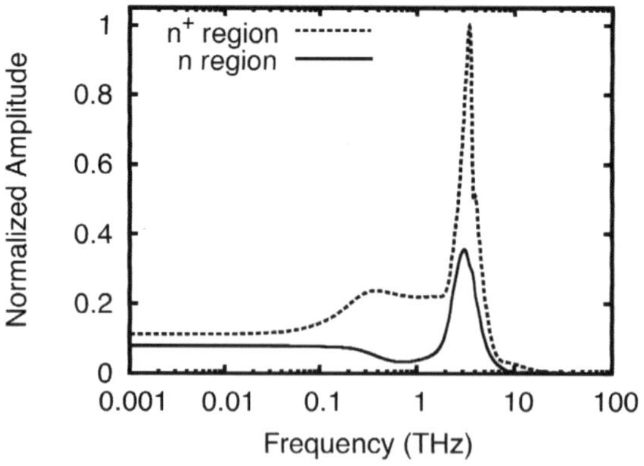


**Figure 1.** Normalized modulus of the electric field response as a function of frequency calculated at the center of the $n^+$ and $n$ regions.

*3.2. Effect of the doping profile on 3D plasma resonance frequencies*

The previous calculations are performed for different doping profiles, and the results are compared with the theoretical 3D plasma frequency relative to the corresponding concentration which verifies the theoretical expression [1]:

$$f_{3D} = \frac{1}{2\pi}\sqrt{\frac{e^2 N_D}{m\varepsilon}} \tag{2}$$

We report in Figure 2 the obtained results as functions of the $n$ region electron density for $n^+/n = 10$ (a) and as functions of the $n^+/n$ ratio for $n = 10^{16}$ cm$^{-3}$ (b). We observe in all the considered cases an excellent agreement between the calculated resonance frequencies for the region $n^+$ and the corresponding analytical 3D plasma frequencies. On the other hand, we observe a discrepancy of the hydrodynamic results for the resonance frequency and the 3D analytical model in the $n$ region of the diode. In fact, the resonance of the diode corresponding to the concentration $n$ is situated between the 3D plasma frequencies corresponding to the concentrations $n$ and $n^+$, so that the resonance $n$ of the diode is shifted towards the 3D plasma resonance corresponding to the concentration $n^+$. This result shows that the main 3D plasma resonance of the diode is that corresponding to the concentration $n^+$.

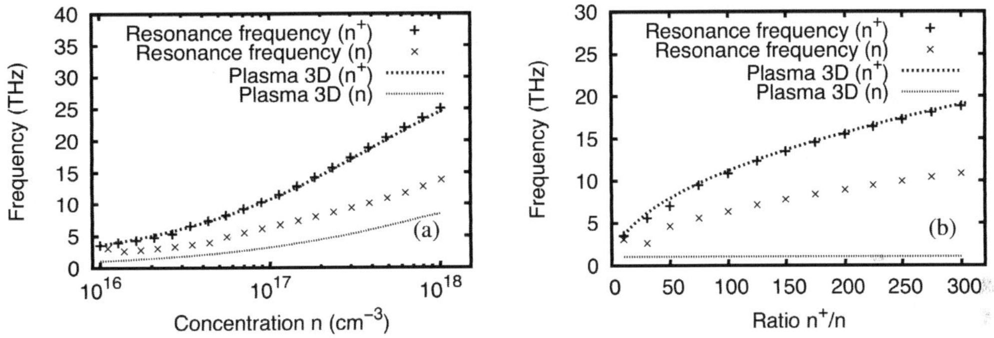


**Figure 2.** Resonance frequencies of the electric field response to an optical beating excitation calculated in the middle of the $n$ and $n^+$ regions of a $n^+ - n - n^+$ diode, as functions of $n$ for $n^+/n = 10$ (a) and as functions of the $n^+/n$ ratio for $n = 10^{16}$ cm$^{-3}$ (b). Symbols refer to the hydrodynamic calculations and lines to the analytical formula (2).

*3.3. Effect of the diode geometry on 3D plasma resonance frequencies*

We have calculated the resonance frequencies for different values of the regions lengths, in order to evaluate the effect of the diode geometry. We have fixed the doping concentrations to $n = 10^{16}$ cm$^{-3}$ and $n^+/n = 10$. Results are shown in Figure 3 as functions of the internal $n$ region length by fixing the $n^+$ regions length to 500 nm (a) and as functions of the external $n^+$ regions length for $n$ region length fixed to 500 nm (b). The results show that, while the effect of the variation of the external regions length is negligible, the increase of the internal region length is crucial to put in evidence the 3D plasma frequency corresponding to the concentration $n$. Indeed, the resonance frequency and amplitude decrease if the length of the internal region increases. In fact, by increasing the length of the $n$ region, the influence of the $n^+$ access region on the dynamic behavior of the internal region decreases. As a result, the resonance frequency calculated in the middle of the $n$ region of the diode decreases and tends more and more towards the 3D plasma resonance frequency corresponding to the concentration $n$. In particular, when the internal region length becomes greater than 2000 nm, the resonance frequency in the region $n$ tends to 1.1 THz which is rigorously close to the 3D plasma resonance frequency corresponding to the concentration $n = 10^{16}$ cm$^{-3}$ (Figure 3). Then, the resonance frequency corresponding to the concentration $n$ in the diode $n^+ - n - n^+$ is evidenced when the length of the internal $n$ region is greater than 2000 nm. On the other hand, we observe that the resonance frequency corresponding to the concentration $n^+$ is evidenced in all cases. Indeed, whatever the lengths of the diode regions, the resonance frequency of the external $n^+$ regions tends to 3.6 THz, that is to the 3D theoretical plasma frequency for the considered doping concentrations (see Figure 3).

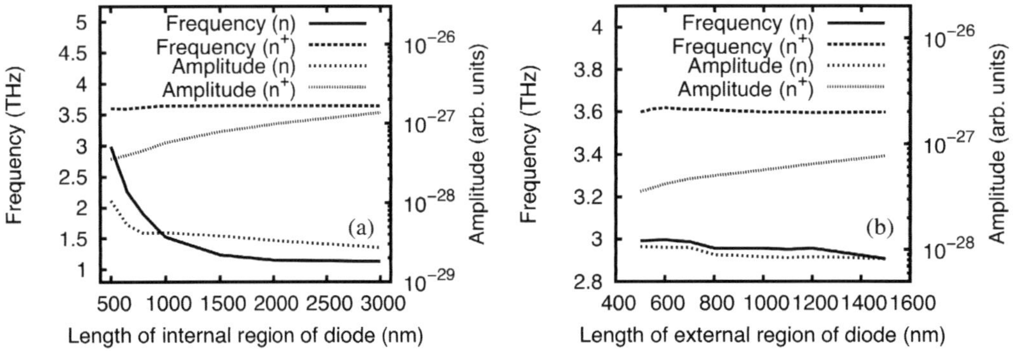


**Figure 3.** Resonance frequency and amplitude in $n$ and $n^+$ regions of the diode as functions of the internal (a) and external (b) region length of the diode.

## 4. Conclusion

We have performed a numerical investigation of the plasma resonances in InGaAs $n^+ - n - n^+$ diodes. We have calculated the frequency and amplitude of the resonance peaks as functions of the doping and the geometry of the devices. In all the considered cases, the presence in the doped access regions of a resonance at the usual 3D plasma frequency of a $n^+$-doped semiconductor has been shown. In the internal $n$ region, we evidenced another resonance whose frequency is shifted from the $n$ 3D plasma frequency toward the $n^+$ frequency. We showed that this shift towards the $n^+$ frequency is reduced and even suppressed for long $n$ regions. Anyway, the results show clearly that the resonances associated with three-dimensional plasma open new possibilities to exploit not only field effect transistors but also diodes in the field of solid-state terahertz devices operating at room-temperature.

## References

[1] Knap W, Lusakowski J, Parenty T, Bollaert S, Coppy A, Popov V V and Shur M S 2005 *J. Appl. Phys.* **97** 64307

[2] El Fatimy A, Teppe F, Dyakonova N, Knap W, Seliuta D, Valusis G, Shchepetov A, Roelens Y, Bollaert S, Coppy A and Rumyantsev S 2006 *Appl. Phys. Lett.* **89** 131926

[3] Torres J, Marinchio H, Nouvel P, Sabatini G, Palermo C, Varani L, Chusseau L, Shiktorov P, Starikov E and Gruzhinskis V 2008 *IEEE J. Sel. Topics Quantul Electron* **14** 491

[4] Kim S, Zimmerman J, Focardi P, Gossard A, Wu D and Sherwin W 2008 *Appl. Phys. Lett.* **92** 253508

Journal of Physics: Conference Series **193** (2009) 012097 doi:10.1088/1742-6596/193/1/012097

# Picosecond electric pulse excitation of three-branch ballistic nanodevices

**Hiroshi Irie and Roman Sobolewski**

Department of Electrical and Computer Engineering and the Laboratory for Laser Energetics, University of Rochester, Rochester, NY 14627-0231 USA

E-mail: roman.sobolewski@rochester.edu

**Abstract**. Picosecond electrical pulse excitation of the ballistic three-branch junctions (TBJs) was performed to investigate the electrical response at terahertz frequencies. The generation of picosecond electrical pulses and the signal detection with sub-picosecond temporal resolution were achieved by a photoconductive switch and the time-resolved electro-optical sampling technique, respectively, with the help of the femtosecond Ti:Sapphire laser system. The electrical response of a TBJ rectifier to the 1.75-ps-wide incident pulse was successfully measured and its frequency spectrum was not degraded in the tested frequency range, as compared to the incident excitation.

## 1. Introduction

Electron devices utilizing ballistic electron transport have been regarded as the most suitable devices for the future high-speed electronics because of their predicted terahertz (THz) bandwidth [1]. One class of such ballistic devices is nanoscale three-branch junctions (TBJs) fabricated in 2-dimensional electron gas (2DEG) heterostructures [2,3]. Because of the extremely small internal capacitance of the TBJ, its frequency response is free from the RC charging time limitations, contrary to the conventional electron devices. This advantage gives rise to a hope of building THz electronics from TBJs. THz operation of the TBJs has been demonstrated by the Monte Carlo simulations [4], however, experimental verifications have been limited to the GHz frequency band [5], due to the lack of proper testing tools. Therefore, developing a testing tool with a bandwidth reaching the THz frequency band is desired to assess the ultimate performance of ballistic devices. In addition, exploring the ultrafast response of ballistic devices is interesting in its own right, because it is expected to observe new physical phenomena such as the kinetic inductive effect [6] in a time scale shorter than the mean scattering time. Although the kinetic inductive effect is negligible in conventional nonsuperconducting electron devices, it could govern the ballistic electron transport in the THz frequency regime.

In this communication, we present an experimental chip consisting of a transmission line structure, which integrates both a photoconductive (PC) switch and a ballistic TBJ rectifier under test. Femtosecond optical pulses were used to excite the PC switch and trigger a single-picosecond voltage transient applied to the TBJ rectifier, while the output of the rectifier was read with the help of an ultrafast electro-optical (EO) sampler. The above arrangement allowed us to investigate the ultrafast electrical response of the TBJ devices and demonstrate their applicability for THz electronics.

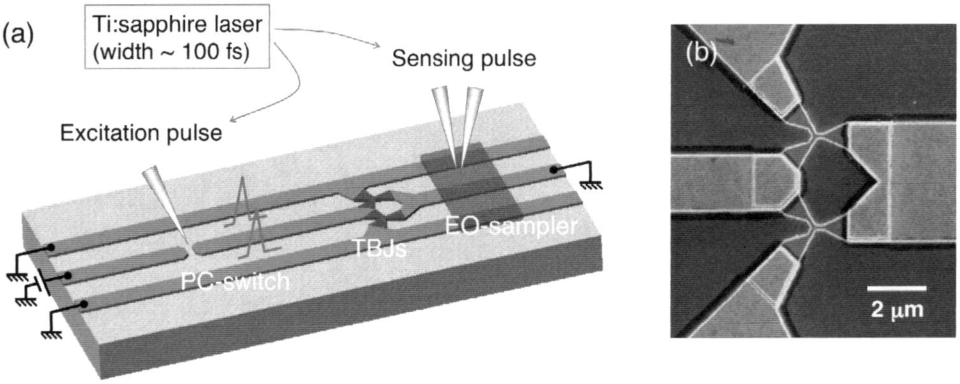


**Figure 1.** (a) Schematics of the picosecond electric pulse excitation setup. The PC switch, and the TBJ rectifier are monolithically integrated in the CPW line. The generation and detection of the electrical pulses are controlled by 100-fs-wide optical pulses from a Ti:Sapphire laser. (b) An SEM image of the TBJ rectifier.

## 2. Integrated system for the generation and detection of the picosecond electrical pulses

In figure 1(a), schematics of our monolithic chip for the picosecond pulse excitation of the TBJ structure and the subsequent read-out by the EO sampler are illustrated. A PC switch is a pulse generator that consists of two metal electrodes deposited on a photoconductive substrate and separated with a micrometer size gap [7]. Since the photoconductive substrate becomes conductive only when it is illuminated by light, one can excite an electrical transient by illuminating the DC-biased switch gap with a short laser pulse. In our experiment, a 50-nm-thick InAlAs layer that was grown on an InP substrate by molecular beam epitaxy (MBE) was used for the photoconductive material and a train of 100-fs-wide optical pulses generated by a commercial, mode-locked Ti:Sapphire laser were employed for the optical excitation.

As is shown in figure 1(a), the PC switch is incorporated into a coplanar waveguide (CPW), which allows the delivery of the electrical pulse to the device under test (DUT), which in our case is a TBJ rectifier consisting of two TBJs in parallel as shown in figure 1(b). The TBJ is a 3-terminal device that exhibits a nonlinear transfer curve originated from the ballistic electron transport [3]. It has been shown that the nonlinear response makes the TBJs work as rectifiers or logic gates [1].

For the detection of the ultrafast transient signal, we have employed an ultrafast EO sampler [8] consisting of an EO transducer (a $LiTaO_3$ crystal overlaying the transmission line) and optics system as a polarization analyzer. The EO transducer converts the electrical field signal, traveling in the CPW line and coupled to the EO crystal, into the change of polarization state of the optical sensing pulses penetrating the EO crystal and reflected from the dielectric mirror at the bottom. By using a linearly polarized incident light and two cross-polarized polarizers, the EO sampler acts as an optical intensity modulator for the sensing beam. It has been verified that EO samplers based on the $LiTaO_3$ crystal exhibit the bandwidth that exceeds 1 THz [9]. Finally, we stress that the EO sensing position can be changed by moving the sensing crystal along the CPW line (e.g., in front of the DUT), so both the incident transients and the ones transmitted by the DUT can be measured with a sub-picosecond temporal resolution.

## 3. Fabrication

The fabrication of the integrated chip started with a modulation-doped $In_{0.53}Ga_{0.47}As/In_{0.52}Al_{0.48}As$ heterostructure, grown by MBE on a semi-insulating InP substrate. The resulting 2DEG layer was approximately 60 nm beneath the surface, and its room temperature mobility $\mu$ and electron sheet density $n_e$ were 1.1 $m^2$/Vs and 4.1x$10^{15}$ $m^{-2}$, respectively. To define the geometry of the Y-shape mesa, electron beam lithography and subsequent ion-mill etching were used, resulting in 130-nm-deep etching and bringing the InAlAs buffer layer that was used for PC switch to the surface. Then, a stack of Ni/Ge/Au was deposited and annealed at 420 °C for 30 seconds in Ar/H forming gas for the ohmic contacts. Finally, the CPW layer was formed by the electron beam lithography and lift-off of a 200-nm-thick Au layer. The width of the CPW's central conductor and its separation from the ground planes were both 30 μm. The gap of the PC switch was chosen to be 25 μm.

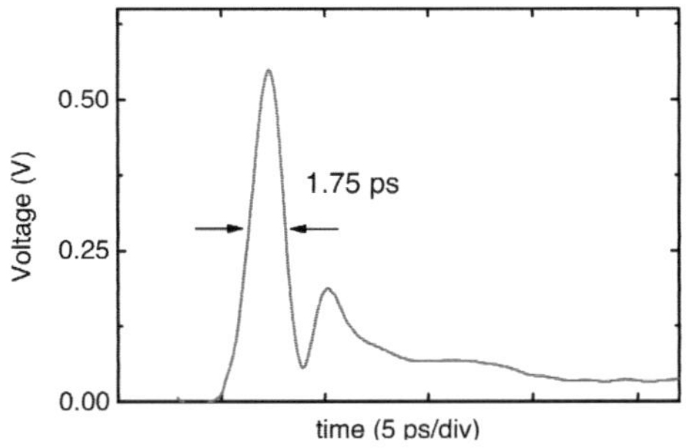


**Figure 2.** EO measurement of a voltage transient generated by the PC switch.

## 4. Picosecond electrical pulse excitation of TBJ rectifier

In the first phase of our studies, we performed EO characterization of the picosecond electrical pulses from the PC switch by fabricating a test structure identical to the one shown in figure 1(a) but without the DUT. Figure 2 presents a transient voltage signal generated by the PC switch, illuminated with 2 mW of optical power and biased at 9 V. The main pulse has the full-width at half-maximum (FWHM) of 1.75 ps and the amplitude exceeding 0.5 V.

In the complete test chip, which included the DUT, the transient waveforms both before (incident signal) and after (transmitted signal) the TBJ rectifier were measured, as is shown in figure 3(a). Looking at the incident waveform, the first pulse is a somewhat dispersed (broadened) switch pulse shown in figure 2 and it is followed by the second pulse, which is a reflection from the DUT. The latter observation was experimentally confirmed by measuring the time separation $\Delta t$ between the two peaks while moving the sensing positions (data not shown). It was observed that $\Delta t$ increased linearly with the distance $\Delta x$ of the sensing position from the TBJs. We also calculated the propagation velocity $v_g$ from the slope of the $\Delta x$ vs. $\Delta t$ dependence, which gave ~1.0x$10^8$ m/s, a value reasonable for our CPW. Another observation is that the amplitude of the reflected pulse, after background subtraction, is close with that of the incident pulse, which indicates that a large portion of the incident signal was actually reflected at the CPW/DUT interface. This large reflection is a consequence of a significant impedance mismatch at the CPW/DUT plane, due to the fact that the characteristic impedance of the CPW was 62 Ω, while the TBJ rectifier input impedance was ~20 kΩ. Thus, the effective excitation signal coupled into the DUT must have been only on the 1 % level of the total incident signal.

The shape of the transmitted signal in the case of unbiased TBJ rectifier is shown in figure 3(a). We note two prominent features: (1) its amplitude is ~22 mV, corresponding to only a few % of the height of the incident pulse; (2) the pulse has an oscillatory shape. The small amplitude is again caused by the impedance mismatch that results in a small power coupled into the TBJs. On the other hand, the negative dip of the transmitted pulse strongly suggests that there exist some reactive elements in the DUT structure. This observation was confirmed by the fact that a time derivative of the incident pulse, shown as a dotted line in figure 3(a), followed almost exactly the transmitted pulse shape. The origin of the reactive elements is most likely the capacitive coupling among the funnel-shape access regions between the nano-channel and the ohmic contacts, as well as external coupling of the TBJ rectifier to the CPW, as it was discussed in the equivalent circuit analysis in [5].

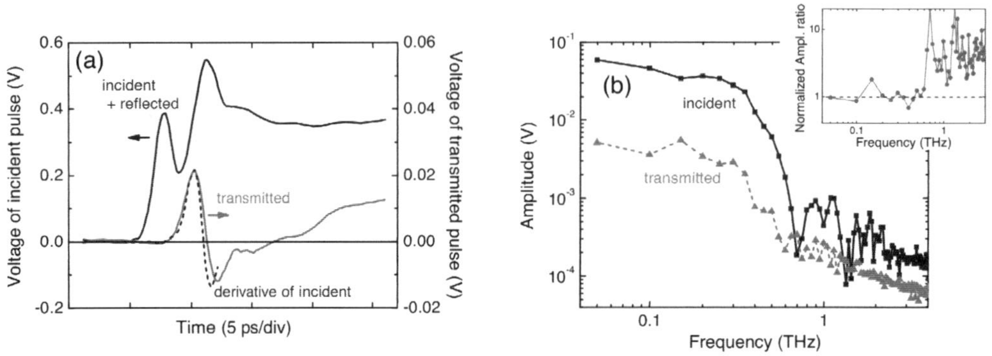


**Figure 3.** (a) EO measurements of the voltage transient signals collected 0.11 mm before (incident) and 0.11 mm after (transmitted) from the DUT. The dotted line represents a normalized time derivative of the incident pulse. (b) Fourier transformed spectra of the incident and transmitted voltage transient signals presented in (a). The inset shows the normalized amplitude ratio of the transmitted spectrum to the incident one.

To show the advantage of the implemented technique, the Fourier transform spectra of the transient signals are shown in figure 3(b). The spectrum of the incident pulse reaches almost 1-THz bandwidth and is limited by the PC switch response time. The spectrum of the transmitted pulse has the frequency dependence similar to the incident pulse, resulting in a constant amplitude ratio within the bandwidth as shown in the inset of figure 3(b). The constant ratio indicates that the TBJ rectifier does not intrinsically degrade the frequency response in the tested sub-THz frequency range. Finally, it is worth to mention that further extension of the system bandwidth is possible by modifying the PC switch geometry into a metal-semiconductor-metal interdigitized structure with a sub-micrometer gap [10] to realize electrical excitations with over 1-THz bandwidth.

## Acknowledgements

This work was supported by the NSF NIRT grant ECS-0609140, ONR grant N00014-08-1-0093, and AFOSR grant FA9550-07-1-0032, and a NYSTAR Grant to the UR CAT-CEIS. The test structures were fabricated at the Cornell NanoScale Science and Technology Facility, a member of NNI Network, which is supported by the NSF grant ECS-0335765.

## References

[1] Song A M 2004 *Encyclopedia of Nanoscience and Nanotechnology* vol 9, ed Nalwa H S (Stevenson Ranch: American Scientific Publishers) pp. 371–379
[2] Xu H Q 2001 *Appl. Phys. Lett.* **78** 2064
[3] Irie H, Diduck Q, Margala M, Sobolewski R and Feldman M J 2008 *Appl. Phys. Lett.* **93** 053502
[4] Mateos J, Vasallo B G, Pardo D, González T, Galloo J- S, Bollaert S, Roelens Y and Capy A 2003 *IEEE Trans. Electron Dev.* **50** 1897
[5] Bednarz L, Rashmi, Hackens B, Farhi G, Bayot V and Huynen I 2005 *IEEE Trans. Nanotech.* **4** 576
[6] Burke P J, Spielman I B, Eisenstein J P, Pfeiffer L N and West K W 2000 *Appl. Phys. Lett.* **76** 745
[7] Auston D H 1983 *IEEE J. of Quant. Electron.* **QE-19**, 639
[8] Valdmanis J A, Mourou G and Gabel C W 1982 *Appl. Phys. Lett.* **41** 211
[9] Zheng X, Xu Y, Sobolewski R, Adam R, Mikulics M, Siegel M and Kordoš P 2003 *Appl. Opt.* **42** 1726
[10] Chou S Y and Liu M Y 1992 *IEEE J. Quantum Electron.* **28** 2358

EDISON 16 IOP Publishing
Journal of Physics: Conference Series **193** (2009) 012098 doi:10.1088/1742-6596/193/1/012098

# Intersubband dynamics of spin polarized carriers

**M Vogel[1], A Vagov[1], V M Axt[1], A Seilmeier[2] and T Kuhn[3]**

[1] Institute for Theoretical Physics III, University of Bayreuth, 95440 Bayreuth, GERMANY,
[2] Institute for Experimental Physics III, University of Bayreuth, 95440 Bayreuth, GERMANY,
[3] Institute for Solid State Theory, University of Münster, 48149 Münster, GERMANY.

E-mail: `martin.axt@uni-bayreuth.de`

**Abstract.** The intersubband (ISB) dynamics of spin orientated carriers generated in undoped quantum wells by circularly polarized interband excitations is analyzed theoretically. It is shown that short THz pulses in resonance with the lowest conduction ISB transition can be used to switch the spin orientation in a given subband without changing the total subband occupation. For longer ISB pulses a periodic modulation of the spin orientation is found. However, when the Coulomb interaction is taken into account the total subband density is typically also modulated which differs from expectations based on calculations in a three band model of non-interacting particles. The dependence on the ratio of the subband masses is discussed as well as the observability of these effects in pump-probe type measurements.

## 1. Introduction

It is a central goal of spintronic devices to manipulate spin and charge degrees of freedom independently. Intersubband (ISB) transitions in quantum wells are not usually discussed in this connection, because in the most common set-ups that test ISB transitions no preferred spin direction is initially prepared and the subsequent laser induced dynamics is typically insensitive to spins. This is because the dominant coupling is to the allowed dipole transitions oriented linearly in growth direction which is not spin selective. Circularly polarized in-plane laser fields couple to ISB transitions only via mechanisms that are typically about three orders of magnitude weaker than the dominant dipole coupling and thus the resulting spin-orientation effects are rather weak [1]. In the present paper we use a combination of interband (IB) and ISB excitations where only the dominant dipole couplings are considered. Circularly polarized laser pulses in resonance with the IB transitions are used to prepare a distribution of spin polarized carriers in the conduction subbands (CSB). The main focus of the paper is the spin dependent ISB dynamics that follows this preparation step.

To be specific, we consider the excitation conditions that are depicted schematically in Fig. 1. First, a $\sigma^+$-polarized IB-pulse in resonance with the transition between the energetically highest heavy-hole type valence subband and the lowest CSB generates an electron-hole pair where the electron is, according to the usual selection rules, spin-down-polarized. Then a linearly polarized ISB pulse performs a Rabi rotation and thus promotes the spin-down electrons to the second CSB. A second IB-pulse, this time $\sigma^-$-polarized, is used to create a population of spin-up electrons in the lowest CSB. This protocol leads to a time evolution through the states (a)-(d) illustrated in the upper part of Fig. 1. The goal of this protocol is to prepare equal occupations of the two lowest CSBs with opposite spins. Once this is achieved the ISB dynamics induced

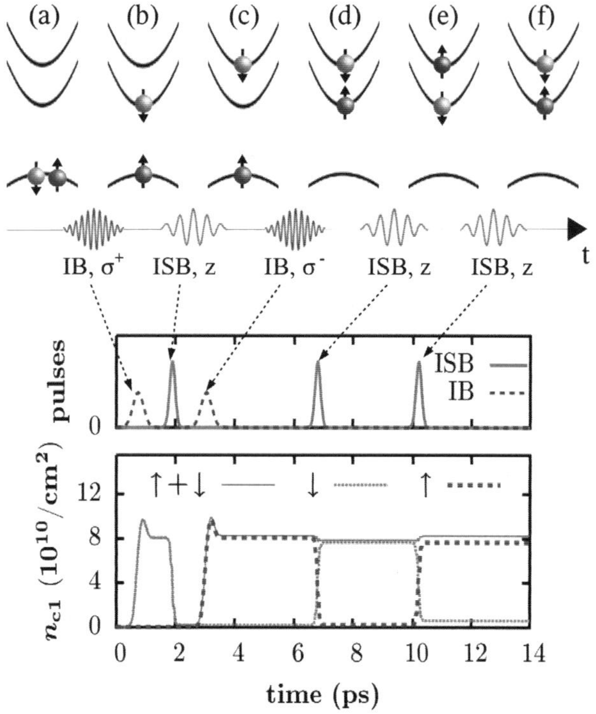


**Figure 1.** Top: preparation and manipulation scheme; Middle: pulse sequence. Bottom: occupation of the lowest conduction subband. The symbol ↑ (↓) specifies the density with spin up (down) while ↑ + ↓ marks the total subband density. The quantization axis for the spin is the $z$-direction, i.e. the growth direction. Parameters: conduction subband masses $m_{c1} = 0.068 m_0$, $m_{c2} = 0.071 m_0$, $m_0$ = free electron mass, decoherence times $T_{IB} = 1$ ps, $T_{ISB} = 3$ ps, well width $d = 25$ nm, separation of conduction subbands $\Delta E_C = E_{c2} - E_{c1} = 25$ meV.

by further pulses in resonance with the transition between these subbands [cf. Fig. 1 (e),(f)] exhibits characteristic spin dependencies which shall be explored in this paper.

## 2. Numerical results

We have simulated the above sketched scenario within a $k$-space density matrix approach where in addition to the band energies the Coulomb interaction is accounted for on the mean-field level [2]. We have modeled a 25 nm wide quantum well with infinitely high barriers using standard GaAs parameters. The separation of the lowest CSBs evaluates for this structure to $\Delta E_C = 25$ meV, i.e.,$\Delta$ $E_C$ is considerably lower than the energy of longitudinal optical phonons of $E_{LO} = 36$ meV. In this case ISB scattering times above 100 ps up to several hundred ps are common at low temperatures [3]. Thus, ISB relaxation can be neglected on the short time scales studied here. Electron spin relaxation times may even reach nanoseconds [4] and are disregarded as well. Hole spin relaxation times may be short (of the order of a few ps) but these processes have only a marginal influence on the ISB dynamics. Corresponding simulations (not shown) revealed that they have some effect on the preparation stages [cf. Fig. 1 (a)-(d)] which involve IB transitions. These effects can, however, easily be compensated by suitably adjusting the intensities of the IB pulses [5] with the result that with or without accounting for hole spin relaxation practically the same final states in the CSBs can be prepared. For simplicity we show here results calculated without hole relaxation. We account, however, for decoherence times of $T_{IB} = 1$ ps for IB and $T_{ISB} = 3$ ps for ISB transitions.

Plotted in the lower part of Fig. 1 are calculated occupations of the lowest CSB. Apart from the total occupation (↑ + ↓, solid) also the occupations of spin-down (↓, dotted) and spin-up (↑, dashed) states are shown. The middle part of the figure illustrates the pulse sequence. We have considered IB pulses of 300 fs duration Full-Width-at-Half-Maximum (FWHM) of the intensity which guarantees a selective excitation of only the lowest subband. The duration of ISB-pulses has been adjusted to 160 fs which for a pulse of central frequency $\hbar\omega = 25$ meV

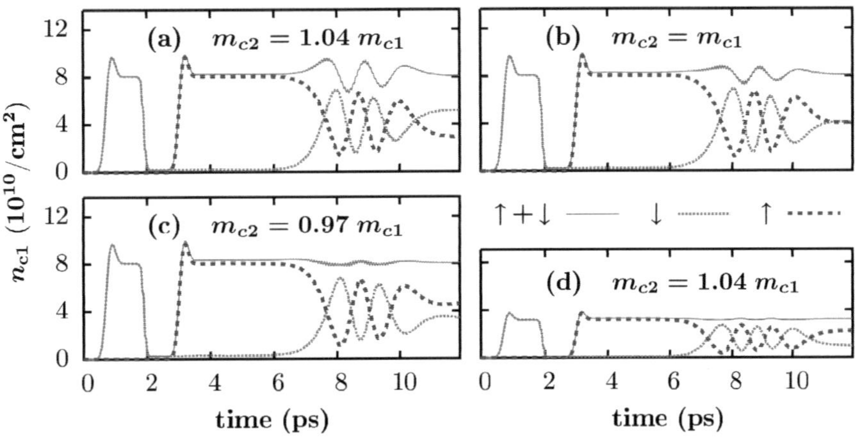


**Figure 2.** Occupation of the lowest conduction subband. (a),(b)(c) calculation with Coulomb interaction and conduction subband mass ratios as indicated. $m_{c1}$ and all other parameters are as in Fig. 1. (d) the same as (a) without Coulomb interaction.

corresponds roughly to a single cycle pulse. Recent advances in laser technology made single cycle THz pulses of high intensity available [6; 7] and it turns out that such short pulses are indeed favorable for the above described protocol (see below). It is clearly seen, that after the second IB pulse practically all electrons in the lowest CSB are spin-up electrons which is the intended result of the preparation. After the preparation two ISB-pulses are used to manipulate the spin orientation of the carriers in the lowest CSB. It turns out that it is indeed possible to switch the spin expectation value in a given subband essentially without changing the total occupation, i.e., the spin has been controlled independent from the charge degree of freedom.

Fig. 2 shows results where after the preparation phase a single $6\pi$ ISB pulse of 2 ps duration has been used to drive the ISB dynamics. Part (a) is calculated with a ratio of the CSB masses of $m_{c2}/m_{c1} = 1.04$ which is a typical value for this separation of subband energies [8] and which has also been used in Fig. 1. We observe a periodic modulation of the spin occupation in the lowest subband. This modulation is, however, accompanied by a similar modulation of the total subband occupation. This is mainly due to renormalizations of the ISB transition energies caused by Coulomb interactions resulting in Rabi rotations that are off-resonant and thus incomplete. Even for initially equal occupations of the two subbands the number of carriers rotated down does not equal the number of carriers rotated up because the renormalizations are different for carriers in different subbands [9]. The amplitude of the modulation of the total subband occupation turns out to be strongly dependent on the CSB mass ratio. Interestingly, the weakest modulation is not found for equal masses [Fig. 2 (b)] but for a slightly lower value of $m_{c2}/m_{c1} = 0.97$ [Fig. 2 (c)] for which the effect is indeed rather weak. It should also be noted that this behavior is not expected in a model where the Coulomb interaction is neglected. Here, the total subband density is only marginally affected by the mass ratio and, for the standard value of $m_{c2}/m_{c1} = 1.04$, practically no modulation is seen [Fig. 2 (d)]. Also note that for the same pulse intensities simulations with Coulomb interaction yield higher electron densities due to the Coulomb enhancement in the absorption process.

It is important that it is possible to decide also experimentally whether spin manipulations have been successfully performed independent from charge manipulations. A standard way to monitor spin-dependent occupations is to compare the absorption of circularly and linearly polarized IB pulses. These signals should reflect the spin dependent densities because of the

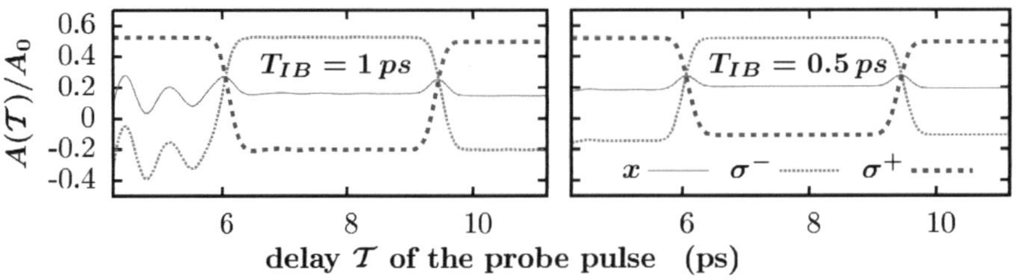


**Figure 3.** Absorption of an IB test pulse with delay $\tau$ with respect to the first IB pulse for different IB decoherence times $T_{IB}$ and different circular or linear polarizations as indicated. All other parameters are as in Fig. 1. Signals are normalized to the absorption of the unexcited sample.

spin dependence of blocking effects. Shown in Fig. 3 is the absorption of weak 300 fs IB test pulses of different polarizations: linear $x$ (solid), circular $\sigma^-$ (dotted) and circular $\sigma^+$ (dashed). The simulations have been performed for the excitation conditions in Fig. 1. The absorption of the test pulse is plotted as a function of the delay $\tau$ from the first IB pulse. The curves in the left panel are calculated as before with $T_{IB} = 1$ ps while for the right panel a value of $T_{IB} = 0.5$ ps has been used. In the latter case the absorption of the circularly polarized pulses clearly reflects the switching dynamics of the respective spin densities and the absorption of the linearly polarized test pulse essentially follows the evolution of the total subband occupation with a time resolution set mainly by the duration of the test pulse. The longer IB decoherence time in the left panel, however, results in coherent signal components that are superimposed on the signals. Even though the main features of the occupation dynamics are still visible, now additional oscillations appear which make such signals harder to interpret.

### 3. Conclusions

We have shown that it is possible to optically prepare CSBs with equal densities of opposite spin orientations. It is demonstrated that after this preparation the spins in a given subband can be switched by ISB pulses on a picosecond time scale. The spin switching can be realized keeping the total subband occupation almost constant when sufficiently short ISB pulses are used. For longer ISB pulses spins and charges in a given subband exhibit a periodic modulation. The oscillation of the total subband density is not expected in a model without Coulomb interaction and strongly depends on the CSB mass ratio when the Coulomb interaction is accounted for. Pump-probe type signals are suitable to monitor the predicted spin-sensitive dynamics, but may exhibit coherent features superimposed on the occupation dynamics.

### References

[1] Khurgin J B 2006 *Appl. Phys. Lett.* **88** 123511
[2] Kaindl R A *et al.* 1998 *Phys. Rev. Lett.* **80** 3575–8
[3] Murdin B N *et al.* 1997 *Phys. Rev.* B **55** 5171–6
[4] Ohno Y, Terauchi R, Adachi T, Matsukura F and Ohno H 1999 *Phys. Rev. Lett.* **83** 4196
[5] Vogel M, Vagov A, Axt V M, Seilmeier A and Kuhn T *Phys. Rev.* B accepted
[6] Bartel T, Gaal P, Reimann K, Woerner M and Elsaesser T 2005 *Opt. Lett.* **30** 2805–7
[7] Sell A, Leitenstorfer A and Huber R 2008 *Opt. Lett.* **33** 2767–9
[8] Ekenberg U 1989 *Phys. Rev.* B **40** 7714–26
[9] Rossi F and Kuhn T 2002 *Rev. Mod. Phys.* **74** 895–950

# Study of the O *K*-edge XANES of Cobalt-doped ZnO diluted magnetic semiconductors

**Tongfei Shi[1†], Yuqi Wang[1], Zhijun Yin[1], Yunpeng Zhang[2] and Shishen Yan[2]**

[1]Key Laboratory of Materials Physics, Institute of Solid State Physics, Chinese Academy of Sciences, Hefei, Anhui 230031, China
[2]The School of Physics and Microelectronics, Shandong University, Jinan, Shandong 250100, China

[†]Email: tongfeishi@gmail.com

**Abstract**. Cobalt-doped ZnO samples were prepared by both sol-gel and magnetron sputtering methods, respectively. Co atoms have incorporated into the ZnO lattice and located at the substitutional sites of the Zn atoms in all samples, while the sol-gel prepared $Zn_{1-x}Co_xO$ samples show paramagnetism and the sputtering prepared $Co_yZn_{1-y}O$ samples appear ferromagnetism. The calculated O *K*-edge X-ray absorption near edge structure (XANES) can commendably reproduce all of the peaks observed in $Co_yZn_{1-y}O$ samples when one Zn interstitial atom ($Zn_i$) nears the central O atom. Compared with the magnetic properties, it is shown that the Zn interstitial atoms play a key role to the observed ferromagnetism in $Co_yZn_{1-y}O$ samples.

## 1. Introduction

ZnO-based diluted magnetic semiconductors (DMSs) had been prepared by both physical and chemical methods, but the observations showed that the magnetic properties of oxide based films are highly sensitive to the preparation conditions. [1-3] Even in systems that appear to be quite pure, reported magnetic behavior can vary widely, ranging from room-temperature ferromagnetism [1] to paramagnetism [2]. This variance in magnetism is attributed to the multitude of synthetic methods used for the production of DMSs materials, which can result in differences in dopant environment, structural disorder or defects. Under the theory of bound magnetic polarons [3], researchers have increasingly found that the defects have remarkable influence to the magnetism of materials, and that a careful control of defects in ZnO and not the doping with magnetic ions might be a better way to obtain reproducible intrinsic high-T*c* ferromagnetism in ZnO [4].

In this letter, we use the X-ray absorption near edge structure (XANES) spectroscopy, which is a powerful technique for the probing of the arrangement of atoms and distributions of impurities, to study differences between the samples prepared by equilibrium and non-equilibrium state methods. FEFF8.2 software [5] is used to simulate the XANES spectra with different defects' models. Compared with experimental spectra, we find directly proof that the shallow donor $Zn_i$ is the key of the ferromagnetism in Cobalt-doped ZnO sample.

## 2. Experimental

The $Zn_{1-x}Co_xO$ DMS samples were prepared by the sol-gel method using $Zn(CH_3COO)_2{\cdot}H_2O$ and $Co(CH_3COO)_2{\cdot}H_2O$ as starting precursors, de-ionized water as a solvent, and polyvinyl alcohol as a stabilizing agent. [6] On the other side, the $Co_yZn_{1-y}O$ DMS films were prepared on glass substrates by alternately sputtering very thin Co layers of nominal thickness of 0.6 nm and ZnO layers of nominal thickness of 1.2 nm for 60 periods at room temperature [7] (the nominal structure is $Co_{0.33}Zn_{0.67}O$). The O *K* edge XANES spectra of Co-doped ZnO samples were measured at the U18 beamline of the National Synchrotron Radiation Laboratory. The beam from a bending magnet was monochromatized with a varied line-spacing plane grating and refocused by a toroidal mirror. The data were collected in a mode of sample drain current under a vacuum better than $5\times10^{-5}$ Pa.

## 3. Results and Discussion

The Co *K*-edge XANES for $Co_yZn_{1-y}O$ samples with different Co doping concentration is shown in figure 1. There is a distinct step during the edge of Co metal's spectrum, and jumping-off point of the *K*-edge of Co metal is 7700eV which is less than 7710 eV for Co-doped ZnO samples. This result implies that Co metal phase doesn't exist in all of our samples. Figure 2(a) shows the O *K*-edge XANES spectra of $Co_yZn_{1-y}O$ samples, and figure 2(b) gives the O *K*-edge XANES spectra of $Zn_{1-x}Co_xO$ samples. The spectra of pure ZnO, CoO and $Co_3O_4$ reference compounds are also exhibited in these figures. It is shown that the spectra of CoO and $Co_3O_4$ both have a shape and strong peak before 530 eV. It is indicated that there are no CoO or $Co_3O_4$ phases in our samples.

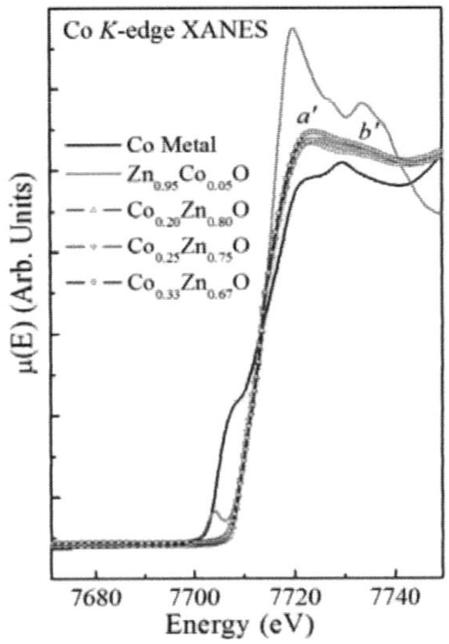


**Figure 1.** The Co *K*-edge XANES spectra for $Co_yZn_{1-y}O$, $Zn_{0.95}Co_{0.05}O$ samples and ZnO, Co metal reference compounds.

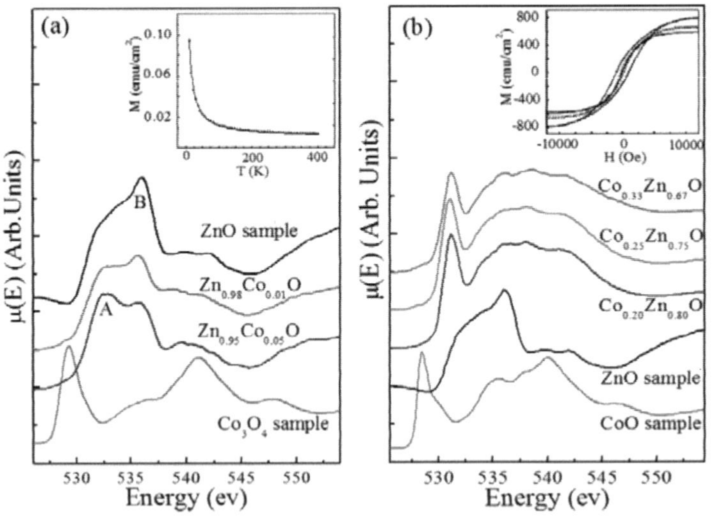


**Figure 2.** (a) The O *K*-edge XANES spectra of $Zn_{1-x}Co_xO$, ZnO and $Co_3O_4$ samples; the insert figure shows paramagnetism of $Zn_{0.95}Co_{0.05}O$; (b) The O *K*-edge XANES spectra of $Co_yZn_{1-y}O$, ZnO and CoO samples; The insert figure shows ferromagnetism of $Co_{0.25}Zn_{0.75}O$ at 5, 80 and 290 K.

In figure 1, it can be found that the spectra of $Co_yZn_{1-y}O$ are similar to the $Zn_{0.95}Co_{0.05}O$ sample's curve. The first peak *a'* and the second peak *b'* both exit in the postedge of $Co_yZn_{1-y}O$ samples, while the intensity has been weakened by the disorder of lattice. These evidences indicate that the local structure of Co atoms in the $Co_yZn_{1-y}O$ and $Zn_{1-x}Co_xO$ samples must be similarly. As we have demonstrated that the Co atoms in $Zn_{1-x}Co_xO$ ($x\leq0.05$) are incorporated into the ZnO lattice and located at the substitutional sites of the Zn atoms [6], the Co atoms have also taken the place of Zn site and entered the $Co_yZn_{1-y}O$ samples lattice.

In figure 2, the insert figures have shown the magnetic properties measured by a superconducting quantum interference device (SQUID). The $Zn_{0.95}Co_{0.05}O$ sample prepared by sol-gel method shows paramagnetic behavior, while the Curie temperature of $Co_{0.25}Zn_{0.75}O$ samples prepared by sputtering is at least near room temperature. As the local structures of these different samples have been investigated that the Co atoms have entered into the ZnO lattice, the related magnetism is intrinsic property. It is puzzle what causes the different magnetic property, when Co atoms have incorporated into the ZnO lattice of both kinds of samples. Then, we turn to the O *K*-edge XANES spectra of the Co-doped ZnO samples. Strikingly, the $Co_yZn_{1-y}O$ films exhibit identical spectral features which quite different from those of $Zn_{1-x}Co_xO$. These distinctly different XANES spectral behaviors demonstrate that the local electronic structure of these two kinds' samples is quite different. In order to find out the differences between these two kinds Co-doped ZnO system, the FEFF8.2 code is used to simulate the XANES spectra with the models containing the different defects in host lattice. For each model structure, 125 atoms are used to achieve convergence of the XANES calculations.

Lee *et al.* have found that the ferromagnetic coupling has a short ranger nature, effective at Co-Co distances of about 3 Å, [8] which implying Co clustering may influence the ferromagnetism of the Co-doped ZnO DMS. Different structural models were considered, by replacing, respectively, one, two, three and four the nearest Zn neighbors of $Co_S$ with Co atoms. Hereafter, these models are named as 1 $Co_S$, 2 $Co_S$, 3 $Co_S$, and 4 $Co_S$. In figure 3(a), the calculated XANES spectra are plotted together with the experimental spectra of $Zn_{0.95}Co_{0.05}O$ sample. It is seen that the main features of the experimental spectra can be reproduced by the 1 $Co_S$ model, including peaks A, B, and C. From the *n* $Co_S$ Models, it can be found that the intensity of peak A increases with *n*. This is because that peak A is associated with the hybridization between Co 3*d* and O 2*p* state. The increased number of Co nearest neighbors around a given O ion enhances the transition probability. These results indicate that the $Co_S$ ions have cluster together via O atoms in $Zn_{0.95}Co_{0.05}O$ sample. As magnetic properties predominantly determined by the nearest-neighbor distance of Co ions [9] and Co clustering randomly distribute in the lattices of $Zn_{1-x}Co_xO$ samples [6], the magnetic moment of these Co clustering should be disorder. Therefore, the sol-gel prepared $Zn_{1-x}Co_xO$ samples show paramagnetism in macroscopy.

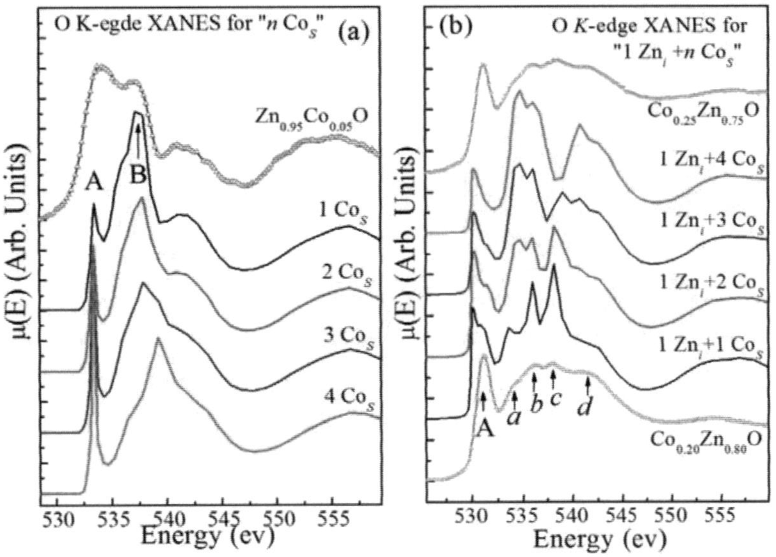


**Figure 3.** (a) O *K*-edge XANES spectra of $Zn_{0.95}Co_{0.05}O$ sample, and the calculated spectra for different structural models by replacing one, two, three, and four the nearest Zn neighbors of $Co_S$ with Co atoms. (b) O *K*-edge XANES spectra of $Co_yZn_{1-y}O$ ($y$=0.20, 0.25) samples, and the calculated spetra for different structural models by placing one $Zn_i$ atom inside the nearest neighbors, and replacing one, two, three and four the nearest Zn neighbors of $Co_S$ with Co atoms.

Gamelin *et al.* have indicated that the $Zn_i$ may introduce an impurity band in Co-doped ZnO, so the native shallow donor interstitial zinc is capable of activating high-T*c* ferromagnetism in Co-doped ZnO [10]. Different structural models with $Zn_i$ atoms have also been considered. It is given only one $Zn_i$ atom inside the nearest-neighbor, and replaced one, two, three and four the nearest Zn neighbors of $Co_S$ with Co atoms, respectively. We named these models as 1 $Zn_i$ +*n* $Co_S$. Experimental O *K*-edge XANES spectra for the $Co_{0.25}Zn_{0.75}O$ and $Co_{0.20}Zn_{0.80}O$ samples and calculated XANES spectra are shown in figure 3(b). Compared with the experimental spectra, peaks (*a*, *b*, *c*, *d*) in samples have been commendably reproduced in the calculated spectra for 1 $Zn_i$ +1 $Co_S$. These peaks can only be found in $Co_yZn_{1-y}O$ samples prepared by sputtering, and not existed in $Zn_{1-x}Co_xO$ samples prepared by the sol-gel method or the pure ZnO sample. It is directly indicated that the spectral features of the O *K*-XANES spectra for $Co_yZn_{1-y}O$ samples is caused by the appearance of $Zn_i$ atoms. In conclusion, the appearance of $Zn_i$ atoms is the difference between $Zn_{1-x}Co_xO$ and $Co_yZn_{1-y}O$ samples.

It can be found that the intensity and FWHM of peak A for $Co_yZn_{1-y}O$ samples is much more enhanced than $Zn_{1-x}Co_xO$, in figure 2(b). Especially, peak A in calculated XANES spectra for 1 $Zn_i$ +1 $Co_S$ model has been divided into two peaks. These indicated that the appearance of $Zn_i$ atoms has strongly enhanced the hybridization of O 2*p*-Co 3*d* state and caused a new impurity band near Fermi lever. Moreover, Chang *et al.* have indicated that the ferromagnetism of Co-doped ZnO samples should arise from the hybridization between Co 3*d* and O 2*p* state around Fermi level. [11] Therefore, the hybridization and charge transfer effect between the $Zn_i$ derived impurity band and the substituting Co ions is the origin of ferromagnetism in $Co_yZn_{1-y}O$ samples. In other words, $Zn_i$ atom is the key of ferromagnetism in $Co_yZn_{1-y}O$ samples.

## 4. Conclusion

In summary, we have employed the XANES technique at O-*K* edge to investigate the influences with different defects in Co-doped ZnO samples prepared by sol-gel and sputtering methods. It is found that Co clustering existed in sol-gel prepared $Zn_{1-x}Co_xO$ samples. As the disorder of magnetic moment, the $Zn_{1-x}Co_xO$ samples show paramagnetism. Compared with $Co_yZn_{1-y}O$ samples, we find the directly proof that $Zn_i$ is the key of the ferromagnetism in $Co_yZn_{1-y}O$ sample. Our results further show that the room temperature ferromagnetism can be achieved by the control of $Zn_i$ in ZnO DMSs.

## References

[1] Hyeon-Jun Lee and Se-Young Jeong, Chae Ryong Cho, Chul, Hong Park 2002 *Appl. Phys. Lett.* **81**, 4020
[2] G. Lawes, A. S. Risbud, A.P. Ramirez, Ram Seshadri 2005 *Phys. Rev. B* **71**, 045201
[3] J. M. D. Coey, M. Venkatesan, and C. B. Fitzgerald 2005 *Nat. Mater.* **4**, 173
[4] Qingyu Xu, Heidemarie Schmidt, Shengqiang Zhou, Kay Pozger, Manfred Helm. Holger Hochmuth, Michael Lorenz, Annette Setzer, Pablo Esquinazi, Christonph Meinecke and Marius Grundmann 2008 *Appl. Phys Lett.* **92**, 082508
[5] E. A. Stern, M. Newville, B. Ravel, D. Haskel and Y. Yacoby 1995 *Physica B* **208**, 117
[6] Tongfei Shi, Sanyuan Zhu, Zhihu Sun, Shiqiang Wei and Wenhan Liu 2007 *Appl. Phys. Lett.* **90**, 102108
[7] Shi-shen Yan, C. Ren, X. Wang, Y. Xin and Z. X. Zhou 2004 *Appl. Phys. Lett.* 84, 2376
[8] Eun-Cheol Lee and K. J. Chang 2004 *Phys. Rev. B* 69, 085205
[9] G. S. Chang, E. Z. Kurmaev, D. W. Boukhvalov, L. D. Finkelstein, S. Colis, T. M. Pedersen, A. Moewes and A. Dinia 2007 *Phys. Rev. B* 75, 195215.
[10] Dana A. Schwartz and Daniel R. Gamelin 2006 *Adv. Mater.* **16**, 2115
[11] G. S. Chang, E. Z. Kurmaev, D. W. Boukhvalov, L. D. Finkelstein, S. Colis, T. M. Pedersen, A. Moewes and A. Dinia 2007 *Phys. Rev. B* 75, 195215.

# Spin-dependent transport in armchair graphene nanoribbon structures with edge roughness effects

**V Hung Nguyen[1,2], V Nam Do[3], A Bournel[1], V Lien Nguyen[2] and P Dollfus[1]**
[1]Institut d'Electronique Fondamentale, UMR 8622, CNRS, Univ. Paris Sud, 91405 Orsay, France
[2]Theoretical Department, Institute of Physics, VAST, P. O. Box 429 Bo Ho, Hanoi 10000, Vietnam
[3]Hanoi Advanced School of Science and Technology, 40 Ta Quang Buu, Hanoi 10000, Vietnam

E-mail: viet-hung.nguyen@u-psud.fr

**Abstract.** We analyze the spin-dependent transport in single ferromagnetic gate structures based on armchair graphene nanoribbon (GNR) using the non-equilibrium Green's function method in a tight binding model. It is shown that the spin polarized current oscillates as a function of the gate-induced barrier height. For perfect GNRs, the larger the energy band gap, the stronger the oscillation of the spin polarization. However, though the edge roughness of the ribbons tends to enlarge the band gap, it also strongly reduces the conductance which finally degrades the spin polarized current.

**Key words**: *Graphene nanoribbon, spin polarized transport, Green's functions method*

## 1. Introduction

Due to unique electronic properties [1], graphene materials may be a key to envision future applications in nano-electronics [2] and spintronics [3]. The charge carriers in 2D-graphene structures behave as relativistic particles and can be described by a Dirac's equation [4]. This property originates to a lot of peculiar phenomena such as the finite conductivity at zero density [5], the unusual integer quantum Hall effect [6], and the Klein tunnelling [7]. In addition, due to very weak spin orbit interaction which leads to a long spin flip length ( $\sim 1\,\mu\text{m}$ ) [8], graphene also offers a good candidate for spintronics applications. Many spin-dependent properties such as the spin field effect transistor [9], the spin Hall effect [10], the spin valve effect [3,11], and the electrical control of spin polarized current [12-14] have been explored in graphene-based structures. Especially, because of their finite energy gap [15-17], graphene nanoribbons (finite width graphene strips) are expected to be a basic material for designing high performance devices. However, their electronic properties have been demonstrated to be drastically affected by the disorder induced by defects [15,17], impurities [18], edge roughness [19], etc... For narrow graphene nanoribbons (GNR) the edge roughness effects are very important since they can govern the whole of transport phenomena of the structures [15,20,21].

In a recent work [14], we have investigated the spin-controllable possibilities of a structure based on armchair-edge ribbons without considering effects of the edge roughness. Our goal here is, therefore, to address this problem. The non-equilibrium Green's functions (NEGF) formalism for an appropriate tight binding model is used as a calculation method. Section 2 is devoted to a brief introduction of the considered structure and the model. Obtained results and discussions are then presented in Section 3. Finally, some conclusions are addressed in Section 4.

## 2. Model and calculation method

The structure under consideration is schematized in Fig. 1: it consists of a GNR sample with rough edges coupled to two semi-infinite perfect GNR leads. The ferromagnetic gate is deposited in the centre part to create a spin-dependent potential barrier which can control electrically the spin polarized current through the structure. The system is described by a tight binding model with one $\pi$ – electron per atom and the calculation procedure is based on the NEGF technique described in Ref. [14].

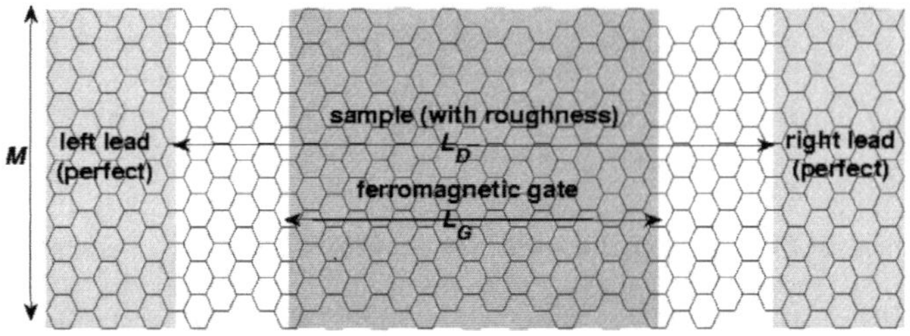


**Figure 1**: (color online) Schematic illustration of the considered structure which is similar to that in Fig. 1(a) of Ref. [14] but the edge roughness is studied in the sample.

To model the rough edges, we assume that each pair of carbon atoms of edge rows has a probability $\Theta_d$ to be removed. Missing atoms are modelled by setting the corresponding hopping elements to zero in the lattice description. The perfect GNR leads have the same width as the central part which is characterized by the number $M$ of carbon atom chains between the two edges. It is worth noting that in the nearest neighbour tight-binding model the perfect GNR behaves as a semiconductor (finite band gap) for the armchair row $M \neq 3n+2$ while is a metal for $M = 3n+2$ (gapless).

## 3. Results and discussions

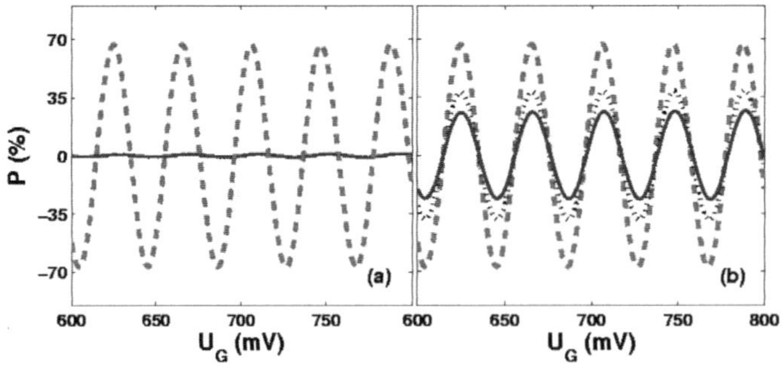


**Figure 2**: (color online) Spin polarization vs the barrier height in perfect structures: role of the band gap. In (a), the solid and dashed curves correspond to metallic ($M = 74$) and semiconducting ($M = 76$) GNRs, respectively. The curves in (b) are for semiconducting ribbons of different widths: $M = 76$ (dashed), 100 (dotted) and 124 (solid line).

As discussed in Refs. [12,13], the conductance and the spin polarization in the 2D-graphene structures have an oscillatory behaviour with respect to the barrier height controlled by the gate voltage $U_G$. This feature has been also observed in the GNRs [14], i. e., as illustrated in Fig. 2. Here, the GNRs have been assumed to be perfect, the Fermi energy $E_F = 75meV$, the gate length $L_G = 42.6nm$ and the exchange energy $h = 10meV$. Actually, it has been shown that the transmission oscillation is stronger when the band gap is larger [14]. Therefore, the transport quantities such as the conductance and the spin polarization oscillate more strongly in semiconducting GNRs than in metallic ones, e. g., as presented in Fig. 2(a) for the cases of $M = 74$ (metallic, solid line) and $M = 76$ (semiconducting,

Journal of Physics: Conference Series **193** (2009) 012100 doi:10.1088/1742-6596/193/1/012100

dashed line). Moreover, it is also found that the amplitude of spin polarization in the semiconducting GNRs is significantly higher than in 2D-graphene, i. e., it is about 67% for $M = 76$ in Fig. 2(a) while its maximum value is only about 30% in the 2D-graphene [12,13]. However, in semiconducting GNRs, as a result of band gap reduction [15], the oscillation of transport quantities gradually weakens when increasing the ribbon's width (see in Fig. 2(b)). For instance, the amplitude of spin polarization reduces from 67% for $M = 76$ (dashed line) to 38.5% for $M = 100$ (dotted line) and 26% for $M = 124$ (solid line). It is worth mentioning that the band gap in graphene can be induced by the interaction of the sample with the substrate [22,23] besides by the effect of finite width. The gap has been predicted to be about 0.26 eV and decreases as the number of graphene layers increases [22].

In Fig. 3, we display the edge roughness effects on the transport quantities. For $W = 9.1nm$, $\Theta_d = 50\%$ and $L_D \approx 85.2\,\text{nm}$, Fig. 3(a) shows that: (i) with rough edges the GNRs are always semiconducting, i.e., they exhibit a finite conduction gap $E_g$ (as a consequence of energy band gap) and (ii) the edge disorder also makes the conductance smaller than in the case of perfect GNRs. The width-dependence of the conduction gap ($E_g \sim 1/W$) and of the conductance plotted in Figs. 3(b) and 3(c), respectively, are in good agreement with experimental data of Ref. [15]. In Figs. 3(d) and 3(e), we also show the dependence of conduction gap and conductance on the intensity of edge disorder which is characterized by the probability $\Theta_d$. Obviously, the conduction gap (conductance) is enlarged (reduced) when increasing $\Theta_d$. The GNR's metallic behavior disappears as soon as a small amount of edge disorder is considered. Here, the conductance is calculated at $E_F = 0.5eV$.

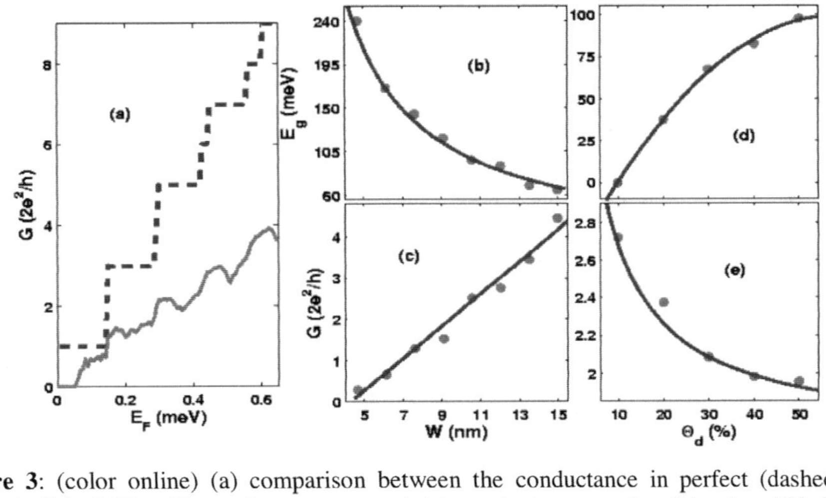


**Figure 3**: (color online) (a) comparison between the conductance in perfect (dashed) and rough (solid) GNRs; (b) conduction gap and (c) conductance vs the ribbon's width W; (d) conduction gap and (e) conductance depending on the disorder probability $\Theta_d$.

Definitely, the edge roughness affects strongly the possibilities of gate controlled spin polarized current. In Figs. 4(a) and 4(b), we plot the spin polarization versus the barrier height obtained when taking into account the edge roughness, with the leads considered to be perfectly metallic and semiconducting GNRs, respectively. On the one hand, due to the enhancement of conduction gap, the spin polarization in the structures with metallic GNR leads is enhanced when increasing the intensity of edge disorder $\Theta_d$. On the other hand, because of the roughness-induced decrease of transport quantities as the conductance, the spin polarized current in the structures with semiconducting leads is reduced when increasing $\Theta_d$. Moreover, the edge disorder makes the gate controllability of spin polarized current blurred. It can be obviously concluded that nearly-perfect semiconducting GNRs with limited edge disorder are desirable to provide the tunability of spin polarized current suitable for designing efficient spintronics devices.

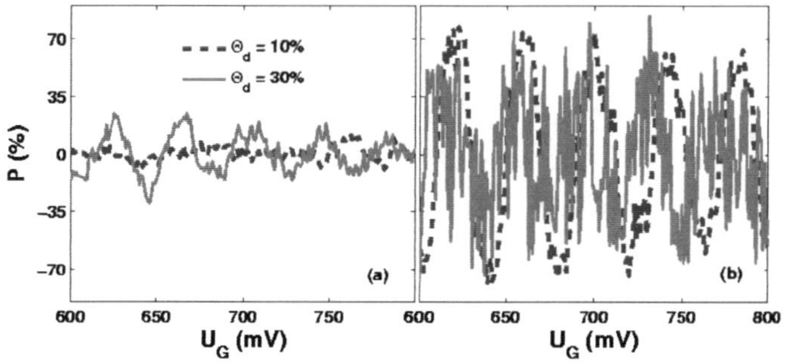


**Figure 4**: (color online) Spin polarization vs the barrier height: edge roughness effects. (a) and (b) are for metallic ( $M = 74$ ) and semiconducting ( $M = 76$ ) leads, respectively.

## 4. Conclusions

Using NEGF method, we have considered the spin-dependent transport in armchair GNRs. It is shown that a highly tunable spin current can be achieved in semiconducting structures. The edge roughness tends to enlarge the conduction gap and to reduce the conductance. These two effects influence differently the spin polarization. Finally, the results suggest that GNRs with nearly perfect edges and large energy band gap should be appropriate for designing efficient spintronics devices.

## References

[1] Castro Neto A H, Guinea F, Peres N M R, Novoselov K S and Geim A K 2009 *Rev. Mod. Phys.* **81** 109
[2] Lemme M C, Echtermeyer T J, Baus M and Kurz H 2007 *IEEE Elec. Dev. Lett.* **28** 282
[3] Hill E W, Geim A K, Novoselov K, Schedin F and Blake P 2006 *IEEE Trans. Magn.* **42** 2694
[4] Katsnelson M I, Novoselov K S and Geim A K 2006 *Nature Phys.* **2** 620
[5] Novoselov K S, Geim A K, Morozov S V, Jiang D, Katsnelson M I, Grigorieva I V, Dubonos S V and Firsov A A 2005 *Nature* **438** 197
[6] Zhang Y, Tan Y -W, Stormer H L and Kim P 2005 *Nature* **438** 201
[7] Stander N, Huard B and Goldhaber-Gordon D 2009 *Phys. Rev. Lett.* **102** 026807
[8] Jozsa C, Popinciuc M, Tombros N, Jonkman H T and van Wees B J 2008 *Phys. Rev. Lett.* **100** 236603
[9] Semenov Y G, Kim K W and Zavada J M 2007 *Appl. Phys. Lett.* **91** 153105
[10] Kane C L and Mele E J 2005 *Phys. Rev. Lett.* **95** 226801
[11] Munoz-Rojas F, Fernandez-Rossier J Palacios J J 2009 *Phys. Rev. Lett.* **102** 136810
[12] Yokoyama T 2008 *Phys. Rev. B* **77** 073413
[13] Nam Do V, Hung Nguyen V, Dollfus P and Bournel A 2008 *J. Appl. Phys.* **104** 63708
[14] Hung Nguyen V, Nam Do V, Bournel A, Lien Nguyen V and Dollfus P 2009 *arXiv*:0905.1868v1
[15] Han M Y, Ozyilmaz B, Zhang Y and Kim P 2007 *Phys. Rev. Lett.* **98** 206805
[16] Li X,Wang X, Zhang L, Lee S and Dai H 2008 *Science* **319** 1229
[17] Chen Z, Lin Y -M, Rooks M J and Avouris P 2007 *Physica E* **40** 228
[18] Lherbier A, Biel B, Niquet Y -M and Roche S 2008 *Phys. Rev. Lett.* **100** 036803
[19] Son Y -M, Cohen M L and Louie S G 2006 *Phys. Rev. Lett.* **97** 216803
[20] Nakada K, Fujita M, Dresselhaus G and Dresselhaus M S 1996 *Phys. Rev. B* **54** 17954
[21] Querlioz D *et al.* 2008 *Appl. Phys. Lett.* **92** 042108, Evaldsson M *et al.* 2008 *Phys. Rev. B* **78** 161407(R), Mucciolo E R *et al.* 2009 *Phys. Rev. B* **79** 075407
[22] Zhou S Y, Gweon G -H, Fedorov A V, First P N, de Heer W A, Lee D -H, Guinea F, Castro Neto A H and Lanzara A *2007 Nature Materials* **6** *770*
[23] Giovannetti G, Khomyakov P A, Brocks G, Kelly P J and van den Brink J *2007 Phys. Rev. B* **76** *073103*

# Dynamics of a single Mn spin in a quantum dot: The role of magnetic fields in Faraday and Voigt geometry

**D E Reiter[1], T Kuhn[1], V M Axt[2] and P Machnikowski[3]**

[1] Institut für Festkörpertheorie, Westfälische Wilhelms-Universität Münster, Wilhelm-Klemm-Straße 10, 48149 Münster, Germany

[2] Institut für Theoretische Physik III, Universität Bayreuth, 95440 Bayreuth, Germany

[3] Institute of Physics, Wrocław University of Technology, 50-370 Wrocław, Poland

E-mail: Doris.Reiter@uni-muenster.de

**Abstract.** A theoretical analysis of the optically driven spin dynamics of a single Mn atom in a quantum dot in the presence of magnetic fields in different geometries is presented. When a magnetic field in Faraday configuration is applied, the Mn spin can be flipped from its initial state into each of its six spin eigenstates via optical excitation and manipulation of excitons. In the case of a magnetic field in Voigt configuration, the presence of an exciton leads to a precession of the Mn spin around the combination of external and exchange-induced magnetic field. By using optical pulses this precession can be switched on and off.

## 1. Introduction

Spin systems are attractive candidates for the implementation of a quantum logic. While usually quantum gates are built from two-level systems, i.e., qubits, it has recently been shown that it might be advantageous to employ higher-dimensional Hilbert spaces [1]. In a solid state environment such a higher-dimensional Hilbert space can be realized by a Mn atom which has a spin of $M = 5/2$ and thus six possible orientations $M_z$. Quantum dot (QD) structures doped with a single Mn atom have recently been fabricated both in II-VI [2] and III-V materials [3]. In both cases, due to the strong exchange interaction between exciton and Mn spin, the exciton line in the photoluminescence (PL) spectra splits into a series of lines. Besides splitting the exciton line, the exchange interaction also gives rise to simultaneous spin flips of electron or hole and Mn. Thus, even if there is no direct coupling of the Mn spin to the light field, an indirect manipulation of the Mn spin state becomes possible via the optical excitation of excitons. This has recently been demonstrated experimentally in single [4] and in double QDs [5]. There the spin transfer was based on incoherent relaxation or transfer processes resulting in typical time scales of the order of several nanoseconds. In contrast, our scheme is based on the coherent light-induced dynamics of the exciton-Mn system resulting in switching times in the picosecond regime [6]. Here we will discuss the role of a magnetic field applied either in Faraday or in Voigt configuration on the Mn spin dynamics.

## 2. Model system

We consider a single CdTe quantum dot doped with a single Mn atom. For the electronic system we take into account the lowest conduction band state ($S_z^e = \pm 1/2$) as well as the highest state

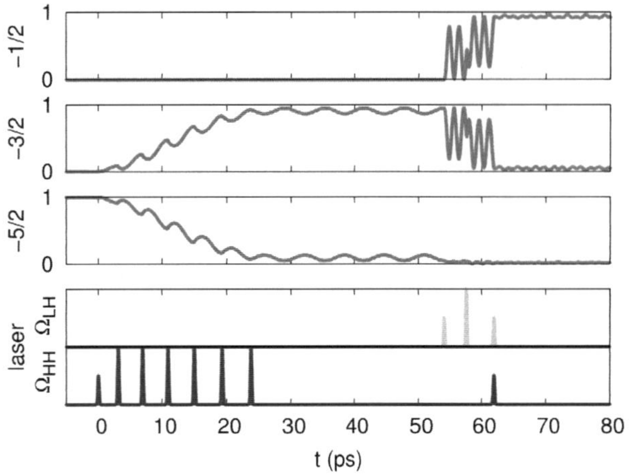


**Figure 1.** Occupation of the Mn spin states $M_z = -5/2$, $-3/2$, and $-1/2$ for the case of excitation with a sequence of laser pulses as shown in the lower part. The pulses are either resonant to the HH exciton ($\Omega_{HH}$) or to the LH exciton ($\Omega_{LH}$) transition.

in the heavy hole (HH, $S_z^h = \pm 3/2$) and in the light hole (LH, $S_z^h = \pm 1/2$) band, respectively. All these electron and hole states are two-fold degenerate. The electronic system thus consists of the ground state, four HH and four LH single exciton states, a HH and a LH biexciton state as well as four mixed HH-LH biexciton states, adding up to 15 states. The coupling of the excitons to the light field is treated in the usual rotating wave and dipole approximation. We assume Gaussian laser pulses with a full width at half maximum of 100 fs; a circularly polarized $\pi$ pulse creates a single spin-polarized exciton in the system. The laser pulse frequency is tuned either to the HH exciton or the LH exciton transition. A HH-LH splitting of 40 meV has been assumed such that both transitions can be selectively addressed. The magnetic field $\vec{B}$ couples via the Zeeman terms $H_Z = \mu_B \left( g_{Mn}\vec{M} + g_e\vec{S}^e + g_h\vec{S}^h \right) \cdot \vec{B}$ with the g-factors $g_{Mn}$, $g_e$, and $g_h$ and $\mu_B$ denoting the Bohr magneton. The exchange interaction is of the usual Heisenberg type $H_{exc} = j_e\vec{M} \cdot \vec{S}^e + j_h\vec{M} \cdot \vec{S}^h + j_{eh}\vec{S}^e \cdot \vec{S}^h$, consisting of electron-Mn, hole-Mn, and electron-hole contribution with the respective coupling constants $j_e$, $j_h$, and $j_{eh}$. With the identity $\vec{M} \cdot \vec{S}^e = M_z S_z^e + \frac{1}{2}(M_+ S_-^e + M_- S_+^e)$ (and analogous expressions for the other contributions) we clearly see the two different types of terms: The Ising type term leading to the energy splitting and the terms leading to correlated spin flips of electron and Mn spin. PL spectra for both Faraday and Voigt configurations have been measured and interpreted in Ref. [7]. A model based on these ingredients has been successfully used to model the observed PL spectra [8]. With this Hamiltonian the Liouville-von Neumann equation for the density matrix of the coupled exciton-Mn system has been set up and solved numerically.

## 3. Spin flip in Faraday configuration

Let us first consider the case of a magnetic field of $B = 6$ T in Faraday configuration. Since the magnetic field lifts the degeneracy of the Mn spin states, the ground state is now given by $M_z = -5/2$. Starting from this initial state we apply a series of laser pulses, first on the HH exciton and then on the LH exciton transition, as plotted in the lower part of Fig. 1. The upper part shows the occupations of the three lowest Mn spin states $M_z = -5/2$, $-3/2$, and $-1/2$. With the first $\sigma^-$ circularly polarized pulse (pulse area $\pi$) a spin polarized $H-1$ exciton is created at time $t = 0$. We denote the exciton states by their valence band type and their angular momentum $S_z^h + S_z^e$. In a pure HH exciton the hole spin, being $S_z^h = -3/2$, is pinned. The electron, however, which is created in the $S_z^e = +1/2$ state can undergo a spin flip by simultaneously flipping the Mn spin from $-5/2$ to $-3/2$. These spin flips manifest themselves as exchange-induced Rabi oscillations between the combined exciton-Mn states $|H-1, M_z = -5/2\rangle$

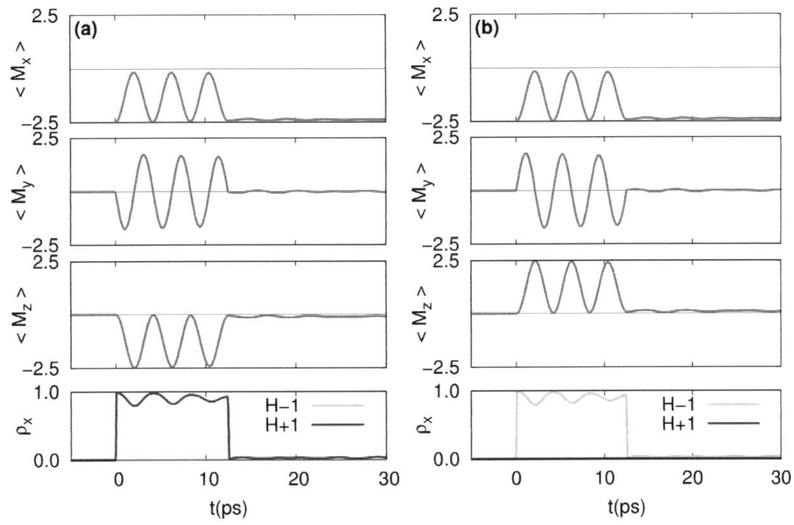


**Figure 2.** Expectation value of the Mn spin in the three directions $\langle M_x \rangle$, $\langle M_y \rangle$ and $\langle M_z \rangle$ as well the occupation of the exciton state as a function of time for the case of a magnetic field of 6 T applied in Voigt configuration. (a) Excitation with $\sigma^+$ pulses; (b) with $\sigma^-$ pulses.

and $|H-2, M_z = -3/2\rangle$, which are off-resonant due to the energy difference between the states and therefore do not lead to a complete flip. With a series of $2\pi$-pulses, however, we can push the system almost completely into the state $|H-2, M_z = -3/2\rangle$, since each $2\pi$ pulse creates a phase jump in the coherence between the involved states. Thus we have effectively switched the Mn spin from $M_z = -5/2$ to $M_z = -3/2$. Indeed, in Fig. 1 we observe that after 7 pulses the occupation of $-5/2$ is nearly zero while the occupation of $-3/2$ has raised to nearly one. Now the exciton is in the dark state $H-2$, which is not accessible anymore by laser pulses resonant to the HH exciton transition. To overcome this bottleneck we apply a $\sigma^+$ polarized $\pi$-pulse on the LH transition and create a combined biexciton $HL-1$. Now the conduction band state is filled with two electrons, the HH spin is still pinned, but the LH can perform a spin flip accompanied by a simultaneous Mn spin flip to switch the Mn spin from $-3/2$ to $-1/2$. Because the coupling of the Mn spin to the holes is four times stronger than to the electrons and the energy splitting between the involved states is much smaller, the exchange-induced Rabi oscillation is much faster and stronger than was the case for the electron spin flip. For this oscillation, as can be seen in Fig. 1, only one $2\pi$ pulse is needed to push the Mn spin state almost completely to $-1/2$. The biexciton now consists of two bright excitons, so two $\pi$ pulses, one resonant to the HH and the other to the LH transition, are applied to bring the electronic system back to its ground state. This scheme can be repeated to switch the Mn spin into all of its spin eigenstates [6]. While this scheme works for any magnetic field strength, the details (number of pulses, total switching time, etc.) strongly depend on the field [9].

## 4. Spin dynamics in Voigt configuration

The situation is very different when a magnetic field in Voigt configuration is applied. We again take a field of $B = 6$ T, but now oriented in $x$-direction. Without exciton the magnetic field defines the quantization axis for the Mn spin, which therefore in the ground state has the spin $M_x = -5/2$. With a single $\sigma^+$ polarized $\pi$-pulse at time $t = 0$ we now create a $H+1$ exciton in the system. In Fig. 2(a) the expectation values of the three components of the Mn spin as well as the occupation $\rho_X$ of the bright exciton states are shown. We start with $\langle M_x \rangle = -5/2$ and

$\langle M_y \rangle = \langle M_z \rangle = 0$. After the excitation all three expectation values of $\vec{M}$ start to oscillate with a period of $T = 4.2$ ps. The reason for this behavior is that the exchange interaction effectively gives rise to an additional magnetic field. The total effective field for the Mn spin is then given by

$$\vec{B}^{Mn}_{eff} = \vec{B} + \frac{j_e}{g_{Mn}\mu_B}\vec{S}^e + \frac{j_h}{g_{Mn}\mu_B}\vec{S}^h. \tag{1}$$

Like in the previous section a HH with spin $S^h_z = \pm 3/2$ cannot flip. Its quantization axis is given by the growth direction, therefore the hole always gives a contribution to the effective field which points in $z$-direction. The electron contribution does not have a fixed direction because the electron spin precesses as well. However, as can be seen in the bottom part of the Figure, the exciton essentially remains in the $H+1$ state, thus also the electron contribution to the effective field is mainly in $z$-direction. Then the effective field is in the $xz$-plane. For our parameters it has a value of $B^{Mn}_{eff} = 8.4$ T and is oriented at an angle of 45° with respect to the $x$-axis. This estimate is completely in agreement with the Mn spin dynamics seen in Fig. 2(a): The $x$- and $z$-components of $\langle \vec{M} \rangle$ oscillate between $-5/2$ and 0 while the $y$-component oscillates symmetrically around zero. Much like the Mn spin also the electron experiences an effective magnetic field. Here, the strongest contribution comes from the electron-hole exchange interaction. The oscillation of the Mn spin gives rise to an additional oscillating contribution in this effective field. The resulting precession of the electron is seen in the oscillations of the occupation of the bright exciton state [bottom part of Fig. 2(a)]. After three periods of the oscillation, i.e. at $t = 12.6$ ps, we apply a second $\pi$ pulse to annihilate the exciton and thus switch off the precession. At this time the Mn spin has rotated back to its initial position. With the annihilation of the exciton, the exchange-induced magnetic field contribution also ends and only the external magnetic field acts on the Mn spin. Thus the Mn spin is now again approximately in an eigenstate and almost no further precession takes place. Figure 2(b) shows that exciting the system with a $\sigma^-$ pulse leads to exactly the same type of dynamics, only $\langle M_y \rangle$ and $\langle M_z \rangle$ have the opposite sign.

## 5. Conclusions

We have analyzed the dynamics of a single Mn spin in a single QD under the influence of a magnetic field in either Faraday or Voigt configuration. In Faraday configuration by starting with an initial spin of $M_z = -5/2$ we were able to switch the Mn spin by 2 into the state $M_z = -1/2$, while the exciton system returned to its ground state. In Voigt configuration the initial Mn spin points perpendicular to the $z$-direction and so in the initial state $\langle M_z \rangle = 0$. If now an exciton is created in the system, $\langle M_z \rangle$ starts to precess. By creation and annihilation of an exciton the effective magnetic field is changed and correspondingly the precession of the Mn spin can be switched on and off on the time scale of the optical pulse.

## References

[1] Lanyon B P *et al.* 2009 *Nature Physics* **5** 134
[2] Besombes L *et al.* 2004 *Phys. Rev. Lett.* **93** 207403
[3] Kudelski A *et al.* 2007 *Phys. Rev. Lett.* **99** 247209
[4] Le Gall C *et al.* 2009 *Phys. Rev. Lett.* **102** 127402
[5] Goryca M *et al.* 2009 *Phys. Rev. Lett.* **103** 087401
[6] Reiter D E, Kuhn T and Axt V M 2009 *Phys. Rev. Lett.* **102** 177403
[7] Léger Y *et al.* 2005 *Phys. Rev.* B **71** 241309(R)
[8] Fernandez-Rossier J 2006 *Phys. Rev.* B **73** 045301
[9] Kuhn T, Reiter D E and Axt V M 2009 *Phys. Status Solidi* B **246** 315

# Transient current in spin blockade condition

**Yasuhiro Tokura**[1], **Keiji Ono**[2], **and Seigo Tarucha**[3]
[1]NTT Basic Research Labs, NTT Corporation, Atsugi-shi, Kanagawa 243-0198, JAPAN
[2]Low Temperature Physics Laboratories, RIKEN, Wako-shi, Saitama 351-0198, JAPAN
[3]Department of Applied Physics, University of Tokyo, Bunkyo-ku, Tokyo 113-8656, JAPAN

E-mail: tokura@nttbrl.jp

**Abstract.** In series double quantum dot with nite bias, strong current suppression by spin-selection rule is expected. Before the current becomes zero starting from a certain initial condition, the transient current ows through the system. We theoretically argue this excess current, under which condition, becomes a simple fraction of charge quantum after being integrated in time. We also propose a scheme to detect this excess current by periodically modulating bias and present the evidence of the fractional excess charge $e/3$.

## 1. Introduction

Spin blockade in the linear transport through a quantum dot (QD) had been proposed by Weinmann *et al.*[1] but had yet not been con rmed experimentally.[2] If one prepares series quantum dots (QDs) with di erent g-factors or di erent but parallel local magnetic elds, we may realize a spin lter in low-bias transport by aligning the energy levels of one of the Zeeman sub-levels in both QDs by gate or bias voltage. However, when one applies large source-drain bias, the spin polarized current ows only transiently. This is because there is a nite probability of occupation of an electron with wrong spin (dark state), which will block the ow of the electrons of correct spin (bright state) by Coulomb blockade e ect as shown in Fig.1.[3, 4] Before the ow stops, we expect small but nite transient charge transfer, $\hat{Q}$, whose expectation value is $\langle\hat{Q}\rangle = \sum_{N=1}^{\infty} eP_N = e$ where $N$ is the number of the transfer processes with the probability $P_N \equiv (\frac{1}{2})^N$ as shown in Fig.1 and $p \equiv \frac{1}{2}$ is the probability that a correct spin is fed to the QDs. Similar spin blocking function had been known for two electron regime in a series QDs where succeeding current ow is blocked when the spin triplet states (dark states) are realized (Pauli spin blockade)[5]. As before, we expect nite charge transfer before the blockade, which is $\langle\hat{Q}\rangle = \sum_{n=1}^{\infty} e(\frac{1}{4})^2 = \frac{e}{3}$ where $p = \frac{1}{4}$ is the probability that the spin singlet state (bright state) is established.[6] It is interesting to notice that the total transient charge $\langle\hat{Q}\rangle$ before the spin blockade completes is a simple fraction of the charge quantum .

The motivation of this work is to clarify the condition when above naive arguments are justi ed and we discuss in detail this transient charge $\langle\hat{Q}\rangle$ using the master equation. Then we argue the experimental setup to observe it using pump-and-probe method and provide an evidence of the fractional transient charge.

## 2. Single electron spin blockade

The Hamiltonian of the series double QDs is $H_{QD} = \sum_{\sigma,\nu=L,R} \epsilon_{\nu\sigma}|\nu\sigma\rangle\langle\nu\sigma| - \Omega_c \sum_\sigma(|L\sigma\rangle\langle R\sigma| +$ h.c.), where $|\nu\sigma\rangle$ stands for the state of $\nu = L, R$ QD with spin $\sigma =\uparrow,\downarrow$ and $\Omega_c$ is the tunnel coupling. The level energies are $\epsilon_{\nu\sigma} = \epsilon_\nu - g_\nu \mu_B B\sigma$ with local g-factor $g_\nu$, Bohr's magneton $\mu_B$ and external magnetic eld $B$. The coupling to the left and right reservoirs are characterized by the line-widths $\gamma_L$ and $\gamma_R$, respectively. We set up the master equation (ME) of the system density matrix $\hat{\rho}$ assuming large intra- and inter-QD Coulomb interaction where the occupation of more than one electron is excluded. For simplicity, we also concentrate to the large bias limit, $\mu_L \gg \epsilon_{\nu\sigma} \gg \mu_R$, where $\mu_\nu$ is the chemical potential of the reservoir $\nu$. The ME for the density matrix elements $\rho_{\sigma 0}$(left QD occupied), $\rho_{0\sigma}$(right QD occupied), and $\rho_{LR\sigma}$(superposition) is[7, 8]

$$\frac{d\rho_{\sigma 0}}{dt} = \gamma_L \rho_0 - i\Omega_c(\rho_{LR\sigma} - \rho_{RL\sigma}), \quad (1)$$

$$\frac{d\rho_{0\sigma}}{dt} = i\Omega_c(\rho_{LR\sigma} - \rho_{RL\sigma}) - \gamma_R \rho_{0\sigma}, \quad (2)$$

$$\frac{d\rho_{LR\sigma}}{dt} = (-i\delta_\sigma - \frac{\gamma_R}{2})\rho_{LR\sigma} - i\Omega_c(\rho_{\sigma 0} - \rho_{0\sigma}), \quad (3)$$

where the energy o set of spin $\sigma$ is $\delta_\sigma \equiv \epsilon_{L\sigma} - \epsilon_{R\sigma}$. The equation for $\rho_{RL\sigma}$ is obtained by taking complex conjugate of the last equation and the probability of empty QDs $\rho_{00}$ is obtained from the sum rule $\rho_{00} + \sum_\sigma(\rho_{\sigma 0} + \rho_{0\sigma}) = 1$. The current is obtained by $I(t) = e\gamma_R \sum_\sigma \rho_{0\sigma}$.

It is convenient to convert this ME into matrix form de ning matrix element vector $\vec{\rho} \equiv (\rho_{\uparrow 0}, \rho_{\downarrow 0}, \rho_{0\uparrow}, \rho_{0\downarrow}, \rho_{LR\uparrow}, \rho_{RL\uparrow}, \rho_{LR\downarrow}, \rho_{RL\downarrow})^t$ and vectors $\vec{u} \equiv (\gamma_L, \gamma_L, 0, 0, 0, 0, 0, 0)^t$ , $\vec{v} \equiv (0, 0, \gamma_R, \gamma_R, 0, 0, 0, 0)^t$ where "$t$" stands for transpose. The ME reads $d\vec{\rho}/dt = \hat{M}\vec{\rho} + \vec{u}$, where $8\times 8$ matrix $\hat{M}$ is obtained from Eqs.(1,2,3). Steady state condition is $d\vec{\rho}_{st}/dt = 0$ and $\vec{\rho}_{st} = -\hat{M}^{-1}\vec{u}$. Then the steady state current and its polarization $P$ using $I_{st\sigma} = e\gamma_R \rho_{st0\sigma}$ are

$$I_{st} = e\vec{v}^t \vec{\rho}_{st} = \frac{2e\gamma_L\gamma_R\Omega_c^2}{\Omega_c^2(\gamma_R + 4\gamma_L) + \gamma_L(\frac{\gamma_R^2}{2} + \delta_\uparrow^2 + \delta_\downarrow^2)}, \quad (4)$$

$$P \equiv \frac{I_{st\uparrow} - I_{st\downarrow}}{I_{st\uparrow} + I_{st\downarrow}} = \frac{\delta_\downarrow^2 - \delta_\uparrow^2}{\delta_\downarrow^2 + \delta_\uparrow^2 + 2\Omega_c^2 + \frac{\gamma_R^2}{2}}. \quad (5)$$

It can be seen that if one of the spin channels is far from the resonant condition: for example $|\delta_\downarrow| \gg \Omega_c, \gamma_\nu$ as shown in Fig.1, the current is suppressed like $2e\gamma_R\Omega_c^2/\delta_\downarrow^2$[3, 4], with almost full polarization $|P| \sim 1$.

Calculation of the transient current is slightly involved. We rst solve the eigenvalue problem of the matrix $\hat{M}$ *s.t.*, $\hat{M}\vec{\xi}_n = \lambda_n \vec{\xi}_n$ for $n = 1 \ldots 8$. All the eigenvalues satisfy $\lambda_n < 0$. We de ne a unitary matrix $\hat{U} \equiv (\vec{\xi}_1, \ldots, \vec{\xi}_8)$ and a diagonal matrix $\hat{\Lambda} \equiv \mathrm{diag}(\lambda_1, \ldots, \lambda_8)$. Further de ning $\vec{\sigma} \equiv \hat{U}^{-1}\vec{\rho}$ and $\vec{\omega} \equiv \hat{U}^{-1}\vec{u}$, we found the ME becomes eight independent di erential equations: $d\vec{\sigma}/dt = \hat{\Lambda}\vec{\sigma} + \vec{\omega}$, which has a general solution with an initial condition $\sigma(0)$, $\vec{\sigma}(t) = (e^{\hat{\Lambda}t} - \hat{1})\hat{\Lambda}^{-1}\vec{\omega} + e^{\hat{\Lambda}t}\vec{\sigma}(0)$. Getting back to the original density matrix and starting from the empty QD ($\rho_{00} = 1$ and others are zero; $\vec{\rho}(0) = \vec{0}$), we have $\vec{\rho}(t) = \hat{U}e^{\hat{\Lambda}t}\hat{\Lambda}^{-1}\hat{U}^{-1}\vec{u} + \rho_{st}$. We plot the current $I(t) = e\vec{v}^t\vec{\rho}(t)$ for two values of $\Omega_c$ in Fig.2. When the tunnel coupling $\Omega_c$ is relatively strong, we can observe sub-peaks corresponding to the number of charge transfer $N$. Transient excess charge is de ned as $Q(t) = e\int_0^t d\tau \vec{v}^t(\vec{\rho}(\tau) - \vec{\rho}_{st})$ and the total transient excess charge is

$$\frac{Q(\infty)}{e} = \vec{v}^t \int_0^\infty d\tau \hat{U}e^{\hat{\Lambda}t}\hat{\Lambda}^{-1}\hat{U}^{-1}\vec{u} = -\vec{v}^t\hat{M}^{-2}\vec{u} \to \frac{2S}{T^2}, \quad (6)$$

where the last expression is obtained by setting $\gamma_L = \gamma_R = \gamma$ and $\delta_\uparrow = 0$ since general formula is too complicated to present, where $S = 2(\frac{\delta_\downarrow}{\gamma})^4 + (2(\frac{\delta_\downarrow}{\gamma})^2 - 7)(\frac{\Omega_c}{\gamma})^2 - 8(\frac{\Omega_c}{\gamma})^4$ and

$T \equiv 2(\frac{\delta_\downarrow}{\gamma})^2 + 1 + 10(\frac{\Omega_c}{\gamma})^2$. It can be seen $Q(\infty)$ is slightly negative for $|\delta_\downarrow| \ll \gamma$ but approaches $e$ for $|\delta_\downarrow| \gg \gamma$ as is argued in Sec.1. Similarly, we can de ne the spin polarization of total excess charge, which approaches unity for $|\delta_\downarrow| \gg \gamma$,

$$P_Q \equiv \frac{Q_\uparrow(\infty) - Q_\downarrow(\infty)}{Q_\uparrow(\infty) + Q_\downarrow(\infty)} \to (\frac{\delta_\downarrow}{\gamma})^2 \frac{T}{S}. \quad (7)$$

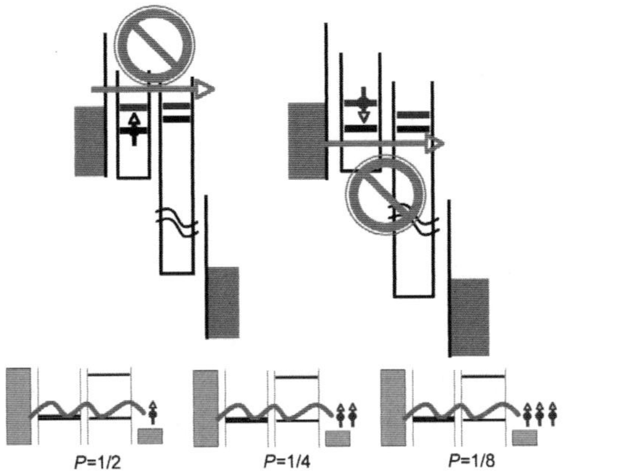


**Figure 1.** Top: The mechanism of the spin blockade in the series double QD under nite bias with di erent g-factors. Bottom: First three series of processes of transport of spin up electrons before the spin down electron is occupied in the left dot.

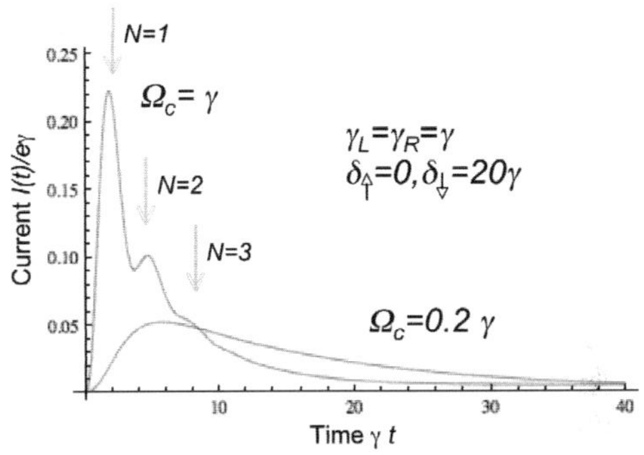


**Figure 2.** Transient current as a function of time starting from empty QD for two di erent inter-dot tunnel couplings. We set $\gamma_L = \gamma_R \equiv \gamma$ and $\delta_\uparrow = 0, \delta_\downarrow = 20\gamma$.

## 3. Pauli spin blockade

ME for the transport involving one and two electron states can be setup up analogously[6, 9] We obtain qualitatively similar behavior of the transient charge as in Sec.2 with a slight modi cation in the number of dark states. Please note that the Pauli spin blockade under zero magnetic eld (also if we can neglect the e ect of Overhauser eld by nuclear spins), the transient current and charge are spin unpolarized.

In the following, we propose an experimental setup to observe this transient charge. In a series double QD with an asymmetry in level energies at zero bias[5], we observe Pauli spin blockade (SB) only for one of the polarities of the bias ($V_s1$) as shown in Fig.3(a). In the other polarity, we observe narrow region of current suppression by Coulomb blockade (CB). When the bias is at the edge of CB condition ($V_s2$), the system is in the one-electron ground state, and hence we can use this condition ($V_s2$) as an initialization stage of the process. We change the bias periodically between $V_s1$ for duration $\tau_{s1}$ and $V_s2$ for duration $\tau_{s2}$ and observe the current in average. We change the duty ratio $\tau_{s2}/\tau_p$ as shown in Fig.3(c) where $\tau_p \equiv \tau_{s1} + \tau_{s2}$ is the period. When the period is much longer than tunneling time, the current as a function of duty ratio is expected simple linear interpolation between the current $I_{V_s1}$ and $I_{V_s2}$ as shown in the dashed line in Fig.4(left). The vertical series double QD device structure is similar to that found in Ref.[5] but we replaced with a semi-insulating substrate to obtain much higher cut-o frequency (> 40GHz). Measurement is performed at 1.5K and the DC pulse from the pulse-generator is applied to the source-drain terminal via bias-Tee and semi-rigid cables. The observed currents as a function of duty ratio for various period shown in Fig.4(left) are clearly deviating from the dashed line. This negative excess current can be attributed the transient current before the SB completed. We plot this excess current as a function of repetition frequency ($f \equiv 1/\tau_p$) in Fig.4(right) with a solid line of ideal excess current obtained by $I = Q(\infty)f$, where $Q(\infty) = \frac{e}{3}$. The deviation of the data points from the ideal line can be understood that the period is too short to transfer all the excess charge, which takes time roughly $\max(\Omega_c^{-1}, \gamma^{-1})$.

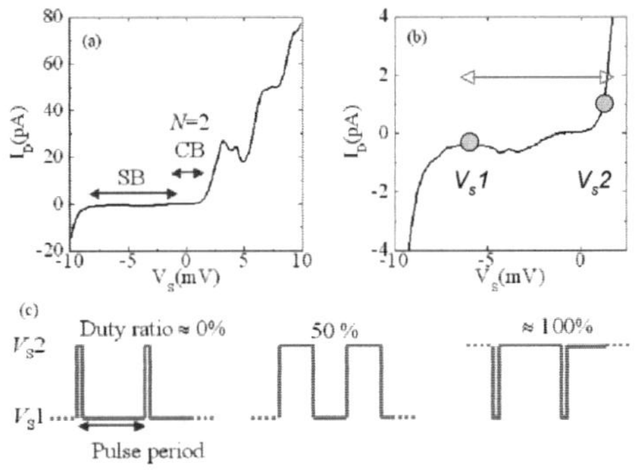


**Figure 3.** (a) Bias dependence of the current through series vertical QD at the Pauli spin blockade condition.(b) Magni ed plot of (a) showing two bias points $Vs1, Vs2$ we set. (c) Schematics of the periodic modulation of bias with three di erent duty ratios.

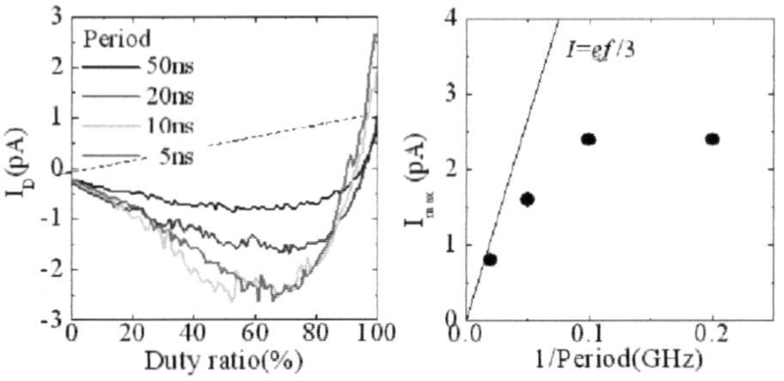


**Figure 4.** (left) Duty ratio dependence of the average current for four di erent periods and the dashed line showing classical interpolation between 0% and 100%. (right) Excess current deviated from the interpolation (dashed line) as a function of repetition frequency. The solid line is the ideal dependence of the excess current on the frequency.

## 4. Conclusion

We argued the excess current/charge theoretically and experimentally before the spin blockade completed in series double quantum dot with nite bias. We clarify the condition of the blockade and polarization of the excess current/charge by using master equation. We also propose a scheme to detect this excess current by periodically modulation of bias and present the evidence of the fractional excess charge $e/3$.

## Acknowledgments

Part of this work is supported nancially by a Grant-in-Aid for Scienti c Research S (No. 19104007), B (No. 18340081), SORST-JST, and special Coordination Funds for Promoting Science and Technology, MEXT.

## References

[1] Weinmann D, Hausler W and Kramer B, 1995 Phys. Rev. Lett. **74** 984
[2] A spin lter realized with a QD connected to reservoirs with di erent g-factor had been proposed in Eugene P R, Sukhorukov V and Loss D, 2000 Phys. Rev. Lett. **85** 1962
[3] Tokura Y, Ono K and Tarucha S, 2009 phys. stat. solidi (b) **246** 740
[4] Huang S M, Tokura Y, Akimoto H, Kono K, Lin J J, Tarucha S and Ono K, Preprint arXiv:0904.1046.
[5] Ono K, Austing D G, Tokura Y and Tarucha S 2002 Science **297** 1313
[6] Qassemi F, Coish W A and Wilhelm F K, 2009 Phys. Rev. Lett. **102** 176806
[7] Gurvitz S A and Prager Ya S, 1996 Phys. Rev. **B 53** 15932
[8] Stoof T H and Nazarov Yu V, 1996 Phys. Rev. **B 53** 1050
[9] Jouravlev O N and Nazarov Yu V, 2006 Phys. Rev. Lett. **96** 176804

# Prediction of giant intrinsic spin-Hall effect in strained p-GaAs quantum wells

**C Schindler and P Vogl**

Walter Schottky Institut, Technische Universitaet Muenchen, Am Coulombwall 3, 85748 Garching, Germany

E-mail: Christoph.Schindler@wsi.tum.de

**Abstract**. We perform spin resolved non-equilibrium Green's function calculations in nanostructured, strained two-dimensional GaAs electron and hole gases. Inelastic scattering is taken into account. We show theoretically that the intrinsic inverse spin-Hall effect provides a simple, sensitive, and purely electrical scheme to measure the spin polarization in nanostructures. We predict large spin polarizations and spin-Hall voltages for several concrete device geometries. We propose tensile strained p-GaAs as being optimally suited for detecting the inverse spin Hall effect and show that the effect is absent in n-GaAs.

## 1. Introduction

The intrinsic spin-Hall effect (SHE) in semiconductors predicted by Murakami *et al.* [1] and Sinova *et al.* [2] is a very promising candidate to manipulate the spin degree of freedom without ferromagnetic elements or external magnetic fields and to generate spin polarized currents, solely caused by the spin-orbit interaction (SOI). The intrinsic SHE has been detected experimentally in semiconductors for two dimensional (2D) hole gases by optical means [3]. However, there are no published reports yet on purely transport based measurements of spin currents in semiconductors, in contrast to metallic systems [4]. In particular, the intrinsic inverse spin-Hall effect that converts a spin polarization into an electric bias, has not yet been observed in semiconductors. In this paper, we predict concrete nanostructured materials and device geometries that exhibit a large intrinsic inverse spin-Hall effect. We provide detailed studies of an H-shaped 2DHG originally proposed by Hankiewicz *et al.* [5] and for several more complex three terminal devices.

## 2. Model and theoretical approach

The investigated 2D geometries are shown schematically in figure 1. For the structure shown in figure 1(a), an applied source-drain bias drives a charge current in the lower leg. Due to the spin-Hall effect, a spin current flows through the vertical connection into the upper leg that, in turn, induces a voltage drop between the two gates due to the inverse spin-Hall effect [5]. For the second, T-shaped, geometry a voltage is applied between the top source contact and the two bottom drain contacts to the right and to the left of the horizontal arm. We have shown previously [6] that large spin polarizations are generated close to the two drain contacts in figure 1(b) that causes the current in the horizontal bar to be spin polarized. To convert this polarization into an electrical bias voltage, we add two analyzing vertical bars into the right and left horizontal arms figure 1(c). The vertical gates exhibit a finite voltage drop in proportion to the spin-polarized current flowing towards the drain contacts. The principle is closely related to the anomalous Hall effect and has been used experimentally in metallic

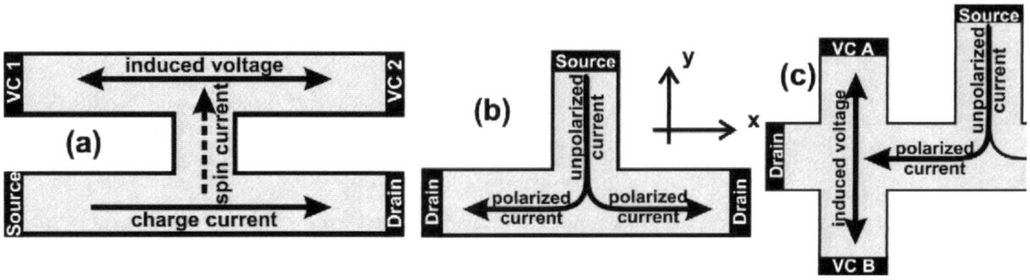


**Figure 1.** Schematic drawing of top gate controlled device geometries investigated in this paper. (a) H-shaped structure, (b) T-shaped structure, and (c) extended T-shaped structure.

systems, where the polarized current has been injected from a ferromagnet [7]. In our case, the spin polarized current is generated directly by the T-shaped geometry due to the intrinsic SHE.

We model the 2D system by a single-band effective mass Hamiltonian including the SO interaction up to cubic order in the in-plane momentum **k**. For a quantum well with $D_{2d}$ symmetry we have [8]

$$H = \frac{\hbar^2 k^2}{2m^*} + V(x,y) + \alpha(k_x\sigma_x - k_y\sigma_y) + c_1(k_y\sigma_x - k_x\sigma_y)k_xk_y + c_2(k_x^3\sigma_x - k_y^3\sigma_y), \quad (1)$$

where $V(x,y)$ is the confinement potential and $\sigma_i$ $(i = x,y,z)$ are the Pauli matrices. The last three terms in equation (1) arise from the SOI and correspond to an effective magnetic field often called the spin-orbit field. The k-linear term proportional to $\alpha$ is the Dresselhaus coupling [9]. There is no Rashba coupling since we only deal with $D_{2d}$ symmetric structures in this paper. The coupling constants $\alpha$, $c_1$, and $c_2$ as well as the effective mass $m^*$ are obtained from atomistic tight binding calculations based on a relativistic $sp^3d^5s^*$ parameterization [10]. For a given band, we map the full atomistic tight-binding matrix onto the $2 \times 2$ effective mass Hamiltonian equation (1). We thereby include the effects of all other bands non-perturbatively.

We employ the non-equilibrium Green's function (NEGF) method for open systems [11], generalized to spin systems [12]. For stationary transport, the two spin resolved Green's functions, the retarded Green function $G^R_{\sigma,\sigma'}(\mathbf{r},\mathbf{r}')$ and the lesser Green function $G^<_{\sigma,\sigma'}(\mathbf{r},\mathbf{r}')$ namely, fully describe the system. Here, $\sigma, \sigma'$ are spin indices. We include scattering by energy, momentum and spin relaxing Büttiker probes [13]. These probes have been modeled in such a way as to mimic realistic mobilities in 2D electron and hole gases. For the hole gases investigated in this paper, we assume a phase breaking mean free path of 200 nm. In this way, known artifacts of purely ballistic transport are eliminated [14]. In the numerical implementation, we solve the NEGF equations in real space by employing a finite difference scheme. Importantly, we apply open boundary conditions, i.e. the device gets attached to two or more semi-infinite leads which remain in thermal equilibrium. In addition, they include the same spin-orbit terms as the active device to suppress artificial reflections that are caused by the abrupt device-lead interfaces.

## 3. Results and discussion

The present atomistic electronic calculations show an exceptionally large k-linear SOI for the top valence band in tensile strained GaAs quantum wells [15]. This large k-linear spin splitting is caused by the strain, which may be generated by growing GaAs on $In_xAl_{1-x}As$ buffer material. Figure 2 shows the calculated Dresselhaus coupling constant within a 14 nm GaAs quantum well as a function of the strain that is generated by embedding the GaAs layer within $In_xAl_{1-x}As$ that sets the lateral lattice constant and acts as barrier material. We note that the SOI in the conduction band remains unaffected by the strain. The tensile strain causes the light hole band to lie above and to be well separated from the heavy hole band. For the other parameters we obtain only a weak dependence on *x*. We find $m^* \approx .375, c_1 \approx -65$ meVnm$^3$, and $c_2 \approx 15$ meVnm$^3$.

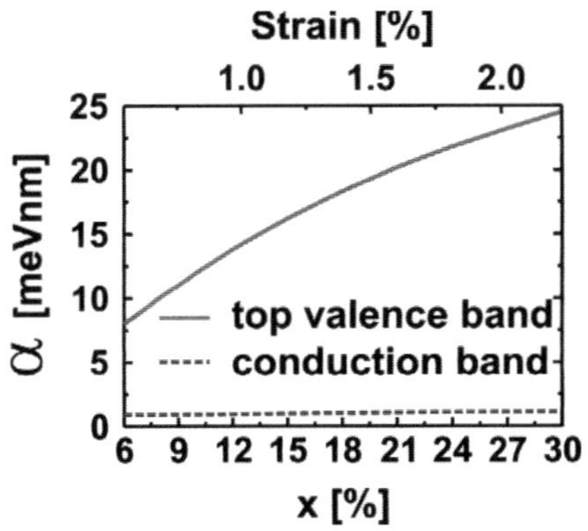


**Figure 2.** Calculated k-linear Dresselhaus coupling constant α in meVnm for a 14 nm wide strained GaAs quantum well plotted versus the Indium content $x$ in % (bottom axis) and tensile strain in % (top axis). The red line (solid) is for the top valance band, while the blue one (dashed) is for the conduction band.

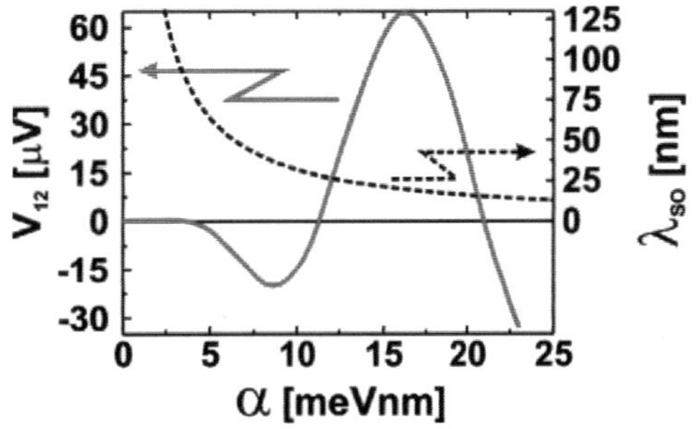


**Figure 3.** Calculated spin-Hall voltage $V_{12}$ in μV and the spin precession length $\lambda_{SO}$ in nm as a function of the Dresselhaus coupling constant α in meVnm. The charge density $n_{2D}$ = $1.75\times10^{11}\text{cm}^{-2}$ and the applied bias $V_{SD}$ = .5 mV are kept fixed.

We now turn to the device with H-bar geometry as shown in figure 1(a) and consider a hole gas in a tensile strained GaAs quantum well. We assume the widths of all channels to be equal to 30 nm. The length of the vertical bar is also taken to be 30 nm. In figure 3, we show the calculated spin-Hall voltage $V_{12}$ between voltage contact (VC) 1 and 2 (cf. figure 1(a)) as a function of the SOI strength for an applied source-drain bias $V_{SD}$ =.5 mV. The hole density amounts to ~1.75×1011 $\text{cm}^{-2}$ so that only the highest hole subband is occupied. The calculations show an oscillatory behavior of the spin-Hall voltage $V_{12}$ as a function of the SOI strength which can be explained as follows. The magnitude and sign of the spin-Hall voltage $V_{12}$ is roughly proportional to the average out-of-plane spin polarization $\langle P_z \rangle$ in the hole gas in the upper leg that has been induced by the charge current in the lower leg. With increasing SOI, the spin precession length $\lambda_{SO}$ changes as depicted in figure 3. Whenever the device width is an even (odd) integer of this length, $\langle P_z \rangle$ becomes a maximum or minimum (zero) which explains the finding in figure 3. For $\alpha < 5$ meVnm, the calculated spin-Hall voltage is negligible. Therefore, the intrinsic inverse SHE is negligible in n-type GaAs, which holds independently of the mobility. For the strained p-GaAs wells, by contrast, the two extrema in figure 3 correspond to a fairly low Indium composition of $x$=7% and $x$=15%. The resistance signal, i.e. the spin-Hall voltage $V_{12}$ divided by the source-drain current $I_{SD}$, is approximately 1 kΩ for $x$=7% and 3 kΩ for $x$=15%, which is markedly larger than those predicted for HgTe quantum wells [16]. Figure 4 shows the spin-Hall voltage $V_{12}$ and the source-drain current $I_{SD}$ as a function of applied source-drain bias $V_{SD}$ for both structures. Since the current is almost equal in both cases, in contrast to the spin-Hall voltage $V_{12}$, we can be sure that the latter is caused by the SOI rather than by a change in the longitudinal conductance.

For the T-shaped device in figure 1(b), we use the same channel widths and hole density as for the H-shaped geometry. Figure 5 shows the three components of the spin polarization at the rightmost drain contact as a function of the length of the side arm. The $y$- and $z$-component of the spin polarization, averaged over the width of the arm, is predicted to oscillate in correspondence with the spin precession length. The $x$-component lies parallel to the spin-orbit field and therefore does not precess. All polarization amplitudes decrease in magnitude due to the phase breaking scattering.

Finally, we attach additional analyzing probes (VC A and VC B) to the T-structure as depicted in figure 1(c). The calculated spin-Hall voltage $V_{AB}$ turns out to be a direct measure for the z-component of the spin polarization. When the probe is attached at a position where positively (negatively)

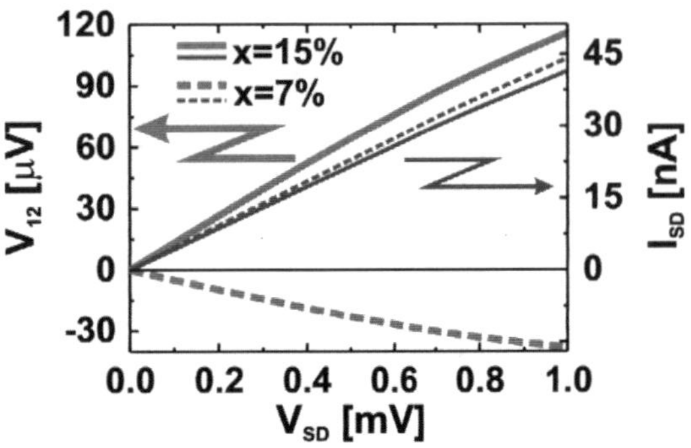


**Figure 4.** Calculated spin-Hall voltage $V_{12}$ (left vertical axis, thick red lines) and the source-drain current $I_{SD}$ in nA (right vertical axis, thin blue lines), as a function of $V_{SD}$. Shown are results for two different strain values. The dashed lines are for an Indium content of $x$=7%, while the solid ones are for $x$=15%.

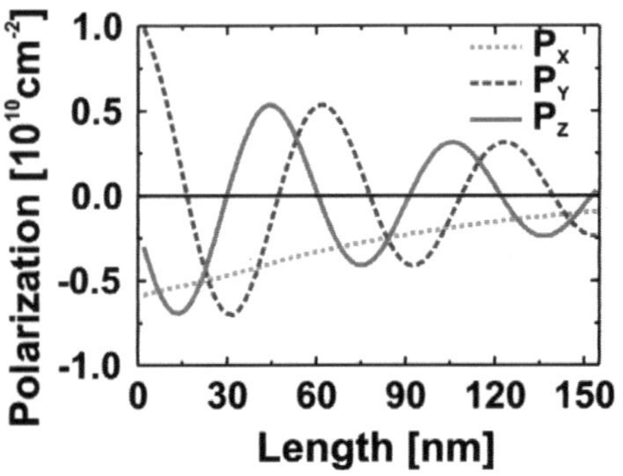


**Figure 5.** Calculated spin polarization averaged along the $y$-direction at the right drain contact in $cm^{-2}$, as a function of the length of the side arm in nm. Red (solid) shows the $z$-component, blue (dashed) the $y$-component, and green (dotted) the $x$-component. The Indium content is 9% and the applied bias equals 1.0 mV.

polarized carriers dominate, we obtain a positive (negative) voltage signal $V_{AB}$. If we attach the probe at 75 nm, for example, we find $V_{AB}$ = -15.6 μV.

**Acknowledgments**
The authors acknowledge financial support from the Deutsche Forschungsgemeinschaft (SPP 1285), and the Nano Initiative Munich.

**References**
[1] Murakami S, Nagaosa N and Zhang S C 2003 *Science* **301** 1348
[2] Sinova J, Culcer D, Niu Q, Sinitsyn N A, Jungwirth T and MacDonald A H 2004 *Phys. Rev. Lett.* **92** 126603
[3] Wunderlich J, Kaestner B, Sinova J and Jungwirth T 2005 *Phys. Rev. Lett.* **94** 047204
[4] Saitoh E, Ueda M, Miyajima H and Tatara G 2006 *Appl. Phys. Lett.* **88** 182509
[5] Hankiewicz E M, Molenkamp L W, Jungwirth T and Sinova J 2004 *Phys. Rev.* B **70** 241301(R)
[6] Kubis T and Vogl P 2008 *phys. stat. sol.* (c) **5** 290
[7] Valenzuela S O and Tinkham M 2007 *J. Appl. Phys.* **101** 09B103
[8] von Allmen P 1992 *Phys. Rev.* B **46** 15382
[9] Fabian J, Matos-Abiague A, Ertler C, Stano P and Zutic I 2007 *acta physica slovaca* **4** 565
[10] Jancu JM, Scholz R, Beltram F and Bassani F 1998 *Phys. Rev.* B **57** 6493
[11] Datta S 1999 *Electronic Transport in Mesoscopic Systems* (Cambridge: Cambridge University Press)
[12] Nicolic B K, Zarbo L P and Souma S 2006 *Phys. Rev.* B **73** 075303
[13] Venugopal R, Paulsson M, Goasguen S, Datta S and Lundstrom M S 2003 *J. Appl. Phys.* **93** 5613
[14] Mamaluy D, Vasileska D, Sabathil M, Zibold T and Vogl P 2005 *Phys. Rev.* B **71** 245321
[15] Majewski J A and Vogl P 2003 *Physics of Semiconductors 2002* (Bristol: Institut of Physics Publ.) p 305
[16] Brune C, Roth A, Novik E G, Konig M, Buhmann H, Hankiewicz E M, Hanke W, Sinova J and Molenkamp L W 2008 (*Preprint* cond-mat/0812.3768v1)

Journal of Physics: Conference Series **193** (2009) 012104 doi:10.1088/1742-6596/193/1/012104

# Tunneling magnetoresistance dependence on the temperature in a ferromagnetic Zener diode

**E. Comesaña, M. Aldegunde and A. García-Loureiro**

Departamento de Electrónica e Computación, Universidade de Santiago de Compostela, 15782 Santiago de Compostela, Spain

E-mail: enrique.comesana@usc.es

**Abstract.** In the present work we focus on the study of the temperature dependence of the tunnelling current in a ferromagnetic Zener diode. We predict the tunneling magnetoresistance dependence on the temperature. Large doping concentrations lead to magnetic semiconductors with Curie temperature $T_C$ near or over room temperature and this will facilitate the introduction of new devices that make use of the ferromagnetism effects. According to our calculations the tunneling magnetoresistance has the form $TMR \propto (T_C^n - T^n)$.

## 1. Introduction

The recent progress in developing new semiconductor materials that show ferromagnetic characteristics at room temperatures is leading to a growing interest to replace charge with spin in signal processing devices. Spintronics is motivated by the belief that spin signal processing may yield advantages in terms of processing speed, power consumption or device density.

Dilute magnetic semiconductors (DMS) are showing good characteristics at rising temperatures and new DMS are being developed to achieve spin control at room temperature [1]. In an optimally doped DMS the density of carriers is approximately half that of the density of magnetic ions which is usually between $x = 5\%$ and $x = 15\%$ [1]. The well established DMS occur with one type of carrier. One of the most investigated materials is the III-V material (GaMn)As in which the Mn ion provides a localized spin $S = 5/2$ and also contributes a hole [1]. The holes are degenerate and strongly polarized at low temperatures [2]. The ferromagnetic III-V doped materials are all p type but there is much effort to find a compatible n type ferromagnetic semiconductor [3]. Magnetic properties have been seen in a number of magnetically doped oxides, particularly $ZnO$, $TiO_2$ and $SnO_2$ [4] all of which occur as n type semiconductors.

A PN junction of two highly degenerated semiconductors makes a Zener tunneling diode which has many applications [5] and adding magnetic functionality would enable more devices such as magnetic switching of microwave devices. Zener tunnelling has been observed in a ferromagnetic-nonmagnetic (GaMn)As/GaAs heterostructure, and a high spin polarization is observed optically [6]. The voltage dependence of the tunnelling current in a spin-polarized Zener diode is well fitted by the theory of a non-magnetic diode [2]. We are developing analytical [7] and numerical models [8] to study the transport in ferromagnetic Zener diodes. Using these models we have predicted the dependence of the tunneling current on the mean magnetization of the system and we have evaluated the tunneling magnetoresistance ($TMR$) for different values of the applied bias and temperatures in a theoretical both-sided-ferromagnetic diode.

The article is structured as it follows: Section 2 sumarizes the effects of the ferromagnetism. Section 3 shows the results for tunnelling current and $TMR$ dependence with the temperature in a ferromagnetic GaAs diode. Conclusions are drawn in Section 4.

## 2. Tunneling transport

The current through a tunnelling diode has three components [5]: the tunnelling current, the excess current and the diffusion current. The tunnelling current rises to a maximum at low voltage, about the lowest of the distances between the Fermi level and the conduction band in the n-side ($\varepsilon_n/q_e$) and the valence band in the p-side ($\varepsilon_p/q_e$). Then it falls to zero when bias increases over $(\varepsilon_n + \varepsilon_p)/q_e$. This fall in the I-V curve is known as negative resistance region and is produced by the overlapping of the density of carrier functions in both P and N sides.

The tunnelling of carriers through the bandgap is an important part of the carrier transport in highly doped PN junctions. There are two tunnelling mechanisms: the direct transition from band to band and the trap assisted tunnelling. The basic principles of the band-to-band tunnelling were explained by Kane [9]. The tunnelling probability, obtained from the WKB approximation, is given by $T_t \simeq \exp\left[-2\int_u^l |\kappa(\mathbf{r})| d\mathbf{r}\right]$, where $\kappa(\mathbf{r})$ is the wave vector associated to the carrier near the barrier and $u$ and $l$ are the classical turning points in the potential barrier. The tunnelling density of current is described as follows [9]:

$$J_t \propto \int_{E_C}^{E_V} [F_C(\epsilon) - F_V(\epsilon)] T_t D_n(\epsilon) D_p(\epsilon) d\epsilon, \tag{1}$$

where $T_t$ is assumed to be equal for both directions, $F_C(\epsilon)$ and $F_V(\epsilon)$ are the Fermi-Dirac distribution functions and $D_n(\epsilon)$ and $D_p(\epsilon)$ are the density of states. Equation (1) shows that the tunnelling current depends on the tunnelling probability $T_t$ as well as a supply function of electrons and holes, described by the density of carriers in each band.

### *2.1. Mean field theory in (III,Mn)V alloys*

The effect of the exchange interaction between the charge carriers and the localized magnetic moments on the energy band structure and the transport properties can be estimated by using a perturbation theory [10]. The total Hamiltonian describing the free carrier and magnetic subsystems in the DMS is given by $H_{tot} = H^{(1)} + H_{exch} + H_m$ where $H^{(1)}$ gives the carrier energies in the spin-polarized band, $H_{exch}$ takes account of the interaction between the free carriers and the magnetic ions and $H_m$ the interaction between magnetic ions and the Zeeman energy when an external magnetic field is applied.

A first order correction of the band energies due to the exchange potential can be obtained, assuming external magnetic field, this is, no paramagnetic behavior of the material. The first order correction can be written as [10]:

$$E_{\mathbf{k}\sigma} = E_{be}^0 + \frac{\hbar^2\mathbf{k}^2}{2m^*} - \frac{\Delta}{2}\left(\delta_{\sigma\uparrow} - \delta_{\sigma\downarrow}\right), \tag{2}$$

where $\Delta = xJ_{exch}\langle S^z\rangle$ is the fist-order band splitting. In the mean field approximation, the average spin polarization of the magnetic moments, $\langle S^z\rangle$, is given by $x\langle S^z\rangle = xSB_S(y)$ where $x$ is the Mn ions concentration and $B_S(y)$ is the Brioullin function for the spin quantum number $S$ with $y = g_L\mu_B SB_{eff}/K_BT$ and $B_{eff}$ is the effective molecular field.

Figure 1 shows the calculated band spliting at $B = 0\ T$ for (GaMn)As where $J_{exch}^{pd} = 1.4eV$, $m_h^* = 0.5m_0$, $T_C = 110\ K$ and $S = 5/2$. We see how the ferromagnetism effects disappear for temperatures over $T_C$ and the splitting rises and saturates to a maximun value for low temperatures. The resulting band diagram at $T = 0\ K$ is shown in figure 2. There are two

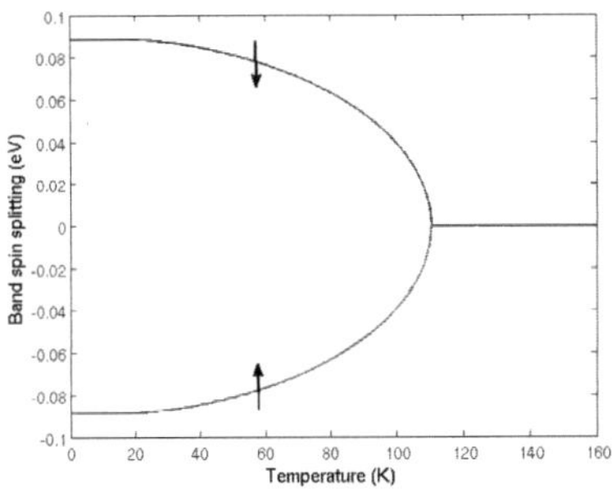


**Figure 1.** Temperature dependence of the band edge in a ferromagnetic semiconductor (GaMn)As at $B = 0\ T$ showing the band splitting when $T_C = 110\ K$.

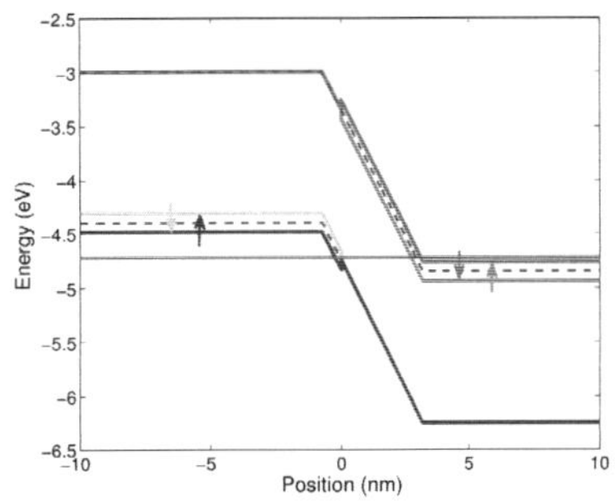


**Figure 2.** Band diagram for a GaAs diode at equilibrium. The dotted lines are the bands when the spin polarization is zero ($p = 3 \cdot 10^{20}\ cm^{-3}$ and $n = 0.7 \cdot 10^{20}\ cm^{-3}$).

conduction/valence band edges, one for electrons/holes with spin up and for electrons/holes with spin down. The tunnelling occurs between the two majority and the two minority bands or, if the relative magnetization of the layer is reversed, between the majority and the minority bands. We call these currents parallel current $I_p(V)$ and antiparallel current $I_{ap}(V)$.

## 3. Results

The tunneling current has been calculated for a theoretical GaAs diode, ferromagnetic on both sides in a range of temperatures between 0 $K$ and $T_C$. Typical characteristic I-V curves are shown in figure 3, where the current through a non spin-polarized diode (solid line) and for the parallel and antiparallel configuration can be seen. The tunneling magnetoresistance ($TMR = \frac{I_p - I_{ap}}{I_p + I_{ap}}$) was evaluated at applied bias $V_a = \frac{min(\varepsilon_n, \varepsilon_p) + \varepsilon_n + \varepsilon_p}{2q_e}$, which is half way between the maximun and zero tunneling current biases for a non spin-polarized diode. This bias value has been chosen because for higher values the TMR is expected to be lower because the excess and thermal current become important and they are not spin dependent. For lower biases the TMR is lower as can be seen in figure 3, since the tunneling current is dominated by the overlapping of high energy states which are less dependent on the splitting.

We sample the value of the $TMR$ at different temperatures to analyze its dependence at low and near $T_C$ temperatures. Results from three of the evaluated configurations are shown in figure 4 as the most significant: a symmetric diode (dashed green line) where the carrier concentration in each side of the diode is the same so they have the same $T_C = 110\ K$, a $P^+N$ diode with $T_C = 110\ K$ in the P-side and $T_C = 27\ K$ in the N-side and a $PN^+$ diode where $T_C = 110\ K$ in the P-side and $T_C = 139\ K$ in the N-side. We have fixed the $T_C$ in the p-side because it is the highest Curie temperature achieved for GaAs at the moment.

Several remarks can be extracted from figure 4. The highest values of the $TMR$ are obtained for temperatures near 0 K, which is a consequence of the dependence of the magnetization and therefore of the splitting on the temperature as shown in figure 1. The most favorable configuration to obtain high $TMR$ values is the $P^+N$ diode, although it is constrained to low temperature values, because for $T > 27\ K$ the N-side loses its ferromagnetic properties and the TMR is no longer observable because it requires that both sides are ferromagnetic. $PN^+$ configuration shows low values of $TMR$, even at 0 K. This can be explained from the low polarization of the carriers in the N-side where the maximum possible splitting is less than 0.1 eV at temperatures near 0 K and the kinetic energy of the higher energy carriers is around 0.4 eV;

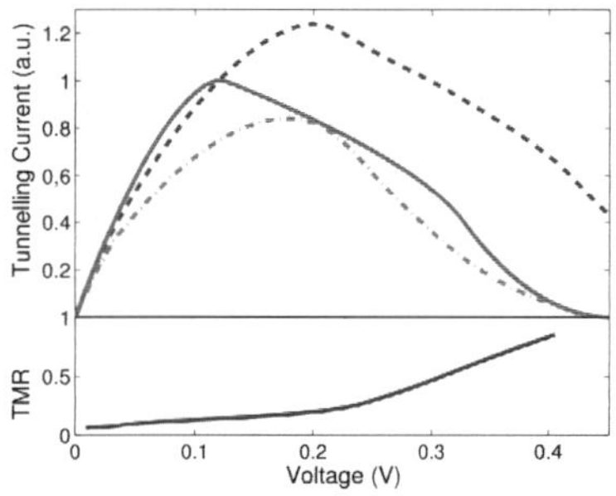


**Figure 3.** Tunnelling current and $TMR$ when the splittings of the p-band and n-band are near to 0.1 eV. The parallel (dashed, blue) and the antiparallel (dot-dashed, red) configurations are given as well as the current for zero spin-splitting (continuous, green).

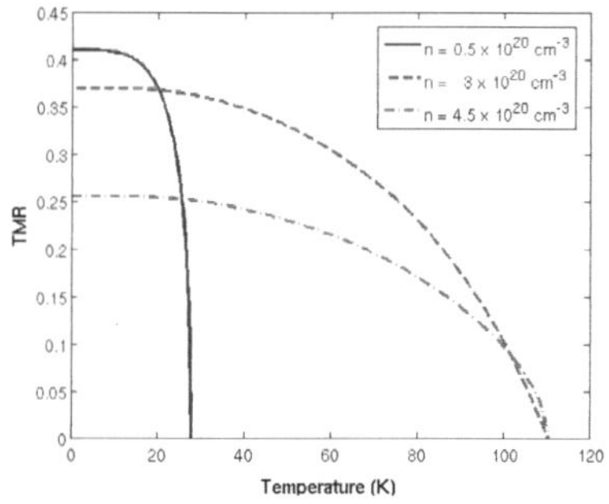


**Figure 4.** $TMR$ dependence on the temperature for different N-side doping levels. The doping in the P-side is fixed to $3 \cdot 10^{20}\ cm^{-3}$ which correspond with $T_C = 110\ K$. The Curie temperatures in the N-side are 27 K (continuous, blue), 110 K (dashed, green) and 139 K (dot-dashed, red)

therefore, only a small part of the carriers are polarized. The symmetric configuration seems to be the best configuration to see good values of $TMR$ in the widest range of temperatures.

Another interesting result from our calculations was that, in a symmetric configuration, the $TMR$ dependence on the temperature has the form $TMR \propto (T_C^n - T^n)$. In particular, for a carrier concentration level that correspond with $T_C = 110\ K$, $n = 3$.

## 4. Conclusions

We extend our previous developed analytical model to add the temperature dependence of the ferromagnetic properties of the DMSs. We found that the $TMR$ has a strong dependence on the temperature as well as on the relative doping values of both diode sides. We conclude that the best configuration to see high $TMR$ values at high temperatures will be a symmetric diode. However, the highest values of $TMR$ can be achieved from non-symmetric configurations that guarantee high polarization level of the carriers in, at least, one of the diode sides; but these values of the $TMR$ are constrained to low temperatures.

## Acknowledgment

This work was supported by Spanish Government (MCYT TIN2007-67537-C03-01), and by Xunta de Galicia (DXIDI07TIC01CT and INCITE08PXIB206094PR).

## References

[1] Jungwirth T, Sinova J, Mašek J, Kučera J and MacDonald A H 2006 *Rev. Mod. Phys.* **78** 809
[2] Holmberg H et all. 2007 *Phys. Stat. Sol. B* **204** 791–804
[3] Mašek J et all. 2007 *Phys. Rev. Lett.* **98** 067202
[4] Chambers S A 2006 *Surface Science Reports* **61** 345–381
[5] Sze S and Ng K K 2006 *Physics of Semiconductor Devices* 3rd ed (John Wiley & Sons, Inc.)
[6] Dorpe P V et all. 2004 *Appl. Phys. Lett.* **84** 3495–3497
[7] Comesaña E and Gehring G 2007 *Appl. Phys. Lett.* **91** 142510
[8] Comesana E et all. *2009 Spanish Conf. on Electron Devices.* pp 160–163
[9] Kane E O 1961 *J. Appl. Phys.* **32** 83–91
[10] Lebedeva N and Kuivalainen P 2003 *J. Appl. Phys.* **93** 9845–9864

Journal of Physics: Conference Series **193** (2009) 012105 doi:10.1088/1742-6596/193/1/012105

# Interface dipole formation between GaMnAs and organic material

**Wenjin Chen[1], Baikui Li[1], Hongtao He[1], Jiannong Wang[1,*], Hoi Lam Tam[2], Kok Wai Cheah[2], Xiancun Cao[3], Yuqi Wang[3], Guijun Lian[4] and Guangcheng Xiong[4]**

[1] Department of Physics, The Hong Kong University of Science and Technology, Clear Water Bay, Kowloon, Hong Kong, China

[2] Department of Physics, Hong Kong Baptist University, Kowloon, Hong Kong, China

[3] Institute of Solid State Physics, Chinese Academy of Sciences, Hefei, Anhui 230031, China

[4] Department of Physics, Peking University, Beijing 100871, China

* Email: phjwang@ust.hk

**Abstract.** The interface band alignment between GaMnAs and organic material was investigated by current transport measurement in hole only devices with GaMnAs/NPB/Al structure. The current density – voltage (*J-V*) curves were measured and numerical simulations performed on this device structure yielded the hole injection barrier between GaMnAs and NPB to be 0.77 eV. A vacuum level shift at GaMnAs/NPB interface was deduced as 0.54 eV, indicating a dipole layer across the interface. We attributed the vacuum level shift to the charge transfer across the interface.

## 1. Introduction

Organic semiconductors have attracted considerable interest for their successful application in organic light emitting diodes and organic thin film transistors [1,2]. Recently, with the development of spintronics, organic semiconductors have been utilized as efficient spin transport layers due to their weaker spin-orbit coupling and hyperfine interaction, which render longer spin lifetime and spin diffusion length than those in inorganic materials [3]. Experimentally, spin valve devices based on tris(8-hydroxyquinoline) aluminium (Alq3) have been studied and spin diffusion length up to hundreds of nanometers at low temperature has been reported [4]. Theoretically, the spin dependent carrier injection and transport with respect to different contact resistances at the metal/organic interface have been calculated [5].

To achieve the spin injection into the organic semiconductor, carriers in the electrode must be highly spin-polarized. Ferromagnetic metals [6] and semimetals [7] have been used in previous studies. However, so far no report has been seen utilizing the ferromagnetic semiconductor GaMnAs as the spin-polarized carrier injection electrode. Ferromagnetic GaMnAs has very high spin polarization for holes in valence band at temperatures below its Curie-temperature [8]. In addition, GaMnAs may provide a better resistance match at the electrode/organic interface in comparison with ferromagnetic metal/organic interface. This in turn leads to higher spin injection efficiency [9]. Thus, GaMnAs is a good candidate for hole spin injection electrode in organic spin based devices. In this paper, we investigated the interface formation between GaMnAs and *N,N'*-diphenyl-*N,N'*-bis(1-naphthyl)(1,1'-biphenyl)-4,4'-diamine (NPB), a prototypical organic hole transporting material. Hole only devices with GaMnAs/NPB/Al sandwiched structure were fabricated. Current density – voltage (*J-V*) measurement were performed on these devices. Interface band alignment between GaMnAs and NPB was extracted by numerical simulation of the measured *J-V* curves to reveal the potential application of GaMnAs as a hole spin injector in NPB based organic devices.

## 2. Experiments

Ferromagnetic GaMnAs thin films with nominal Mn concentration of 5% and thickness of 20 nm were grown at 250 °C in a Riber 32 MBE system on (001) semi-insulating GaAs (SI-GaAs) substrates [10]. Post-grown annealing was performed in nitrogen atmosphere at 260 °C for 1 hour to enhance the hole concentration and the ferromagnetism of GaMnAs [11]. The curie temperature of the annealed GaMnAs was 103 K.

The device structure for *J-V* measurement is schematically shown in figure 1. The GaMnAs thin films were patterned into stripes with standard photolithography process, and etched in diluted HCl solution to produce Ga-Cl terminated surface [12], which is stable in air for at least 1 hour, long enough for the sample transfer [13]. NPB films of 70nm and Al of 60 nm were thermally evaporated onto the patterned substrates using shadow masks in an Edwards Auto 306 system. The device area was 400μm x 500μm. After deposition, the devices were transferred into a vacuum chamber where *J-V* curves were measured at room temperature using an HP 4155A semiconductor parameter analyzer.

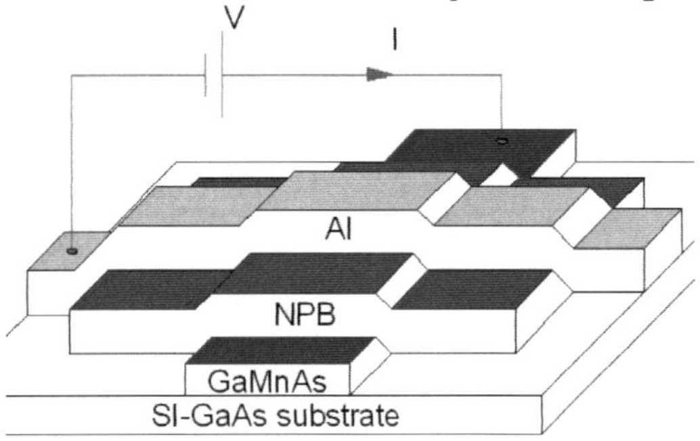


**Figure 1**. (Color online) a schematic device structure for *J-V* measurements; GaMnAs is the anode and Al is the cathode.

## 3. Numerical model

Numerical simulations of the *J-V* curves have been performed using the hole injection barrier between GaMnAs and NPB interface as the fitting parameter. The theoretical models are based on the injection limited current (ILC) model [14,15] and the space charge limited current (SCLC) model [16,17]. The current continuity equation, Poisson's equation, and the drift-diffusion equation are used to describe the charge transport.

We note that the electron injection barrier at the cathode interface (Al/NPB) is as high as 1.75 eV which is too large for an effective electron injection via Al/NPB interface within the bias range studied [18]. Unipolar device model is therefore constructed.

The vacuum level shift at interfaces is included in the simulation. At NPB/Al interface, the vacuum level shift $\Delta_2$ is taken to be 0.45 eV [18]. The vacuum level shift at the GaMnAs/NPB interface is related to the hole injection barrier $\phi_b$ (at zero bias) by:

$$\Delta_1 = I_{GaMnAs} + \phi_b - I_{NPB} \qquad (1)$$

where $I_{GaMnAs}$ and $I_{NPB}$ are the ionization energy of GaMnAs and NPB, respectively. The ionization energy of GaMnAs $I_{GaMnAs}$ is 5.27 eV, taking into account the broadening of the Mn impurity band and the merge of it with GaMnAs valence band [19]. The ionization of NPB is taken as 5.5 eV [20,21]. GaMnAs Fermi energy is assumed to be aligned with its valence band top due to its high doping level.

In the space charge limited conduction, Ohmic contact is assumed with the boundary condition [16]:

$$p(0) = p_0 \quad (2)$$

where $p(0)$ is the hole carrier density at the GaMnAs/NPB interface, and $p_0$ is the effective hole density of states in NPB highest occupied molecular orbital (HOMO).

## 4. Results and discussion

A typical *J-V* curve measured on GaMnAs/NPB/Al device is shown in figure 2 in semi-logarithmic scale, together with the simulated results by different models. By comparison, the injection limited model with the hole injection barrier of 0.77 eV at GaMnAs/NPB interface best fits the measured curve. The fact that the measured current density is much less than the current density predicted by the space charge limited model implies that the current transport of GaMnAs/NPB/Al device is limited by the interface injection. The discrepancy between the measured and the calculated *J-V* curve at lower voltage range (0 to 6 V) is possibly due to the Ohmic conduction before effective injection starts [16]. With the extracted hole injection barrier at GaMnAs/NPB interface, the vacuum level shift can be further calculated by equation (1) to be 0.54 eV. The calculated injection barrier and the vacuum level shift agree well with our UPS results discussed elsewhere [22].

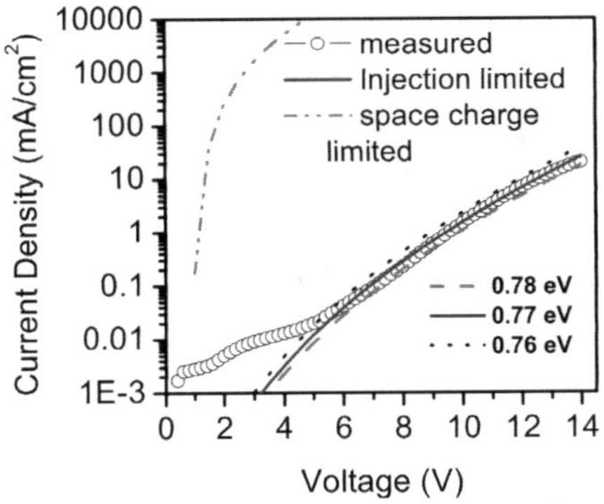


**Figure 2.** (Color online) *J-V* curve measured (black open circle) and simulated by space charge limited model (red dash dot-dotted) and by injection limited model with 0.78 eV, 0.77 eV and 0.76 eV hole injection barriers at the interface (red dashed, blue solid, and black dotted, respectively).

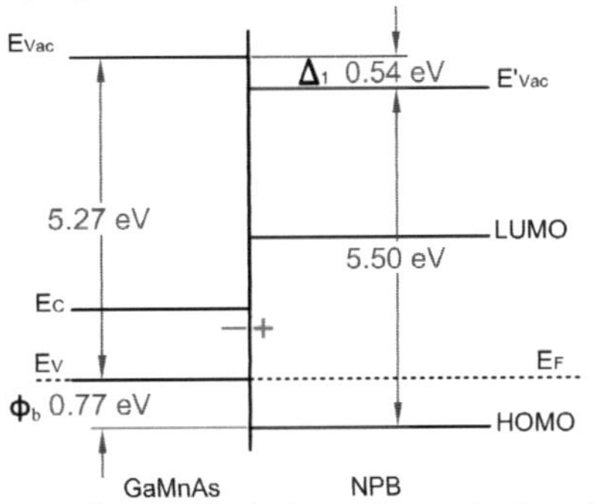


**Figure 3.** (Color online) the schematic band diagram at GaMnAs/NPB interface deduced from the *J-V* measurement and simulation. $E_V$ ($Ec$) is the valence (conduction) band edge of GaMnAs. $E_{Vac}$ ($E'_{Vac}$) is the vacuum level on GaMnAs (NPB) side. $E_F$ is the Fermi energy, $\Delta_I$ is the vacuum level shift at the interface, and $\phi_b$ is the band edge offset across the interface. The red markers indicate the charges forming the interface dipole.

Based on the simulation of the *J-V* measurement, the band alignment diagram at GaMnAs/NPB interface is schematically shown in figure 3, where the injection barrier calculated is represented as the band offset between the valence edge of GaMnAs and the HOMO edge of NPB, which is actually independent of the Fermi level position. The band bending at the surface of GaMnAs [22] is not shown for simplicity. The downward shift of the NPB vacuum level indicates a dipole layer as shown by red mark. This vacuum level shift is generally observed at metal/organic and semiconductor/organic interface, though the exact origin of the shift is not conclusive so far [23-26].However, one theory based on the metal induced gap states have recently been used extensively to explain metal/organic interface dipole formation [24,25]. These gap states can be either donor-like (close to the valence band) or acceptor-like (close to the conduction band). A charge neutrality level is then defined as the point separating the donor-like and acceptor-like levels. The charge neutrality level tends to align with the Fermi energy of the electrode as a result of charge transfer between the electrode and the induced gap states, forming an interface dipole. Based on this mechanism GaMnAs with very high hole concentration (~$10^{20}/cm^3$) induces the gap states at GaMnAs/NPB interface. As the charge neutrality level of NPB (4.1 eV – 4.2 eV [24,25]) is much smaller than the work function of

Journal of Physics: Conference Series **193** (2009) 012105 doi:10.1088/1742-6596/193/1/012105

GaMnAs (5.27 eV), it is therefore believed that charge transfer occurs between GaMnAs and NPB interface gap states, and an interface dipole is formed, as is shown in figure 3.

## 5. Conclusion

Interface formation between GaMnAs and NPB is studied by the *J-V* measurements. The hole injection barrier and vacuum level shift at GaMnAs and NPB interface were estimated to be 0.77 eV and 0.54 eV, respectively, by fitting the measured *J-V* curves. The band alignment diagram between GaMnAs and NPB is constructed based on the measurement and the calculation. We ascribed the vacuum level shift to charge transfer between GaMnAs and NPB interface gap states. This study should serve as a starting point for the investigation of GaMnAs based organic spin devices.

**Acknowledgement:** The authors are grateful to the financial support from Research Grant Council of Hong Kong SAR government via grant number N_HKUST611/05.

## References

[1] Tang C W and VanSlyke S A 1987 *Appl. Phys. Lett.* **51** 913
[2] Dodabalapur A, Torsi L, and Katz H E 1995 *Science* **268** 270
[3] Naber W J M, Faez S, and van de Wiel W G 2007 *J. Phys. D: Appl. Phys.* **40** R205
[4] Xiong Z H, Wu D, Valy Vardeny Z, and Shi J 2004 *Nature* **427** 821
[5] Yunus M, Ruden P P, and Smith D L 2008 *J. Appl.Phys* **103** 103714
[6] Wang F J, Yang C G, Valy Vardeny Z, and Li X G 2007 *Phys.Rev.B* **75** 245324
[7] Dediu V, Murgia M, Matacotta F C, Taliani C, and Barbanera S 2002 *Solid State Commun.* **122** 181
[8] Sapega V F, Moreno M, Ramsteiner M, Däweritz L, and Ploog K H 2005 *Phys. Rev. Lett.* **94** 137401
[9] Ohno Y, Young D K, Beschoten B, Matsukura F, Ohno H, and Awschalom D D 1999 *Nature* **402** 790
[10] Ji C J, Cao X C, Han Q F, Qiu K, Zhong F, Li X H, He H T, Wang J N, and Wang Y Q 2007 *Appl. Phys. Lett.* **90** 232501
[11] Ji C J, He H T, Cao X C, Qiu K, Zhong F, Li X H, Han Q F, Xu F Q, Wang J N, and Wang Y Q 2007 *Europhys. Lett.* **78** 57006
[12] Kang M G, Sa S H, Park H H, Suh K S, and Oh K H 1997 *Thin Solid Films* **308-309** 634
[13] Lu Z H, Chatenoud F, Dion M M, Graham M J, Ruda H E, Koutzarov I, Liu Q, Mitchell C E J, Hill I G, and Mclean A B 1965 *Appl. Phys Lett.* **67** 670.
[14] Walker A B, Kambili A, and Martin S J 2002 *J.Phys.: Condens.Matter* **14** 9825
[15] Davids P S, Campbell I H, and Smith D L 1997 *J.Appl.Phys.* **82** 6319
[16] Blom P W M, de Jong M J M, and van Munster M G 1997 *Phys.Rev.B* **55** R656
[17] Schwartz G, Ke T H, Wu C C, Walzer K, and Leo K 2008 *Appl Phys Lett.* **93** 073304
[18] Xie Z T, Ding B F, Gao X D, You Y T, Sun Z Y, Zhang W H, Ding X M, and Hou X Y 2009 *J.Appl.Phys.* **105** 106105
[19] Jungwirth T, Sinova J, Macdonald A H, Gallagher B L, Novák V, Edmonds K W, Rushforth A W, Campion R P, Foxon C T, Eaves L, Olejník E, Mašek J, Yang S -R Eric, Wunderlich J, Gould C, Molenkamp L W, Dietl T, and Ohno H, 2007 *Phys. Rev. B* **76** 125206
[20] Chassé T, Wu C –I, Hill I G, and Kahn A 1999 *J Appl. Phys.* **85** 6589
[21] Kahn A, Koch N, and Gao W 2003 *J. Polym. Sci., Part B: Polym. Phys.* **41** 2529
[22] Unpublished results
[23] Ishii H, Sugiyama K, Ito E, and Seki K 1999 *Adv. Mater.* **11** 605
[24] Vázquez H, Flores F, and Kahn A 2007 *Organic Electronics* **8** 241
[25] Helander M G, Wang Z B, Qiu J, and Lu Z H 2008 *Appl. Phys. Lett.* **93** 193310
[26] Hirose Y, Chen W, Haskal E I, Forrest S R, and Kahn A 1994 *Appl. Phys. Lett.* **64** 3482

# Improvement of carrier transport and luminous efficiency of organic light emitting diodes by introducing a co-deposited active layer

**Naoki Ohtani, Masaya Murata, Keiichiro Kashiwabara and Kazunori Kurata**
Department of Electronics, Doshisha University,
3-1 Tatara-Miyakodani, Kyotanabe-shi, Kyoto 610-0321, Japan

E-mail: ohtani@mail.doshisha.ac.jp

**Abstract.** We evaluated carrier transport and luminous efficiency of organic light-emitting diodes (OLEDs) whose active regions consist of a single co-deposited layer. One organic material is a hole transport material N,N'-Bis(3-methylphenyl)-N,N'-diphenylbenzidine (TPD), while the other is an electron transport/emissive material Tris(8-hydroxyquinolinato)-aluminum (Alq3). It was found that the luminous efficiency strongly depends on the thickness and the ratio of the TPD:Alq3 co-deposited layer. This indicates that the carrier balance in the active region can be improved by changing the co-deposited layers. In addition, we performed the dye-doping method to clarify the recombination region. As a result, we found that the radiative recombination is caused in the whole TPD:Alq3 co-deposited layer.

## 1. Introduction

Recently, organic light-emitting diodes (OLEDs) have attracted much interest due to their application in flat-panel displays and illumination light sources [1,2]. OLEDs generally consist of hetero-structures involving three or more organic layers, which work as electron/hole transport and emissive layers [3], and the characteristics of OLEDs are controllable by changing the layer construction [4]. We have investigated the influence of introducing co-deposited layers in the active region on the carrier transport and luminous efficiency of OLEDs. We found that luminous efficiency of OLEDs can be improved by introducing the co-deposited layer in the active region [5].

In this paper, we investigated the current-voltage (I-V) characteristics and electroluminescence (EL) properties of various samples in which the thicknesses and compound ratios of the co-deposited layers are different. In addition, we performed δ-function-like dye-doping method to clarify the radiative recombination region [6]. This research leads to the answer the relation between the luminous efficiency and the recombination region in OLEDs.

## 2. Experimental and sample structures

We fabricated OLED samples consisting of a single co-deposited layer consisting of two organic materials; one is a hole transport material N,N'-Bis(3-methylphenyl)-N,N'-diphenylbenzidine (TPD) and the other is an electron transport/emissive material Tris(8-hydroxyquinolinato)-aluminum (Alq3). The indium-tin-oxide (ITO) coated glass substrates were ultrasonically cleaned in acetone, ethanol and isopropyl alcohol for 10 minutes, respectively. The organic materials were vacuum deposited under a

pressure of $1\times10^{-4}$ Pa on ITO-coated glass substrates. The organic layers consist of two organic materials; one is TPD as a hole transport material, and the other is as an electron transport/emissive material Alp3. In the co-deposition process, two organic sources were independently controlled. The compound ratio of TPD:Alq3 co-deposited layers were precisely controlled by the deposition rates using a thickness monitor. Rubrene was additionally doped in some samples for the dye-doping method. Finally, the aluminum (Al) layer was vacuum deposited as cathode metal. The active area of sample was 2×2 $mm^2$. We fabricated six types of samples as follows;
A) ITO / TPD (50 nm) / Alq3 (50 nm) / Al,
B) ITO / TPD:Alq3=x:y (100 nm) / Al,
C) ITO / TPD (x nm) / TPD:Alq3=1:1 (100-2x nm) / Alq3 (x nm) / Al,
D) ITO / TPD (50 nm) / Alq3+Rubrene [2 %] (50 nm) / Al,
E) ITO / TPD (30 nm) / TPD:Alq3=1:1+Rubrene [2 %] (40 nm) / Alq3 (30 nm) / Al,
F) ITO / TPD (30 nm) / TPD:Alq3=1:1 (x nm) / TPD:Alq3+Rubrene [2 %] (5 nm) / TPD:Alq3=1:1 (35-x nm) / Alq3 (30 nm) / Al.
Note that the thicknesses of the total organic layers are the same as to be 100 nm. Figure 1 shows schematic illustrations of Rubrene-doped three sample structures; samples D, E, and F. Please note that we fabricated three different type of sample F, in which the position of the δ-function-like Rubrene doping is different to clarify the radiative recombination position in TPD:Alq3 co-deposited layer. Each δ-doping layer thickness is 5 nm.

I-V characteristics were measured by using a semiconductor parameter analyzer (4155C, Agilent Technologies). EL spectra were measured by using a CCD spectrometer (USB2000, Ocean Optics). Luminous efficiencies were measured by using an integrating sphere. All measurements were performed in air and at room temperature.

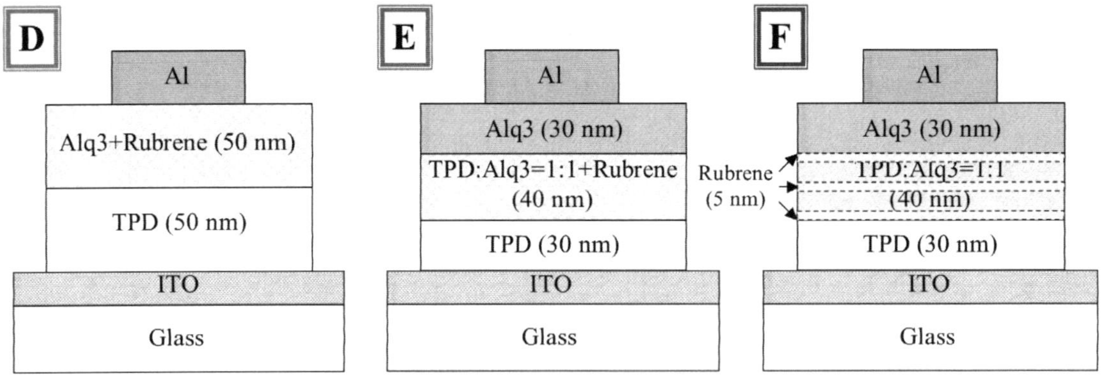


**Figure 1.** Schematic illustrations of Rubrene-doped three samples.

## 3. Results and discussion

### *3.1 External luminous efficiencies in samples A, B, and C*

In the comparison of samples A, B and C, we have found that the brightest EL signal was observed in sample C. Thus, we concluded that the carrier balance was improved in sample C whose active region consists of TPD:Alq3 co-deposited layer sandwiched between TPD and Alq3 layers [5]. In this paper, we evaluated the external luminous efficiencies $\eta_{ex}$ of samples A and C by using an integral sphere. The external luminous efficiency of sample A is about 0.89 %. On the other hand, $\eta_{ex}$ of sample C strongly depends on the layer thickness of TPD:Alq3 co-deposited layer. We found that $\eta_{ex}$ of sample C becomes greater by narrowing the TPD:Alq3 co-deposited layer. The maximum $\eta_{ex}$ of sample C is

2.02 % when the thickness of TPD:Alq3 co-deposited layer is 5 nm. However, $\eta_{ex}$ of sample C becomes smaller than that of sample A, when the thickness of TPD:Alq3 co-deposited layer is wider than 50 nm. These results indicate that the mobility of carriers in the TPD:Alq3 co-deposited layer is much slower than that in TPD layer and Alq3 layer, resulting in the smaller luminous efficiency in wider TPD:Alq3 co-deposited layer.

*3.2 Identification of the radiative recombination position in the co-deposited layer*
The starting voltage of light emission of sample C is about 12 V, which is independent of the thickness of TPD:Alq3 co-deposited layer. This means that the radiative recombination is most likely caused at the same position in the co-deposited layer. Thus, we have evaluated the position of the radiative recombination by using the dye-doping method [6].

Figure 2 shows normalized EL spectra of samples D, E, and F. Note that EL spectra of samples D and E are almost equal. In addition, EL spectra of samples F [x=17.5] and [x=35] are also equal. The peaking EL wavelengths of Alq3 and Rubrene are 520 and 560 nm, respectively. Thus, the EL spectra of samples D and E originate from only Rubrene. This indicates that the Rubrene was uniformly doped in the Alq3 layers of samples D and E, resulting in the carrier relaxation into the lower bandgap material Rubrene. On the other hand, the three EL spectra of sample F originate from the combination of Alq3 and Rubrene. This means that the contribution of Rubrene to EL emission is not independent of the position of δ-function-like Rubrene doping in the TPD:Alq3 co-deposited layer. This result clearly indicates that the radiative recombination occurs in the whole TPD:Alq3 co-deposited layer.

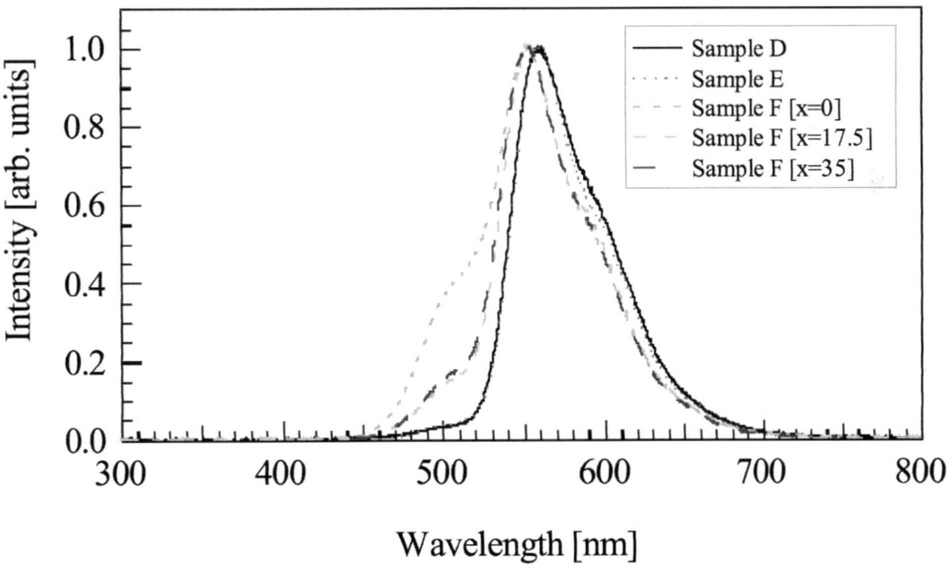


**Figure 2.** EL spectra of samples D, E, and F

*3.3 Radiative recombination and carrier transport process in sample C*
Figure 3 shows schematic energy band diagrams and carrier transport processes at the interface of TPD and Alq3: one has TPD:Alq3 co-deposited layer corresponding to sample C, the other has TPD:Alq3 interface corresponding to sample A. In the case of sample A, the interface of TPD and Alq3 plays a role of the obstruction of hole transport, resulting in the accumulation of holes at the interface. In the TPD and Alq3 heterostructure system, hole transport is more important because the mobility of hole in TPD is much faster than that of electron in Alq3.

Journal of Physics: Conference Series **193** (2009) 012106 doi:10.1088/1742-6596/193/1/012106

In the case of sample C, on the other hand, the TPD:Alq3 co-deposited layer makes the interface unclear, resulting in the improvement of hole transport. Holes can transfer into the whole TPD:Alq3 co-deposited layer then radiatively recombine with electrons. This leads to the lower starting voltage of light emission as well as the higher external luminous efficiency of sample C [5]. This carrier transport model can easily explain the TPD:Alq3 co-deposited layer thickness dependence on the external luminous efficiency of sample C. The narrower thickness of TPD:Alq3 co-deposited layer can increase the carrier concentration, thus the external luminous efficiency is improved.

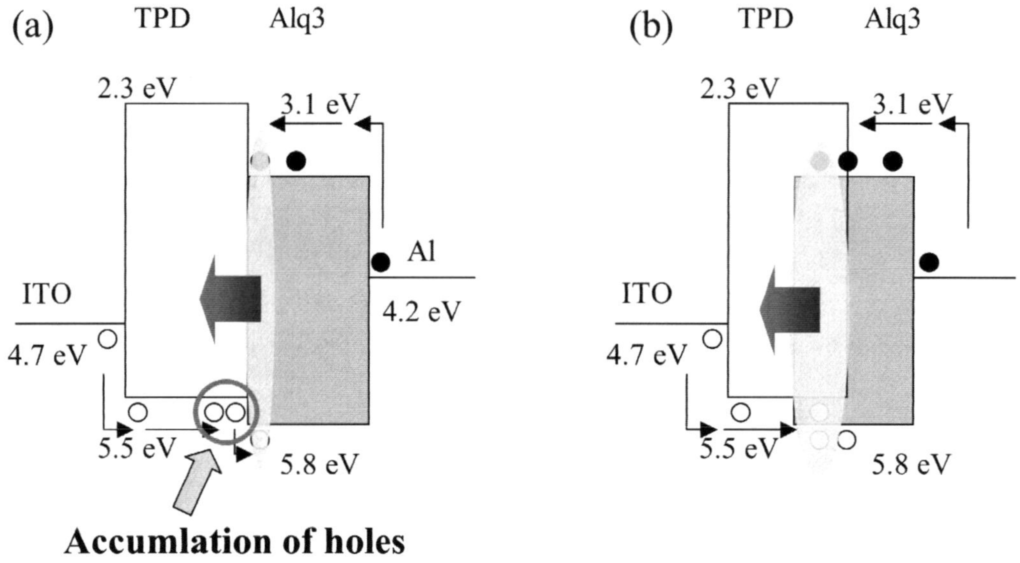


**Figure 3.** Schemetic energy band diagrams and carrier transport processes. (a) Sample A having TPD:Alq3 interface, (b) sample C having TPD:Alq3 co-deposited layer.

**4. Conclusion**

We have investigated the radiative mechanism in organic light-emitting diodes consisting a co-deposited layer in the active region by using the dye-doping method. We found that the radiative recombination occurs in the whole co-deposited layer. The external luminous efficiency becomes larger in the sample consisting the narrower co-deposited layer. This is most likely caused by the increased carrier concentration due to the narrower co-deposited layer.

**References**

[1] Tang C. W and VanSlyke S. A 1987 *Appl. Phys. Lett.* **51** 913
[2] Xu Q, et al., *Appl. Phys. Lett.* 2003 **83** 4695
[3] Tang C.W VanSlyke S. A and Chen C. H 1989 *J. Appl. Phys.* **65** 3610
[4] Matsushima T and Adachi C 2007 *Jpn. J. Appl. Phys.* **46** L861
[5] Murata M, Yamamoto T, Haishi M, Ando T and Ohtani N 2009 *phys. stat. sol. (c)* **6** 330
[6] Aziz H and Popovic Z. D 2002 *Appl. Phys. Lett.* **80** 2180

# Detecting conformational change by current transport in proteins: the case of bacteriorhodopsin monolayers

**E. Alfinito and L. Reggiani**

Dipartimento di Ingegneria dell'innovazione. Università del Salento, via Monteroni, I-73100 Lecce, Italy, EU
CNISM - via della Vasca Navale, 84, I-00146, Roma, Italy, EU

E-mail: eleonora.alfinito@unisalento.it
lino.reggiani@unisalento.it

**Abstract.** Recent experiments on the light receptor bacteriorhodopsin have revealed the protein conductive properties and connected them to its sensing action. In particular it was shown that the super-Ohmic I-V characteristic acquired in dark, changes in the presence of green light, with an enhancement of current at increasing bias values. Here we propose a current transport model for proteins able to reproduce experimental data, mainly the dependence of current on their three dimensional (tertiary) structure. The model is based on a resistance network model and implements a tunneling mechanism of charge transfer between the amino-acids constituting the protein.

## 1. Introduction

Charge transport in biological materials is a topic receiving much attention both from a theoretical and an experimental point of view [1]. In doing so, new techniques have been developed, both to perform measurements and implement theoretical models [2, 3, 4]. In this respect, particular attention was devoted to proteins, which are quite interesting for their function-topology correlation. Among proteins, recently some experiments have investigated the charge transport characteristics of the light receptor bacteriorhodopsin (bR). This protein is present in the purple membrane of *Halobacterium salinarum*, which is organized as a 2D crystal lattice with a thickness of about 5 nm (the protein size). Furthermore, monolayers of bR are enough easy to be manipulated for experimental testing [3] being very stable under external stress. In particular, current voltage (I-V) characteristics were explored with two different techniques: first with an *electrode-bilayer-electrode* structure [5], and second within a conductive AFM device [6]. In the former measurements, the light-dependent conductivity was also explored. In the latter measurements, the transition between a direct and an injection tunneling regime was analyzed. Both experiments provided interesting hints on the protein conductivity. First, the main result of both the experiments is that a charge transport is detectable and that the current voltgae characteristic exhibits a significant super-linear behaviour. The common interpretation of this behaviour is that charge transport is controlled by a tunneling mechanism [5]. Furthermore, the light dependence of the I-V characteristics reveals an enhancement of current in the presence of light, with respect to the dark case, for a fixed bias value. These

Journal of Physics: Conference Series **193** (2009) 012107 doi:10.1088/1742-6596/193/1/012107

results make bR a very intriguing material for investigation also from the theoretical point of view. In this paper we present a theoretical model able to reproduce the main features of the experiments with the aim of capturing the essence of the conduction-topology correlation in proteins.

## 2. Model

We produce a protein backbone by mapping its 3D structure into a graph whose nodes correspond to the protein amino-acids. The position of each amino-acid is taken as that of its C$\alpha$ atom, as given by the protein data bank [7]. Finally, two nodes are connected with a link only if their distance is less than an assigned cut-offradius $R$. The value of $R$ has the role of an interaction radius, and the graph is an interaction map, naturally reflecting the protein topology. Finally we fix the kind of interaction by interpreting each link as a privileged channel of charge transfer [4, 8], thus attributing to it a finite resistance $r_{i,j} = \rho\, l_{i,j}/\mathcal{A}_{i,j}$, where $\rho$ indicates the resistivity, taken the same for all the amino-acids, the pedices $i, j$ refer to the amino-acids between which the link is stretched, $l_{i,j}$ is the link length and $\mathcal{A}_{i,j}$ is the cross-sectional area shared by the amino-acids: $\mathcal{A}_{i,j} = \pi\left( R^2 - l_{i,j}^2/4\right)$. Accordingly, the graph becomes a resistance network. By taking appropriate contacts and applying an external bias, the electrical network is solved within a linear circuit analysis.

### *2.1. Direct tunneling*

To account for the super-Ohmic I-V characteristics we implement a tunneling mechanism of charge transfer between amino-acids as follows. The value of the link resistivity is taken to assume two different values, $\rho_{MAX}$ which corresponds to a nearly insulating material, and $\rho_{min} \ll \rho_{MAX}$. The possibility for the link to switch from the high to the low resistivity is linked to a voltage driven tunneling mechanism through a rectangular potential barrier. For low applied voltages, we adopt the JWKB approximation for quantum tunneling, i.e. the transition probability $\mathcal{P}_{i,j}$

$$\mathcal{P}_{i,j} = \exp\left[-\frac{2l_{i,j}}{\hbar}\sqrt{2m(\Phi - \beta e V_{i,j})}\right] \quad , \qquad \Phi > eV_{i,j} \; ; \tag{1}$$

where $V_{i,j}$ is the potential drop between the couple of $i, j$ amino-acids, $\Phi$ is the potential barrier and $\beta$ a numerical parameter accounting for the geometrical shape of the barrier, here taken as $\beta = 1$. For$\Phi$ $\leq eV_{i,j}$ $\mathcal{P}_{i,j} = 1$, i.e. all the charges are transmitted through the barrier, and this condition determines the value of$\Phi$.

In this way, charge transport across the protein is modelled as a *sequential* direct tunneling mechanism over distances of a few Å, which we believe to be more feasible than a single tunneling across the 5 nm entire protein. Furthermore, the model is able to produce different responses with respect to different conformations, i.e. different illumination conditions.

### *2.2. Fowler-Nordheim tunneling*

In the presence of intense applied voltages, the approximation of rectangular potential barrier is no longer valid, and the barrier is better described by a triangular or round triangular shape. The semi-classical approach to this problem furnishes the Fowler-Nordheim (FN) tunneling probability [9, 6]:

$$\mathcal{P}_{ij}^{FN} = \exp\left[-\left(\frac{2l_{i,j}\sqrt{2m}}{\hbar}\right)\frac{\Phi}{eV_{i,j}}\sqrt{\frac{\Phi}{2}}\right] \quad , \qquad \Phi \leq eV_{i,j} \; ; \tag{2}$$

which leads to a $\ln(I/V^2) \propto 1/V$ behaviour. At low bias, the direct tunneling probability, Eq. (1) can be also used [9]. In this case the value of $\beta$ which provides the two regime matching is $\beta = 1/2$ (triangular barrier). Notice that this choice leads to an overestimation of $\Phi$with respect to the rectangular shape. As a matter of fact, the condition $eV_{i,j} = \Phi$ only separates the direct from the FN regime, while the direct probability goes to unity for $eV_{i,j} = 2\Phi$. In this model, the continuous transition from direct to FN tunneling is reproduced by the relation:

$$\rho(V) = \begin{cases} \rho_{MAX} & \Phi > eV \\ \rho_{MAX}(\frac{\Phi}{eV}) + \rho_{min}(1 - \frac{\Phi}{eV}) & \Phi \leq eV \end{cases} \tag{3}$$

Accordingly, in the presence of a FN tunneling, the resistance link takes the $\rho_{min}$ value, otherwise, it takes the $\rho(V)$ value.

### *2.3. Materials*

To model bacteriorhodopsin we have taken the protein data base (PDB) entries 2NTU and 2NTW which label the native (in dark) and activated (in green light) states, respectively [7]. These are not the only protein representations given in PDB but, being obtained under the same experimental conditions [8], they are considered as the most reliable. In any case, the model was tested also on protein representations in different activation states, like those labelled 1M0K-1M0M, and the results were found to be compatible with those reported here.

## 3. Results

Figure (1) reports the comparison between experimental data [5] and theoretical calculations performed with two different barrier heights $\Phi$ = 59 meV and 53 meV, respectively. Because of the symmetry of simulated data with respect to the transformation V→ -V, we report the 59 meV data for positive bias and the 53 meV data for negative bias. Despite of the small difference in the barrier values, calculations produce significant different resolutions at V=1 V, which the maximal current response given by experiments corresponds to. In particular, Fig. (1) shows a larger resolution for the smaller $\Phi$ value, while the behaviour of the native state is better reproduced by the larger $\Phi$ value. Furthermore, data reported in Ref. [5] suggest large error bars on experimental curves (up to 50 %), so that our estimation of the barrier height has to be considered as indicative [4].

Since charge transport in disordered organic materials can be described as due not to a single but to a Gaussian distribution of potential barriers [1, 10], we have calculated the I-V characteristics by using a Gaussian distributed $\Phi$ values within an average value of 59 meV and two dispersion $\sigma$ values: $\sigma$ = 19 meV, 94 meV, respectively. The results are reported in Fig. (2) where we can appreciate an effective improvement of the fit for the large dispersion. Finally, we have also considered the case of a high voltage where the FN tunneling becomes dominant, as reported in Ref. [6]. In doing so, as fitting parameter we have found a potential barrier higher than in the previous case, $\Phi = 194 \pm 25$ meV. This larger value of $\Phi$ is mainly due to the choice of a triangular barrier and also the difficulty to compare the results of two, so intrinsically different, experiments [5, 6]. Figure 3 reports the fitting data of experiments [6] and also the calculated values for the representation of the activated state 2NTW. In the inset we can appreciate the current enhancement in the range 0-1 V, which is the same of that observed in ref.[5], although the absolute value of the currents is lowered by a factor of three. The origin of this discrepancy is attributed to the different experimental procedures.

## 4. Conclusions

We have presented a resistor network model to investigate the correlation between a conformational change and the variation in electrical transport of a sensing protein. To this purpose we have investigated the change of the static I-V characteristics. The theory is validated

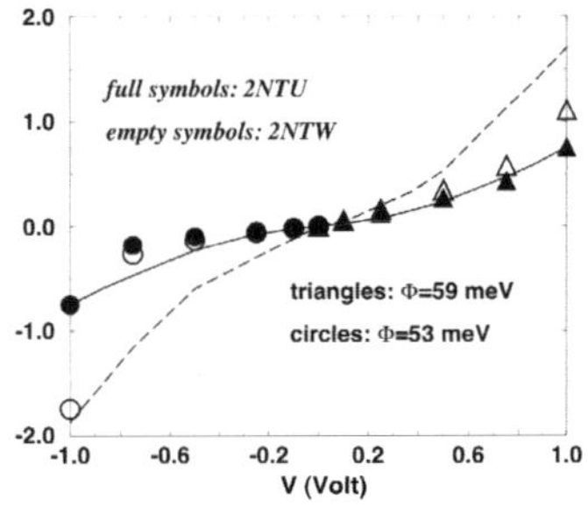


**Figure 1.** I-V characteristic for bR in the native (2NTU) and activated (2NTW) states. Calculations are carried out with two different barrier heights,$\Phi$ =59 meV (triangles) and$\Phi$ =53 meV (circles). The continuous and dashed lines report the experimental data in the native and activated state, respectively.

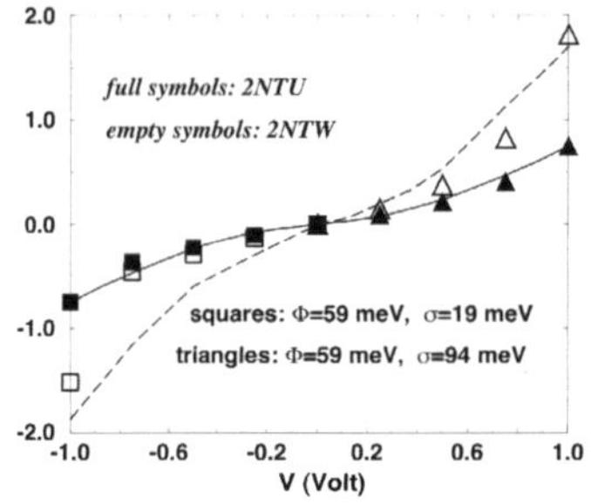


**Figure 2.** I-V characteristic for bR in the native (2NTU) and activated (2NTW) states. Data were calculated by using a Gaussian distribution of barrier heights. The continuous and dashed lines report the experimental data in the native and activated state, respectively.

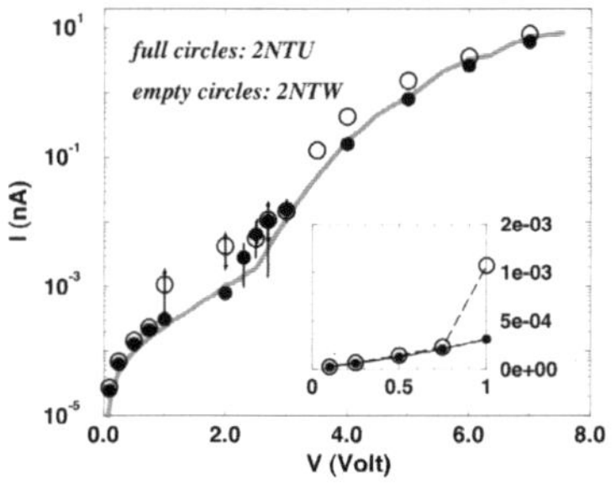


**Figure 3.** I-V characteristic for bR in a wide bias range. Calculations include the FN tunneling regime with a barrier height $\Phi = 194 \pm 25$ meV. Full circles refer to the native state, while empty circles refer to the activated state.

on recent experiments carried out on layered structures of bR, a light-activated protein, which evidences a substantial increase of the current, at a given voltage, when passing from dark to the presence of a green light. Even if remaining mostly qualitative, the best agreement with experimental data is obtained when considering a maximum interaction radius of 6 Å and an average barrier energy of 59 meV with a Gaussian distribution. Furthermore, taking advantage of its flexibility, the model is implemented by introducing the FN tunnelling and shown able to reproduce satisfactorily the experimental data of [6], but with a barrier energy of about 194 meV, significantly higher than that of direct tunneling through a rectangular barrier. At present this discrepancy remains an open problem which needs further investigation.

## Acknowledgments
This research is carried out within the BOND project under evaluation at the CE within the 7th Program, grant agreement:228685-2.

[1] Baranovski S (Ed) 2006 *Charge transport in disordered solids* ( The Atrium, Southern Gate, Chichester, West Sussex PO19 8SQ, England, John Wiley & Sons Ltd.)
[2] Andolfi L, Bizzarri A R and Cannistraro S 2006 *Appl. Phys. Lett.* **89** 183125
[3] Hou Y, *et al* 2007 *Biosensors & Bioelectronics* **22** 1550
[4] Alfinito E and Reggiani L 2009 *Europhys. Lett.* **85** 68002
[5] Jin Y D, Friedman N, Sheves M, He T and Cahen D 2006 *PNAS* **103** 8601; Jin Y D, Friedman N, Sheves M, and Cahen D 2007 *Adv. Funct. Mater.* **17** 1417
[6] Casuso I, Fumagalli L, Samitier J, Padrós E, Reggiani L, Akimov V and Gomila G 2007 *Nanotechnology* **18** 465503
[7] Berman H M, Westbrook J, Feng Z, Gilliland G, Bhat T N, Weissing H, Shindyalov I N and Bourne P 2000 *Nucleic Acids Research* **28** 235
[8] Alfinito E, Pennetta C and Reggiani L 2008 *Nanotechnology* **19** 065202
[9] Simmons J G 1963 *J. Appl. Phys.* **34** 1793
[10] Bässler H 1993 *Phys. Stat. Sol.(b)* **175** 15

Journal of Physics: Conference Series **193** (2009) 012108 doi:10.1088/1742-6596/193/1/012108

# Polaronic transport in polysilanes

**S Nešpůrek[1], J Nožár[1], A Kadashchuk[2] and I I Fishchuk[2]**
[1]Institute of Macromolecular Chemistry, Academy of Sciences of the Czech Republic, v.v.i., Heyrovský Sq. 2, 162 06 Prague 6, Czech Republic
[2] Institute of Physics, Academy of Sciences of Ukraine, Prospect Nauki 46, 03028 Kiev, Ukraine

E-mail: nespurek@imc.cas.cz

**Abstract.** Analytical theory based on the effective medium approach is used to explain electric field dependences of charge carrier mobility in poly[methyl(phenyl)silylene]. It is shown that the charge transport can be described by a small polaron movement in the presence of correlated energetic disorder. Polaron binding energy is the main part of the activation energy of the mobility.

## 1. Introduction

Amorphous silicon is a member of the larger class of silicon backbone solids which, in addition to crystalline phase, also contain silicon clusters, silane oligomers, poly(siloxanes), and poly(organylsilanes). The last mentioned backbone polymers are under considerable research interest because of their unusual optical and electrical properties and possible models for the charge carrier transport in one-dimmensional silicon structures, where quantum dots can be created [1,2].

Charge carrier transport in silicon backbone polymer, poly[methyl(phenyl)silylene] (PMPSi), was discussed by Kepler et al. [3] and by Abkowitz et al. [4]. The experimental results were interpreted in the frame of charge hopping between backbone-derived localized states. These states were associated with domain-like suborganization of the silicon backbone into all-trans segments of approximately 10 – 20 repeat units separated by conformationally disordered regions. An alternative explanation [5] was done on the basis of intragap states which spontaneously arise, even in a perfectly ordered chain, from bond order polaron formation. This is a consequence of the coupling of charge carriers to the quasi-one-dimensional silicon backbone.

Several values of the polaron binding energy is mentioned in literature for PMPSi. In this paper we summarize the published methods suitable for the determination of this energy with some criticism of their use. The second part of the paper deals with our new methods for the polaron detection and the determination of the polaron binding energy.

## 2. Experimental

Poly[methyl(phenyl)silylene] (PMPSi) was prepared by Wurtz coupling polymerization, as described by Zhang and West [6]. The low-molecular-weight fractions were extracted with boiling diethyl ether. The residual polymer, obtained in c. 17 % yield, possessed a unimodal but broad molar mass distribution, $M_W = 4 \times 10^4$ g mol$^{-1}$.

Journal of Physics: Conference Series **193** (2009) 012108 doi:10.1088/1742-6596/193/1/012108

Thin films (thickness from 1 to 3 μm) were prepared from a toluene solution by casting on conductive ITO glasses. The top Al or Au electrodes, 40 – 60 nm thick, were prepared by the vacuum evaporation. Before the deposition, the polymer was three times purified by precipitation in methanol and toluene solution and centrifuged (12 000 rpm, 15 min). After the deposition the films were dried in the vacuum of 0.1 Pa at 330 K for at least 4 hours. The film thicknesses were determined by surfometric measurements.

Charge carrier mobility was measured by a time-of-flight method using an electrical circuit consisting of voltage source, sample and oscilloscope (HP 54510A, 50 Ω input impedance) connected in series. The samples were illuminated by the 347 nm laser pulses (duration 20 ns) generated by a ruby laser (Korad model K1QS2) in conjunction with a frequency doubler through the transparent ITO electrode. The penetration depth of the light was 0.13 μm which represents less than 4 % of the sample thickness. The samples were kept in a vacuum cryostat ($10^{-4}$ Pa) or under argon during the measurement. In both cases the same experimental results were obtained.

## 3. Charge carrier transport

The previous experimental results mentioned in literature [4] suggest that in 3D samples the charge carrier transport is realized by hopping through sites distributed both in energy and space [7]. Tails, observed in the current transients (time-of-flight method), suggest the dispersive character of the transport [8].

Recently, Kim et al. [9], from the measurements of the velocity of the excitation energy along the PMPSi chain, have found the excitations are carrying significant distortions of the backbone. One possibility of explaining this behaviour is to use disordered polaronic theory [10]. In this case the activation energy of the charge mobility has contributions both from the dynamic disorder, i.e., the polaronic barrier, and from the static disorder, i.e., the variation of the energy as a result of the environment. Thus, in the first approximation, one can write for the activation energy of the charge carrier mobility the relation [11]

$$E_a^{eff} = E_a^{pol} + E_a^{dis} = \frac{E_p}{2} + \frac{8}{9}\frac{\sigma^2}{kT} \tag{1}$$

where $E_a^{pol}$ and $E_a^{dis}$ represents polaronic and disorder contributions and $E_p$ is the polaron binding energy.

We proposed two methods for the polaron detection in polymeric materials:
(i) The first method, published recently [12], is based on the shift of the low temperature thermostimulated photoluminescence (TSPL) peak to higher temperatures after the additional infrared illumination. The idea is based on the following principle. The initial energy distribution of polaron was formed in the course of low-temperature energy relaxation. The polaron formation strongly affects both low-temperature energy relaxation and thermally assisted hopping of charge carriers in disordered material. If the polaron binding energy is comparable with or larger than the density-of-states (DOS) width, the energetically downward hopping is strongly suppressed by the lack of deeper vacant sites, which are only accessible for further jumps at low temperatures. Therefore, after a fast initial polaron relaxation, the energy distribution of polarons remains almost "frozen in" over the entire dwell time between low-temperature photoexcitation of the sample and the onset of the TSPL heating run. The polaronic contribution to the activation energy of every individual upward jump is equal to a half of the polaron binding energy, $E_p/2$, while the low-temperature energy relaxation of polarons within the DOS distribution is slowed down because the polaron energy distribution is shifted down by $E_p$. IR irradiation causes a local heating in the environment of charged molecules and at low temperatures efficiently stimulates energy relaxation towards the deeper tail states of polarons. Concomitantly, charge carriers get a possibility to occupy those tail states, which would be accessible

for them when there were no polaron effects. This results in the shift of the TSPL peak towards higher temperatures and in a decrease in its intensity.
(ii) We present here a new method based on the electric field ($F$) and temperature ($T$) dependencies of the charge carrier mobility ($\mu$). The $\ln(\mu)$ versus $\sqrt{F}$ dependence is linear. From detail calculations [13] follows that Poole-Frenkel factor $\alpha = \partial \ln(\mu/\mu_0)/\partial \sqrt{F}$ varies almost linearly with $T^{-3/2}$ in the case of polaron-free transport. Alternatively, this dependence deviates significantly from a straight line in the case of polaron model.

The polaron binding energy in PMPSi was estimated by Bässler et al. as $E_p = 0.16$ eV [11] by the linear extrapolation of the temperature dependence of the activation energy of the charge carrier mobility to zero temperature. Pan et al. [14] obtained the value of $E_p = 0.08$ eV for exciton-polarons. Regarding to the fact that in molecular crystals the defects of the same feature form about 3 times deeper traps for charge carriers than for singlet excitons [15], one can write for the charge-polaron binding energy $E_p = 0.24$ eV. Even if the polaron binding energy is higher than the width of the density of states, the disorder seems to be still an important contributor to the temperature and field dependencies of the mobility in macroscopic systems where inter- and intramolecular hopping takes place.

Let us put some comments to the facts mentioned above:
(i) The plot of the dependence of the charge carrier mobility on temperature is not usually linear. Thus, $E_p$ value is only the rough estimation.
(ii) The simplified jump rate equation used in Ref. [11] for polaron hopping transport has no theoretical justification. Paris et al [16] demonstrated that the problem related to physical parameters can be solved assuming the small polaron transport occurs in the presence of correlated energetic disorder. In this case the coefficient (8/9) - see Eq. (1) - in the disorder term must be changed to (1/4) [13]. Discussion of the problem for general case of the Marcus jump rate is discussed in Ref. [13].
According to our effective medium approach for the symmetrical jump rate the mobility can be approximated by the expression [13]

$$\mu \approx \exp\left[-\frac{E_p}{2kT} - \frac{1}{8}\left(\frac{\sigma}{kT}\right)^2 + \left(\frac{\sigma}{kT} - 1\right)\sqrt{\frac{erF}{8kT}}\right], \tag{2}$$

where $e$ is the unit change, $F$ is the electric field strength and $r$ is the average distance between neighboring hopping sites of the disordered polaron transport. Equation 2 represents a good approximation for the electric field dependences of the charge carrier mobility.

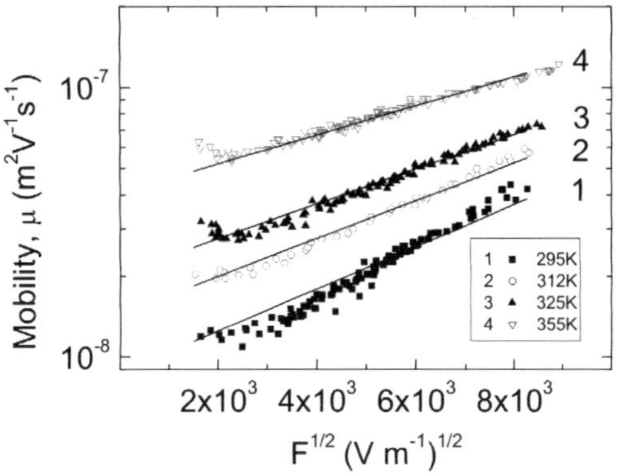


**Figure 1** Electric field dependences of the charge carrier mobility of PMPSi parametric in temperature; $T = 295$ K (curve 1), 312 K (curve 2), 325 K (curve 3), 355 (K) (curve 4). Lines represent the best common fit according to Eq. 2.

After the fitting the electric field dependencies (see Fig. 1) using Eq. 2, the following parameters were obtained: energy distribution parameter of hopping states $\sigma = 0.09$ eV, $E_p = 0.29$ eV, the most probable hopping distance $r = 1.2$ nm. The common fit for all dependencies is represented by straight lines in Fig. 1. From this result follows that the polaron binding energy is about two times higher (compare $E_p = 0.16$ eV vs. 0.28 eV). In the latter case the small polaron transport in the presence of correlated energetic disorder was assumed. Thus, the activation energy of the charge mobility is not only a simple sum of the contributions due to the polaron formation and energetic disorder. The contribution of the disorder to the activation energy of the mobility is only 26 % (0.076 eV).

**Acknowledgements**
This work was supported by the grants No. IAA100100622 and KAN400720701 from the Grant Agency of the Academy of Sciences of the Czech Republic.

[1] Nešpůrek S, Toman P and Sworakowski J 2003 *Thin Solid Films* **438-439** 268
[2] Nešpůrek S, Sworakowski J and Kadashchuk A 2001 *IEEE Trans. Diel. Electr. Insul.* **3** 432
[3] Kepler R G, Zeigler J M, Harrah L A and Kurtz S R 1982 *Phys. Rev. B* **35** 2819
[4] Abkowitz M A, Rice M J and Stolka M 1990 *J. Philos. Mag. B* **61** 25
[5] Rice M J and Phillpot S R 1987 *Phys. Rev. Lett.* **58** 937
[6] Zhang X-H and West R 1984 *J. Polym. Sci., Polym. Chem. Ed.* **22** 159
[7] Bässler H 1991 *J. Philos. Mag. B* **63** 587
[8] Nešpůrek S, Eckhardt A 2001 *Polym. Adv. Technol.* **12** 427
[9] Kim Y R, Lee M, Thorne J R , Hochstrasser R M and Zeigler J M 1988 *Chem. Phys. Lett.* **145** 75
[10] Kenkre V M and Dunlop D H 1992 *J. Philos. Mag. B* **65** 831
[11] Bässler H, Borsenberger P M and Perry R J 1994 *J. Polym. Sci. Polym. Phys.* **32** 1677
[12] Arkhipov V I, Emelianova E V, Kadashchuk A, Blonsky I, Nešpůrek S, Weiss D S and Bässler H 2002 *Phys. Rev. B* **65** 165218
[13] Fishchuk I I, Kadashchuk A, Bässler H and Nešpůrek S 2003 *Phys. Rev. B* **67** 224303
[14] Pan L, Zhang M and Nakayama Y 1999 *J. Chem. Phys.* **110** 10509
[15] Silinsh E A 1980 *Organic Molecular Crystals: Their Electronic States* (Berlin: Springer-Verlag)
[16] Paris P E, Kenkre V M and Dunlap D H 2001 *Phys. Rev. Lett.* **87** 126601

# Magnetic field effect on photocurrent in *N,N′*-di(naphthalene-1-yl)-*N,N′*-diphenyl-benzidine (NPB)

**B K Li[1], W J Chen[1], M K Lam[2], K W Cheah[2] and J N Wang[1]**
[1]Department of Physics, the Hong Kong University of Science and Technology, Clear Water Bay, Kowloon, Hong Kong, China
[2]Center of Advanced Luminance Materials, Department of Physics, Hong Kong Baptist University, Hong Kong, China

E-mail: libk@ust.hk, wenjin@ust.hk, s164367@hkbu.edu.hk, kwcheah@hkbu.edu.hk, phjwang@ust.hk

**Abstract.** We studied organic magnetoresistance (OMAR) and magneto-photocurrent (MPC) in ITO/*N,N′*-di(naphthalene-1-yl)-*N,N′*-diphenyl-benzidine (NPB)/Al sandwich devices with NPB thicknesses of 50 nm. Without illumination, no OMAR is observable. While the MPC curves are found to have different line shapes containing a low field and a high field component, depending on excitation photon energy and applied bias. The analysis of experimental results indicate that: (i) the magnetic field effects in NPB is associated with excitonic states, and (ii) the dissociation of singlet excitons is associated to the high field component and triplet-exciton-charge interaction is to the low field component in MPC curves.

## 1. Introduction

Organic magnetoresistance (OMAR), observed in organic semiconducting materials (OSE), is receiving increasing interest in recent years [1-9]. Due to its large value and high sensitivity, up to 10% in a relatively small magnetic field (~50 mT) at room temperature, OMAR possesses potential utilization in "plastic" electronics, such as optoelectronic devices controlled by magnetic field, magnetic sensors, and even in organic spintronics [10]. Besides helping in the development of potential applications, understanding the mechanism the OMAR will not only improve our understanding of charge transport in OSE, but also, more generally, magnetism, since all the components of the device are nonmagnetic.

OMAR can be positive or negative depending on the temperature, applied bias and organic layer thickness [5, 6], and it can also be enhanced by electrical conditioning [7]. Most of the reported OMAR versus field follow empirical laws given by either the non-Lorentzian line shape ($B^2/(|B|+B_0)^2$) or a Lorentzian ($B^2/(B^2+B_0^2)$), depending on materials [1], where characteristic field width $B_0$ is small ~10 mT. Three models based on hyperfine interaction have been proposed to explain the observed OMAR. Two of them are excitonic, in which the magnetic field affects the inter-conversion between singlet and triplet excitons [2, 6], or the triplet-exciton-charge interaction [3]. The third one relies on magnetic field effects on bipolaron formation [4]. Obviously the excitonic model is based on the existence of excitons and the bipolaron model predicts that OMAR to be observable in electron-only or hole-only devices as well.

Up to now, most of the OMAR studies are based on the current 'injection' from electrodes. However, the studies of magnetic field effect on current 'generation' from illumination (photocurrent) in organic semiconductors may provide some insights into the origin of OMAR from another point of view. In this work, we investigated both the OMAR and magneto-photocurrent (MPC) in ITO/*N,N'*-di(naphthalene-1-yl)-*N,N'*-diphenyl-benzidine (NPB)/Al sandwiched devices at room temperature with NPB thickness of 50 nm. Without illumination, no OMAR can be observed. In contrast, under illumination the magnetic field effect on conductance is significant. Excitation photon energy ($E_{ph}$), and applied bias ($V$) dependent MPC curves were measured and two types of MPC line shapes were found. We identify that the dissociation of photogenerated singlet excitons is a high field ($B_0 \sim 140$ mT) component, while the triplet-exciton-charge interaction is a low field ($B_0 \sim 8$ mT) one.

## 2. Experiments

The 3mm X 3mm devices were fabricated on ITO patterned glass substrates. The substrate was rigorously cleaned before evaporation of 50 nm NPB active layer and the 120 nm Al electrode in a standard vacuum chamber with a base vacuum of $10^{-6}$ Torr. After fabrication of the sandwiched structures, the devices were transferred to an inert atmosphere glove box and packaged. A xenon lamp together with a Spex1681 monochromator was used as light source to study the photocurrent spectra and MPC using a standard lock-in technique. The current-voltage ($I$-$V$) characteristics were measured by a HP 4155A semiconductor parameter analyzer, which was also used as a voltage source to measure the OMAR and MPC under different applied bias. MPC is defined as the change in photocurrent as a function of magnetic field, MPC = $[PC(B) - PC(0)]/PC(0)$, where $PC$ is the photocurrent. All the measurements are conducted at room temperature and no magnetic field direction dependence are observed.

## 3. Results and discussion

NPB, usually used as a hole transport layer in organic light emitting diode, has high hole mobility and low electron mobility [11]. And taking into account of the energy difference between the Fermi levels of electrodes and the highest occupied molecular orbital (HOMO)/lowest unoccupied molecular orbital (LUMO) of NPB, the transport in the device studied in this work is dominated by holes. Figure 1(a) shows the $I$-$V$ characteristic and Fig. 1(b) shows the short circuit photocurrent spectrum of the device measured at room temperature.

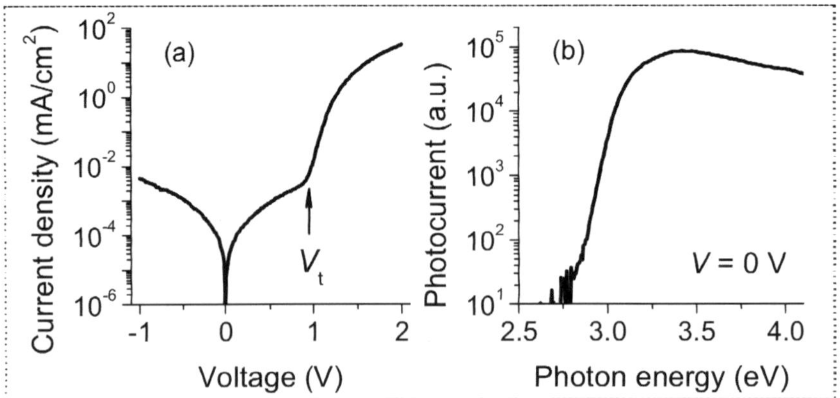


**Figure 1.** (a) $I$-$V$ characteristics and (b) short circuit photocurrent spectrum of the device measured at room temperature.

Without illumination, no magnetic field effect can be observed under different bias conditions, varying from -1 V to 2 V, with a magnetic field up to 800 mT. This absence of measurable OMAR indicates that bipolaron formation is negligible in NPB. In contrast, when the device is illuminated by a light with photon energy larger than the "bandgap" of NPB, significant OMAR can be measured. The OMAR observed under illumination actually originates from the magnetic field effect on photocurrent (MPC). Both bias dependent and excitation photon energy dependent MPC were studied.

Figure 2 presents the excitation photon energy dependent MPC curves measured at zero bias, with $E_{ph}$ ranging from 2.974 eV to 4.039 eV as indicated. It can be seen that the MPC does not show significant excitation photon energy dependence. This MPC independence on $E_{ph}$ is resulted from the small thickness of NPB layer in our device. For a thin NPB layer with thickness comparable to the shortest photon penetration depth the effect of photon energy is negligible. The line shape of the MPC curves at zero bias can be fitted with the empirical law, MPC ($B$) = $MPC_{\infty}\cdot B^2/(|B|+B_0)^2$, where $B_0$ is the characteristic field width and $MPC_{\infty}$ is the MPC at an infinite $B$ field [1]. The fitting result of $B_0$ is 140 mT, which is significantly larger than those reported ones in OMAR [1, 5]. We refer this MPC behavior as the high field component.

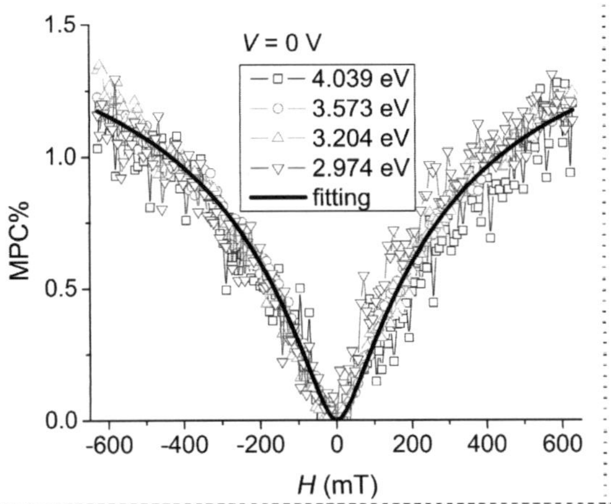


**Figure 2.** MPC curves measured with excitation photon energy varying from 2.974 eV to 4.039 eV at zero bias. The solid line is fitting results using the empirical law, $B^2/(|B|+B_0)^2$, with $B_0$ = 140 mT.

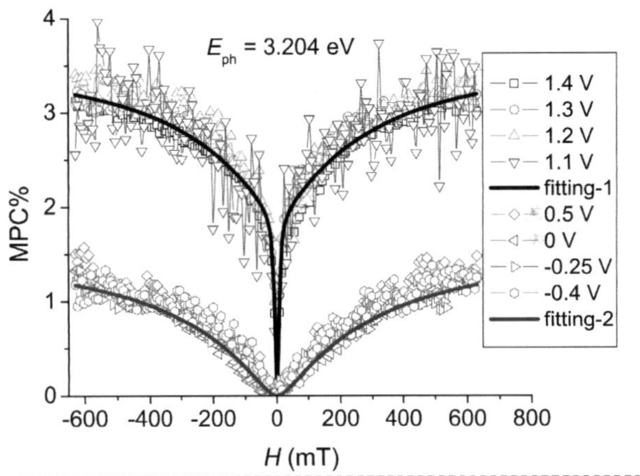


**Figure 3.** MPC curves measured at different bias with excitation photon energy fixed at 3.204 eV. The solid blue line is the fitting curve using empirical law $B^2/(|B|+B_{0HF})^2$ for $V < V_t$ with $B_0 \sim 140$ mT, and the solid black line using $B^2/(|B|+B_{0HF})^2 + B^2/(B^2+B_{0LF}^2)$, for $V > V_t$ with $B_{0HF} \sim 140$ mT and $B_{0LF} \sim 8$ mT.

In Fig. 3, measured MPC curves with applied biases varying from -0.4 V to 1.4 V and excitation photon energy fixed at 3.204 eV are illustrated. As it can be seen, for applied biases smaller than the turn-on voltage ($V_t$), approximately 0.9 V (see Fig. 1(a)), the MPC curves can be fitted with the high field component, i.e. MPC ($B$) = $MPC_{\infty HF}\cdot B^2/(|B|+B_{0HF})^2$ with $B_{0HF} \sim 140$ mT. When the applied bias is larger than $V_t$, a new low field MPC component is emerged while the high field MPC component remains. The MPC curves can be fitted with

$$MPC(B) = \frac{MPC_{\infty HF} \cdot B^2}{\left(|B| + B_{0HF}\right)^2} + \frac{MPC_{\infty LF} \cdot B^2}{B^2 + B_{0LF}{}^2}$$

with $B_{0HF} \sim 140$ mT and $B_{0LF} \sim 8$ mT, respectively. The line shapes of MPC do not show significant bias dependence within each bias range.

Photon absorption can only generate singlet excitons by exciting an electron from the highest occupied molecular orbital to the lowest unoccupied molecular orbital in a single molecule [12]. The photon generated singlet excitons will either recombine to emit photons or dissociate into free electrons and holes. The electrons and holes will either drift towards opposite electrodes because of the built-in electric field to generate photocurrent, or will recombine to form singlet/triplet excitons with the ratio of 1:3. In a steady state both singlet and triplet excitons exist in the device. There are two dominating photocurrent generation channels in organic semiconducting materials: singlet exciton dissociation and triplet-exciton-charge reaction [9]. These two processes are both magnetic field sensitive, but the magnetic field dependence of these two processes is unclear [3, 9]. From the above mentioned bias dependent MPC study, we can see when applied bias is below $V_t$, there is no effective carrier injection and the free carrier density is very low in the NPB layer. As a result, triplet-exciton-charge reaction process is unlikely to occur and MPC is dominated by the magnetic field effect on singlet exciton dissociation. However, when the bias is above $V_t$, an effective carrier injection produces a higher free carrier density in the NPB layer. This large amount of free carrier enhances triplet-exciton-charge reaction process and leads the emergence of a low field MPC component. The MPC is consisted of magnetic field effects on both singlet exciton dissociation and triplet-exciton-charge reaction process. Based on above analysis, we can be attribute high field and low field MPC component to singlet exciton dissociation and triplet-exciton-charge reaction, respectively.

## 4. Conclusions

In conclusion, no OMAR is observed without illumination in NPB, which means bipolaron formation is negligible in NPB. The analysis of bias dependent MPC lead us to identify that magnetic field dependence of singlet exciton dissociation and triplet-exciton-charge reaction are associated to a high field component and a low filed component with characteristic field width of 140 mT and 8 mT, respectively.

## Acknowledgments

This work was supported by the Research Grant Council of Hong Kong SAR government via grant number HKUST16/CRF/08.

## References

[1] Ö. Mermer, G. Veeraraghavan, T.L. Francis, Y. Sheng, D.T. Nguyen, M. Wohlgenannt, A. Köhler, M.K. Al-Suti, and M.S. Khan 2005 *Phys. Rev. B* **75** 094423.
[2] V.N. Prigodin, J.D. Bergeson, D.M. Lincoln, and A.J. Epstein 2006 *Synth. Met.* **156** 757.
[3] P. Desai, P.Shakya, T. Kreouzis, W.P. Gillin, N.A. Morley, and M.R.J. Gibbs 2007 *Phys. Rev. B* **75** 094423.
[4] P.A. Bobbert, T.D. Nguyen, F.W.A. van Oost, B. Koopmans, and M. Wohlgenannt 2007 *Phys. Rev. Lett.* **99** 216801.
[5] F.L. Bloom, W. Wagemans, M. Kemerink, and B. Koopmans 2007 *Phys. Rev. Lett.* **99** 257201.
[6] J.D. Bergeson, V.N. Prigodin, D.M. Lincoln, and A.J. Epstein 2008 *Phys. Rev. Lett.* **100** 067201.
[7] U. Niedermeier, M. Vieth, R. Pätzold, W. Sarfert, and H. von Seggern 2008 *Appl. Phys. Lett.* **92** 193309.
[8] F.J. Wang, Heinz Bässler, and Z. Valy Vardeny 2008 *Phys. Rev. Lett.* **101** 236805.
[9] B. Hu, L. Yan, and M. Shao 2009 *Adv. Mater.* **21** 1500.
[10] W.J.M. Naber, S. Faez and W.G. van der Wiel 2007 *J. Phys. D* **40** R205.
[11] S.W. Liu and J. K. Wang 2006 *Proc. SPIE* **6333** 63331R-1.
[12] A. Köhler and D. Beljonne 2004 *Adv. Funct. Mater.* **14** 11.

Journal of Physics: Conference Series **193** (2009) 012110 doi:10.1088/1742-6596/193/1/012110

# New methods for enhancing $ZT$ value of nanowires

**G. R. Zhang and Z. Zeng***

Key Laboratory of Materials Physics, Institute of Solid State Physics, Chinese Academy of Sciences, Hefei 230031, P.R. China

E-mail: zzeng@theory.issp.ac.cn

**Abstract.** We report an important modulating factor, degeneracy, for $ZT$ value enhancement of nanowire. Three approaches to design nanowire of large $ZT$ value by degeneracy effects are proposed: (1) controlling growth direction, (2) increasing the number of near degenerate conduction (valence) subbands induced by quantum confinement and (3) searching for bulk materials with two (or more) sets of near degenerate conduction (or valence) bands to fabricate nanowires.

## 1. Introduction

Thermoelectricity is one of the simplest technologies applicable to energy conversion. The efficiency of thermoelectricity is governed by a basic property of thermoelectrical material, dimensionless figure of merit $ZT = \frac{S^2\sigma}{\kappa}T$, where $S$ is the Seebeck coefficient, $\sigma$ is the electrical conductivity, $\kappa$ is the thermal conductivity which consists of electron and lattice thermal conductivity, $\kappa_e$ and $\kappa_L$, and $T$ is the absolute temperature. In the 1990's Hicks and Dresselhaus pointed out that low dimensional materials should have better efficiency than bulk ones due to low-dimensional effects on both charge carriers and lattice waves[1, 2]. The validity of attaining higher $ZT$ value in low dimension systems has been experimentally demonstrated on $Bi_2Te_3/Sb_2Te_3$ superlattices [3] and on PbTe/PbSeTe quantum-dots [4]with $ZT \sim 2.4$ and 1.6 respectively at 300K.

Therefore, nanowires are potentially good thermoelectrical systems for application. In this paper, we report another important modulating factor in $ZT$ value enhancement of nanowire, degeneracy. Three approaches to design nanowire of large $ZT$ value by degeneracy effects are proposed. The degeneracy effects on $ZT$ value enhancement of nanowire are demonstrated by the calculated $ZT$ values of $Bi_2Te_3$ and $Sb_2Te_3$ wires with diameter below 10 nm along trigonal and binary axes.

## 2. Theoretical approaches

The band structure and the thermoelectric transport properties of nanowires are calculated via the theoretical approach from Ref.[5]. In such an approach, the cylindrical wire boundary conditions, the anisotropic carrier effective mass tensor and the multiple carrier pockets are taken into account to calculate the band structure, and the thermoelectric transport coefficients are obtained from a semiclassical model based on the Boltzmann transport equation. The effective mass parameters are from Refs.[6, 7] and the parameters for calculating thermoelectric transport properties are from Refs. [1, 8, 9, 10, 11, 12, 13, 14].

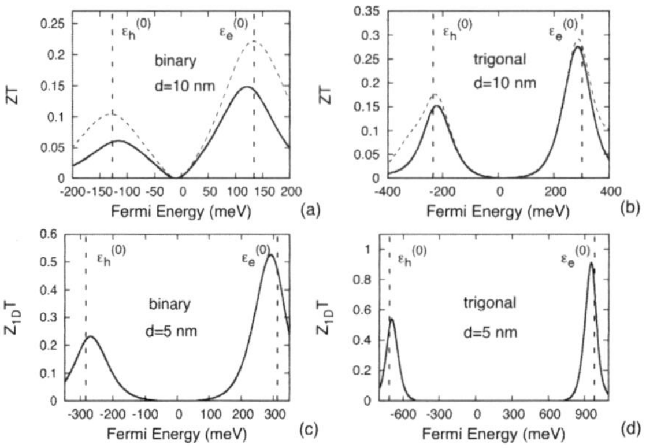


**Figure 1.** The Fermi energy dependence of calculated $ZT$ value at 300K for $Bi_2Te_3$ 5 nm and 10 nm nanowires orientated along binary and trigonal axes. Solid line: only the first conduction or valence subband is taken into account. Dashed line: two subbands are taken into account. Energies are respected to the center of bulk band gap. $\varepsilon_e^{(0)}$ denotes the lowest subband energy edge of electrons and $\varepsilon_h^{(0)}$ the highest one of holes.

## 3. Results and discussions

The obtained $ZT$ value at 300 K as a function of Fermi energy for $Bi_2Te_3$(n-type and p-type) and $Sb_2Te_3$ (p-type) nanowires with different diameters (10nm and 5nm) are presented in Fig. 1 and Fig. 2. The nanowires are orientated along the two principal crystallographic axes ( trigonal and binary directions). Zero energy level is selected at the center of band gap of bulk material for $Bi_2Te_3$ wire and valence band edge of bulk material for $Sb_2Te_3$ wire, respectively. The lowest subband edge of electrons is labeled by $\varepsilon_e^{(0)}$ and the highest one of holes is denoted by $\varepsilon_h^{(0)}$. As shown in the two figures, the $ZT$ values are quite small for the nanowires with a diameter of 10 nm when only one subband is taken into account, due to low mobility of carriers in $Bi_2Te_3$ and $Sb_2Te_3$. As the diameter reduces, $ZT$ value increases significantly owing to the quantum confinement. However, the increment ratio is not so large as in Bi nanowire [5] since the thermal conductivity is affected little by this quantum effect for $Bi_2Te_3$ and $Sb_2Te_3$ wires. Another general feature is that all the Fermi energies are near the subband edges but not exactly at these points when largest $ZT$ values occur. Comparatively, we find that, for $Bi_2Te_3$, n-type wire may be more favorable in application than p-type owing to the relatively larger mobility and smaller effective mass. We also notice that the maximal $ZT$ value of $Sb_2Te_3$ wire are smaller than that of $Bi_2Te_3$ wire with the same reasons.

Now we look into the importance of degeneracy. For bulk $Bi_2Te_3$ and $Sb_2Te_3$ materials, most attentions have been paid to the cleavage plane in that the mobility in this plane is much higher than that along trigonal direction. However, when such materials are fabricated as slim nanowires, the trigonal direction becomes important. As demonstrated in Fig. 1, the maximal $ZT$ values of trigonal wires are larger than those of binary wires even though the mobility in binary axis is the highest. We attribute this to degeneracy. Due to the structural difference between nanowire and bulk material, the degeneracy of carrier pocket varies significantly when a wire is orientated along different direction. For a trigonal nanowire made from $Bi_2Te_3$ or $Sb_2Te_3$, the electron and hole pockets are six-fold degenerate. However, for wires orientated along the binary axis, the carrier pockets are divided into two groups: one denoted as binary group I is four-fold degenerate and another denoted as binary group II is two-fold degenerate. From our results, the lowest conduction (uppermost valence) subband of binary nanowire originates from carrier pockets of binary group II. Therefore, the degeneracy of lowest conduction (uppermost

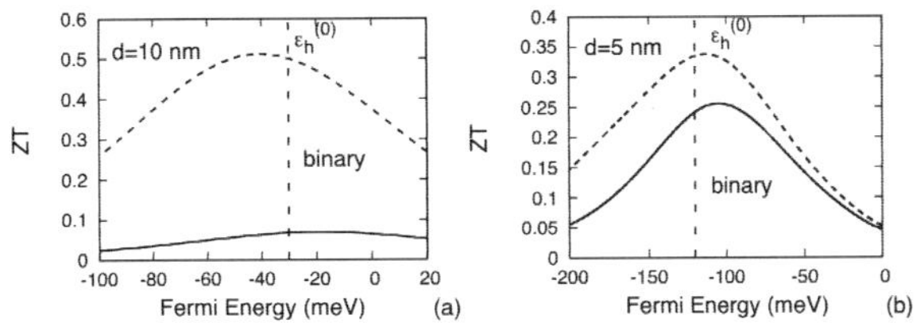


**Figure 2.** The Fermi energy dependence of calculated $ZT$ value at 300K for 5nm (a) and 10 nm (b) $Sb_2Te_3$ nanowires orientated along binary axis. Energies are respected to the valence band edge of bulk material. $\varepsilon_h^{(0)}$ denotes the highest subband energy edge of holes.

valence) subband of binary nanowire is only one third of that of trigonal wire. Such a difference of degeneracy results in the maximal $ZT$ values of trigonal wires being larger than those of binary wires with small diameter (bellow 10 nm for $Bi_2Te_3$). This phenomena can also be found in the theoretical results for Bi nanowire [5]. As a result, we infer that the degeneracy is important in nanowires and favorable thermoelectrical properties can be achieved through controlling the wire direction.

The second interesting phenomenon is exhibited in Fig. 2 which shows the calculated $ZT$ values as a function of Fermi level for $Sb_2Te_3$ nanowires with diameter 10 nm and 5 nm in binary direction. We find a significant enhancement of $ZT$ value comparing the multi-subband contribution (dashed line in Fig. 2(a)) with single subband contribution(solid line in Fig.2(a)). This phenomenon can also be found in $Bi_2Te_3$ wire orientated along binary axis (see Fig. 1). Moreover, it is very interesting that even the maximum of $ZT$ value for a 10 nm $Sb_2Te_3$ wire is larger than that for a 5 nm wire (see Fig. 2(b)). Such a phenomenon originates from another degeneracy effect. For a binary $Sb_2Te_3$ nanowire with a diameter of 10 nm, our results show that the edges of first three subbands of binary group II (30.14, 45.95, 66.89 meV) and the first subband of binary group I (65.97 meV) are very vicinal. Hence each of these subbands makes a contribution to the $ZT$ value. If we define vicinal degeneracy as the number of subbands whose energy edges are within 40 meV above (below) the lowest conduction (upermost valence) subband, a vicinal degeneracy with a value of 10 is obtained since the subband of binary II group is two-fold degenerate and the subband of binary group I is four-fold degenerate. However, the vicinal degeneracy decreases when a wire becomes slimmer due to the quantum confinement effect. This is the reason of that 10 nm $Sb_2Te_3$ wire has a larger maximal $ZT$ value than 5 nm wire. This vicinal degeneracy effect provides an useful guidance and an easy way to design new favorable thermoelectrical materials since it does not need to fabricate quite slim wire for a large $ZT$ value. Moreover, at a larger diameter, the electrical conductivity is not reduced strongly by localized effect.

The other way to achieve large $ZT$ value through degeneracy effect is searching for (or designing) materials whose bulk band structure has the shape as sketched in Fig. 3. Materials of this type have two conduction (or valence) bands and the difference between the two energy edges is quit small. Moreover, the effective mass of upper conduction (lower valence) band is larger than that of the lower conduction (upper valence) band. For nanowires made from such materials, the energy shifts of the subband edges with respect to the upper bulk conduction (lower valence) band increases more slowly, compared with these respect to the lower conduction(upper valence) band. So, at some critical diameter, some subbands induced from upper bulk conduction (lower valence) band will be close to those induced from lower conduction(upper valence) band. If each subband is multi-fold degenerate, larger vicinal degeneracy can easily been obtained, and it is possible to enhance $ZT$ value significantly. However, an important condition that mobility of upper conduction (or lower valence) band is not too low should be satisfied. In fact, the band

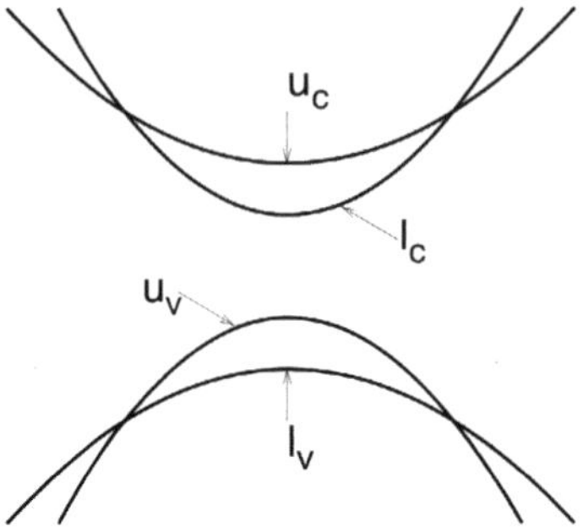


**Figure 3.** The schematic diagram of bulk band structure for the favorable thermoelectrical material from which nanowire of large $ZT$ value is made. $u_c$, $l_c$, $u_v$, and $l_v$ denote upper conduction band, lower conduction band, upper valence band and lower valence band, respectively.

structures of $Bi_2Te_3$ and $Sb_2Te_3$ belong to this type. But, unfortunately, the second conduction (valence) band of bulk material has quite low mobility. If this disadvantage is removed, $Bi_2Te_3$ and $Sb_2Te_3$ nanowires will be favorable in thermoelectrical application.

## 4. Conclusions

Enhancement of $ZT$ value of nanowire has the potential to be achieved by the following approaches: (1) controlling growth direction, (2) increasing the number of near degenerate conduction (valence) subbands induced by quantum confinement and (3) searching for bulk materials with two (or more) sets of near degenerate conduction (or valence) bands to fabricate nanowires.

## 5. Acknowledgements

This work was supported by the special Funds for Major State Basic Research Project of China(973) under grant no. 2005CB623603, Knowledge Innovation Program of Chinese Academy of Sciences, and Director Grants of Hefei Institutes of Physical Sciences. Part of the calculations were performed in Center for Computational Science, Hefei Institutes of Physical Sciences.

## References

[1] L. D. Hicks, M. S. Dresselhaus, Phys. Rev. B **47**, 12727 (1993).
[2] L. D. Hicks, M. S. Dresselhaus, Phys. Rev. B **47**, 16631 (1993).
[3] R. Venkatasubramanian, E. Siivola, T. Colpitts, B. O'Quinn, Nature. **413**, 597 (2001).
[4] T. C. Harman, P. J. Taylor, M. P. Walsh, B. E. LaForge, Science. **297**, 2229 (2002).
[5] Yu-Ming Lin, Xiangzhong Sun, and M. S. Dresselhaus, Phys. Rev. B **62**, 4610 (2000).
[6] M. Stordeur, M. Stölzer, H.Sobotta and V. Riede, Phys. Status Solidi(b). **150**, 165 (1988).
[7] H. Kohler, Phys. Status Solidi B **73**, 95 (1976);
H. Kohler, Phys. Status solidi B **74**, 591 (1976).
[8] Rama Venkatasubramanian, Phys. Rev. B **61**, 3091 (2000).
[9] V. M. Zhdanov, Zh. Fiz. Khim. **45**, 2388 (1971).
[10] Th. L. Anderson, H. Brigitte. Krause, Acta Crystallogr. B **30**, 1307 (1974).
[11] J. S. Dyck, W. Chen, and C. Uher, Č. Drašar and P. Lošt'ák, Phys. Rev. B **66**, 125206 (2002).
[12] L. P. Caywood, Jr. and G. R. Miller. Phys. Rev. **30**, 1002 (1968);
L. R. Testardi and E. Burstein, Phys. Rev. B **6**, 460 (1972).
[13] Saji Augustine and Elizabeth Mathai, Semicond. Sci. Technol. **18**, 745 (2003).
[14] Black, Conwell, Seigle, and Spencer, J. Phys. Chem. Solids. **2**, 240 (1957);
Harman, Paris, Miller, and Goering, J. Phys. Chem. Solids. **2**, 181 (1957).

Journal of Physics: Conference Series **193** (2009) 012111 doi:10.1088/1742-6596/193/1/012111

# Zener Tunnelling in Graphene Based Semiconductors - the k·p Method

**William Vandenberghe**[1,2], **Bart Sorée**[1], **Wim Magnus**[1,3] **and Guido Groeseneken**[1,2]
[1] *IMEC, Kapeldreef 75, 3001 Leuven, Belgium*
`william.vandenberghe@imec.be`

[2] *Dept. of Electrical Engineering, Katholieke Universiteit Leuven*
*Kasteelpark Arenberg 10, 3001 Leuven, Belgium*

[3] *Department of Physics, Universiteit Antwerpen*
*Groenenborgerlaan 171, 2020 Antwerpen, Belgium*

**Abstract.**
The carbon nanotube and graphene nanoribbon band structure is derived using the two-band $\mathbf{k}\cdot\mathbf{p}$ method and shown to have a similar band structure as a III-V semiconductor. Contrary to a previous claim, it is shown that the tunnelling probability is lower for a graphene based semiconductor than for a III-V semiconductor with the same bandgap. Considering the relation between the bandgap and the effective mass we conclude that a graphene based semiconductor is not well-suited for a classical device which suffers from Zener tunnelling, but is rather promising for a device which has its working principle based on Zener tunnelling.

## 1. Introduction

Graphene based materials such as carbon nanotubes and graphene nanoribbons provide an interesting research topic since they have a band structure which is different from traditional semiconductors. Graphene has a zero bandgap at the K-points in the Brillouin zone, but some carbon nanotubes or graphene nanoribbons exhibit a non-zero bandgap which, in principle, enables the fabrication of semiconductor devices. These semiconducting carbon nanotubes and graphene nanoribbons will be referred to as graphene based semiconductors throughout this paper.

Graphene based semiconductors generally have a small bandgap and hence Zener tunnelling, also called band-to-band tunnelling (BTBT), cannot be neglected in possible graphene based semiconductor devices. A recent publication [1] shows a calculation of the Zener tunnelling current using the WKB method and concludes that it exceeds the current of group IV and III-V semiconductors with identical bandgaps.

In this paper, we first develop the band structure within the $\mathbf{k}\cdot\mathbf{p}$ scheme for bulk graphene as well as for nanoribbons or nanotubes with a non-zero bandgap. Next, we consider the tunnelling probability and refute the claim that tunnelling is more efficient in a graphene based semiconductor compared to a III-V semiconductor with the same bandgap, and we explain the discrepancy with the previous publication [1].

## 2. Graphene based semiconductors: band structure

For the $\mathbf{k}\cdot\mathbf{p}$ expansion of the bulk graphene band structure around the $K$-point we use two bands: the $\pi^*$ conduction band and the $\pi$ valence band which are degenerate at the $K$-point due to symmetry [2]. Using these states to construct the $\mathbf{k}\cdot\mathbf{p}$ Hamiltonian, we need to determine the momentum matrix elements

$$\mathbf{p_{11}} = \int u_1^*(\mathbf{r},z)\mathbf{p}\,u_1(\mathbf{r},z)\mathrm{d}^3r \quad , \quad \mathbf{p_{12}} = \int u_1^*(\mathbf{r},z)\mathbf{p}\,u_2(\mathbf{r},z)\mathrm{d}^3r \quad , \quad \mathbf{p_{22}} = \int u_2^*(\mathbf{r},z)\mathbf{p}\,u_2(\mathbf{r},z)\mathrm{d}^3r, \tag{1}$$

appearing as integrals over the 2D graphene unit cell in the plane and over the entire space in the $z$-direction perpendicular to the plane,

$$\int \mathrm{d}^3 r = \int_{\text{unitcell}} \mathrm{d}^2 r \int_{-\infty}^{\infty} \mathrm{d}z \,. \tag{2}$$

Here, $u_1(\mathbf{r}, z)$ and $u_2(\mathbf{r}, z)$ are two Bloch functions which are orthogonal solutions to the Schrödinger equation with eigen energy equal to zero:

$$\left[-\frac{\hbar^2 \nabla^2}{2m} + U_{\text{lat}}(\mathbf{r}, z)\right] u_{1,2}(\mathbf{r}, z) = 0 \tag{3}$$

$U_{\text{lat}}(\mathbf{r}, z)$ is the potential due to the graphene lattice which is periodic with respect to $\mathbf{r}$, $m$ is the free electron mass and $\hbar$ is the reduced Planck constant.

As the $z$-axis stands perpendicular to the graphene plane, both the $\pi$ and $\pi^*$ bands are odd under a reflection in the $xy$-plane and as a result the 3 selected matrix elements have no $z$ component. The $y$ direction is chosen such that $u_1$ and $u_2$ are respectively odd and even under reflection in the $yz$-plane. This requires $\mathbf{p_{11}}$ and $\mathbf{p_{22}}$ to be in the $y$ direction and $\mathbf{p_{12}}$ in the $x$ direction. At the $K$-point the valence and conduction band are degenerate due to an additional rotational symmetry over 120°. This adds the requirement $|\mathbf{p_{11}}| = |\mathbf{p_{12}}| = |\mathbf{p_{22}}| = p$ and $\mathbf{p_{11}} = -\mathbf{p_{22}}$.

According to the $\mathbf{k} \cdot \mathbf{p}$ method a state with wavevector $\mathbf{k}$ relative to a $K$-point can be written as

$$\psi(\mathbf{r}, z) = e^{\mathrm{i}\mathbf{k}\cdot\mathbf{r}} \left(a_1 u_1(\mathbf{r}, z) + a_2 u_2(\mathbf{r}, z)\right) \tag{4}$$

and its energy eigenvalues $E_\pm(\mathbf{k})$ are determined by the 2-band $\mathbf{k} \cdot \mathbf{p}$ Hamiltonian

$$H = \begin{pmatrix} \frac{\hbar^2}{2m}(k_x^2 + k_y^2) + \frac{\hbar p}{m} k_x & \frac{\hbar p}{m} k_y \\ \frac{\hbar p}{m} k_y & \frac{\hbar^2}{2m}(k_x^2 + k_y^2) - \frac{\hbar p}{m} k_x \end{pmatrix} \tag{5}$$

yielding

$$E_\pm(\mathbf{k}) = \frac{\hbar^2 |\mathbf{k}|^2}{2m} \pm \frac{\hbar p}{m} |\mathbf{k}|, \quad |\mathbf{k}| = \sqrt{k_x^2 + k_y^2}. \tag{6}$$

To first order, this corresponds to the nearest neighbour tight binding expression

$$E_\pm(\mathbf{k}) \approx \pm \hbar v_F |\mathbf{k}| \tag{7}$$

and the matrix element $p$ can be determined as $p = m v_F$ with $v_F$ the graphene Fermi velocity.

Note that an alternative choice of basis functions $u_1$ and $u_2$ and omission of the second derivative transforms Eq. (5) into a more familiar formulation of the Hamiltonian

$$H = \frac{\hbar}{m} p \begin{pmatrix} 0 & k_x - \mathrm{i}k_y \\ k_x + \mathrm{i}k_y & 0 \end{pmatrix}, \tag{8}$$

the resulting Schrödinger equation being algebraically identical to the 2D Weyl equation [3].

Returning to Eq. (5) for the case of a semiconducting nanoribbon or nanotube we use the zone folding technique to determine the band structure. The $\mathbf{k}$ vector is continuous in one direction and quantized in the perpendicular direction. For the sake of convenience we select a zig-zag nanoribbon or nanotube such that $k_x$ is quantized and $k_y$ is continuous. The following derivation of the band structure is valid however for an arbitrary orientation of the tube or ribbon since Eq. (6) is invariant under rotations.

Denoting $k_0$ by the smallest allowed value of $k_x$ we obtain the bandgap of the graphene based semiconductor

$$E_{\text{g}} = 2\frac{\hbar p}{m} k_0 \tag{9}$$

where size quantization allows for $k_0 \neq 0$ in the case of a semiconducting nanotube or ribbon, while Eq. (5) turns into

$$H = \begin{pmatrix} E_0 + \frac{\hbar^2 k_y^2}{2m} + \frac{E_{\text{g}}}{2} & \frac{\hbar p}{m} k_y \\ \frac{\hbar p}{m} k_y & E_0 + \frac{\hbar^2 k_y^2}{2m} - \frac{E_{\text{g}}}{2} \end{pmatrix} \tag{10}$$

Journal of Physics: Conference Series **193** (2009) 012111 doi:10.1088/1742-6596/193/1/012111

with $E_0 = \hbar^2 k_0^2/2m$ and the energy dispersion relation for the lowest subband reads

$$E_{\pm} = E_0 + \frac{\hbar^2 k_y^2}{2m} \pm \sqrt{\frac{E_g^2}{4} + \frac{\hbar^2 p^2}{m^2} k_y^2}. \tag{11}$$

Within the 2 band approximation Eq. (10) is formally identical to the $\mathbf{k} \cdot \mathbf{p}$ Hamiltonian of a III-V semiconductor with a bandgap $E_g$, a matrix element $p$ and an energy offset $E_0$ [4, 5]. More specifically, in a III-V semiconductor, the two-band Hamiltonian describes the interaction between the light valence band and the conduction band. The energy dispersion relation of a graphene based semiconductor is therefore the same as that of a III-V semiconductor apart from the absence of a heavy and a split-off valence band.

### 3. Tunnelling probability

As a graphene based semiconductor and a III-V semiconductor share a similar band structure, the expression of the Zener tunnelling probability for an electron in a uniform field will also be the same:

$$T = \exp\left(\frac{-\pi E_g^2 m}{4\hbar p q F}\right) \tag{12}$$

with $F$ the applied uniform electric field and $q$ the unit charge [5]. To compare different materials, we consider the matrix element $p$ which is generally expressed as the energy unit $E_p = 2p^2/m$ taking the value $E_p = 2mv_F^2 = 11.4$eV for graphene and $E_p \approx 20$eV for III-V semiconductors, such as GaAs, InAs, InSb, ...[6].

For conventional devices, such as the metal-oxide-semiconductor field-effect transistor (MOSFET), BTBT is a detrimental effect since it increases the off-state current. To make a fair comparison concerning the amount of BTBT we must consider MOSFETs with a common figure of merit such as the transport effective mass. Making use of the relation between the effective mass and the bandgap, $m^* \approx mE_g/E_p$ and the definition of $E_p$, the transmission coefficient (12) can be written as

$$T \approx \exp\left(\frac{-\pi E_p^{3/2} m^{*2}}{2\sqrt{2} m^{3/2} \hbar q F}\right). \tag{13}$$

Since $E_{p,\mathrm{graphene}} < E_{p,\mathrm{III-V}}$, Eq. (13) shows that for the same effective mass, the tunnelling probability will be much greater for a graphene based MOSFET compared to a III-V based MOSFET as shown in Fig. 1. Or, in other words a graphene based MOSFET with the same on-current characteristics as a III-V based MOSFET will show a higher leakage current due to increased BTBT.

In the case of a III-V semiconductor, the effective mass ($m^*$) refers to the conduction band mass or the light hole mass. In the case of a graphene based semiconductor, there is only one direction with a continuous wavevector defining single valence and conduction bands. The absence of a heavy hole and a split-off band in graphene band structure could still allow graphene to outperform a similar III-V device but certainly for nMOS devices we do not expect an improvement over III-V MOSFETs.

For devices operating on the basis of BTBT such as the tunnel FET [7], large tunnelling probabilities are required. Since the matrix element $p \propto \sqrt{E_p}$ is smaller for graphene than for III-V semiconductors, graphene based semiconductors have a smaller tunnel probability for the same bandgap as shown in Fig. 1. To obtain the same performance as a III-V tunnel FET, a graphene based semiconductor tunnel FET must have a smaller bandgap. But since for graphene based semiconductors the bandgap can be made smaller by selecting a different size of nanoribbon or nanotube, this needs not be a disadvantage if graphene based semiconductors can be fabricated reliably.

Finally, we report on a discrepancy between our results and the work by Jena et al. [1] who came to the opposite conclusion concerning the tunnelling probability. To our opinion, the disagreement stems from the assumption by Jena et al. that traditional semiconductors have a "parabolic bandgap". Obviously, the parabolic nature of the band structure in any semiconductor only holds in a limited energy range and does not cover the entire Brillouin zone. As we have shown in our previous section, there is no formal difference in the band structure between a graphene based semiconductor and a III-V semiconductor except for the presence of an additional heavy hole and split-off band in the latter.

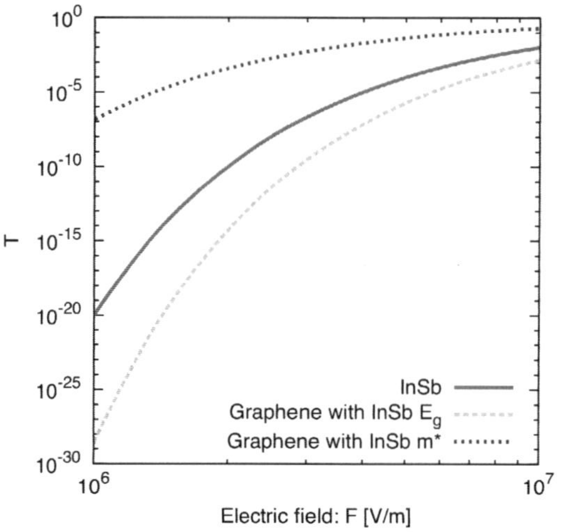


**Figure 1.** Comparison between the Zener tunnelling probability of InSb and a graphene based semiconductor with the InSb bandgap (Eq. 12) and a graphene based semiconductor with the InSb effective mass (Eq. 13).

The direct cause of the disagreement lies in the use of the transmission coefficient [8]

$$T = \exp\left(-\frac{4\sqrt{2m^*}E_g^{3/2}}{3\hbar F}\right) \tag{14}$$

by Jena et al. In the case of Zener tunnelling, the "parabolic bandgap" approximation is too crude, comparing Eq. (14) with Eq. (12) and using $p = m/\sqrt{2}\sqrt{E_g/m^*}$ we observe a deviation of a factor $3\pi/16 = 0.59$ in the exponential. Correcting for this factor the conclusion from the previous publication that graphene has a larger tunnel probability compared to a III-V with the same bandgap is reversed and agrees with our current publication, claiming that the III-V material will show the larger tunnel probability for the same bandgap.

## 4. Conclusions

The band structure of the valence and the conduction band in a graphene based semiconductor can be described in the same way as the light valence and the conduction band structure in a III-V semiconductor.

Graphene based semiconductors have a lower Zener tunnelling probability than III-V semiconductors for the same bandgap.

Semiconductor devices where the effective mass is a figure of merit and where leakage (off-current) due to Zener tunnelling is unwanted such as the MOSFET will show higher leakage for the same on-current in the case of graphene based semiconductors compared to III-V semiconductors. On the other hand, graphene based semiconductors are well suited for devices which operate on the basis of Zener tunnelling such as the tunnel FET.

## Acknowledgments

William Vandenberghe gratefully acknowledges the support of a Ph.D. stipend from the Institute for the Promotion of Innovation through Science and Technology in Flanders (IWT-Vlaanderen).

## References

[1] Jena D, Fang T, Zhang Q and Xing H 2008 *Applied Physics Letters* **93** 112106
[2] Stéphanie Reich Christian Thomsen J M 2004 *Carbon nanotubes: basic concepts and physical properties* (Wiley-VCH)
[3] DiVincenzo D P and Mele E J 1984 *Phys. Rev. B* **29** 1685–1694
[4] Vandenberghe W, Sorée B, Magnus W and Groeseneken G 2009 Submitted for publication to Phys. Rev. B
[5] Kane E O 1959 *Journal of Physics and Chemistry of Solids* **12** 181–188
[6] Vurgaftman I, Meyer J R and Ram-Mohan L R 2001 *Journal of Applied Physics* **89** 5815–5875
[7] Vandenberghe W, Verhulst A, Groeseneken G, Sorée B and Magnus W 2008 *SISPAD 2008*. pp 137–140
[8] Sze S 1981 *Physics of Semiconductor Devices* (2nd ed. John Wiley & Sons)

Journal of Physics: Conference Series **193** (2009) 012112 doi:10.1088/1742-6596/193/1/012112

# Full-band and atomistic study of electron-phonon interaction in graphene nanoribbons

**Tatsuro Kitayama, Hideki Minari, and Nobuya Mori**
Division of Electrical, Electronic and Information Engineering, Osaka University
2-1 Yamada-oka, Suita, Osaka 565-0871, Japan

E-mail: kitayama@si.eei.eng.osaka-u.ac.jp

**Abstract.** Transmission function through metallic armchair-edge graphene nanoribbons is calculated within non-equilibrium Green's function method including full-band electron–optical-phonon interaction. Structures appear in transmission function, which are originated in both zone-center and zone-edge optical-phonon scattering.

## 1. Introduction

Recently, Novoselov *et al.* successfully fabricated field-effect transistors (FETs) of graphene [1]. Graphene sheets have also been patterned into narrow nanoribbons to produce graphene nanoribbons (GNRs) with a width of several tens of nanometers down to a few nanometers. One can control the band-structure of GNRs by changing their width and orientation relative to the graphene crystal structure. Graphene-based FETs are now considered to be one of most promising device structures for future integrated circuits. For improvements in the transport properties of GNRs, it is essential to identify the major sources of carrier scattering. Since graphene-based FETs operate at room temperature, electron-phonon interaction is considered to play an important role as a dominant scatterer. In the present work, we have numerically studied effects of electron–optical-phonon interaction in metallic armchair-edge GNRs (A-GNRs). We have calculated transmission function through A-GNRs using the method developed by Frederiksen *et al.* [2] based on non-equilibrium Green's function (NEGF) [3–5]. The electron and phonon dispersion curves are obtained within tight-binding approximation, and optical-phonon scattering are taken into account within full-band and atomistic NEGF formalism.

## 2. Calculation method

### *2.1. Electronic states*

We use a tight-binding model to describe the electronic states of GNRs. The wave function is written as

$$\Psi(\boldsymbol{r}) = \sum_{\boldsymbol{R}_n} \Psi(\boldsymbol{R}_n)\, \phi(\boldsymbol{r} - \boldsymbol{R}_n), \tag{1}$$

where $\phi(\boldsymbol{r})$ is the wave function of the $p_z$ orbital of a carbon atom located at the origin and $\boldsymbol{R}_n$ represents the position of a carbon atom. The Hamiltonian is written as

$$\boldsymbol{H} = \sum_{nm} E_{nm} |\phi_{nm}\rangle\langle\phi_{nm}| + \sum_{nm \neq n'm'} \gamma_{nm,n'm'} |\phi_{nm}\rangle\langle\phi_{n'm'}|, \tag{2}$$

where $E_{nm}$ is the on-site energy, $\gamma_{nm,n'm'}$ is the transfer integral between orbitals, $n$ is a unit cell number, and $m$ is a site number in a unit cell (see inset of Fig.1). We consider nearest-neighbor interaction with the transfer integral of $-\gamma_0$ and the case without potential ($E_{nm} = 0$). Although the case with a non-zero potential would be important, we only consider the zero potential case in the present study for investigating fundamental properties of GNRs with phonon scattering.

### *2.2. Optical-phonon modes*

We describe the optical-mode lattice vibration by considering two types of restoring forces between carbon atoms [6]. One is restoring force for the bond lengths and the other is that for the bond angles. In the present study, we ignore edge effects on phonon modes and assume bulk phonons for simplicity.

### *2.3. Electron-phonon interaction*

We include the full-band electron-phonon interaction in NEGF formalism by using the method developed by Frederiksen *et al.* [2]. The transfer integral between an atom labeled $i$ and an atom labeled $j$ depends on the change of the lattice displacement as

$$\gamma_{i,j}(|\boldsymbol{d}_l + \boldsymbol{u}_i(\boldsymbol{R}_i) - \boldsymbol{u}_j(\boldsymbol{R}_j)|) = \gamma_0 + \frac{1}{b}\frac{\mathrm{d}\gamma_0}{\mathrm{d}b}\boldsymbol{d}_l \cdot [\boldsymbol{u}_i(\boldsymbol{R}_i) - \boldsymbol{u}_j(\boldsymbol{R}_j)], \tag{3}$$

where $\boldsymbol{u}(\boldsymbol{r})$ is the lattice displacement, $\boldsymbol{d}_l = \boldsymbol{R}_i - \boldsymbol{R}_j$ are the vectors connecting nearest neighbor carbon atoms ($l = 1, 2, 3$) [7], and $b$ is the bond length without the lattice displacement [8–10].

By solving the dynamical equation of the two restoring forces between carbon atoms, we obtain the normal mode vector $\boldsymbol{e}_\mu$ (normalized $\boldsymbol{e}_\mu \cdot \boldsymbol{e}_\mu = 1$) and the eigen-frequency $\omega_\mu(\boldsymbol{q})$. Here $\mu$ is the mode number and $\boldsymbol{q}$ is the phonon wavevector. The quantized expression for $\boldsymbol{u}_i$ is then written as

$$\boldsymbol{u}_i(\boldsymbol{R}) = \sum_\mu \sum_{\boldsymbol{q}} (\boldsymbol{e}_{\mu i}) \mathrm{e}^{\mathrm{i}\boldsymbol{q}\cdot\boldsymbol{R}} \sqrt{\frac{\hbar}{2MN\omega_\mu(\boldsymbol{q})}} (b^\dagger_{\mu,-\boldsymbol{q}} + b_{\mu,\boldsymbol{q}}), \tag{4}$$

$$\boldsymbol{u}_j(\boldsymbol{R} - \boldsymbol{d}_l) = \sum_\mu \sum_{\boldsymbol{q}} (\boldsymbol{e}_{\mu j}) \mathrm{e}^{\mathrm{i}\boldsymbol{q}\cdot\boldsymbol{R}} \sqrt{\frac{\hbar}{2MN\omega_\mu(\boldsymbol{q})}} (b^\dagger_{\mu,-\boldsymbol{q}} + b_{\mu,\boldsymbol{q}}), \tag{5}$$

where $N$ is the number of unit cells, $M$ is the mass of a carbon atom, and $b^\dagger_{\mu,\boldsymbol{q}}$ and $b_{\mu,\boldsymbol{q}}$ are the creation and destruction operators, respectively. Inserting Eqs. (4) and (5) into Eq. (3), we obtain

$$\gamma_{i,j} = \gamma_0 + \sum_\mu \sum_{\boldsymbol{q}} (\boldsymbol{M}_{\mu,\boldsymbol{q}})_{i,j} (b^\dagger_{\mu,-\boldsymbol{q}} + b_{\mu,\boldsymbol{q}}), \tag{6}$$

where $\boldsymbol{M}_{\mu,\boldsymbol{q}}$ is the electron–optical-phonon coupling defined as

$$(\boldsymbol{M}_{\mu,\boldsymbol{q}})_{i,j} = \frac{1}{b}\frac{\mathrm{d}\gamma_0}{\mathrm{d}b}\boldsymbol{d}_l \cdot (\boldsymbol{e}_{\mu i} - \boldsymbol{e}_{\mu j}) \mathrm{e}^{\mathrm{i}\boldsymbol{q}\cdot\boldsymbol{R}} \sqrt{\frac{\hbar}{2MN\omega_\mu(\boldsymbol{q})}}. \tag{7}$$

### *2.4. NEGF formalism*

The electron-phonon interaction is included into NEGF formalism through electronic self-energies caused by the optical-phonon scattering$\Sigma^{r,\lessgtr}_{\mathrm{ph}}(\varepsilon)$ as (we follow the notation as in Ref. [2])

$$\Sigma^r_{\mathrm{ph}}(\varepsilon) = \mathrm{i}\sum_\mu \sum_{\boldsymbol{q}} \int_{-\infty}^{\infty} \frac{\mathrm{d}\varepsilon'}{2\pi} \boldsymbol{M}^\dagger_{\mu,\boldsymbol{q}} \Big\{ \frac{4}{\hbar\omega_\mu(\boldsymbol{q})} \mathrm{Tr}[\boldsymbol{G}^<(\varepsilon')\boldsymbol{M}_{\mu,\boldsymbol{q}}]$$
$$+ D^r_\mu(\boldsymbol{q}, \varepsilon - \varepsilon')[\boldsymbol{G}^<(\varepsilon') + \boldsymbol{G}^r(\varepsilon')]\boldsymbol{M}_{\mu,\boldsymbol{q}} + D^<_\mu(\boldsymbol{q}, \varepsilon - \varepsilon')\boldsymbol{G}^r(\varepsilon')\boldsymbol{M}_{\mu,\boldsymbol{q}} \Big\}, \tag{8}$$

$$\Sigma^{\lessgtr}_{\mathrm{ph}}(\varepsilon) = \mathrm{i}\sum_{\mu}\sum_{\boldsymbol{q}} \int_{-\infty}^{\infty} \frac{\mathrm{d}\varepsilon'}{2\pi} \boldsymbol{M}^{\dagger}_{\mu,\boldsymbol{q}} D^{\lessgtr}_{\mu}(\boldsymbol{q}, \varepsilon - \varepsilon') \boldsymbol{G}^{\lessgtr}(\varepsilon') \boldsymbol{M}_{\mu,\boldsymbol{q}}, \tag{9}$$

where $G^{\lessgtr}$ are the lesser and greater electron Green function and $D^{r,\lessgtr}_{\mu}$ are the noninteracting phonon Green functions. The retarded electron Green function is given by

$$\boldsymbol{G}^{r}(\varepsilon) = \left[(\varepsilon + \mathrm{i}0)\boldsymbol{I} - \boldsymbol{H} - \Sigma^{r}_{\mathrm{L}}(\varepsilon) - \Sigma^{r}_{\mathrm{R}}(\varepsilon) - \Sigma^{r}_{\mathrm{ph}}(\varepsilon)\right]^{-1}, \tag{10}$$

where$\Sigma^{\,r}_{\mathrm{L,R}}$ are the contact self-energies. We calculated the retarded Green function $G^{r}$ self-consistently using Eqs. (8) and (10). Transmission function is then obtained by

$$T(\varepsilon) = \mathrm{Tr}\left[\Sigma^{<}_{L}\boldsymbol{G}^{>}(\varepsilon) - \Sigma^{>}_{L}\boldsymbol{G}^{<}(\varepsilon)\right]. \tag{11}$$

## 3. Result and Discussion

Figure 1 shows the band-structure of an A-GNR with a width of $W = (7/2)a$ ($a$ is the lattice constant). We set the transfer integral to $\gamma_0 = 3.03\,\mathrm{eV}$. Figure 2 shows optical-phonon scattering rate at an edge atom (marked in the inset of Fig. 1) as a function of electron energy. We set the phonon occupation factor to zero and consider the emission process only. We find high scattering rate $\gtrsim 10\,\mathrm{ps}^{-1}$ for $2\,\mathrm{eV} \lesssim \varepsilon \lesssim 9\,\mathrm{eV}$. This can be attributed to the high density-of-states for that energy region.

Figure 3 shows the transmission function through A-GNRs with a width of $W = (7/2)a$ for $L = (5\sqrt{3}/6)a$, $(11\sqrt{3}/6)a$, and $(17\sqrt{3}/6)a$. In Fig. 3, we also plot the transmission function without electron-phonon interaction for comparison. The electron-phonon interaction has strong impact on the transmission for electron energies near the band center ($\varepsilon \approx 3\,\mathrm{eV}$), while weak impact near the band edges ($\varepsilon \approx 0\,\mathrm{eV}$ and $9\,\mathrm{eV}$). We find several structures appearing on the transmission function, which can be attributed to not only zone-center phonons ($\hbar\omega = 196\,\mathrm{meV}$) but also zone-edge phonons ($\hbar\omega = 162\,\mathrm{meV}$) (see Fig. 4). When electron kinetic energy

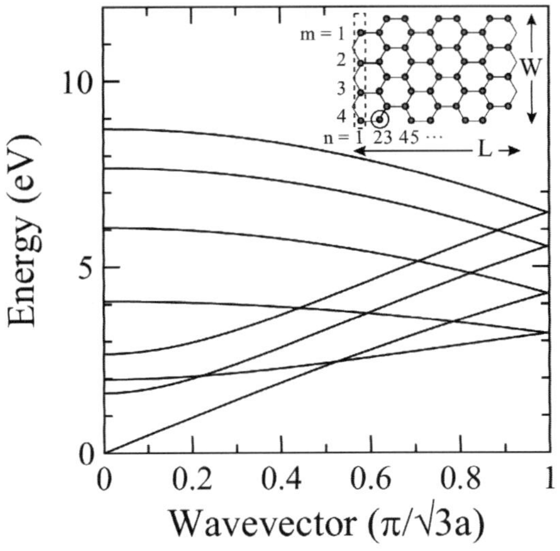


**Figure 1.** Band-structure of an armchair-edge graphene nanoribbon with a width of $W = (7/2)a$. Inset shows a schematic of the armchair-edge graphene nanoribbon.

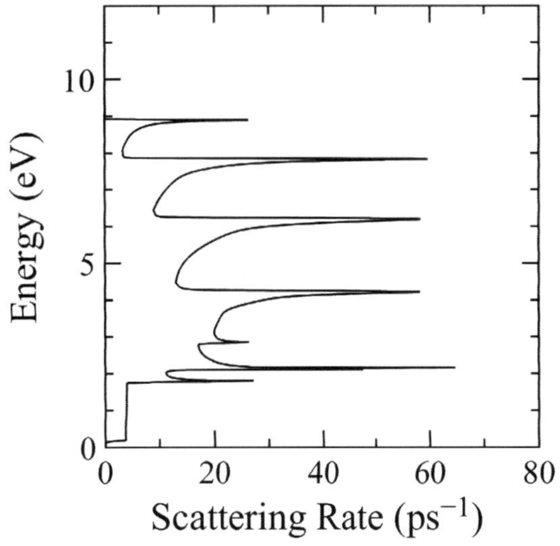


**Figure 2.** Optical-phonon scattering rate of an armchair-edge graphene nanoribbon with a width of $W = (7/2)a$ and a length of $L = (17\sqrt{3}/6)a$ at an edge atom (marked in the inset of Fig. 1).

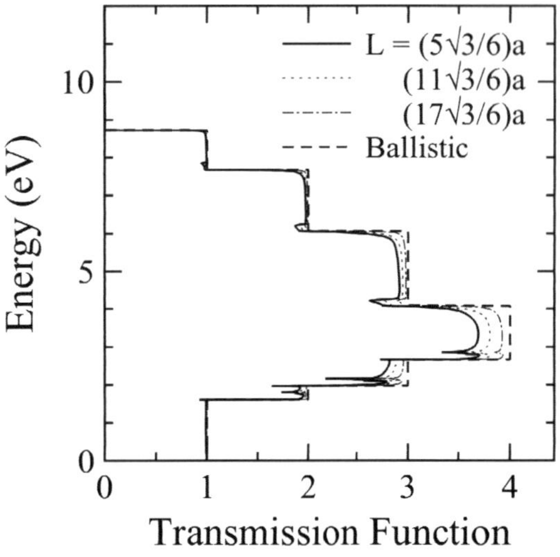


**Figure 3.** Transmission function through armchair-edge graphene nanoribbons with a width of $W = (7/2)a$ for $L = (5\sqrt{3}/6)a$, $(11\sqrt{3}/6)a$, and $(17\sqrt{3}/6)a$. Thick dashed line shows the results without optical-phonon scattering.

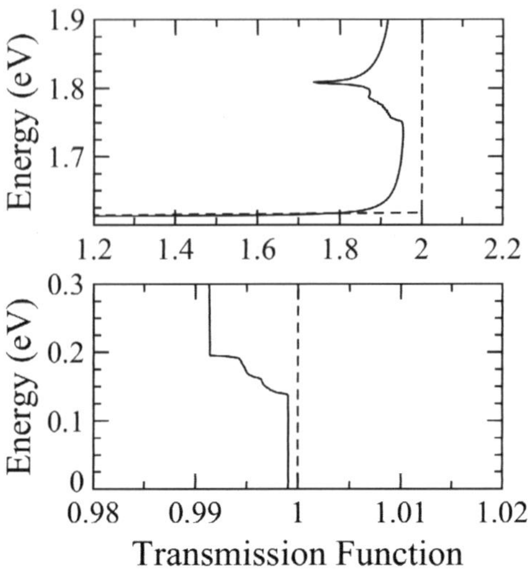


**Figure 4.** The same as Fig. 3 of $L = (17\sqrt{3}/6)a$ but on enlarged scales.

becomes higher than optical phonon energy, phonon emission occurs with the probability being proportional to the final density-of-states. This results in the structures on the transmission function.

## 4. Summary

We have calculated transmission function through metallic A-GNRs within NEGF method including full-band electron–optical-phonon interaction. We find structures appearing in transmission function, which are originated in both the zone-center and zone-edge optical-phonon scattering. We also find that the electron-phonon interaction has strong impact on the transmission for electron energies near the band center, while weak impact near the band edges.

## Acknowledgments

This work was supported in part by a Global COE Program at Osaka University "Center for Electronic Devices Innovation" from the Ministry of Education, Culture, Sports, Science and Technology.

[1] Novoselov K S, Geim A K, Morozov S V, Jiang D, Zhang Y, Dubonos S V, Grigorieva I V and Firsov A A 2004 *Science* **306** 666
[2] Frederiksen T, Brandbyge M, Jauho A P and Lorente N 2004 *J. Comput. Electron.* **3** 423
[3] Datta S 1995 *Electronic Transport in Mesoscopic Systems*, Cambridge University Press, Cambridge, UK
[4] Haug H and Jauho A P 1996 *Quantum Kinetics in Transport and Optics of Semiconductors*, Springer, Berlin
[5] Datta S 2000 *Superlatt. & Microstruct.* **28** 253
[6] Yoshimori A and Kitano Y 1956 *J. Phys. Soc. Jpn.* **11** 352
[7] Ando T 2005 *J. Phys. Soc. Jpn.* **74** 777
[8] Su W P, Schrieffer J R and Heeger A J 1979 *Phys. Rev. Lett.* **42** 1698
[9] Ishikawa K and Ando T 2006 *J. Phys. Soc. Jpn.* **75** 084713
[10] Suzuura H and Ando T 2008 *J. Phys. Soc. Jpn.* **77** 044703

Journal of Physics: Conference Series **193** (2009) 012113 doi:10.1088/1742-6596/193/1/012113

# Electron states near graphene edge

**V.A. Volkov and I.V. Zagorodnev**

V.A. Kotelnikov Institute of RadioEngineering and Electronics, Russian Academy of Sciences, Mokhovaya 11-7, Moscow, 125009, Russia

E-mail: vova@cplire.ru

**Abstract.** The band structure of graphene near the Fermi surface is often well described by the effective mass approximation. In this approach electron wave function obeys the Dirac-type equation. To describe a number of effects it is necessary to supplement the equation with boundary conditions. Using the Hermiticity and the time reversal symmetry we derive the phenomenological boundary conditions for a multicomponent effective wave function. The boundary conditions in the case of smooth crystalline potential near the edge are parameterized in terms of a real phenomenological constant that describes a general type edge. The edge states spectra (i.e. Tamm states spectra) for semi-infinite graphene are analyzed.

## 1. Introduction

Graphene, a monoatomic plane layer of carbon atoms at sites of a 2D hexagonal lattice, is a building block for the formation of 3D graphite, 1D nanotubes, and 0D fullerenes. Its unique properties attract increasing interest of physical community [1].

An electron in graphene has an ultrarelativistic spectrum near each corner of the hexagonal Brillouin zone. A charged two-dimensional (2D) massless Weyl neutrino would have had the same spectrum, if it existed. This spectrum is twice valley-degenerated. The centers of valleys $K$ and $K'$ are being spaced by vector $2\vec{p}_0$ in Brillouin zone. Such band structure is described in the effective mass approximation by the four-component Dirac-type equation for bispinor $(\psi,\psi')^T$:

$$v\begin{pmatrix} \boldsymbol{\sigma}(\hat{\boldsymbol{p}}-\boldsymbol{p}_0) & 0 \\ 0 & -\boldsymbol{\sigma}(\hat{\boldsymbol{p}}+\boldsymbol{p}_0) \end{pmatrix}\begin{pmatrix} \psi \\ \psi' \end{pmatrix} = E\begin{pmatrix} \psi \\ \psi' \end{pmatrix}. \tag{1}$$

Here $\hat{\boldsymbol{p}} = (\hat{p}_x, \hat{p}_y, 0)$ is the momentum operator in 2D system, $\boldsymbol{\sigma} = (\sigma_1, \sigma_2, \sigma_3)$ are Pauli matrices in the standard representation, $v = 10^6$ m/s is effective speed of light. The Hamiltonian acts on the four-component effective wave function $(\psi_A, \psi_B, \psi'_A, \psi'_B)^T$ [2], [3]. The difference of the Hamiltonian (1) from the traditionally used one is the shift by the quasi-momentum $\boldsymbol{p}_0$ been included in (1). This is necessary in order to take into account intervalley scattering of electrons.

To describe a number of effects, including the abnormal quantum Hall effect, size quantization in nanoribbons and quantum dots, it is necessary to understand the behavior of electrons or holes near the graphene edge. To do this, one should derive boundary conditions (BCs) for envelope wave functions. In general case the BCs have rather complicated form [4]. That is why the model BCs [1], [3] are often in use. In the tight-binding approximation armchair and zigzag edges are usually considered [5], [6] and the BCs are taken from the simplest consideration, that is however not well-founded. The model of infinitely-massive wall is also in use [7], [8]. To find edge states (ESs) spectra (an analog of Tamm states) one should solve the pseudorelativistic equations for the envelope wave functions with the derived BCs. It was found that the ESs spectra are sensitive to the choice of the model of an edge [7], [9]-[11].

The goal of this paper is to derive the BCs for the equation (1) satisfying general physical requirements without any assumptions about microscopic structure of an edge of graphene. Peculiarity of the boundary problem is the following: though there are no intervalley terms in the equation (1) such terms could be in BCs. It is because the equation (1) is applicable only for slowly varying potential. Crystalline potential at the edge may be sharp that leads to intervalley scattering.

To derive the BCs we assume that the Hamiltonian is Hermitian in restricted area, i.e. we construct self-adjoint extension of the Hamiltonian. We restrict ourselves to BCs that do not break time reversal symmetry. Similar approach was used before [4], [8], [12]-[14]. We concentrate on the simplest case when the intervalley scattering is absent. In this case the Dirac-type equation (1) is reduced to a couple of Weyl equations.

The outline of this paper is as follows. In Sec. 2 we consider BCs for the Weyl equation and find the ESs for Weyl neutrino on half-plane. In Sec. 3 the problem of possible boundary conditions for the equation (1) in the case of absence of intervalley scattering at the boundary is reduced to the problem of finding boundary conditions for the Weyl equation. Then, using symmetry arguments we can find ESs spectra for semi-infinite graphene sample from ESs spectra for the Weyl neutrino on semi-infinite plane. Finally, we discuss effects of intervalley scattering qualitatively.

## 2. The edge states for the Weyl equation on semi-infinite plane

We start with finding BCs for the 2D Weyl equation for two-component wave function $(\psi_1, \psi_2)^T$:

$$v\boldsymbol{\sigma}\hat{\boldsymbol{p}}\psi = E\psi. \tag{2}$$

We assume only the Hermiticity of the equation (2) in restricted area $S$. Performing partial integrations of $< \psi|\boldsymbol{\sigma}\hat{\boldsymbol{p}}|\phi >$ and equating the surface term to zero for a restricted domain in the $(x, y)$ plane we obtain the BCs:

$$\left(\psi_1 + i\cot\left(w/2\right)e^{-i\alpha}\psi_2\right)\big|_S = 0. \tag{3}$$

$w$ is a phenomenological real parameter, that characterizes the edge of the domain, $\alpha$ is the angle between the normal $\boldsymbol{n}$ and $X$-axis.

We solve the 2D Weyl equation (2) with the BC (3) on a half-plane for a particle with specified momentum $k$ along the boundary. The parameter $w$ is a constant on the edge. At an infinite distance from the boundary we assume that the wave function must be finite. Among the solutions there are those corresponding to the bulk spectrum and solutions corresponding to the ESs. For the latter we obtain the following solutions:

$$E = vk\sin w, \quad k\cos w \geq 0. \tag{4}$$

The ESs band has a linear dependence on the momentum $k$, Fig. 1. Depending on the value of the parameter $w$, it exists in different quadrants of the $(E, k)$ plane. The ESs band is chiral. It is not symmetric about $k = 0$. This is because of the Weyl Hamiltonian is not invariant under space inversion $\boldsymbol{r} \rightarrow -\boldsymbol{r}$.

The time reversal operator for the Weyl equation is $\hat{T} = \sigma_2\hat{K}_0$ ($\hat{K}_0$ is the complex conjugation operator). Note that the BC (3) breaks the time reversal symmetry. If we assume that the BC does not break time reversal symmetry, we will obtain only zeros BC.

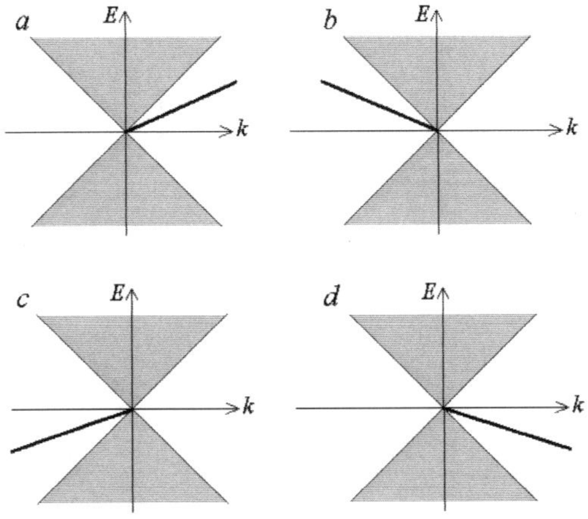


**Figure 1.** Electron spectrum $E(k)$ for the 2D Weyl equation on a half-plane. The shaded region corresponds to the continuous spectrum (bulk solutions), the thick lines to the edge states, which decay exponentially away from the boundary. The edge states exist in different quadrants, depending on the parameter $w$: $w \in (\pi/2, \pi)$ (b), $w \in (0, \pi/2)$ (a), $w \in (-\pi/2, 0)$ (d), $w \in (-\pi, -\pi/2)$ (c).

## 3. The edge states for semi-infinite graphene without intervalley scattering

We are now ready to consider BCs for the equation (1). If the crystalline potential near the edge of graphene is smooth enough we can neglect the intervalley scattering. It means that the BCs have no intervalley terms, i.e. components of the spinors from different valleys do not mix. In this case electron in graphene can be considered as a couple of the left and the right Weyl neutrino that is massless limit of the Dirac electron:

$$v \begin{pmatrix} \boldsymbol{\sigma}\hat{\boldsymbol{p}} & 0 \\ 0 & -\boldsymbol{\sigma}\hat{\boldsymbol{p}} \end{pmatrix} \begin{pmatrix} \psi \\ \psi' \end{pmatrix} = E \begin{pmatrix} \psi \\ \psi' \end{pmatrix}. \tag{5}$$

The time reversal operator (corresponding to complex conjugation of the total wave function) in our representation is [3]:

$$\hat{T} = \begin{pmatrix} 0 & \sigma_1 \\ \sigma_1 & 0 \end{pmatrix} \hat{K}_0, \tag{6}$$

where $\hat{K}_0$ is the complex conjugation operator. Following the procedure described in Sec. 2 and using T-invariance in addition to the Hermiticity of the problem we find the following BCs:

$$(\psi_A + i\cot(a/2)e^{-i\alpha}\psi_B)|_S = 0, \quad (\psi'_A + i\tan(a/2)e^{-i\alpha}\psi'_B)|_S = 0, \tag{7}$$

here $\alpha$ is the angle between armchair orientation and the edge of graphene. This BCs depend on a real parameter $a$ as for the Weyl case (2), (3).

We solve the equation (1) with derived BCs on a graphene half-plane the lateral boundary of which is specified by a 2D vector normal $(\cos\alpha, \sin\alpha)$ for a particle with specified momentum $k_y$ along the boundary ($x = 0$). As in Sec. 2, among the solutions there are solutions corresponding to the bulk spectrum and solutions corresponding to the ESs, see Fig. 2. We find that the ESs bands in $E(k_y)$ plane are rays starting in the projection of the valley centers on the edge, $\pm p_{0y}$, where $p_{0y} = p_0 \sin\alpha$. The ESs bands are described by the dispersion equations:

$$E_s(k_y) = sv(k_y + sp_{0y})\sin a, \quad s(k_y + sp_{0y})\cos a > 0. \tag{8}$$

$s = +1$ in the left valley and $s = -1$ in the right valley.

Solution for the ES envelope wave function is:

$$\psi(x) \sim e^{-x(k_y + p_{0y})\cos a}, \quad \psi'(x) \sim e^{x(k_y - p_{0y})\cos a}. \tag{9}$$

The ESs band is not symmetric about the center of the valleys $k_y = \pm p_{0y}$, but it is symmetric about the center of the Brillouin zone $k_y = 0$.

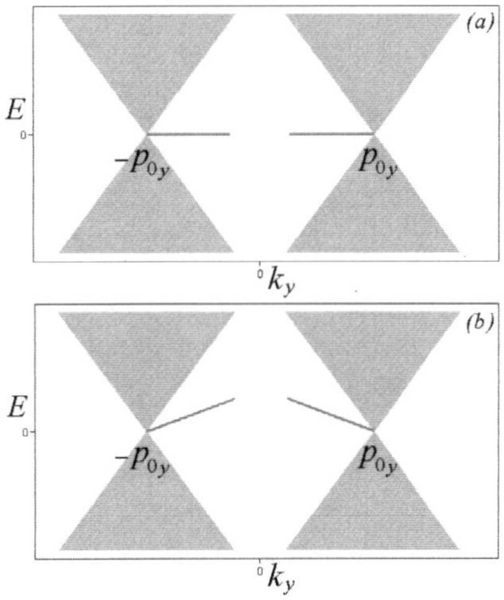


**Figure 2.** Electron spectrum $E(k_y)$ for a graphene half-plane (boundary x = 0) for two meaning of the parameter $a$. $p_{0y} = p_0 \sin\alpha$ is the projection of the valley centers on the $y$-axis. The shaded region corresponds to the continuous spectrum, and the lines to the edge states. The slope of the spectrum of the edge states with respect to the $x$-axis is determined by the boundary parameter $a$: (a) $\sin a = 0$, (b) $\sin a = 0.2$

## 4. Discussion

In this paper we have studied the edge states of graphene in the case of smooth crystalline potential near the edge. We found that a continuum description of electron wave function in the form of the Dirac equation must be supplemented with certain BCs. The BCs were derived from general physical requirements: the Hermiticity and the time-reversal invariance of the boundary problem. In this simplest case the graphene edge appears to be described by a real parameter $a$. We revealed that without intervalley scattering the Tamm band of the edge states for semi-infinite sample of graphene was easily obtained from Tamm states for the Weyl neutrino.

The ESs band is represented with the rays, Fig. 2, beginning at the projection of the valley centers on the edge and located in one of the quadrants of $(E, k_y)$ plane, depending on the parameter $a$. For the value $a = 0$ the known dispersionless spectrum for zigzag edge is obtained [6], [5]. For $\sin a \approx 0.2$ the result, Fig. 2b, agrees with the next-nearest-neighbors tight-binding calculations [9], [11].

In the case of arbitrary crystalline potential near the edge of graphene we derived more complicated BCs taking the intervalley edge scattering into account. The same phenomenological approach was used. We found that in this case left and right ESs rays instead of crossing near $k_y = 0$ have an anticrossing with a gap depending on the intervalley edge scattering.

## References

[1] Castro Neto A H, Guinea F, Peres N M R, Novoselov K S and Geim A K 2009 *Rev. Mod. Phys.* **81** 109
[2] Ando T 2005 *J. Phys. Soc. Jpn* **74** 777
[3] Beenakker C W J 2008 *Rev. Mod. Phys.* **80** 1337
[4] McCann E and Fal'ko V I 2004 *J. Phys.: Condens. Matter* **16** 2371
[5] Nakada K, Fujita M, Dresselhaus G and Dresselhaus M S 1996 *Phys. Rev.* B **54** 17954
[6] Fujita M, Wakabayashi K, Nakada K and Kusakabe K 1996 *J. Phys. Soc. Jpn.* **65** 1920
[7] Peres N M R, Rodrigues J N B, Stauber T and Lopes dos Santos J M B 2008 arXiv:0810.4768v1
[8] Berry M V and Mondragon R J 1987 *Proc. R. Soc. Lond.* A **412** 53
[9] Peres N M R, Guinea F and Castro Neto A H 2006 *Phys. Rev.* B **73** 125411
[10] Wakabayashi K, Takane Y, Yamamoto M and Sigrist M 2009 *CARBON* **47** 124
[11] Sasaki K, Shimomura Y, Takane Y and Wakabayashi K 2009 *Phys. Rev. Lett.* **102** 146806
[12] Akhmerov A R and Beenakker C W J 2008 *Phys. Rev.* B **77** 085423
[13] Volkov V A and Pinsker T N 1979 *Surf. Sci.* **81** 181
[14] Volkov V A and Pinsker T N 1981 *Sov. Phys. Solid State* **23** 1022

# Organic thin-film phototransistors based on poly(3-hexylthiophene)

**G Scarpa, E Martin, S Locci, B Fabel and P Lugli**

*Institute for Nanoelectronics, Technische Universität München, D-80333 Munich, Germany*

E-mail: scarpa@tum.de

**Abstract**. We investigated visible and UV light response of OTFTs based on the solution processable poly(3-hexylthiophene) (P3HT) semiconducting polymer. The transistors show a photoresponse in channel current under white and UV light illumination. Stabilization time after illumination and recovery time after turning off the light sources are also studied. A transport analysis based on a modified drift-diffusion model is presented.

## 1. Introduction

Organic phototransistors (OPTs) are one of the best feasible applications of organic thin film transistors (OTFTs) because of the large absorption and the excellent photocurrent generation efficiency of organic semiconductors both in the ultraviolet (UV) and visible range [1]. Solution-processable OPTs [2] can be easily integrated in smart clothes, packages, and biological systems as light sensors, biosensors and/or multifunctional sensors. OPTs fabricated by thermal evaporation of some oligomers as pentacene [3] and rubrene [4] have been found to have a very high photoresponse. Also solution processible OPTs based on conjugated polymers, such as poly(9,9-dioctylfluorene-co-bithiophene) [5] and methano [6,6]-phenyl C61-butyric acid methyl ester (PCBM) [6] have been reported.

Poly(3-hexylthiophene) (P3HT) is a solution processable semiconducting polymer, which is suitable for roll-to-roll process applications. It has reported field effect mobility up to ~0.1 $cm^2/V$ s [7], close to the value of amorphous silicon and it is one of the most promising materials for applications in OTFTs and solar cells [8, 9]. However, the use of P3HT for phototransistor applications is only reported in combination with other compounds. Meixner et. al. reported the use of dye-doped P3HT OPTs using dyes for wavelength selection [10], while Mok et. al. adopted P3HT in composite form combined with $TiO_2$ nanoparticles demonstrating good photocurrent generation under white light illumination [11]. Finally, the UV light sensitivity of OTFTs based on P3HT has not yet been analyzed.

In this work, a good photosensitive behavior due to response to UV and white light is reported for OTFTs based on P3HT processed from solution without any further treatment. The results show a fast response of transistors under light illumination, constituting a promising performance for further application of OPTs.

## 2. P3HT-transistor

### 2.1. Device fabrication

Bottom-gate OTFTs on boron-doped silicon wafers (p+-Si) as substrate were fabricated. The doping level is around $10^{18}$ cm$^{-3}$ and the conductivity is equal to 10,000 S/m (purchased from Si-Mat, Landsberg). On top of the p+-Si surface an 80 nm thick silicon dioxide ($SiO_2$) layer was grown in a dry oxidation process at 1000ºC. This insulator separates the p+-Si gate and the contact electrodes titanium/gold (2 nm Ti, 45 nm Au), patterned by lift-off in an optical lithographic step. The channel has a large cross-section area (W/L) because of the interdigited electrode structure used for the contacts [12]. In the inset of Figure 1 (left), a schematic cross-section of the transistor structure is shown. The values for the channel length (L) were chosen between 10 µm and 100 µm. In particular, three transistor structures with 10 µm, 50 µm and 100 µm channel lengths are investigated here. The conducting polymer (regio-regular poly-3-hexyl-thiophene, rr-P3HT, Sigma-Aldrich) was used as received and dissolved in toluene solvent with a mass relation of 1 to 100. The active layer was deposited by spin coating (1000 rpm for 20 s) under nitrogen atmosphere. The polymer and the dielectric surface were not modified. A semiconductor characterization system (Keithley 4200-SCS) was used for all measurements. The analysis has been carried out at room temperature under ambient conditions.

### 2.2. Device physics and simulations

Our simulation is based on the drift-diffusion approach implemented in the commercial software tool SENTAURUS™, which is a widely used and well-known software package designed for inorganic semiconductors [13]. When the organic phototransistor is illuminated, excitons are generated in the semiconductor [14]. Those excitons which do not recombine lead to an electron-hole pair. In the device, these carriers, as well as the others that are injected by the contacts, will move according to the drift-diffusion model, which means that they will drift, driven by an applied electric field, or diffuse, due to the presence of a gradient of the carrier density [15]. Although we were unable to simulate the dynamics of exciton generation and recombination, we simulated the phototransistors using the SENTAURUS drift-diffusion simulator adding an optical generation term. Optical generation is considered by adding a generation rate G(**r**,*t*) (2) to the continuity equations (1):

$$\nabla \cdot \mathbf{J}_n = -qG(\mathbf{r},t) + \frac{\partial n}{\partial t}, \quad -\nabla \cdot \mathbf{J}_p = -qG(\mathbf{r},t) + \frac{\partial p}{\partial t} \tag{1}$$

$$G(\mathbf{r},t) = \frac{P_0(\mathbf{r},t)}{E_{\mathrm{Ph}}}\alpha \tag{2}$$

where $P_0(\mathbf{r},t)$ is the incident power of the wave beam in the position **r** at the time *t*, $E_{\mathrm{Ph}}$ is the photon energy and $\alpha$ is the absorption coefficient. We assume a constant light absorption throughout the entire device. In this framework excitons are not considered and the photons directly generate an electron-hole pair. P3HT parameters used for the simulations are electron mobility $\mu = 1\cdot 10^{-6}$ cm$^2$/Vs, hole mobility $\mu = 1.28\cdot 10^{-3}$ cm$^2$/Vs, band gap 2.13 eV, electron affinity 3.2 eV, dielectric permittivity 3 and density of states $N_c = N_v = 2\cdot 10^{21}$ cm$^{-3}$.

### 2.3. Electrical characteristics

Figure 1 (left) shows the measured and simulated output characteristics under dark conditions of a sample device with 20 µm channel length and a W/L= 13,350. As the surface of the silicon dioxide was not treated separately, nearly all transistors showed a positive threshold voltage of about +5 V, thus displaying a clearly conducting channel already at a zero gate voltage, as can be seen from the picture. The field-effect mobility extracted from the transfer characteristics of the transistors was in the order of $10^{-4}$ cm$^2$/Vs, which was consistent with previous findings and experiments [12]. Figure 1 (right) shows the space charge distribution in the device. The transverse (gate induced) electric field attracts holes at the interface forming a layer of charges that acts as the channel.

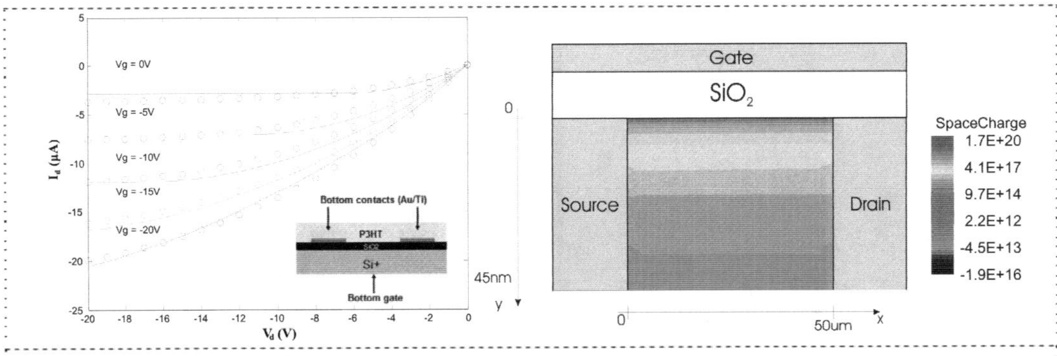

**Figure 1.** Left: Measured (circles) and simulated (lines) output characteristics of an OFET with P3HT spun out of toluene with a channel length of 20 μm and W/L = 13,350 on 80 nm silicondioxide. Inset: schematic picture of the transistor structure (cross-section) adopted in our studies. Right: Space charge distribution at $V_g$ = -10V, $V_d$ = -15V

## 3. Photoresponse of the devices

Several measurements for evaluating the photoresponse of the transistors to white and UV light were performed. A white light source with an intensity of 7.9 mW/cm$^2$ and a UV diode arrays-source with an intensity of 8.6 mW/cm$^2$ (with a peak wavelength of 375 nm) has been used. All experiments were done with the light illuminating the entire surface of the transistors. The distance between the focus and the sample for white light was fixed at 10 cm and for the UV measurements 2 cm was used. A fast response under illumination and the capability of the transistors to recover to the initial state are important factors for further applications of OPTs. Therefore, measurements of the channel current, under light illumination, versus time have been performed. Figure 2 shows a channel current change of 1 μA and 0.8 μA under UV and white light illumination, respectively; by fixed values for $V_{ds}$ (-15 V) and $V_{gs}$ (-10 V). In spite of the small values, this current increment could be high enough for photosensors applications. The recovery time after turning off the light, is about 90 s, comparable with recovery times reported for phototransistors [1]. However, it is so far a quite high value and constitutes an important disadvantage.

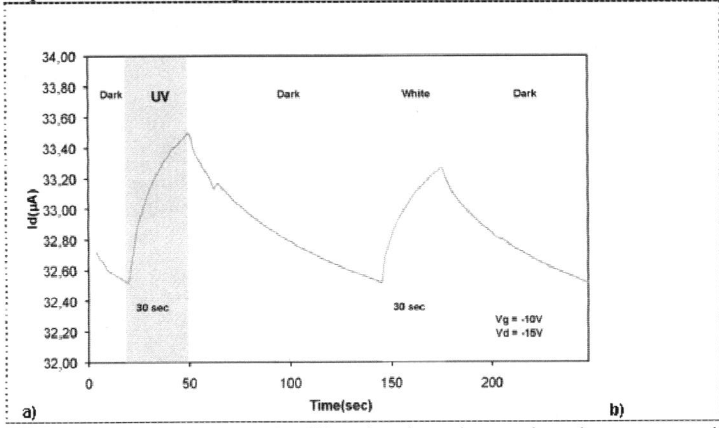

**Figure 2.** Channel current responses under white and UV light illumination is presented. 30 s UV and 30 s white light has been applied. $V_{ds}$ = -15 V, $V_g$ = -10 V.

The simulated response to the illumination shows that the generated carriers determine a space charge distribution which is almost uniform along the channel, except for the vicinity of the contacts. The potential being also uniformly flat in the device (most of the voltage drop occurs in the vicinity of the source contact), very long times are required to relax the exceeding carriers once the light is turned off, because both the electric field and the carrier gradient are low. The proposed simulation (Figure 3 left) qualitatively agrees with the experimental data, but still exciton dynamics should be included to provide a better fitting.

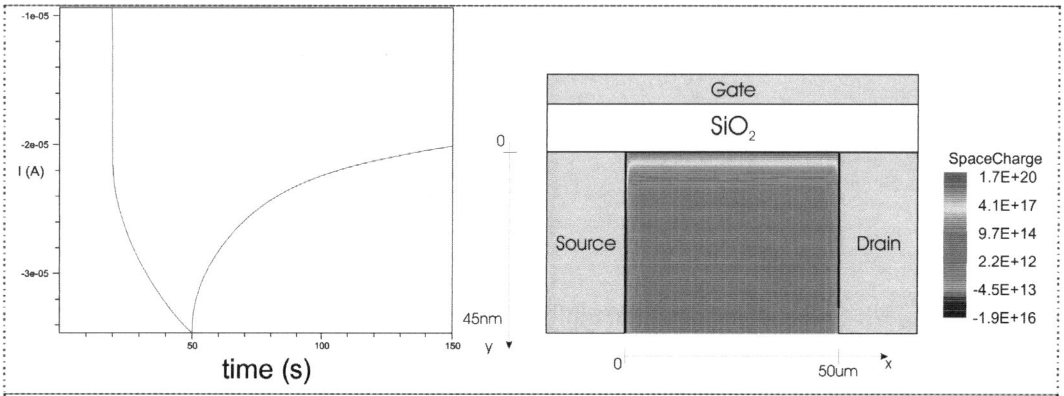


**Figure 3.** Left: Simulated response to illumination. Right: Space charge under illumination.

## 4. Conclusion

The photosensitivity of OTFTs based on solution-processable P3HT p-type organic semiconductor has been tested. All measurements were done at room temperature ambient conditions and the polymer was used as received. The transistors show a good response to white and a higher one to UV light. The time analysis shows a fast response after the illumination and a slow recovery time after turning off the light source. This delay for recovering the initial value in channel current is a big disadvantage of the phototransistors. The recovery time depends also on the measurement starting point. In other words, the initial polarization state and the density of charges trapped are important factors before starting the measurement. Here, the gate voltage plays an important role for controlling the initial conditions. Despite these disadvantages, low-cost, large-area applications of solution-processable P3HT-based phototransistors are feasible.

**References**

[1] Hamilton MC, Martin S, Kanicki J, IEEE Trans. Elect. Dev. **51**, 877 (2004).
[2] Marjanovic N, et al., Organic Electronics, **7**, 188 (2007).
[3] Hu Y, Dong G, Liu C, Wang L, Qiu Y, Appl. Phys. Lett. **89**, 072108 (2006).
[4] Podzorov MEGV, Physical Review Letter **95**, 016602 (2005).
[5] Saragi TPI, Pudzich R, Fuhrmann T, Salbeck J, Appl. Phys. Lett. **84**, 2334 (2004).
[6] Narayan KS, Kumar N, Appl. Phys. Lett. **79**, 1891 (2001).
[7] Sirringhaus H, Tessler N, Friend RH, Science **280,** 1741 (1998).
[8] Sirringhaus H, et al., Synthetic Metals **111–112,** 129 (2000).
[9] Mihailetchi VD, et al., Adv. Funct. Mat. **16**, 699 (2006).
[10] Meixner RM, et al., Appl. Phys. Lett. **89**, 092110 (2006).
[11] Mok SM, Yan F, Chan HLW, Appl. Phys. Lett. **93**, 023310 (2008).
[12] Goetz SM, et al., Organic Electronics, **10**(4), 573 (2009).
[13] www.synopsys.com
[14] Schäfer W. & Wegener M. 2002 *Semiconductor Optics and Transport Phenomena* (Berlin: Springer).
[15] Sze S M 1985 *Semiconductor Devices: Physics and Technology* (New York: Wiley)

# Design and fabrication of organic solar cells structured via nanoimprint lithography

**O Wiedenmann, A Abdellah, G Scarpa and P Lugli**

*Institute for Nanoelectronics, Technische Universität München, D-80333 Munich, Germany*

E-mail: lugli@tum.de

**Abstract**. A theoretical analysis based on a drift-diffusion simulation method is performed for investigating device performances of nanostructured organic solar cells. The results show an increase of the short-circuit current density, which can be attributed to the increased interface area. Patterning via nanoimprint techniques is demonstrated, which constitutes the first step towards the realization of bulk-heterojunction organic solar cells with controlled morphology.

## 1. Introduction

In recent years, research and development in the field of organic semiconducting materials has advanced tremendously, driven by the potential for a variety of low cost, large area devices with attractive market perspectives. Ever since the discovery of conductivity and semi-conductivity in polymers [1] as well as the discovery of metallic conductivity in poly-acetylene in 1977 [2], the electrical properties of various small molecules and polymers have been investigated. This paved the way for the realization of organic light-emitting diodes (OLED), photodiodes, transistors and sensors. The field of organic optoelectronics has become one of the most promising for research and for industry. It is therefore not surprising that it is in this field where the first commercial products like OLED displays have emerged. Utilizing the full range of advantages offered by organic semiconductor technology (e.g. large-area, cheap manufacturing, flexible substrates), solar cells based on organic materials can potentially be manufactured in mass production at lower costs with respect, e.g., to silicon [3] since they do not require high deposition temperatures, complex processing or clean room environments. However, up to now the efficiencies of organic solar cells cannot compete with conventional photovoltaics. One major challenge is the control of the interfacial area between donor-acceptor materials. In fully organic solar cells two organic components with different electron and hole affinities are mixed [3]. Generally the fabrication methods (such as solution processing and vacuum sublimation) used for preparing the active films of these components do not allow high control over the phase separation of the organic layers. A promising alternative is provided by nanopatterning techniques and in particular by nanoimprint lithography [4-7]. This method allows one to structure organic layers in the nanometer scale and enables the tuning of the interfacial area between the two organic materials. In this paper, we discuss the influence of morphology on the performance of organic solar cells via a two dimensional simulation. In addition, we show how nanoimprinting can realize patterning of the active layers down to the required nanometer scale.

## 2. Methods

### 2.1. Nanoimprint lithography

Nanoimprint lithography (NIL) can be used for structuring thin P3HT layers. This technique would allow the realization of ordered heterostructure solar cells [6,7] using P3HT as donor material, for further improving the efficiency of these devices. In order to fabricate heterostructures based on the P3HT/PCBM material system, one should be able to deposit PCBM on top of P3HT layers. However, the use of spin-coating techniques could be difficult, since successively deposited layers would eventually destroy the layer underneath. Our experiments revealed that by using different solvents for the two polymers, a bilayer structure of P3HT and PCBM can be realized. Our films were spin coated at 1500 rpm for 20 s from a chloroform solution (with a concentration of 0.5 % by weight) onto 0.5 μm-thick polished glass covered by a PEDOT:PSS (Baytron P purchased by H. C. Starck) layer, which was spun on the substrates prior to deposition of the P3HT layer (2000 rpm for 20 s baked at 150°C for 15 min). All films were deposited under nitrogen atmosphere and no other thermal treatment was applied after depositing the P3HT.

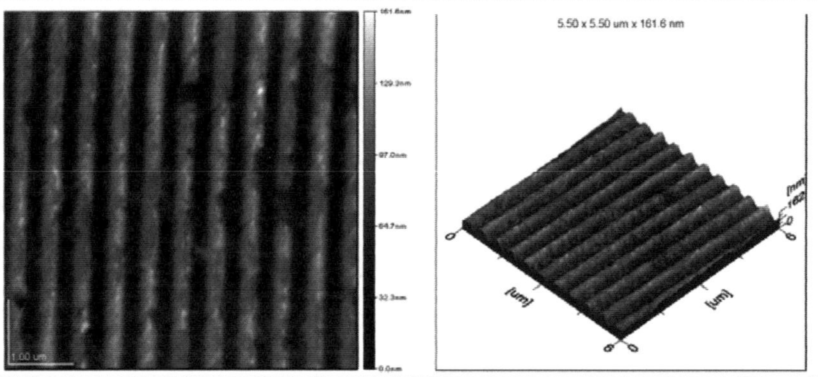


**Figure 1.** AFM topographic top view images of 200-nm-wide lines imprinted into P3HT.

After deposition, the samples were placed on the chamber of a commercial (Obducat 2.5 inch) NIL tool with the mold manually placed upside down onto the imprint polymer layer. Imprint duration, temperature and pressure were computer-controlled, in-situ monitored, and can be preset. Figure 1 shows topographic AFM images of imprinted P3HT polymer surfaces structured with the NIL process described above, with grating patterns comprising 200-nm-wide lines. The imprint temperature was 180°C. The surface structure of the P3HT film layer shows good feature transfer quality with an imprint depth of approximately 80 nm and rounded sidewalls.

### 2.2. Simulation method

Our simulation is based on the drift diffusion approach implemented in the commercial software tool SENTAURUS$^{TM}$, which is a widely used and well-known software package developed for inorganic semiconductors [8]. It is a multidimensional, electro-thermal, mixed-mode device and circuit simulator designed to simulate inorganic one-dimensional, two-dimensional and three-dimensional semiconductor devices. The self-consistent numeric scheme solves the Poisson equation coupled to the drift-diffusion and continuity equations while considering van-Neumann type boundary conditions for contacts and geometry. This approach showed already good agreement for organic bulk heterojunction solar cells [9]. In order to account for the special nature of organic semiconductors, input parameters such as the energy of the transport levels and the mobility needed to be adjusted [10]. Furthermore, we have implemented a number of physical models (e.g. field-dependent and density-dependent mobility, Langevin-type recombination), which are essential in organic materials. The material parameters used for the donor material (P3HT), the acceptor material (PBCM) and the blend structure (P3HT:PCBM) are shown in Table 1. To introduce the blend into our simulations, we have combined the properties of both P3HT and PCBM in a virtual semiconductor. The LUMO and HOMO levels are taken from PCBM and P3HT, respectively, as illustrated in the inset of Figure 2. Figure 2 demonstrates the precision that we can obtain in the reproduction of experimental current-voltage characteristics of BHJ photodiodes. In the implementation of the simulations, it has been a key success factor to derive a suitable material model that would describe the polymer bulk heterojunction in the blend material. The analyzed device has a 200 nm active layer of P3HT:PCBM blend contacted by ITO and Calcium electrodes. In the case of nanostructured solar cells, in order to take into account the effects of exciton-dissociation at the interface between donor and acceptor material, a 10 nm region along the interface has been defined as virtual semiconductor (as explained above). The absorbing region defined as a virtual semiconductor is indicated by the dotted area along the donor/acceptor interface in the inset of Figure 3. This allowed us to mimic exciton processes (generation, diffusion and dissociation) and to properly account for the separation of the charges generated from exciton-dissociation and their further contribution to the carrier density at the respective energy levels of the materials (inset of Figure 2). In other words, incident photons generate excitons, which start to diffuse in the material. As assumption, only those excitons generated within the exciton diffusion length from the interface can dissociate. After dissociation, a free electron can move in the LUMO of the acceptor (PCBM), while a free hole can travel in the HOMO of the donor (P3HT). Finally, the carriers are collected by the respective metal contacts.

**Table 1.** Virtual semiconductor parameters used for the P3HT:PCBM blend.

| | P3HT | virtual semiconductor (blend structure) | PCBM |
|---|---|---|---|
| **LUMO [eV]** | 3.0 | 3.7 | 3.7 |
| **HOMO [eV]** | 5.0 | 4.86 | 6.1 |
| **Band gap [eV]** | 2.0 | 1.16 | 2.4 |
| **Hole mobility [$cm^2$/Vs]** | 1.0e-3 | 1.0e-3 | 1.0e-5 |
| **Electron mobility [$cm^2$/Vs]** | 1.0e-5 | 1.0e-3 | 1.0e-3 |
| **Effective absorption [$cm^{-1}$]** | 2.0e4 | 2.0e4 | 2.0e4 |
| **Density of states [$cm^{-3}$]** | 5.0e20 | 5.0e20 | 5.0e20 |
| **Carrier lifetime [s]** | 5.0e-7 | 5.0e-7 | 5.0e-7 |

## 3. Results

The current-voltage characteristics of organic solar cells with different interface geometries are shown in Figure 3, while the insets show the schematic cross-section of the devices. A bilayer structrure of 100 nm P3HT and 100 nm PCBM is simulated as reference cell (Table 2). The nanostructured devices have a total active layer thickness of 200 nm, which is divided into a structured region of 100 nm height (parameter h in the inset of Figure 3) embedded in a 50 nm thick donor and acceptor layer, respectively. The simulated widths (parameter w in the inset of Figure 3) are indicated in Table 2.

Whereas the open circuit voltage $V_{OC}$ of the devices remains constant in agreement with previous findings [5], the short circuit current density $J_{SC}$ is rising with larger interface areas. This fact can be explained with the increasing number of excitons reaching the donor/acceptor interface in order to dissociate, thus generating more carriers.

Journal of Physics: Conference Series **193** (2009) 012115 doi:10.1088/1742-6596/193/1/012115

Table 2 shows a comparison between the sizes of the interface area of the active region (where exciton splitting is assumed) and the respective short circuit current density values of the devices extracted from our simulations. It is worth to observe that the correlation between interface area and short circuit current density is quite accurate.

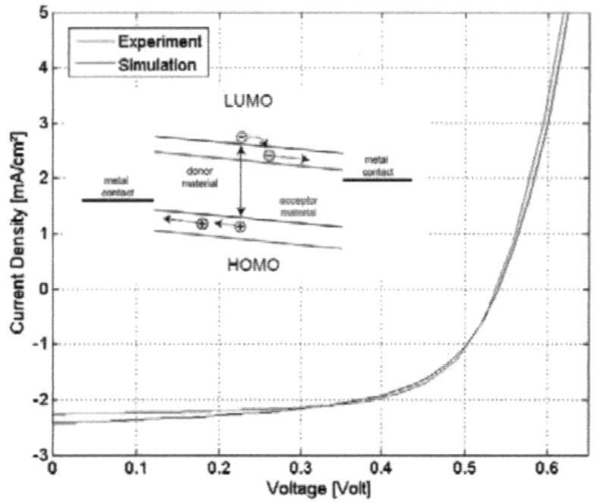


**Figure 2.** Simulation results compared to experimental reference values. Inset: transport levels in bulk heterojunction solar cells based on acceptor and donor materials.

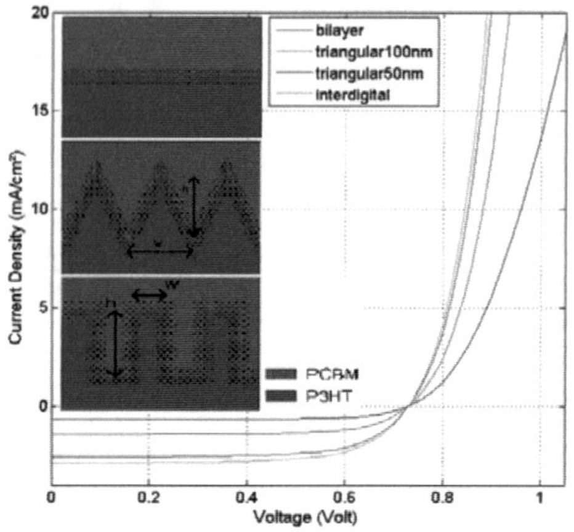


**Figure 3.** Simulation results for various device geometries. The short circuit current values are reported in Table 2. Inset: schematic cross-sections of the structures. The dotted area represents the absorbing region.

**Table 2.** Comparison of the different geometries with the calculated interface areas.

| | $A/A_0$ | $J_{SC}$ | $J_{SC}/J_{SC0}$ | $V_{OC}$ |
|---|---|---|---|---|
| **Bilayer (reference cell)** | 1 | 0.65 | 1 | 0.73 |
| **Triangular (w=50 nm)** | 2.2 | 1.41 | 2.17 | 0.73 |
| **Triangular (w=25 nm)** | 4.0 | 2.58 | 3.97 | 0.73 |
| **Interdigital (w=25 nm)** | 4.15 | 2.90 | 4.46 | 0.73 |

## 4. Conclusions

The simulation will help understand the physical models and the principal effects of device specific parameters, as simulation variables for the materials and their impact on the results can be illustrated. The ability to accurately simulate the electrical and optical properties of solar cells based on organic semiconducting materials will aid in the design process.

## References

[1] Shirakawa H, Louis EJ, Macdiarmid AG, Chwan K, Heeger AJ, Chem. Comm. **16**, 578 (1977).
[2] Chiang CK, et al., Phys. Rev. Lett. **39**, 1098 (1977).
[3] Thompson BC, Frechet JMJ, Angew. Chem. Int. Ed. **47**, 58 (2008).
[4] Chou SY, Appl. Phys. Lett. **67**, 3114 (1995).
[5] Kim M, et al., Appl. Phys. Lett. **90**, 123113 (2007).
[6] Na S, et al., Appl. Phys. Lett. **91**, 173509 (2007).
[7] Cheyns D, et al., Nanotechnology **19**, 424016 (2008).
[8] www.synopsys.com
[9] Waldauf C, Schilinsky P, Hauch J, Brabec CJ, Thin Solid Films **451–452**, 503 (2004).
[10] Bolognesi A, Di Carlo A, J. of Comp. Electr. **2(2-4)**, 297 (2003).

Journal of Physics: Conference Series **193** (2009) 012116 doi:10.1088/1742-6596/193/1/012116

# Crossover from Weakly-Disordered Fermi Liquid to Luttinger Liquid in quasi-1D Nanostructures

**V Mitin, A Sergeev, M Bell, J Bird and A Verevkin**
Dept. of Electrical Engineering. State University of New York at Buffalo, 14260, USA
E-mail: vmitin@buffalo.edu

**Abstract.** We investigated the temperature–dependent conductance of a long and narrow high-mobility channel, which is formed from an AlGaAs/GaAs heterostructure using the split-gate technique. When the channel is wider than 100 nm, in the range of 1-10 K we observe logarithmic temperature dependence, which is a key signature of the interaction correction in the quasi-one-dimensional weakly-disordered Fermi liquid. For the narrowest channels with width smaller than 100 nm, we observe power-law dependences, which are typical for the Luttinger liquid. The crossover from the weakly-disordered Fermi liquid to the Luttinger liquid takes place when the channel width corresponds to 2-3 one-dimensional subbands.

## 1. Introduction

Interference of electron-electron and electron-impurity scattering leads to nontrivial transport properties of the Fermi liquid. Interference terms in electrical conductivity have specific temperature dependencies. In the diffusive limit, $T\tau << 1$ (τ is the elastic mean free time), the Al'tshuler-Aronov corrections to conductivity have been intensely studied in 3D conductors ($T^{1/2}$-term), in quasi-2D metallic films and disordered semiconductor structures (ln$T$-term), and quasi-1D metallic and semiconductor nanowires ($T^{-1/2}$-term) [1]. In the opposite, i.e. quasi-ballistic limit, $T\tau >> 1$, the interference 2D corrections ($T$-term) [2] have been also investigated in various semiconductor heterostructures. For high-mobility quasi-1D conductors, i.e. for the weakly-disordered 1D multi-channel Fermi liquid, a corresponding theory of many-body effects has been recently developed by Sergeev et al. [3]. It predicts temperature-dependent corrections proportional to ln$T$.

For even narrower channels with essentially one subband, the electron-electron interaction leads to formation of the Luttinger Liquid (LL) [4]. For sufficiently long channels in the presence of disorder, suppression of the conductance is expected. This negative correction to the conductance versus temperature has been predicted to be proportional to $T^{g-1}$, where $g < 1$ is an interaction parameter [5,6]. In the current paper we investigate transition from the weakly-disordered Fermi liquid to the Luttinger liquid behaviour.

## 2. Experiment

In this paper we investigate corrections to the conductivity in weakly disordered, quasi-one-dimensional conductors, where the conductor width is substantially smaller than the characteristic length $L_T = v_F/T$, $v_F$ is the Fermi velocity. The device measured consisted of a 6 voltage-probe Hall bar made from an AlGaAs/GaAs two-dimensional electron gas (2DEG). The measured mobility $\mu$ and electron concentration $n_{2D}$ of the 2DEG is $5\times10^5$ cm$^2$/Vs and $3\times10^{11}$ cm$^{-2}$ at 4.2 K respectively. Gate electrodes were deposited on top of the Hall bar to form a split-gate to electrostatically define an adjustable width channel with lithographic dimensions of 0.5 μm x 100 μm. A schematic of the device is shown in Figure 1(a).

Magnetoresistance measurements were used to determine the effective width $w$ of the channel for different gate voltages $V_g$ [7,8]. The low-field magnetoresistance of the channel for a gate voltage of -1150 mV is shown in Figure 2 inset. The negative magnetoresistance which exhibits a sharp peak at low fields is due to weak-localization (WL) effects in the channel. At higher fields above ~100 mT effects of WL are suppressed, at these fields the resistance vs. temperature $R(T)$ measurements were performed. From the mobility of the 2DEG the low temperature limit for the quasi-ballistic regime $T\tau >> 1$ was determined to be above ~700 mK.

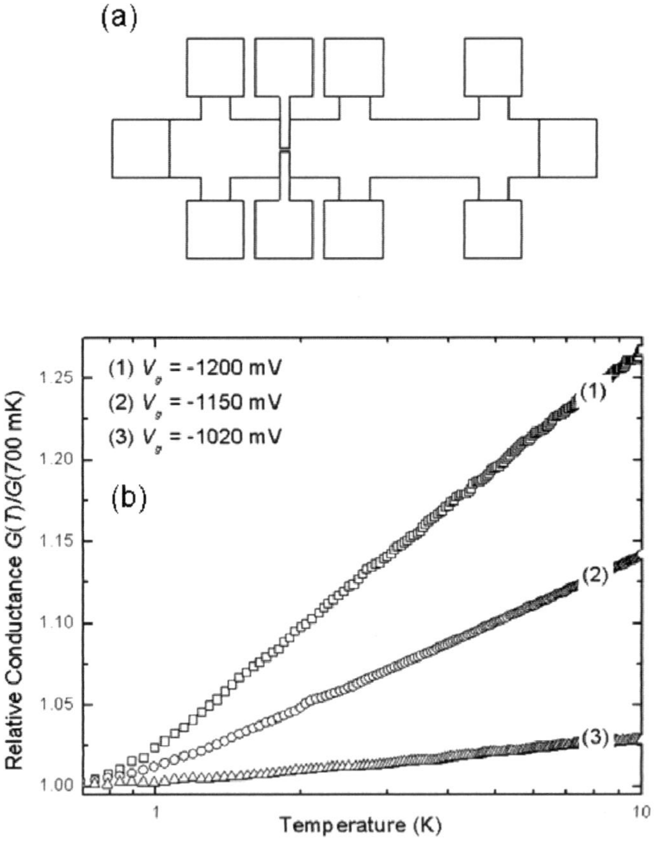


**Figure 1**. (a) Schematic of the device measured. (b) Conductance normalized to the residual conductance at 700 mK vs. temperature for three different gate voltages ($V_g$ = -1.2, -1.15, and -1.02 V).

## 3. Discussion

Plotted in Fig. 1(b) is the dimensionless conductance normalized to the conductance measured at 700 mK vs. temperature for three different gate voltages $V_g$ = -1.2 ($w \approx$ 118 nm), -1.15 ($w \approx$ 125 nm), and -1.02 V ($w \approx$ 160 nm). For these channels the conductance shows a ln($T$) dependence, which was predicted for the quasi-one dimensional weakly-disordered Fermi liquid [3]. In weakly disordered conductors, characteristic momentum transfers are of the order of $T/v_F$ , which is significantly smaller than the Fermi momentum. Therefore, the Landau Fermi-Liquid theory is applicable and all processes with large momentum transfer are taken into account by the effective Fermi-liquid constants. Both singlet and triplet channels of the electron-electron interaction give the interference corrections to conductivity [3],

$$\frac{\delta_{ee}\sigma}{\sigma_0} = \left[1 + \frac{3\left[F_0^\sigma + G\left(F_0^\sigma\right)\right]}{1+F_0^\sigma}\right]\frac{1}{wp_F}\ln\left(\frac{2T}{\varepsilon_F}\right), \quad (1)$$

where $p_F$, $\varepsilon_F$, and $F_0^\sigma$ are the Fermi momentum, Fermi energy, and Fermi liquid parameter respectively. The function $G(F_0^\sigma)$ was calculated in Ref. 3. Due to the Coulomb potential divergence at small momenta, the single-channel interaction corresponds to the unitary limit and the corresponding corrections are independent of the Fermi-liquid parameters. The triplet-channel corrections have the same temperature dependence as the singlet-channel corrections. Contrary to the singlet channel, the triplet channel corrections are not universal. Therefore, the value and sign of the total correction depend on the Fermi-liquid parameter in the triplet channel. In the weak coupling limit, $F_0^\sigma << 1$, the singlet-channel dominates over the triplet one and the corrections to conductivity are positive. The temperature dependant normalized conductance data in Fig. 1(b) was fitted to Eq. 1 for channels of various widths and show that the normalized conductance has a ln($T$) dependence and the pre-logarithm of the data shows $w^{-1}$ dependence consistent with the theory.

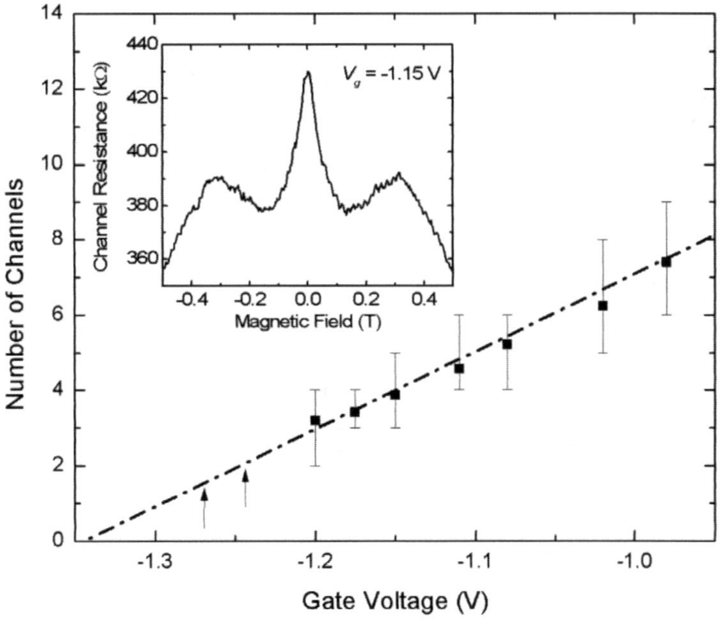


**Figure 2.** Number of subbands in the channel determined from the width of the channel as a function of gate voltage. Solid line is a guide to the eye. Arrows indicate the gate voltages of the channel which show a power-law conductance dependence on temperature. Inset low-field magnetoresistance of the channel with a gate voltage of $V_g$ = -1150 mV.

Shown in Figure 2 is the number of subbands determined from the width of the channel vs. gate voltage. As can be seen, for the range of gate voltages where the ln($T$) dependence of conductance was observed $V_g$ = -1020 mV to -1200 mV, the channel consists of multiple subbands, wich corresponds to the weakly disordered multi-channel Fermi liquid. As the gate voltage is made more negative, $V_g$ < -

1245 mV, which results in a narrowing of the channel (indicated by the arrow in Figure 2), there is only one – two subbands in the channel. For $V_g$ = -1245 mV and -1265 mV a power-law dependence of conductance on temperature is observed and is shown in Figure 3. This power-law dependence is a key signature of Luttinger Liquid behaviour [5,6]

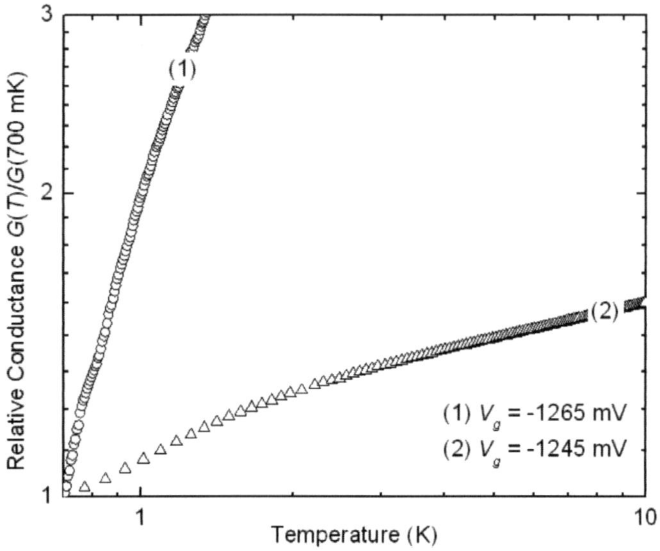


**Figure 3.** Conductance normalized to the residual conductance at 700 mK vs. temperature for $V_g$ = -1245 mV and -1265 mV.

In summary, we experimentally investigated the conductance of a long and narrow high-mobility channel and observed crossover from weakly-disordered multi-channel Fermi liquid to the Luttinger liquid when decreasing channel width to ~ 100nm. In the range of channel widths 118 – 160 nm, at temperatures 1- 10 K we clearly observe the logarithmic temperature dependences of the conductance. These dependences and corresponding pre-logarithmic factors are adequately explained by effects of electron-electron interaction in weakly-disordered quasi-one dimensional (1D multi-channel) Fermi liquid. When the width further decreases, the logarithmic dependences in conductance change to power-law dependences, which are typical for Luttinger liquid. This crossover takes place when the channel width corresponds to 2-3 one-dimensional subbands.

## References

[1] B. L. Altshuler and A. G. Aronov, *Electron-Electron Interaction in Disordered Systems,* edited by A. L. Efros and M. Polak (North-Holand, Amsterdam, 1985).
[2] G. Zala, et. al., *Phys. Rev. B* 64, 214204 (2001).
[3] A. Sergeev, M. Yu. Reizer, and V. Mitin, *Phys. Rev. B* 69, 075310 (2004).
[4] S. Tomonaga, *Prog. Theor. Phys.* 5, 544 (1950).
[5] M. Ogata and H. Fukuyama, *Phys. Rev. Lett.* 73, 468 (1994).
[6] D. L. Maslov, *Phys. Rev. B* 52, R14368 (1995).
[7] T. J. Thornton, *et. al., Phys. Rev. Lett.* 63, 2128 (1989).
[8] K.-F. Berggren, T. J. Thornton, D. J. Newson, and M. Pepper, *Phys. Rev. Lett.* 57, 1769 (1986).

# Simulation of transport and 1/f noise in carbon nanotube films

**S. Soliveres, F. Martinez, A. Hoffmann, F. Pascal**
IES, University of Montpellier 2, place E. Bataillon, 34095 Montpellier, France

E-mail: soliveres@ies.univ-montp2.fr

**Abstract.** In this paper we present a simulator for electrical properties of carbon nanotube film field-effect transistors. The simulator, based on carbon nanotube physics uses Landauer formalism and Poisson equation. The total film is described as an electrical network. A modified nodal analysis provides DC and noise characteristics. These simulations are in good agreement with experimental results.

## 1. Introduction

The properties of an individual nanotube make it interesting to use in microelectronic as transistor (CNFET). If the fabrication of high quality and reproducible CNFETs remains a difficult challenge, sensor applications using large quantities of CNTs are already available. The sensors consist of field-effect transistors based on films of CNTs arranged randomly. Although transport properties in individual nanotube have been intensively studied, little is known about structures with large quantities of CNTs.

The material consists of a large quantity of nanotubes arranged randomly. The first studies have shown that nanotube devices exhibit linear current-voltage characteristics and a large 1/f noise component. Collins et al have observed that this 1/f noise level appears to be proportional to the device resistance for individual NTs, films and mats that are CNT random networks [1]. Snow et al have measured the noise in films and have shown empirically that the noise appears to be proportional to the resistance and nearly inversely proportional to the film's lengths [2]. This length dependence has not been observed in individual NTs, whether in the case of diffusive or ballistic transport, and appears to be specific to NT macrostructures. We have shown in an experimental paper that this length dependence is explained if we consider the structures as percolating networks [3]. In such a structure, CNT are links which are connected to form a path between two electrodes.

In this paper, we present an electrical simulator which allows to reproduce experiment and to predict the behaviour of noise and conduction in CNTs films and transistors.

## 2. Theory and modelling

### *2.1. Theory*

Experiments have shown that nanotube films seem to behave as a small band gap semiconductor (sc) with p-type conduction. However, it's not possible to describe a NT film as a crystalline

semiconductor compound. Due to the absence of covalent bond and to the cylinder nature of NTs, there is a weak coupling between NTs. Moreover the fact that CNTs can be described by links in a percolating network shows that NTs conserve their individual properties in the network.
Two microscopic charge transport mechanisms occur in the film: the transport along nanotube themselves and the transport between crossed nanotubes. Considering the large mean free path in CNTs and the weak coupling between NTs, we assume that the contacts between NTs dominate the transport through the film.

### *2.2. Model*

In order to get a microscopic understanding of the transport phenomenon, we have developed a simulator for the electrical properties of CNTs films. Nanotubes are set randomly on a surface and wide electrodes are defined to form source and drain with a low contact resistance (figure 1). In the network, we assume that the resistance of the intrinsic NT is negligible compare to NT junction. Thus we can draw an electrical equivalent circuit where nanotubes are wire of constant potential and resistance between tubes are due to junction resistances (figure 1). The tubes potential are computed using a modified nodal analysis (MNA). By this way the computational effort is hugely reduced for high tube densities because the matrix dimension depends on the number of tubes and not on the number of junctions and segments.

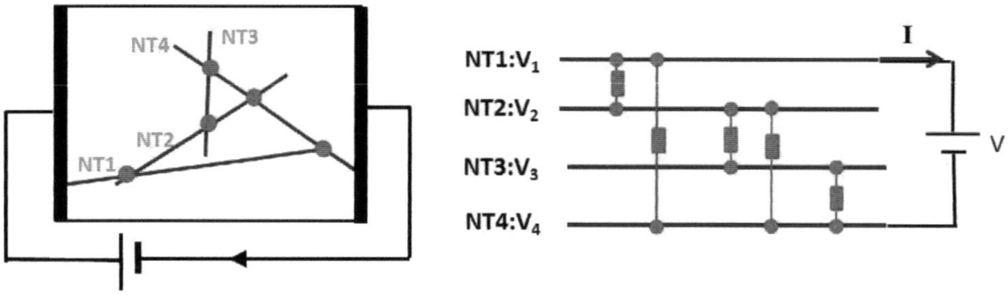


**Figure 1.** A simple network and his circuit representation.

Each intersection between tubes (metal/metal, metal/semiconductor or sc/sc) is modeled as a dipole with a dynamic resistance rc. The hopping from a tube to another is modeled as a perturbation in the transmission probabilities in the framework of Landauer formalism. Transmission probability *T(E)* depends on the nanotube charge $\rho$ and on the energy barriers between nanotubes. Charge inside a tube is calculated self consistently with potential (Poisson equation) taking into account the gate voltage and the density of states *D(E)*. The energy barrier is obtained by a first neighbor tight-binding calculation. From these values, the transmission probability is obtained by WKB approximation (to reduce computational time). In the considered experimental conditions each junction exhibits only 1/f noise source. The current noise spectral density is expressed as $S_I(f)=K\,I^2/f$ with $K=10^{-10}r_c$ [4], and the total noise is calculated using the adjoint network method. The general algorithm is presented figure 2.a and the band diagram for a metal-sc junction is presented figure 2.b.

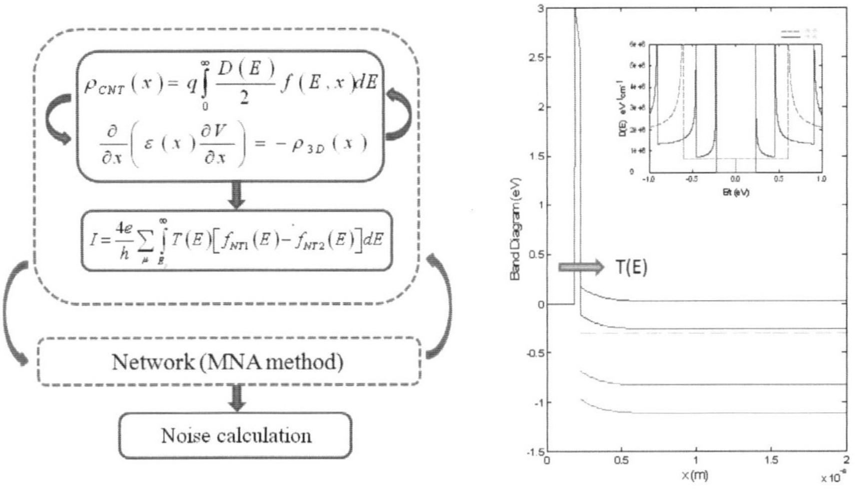


**Figure 2.** (a) General algorithm. (b) Band diagram for subbands around Fermi level for a M-SC junction. In inset the density of states of nanotubes.

## 3. Simulation results

Device simulations are performed using the model presented previously. Figure 3 shows the simulated potential along a device for $V_{DS}$=1 V. For high CNTs densities the potential decreases linearly. For low CNTs densities, the potential drop is not homogeneous and there is dispersion in nanotube potentials. 1/f noise is more important in the non homogeneous parts and contributes mainly to the total noise.

Figure 4.a and 4.b present simulated conductivity and the noise coefficient K for the different densities. The curves clearly exhibit the same behavior than experimental results [3], with similar exponents. Figure 5 presents a simulation of the drain current versus gate voltage for a CNT film composed of semiconductor nanotubes only. It indicates that transistors with very good $I_{ON}/I_{OFF}$ ratio could be competitive if it were possible to sort metallic and semiconductors nanotubes.

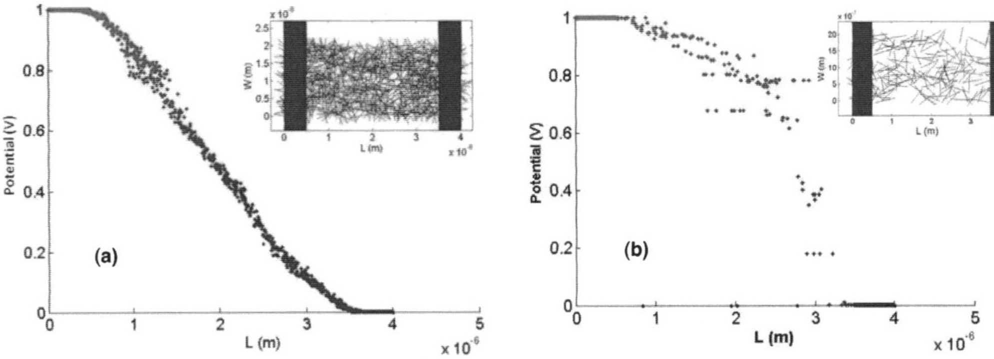


**Figure 3.** NT potentials along the film for a density (a) 125 NT/µm², (b) 25 NT/µm².

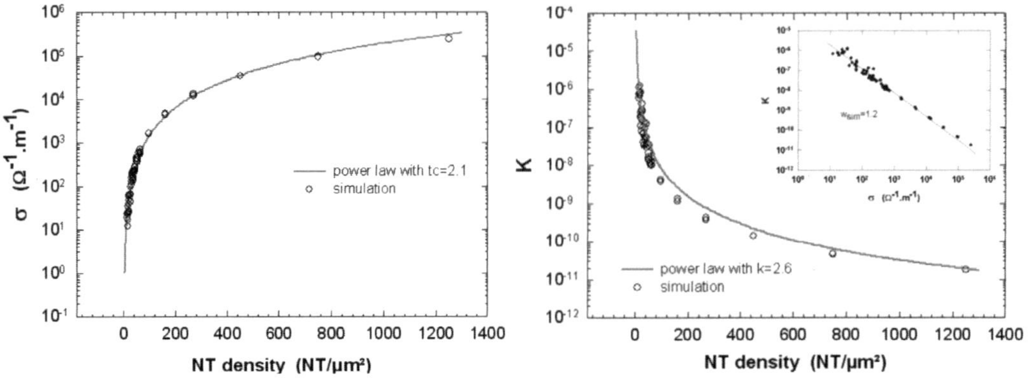


**Figure 4.** a: Simulated conductivity (o) and power law (-) versus the NT density. b: Simulated noise (o) coefficient and power law (-) versus the NT density. W/L=1µm/4µm.

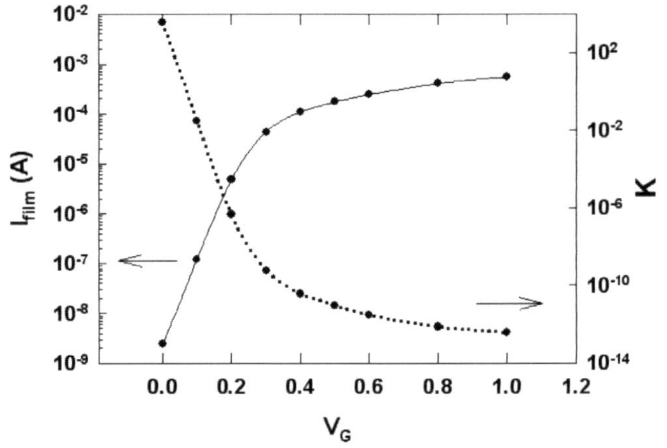


**Figure 5.** Simulated current and noise coefficient versus the gate bias for a CNT film field-effect transistor, for 60 NT/µm². W/L=1/4µm.

## 4. Conclusion

We have developed a simulator based on random generated CNT film. The physical description of NT-NT junction allows to simulate CNT-FET based on carbon nanotube film. The simulator describes the conduction and noise measurements and their variations with the CNT density. These simulations show that this type of transistors can present good performances for low cost electronic applications.

[1] Collins P G, Fhurer M S and Zettl A 2000 Appl. Phys. Lett. **76** 894
[2] Snow E S, Novak J P, Lay M D and Perkins F K 2004 Appl. Phys. Lett. **85** 4172
[3] Soliveres S, Gyani J, Delseny C, Hoffmann A, Pascal F 2007 Appl. Phys. Lett. **95** 082107
[4] Benham A, Bosman G, Ural A 2008 Phys. Rev. B **78** 085431

Journal of Physics: Conference Series **193** (2009) 012118 doi:10.1088/1742-6596/193/1/012118

# Room temperature velocity saturation in intrinsic graphene

**R S Shishir, D K Ferry[1] and S M Goodnick**
School of Electrical, Computer, and Energy Engineering, and Center for Solid State Electronics Research, Arizona State University, Tempe, AZ 85287

E-mail: ferry@asu.edu

**Abstract.** We report ensemble Monte Carlo (EMC) studies of velocity saturation in intrinsic graphene at room temperature. The parameters for the phonon scatterings were obtained from fitting experimental data of graphene sheets in a variety of dielectric media, in which mobilities as high as 44,000 $cm^2/Vs$ were observed at 300 K. In our work, velocity saturation was clearly observed at low carrier density ($\leq 5\times10^{12}$ $cm^{-2}$). Saturation velocity as high as $5.4\times10^7$ cm/s was achieved at the lowest density ($5\times10^{11}$ $cm^{-2}$). Only scatterings due to the acoustic phonon and the *K*-point intervalley optical phonon were considered in the simulation. As the density increases, the onset value increases and the saturation velocity decreases.

## 1. Introduction

Graphene is one of the most notable discoveries made in last century. Although scientists have been investigating graphene for nearly seven decades, it became a serious topic after 2004 when A. Geim and his group were successfully able to extract graphene in its two-dimensional form by mechanical exfoliation [4]. After that a great deal of interest both in theoretical and experimental fields initiated for investigating the electronic properties of two dimensional graphene [5]. Graphene is a strictly two-dimensional material which has excellent high crystal and electronic quality and despite being a new comer, it presented itself as a promising material for future nanotechnology. Graphene is indeed an extraordinary material having unusual behavior which distinguishes it from other two dimensional semiconductor systems. In its single layer form, graphene is only one atom thick sheet of carbon atoms where the atoms are arranged in a honeycomb shape. It has unique bandstructure where the bands and electron transport are mainly influenced by the linear energy bands near the Dirac point [6]. The band structure itself is governed by the Dirac equation, in which charge carriers imitate relativistic particles with zero rest mass.

In order to utilize graphene in coming nanoscale devices, its high field behavior must be investigated thoroughly. Since the first studies into the breakdown and current saturation in semiconductors, the hot electron behavior is of great interest among the researchers [7,8]. It appears that at high electric fields, only the strong metals are able to maintain linear behavior [9]. But, graphene is a strange metallic material, with somewhat conflicting device characteristics. Although intense theoretical studies are carried on graphene's transport [10,11,12,13,14], however, there have been apparently no reports of high electric field behavior in intrinsic graphene to date. Lin *et al.*, observed little effect of the gate in graphene field effect transistors [15] which was expected from the Klein paradox where electrons tunnel through finite length barriers without reflection [16]. But Meric

[1] To whom any correspondence should be addressed.

*et al.* found contradictory results in their experiments. They observed current saturation in similar transistors as expected by electrostatic depletion from the gate [17]. While they considered the possibility of velocity saturation, which is present in carbon nanotube, not much is presently known about this phenomenon in graphene [18].

In this paper, we conduct a systematic study of velocity saturation in graphene as a function of electron density and electric field. For this purpose we employed an ensemble Monte Carlo procedure to solve the Boltzmann transport equation with the unique graphene band structure. In order to determine the drift velocity in any material, usually the value of mass is required. For graphene, away from the Dirac point, the mass is found easily by relating the particle momentum to the crystal momentum. This rather unusual way of determining mass yields a result that is the same as the cyclotron mass where it varies linearly with momentum [19]. Therefore, we calculate the drift velocity in graphene from the following relationship

$$v_d = v_F \left\langle \frac{k_y}{k} \right\rangle , \tag{1}$$

where $v_F$ is the Fermi velocity and the angular brackets denote the ensemble average.

## 2. The transport model

For this paper, our interest lies only in the velocity behavior which is limited by intrinsic phonon in graphene at room temperature. So far most of the room temperature transport studies addressed limited mobility due to the impact of impurities in the $SiO_2$ upon which the exfoliated graphene is deposited. Recently, this situation changed when single layer graphene samples were immersed in various high K dielectric solvent in order to screen the effect of charged impurities which resulted in very high mobilities at room temperature [20].Charged impurity scattering is the most dominating scattering mechanism in graphene at low density. When its effect was drastically reduced by these strong polar solvents, graphene's mobility was basically determined by the intrinsic phonon scattering. This helped

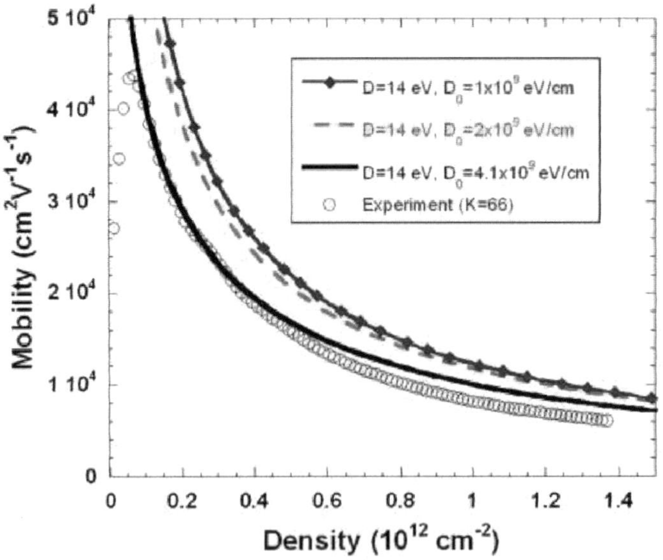


**Figure 1**. The mobility extracted from experimental data with high K dielectric solvents, K=66 (blue circles). Theoretical mobilities with different coupling constants are superimposed on the measured plot to obtain the best fit.

us to separate the intrinsic mobility at various densities and moreover, allowed us to determine the coupling constant for the acoustic phonons as well as to estimate the coupling constant for the optical phonons. Here, we include only these scattering mechanisms in the calculation. The optical phonon scattering is by *K* point phonons which couple the two equivalent valleys of the conduction band. From studies of the phonon structure, this is the LA+LO mode with an energy of 150 meV [21]. Detailed derivations for the scattring rates are given elsewhere [22].

### 3. Results

Fig 1 illustrates the mobility for high K dielectric solvent (K=47 and K=66) along with the theoretical intrinsic mobilities calculated using different coupling constants. We can see the best set of values to fit the experimental results are $D$ = 14 eV for the acoustic modes and $D_0 = 4.1\times10^9$ eV/cm for the optical mode. As mentioned above, these values lead to a room temperature mobility above 40,000 $cm^2$/Vs in what is believed to be intrinsic graphene. Although these coupling constants provide very good fit for the low density regime of the data, the theoretical curve lies above the measured data at high density region. Although we currently do not have full understanding about this discrepancy, but low energy phonon modes that are accompanied by high dielectric constant may be the reason for mobility degradation at higher density. However more work need to be done to confirm this statement. It is important to note that these coupling constants are not unique. Calculation with different coupling constants obtained from the same set of data has also been published [23].

In Fig. 2, we plot the steady state drift velocity as a function of applied field for different carrier densities. The onset of saturated behavior is found to occur at lower field but the velocity gets higher with lower carrier density which is indeed remarkable. Because of graphene's unusual linear dispersion relation near the Dirac point, both the energy and the mass increase linearly with total momentum. At lower carrier density, scattering increases but mass decreases and both of them contribute for velocity increment. The ratio of average energy to the Fermi energy is plotted against the electric field in Fig 3 for different densities. At low electric field and high density, the ration is very close to 0.67 which is expected for degenerate electron gas in graphene at equilibrium. In Fig. 4, the velocity at 30 kV/cm is plotted as a function of the average energy attained at the field. It is easily

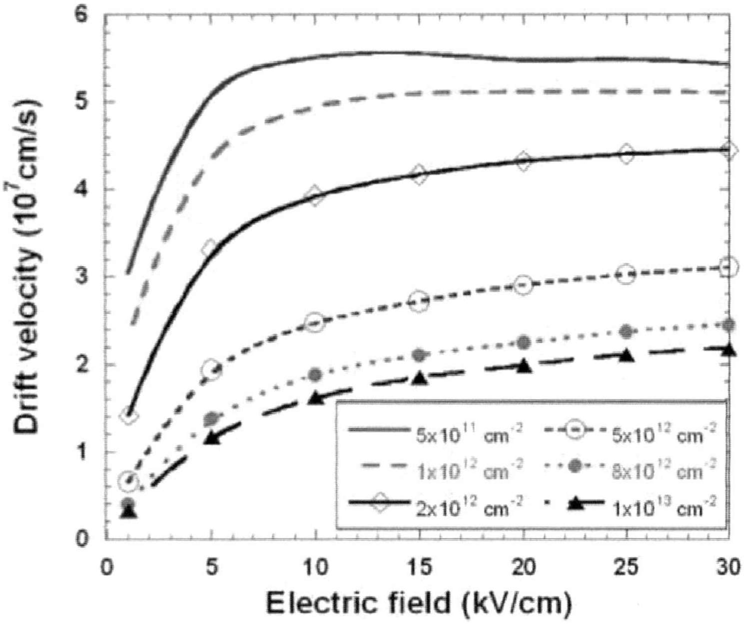


**Figure 2.** The Steady state velocity as a function of applied electric field for different carrier densities.

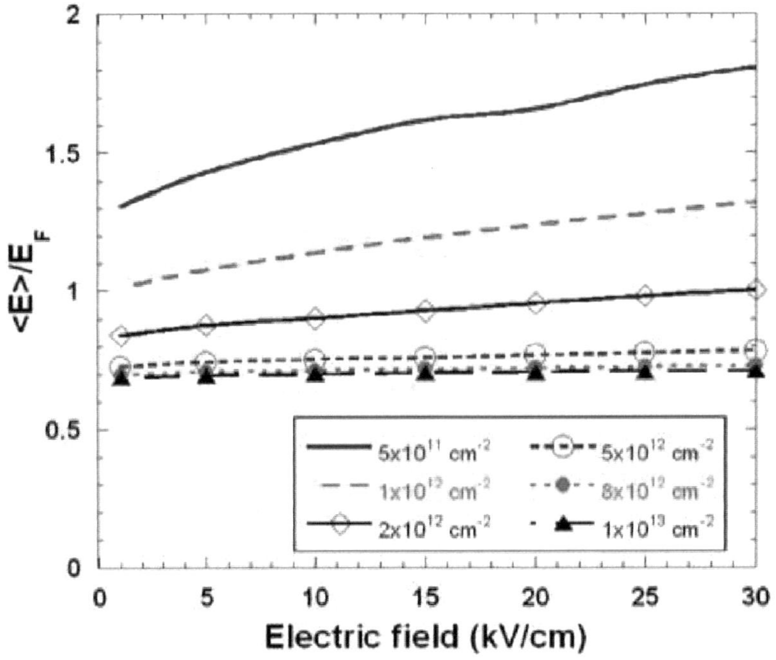


**Figure 3**. The ratio of average energy to the Fermi energy as a function applied electric field for different carrier densities.

observable that there is an inverse square relationship present between the average energy and the velocity. Incidentally, this is not unexpected at all for graphene. At any value of the field, there is an effective mobility relating the velocity to this field, as

$$v = \mu F = \frac{e\tau}{m^*} F \quad . \tag{2}$$

Here $\tau$, the relaxation time, is inversely proportional to the density of states and also to energy. Again, the effective mass increases linearly with energy for graphene. Because of these, it is expected, to first order, that the velocity will decrease with the square of the energy, and this is very nearly the situation found in Fig. 4.

Finally, in Fig. 5, we illustrate how the initial Fermi-Dirac distribution is broadened and shifted in the high electric field. This confirms that there is a change in both the average energy (change in the Fermi energy) and in the temperature (broadening of the distribution) as the field is increased.

## 4. Conclusions

We have studied the transport of energetic electrons in graphene at room temperature. anomalous velocity saturation, that is velocity saturation behavior different than observed in normal semiconductors, is observed in intrinsic graphene at room temperature when only phonon scatterings are considered. The reason behind this anomaly lies in graphene's unique bandstructure which has a linear dispersion relation near the Dirac point. The saturated velocities showed systematic variation with both the average energy and the carrier density. We also find that the value of the saturation velocity is a function of the electron density itself, with strong saturation being observed only at low density.

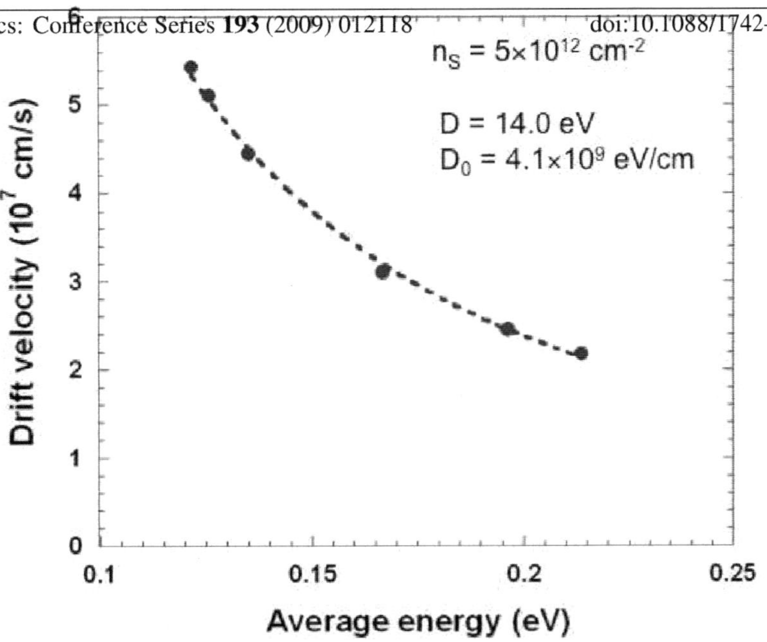


**Figure 4.** The Steady state drift velocity at 30 kV/cm applied field as a function of average energy for carrier density, $n_s=5\times10^{12}cm^{-2}$.

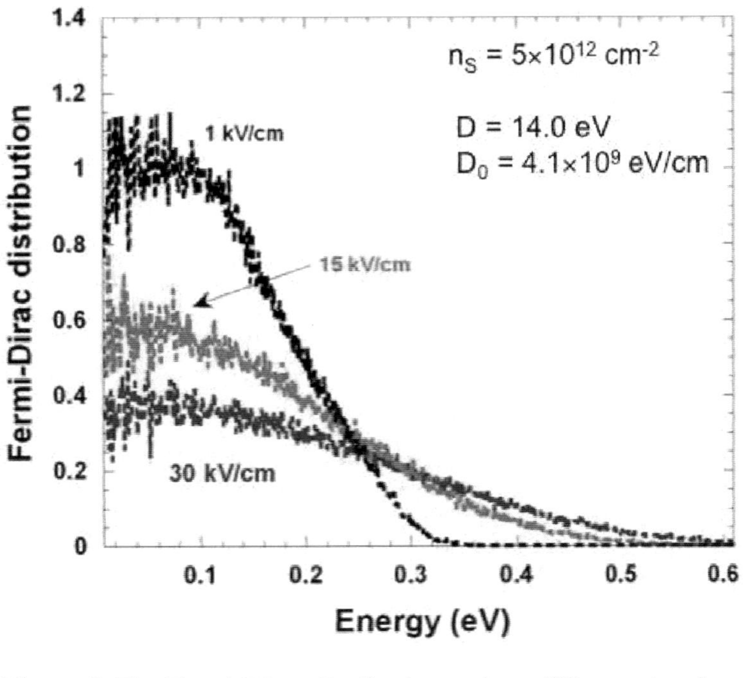


**Figure 5**. The Fermi-Dirac distribution at three different electric fields for carrier density, $n_s=5\times10^{12}cm^{-2}$.

[4] Novoselov, K S *et al*. 2004 *Science* 306 666-669.
[5] Castro Neto A H et al. 2009 *Rev. Mod. Phys.* 81 109.

[6] Wallace P R 1947 *Phys. Rev.* 71 622.
[7] Fröhlich H & Seitz F 1950 *Phys. Rev.* 79 526.
[8] Ryder E J 1953 *Phys. Rev.* 90 766.
[9] Heinrich H & Jantsch W 1969 *Sol. State Commun.* 7 377.
[10] Hwang E H, Adam S & Das Sarma S 2007 *Phys.Rev. Lett.* 98 186806.
[11] Adam S *et al.*2007 *Proc. Natl. Acad. Sci. USA* 104 18392.
[12] Ando T 2006 *J. Phys. Soc. Jpn* 75 074716.
[13] Aleiner I and Efetov K 2006 *Phys. Rev. Lett.* 97 236801.
[14] Nomura K and MacDonald A H 2007 *Phys. Rev. Lett.* 98 076602.
[15] Lin Y M *et al.* 2009 *Nano Lett.* 9 422.
[16] Katsnelson M *et al.* 2006 *Nature Phys.* 2 620.
[17] Meric I *et al.* 2008 *Nature Nanotech.* 3 654.
[18] Verma A, Kauser M Z and Ruden P P 2005 *J. Appl. Phys.* 97 114319.
[19] Novoselov K S *et al.*2005 *Nature* 438 197.
[20] F. Chen *et al.* 2009 *Nano Lett.*in press.
[21] Y anagisawa H *et al.2005 Surf. Interf. Anal.* 37 133.
[22] Shishir R S *et al. J. Comp. Electr., submitted for publication.*
[23] Shishir R S *et al.* 2009 *J. Phys.: Condens. Matter* in press.

EDISON 16 IOP Publishing
Journal of Physics: Conference Series **193** (2009) 012119 doi:10.1088/1742-6596/193/1/012119

# Impurity induced current oscillations in one-dimensional conductors

**S N Artemenko, D S Shapiro, R R Vakhitov, S V Remizov**

V.A. Kotelnikov Institute for Radioengineering and Electronics of the RAS, 125009 Moscow, Russia

E-mail: art@cplire.ru

**Abstract.** We study theoretically electronic transport through an isolated local defect in a 1D conductor described in terms of the Luttinger liquid, and show that the well-known tunneling regime of electronic transport leading to power-law I-V curves takes place only in the limit of small voltage. At voltages exceeding a threshold value a new dynamic regime of transport starts in which the DC current $\bar{I}$ induces AC oscillations of frequency $f = \bar{I}/e$. In gated quantum wires where interaction between electrons is short-ranged, generation linewidth is small provided the inter-electronic repulsion is strong enough, otherwise a wide-band noise is generated. In case of long-range Coulomb interaction generation is coherent at any interaction strength. The effect is related to interaction of the current with Friedel oscillations of the electronic density around the impurity. Manifestations of the effect resemble the Coulomb blockade and the Josephson effect. Oscillations of the electric current are accompanied by spin current oscillations. The results are related to semiconducting quantum wires, metallic atomic chains, carbon nanotubes, graphene nanoribbons and others.

Basic electronic properties of three-dimensional (3D) solids are usually well described within Landau's Fermi-liquid picture. This is not the case in 1D systems where interaction is always strong. In 1D systems of interacting electrons single-electron quasi-particles (electrons and holes) do not exist, the only low energy excitations are charge and spin collective modes with a sound-wave-like spectrum. These excitations form the Luttinger liquid (LL) that is an alternative to Fermi liquid in 1D (for a review see Ref. [1]). Inter-electronic interaction greatly affects electronic transport in conducting 1D systems. In particular, isolated impurities form effectively large barriers in 1D systems with repulsive inter-electronic interaction and strongly suppress conductivity which is reflected in power-law dependences of conductivity on voltage and/or temperature [1]. This was confirmed experimentally in various 1D systems, including carbon nanotubes [2] and semiconductor quantum wires [3]. This effect was described in terms of macroscopic tunneling between different minima of an inclined by external bias periodic potential of interaction of the electronic system with the impurity. The periodic potential is associated with Friedel oscillations induced by impurity. We have shown recently [4] that the power-law regime takes place only in the limit of small enough currents/voltages, while above a threshold voltage a new dynamical regime starts in which current increases rapidly and the DC current is accompanied by oscillations with the washboard frequency $f = \bar{I}/e$. This regime takes place when applied voltage exceeds a threshold value corresponding to the slope at which the system can roll out from the minimum of the washboard potential renormalized by fluctuations. The effect is induced by the inter-electronic repulsion and cannot be found by means of an approach based on non-interacting electrons. Theoretical study of the effect is complicated by strong fluctuations that in many respects determine behavior of 1D systems. The approach of ref. [4] is strict in the limit of strong inter-electronic repulsion, when fluctuations at the impurity site are not large, while for the general case the Gaussian model has been used that is not justified strictly. Here we study the effect in the limit of large voltages/currents, where non-Gaussian part of fluctuations is small. This enabled us to calculate I-V curves and generation linewidth strictly.

For brevity we consider first the response of a spinless repulsive LL to an external DC voltage at zero temperature using the Tomonaga-Luttinger model [1] with short range interaction between electrons characterized by a constant $K_\rho < 1$. The short-range interaction describes gated quantum wires where the long-range part of the interaction is screened by 3D gate electrodes. At the end we will discuss essential modifications induced by spin, the long-range Coulomb interaction, and finite temperature.

We consider a 1D conductor with impurity at $x = 0$ and apply boundary conditions [5] derived for a wire adiabatically connected to ideal Fermi-liquid reservoirs at $x = \pm L/2$ with voltage difference $V$. We start with the Hamiltonian

$$\hat{H} = \int dx \left\{ \frac{\hbar \pi v_F}{2} \left[ \hat{\Pi}^2 + \frac{1}{\pi^2 K_\rho^2} (\partial_x \hat{\Phi}_\rho)^2 \right] - \frac{1}{\pi} W_i \delta(x) \cos 2\hat{\Phi}_\rho \right\} \tag{1}$$

that includes the standard bozonized Tomonaga-Luttinger (TL) Hamiltonianin written in terms of the bosonic (plasmon) displacement field $\hat{\Phi}_\rho(t,x)$, and the impurity term with strength $W_i$ related to $2k_F$-matrix element of the impurity potential [1]. Time and coordinate derivatives of $\hat{\Phi}_\rho(t,x)$ determine the operators of current, $\hat{I} = (e/\pi)\partial_t \hat{\Phi}_\rho$, and charge density, $\hat{\rho} = (e/\pi)\partial_x \hat{\Phi}_\rho$.

Commuting $\hat{\Phi}_\rho$ with the Hamiltonian we derive the equation of motion for the Heisenberg operator

$$\left( v^2 \partial_x^2 - \partial_t^2 \right) \hat{\Phi}_\rho(t,x) = \frac{2v_F}{\hbar} W_i \sin 2\hat{\Phi}_{\rho 0}(t)\, \delta(x), \tag{2}$$

where $\hat{\Phi}_{\rho 0}(t) \equiv \hat{\Phi}_\rho(t, x=0)$, and $v = v_F/K_\rho$ is the velocity of plasmons.

In ref. [4] we considered the limit of a long conducting channel with the length much larger than the damping length of plasmon excitations. Now we consider a short channel, so we cannot neglect reflections of current pulses generated at the impurity from the contacts.

Using Eq. (2) with boundary conditions at the contacts we express operator $\hat{\Phi}_\rho(t,x)$ in terms of $\hat{\Phi}_{\rho 0}(t)$ and find equation of motion for the expectation value$\Phi_{\rho 0}(t) = \langle \hat{\Phi}_{\rho 0}(t) \rangle$:

$$\partial_t \Phi_{\rho 0}(t) + \int_0^\infty dt_1 Z(t-t_1) \langle \sin 2\hat{\Phi}_{\rho 0}(t_1) \rangle = \frac{eV}{2\hbar}, \quad Z(t) = \int e^{-i\omega t} d\omega \frac{W_i K_\rho (1 - i K_\rho \tan \frac{\omega L}{2v})}{h(K_\rho - i \tan \frac{\omega L}{2v})}. \tag{3}$$

The non-linear term in Eq. (3) depends on fluctuations, $\langle \sin 2\hat{\Phi}_{\rho 0}(t_1) \rangle = \langle \cos 2\delta\hat{\Phi}_\rho \rangle \sin 2\Phi_{\rho 0} + \langle \sin 2\delta\hat{\Phi}_\rho \rangle \cos 2\Phi_{\rho 0}$, $\delta\hat{\Phi}_\rho(t) = \hat{\Phi}_{\rho 0}(t) - \Phi_{\rho 0}(t)$. Correlation function of fluctuations can be calculated from Keldysh Green's function $D^K = -i\langle \{\delta\hat{\Phi}_\rho(t), \delta\hat{\Phi}_\rho(t')\}_+ \rangle$. This function can be expressed via retarded and advanced Green's functions $D^{R(A)} = \pm i\theta[\pm(t-t')]\langle [\delta\hat{\Phi}_\rho(t), \delta\hat{\Phi}_\rho(t')]_- \rangle$ by relation of their Fourier transforms, $D^K(\omega,\omega') = D^R(\omega,\omega') f(\omega') - f(\omega) D^A(\omega,\omega')$, where $f$ is related to the distribution function of bosonic excitations $N(\omega)$, $f(\omega) = 1 + 2N(\omega)$. In the equilibrium state $N(\omega)$ is the Planck distribution function. At low temperatures, smaller than all characteristic energies of the system, one can neglect contribution of thermally excited excitations and $D^K$ can be expressed via the retarded and advanced functions.

Now we derive equations of motion for the retarded and advanced Greens functions of fluctuations. This can be done in a standard way multiplying Eq. 2 by $\hat{\Phi}_\rho(t', x')$ from the left and from the right, and combining them in order to obtain corresponding Green's function after averaging. Then using the Fourier transformation we express Greens functions at the impurity site and get rid of the coordinate dependence of Green's function. After that we subtract expectation values and obtain equations of motion for $D^{R(A)}$. The equation for $D^R$ reads

$$\partial_t D^R(t,t') + \frac{1}{\hbar} W_i K_\rho D_s^R(t,t') = -\frac{\pi K_\rho}{2}\delta(t-t'), \tag{4}$$

where $D_s^R(t,t') \equiv i\theta(t-t')\left\langle\left[\sin 2\hat{\Phi}_\rho(t), \delta\hat{\Phi}_\rho(t')\right]\right\rangle$. Eq. (4) is not close since it contains $D_s^R$. The latter can be easily related to $D^R$ when fluctuations are Gaussian. In the Gaussian model the cosine term is substituted by self-consistent quadratic term $\cos 2\delta\hat{\Phi}_\rho \to e^{-2\langle\delta\hat{\Phi}_\rho^2\rangle}(1 - 2\delta\hat{\Phi}_\rho^2 + 2\langle\delta\hat{\Phi}_\rho^2\rangle)$. We calculate $D_s^R$ with full Hamiltonian (1), considering difference between cosine terms and its self-consistent harmonic approximation perturbatively, and find that non-Gaussian terms decay at large voltages, $V \gg V_T \sim (W_i/e)\left(\frac{K_\rho^{3/2} W_i}{\Lambda}\right)^{\frac{K_\rho}{1-K_\rho}}\sqrt{1-K_\rho}$ (here$\Lambda$ $\sim p_F v$ is a large cut-off energy of the TL model), and become a small correction to the Gaussian part of correlation functions. Here $V_T$ is the threshold voltage above which the dynamical regime of transport starts, it is smaller than impurity potential $W_i$ due to quantum fluctuations [4].

Thus considering the limit of large voltages we can use the Gaussian approximation for fluctuations and relating $D_s^R$ to $D^R$ we obtain close equation for $D^R$. Then this equation and similar equation for the advanced Green's function can be solved analytically at $V \gg V_T$. The solutions are

$$D^{R(A)} = -\frac{\pi K_\rho}{2}\theta[\pm(t-t')]e^{\mp\frac{2}{\hbar}W_i K_\rho \int_{t'}^{t} C(t_1)dt_1}, \quad C(t) \equiv e^{-2\langle\delta\hat{\Phi}_\rho^2(t)\rangle}\cos 2\Phi_{\rho 0}(t). \tag{5}$$

This gives us

$$\langle\delta\hat{\Phi}(t)\delta\hat{\Phi}(t')\rangle = \frac{K_\rho}{4}\left[\int_{t-t'}^{\infty}\frac{dt_1}{t_1}e^{-\frac{2}{\hbar}W_i K_\rho \int_{t-t'}^{t_1} C(t-t_2)dt_2} + \int_{t'-t}^{\infty}\frac{dt_1}{t_1}e^{-\frac{2}{\hbar}W_i K_\rho \int_{t'-t}^{t_1} C(t-t_2)dt_2}\right]. \tag{6}$$

Eq. (6) must be solved self-consistently with (3) in order to find DC and AC currents and spectrum of fluctuations. For large voltages, $V \gg V_T$, it is not difficult to find analytical solutions perturbatively and to calculate current $I = (e/\pi)\langle\partial_t\hat{\Phi}_\rho\rangle$ and its correlation function.

$$I = \bar{I} + I_{ac}\sin\omega_0 t, \quad \bar{I} = VG_0 - I_{nl}, \quad \omega_0 = 2\pi f \equiv \frac{2\pi\bar{I}}{e} \approx \frac{eV}{\hbar}, \tag{7}$$

$$I_{ac} \propto \frac{G_0 V_T\left(\frac{K_\rho^{3/2}V_T}{V}\right)^{\frac{1}{1-2K_\rho}}}{\sqrt{\sin^2\frac{\omega_0 L}{2v} + K_\rho^2\cos^2\frac{\omega_0 L}{2v}}}, \quad I_{nl} \propto K_\rho V_T G_0\left(\frac{K_\rho^{3K_\rho}V_T}{V}\right)^{\frac{1}{1-2K_\rho}}\frac{\tan^2\frac{\omega_0 L}{2v} + K_\rho}{\tan^2\frac{\omega_0 L}{2v} + K_\rho^2}, \tag{8}$$

$$\langle\hat{I}(\omega)\hat{I}(-\omega)\rangle \propto \frac{eG_0 V_T K_\rho^{3(1-K_\rho)}}{\sin^2\frac{\omega_0 L}{2v} + K_\rho^2\cos^2\frac{\omega_0 L}{2v}}\left(\frac{eV_T}{\hbar|\omega-\omega_0|}\right)^{1-2K_\rho} \tag{9}$$

where $G_0 = e^2/h$ is the conductance quantum. These results are valid at $K_\rho < 1/2$, that is for strong enough inter-electronic repulsion. In the last expression we skipped a small term proportional to $\delta(\omega-\omega_0)$ that gives a negligible contribution to the fluctuation spectrum. At larger values of $K_\rho$ the Gaussian model yields that the pronounced maximum at $\omega = \omega_0$ in the fluctuation spectrum is absent and a wide-band noise is generated. For gated quantum wires the interaction parameter $K_\rho$ can be estimated roughly as

$$K_\rho \sim \sqrt{\frac{\hbar v_F \epsilon}{e^2}} \approx 0.2\sqrt{\frac{\epsilon v_F(\mathrm{cm/s})}{10^7}},$$

where $\epsilon$ is a background dielectric constant.

Oscillations of currents in Eqs. (7-9) as function of voltage are induced by reflections of current pulses from the contacts. The oscillatory dependence of conductivity versus voltage was, presumably, observed in [6]. Note that the shape of the oscillations will be different if the impurity is not situated at the middle of the wire. Note also that we assumed the regime of the fixed voltage bias at the contacts. Such a regime takes place if a small impedance $ZG_0 \ll 1$ is connected to the wire either in parallel or in series. In general case the results (7-9) would be modified.

For the case of long-range Coulomb interaction the results are qualitatively similar, the most important difference is that the maximum maximum at $\omega = \omega_0$ in the fluctuation spectrum is always present, that is at any values of parameter $e^2/(\hbar v_F \epsilon)$ measuring the strength of the Coulomb interaction.

In case of finite temperatures, modifications of the results (7-9) are negligible at temperatures $T \ll T_c \sim \frac{eV_T}{k_B}\left(\frac{V_T}{V}\right)^{1/(1-2K_\rho)}$. The main modification is that thermal fluctuations change shape of the line at $|\omega - \omega_0| \ll T$, singularity of the peak being washed out and becoming approximately of Lorentzian form $\sim (G_0 W_i)^2 K_\rho^3 \omega_T / [(\pi K_\rho \omega_T)^2 + (\omega - \omega_0)^2]$, where $\omega_T = k_B T/\hbar$. At temperatures $T > T_c$ the effect of generation disappears because of thermal destruction of the impurity pinning.

Now we discuss what happens in the system of electrons with spin. In the spinful LL the impurity Hamiltonian contains contribution due to the spin phase field, $\hat{\Phi}_\sigma$, and $\cos 2\hat{\Phi}_\rho$ in (1) must be substituted for $\cos\sqrt{2}\hat{\Phi}_\rho \cos\sqrt{2}\hat{\Phi}_\sigma$ [1]. This leads to partial violation of the spin-charge separation and modifies the results. The threshold voltage becomes smaller than in the spinless case due to spin fluctuations, for a spin-independent interaction ($K_\sigma = 1$) it reads $V_T \sim W_i \left(\frac{W_i}{\Lambda}\right)^{\frac{1+K_\rho}{1-K_\rho}} K_\rho^{\frac{1+K_\rho}{3K_\rho}}$. Non-linear term in equation (2) for the charge phase is also changed, $\sin 2\hat{\Phi}_{\rho 0} \to \sin\sqrt{2}\hat{\Phi}_{\rho 0} \cos\sqrt{2}\hat{\Phi}_{\sigma 0}$. Further, we have to solve equation of motion for the spin phase that can be obtained from (2) and (3) by exchanging subscripts $\sigma$ and $\rho$ and putting $K_\sigma = 1$. In the limit of small $K_\rho$ the characteristic time of variations of the spin phase is much smaller than that for the charge phase. Using this fact we can study phase evolution in the limit $V_T \ll V \ll V_T/K_\rho$. Then the phase$\Phi_{\rho 0}$ increases almost linearly,$\Phi_{\rho 0} \approx \frac{\pi \bar{I}}{\sqrt{2}e} t$, while$\Phi_{\sigma 0}$ stays, firstly, practically constant and equal to $2\pi n/\sqrt{2}$, and then, when $\cos\sqrt{2}\Phi_{\rho 0}$ changes its sign, $\Phi_{\sigma 0}$ rapidly changes by $\pm\pi/\sqrt{2}$ since the previous state of$\Phi_{\sigma 0}$ becomes unstable. This variations of the spin phase means generation of pulses of spin current corresponding to transport of one spin through the impurity, the time average of the spin current being equal to zero. However, if the wire is placed in a gradient of the magnetic field then a preferential direction for the spin current would appear resulting in a DC component of the spin current.

In conclusion, we predict the new regime of electronic transport through an impurity or a local defect in 1D conducting channel. The effect is induced by the inter-electronic repulsion. Manifestations of the effect resemble the AC Josephson effect and Coulomb blockade. The results are applicable to semiconducting quantum wires, metallic atomic chains, carbon nanotubes, graphene nanoribbons and other 1D conducting systems.

## References

[1] Giamarchi T 2003, *Quantum Physics in One Dimension*, (Oxford: Clarendon Press)
[2] Ishii H, Kataura H, Shiozawa H et al 2003, *Nature* **426**, 540
[3] Auslaender O M, Steinberg H, Yacoby A et al 2005, *Science* **308**, 88
[4] Artemenko S N, Remizov S V, Shapiro D S 2008 *Pis'ma v ZhETF* **87**, 792 [*JETP Lett.* **87**, 692].
[5] Egger R, Grabert H 1998, *Phys. Rev. B* **58**, 10761
[6] Al Ahmad M, Dragoman D, Dragoman M, Plana R, Ting J-H, Huang F-Y, Li T-L, *Preprint* arXiv:0901.0363v1

# Wide frequency band detection on laterally constricted GaAs structures having nanometric heavily doped layers

**S Ašmontas[1], V Kazlauskaitė[1], A Sužiedėlis[1], J Gradauskas[1] and V Derkach[2]**

[1]Semiconductor Physics Institute, A.Goštauto 11, Vilnius LT-01108, Lithuania

[2] Institute of Radiophysics and Electronics, Akademika Proskury St. 12, Kharkov 61085, Ukraine

E-mail: asmontas@pfi.lt

**Abstract.** Experimental results of planar asymmetrically and symmetrically shaped hot carrier microwave detectors having five delta-doped GaAs layers are presented. The results are compared with the detectors having the same geometrical shape but only homogeneously doped GaAs layer of submicrometric thickness. DC measurements showed that the structure with delta-doped layers had slightly lower resistance than the detector with homogeneously doped layer due to higher carrier mobility in the delta-doped structure. Increase in carrier mobility corresponded to higher voltage sensitivity, which is essential parameter of a microwave detector. Voltage power characteristics of the detectors of both types were linear up to 10 mW of microwave power, whereas at higher powers they became super-linear or sub-linear with the following change in polarity of the detected voltage. Distinctive dependence of the detected voltage on microwave power at strong electric field is attributed to intervalley transfer of charge carriers. Experimental results of voltage sensitivity in microwave frequency range showed that drop-off of sensitivity going from $K_a$ range to W range for both types of detectors is not larger than 3 dB, suggesting their use for wide frequency detection.

## 1. Introduction

Schottky barrier diode is a device widely used in microwave electronics and demonstrating high voltage sensitivity in the detection of electromagnetic radiation from microwaves up to terahertzes [1]. However, some drawbacks of the device, such as poor sustainability against electrical overload and atmospheric influence, considerable frequency dependence of detected voltage stimulate research and creation of alternative detectors, preferably having active region in the bulk of semiconductor to avoid destructive influence of it surface onto device properties. Several decades ago hot-carrier microwave (MW) detector was proposed [2]. Its operation was based on thermoelectric effect of hot charge carriers in a semiconductor. However, the design of such a detector had point-contact diode configuration with metallic whisker serving as an antenna and a terminal for contacting to semiconductor point contact. This design suffered from poor robustness of the detector as well as inconvenience of its use at higher frequencies. Therefore, we proposed a planar design of the hot-carrier MW detector on the base of asymmetrically shaped $n^+$-$n$-GaAs junction [3]. Voltage sensitivity of this detector was inversely proportional to the thickness of the semiconductor epitaxial layer. Planar asymmetrically shaped microwave diodes fabricated on the base of MBE grown heavily doped thin GaAs layer onto semiinsulating substrate (structure of submicrometric thickness) had good detection properties in millimeter wavelength range [4]. However, the diodes were not optimal from the point of view of both thickness of the active layer and voltage sensitivity. In this paper, MW detection properties of semiconductor structures on the base of heavily doped delta layers are investigated and compared with the ones of MW diodes having submicrometric heavily doped GaAs layer.

## 2. Samples

Two types of GaAs structures were grown by molecular beam epitaxy (MBE) for MW diodes' fabrication: *i)* submicrometric structure (SS) was heavily doped GaAs layer of 100 nm thickness on semiinsulating substrate, and *ii)* nanometric structure (NS) was the structure of 5 delta-doped layers with 25 nm spacing between them. Donor density in the SS heavily doped layer was $1\times10^{18}$ cm$^{-3}$ and in the delta doped layer of the NS it was $1\times10^{12}$ cm$^{-2}$. First step of fabrication process of the planar diodes was creation of asymmetrically and symmetrically shaped semiconductor mesas using photolithography and wet chemical etching techniques. Etching depth exceeded 100 nm in the SS case and 150 nm in the NS case, both with the aim to reach semiinsulating substrate of the MBE grown wafers. Metal contacts were thermally evaporated through the windows opened in a negative photoresist during the second photolithography process. Then trenches of 5 $\mu$m depth were etched around the diodes. These trenches served as guide-marks for the subsequent substrate thinning process and as alignment marks for the last photolithography step. Use of microwave diodes at high MW frequencies requires their planar design with micrometric thickness of a semiconductor structure. Therefore several micrometer-thick polyimide film was used to cover the face side of the wafer, and then thinning of the GaAs substrate was performed having the wafer face down stuck onto silicon substrate. The thinning of the substrate was performed mechanically up to ~100 $\mu$m thickness and then was finished in GaAs etchant till the frame of the guide-marks filled with polyimide appeared. The last photolithography was performed onto the back-side of the thinned wafer to strip metallic contacts of the diodes. After etching GaAs around the shielded mid-part of the diode the matrix of the planar diodes' mesas onto polyimide film was achieved. Schematic view of the asymmetrically and symmetrically shaped planar microwave diodes from the polyimide film side is presented in figure 1 (the polyimide film is not presented in the figure for simplicity). The matrix of the diodes was divided into separate diodes mechanically by cutting the polyimide film. The MW diodes were mounted into a lowered rectangular waveguide head and in a strip-line head for detection measurements in $K_a$ (26÷37.5 GHz) and W (75÷110 GHz) frequency ranges, respectively.

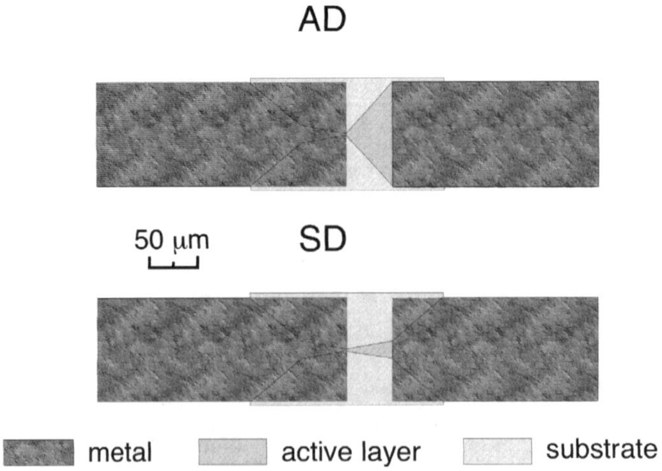


**Figure 1**. Schematic view of the asymmetrically (AD) and symmetrically (SD) shaped diodes.

Journal of Physics: Conference Series **193** (2009) 012120 doi:10.1088/1742-6596/193/1/012120

### 3. Experimental results and discussion

Experimental values of electrical resistance of symmetrically and asymmetrically shaped diodes obtained from their current-voltage (I-V) characteristics turned out to be slightly higher than the resistance calculated from the diodes' geometry. This difference could not be explained by the influence of contact resistance of the diodes because the value of specific contact resistance of the samples did not exceed 0.1 Ω·mm. The difference was greater for the structures with narrower "neck". Therefore, higher experimental value of the resistance of the "necked" structures could be explained by charge carrier trapping in the narrowest part of the structure. This presumption was also confirmed by capacitance measurements: 1) the dependence of capacitance on applied voltage was more pronounced for the more "necked" structures; 2) the increase of capacitance with bias voltage was higher at higher frequencies. Carrier trapping in the "neck" of the structure is responsible for the asymmetry sign of I-V characteristic which is usually opposite to the expected asymmetry of the I-V characteristic of an ideal $n^+$-$n$ junction.

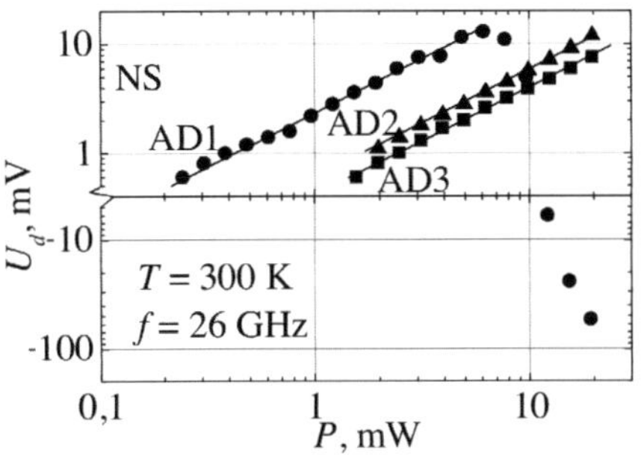


**Figure 2.** Voltage power characteristics of the asymmetrically shaped NS diodes. Lines are the guides for eye of linear dependence.

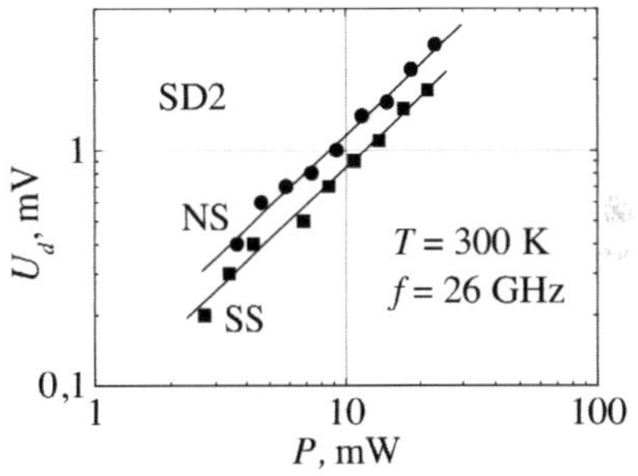


**Figure 3.** Voltage power characteristics of the symmetrically shaped NS and SS diodes. Lines are the guides for eye of linear dependence.

Polarity of voltage signal induced on the diodes placed in microwave waveguide corresponded to that of thermoelectric electromotive force of hot carriers: negative potential arose on the lightly doped side of the $n$-$n^+$ junction (right side of the diode in figure 1). Dependences of the detected voltage on 26 GHz frequency microwave power for asymmetrically shaped NS diodes measured at room temperature are shown in figure 2. The presented characteristics are for different width of the "neck", i.e. 1, 2 and 3 micrometers for the diodes AD1, AD2 and AD3, respectively. Voltage sensitivity $S = U_d/P$ (here $U_d$ is detected voltage, $P$ stands for the microwave power in a waveguide) of the diodes depends on the width of the "neck" of the structure: the sensitivity decreases with wider "neck". Change of polarity of the detected voltage in case of the AD1 diode should be related to Gunn's domains formation in $n$ region of the $n$-$n^+$ junction in a strong microwave electric field [5] in the vicinity of the "neck" of the structure. Figure 3 depicts voltage power characteristics of the symmetrically shaped semiconductor structures on the base of NS and SS structures. Voltage sensitivity of the symmetrically shaped diodes was lower than that of the asymmetrically shaped ones since the asymmetric shape of the semiconductor structure conditions additional mechanism of microwave detection, namely, bigradient electromotive force arises in such a structure [4]. Polarity of the bigradient emf has the same sign as that of thermoelectric electromotive force of hot carries. The microwave diodes on the base of nanometric structures revealed slightly higher voltage sensitivity than the diodes on the base of submicrometric structure. This experimental fact could be explained by narrow peaks of delta-doping cutout in the NS case. However, the difference in the sensitivity is not very pronounced due to relatively high level of electron density background in the delta-doped structure.

Frequency dependence of voltage sensitivity of the asymmetrically shaped NS diodes in W frequency range at room temperature are shown in figure 4. In the whole investigated frequency range MW diodes with narrower "neck" prevail over the others in sense of voltage sensitivity. Positive feature of the MW diodes on the base of delta-doped structure is almost flat frequency characteristic of the voltage sensitivity exhibiting no trend of decrease at higher frequencies. Figure 5 depicts frequency dependence of voltage sensitivity of the symmetrically shaped diodes with 3 $\mu$m width of the "neck"on the base of nanometric and submicrometric structures in $K_a$ and W frequency ranges. The MW diodes on the base of NS exhibited weak frequency dependence of the voltage sensitivity in the whole investigated frequency range. The diodes on the base of SS had significant fall of the sensitivity values when going from $K_a$ to W frequency range, whereas they demonstrated sensitivity increase when going to lower temperatures.

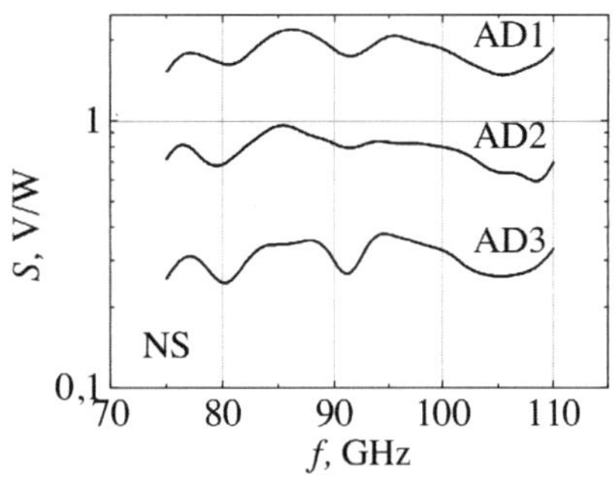


**Figure 4.** Frequency dependence of voltage sensitivity of the asymmetrically shaped NS diodes in W frequency range at room temperature.

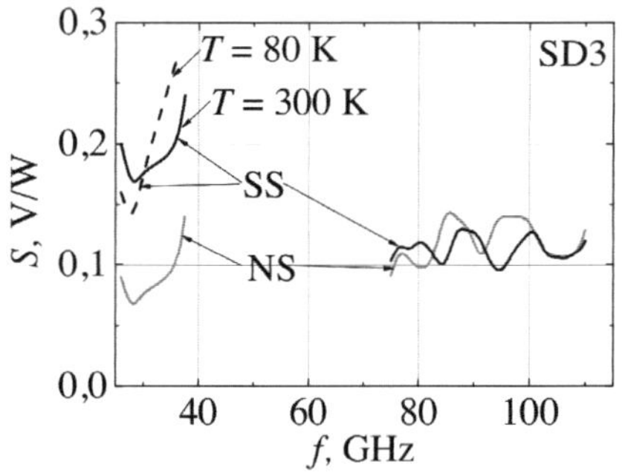


**Figure 5.** Frequency dependence of voltage sensitivity of the symmetrically shaped NS and SS diodes in $K_a$ and W frequency ranges at room and liquid nitrogen temperatures.

In conclusion, properties of all types of diodes investigated in this work revealed that the asymmetrically shaped diode on the base of nanometric structure having delta-doped layers is the best candidate for wide frequency range MW detection. Narrowing of the "neck" of the diode and decrease of lattice temperature improves voltage sensitivity of the detector.

## 4. References

[1] Siegel P H *IEEE Trans. Microwave Theory Tech.* 2002 **50** 910
[2] Harrison R I, Zucker J *Proceedings of the IEEE* 1966 **54** 588
[3] Sužiedėlis A, Gradauskas J, Ašmontas S, Valušis G, Roskos H G 2003 *J. Appl. Phys.* **93** 3034
[4] Ašmontas S, Gradauskas J, Kozič A, Shtrikmann H and Sužiedėlis A 2005 *Acta Physica Polonica A* **107** 147
[5] Ašmontas S, Gradauskas J, Kundrotas J, Sužiedėlis A, Šilėnas A, Valušis G 1998 *Proc. Int. Conf. on Physics of Semiconductors (Jerusalem)* CD-ROM (World Scientific) 1193.pdf

## Acknowledgments

This work was supported in part by Lithuanian State Science and Studies Foundation under the contracts No. P-01/2009 and No. V-21/2009 as well as by Agency of International Science and Technology Development Programmes in Lithuania under the contract No. 31V-149. The samples were fabricated at Braun Submicron Research Centre under contract No. RITA-2003-506095.

Journal of Physics: Conference Series **193** (2009) 012121 doi:10.1088/1742-6596/193/1/012121

# Generation of squeezed phonon states by optical excitation of a quantum dot

**D E Reiter[1], S Sauer[1], J Huneke[1], T Papenkort[1], T Kuhn[1], A Vagov[2] and V M Axt[2]**

[1] Institut für Festkörpertheorie, Westfälische Wilhelms-Universität Münster, Wilhelm-Klemm-Straße 10, 48149 Münster, Germany

[2] Institut für Theoretische Physik III, Universität Bayreuth, 95440 Bayreuth, Germany

E-mail: Doris.Reiter@uni-muenster.de

**Abstract.** A theoretical analysis of the fluctuation properties of optical phonons generated by the optical excitation of a quantum dot is presented. In particular we study whether squeezed states may be generated where either the fluctuations of the displacement or of the momentum fall below the vacuum level. The calculations are based on a generating functions formalism which provides analytical results for the case of excitation by ultrashort optical pulses. While after a single pulse excitation no squeezing occurs we find that after an excitation with two phase-locked pulses phonon states can be created where the uncertainties of displacement or momentum temporarily fall below their vacuum values.

## 1. Introduction

Squeezed states of light have a long tradition in the field of quantum optics [1] and there are various applications, e.g., in gravitational wave detection or quantum information processing [2]. In analogy with photon states in quantum optics the generation and manipulation of specific phonon states has become a major issue in solid state physics [3]. While in most cases coherent phonon amplitudes and nonequilibrium phonon occupations have been investigated, in recent years there has been a growing interest in the study of the fluctuation properties of the generated phonons. The generation of squeezed phonons by second-order Raman scattering has been predicted [4] and signatures of phonon squeezing have been observed by pump-probe spectroscopy in transmission [5] or reflection [6] geometry as well as recently in femtosecond X-ray diffraction [7].

The impulsive optical excitation of a quantum dot (QD) gives rise to the formation of a lattice distortion in the region of the dot, i.e., to the creation of a polaron. This distortion is characterized by a non-vanishing mean value of the lattice displacement. However, it is known that only under specific excitation conditions the phonon system is in a fully coherent state [8], in which the fluctuations of the displacement and the momentum are time-independent and agree with the respective vacuum fluctuations. For other excitation conditions not much is known about the fluctuation properties of the phonons. In this contribution we will study whether the fluctuations of either the displacement or the momentum can be reduced below their vacuum values, i.e., whether squeezed phonon states can be optically generated in a QD system.

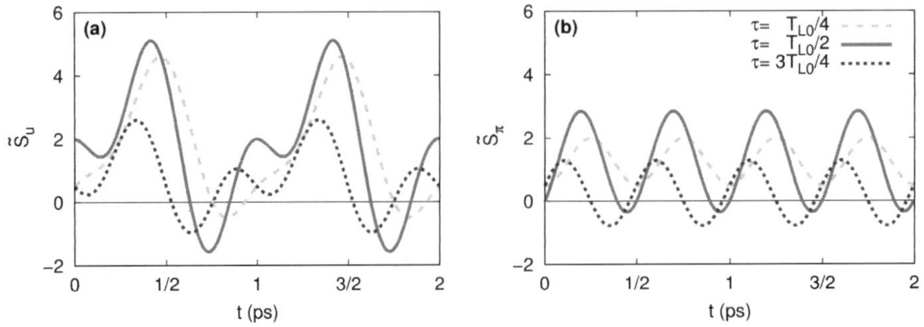


**Figure 1.** Time dependent fluctuations (a) $\tilde{S}_u$ and (b) $\tilde{S}_\pi$ for fixed pulse areas $A_{1,2} = \pi/2$ and phase difference $\Phi = \pi/2$ at different delay times $\tau$.

## 2. Theoretical background

We consider a QD in the strong confinement limit excited by circularly polarized light, such that we can restrict ourselves to two electronic states, the ground state and the one-exciton state. Due to the different charge densities in ground and excited state, the exciton is coupled to longitudinal optical (LO) phonons with frequency $\omega_{LO}$ via the Fröhlich coupling matrix element $g_\mathbf{q}$ [9]. Excitons can be created and manipulated in the system via ultra short laser pulses applied at times $t_j$ with pulse areas $A_j$ and phases $\phi_j$. For this system analytical formulas for all electron and phonon variables are provided by a generating function formalism [10].

For a phonon system the two conjugate variables are the lattice displacement $\mathbf{u}(\mathbf{r},t)$ and the momentum $\pi(\mathbf{r},t)$. In the case of LO phonons the corresponding operators are calculated from the creation (annihilation) operators $b^\dagger_\mathbf{q}$ ($b_\mathbf{q}$) for a phonon with wave vector $\mathbf{q}$ according to

$$\hat{\mathbf{u}}(\mathbf{r},t) = -i\frac{u_0}{\sqrt{N}}\sum_\mathbf{q}\frac{\mathbf{q}}{q}\left(b_\mathbf{q}e^{i\mathbf{q}\cdot\mathbf{r}} - b^\dagger_\mathbf{q}e^{-i\mathbf{q}\cdot\mathbf{r}}\right), \quad \hat{\pi}(\mathbf{r},t) = -\frac{\pi_0}{\sqrt{N}}\sum_\mathbf{q}\frac{\mathbf{q}}{q}\left(b_\mathbf{q}e^{i\mathbf{q}\cdot\mathbf{r}} + b^\dagger_\mathbf{q}e^{-i\mathbf{q}\cdot\mathbf{r}}\right), \tag{1}$$

with $u_0 = \sqrt{\hbar/(2M_r\omega_{LO})}$ and $\pi_0 = \sqrt{\hbar M_r \omega_{LO}/2}$; $M_r$ is the reduced mass of the lattice ions and $N$ is the number of unit cells. In a thermal state at temperature $T$ the uncertainties are given by $(\Delta\mathbf{u}(\mathbf{r},t))^2 = u_0^2(2n_{LO}+1)$ and $(\Delta\pi(\mathbf{r},t))^2 = \pi_0^2(2n_{LO}+1)$ with $n_{LO} = [\exp(\hbar\omega_{LO}/k_BT)-1]^{-1}$. $u_0^2$ and $\pi_0^2$ are thus the corresponding vacuum fluctuations. We assume sufficiently low temperatures such that $n_{LO}$ is negligible. It is convenient to define dimensionless fluctuations as $S_u = (\Delta\mathbf{u})^2/u_0^2 - 1$ and $S_\pi = (\Delta\pi)^2/\pi_0^2 - 1$, which vanish in the vacuum state. Negative values of one of these quantities thus correspond to squeezing of the corresponding variable $\mathbf{u}$ or $\pi$. The Heisenberg uncertainty relation rewritten in these variables reads $S_u + S_\pi + S_uS_\pi \geq 0$. In our case, where typically $|S_u|, |S_\pi| \ll 1$, this effectively reduces to $S_u + S_\pi \geq 0$.

From the generating function approach we obtain closed but in general lengthy formulas for $\langle b_\mathbf{q}\rangle$, $\langle b_\mathbf{q}b_{\mathbf{q}'}\rangle$, and $\langle b^\dagger_\mathbf{q}b_{\mathbf{q}'}\rangle$ [8, 10] and thus for $S_u$ and $S_\pi$. Because of the vanishing dispersion of LO phonons there is no spatial transport and the space dependence of both $S_u$ and $S_\pi$ is given by the time-independent function $|I_0(\mathbf{r})|^2$ with $I_0(\mathbf{r}) = \sum_\mathbf{q}\frac{\mathbf{q}\cdot\mathbf{r}}{qr}(g^*_\mathbf{q}/\omega_{LO})e^{i\mathbf{q}\cdot\mathbf{r}}$. In the Figures we therefore plot the space-independent quantities $\tilde{S}_u = S_u/|I_0|^2$ and $\tilde{S}_\pi = S_\pi/|I_0|^2$. All calculations have been performed for a spherical GaAs quantum dot with diameter 5 nm.

## 3. Results

From the general formulas it is easily seen that after a single pulse excitation, although $\tilde{S}_u$ and $\tilde{S}_\pi$ are in general time-dependent, they never assume negative values. Thus, there is no squeezing in this case. Therefore we will concentrate on the case of excitation by two laser

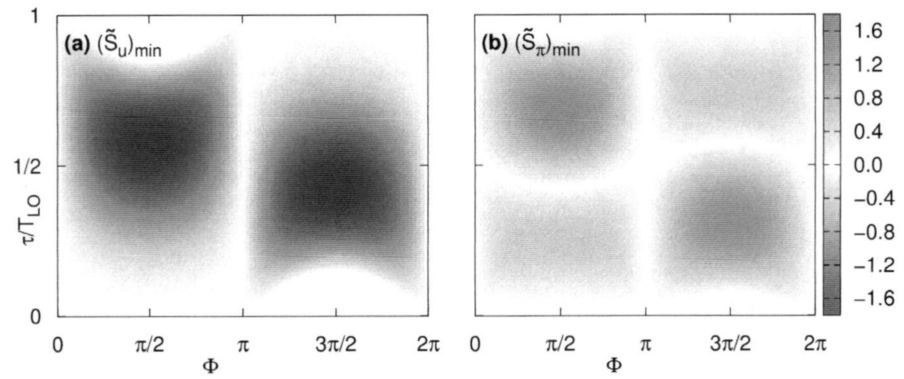


**Figure 2.** Minima of the time dependent fluctuations (a) $(\tilde{S}_u)_{min}$ and (b) $(\tilde{S}_\pi)_{min}$ as functions of the delay time $\tau$ and phase difference $\Phi$ for fixed pulse areas $A_{1,2} = \pi/2$.

pulses at $t_1 = -\tau$ and $t_2 = 0$. The fluctuation properties of the phonons generated by such an excitation depend on the pulse areas $A_{1,2}$, the delay time $\tau$ between the pulses and the phase difference $\Phi = \phi_2 - \phi_1 - \bar{\omega}\tau$, $\bar{\omega}$ being the resonance frequency of the exciton.

Figure 1 shows the time-dependent fluctuations $\tilde{S}_u(t)$ and $\tilde{S}_\pi(t)$ for fixed pulse areas $A_{1,2} = \pi/2$ and fixed phase difference $\Phi = \pi/2$ at different delay times $\tau = T_{LO}/4, T_{LO}/2$ and $3T_{LO}/4$, with $T_{LO} = 2\pi/\omega_{LO}$ being the phonon oscillation period. We find that the fluctuations may become negative in certain time windows, i.e., that now squeezing may occur. The uncertainty relation requires that at any time $\tilde{S}_u + \tilde{S}_\pi \geq 0$, which is indeed fulfilled. The fluctuation of the displacement $\tilde{S}_u$ is periodic with a period $T_{LO}$. For all delay times $\tilde{S}_u$ becomes negative in some time intervals, i.e., all phonon states generated by these pulse sequences are displacement-squeezed. For $\tau = T_{LO}/2$ the most pronounced negative values can be seen, while for $\tau = T_{LO}/4$ and $3T_{LO}/4$ the fluctuation only slightly goes below zero. In contrast, the fluctuation of the momentum $\tilde{S}_\pi$ oscillates with a period of $T_{LO}/2$. Here the most negative values appear for $\tau = 3T_{LO}/4$. For $\tau = T_{LO}/2$ the fluctuation still becomes slightly negative, while for $\tau = T_{LO}/4$ it remains positive, thus there is no momentum squeezing in this case.

A more detailed analysis of the conditions for squeezing is shown in Fig. 2. Here we have plotted the minima of the time dependent fluctuations $(\tilde{S}_u(t))_{min}$ and $(\tilde{S}_\pi(t))_{min}$ as functions of the delay time $\tau$ and the phase difference$\Phi$ . The pulse areas again are kept at $A_{1,2} = \pi/2$. Blue areas indicate that the fluctuation has a minimum below zero, i.e., the quantity is squeezed. Red areas indicate fluctuations which always exceed the vacuum value. We notice that there is a symmetry when exchanging$\Phi \rightarrow 2\pi - \Phi$and $\tau \rightarrow T_{LO} - \tau$. The most pronounced squeezing occurs for $\Phi = (2n-1)\pi/2$. For the displacement the minimum values close to $T_{LO}/2$ are strongly negative, while for the momentum the most negative values appear close to $\tau = 3T_{LO}/4$ for $\Phi = \pi/2$ and $\tau = T_{LO}/4$ for $\Phi = 3\pi/2$. But not all generated states are squeezed: at $\tau = nT_{LO}$ or $\Phi = n\pi$ no squeezing occurs. For delay times around $T_{LO}/4$ and phase differences around $\pi/2$ or delay times around $3T_{LO}/4$ and phase differences around $3\pi/2$ the momentum fluctuations are enhanced compared to the vacuum level while the lattice displacement still exhibits squeezing.

The absolute strength of fluctuations is typically hard to measure in solid state systems and therefore a unique proof of squeezing is difficult. Therefore, the appearance of a Fourier component at a frequency $2\omega_{LO}$ in the spectrum of the fluctuations has been taken as an indication for the presence of a squeezed state [4, 5]. To analyze the spectral properties we have plotted in Fig. 3 the absolute value of the Fourier coefficients $\tilde{S}_u(\omega)$ and $\tilde{S}_\pi(\omega)$ at the

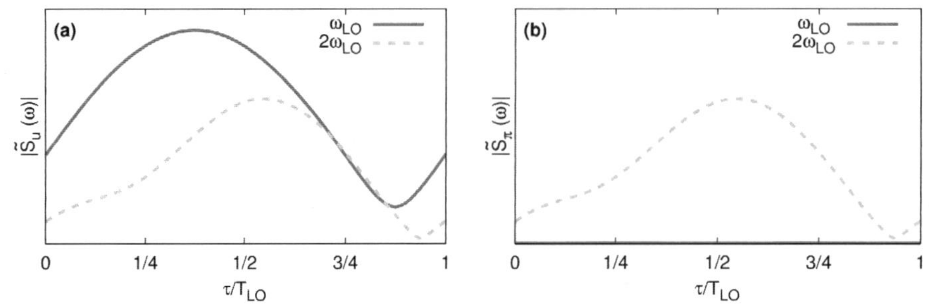


**Figure 3.** Absolute value of the Fourier coefficients (a) $|\tilde{S}_u(\omega)|$ and (b) $|\tilde{S}_\pi(\omega)|$ at the single ($\omega = \omega_{LO}$) and double ($\omega = 2\omega_{LO}$) phonon frequency as functions of the delay time $\tau$ for fixed pulse areas $A_{1,2} = \pi/2$ and phase difference $\Phi= \pi/2$.

single ($\omega = \omega_{LO}$) and double ($\omega = 2\omega_{LO}$) phonon frequency. Here the delay time $\tau$ has been varied while pulse areas are kept at $A_{1,2} = \pi/2$ and the phase difference is $\Phi= \pi/2$. In the case of $\tilde{S}_u$ the Fourier component at $\omega_{LO}$ is always stronger than the component at $2\omega_{LO}$. If we compare the Fourier coefficient with the minimum value $(\tilde{S}_u)_{min}$ shown in Fig. 2, we notice that the coefficient $2\omega_{LO}$ roughly follows the shape of the curve of the minima. The largest Fourier coefficient at $2\omega_{LO}$ corresponds to the most negative value of $(\tilde{S}_u)_{min}$. For the momentum no oscillation with the single phonon frequency $\omega_{LO}$ occurs and its Fourier coefficient is zero for all delay times $\tau$. The behavior of the component at $2\omega_{LO}$ is exactly the same as for $\tilde{S}_u$. When comparing this curve with $(\tilde{S}_\pi)_{min}$ shown in Fig. 2, however, we find that there is no correspondence between the strength of the Fourier coefficient and the appearance of squeezing.

## 4. Conclusions

We have studied the conditions for the generation of squeezed LO phonon states in a QD system excited optically with one or two laser pulses. While for single pulse excitation no squeezing occurs, two pulse excitations may lead to squeezing of the displacement and of the momentum. The details strongly depend on the delay time and the relative phase of the two laser pulses. However, we have found that there is no direct relation between the presence of squeezing and the appearance of a Fourier component in the fluctuations at twice the phonon frequency.

This work has been partially supported by the DFG (grant KU 697/11-1).

## References

[1] Walls D F 1986 *Nature* **324** 210
[2] Polzik E 2008 *Nature* **453** 45
[3] Först M and Dekorsy T 2008 Coherent phonons in bulk and low-dimensional semiconductors *Coherent vibrational dynamics* ed De Silvestri S, Cerullo G and Lanzani G (Boca Raton: CBC press) p 129
[4] Hu X and Nori F 1997 *Phys. Rev. Lett.* **79** 4605
[5] Garrett G A, Rojo A G, Sood A K, Whitaker J F and Merlin R 1997 *Science* **275** 1638
[6] Misochko O V, Sakai K and Nakashima S 2000 *Phys. Rev. B* **61** 11225
[7] Johnson S L *et al.* 2009 *Phys. Rev. Lett.* **102** 175503
[8] Axt V M, Kuhn T, Vagov A and Peeters F M 2005 *Phys. Rev. B* **72** 125309
[9] Krummheuer B, Axt V M and Kuhn T 2002 *Phys. Rev. B* **65** 195313
[10] Vagov A, Axt V M and Kuhn T 2002 *Phys. Rev. B* **66** 165312

# Effect of Interface State Trap Density on the Performance of Scaled Surface Channel $In_{0.3}Ga_{0.7}As$ MOSFETs

**B. Benbakhti*, J. S. Ayubi-Moak, K. Kalna and A. Asenov**

Device Modelling Group, Dept. Electronics & Electrical Engineering, University of Glasgow, Glasgow, G12 8LT, United Kingdom

E-mail: *brahim@elec.gla.ac.uk Tel +44 141 330 2964 Fax +44 141 330 4907

**Abstract.** The effect of interface state trap density, $D_{it}$, on the $I_D$-$V_G$ characteristics of scaled surface channel MOSFETs based on $In_{0.3}Ga_{0.7}As$ channel has been investigated using drift-diffusion simulations. We have developed a methodology to include arbitrary energy distributions of interface states into the input simulation decks and analysed their impact on subthreshold characteristics and drive current when these devices are scaled from a gate length of 65 nm to 35 nm, 25 nm and 18 nm. The distributions of interface states having high density tails that extend into the conduction band can significantly impact the subthreshold performance of the larger gate length device. Furthermore, the same distributions have smaller impact on the performance of shorter channel devices which were designed with smaller high-$\kappa$ thickness.

## 1. Introduction

MOSFETs based on III-V semiconductors are currently the most promising devices for the near-end of the ITRS [1]. High electron mobility and low effective mass resulting in a very high injection velocity and low backscattering promise very high device performance [2]. The development of a suitable $Ga_2O_3$/GdGaO (GGO) gate stacks for GaAs with low interface state density ($\sim 1\times10^{11}/cm^2$) have been already reported resulting in an unpinned oxide/semiconductor interface [3][4]. This is an essential step enabling the realisation of III-V transistors in future CMOS circuits and systems.

However, the introduction of such novel III-V materials into MOSFETs, requires transistor architectures that can take a full advantage of the high mobility and high injection velocity in the channels [2]. The presence of interface state density distributions in these material systems can still impact the overall performance of the scaled transistors. Herein, we investigate the potential device performance and the impact of interface states ($D_{it}$) on the output characteristics of scaled surface channel (SC) $In_{0.3}Ga_{0.7}As$ MOSFETs using a commercially available Drift-Diffusion (DD) TCAD tool [5].

## 2. Surface Channel $In_{0.3}Ga_{0.7}As$ MOSFET architecture

We have simulated scaled SC $In_{0.3}Ga_{0.7}A$ MOSFETs from a gate length of 65 nm to 35 nm, 25 nm and 18 nm. A cross section of the investigated SC devices is illustrated in Fig. 1. This design is based on a traditional bulk MOSFET device structure with $In_{0.3}Ga_{0.7}As$ channel growth on $p$-type doped GaAs buffer substrate. On top of this sits a high-$\kappa$ GGO layer ($\kappa = 20$) separating the $In_{0.3}Ga_{0.7}A$ channel from the metal gate that is modelled using an assumed work function of 4.25 eV. Highly doped implanted $n$-type regions are simulated using a Gaussian-type distribution with a peak doping dose reported in table of Fig. 1. The lateral extensions are similarly defined and self-aligned to the gate.

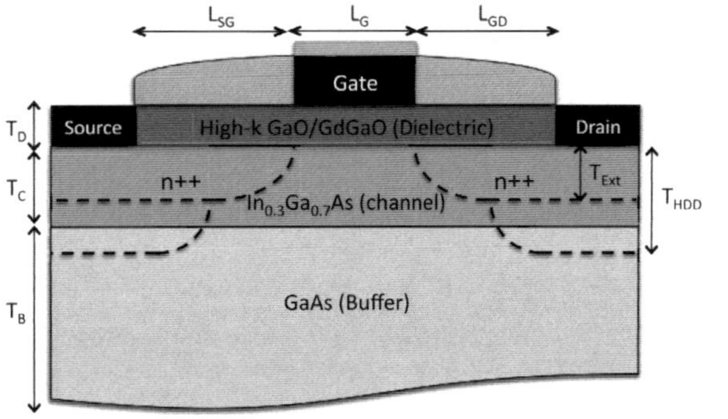


| Thickness [nm] | Gate length [nm] | | | |
|---|---|---|---|---|
| | 65 | 35 | 25 | 18 |
| Source-to-gate spacer ($L_{SG}$) | 80 | 37 | 26 | 20 |
| Gate-to-drain spacer ($L_{GD}$) | 80 | 37 | 26 | 20 |
| High-$\kappa$ GGO ($T_D$) | 10 | 5.1 | 4.6 | 3.6 |
| $In_{0.3}Ga_{0.7}As$ channel ($T_C$) | 10 | 10 | 7 | 5 |
| GaAs buffer ($T_B$) | 300 | 300 | 300 | 300 |
| HDD depth ($T_{HDD}$) | 70 | 35 | 25 | 18 |
| Extension depth ($T_{Ext}$) | 22 | 20 | 14 | 10 |
| EOT | 1.95 | 1.0 | 0.9 | 0.7 |
| Doping concentrations [$\times 10^{17}/cm^3$] | Gate length [nm] | | | |
| | 65 | 35 | 25 | 18 |
| Peak S/D n-type | 100 | 100 | 200 | 400 |
| Uniform p-type | 10 | 5 | 10 | 24 |

**Figure 1.** Cross-section of scaled SC $In_{0.3}Ga_{0.7}As$ transistors and the corresponding dimensions, doping concentrations and effective oxide thickness (EOT).

## 3. Results and discussion

The layer structures of the scaled SC transistors as well as dimensions and doping concentrations are summarised in Fig. 1. The high-$\kappa$ gate oxide is assumed to be GGO with a dielectric constant of 20. The dimensions and doping profiles are optimised using a DD simulations and account for limitations of doping concentrations in III-V materials. The simulated $I_D$-$V_G$ characteristics at low and high drain voltages for each SC gate length MOSFET are plotted in Fig. 2. A proper scaling of critical dimensions and doping profiles is made following the ITRS prescriptions and in order to minimise subthreshold slope and DIBL deterioration (Fig. 3). The effect that the $D_{it}$ has on the performance characteristics is then studied by developing a methodology to include an arbitrary $D_{it}$ distributions into the device simulation procedure. Due to the use of the drift-diffusion approach, the simulations are restrained only in the subthreshold region at low drain voltage aiming mainly to illustrate the impact of the $D_{it}$ on subthreshold slope and mobile charge control in the channel.

The measured and used $D_{it}$ distributions in the device simulations are plotted in Fig. 4. In each case, zero corresponds to the position of the conduction band edge ($E_C$). The measured distribution in Fig. 4(A) was recently reported in [6] and is that for a $GdGaO/Ga_2O_3$ dielectric stack on heterostructure with an $In_{0.3}Ga_{0.7}As$ channel. The $D_{it}$ distribution plotted in Fig. 4(B) is the same as that shown in Fig. 4(A) but includes an extrapolated extension of the $D_{it}$ distribution into the conduction band. The $D_{it}$ of Fig. 4(C) is 2.5 times higher than the measured one. In the device simulations, only acceptor states in the upper half of the conduction band are considered. Furthermore, each one of the plotted distributions in Fig. 4 is then aligned to the conduction band edge of $In_{0.3}Ga_{0.7}As$ which is immediately below the high-$\kappa$ layer. We assume that the experimentally determined $D_{it}$ distributions will not dramatically change when the devices are scaled down.

The simulated $I_D$-$V_G$ characteristics at low drain bias ($V_{DS}$=50 mV) for 65 nm gate length $In_{0.3}Ga_{0.7}$ SC MOSFET excluding and including $D_{it}$ from Fig. 4(A) are plotted in Fig. 5. The interface density of states which has a large tail density at the conduction band will not affect significantly the deep subthreshold region of the $In_{0.3}Ga_{0.7}As$ SC MOSFET. The subthreshold

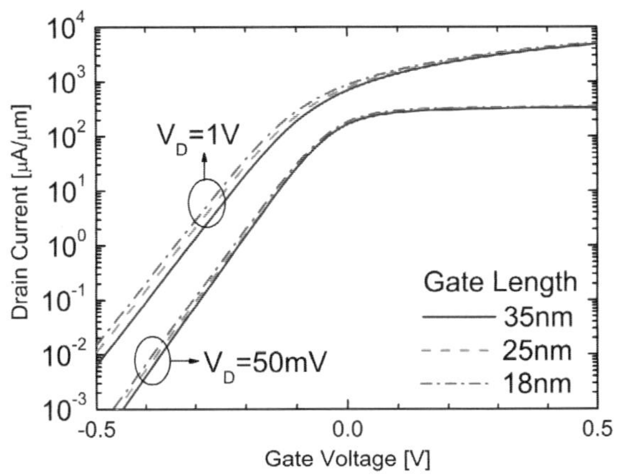


**Figure 2.** $I_D$-$V_G$ characteristics at low ($V_D$=50 mV) and high ($V_D$=1 V) drain voltages of scaled SC $In_{0.3}Ga_{0.7}As$ device.

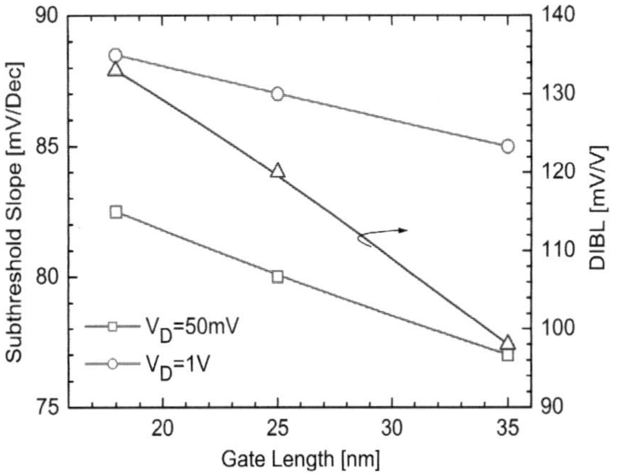


**Figure 3.** Subthreshold slope (SS) and DIBL versus gate length of the scaled SC MOSFETs at low drain voltage (squares); at high drain voltage (circles) and DIBL (triangles).

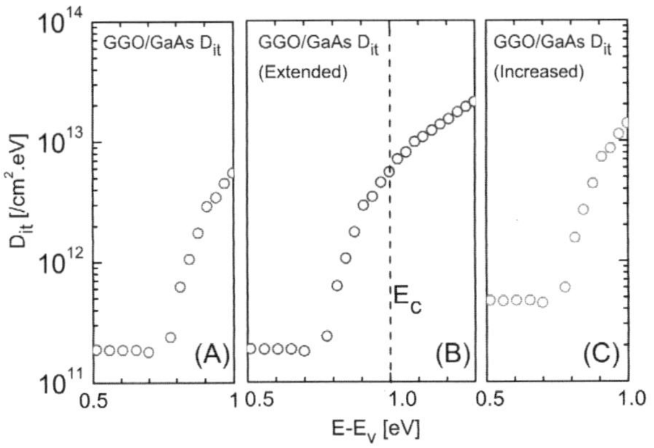


**Figure 4.** Measured and modified GGO/GaAs $D_{it}$ distributions used in simulations of the scaled SC device.

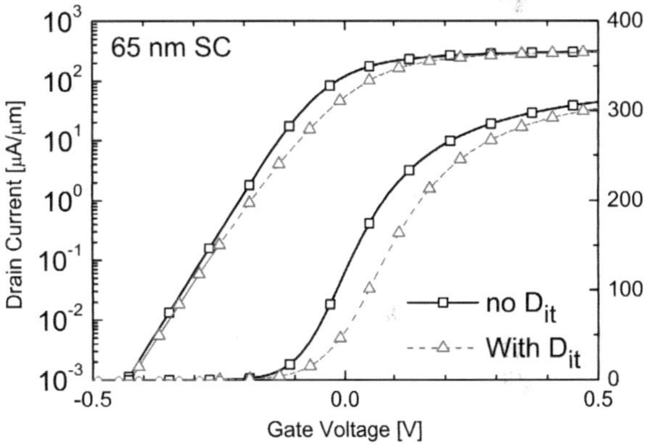


**Figure 5.** $I_D$-$V_G$ characteristics at $V_D$=50 mV of 65 nm gate length SC MOSFET, with $D_{it}$ (triangles) and without $D_{it}$ (squares).

slope (SS) deteriorates close to the threshold voltage and will cause its shift. The on-current is lowered by nearly 40% just above the threshold but will recover at the large gate bias overdrive.

The simulated $I_D$-$V_G$ characteristics at low drain bias of ($V_{DS}$=50 mV) for the scaled 35 nm, 25 nm and 18 nm gate length devices are plotted in Fig. 6. The inclusion of the measured GGO/GaAs $D_{it}$ distribution from Fig. 4(A) into the simulation has little affect on the deep subthreshold behaviour for each of the scaled device structures. This is because the Fermi level sweeps through a relatively low $D_{it}$ ($\sim 2\times10^{11}$/cm$^2$.eV) until approximately 0.2 eV below the conduction band edge as shown in Fig. 4 and demonstrated in Figs. 5 and 6. Above this energy $D_{it}$ dramatically increases to approximately $4\times10^{12}$/cm$^2$.eV resulting in significant deterioration of the subthreshold slope and a shift in $V_T$. Due to the scaling process, the high-$\kappa$ oxide thickness is reduced as well as channel thickness which both affect the electrostatic integrity. Therefore, the drive current increases for each shorter channel SC MOSFET. The same $D_{it}$ in each case affects the drain current with a different ratio due to the increasing values of current for each scaled gate length device. The resulting degradation in the SS due to the measured $D_{it}$ distributions is less significant for shorter channel length devices. Fig. 6(C) for example shows a smaller degradation in the SS and corresponding smaller shift in the $V_T$ when compared with the results in Fig. 6(A). However, the possible extension of the $D_{it}$ in the conduction band (Fig. 4(B)) or larger magnitude of $D_{it}$ (Fig. 4(C)) due to different fabrication processes for smaller device (below 65 nm gate length) can affect the SS and reduces the mobile charge in the channel above threshold (Fig. 6).

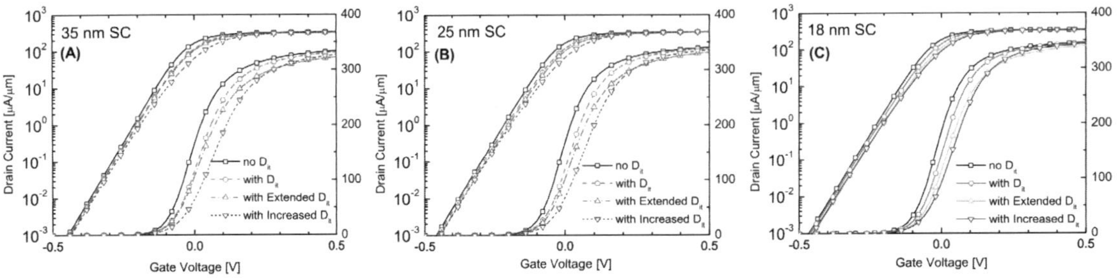


**Figure 6.** $I_D$-$V_G$ characteristics of scaled $In_{0.3}Ga_{0.7}As$ SC MOSFET with gate length of (A) 35 nm, (B) 25 nm and (C) 18 nm at $V_D$=50 mV. The impact of GGO/GaAs $D_{it}$ from Fig. 4 is also shown. Current is plotted on both log (left) and linear scales (right) in each case.

## 4. Conclusion

The effect of $D_{it}$ on the subthreshold characteristics and performance of scaled $In_{0.3}G_{0.7}As$ surface channel has been investigated. Numerical investigations demonstrate that specific distributions of interface states can significantly affect the subthreshold performance of these device. Surface channel designs are more sensitive to interface states in the upper half of the bandgap. This is due to the manner in which the Fermi level sweeps through this particular energy region during device operation.

The results herein show that smaller gate length devices are less affected by the interface state density distributions. For smaller device lengths, the high-$\kappa$ oxide thickness is reduced as well as the channel thickness which both affect the electrostatic integrity. Therefore, the drive current is increased for the shorter channel SC MOSFET and the applied $D_{it}$ affects the drain current with different ratios due to the larger values of the drive current for smaller gate length devices. The larger magnitude of $D_{it}$, possibly due to different fabrication processes for smaller devices, can affect both the subthreshold characteristics and $I_{on}$ current. This limits the range of distribution of interface states that can contribute to the trapping of charge and results in performance degradation. We would like to point out that the simulation results herein are strictly for low drain bias assuming a constant mobility in the channel. A more accurate study at a high drain voltage will require drift-diffusion simulations calibrated to the results obtained from ensemble Monte Carlo simulations [2][7] in order to accurately adjust both the low field mobilities and saturation velocities in each case.

## References

[1] R. Chau et al., IEEE Transactions on Nanotechnology, vol. 4, no. 2, pp. 153-158, 2005.
[2] K. Kalna et al., IEEE Transactions on Electron Devices, vol. 55, no. 9, pp. 2297-2306, 2008.
[3] M. Passlack, O. Hartin, M. Ray, and N. Medendorp, US Patent 6 963 090, Nov. 8, 2005.
[4] M. Passlack et al., IEEE Transactions on Electron Devices, vol. 53, no. 10, pp. 2454-2459, 2006.
[5] Sentaurus sdevice Z-2007.03, Synopsys, Inc.
[6] M. Passlack et al., IEEE Electron Device Letters, vol. 30, no. 1, pp. 2-4, 2009.
[7] K. Kalna et al., IEEE Transactions on Nanotechnology, vol. 6, no. 1, pp. 106-112, 2007.

# Floquet projections of a Gaussian Wigner function in a Kronig-Penney potential

**L. Maccari[1], L. Demeio[1] and P. Bordone[2,3]**

[1] Dipartimento di Scienze Matematiche, Via Brecce Bianche, Università Politecnica delle Marche, 60131 Ancona – Italy
[2] Physics Department, University of Modena and Reggio Emilia, Via Campi 213/A, I-41025 Modena, Italy
[3] S3 Research Center, CNR-INFM, Via Campi 213/A, Modena I-41125, Italy

E-mail: maccari@dipmat.univpm.it

**Abstract.** In this contribution, we apply the multiband model developed in [1] to a Kronig-Penney potential and calculate the Floquet projections of a Gaussian Wigner function in phase space. The Bloch functions are calculated numerically and then used in the expressions of the Wigner projections.

## 1. Introduction

The Wigner-function approach is a commonly used method in the study of electron transport properties in semiconductors and in the description of their physical properties. [2, 3, 4]. All the relevant and realistic applications of this approach, however, have considered until now only those processes which can be adequately described within the single-parabolic-band approximation. This formalism cannot be applied to those processes which involve non-parabolicity of the band profiles or multi-band transitions.The latter ones, for example, play a significant role in determining the flow of current in interband devices (e.g., the Resonant Interband Tunneling Diode [5]). Therefore, the Wigner-function approach needs to be extended; the definition of the Wigner function should include the populations of all bands involved in the transport processes and the evolution equation should take into account possible non-parabolicity effects. A general model of multiband, non-parabolic transport by the Wigner-function approach was developed in [1], where a multi-band Wigner function was introduced by using a straightforward expansion of the wave functions in Bloch states and a Bloch-state representation of the density matrix. This multi-band model is probably the most general way of using the Wigner-function approach in the context of multi-band model problems; other formulations, such as the envelope function model and the Kane model, have recently been explored in the literature [6, 7, 8, 9].

While the effects of non-parabolicity are easy to include in Wigner-function models (see [10]), the description of multiband transitions requires the knowledge of the energy bands and of the Bloch functions of the periodic potential, which enter the expressions of the Floquet projections of the Wigner function. The calculation of these projections is an essential intermediate step towards the numerical solution of the equation which governes the time evolution of the Wigner function, and have not been obtained until now. In this work, we show the Floquet projections of a Gaussian Wigner function in a one-dimensional Kronig-Penney potential.

## 2. Multiband Wigner function

By using the completeness of the Bloch functions, we express the Wigner function as a sum over the components onto the Floquet band subspaces:

$$f(x,p) = \sum_{mn} f_{mn}(x,p) \tag{1}$$

with

$$f_{mn}(x,p) = \frac{1}{2\pi} \int\int_{B^2} \frac{dk'\,dk'}{|B|^2} \rho_{mn}(k,k')\phi_{mn}(k,k',x,p) \tag{2}$$

where $B$ is the Brillouin zone and $|B|$ its volume, the $\rho_{mn}$'s are the elements of the density matrix in the Bloch-state representation, the coefficients $\phi_{mn}$ are given by

$$\phi_{mn}(k,k',x,p) = \int_{\mathbb{R}} \Psi_{mk}(x+\eta/2)\Psi^{\,*}_{nk'}(x-\eta/2)\, e^{-ip\eta}\, d\eta$$

and$\Psi_{\,mk}(x) = u_{mk}(x)e^{i\,k\,x}$ is the Bloch function of the $m-$th band of the periodic potential. The diagonal projections $f_{mm}$ are real, while the off-diagonal ones are complex conjugates, $f_{mn} = f^*_{nm}$. After some steps, we find

$$\begin{aligned} f_{mn}(x,p) = \sum_{m'n'} \sum_{m''n''} & e^{i\,(m'-n')\,x}\, \Theta\left(\alpha_{m'n'}(p)\right) \int_{-\alpha_{m'n'}(p)}^{\alpha_{m'n'}(p)} e^{i\,\tau x} \\ & U_{mm'}\left(K_{m'n'}+\frac{\tau}{2}\right) U^*_{nn'}\left(K_{m'n'}-\frac{\tau}{2}\right) U^*_{mm''}\left(K_{m'n'}+\frac{\tau}{2}\right) U_{nn''}\left(K_{m'n'}-\frac{\tau}{2}\right) \\ & \widehat{f}\left(\tau+m''-n'',\, K_{m'n'}+\frac{m''+n''}{2}\right) d\tau \end{aligned} \tag{3}$$

where we have indicated by $\widehat{f}(\lambda,p)$ the Fourier transform of $f(x,p)$ with respect to $x$. Here, the $U_{mm'}(k)$ are the $m'-$th Fourier coefficients of the periodic part of the Bloch functions. Also, $K_{mn}(p) = p-(m+n)/2$, $\alpha_{mn}(p) = 1-2|K_{mn}(p)|$ and $\Theta$ is the Heaviside function. In order to obtain analytical or numerical data for the $U_{mm'}(k)$'s, we consider a Kronig-Penney potential of period $2\pi$, given by

$$V(x) = \begin{cases} V_0, & -b \le x < 0 \\ 0, & 0 \le x < c, \end{cases}$$

with $b+c = 2\pi$. In the equations above, we have used dimensionless variables. If $a$ is the lattice period, we have scaled the space variable $x$, the energy $E$, the potential $V$, the Wigner momentum $p$ and the wave function $\Psi$ according to

$$2\pi\frac{x}{a} \to x \qquad \frac{2\,m\,a^2}{(2\pi\hbar)^2}\,E \to E \qquad \frac{2\,m\,a^2}{(2\pi\hbar)^2}\,V(x) \to V(x) \qquad 2\pi\frac{a}{\hbar}\,p \to p \qquad \sqrt{\frac{a}{2\pi}}\,\psi \to \psi.$$

For given values of the parameters $b$ and $V_0$, the eigenvalue problem for the Kronig-Penney potential can be solved completely, the Fourier coefficients $U_{mm'}$ can be evaluated and the Floquet projections (3) of the Wigner function can be calculated.

## 3. Numerical results

In this section, we calculate the projections of a Gaussian shaped Wigner function which, as is well known [11], results from the Wigner-Weyl transform of a Gaussian wave packet,

$$f(x,p) = \frac{1}{2\pi\sqrt{M}}\, e^{-\frac{\sigma_{02}\,(x-x_0)^2}{2M} - \frac{\sigma_{20}\,(p-p_0)^2}{2M} + \frac{\sigma_{11}\,(x-x_0)\,(p-p_0)}{M}}, \tag{4}$$

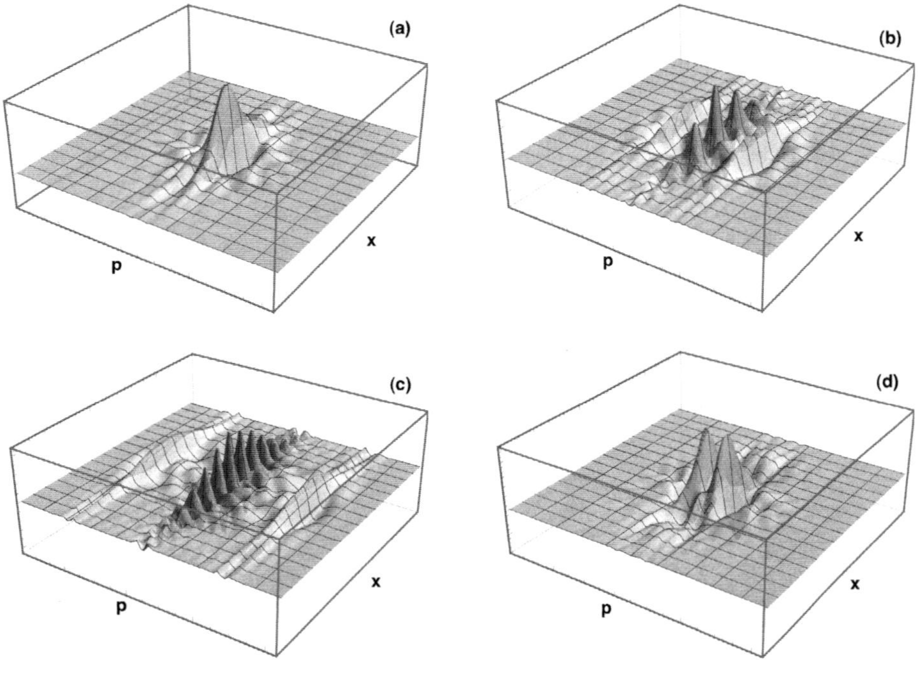


**Figure 1.** (a) $f_{00}(x,p)$, (b) $f_{22}(x,p)$, (c) $f_{44}(x,p)$ and (d) real part of $f_{01}(x,p)$ projections of the Gaussian Wigner function (4) for $p_0 = 0$.

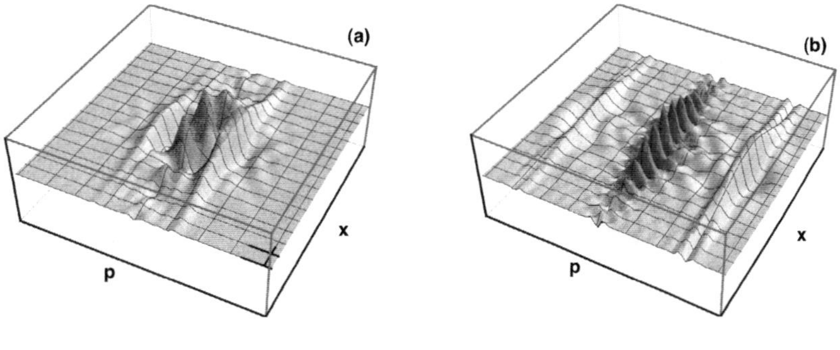


**Figure 2.** (a) $f_{11}(x,p)$ and (b) $f_{44}(x,p)$ projections of the Gaussian Wigner function (4) for $p_0 = 1$.

(with $M = \sigma_{20}\,\sigma_{02} - \sigma_{11}^2$), in the Kronig-Penney potential introduced above. The results are shown in Figures 1 and 2. We have chosen $b = 0.2$ and $V_0 = 10$ as characteristic parameters of the Kronig-Penney potential, and $x_0 = 0$, $\sigma_{02} = 0.2$, $\sigma_{11} = 0$ and $\sigma_{20} = 0.3$, with $p_0 = 0$ (Figure 1) and $p_0 = 1$ (Figure 2).

The projection onto the $m = 0$ band space shows a maximum centered at $x = 0, p = 0$ surrounded by lower maxima and negative minima (not seen), due to interference effects. The projections onto the higher band subspaces show an oscillating structure along the $p = 0$ line, with two other peaks located symmetrically about $p = 0$; they are of equal height in the $p_0 = 0$ case (Figure 1), while in the $p_0 = 1$ case (Figure 2) the one at positive $p$ is higher. The oscillating

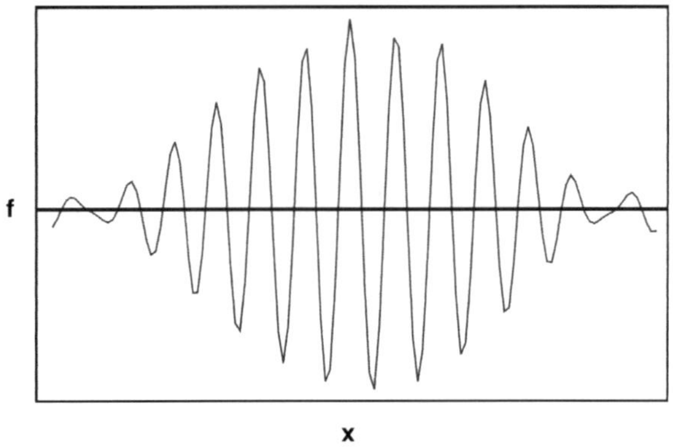


**Figure 3.** $f_{44}(x,p)$ at $p=0$ as a function of $x$, for $p_0=1$.

structure at $p=0$, shown in Figure 3, is due to the interference of different contributions to the sum in equation (3), which have an almost gaussian profile. This effect is similar the one observed in Schrödinger cat states.

## 4. Conclusions and outlook

We have calculated the Floquet projections of a Gaussian shaped Wigner function in a Kronig-Penney periodic potential. These are important intermediate results, needed in the study of the multiband dynamics of a group of carriers in a semiconductor by the Wigner-function approach. The algorithms and techniques for the calculation of these projections will be included in a large-scale numerical computation, which is underway, with the aim of describing a coherent interband transition by the Wigner-function approach.

## References

[1] L. Demeio, P. Bordone and C. Jacoboni, *Multi-band, non-parabolic Wigner-function approach to electron transport in semiconductors*, Transport Theory and Statistical Physics, **34**(7), 499-522 (2005).
[2] E. Wigner, *On the Quantum Correction for Thermodynamic Equilibrium*, Phys. Rev **40**, 749-759 (1932).
[3] W. R. Frensley, *Boundary conditions for open quantum systems driven far from equilibrium*, Rev. Modern Phys. **62**(3), 745-791 (1990).
[4] C. Jacoboni, R. Brunetti, P. Bordone, and A. Bertoni, *Quantum transport and its simulation with the Wigner-function approach*, International journal of high speed electronics and systems, **11**(2), 387-423 (2001).
[5] M. Sweeney and J. M. Xu, *Resonant Interband Tunneling Diodes*. Appl. Phys. Lett. **54**, 546-548 (1989).
[6] O. Morandi and M. Modugno, *A multiband envelope function model for quantum transport in a tunneling diode*, Phys. Rev. B **71**, 235331(1-8) (2005).
[7] O. Morandi and L. Demeio, *A Wigner-function approach to interband transitions based on the multiband-envelope-function model*, Transport Theory and Stat. Phys., **37**, 437-459 (2008).
[8] L. Barletti, *Wigner Envelope Functions for Electron Transport in Semiconductor Devices*, Transport Theory Stat. Phys. **32**, 253-278 (2003).
[9] G. Borgioli, G. Frosali, and P. F. Zweifel, *Wigner Approach to the Two-Band Kane Model for a Tunneling Diode*, Transport Theory Stat. Phys. **32**, 347-366 (2003).
[10] L. Demeio, P. Bordone and C. Jacoboni, *Numerical simulation of an intervalley transition by the Wigner-function approach*, Semiconductor Science and Technology, **19**, 1-3 (2004).
[11] W. Tatarskii, *The Wigner picture of quantum mechanics*, Sov. Journal Uspehi Phys. Sci. **139**, 587-619 (1983).

# Monte Carlo model for the photoluminescence kinetics of a quantum dot embedded in a nanocavity

**G. Tarel and V. Savona**
Institute of Theoretical Physics, Ecole Polytechnique Fédérale de Lausanne EPFL, CH-1015 Lausanne, Switzerland
E-mail: guillaume.tarel@epfl.ch

**A. Badolato**
Department of Physics and Astronomy, University of Rochester, Rochester, New York 14627, USA

**M. Winger, T. Volz, and A. Imamoglu**
Institute of Quantum Electronics, ETH Zürich, CH-8093 Zürich, Switzerland

**Abstract.** We address the problem of the photoluminescence of a quantum (QD) dot in a nanocavity, with focus on the case of nonzero detuning. In this regime, experiments have shown that strong emission from the cavity-like peak is still present for dot-cavity detuning exceeding 10 meV, which seems puzzling. We will discuss the general theory of *cavity feeding*, due to the relaxation and recombination kinetics of a multiply excited QD. We first compute the multi-exciton manifolds using a configuration-interaction scheme, starting from a truncated single-particle basis. We then run Monte-Carlo paths of excitation-emission kinetics on these states. This allows to extract photoluminescence spectra and two-photon correlation curves. The agreement with experimental data[1, 2] is very good. Our result shows unambiguously that the cavity feeding mechanism at large detunings can be attributed to excited-state multiexciton radiative decay (mostly biexcitons), also involving states in the wetting layer continuum.

## 1. Introduction

Once the strong coupling was achieved for semi-conductor quantum dots (QDs) in micro/nano-cavities [1, 3, 4, 5, 6, 2], an interesting parallel was drawn between cavity quantum electrodynamics (CQED) using atoms or semiconductor QDs. The question however arises, in which terms the solid state system differs from its atomic counterpart. Indeed, as it is now well established, a QD can not be considered as a perfect two level system, due to its efficient coupling to the surrounding semi-conductor (SC) matrix. In particular, this non-ideal behavior leads to an unexpected persistence of a PL peak at the cavity frequency, even for large QD-cavity detuning. Here, we develop a model of excitation-emission kinetics based on multi-exciton manifolds, that we compute from an effective mass model. We show that the emission at the cavity frequency is produced by decays from N- to N-1-exciton manifolds, with the initial state mostly involving continuum states above the QD barrier. The agreement of our results with recent experiments is excellent, especially in predicting two-photon correlations. In this respect, contrary to many existing models [14, 15, 13], our picture is the only one providing a full quantitative account of spectra and time-correlation functions.

## 2. Eigenstates of the many-body Hamiltonian

We compute the single particle states in the QD by using an effective mass description, restricting to one electron and one heavy hole band. To account for the finite height of the QD barriers – giving rise to the 2D extended states of the wetting layer – we use a truncated parabolic potential, with frozen z-motion. Features such as the precise QD shape, strain, etc., are purposely not accounted for, as they would only produce a quantitative effect on the principial mechanism that we are describing. Single-particle states are sketched in Fig. 1. The sets of single-particle states are truncated, including 18 electron, 32 holes states, and their spin degrees of freedom. Most of these single-particle states are extended states in the wetting layer. Accounting for these states is essential, as they will result in a quasi-continuum of excited multi-exciton states, allowing to match the transition frequency corresponding to the cavity mode. This feature has no counterpart in CQED of atoms, and rules out simple two-level models often adopted for modeling QD spectra.

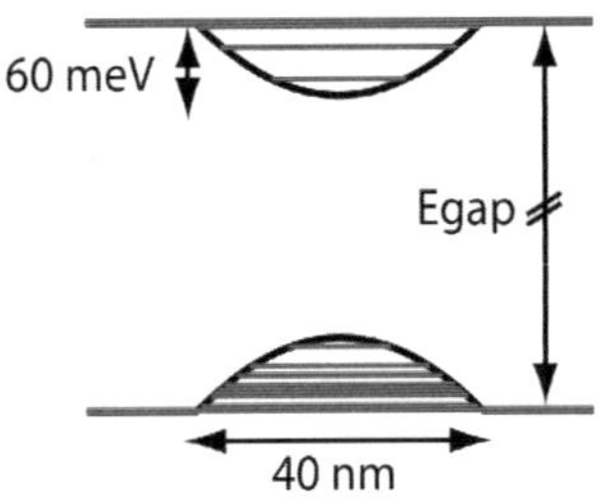


**Figure 1.** Single electron/holes states, calculated in a truncated parabolic potential.

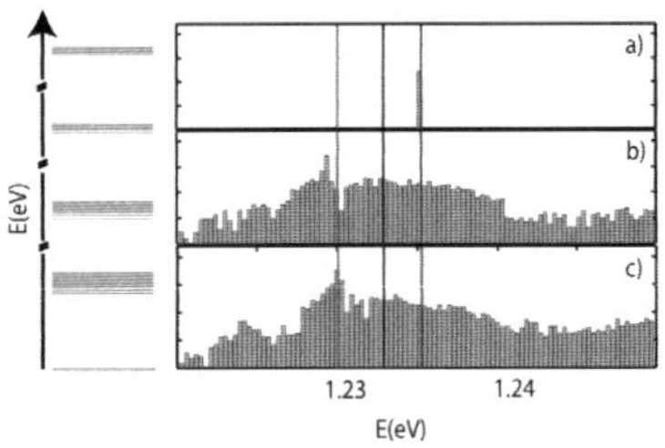


**Figure 2.** Left: Multiexcitonic states – Right: joint density of states for a) 1X manifold to ground state b) 2X manifold to 1X manifold c) 3X to 2X

We now compute eigenstates of the Coulomb hamiltonian including extended multiexcitonic states. For this, we start from the Hatree-Fock configurations of $n$ electron-hole pairs, thus neglecting charged states. We will use these configurations as a basis for the configuration interaction calculation of the Coulomb correlated eigenstates [9, 10, 16]. The Coulomb Hamiltonian is

$$H = \sum_i \epsilon_i \hat{c}_i^\dagger \hat{c}_i + \sum_j \epsilon_i \hat{d}_i^\dagger \hat{d}_i + \frac{1}{2} \sum_{i_1,i_2,i_3,i_4} V_{i_1,i_2,i_3,i_4} \hat{c}_{i_1}^\dagger \hat{c}_{i_2}^\dagger \hat{c}_{i_3} \hat{c}_{i_4} + \frac{1}{2} \sum_{j_1,j_2,j_3,j_4} V_{j_1,j_2,j_3,j_4} \hat{d}_{j_1}^\dagger \hat{d}_{j_2}^\dagger \hat{d}_{j_3} \hat{d}_{j_4} - \sum_{i_1,i_2,j_1,j_2} V_{i_1,j_1,j_2,i_2} \hat{c}_{i_1}^\dagger \hat{d}_{j_1}^\dagger \hat{d}_{j_2} \hat{c}_{i_2} . \quad (1)$$

We compute more than 2000 multiexcitonic states, including spin, up to the four-exciton multiplet (see Fig. 2, left part). In the right panel of this figure is the computed joint density of states for the QD. While for the one electron-hole pair multiplet (excitonic multiplet), the only accessible transition is the X0, there is a large DOS for transitions from multiexcitonic manifolds (at least 2nd multiplet) to the next downwards. For a QD in a uniform medium, these transitions will only provide a broad background to the PL spectrum: a PL measurement would show an ensemble of distinct peaks originating from X0, XX0 etc., superposed to a broad background. In presence of a resonant cavity, the transitions resonant with the cavity mode will be enhanced by Purcell effect, thus feeding the emission at the cavity frequency.

## 3. Monte-Carlo simulation

To prove that the emission background is actually able to feed the cavity efficiently, we use a Monte-Carlo method to simulate the continuous-wave PL under non-resonant excitation. We compute the rates of all processes starting from any level and ending in any other level. The first process we account for, is the capture process from a multiexciton manifold N-X to the next manifold (N+1)-X, by absorption of an additional electron hole pair. This process models the capture, in the multi-exciton states under study, of an electron-hole pair produced by the excitation laser in higher-lying states of the semiconductor. The second process is the radiative decay, that we model including the Purcell factor implied by the cavity resonance. Finally, inside each multiplet, relaxation mechanism through optical phonon are modeled as in Ref. [8]. In addition, we have to account for a spin-flip process, which we treat as a quasi resonant process with a lifetime of $\approx 100ps$. For all these process we compute the transition rates, starting from appropriate Hamiltonians in the electron-hole basis.

## 4. Results

The Monte-Carlo simulation gives access to both PL spectra and two-time correlation functions, as measured in a Hanbury-Brown and Twiss experiment. The simulated PL spectrum is shown in Fig. 3. For this 5nm red detuned device, a cavity peak (C) is always visible together with the usuals X0 and XX0 peaks (blue and black arrow). This is a first evidence that our model can explain the cavity feeding mechanism. In addition, we can simulate two time correlations ($g^2(\tau)$) curves. Fig. 4 shows the simulated curves for the exciton-exciton, cavity-cavity, and exciton-biexciton correlations, and the corresponding measured ones. For the cavity-cavity correlation, we observe that the measured $g^2(0)$ is significantly lower than in the calculations. We attribute this difference to the absence of charged excitons in our model. In any case, our explanation in terms of multiexcitons is perfectly compatible with the sub-poissonian nature of this curve, due the the possibility of cavity photon cascades. Further validation of the cavity feeding mechanism

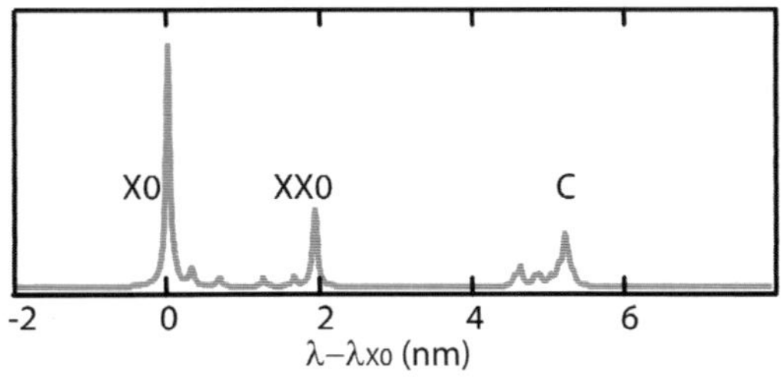


**Figure 3.** Computed PL spectrum for a red detuned device. For this detuning of 5nm, a peak at the cavity frequency (C) is visible.

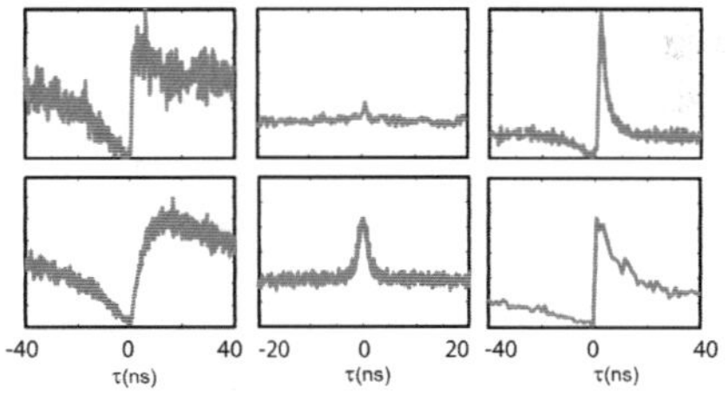


**Figure 4.** Measured (top) and calculated (bottom) $g^2(\tau)$ time correlations. From left to right: cavity-exciton, cavity-cavity, and exciton-biexciton correlations.

comes from the analysis of the power dependence of the spectrum (see Fig. 5). While the X0 and XX0 peaks saturate at high excitation power, as expected, the cavity peak does not saturate within the considered power range. This feature is explained as the cavity feeding involves transitions from high multi-exciton manifolds, that are expected to saturate at higher power. In addition, for low pump power the cavity peak shows a superlinear power dependence, again due to the multi-excitonic origin of the cavity feeding. Both the power dependence of the cavity PL and the two-photon correlations have been demonstrated in experiments. They are a natural consequence of the assumptions of our model. On the contrary, they cannot be explained by any model that assumes PL originating from a single exciton level [13, 14, 15] – no matter how spectrally shaped – that would imply linear power dependence, saturation similar to the exciton line, and antibunched emission at the cavity frequency whatever the excitation power.

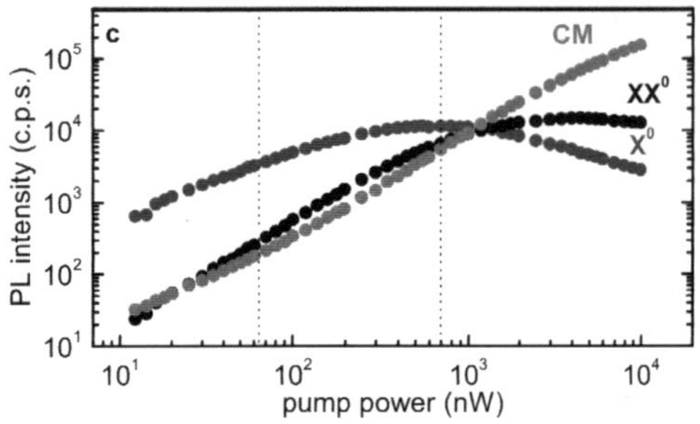


**Figure 5.** Measured PL intensity as a function of the pump power X0, XX0 and cavity peak.

## 5. Conclusion

Our simulations clearly show that multiexcitonic manifolds are an unavoidable ingredient for the explanation of the cavity feeding mechanism at large detuning. In a typical experiment, given the small QD size and the minimum power required to detect any PL signal, the QD is always in a highly excited state, where multiexcitons are excited during a considerable fraction of time. In computing multiexcitonic states and including them in our dynamic model, we are able to explain all the observations related to the cavity feeding mechanism: the surviving cavity PL at large detuning, the second order time correlation curves and the power dependence.

## Acknowledgments

We thank K. Hennessy and E. Hu for their contribution in the samples processing, J. Finley and A. Beveratos for their contribution to the elaboration of the model and S. Portolan for support regarding the Monte-Carlo simulations. We would also like to thank Wolfgang Langbein for enlightening discussions. This work is supported by NCCR Quantum Photonics (NCCR QP), research instrument of the Swiss National Science Foundation (SNSF).

## References

[1] K. Hennessy et al., *Nature*, vol. 445, p.896-899 (2007).
[2] M. Kaniber et al., *Phys. Rev. B.*, vol. 77, p.161303-+ (2008).
[3] T. Yoshie et al., *Nature*, vol. 432, p.200-203 (2004).
[4] E. Peter et al., *Phys. Rev. Lett.*, vol. 95, 067401 (2005).
[5] J.P. Reithmaier et al., *Nature*, vol. 432, p.197-200 (2004).
[6] K. Srinivasan and O. Painter, *Nature*, vol.450, p.862-866 (2007).
[7] S. Strauf et al., *Phys. Rev. Lett.*, vol. 96, 127404 (2006).
[8] T. Grange et al., *Phys. Rev. B*, vol. 96, 241304(R) (2007).
[9] E. Biolatti et al., *Phys. Rev. B.*, vol. 65, 075306 (2002).
[10] A. Barenco et M. A. Dupertuis, *Phys. Rev. B*, vol. 52, 2766 (1995).
[11] C. Jacoboni and L. Reggiani, *Rev. Mod. Phys.*, vol. 55, 645 (2003).
[12] M. Grundmann and D. Bimberg, *Phys. Rev. B*, vol. 55, 9740 (1997).
[13] A. Naesby et al., *Phys. Rev. A*, vol. 78, 045802 (2008).
[14] A. Auffèves et al., *Phys. Rev. A*, vol. 79, 053838 (2009).
[15] J. Suffczynski et al., *Arxiv*, arXiv: 0904.0271 (2009).
[16] W. Sheng and P. Hawrylak, *Phys. Rev. B.*, vol. 72, 035326 (2005).

# A new method for coherent control in a quantum dot ensemble

**M. Kujiraoka[1,2], J. Ishi-Hayase[1,3,4], K. Akahane[1], N. Yamamoto[1], K. Ema[2], and M. Sasaki[1]**

[1] National Institute of Information and Communications Technology 4-2-1, Nukui-Kitamachi, Koganei, Tokyo 184-8795, Japan

[2] Department of Physics, Sophia University, 7-1 Kioi-cho, Chiyoda-ku, Tokyo 102-8554, Japan

[3] The Center for Frontier Science and Engineering, The University of Electro-Communications, 1-5-1 Chofugaoka, Chofu, Tokyo 182-8585, Japan

[4] PRESTO, Japan Science and Technology Agency (JST), 4-1-8 Honcho, Kawaguchi, Saitama 332-0012, Japan

E-mail: m-kujira@nict.go.jp

**Abstract.** We propose a new method for controlling macroscopic coherent phenomena by taking advantage of the characteristics of quantum dot (QD) ensembles. We demonstrate that the ensemble average of Rabi oscillations of excitonic polarization can be changed drastically by selecting the spatial distributions of input electric fields. We show that the changes in ensemble averages are caused by polarization interference; the results of our previous experiments agree with our calculations. This result demonstrates that a macroscopic coherent response from a QD ensemble can be controlled by changing the spatial distribution of excitation beams.

## 1. Introduction

Optical Rabi oscillations (ROs) play crucial roles in coherent control of excitons in semiconductor quantum dots (QDs). Most reported ROs of QD excitons were observed in a single QD [1-3] because a large inhomogeneous distribution of Rabi frequencies results in smearing out ROs in a QD ensemble. Thus, QD ensembles have been considered unsuitable for coherent manipulation.

In a previous study, we demonstrated an "ensemble effect" on the ROs of excitonic polarization and population by measuring four-wave mixing signals and differential transmission. The ensemble effect is defined as the pulse area distribution for each QD with an exciton; this distribution is attributed to the inhomogeneous distribution of exciton transition dipole moments and the spatial distribution of input electric fields [4]. The ensemble effect mainly causes strong damping of ROs. However, the effects on ROs of excitonic polarization are different from those on ROs of excitonic population. We first showed that part of the ensemble effect is cancelled out only in the ROs of excitonic polarization. This is explained well by polarization interference.

Journal of Physics: Conference Series **193** (2009) 012125 doi:10.1088/1742-6596/193/1/012125

In this study, we propose a more positive way to control macroscopic coherent phenomena by taking advantage of the characteristics of QD ensembles. We demonstrate that the ensemble average of ROs of excitonic polarization can be changed drastically by selecting the spatial distributions of input electric fields. We show that the changes in ensemble averages are caused by polarization interference.

## 2. Calculation

### 2.1. Calculation Models

We assume that an excitonic system is a two-level system. According to the optical Bloch equations, in an ideal two-level system, the Bloch vector rotates from the lower to the upper state with an increase in the pulse area, as shown in Fig. 1 (a). In the case of excitonic ROs, the upper state is equivalent to the exciton ground state, and the lower state is equivalent to the crystal ground state, where no exciton exists in a QD. The excitonic polarization corresponding to the value of $\boldsymbol{v}$ is proportional to $\sin\Theta_1$, while the excitonic population corresponding to the value of $\boldsymbol{w}$ is proportional to $\sin^2(\Theta_1/2)$, where $\Theta_1$ is the pulse area of pulse 1. By changing $\Theta_1$, we expect to observe ROs of excitonic polarization and population consistent with the abovementioned equations. Two-pulse four-wave mixing (FWM) and differential transmission (DT) techniques were used to investigate the ROs of excitonic polarization and population, respectively. This is because that the amplitude of the FWM signals observed in the arrangement shown in Fig. 1 (b) is described by $\boldsymbol{P}_{\mathrm{FWM}} \propto \sin\Theta_1$ [5], while the DT observed in the arrangement shown in Fig. 1 (c) is described by $\mathrm{DT} \propto \sin^2(\Theta_1/2)$.

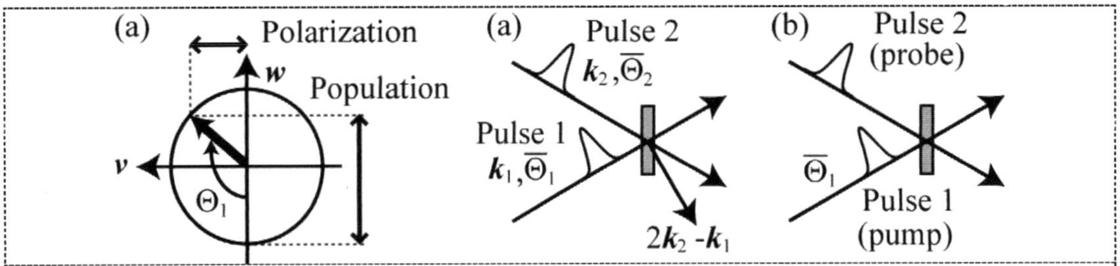


**Figure 1.** (a) Bloch sphere for a two-level system. Excitonic population is proportional to $\sin^2(\Theta_1/2)$, and excitonic polarization is proportional to $\sin\Theta_1$. (b) Experimental arrangement of FWM technique. (c) Experimental arrangement of pump-probe technique.

We calculated the ensemble average of FWM signals and DT, taking into accounts the spatial distributions of input pulse areas. The average pulse area $\overline{\Theta}_i$ of the $\boldsymbol{k}_i$ pulse is changed by varying the spatially averaged electric field $\overline{E}_i$ according to the equation $\overline{\Theta}_i = \overline{\mu}\,\overline{E}_i \Delta T/\hbar$, while maintaining a constant temporal duration $\Delta T$ for the a rectangular pulse. Here, we leave the dephasing process out of consideration. $\overline{\mu}$ and $\overline{E}_i$ represent the average transition dipole moment of the excitons and the average electric field, respectively. If the focused beam has a Gaussian spatial distribution, the QDs in the beam spot are excited not by the same electric field strength $\overline{E}_i$ but by different field strengths depending on their positions $E_i(x, y)$. This results in a spatial distribution of the pulse area for each QD at each position $\Theta_i(x, y)$. In this study, we consider that the spectral widths of the incident pulses are much smaller than the inhomogeneous broadening of the excitonic energies in QDs. The transition dipole moment μ is thus assumed to be homogeneous compared to $E_i(x, y)$.

Figure 2 shows the distribution functions of the excitation beam used in our calculations, i.e., (a) a flat-top beam, (b) a Gaussian TEM00 beam, and (c) a Gaussian TEM10 beam. The average electric field $\overline{E}_i$ is defined as $\iint E_i(x, y)dxdy/\pi R^2$, where $R$ is the beam radius. Since it's reasonable to assume that $R$ (larger than several μm) is much greater than lateral QD size and the distance between QDs (~typically several tens nm), we consider that a large number of QDs are excited resonantly in a beam spot.

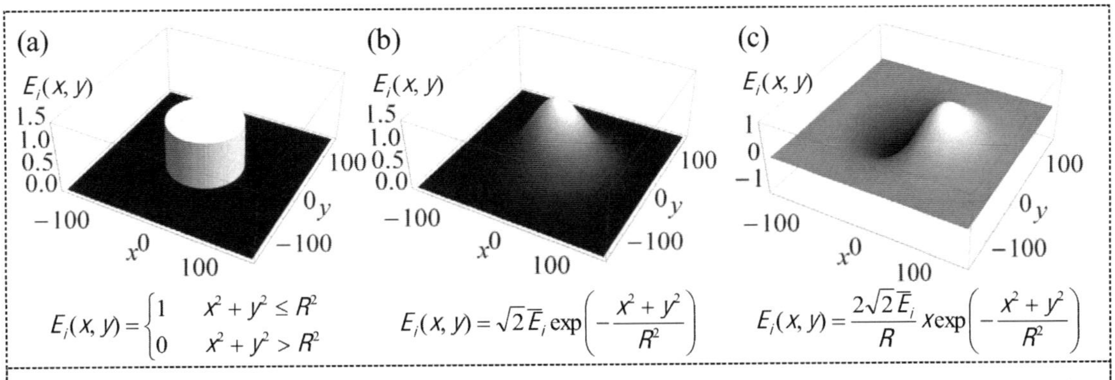


**Figure 2.** Spatial distribution of incident electric fields: (a) a flat-top beam, (b) a Gaussian TEM00 beam, and (c) a Gaussian TEM10 beam.

### 2.2. Calculation Results

Figure 3 shows the calculated FWM signals (a) and DT (b) as a function of $\overline{\Theta}_1$. Note that the horizontal axis shows the "average" pulse area $\overline{\Theta}_1$ ; the pulse areas for different QDs are distributed even at the same $\overline{\Theta}_1$ according to the distribution function of the electric field shown in Fig. 2. The spatial distribution of pulse 2 is assumed to be the same as that of pulse 1. $\overline{\Theta}_2$ was fixed at $\pi$ in this study. As shown in Fig. 3(a), the ensemble average of ROs of excitonic polarization can be changed drastically by selecting the spatial distributions of input electric fields. The FWM signals excited by flat-top beams coincide with ROs in an ideal two-level system represented as $\sin\overline{\Theta}_1$. This is because the inhomogeneity of transition dipole moment $\mu$ is assumed to be negligible compared to the distribution of $E_i(x,y)$ in this study. In practice, the inhomogeneity of transition dipole moment slightly influences the RO; in other words, the ensemble of ROs excited by a flat-top beam will provide direct information on the inhomogeneity of $\mu$. The FWM excited by Gaussian TEM00 beams coincides with $\sin\overline{\Theta}_1$ at $\overline{\Theta}_1 <\pi$ and shows the second peak at around $\overline{\Theta}_1 =1.5\pi$ although the amplitude is strongly damped. This is because of interference between excitonic polarizations with various amplitudes and directions [4]. The FWM excited by Gaussian TEM10 beams vanishes because negative and positive polarizations cancel out completely, which is much different from the calculated result for a TEM00 beam. As shown in Fig. 3(b), the behavior of DT is quite different from that of FWM signal. Though DT excited by flat-top beams coincides with ROs in an ideal two-level system represented as $\sin^2(\overline{\Theta}_1/2)$, DT excited by Gaussian TEM00 does not decrease with an increase in $\overline{\Theta}_1$ and there are few differences in DT between TEM00 and TEM10 beams. The reasons for this difference between FWM signals and DT are as follows: since the integration in the FWM signals involves the superposition of $\sin\overline{\Theta}_1$, which have both positive and negative values, the distribution of pulse area is partly cancelled out in the integration. On the other hand, $\sin^2(\overline{\Theta}_1/2)$ involved in DT have always positive values and the integration just smears out oscillation.

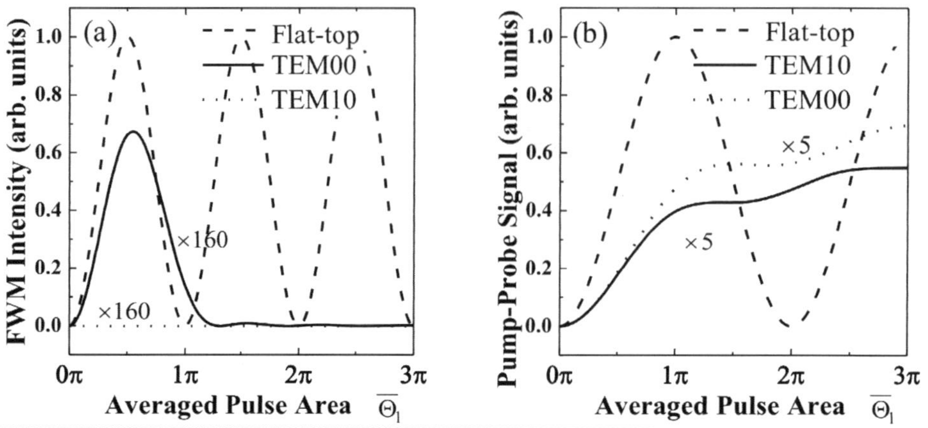


**Figure 3.** Calculations of Rabi oscillations of excitonic polarization (a) and population (b) excited by a flat-top beam (dashed line), a Gaussian TEM00 beam (solid line), and a Gaussian TEM10 beam (dot line).

## 3. Conclusion

We demonstrated that the ensemble average of ROs of excitonic polarization can be changed drastically by selecting the spatial distributions of input pulses due to the polarization interference. The effects of the spatial distributions are different between ROs of excitonic polarization and ROs of excitonic population. These results demonstrate that a macroscopic coherent response from a QD ensemble can be controlled by changing the spatial distribution of excitation beams.

## Acknowledgment

This work was partially supported by a Grant-in-Aid for JSPS Fellows and for Young Scientists (A).

## References

[1] Borri P, Langbein W, Schneider S, Woggon U, Sellin R. L, Ouyang D and Bimberg D 2002 Phys. Rev. B **66** 081306R

[2] Stievater T H, Li X, Steel D G, Gammon D, Katzer D S, Park D, Piermarocchi C and Sham L J 2001 Phys. Rev. Lett. **87** 133603

[3] Kamada H, Gotoh H, Temmyo J, Takagahara T and Ando H 2001 Phys. Rev. Lett. **87** 246401

[4] Kujiraoka M, Ishi-Hayase J, Akahane K, Yamamoto Y, Ema K and Sasaki M 2009 Phys. Status Solidi (a) **206** 952

[5] Abram I 1989 Phys. Rev. B **40** 5460

Journal of Physics: Conference Series **193** (2009) 012126 doi:10.1088/1742-6596/193/1/012126

# Electron Dynamics in Silicon Nanowire using a Monte-Carlo Method

**E. Sarrazin[1], S. Barraud[1], A. Bournel[2] and F. Triozon[1]**

[1] CEA-LETI-MINATEC, 38 054 Grenoble, France

[2] Institut d'Electronique Fondamentale, Univ. Paris-Sud, CNRS UMR 8622, 91 405 Orsay, France

E-mail: `sylvain.barraud@cea.fr`

**Abstract.** We present a theoretical study of electron transport in silicon nanowire (SNW). A self-consistent 2D-Poisson-Schrödinger solver provides the band structure. Then, both electron velocity and low-field electron mobility along the SNW axis are computed with an ensemble Monte-Carlo method. Scattering mechanisms due to phonons (acoustic phonons, zero-order and first-order intervalley phonons) and surface roughness are taken into account. We investigate the effect of cross section size and transverse electric field on electron mobility in SNW.

## 1. Introduction

Semiconductor nanowires represent an important and broad class of one-dimensional nanostructures at the forefront of nanoscience and nanotechnology. Recent advances in growth, characterization, assembly and integration of nanowires have enabled the emergence of a large number of applications in nanoelectronics, nano-optoelectronics, nanosensors, nanobiotechnology and energy harvesting [1]. Most of these applications, like transistors, sensors and thermoelectric devices are based on charge transport. Consequently, the knowledge of low-field electron mobility is essential to determine the performance of such devices. In this work, we computed electron velocity and low-field electron mobility at room temperature for a [100]-oriented silicon nanowire (SNW) with a square cross section surrounded by oxide materials ($SiO_2/HfO_2$) and metal gate.

This paper is organized as follows: sec. 2 explains the self-consistent calculation of the band structure. Sec. 3 contains a description of scattering mechanisms. Sec. 4 presents both electron velocity and low-field electron mobility in SNW obtained by Monte-Carlo simulation.

## 2. Band structure

The band structure of the SNW is computed with a 2D-Poisson-Schrödinger solver using the effective mass approximation (EMA) model. We obtain electronic states and corresponding wavefunctions as well as the potential distribution along the cross section. These results constitute the basis of our transport simulation. Comparisons with tight-binding simulations have shown that simulations with EMA are valid for large SNWs (over 5 nm in width) [2].

Because of 2D quantum confinement, the reciprocal space is restricted to 1D and the first Brillouin zone extends over $[-\pi/a_0; \pi/a_0]$, where $a_0$ is the silicon lattice parameter. The six $\Delta$ conduction valleys in the bulk split into two groups: one group of four valleys ([100], $[\bar{1}00]$, [010] and $[0\bar{1}0]$) is projected at $k = 0$ and the other group ($[001]$, $[00\bar{1}]$) is projected at $k \simeq 0.33\pi/a_0$ (for narrow nanowires, subband minima shift towards $k = 0$) [3].

## 3. Scattering mechanisms

Using first-order perturbation theory, the scattering rates of electrons with phonons and surface roughness are computed.

### *3.1. Electron-phonon*

In this study, acoustic phonon, zero-order and first-order intervalley phonon scattering rates are computed using the deformation potential approximation.

For acoustic phonons, we analyze phonon quantum confinement due to the reduction of nanowire width and we compare elastic bulk phonon scattering with inelastic confined phonon scattering. In order to compute the scattering rates, the phonon dispersion is required. For bulk phonons, the phonon dispersion is approximated with a linear relation. Considering the dimensions of the SNWs in this study, confined phonon modes are obtained by solving elastic continuum equation thanks to the $xyz$ algorithm with free-standing boundary conditions [4, 5]. The figure 1 shows the lowest subbands of confined acoustic phonon dispersion for a SNW of 6 nm in width. The figure 2 shows the acoustic phonon scattering rates for the first subband of a SNW of 6 nm in width. Elastic bulk acoustic phonon scattering rates are proportional to the electron density of states. An electron is able to scatter into a subband, if its energy reaches the subband energy minimum. For inelastic confined acoustic phonon scattering, scattering rates is composed of two groups of peaks, which are localized on both sides of elastic bulk acoustic phonon scattering peaks. The group below the subband energy levels corresponds to phonon absorption and the group above the subband energy levels corresponds to phonon emission. Inelastic confined acoustic phonons scattering rate is higher than the acoustic scattering rate calculated using bulk phonons, because confined phonon group velocity is smaller than bulk phonon group velocity and scattering rates are inversely proportional to group velocity [6].

Zero-order and first-order intervalley are composed of long wavevector acoustic or nonpolar optical phonons. Intervalley phonons are considered as bulk phonons and the phonon dispersion is assumed to be constant. Zero-order intervalley scattering is modeled using the nonpolar optical model and first-order intervalley phonon following the approach developed by Monsef *et al.* [7].

### *3.2. Surface roughness*

The imperfections of $Si/SiO_2$ interface induce fluctuations of electrostatic potential, energy levels, wavefunctions and also a redistribution of electron density within the cross section. Our model, based on Jin's model, will be restricted only to the fluctuations of potential, energy levels and wavefunctions [6, 8].

## 4. Transport

The electron transport along the SNW axis is simulated with an ensemble Monte-Carlo method. Both electron velocity and mobility are extracted at room temperature under a constant electric field applied along the transport direction.

The deformation potential values of electron-phonon scattering have been calibrated in order to recover bulk mobility for a large cross section size (typically 30 nm) in the flat band regime, because it is assumed that the electronic properties of a large SNW are quasi-identical to those of bulk silicon and consequently, mobility too. The parameters for surface roughness (rms height $\Delta = 0.35$ nm and correlation length $\Lambda = 1.3$ nm) are adjusted in the high inversion density $N_{inv}$ range in order to reproduce experimental mobility variations measured in large SNWs [10].

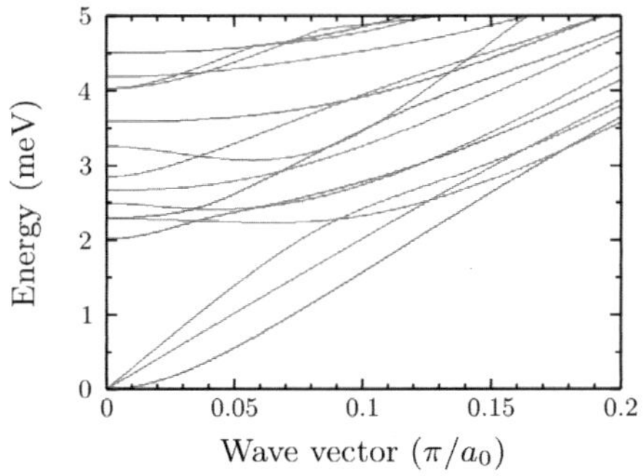


**Figure 1.** Confined acoustic phonon dispersion calculated using the $xyz$ algorithm for a SNW of 6 nm in width. Only the lowest phononic subbands are shown.

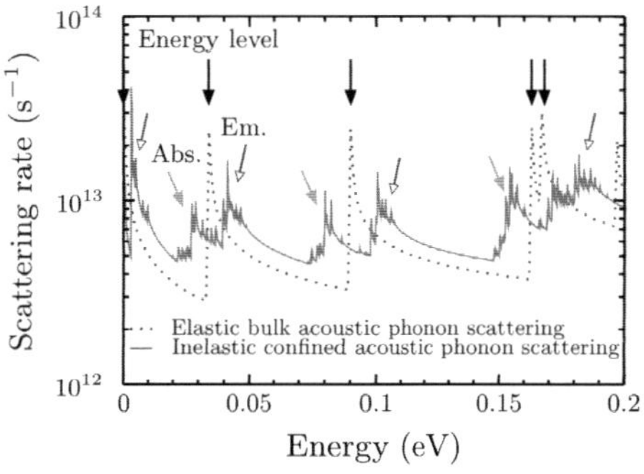


**Figure 2.** Acoustic phonon scattering rate in relation with electron energy for a SNW of 6 nm in width. Only the first subband is shown.

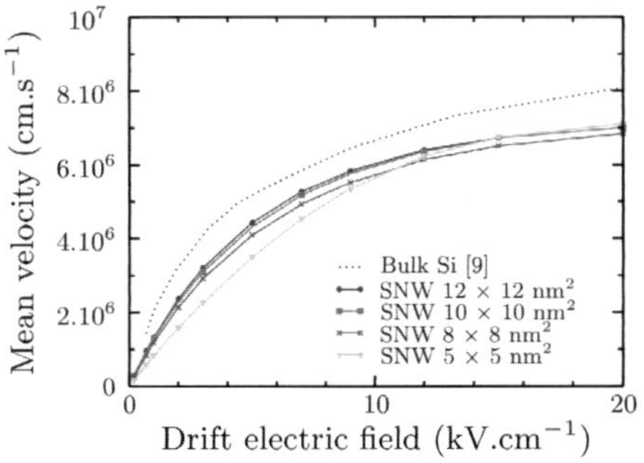


**Figure 3.** Electron mean velocity limited by bulk phonon as a function of the longitudinal electric field for different SNW widths. The bulk values are provided by ref. [9].

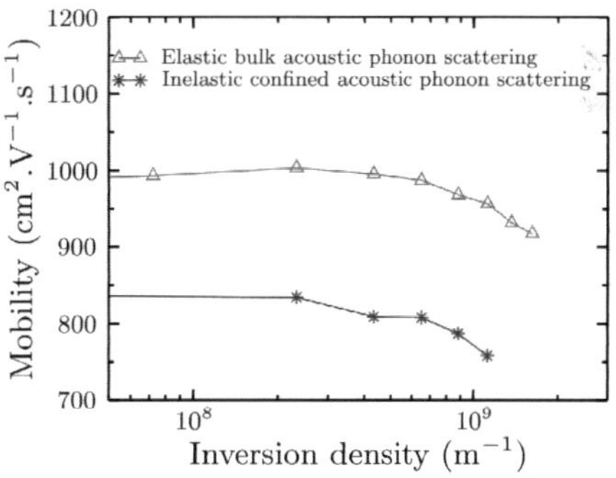


**Figure 4.** Phonon limited low-field mobility for a SNW of 6 nm in width assuming bulk and confined acoustic phonons.

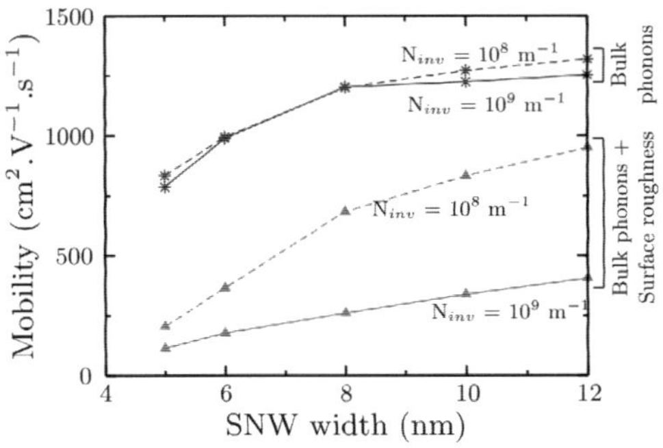


**Figure 5.** Low-field mobility as a function of SNW width at weak ($N_{inv} = 10^8\ m^{-1}$) and high ($N_{inv} = 10^9\ m^{-1}$) inversion densities.

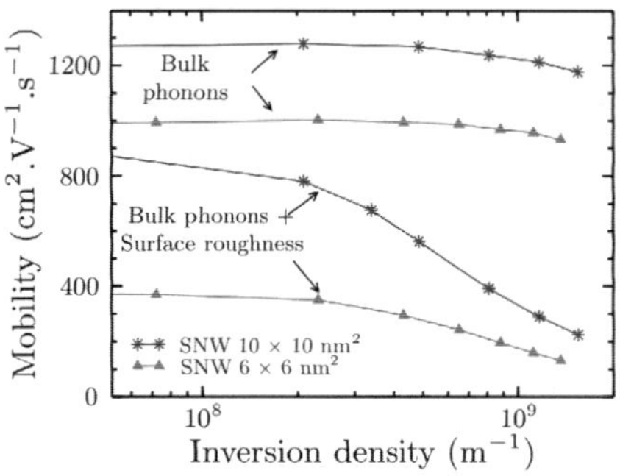


**Figure 6.** Low-field mobility as a function of inversion density for two SNW widths (6 nm and 10 nm).

First, the velocity-field characteristics in undoped SNW are calculated for various cross-section sizes (cf. figure 3). Only the electron-bulk phonon scattering rates are taken into account. The behavior in presence of longitudinal electric field is different for large and thin SNWs due to the modification of valley populating between large and thin SNWs.

The figure 4 shows the impact of phonon confinement. In this figure, low-field phonon limited mobility is calculated using both bulk and confined acoustic phonon scattering rates for a SNW of 6 nm in width. Low-field phonon limited mobility is smaller with inelastic confined acoustic phonon scattering than with elastic bulk acoustic phonon scattering. The next calculations are done using only bulk phonons.

In figure 5, low-field mobility is plotted as a function of the SNW width for various inversion densities. Regardless the scattering mechanisms considered, the mobility is degraded with the reduction of SNW width. With the decrease of SNW width, subband modulation tends to suppress intersubband and intervalley transitions. This should induce an increase of mobility. Unfortunately, the scattering rates increase with decreasing the SNW width. Indeed, the electron-phonon scattering depends both on electron density of states and overlap between wavefunctions of initial and final states. When SNW width is reduced, electron density of states decreases, whereas the overlap of wavefunctions increases. So, the reduction of low-field phonon limited mobility is mainly due to the increase of wavefunction overlap. The decrease of mobility observed when the surface roughness scattering is taken into account is mainly due to the wavefunction fluctuations, which becomes the dominant term.

The figure 6 presents the low-field mobility as a function of inversion density for a large SNW (10 nm in width) and a small SNW (6 nm in width). We can notice that the mobility is reduced as the inversion density increases. Phonon scattering rates increase with increasing $N_{inv}$ because of the wavefunction overlap which is enhanced by the confinement induced by the transverse field. The contribution of potential fluctuations, strongly dependent on $N_{inv}$, become more important for large SNWs than for small SNWs. This is due to the localization of electrons close to the $Si/SiO_2$ interface for large SNWs, when $N_{inv}$ is high. This explains why mobility for large SNWs is more impacted by a transverse field. This figure also shows that surface roughness scattering has a strong impact on low-field electron mobility and appears as the major scattering mechanism.

## 5. Conclusion

We have performed a complete study of low-field electron mobility in SNW using an ensemble Monte-Carlo method. To do that, the band structure with the effective mass approximation and scattering rates have been calculated. From these simulations, we can conclude that: (i) low-field mobility decreases both with the reduction of SNW width and with the increase of inversion density, and (ii) surface roughness appears to be the major scattering mechanism.

## References

[1] Lieber C M and Wang Z L 2007 *MRS Bulletin* **32** 99–104
[2] Sarrazin E, Barraud S, Triozon F and Bournel A 2008 *International Conference on Simulation of Semiconductor Processes and Devices* pp 349–352
[3] Niquet Y M, Lherbier A, Quang N H, Fernández-Serra M V, Blase X and Delerue C 2006 *Phys. Rev. B* **73** 165319
[4] Visscher W M, Migliori A, Bell T M and Reinert R A 1991 *J. Acous. Soc. Am.* **90** 2154–2162
[5] Nishiguchi N, Ando Y and Wybourne M N 1997 *J. Phys.: Condens. Matter* **9** 5751–5764
[6] Ramayya E B, Vasileska D, Goodnick S M and Knezevic I 2008 *J. Appl. Phys.* **104** 063711
[7] Monsef F, Dollfus P, Galdin S and Bournel A 2002 *Phys. Rev. B* **65** 212304
[8] Jin S, Fischetti M and Tang T W 2007 *IEEE Transactions on Electron Devices* **54** 2191–2203
[9] Canali C, Jacoboni C, Nava F, Ottaviani G and Alberigi-Quaranta A 1975 *Phys. Rev. B* **12** 2265–2284
[10] Chen J, Saraya T, Miyaji K, Shimizu K and Hiramoto T 2008 *Symposium on VLSI Technology* pp 32–33

Journal of Physics: Conference Series **193** (2009) 012127 doi:10.1088/1742-6596/193/1/012127

# Electron-electron interaction and tunnel density of states at the Fermi level in high-density two-dimensional electron system

**E.N. Morozova, I.N. Kotel'nikov, S.E. Dizhur**
Institute of Radio Engineering and Electronics RAS,125009 Moscow, Mokhovaya 11/7, Russia

**E.V. Deviatov, V.T. Dolgopolov**
Institute of Solid State Physics RAS, 142432 Chernogolovka Moscow region, Russia

E-mail: elena.morozova@gmail.com

**Abstract.**
The zero bias anomaly (ZBA) of the tunnel conductivity are investigated in the $Al/\delta-GaAs$ tunnel structure with 2D electron concentration in $\delta$-layer to be equal to $3.6{*}10^{12}$ cm$^{-2}$. We find that the dip (ZBA) in the tunnel density of states near the Fermi level $E_F$ reveals logarithmic dependence on the energy $\varepsilon$ in the range $kT < \varepsilon < \hbar/\tau$. Here $\varepsilon$ is the energy relative to $E_F$ and $\tau$ is the elastic relaxation time of 2D electrons. The depth of the ZBA is proportional to $ln(T/T_0)$ in the range $T = 0.1 - 20$ K. These results are in agreement with Aronov-Altshuler theory in which the correction to the 2D density of states is due to the effects of the electron-electron interaction in diffusion channel. The unusual behavior of the tunnel magnetoconductivity is observed.

## 1. Introduction

Localization effects in the 2D electron system (2DES) in semiconductor structures are intensively investigated, see for review [1,2]. The main interest in such structures is study of the tunneling density of states (TDOS) near the Fermi level of 2DES. Theory [3] predicts manifestations of inter-electronic interaction in the regime of weak localization of carries when electrons concentration sufficiently high ($k_F l \gg 1$). In this case for the diffusion 2D channel in the TDOS $\rho(\varepsilon)$ at the Fermi level $E_F$ of 2DES should be observed a considerable increase of the dip $\Delta\rho/\rho \sim \ln(kT\tau/\hbar)$, when energy $|\varepsilon| = |E - E_F| < kT$ and $L_{ee} = V_F(\hbar\tau/kT)^{1/2} >> l$. Here $\tau$ is the elastic relaxation time and $l = V_F\tau$, $L_{ee}$ is the phase relaxation length due to electron-electron collisions. Measurements of the TDOS, which confirmed results of [3], were made on the metal-films (see, for example [4,5]), where conditions of the "dirty limit" for metal conductivity is easy to realize. In the semiconductor 2D structures the corrections to the 2DES conductivity had been studied until now, see for example [6,7]. The "dirty limit" in the semiconductor structures with 2DES is well-satisfied in $GaAs$ structures in which 2DES with Fermi energy $\simeq 100$ meV is situated in the self-consistent potential of the $\delta$-layer [8-10]. In this work we present studies of the first investigations of TDOS for the semiconducting structure $Al/\delta$-$GaAs$ ($k_F l \simeq 10$) in the temperature range $0.03 - 30$ K and magnetic field up to 14 Tesla.

## 2. Samples

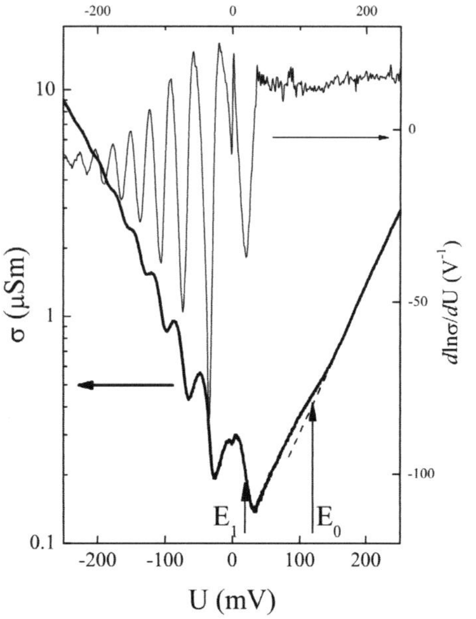


**Figure 1.** Tunnel conductivity ($\sigma = dI/dU$) and tunnel spectrum ($dln\sigma/dU$) of the sample "d" with short gate measured at 4.2 K. The arrows ($E_0$ and $E_1$) indicate the steps in the tunnel conductivity for occupied 2D subbands or the dip in tunnel spectrum.

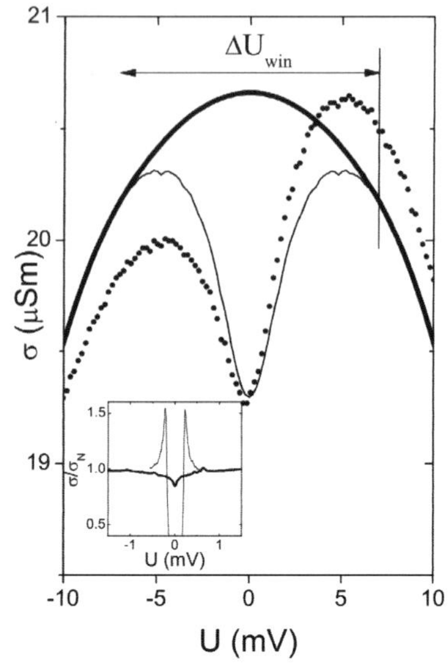


**Figure 2.** Tunnel conductivity of the sample "e" with long gate measured at 4.2 K. Points - conductivity data $\sigma$ (U), thin line - symmetrized conductivity $\sigma_{sym}$(U)=$(\sigma(U) + \sigma(-U))/2$, thick line is the background for $\sigma_{sym}(U)$. The inset shows curves $\sigma$ (U)/$\sigma\,(-2$mV) for two cases: at B=0 - superconducting gap in Al gate and at B=0.16 T - tunnel zero bias anomaly, T=0.03K.

In our experiments, we were using the MBE grown $Al\ /\delta\text{-}GaAs$ tunnel structure. $Al$ gate was grown *in situ* to achieve high quality of the $Al/GaAs$ interface. $\delta$-layer was grown at the distance of $\simeq$ 20 nm from the$Al/GaAs$ interface and the concentration of $Si$ $\delta$-doping was $\simeq 7{*}10^{12}$ cm$^{-2}$. Epitaxial GaAs was slightly unintentionally p-doped at the level of $\sim 10^{16}$ cm$^{-3}$. The samples were prepared in the form of the Hall bars with the width of the 2D channel of 0.5 mm and the full length of 1.5 mm. We studied two types of samples. One of them had two tunnel gates of the length of 10 $\mu$m close to the ends of the Hall bar (sample "d"), another one (sample "e") had the gate length 1.3 mm. The $\rho$-contacts were between the gates in the first case and under the gate in the second one. The tunnel characteristics for the sample "d" are shown in Fig.1. It is known, that tunnel conductivity is $\sigma \propto D * \rho_{2D}$, where D is the tunnel barrier transparency and $\rho_{2D}$ is 2D density of states in the 2DES electrode. So the steps of $\rho_{2D}$ ($\varepsilon$=-eU) can be observed in $\sigma$(U) dependence, as in Fig.1 [11]. These steps become the dips in tunnel spectrum. So the positions of the dips at positive bias gives us the values of energies (with respect to Fermi level or U=0) for the occupied subband bottoms of $E_0 \simeq$ 120meV and $E_1 \simeq$ 20meV in our samples. The Hall measurements and the tunneling data permit to determine the electron concentrations and mobilities in the 2D channel under the gate. These values are $n_0 = 3.1 * 10^{12}$cm$^{-2}$ and

$n_1 = 5.1 * 10^{11}\mathrm{cm}^{-2}$ and $\mu_0 = 400\,\mathrm{cm}^2/\mathrm{Vs}$ and $\mu_1 = 1000\,\mathrm{cm}^2/\mathrm{Vs}$ for the ground and the first-excited subbands, respectively.

We are interested in the ZBA that is seen in Fig.1 as weak feature in $\sigma$(U) at U=0. The region of the ZBA, which associated with the TDOS near the Fermi level, is shown on Fig. 2. The inset in Fig.2 shows the superconducting gap in Al gate of $Al/\delta$-$GaAs$. The gap is observed at $T < 1.1$ K and magnetic field B=0 that confirms the high quality of our $Al/\delta$-$GaAs$ structures and the possibility to measure effects in density of states at Fermi level in electrodes. Since TDOS near the Fermi level is symmetrical function of energy $\varepsilon = E - E_F$, we will used symmetrized tunnel conductivity $\sigma_{sym}$ (Fig.2) to investigate the ZBA behavior.

## 3. Results and discussions

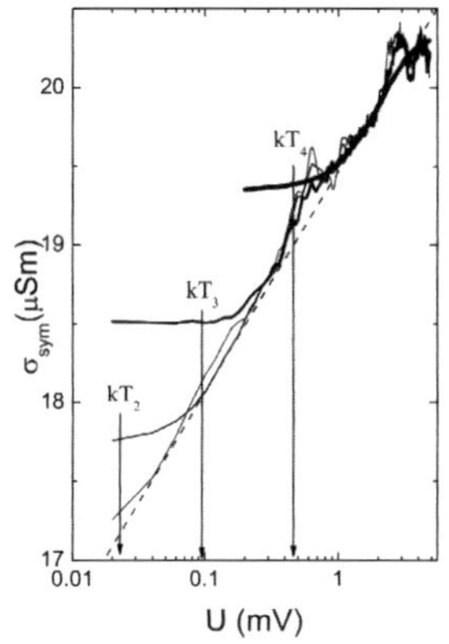

**Figure 3.** Symmetrized tunnel conductivity dependences near zero bias at temperatures $0.03 - 0.3 - 1.2$–$4.2\,\mathrm{K}$ (width of lines increases with T). Dashed line shows $lnU$ dependence.

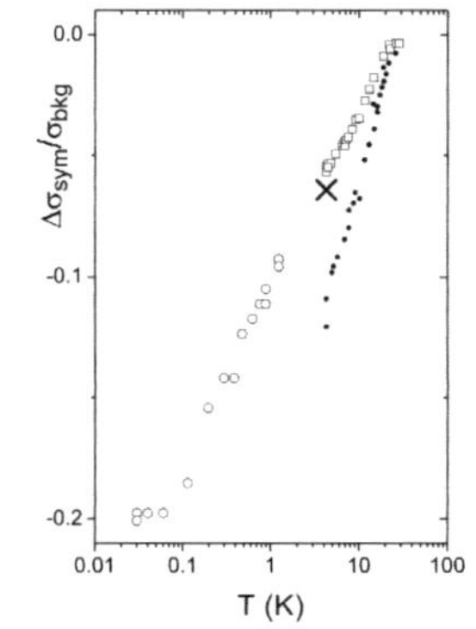

**Figure 4.** Dependence of the$\Delta$ $\sigma_{sym}/\sigma_{bkg} \propto \Delta\rho/\rho$ on temperature for the sample "e" (open circles and cross). Points - another sample with $n_{2D} \simeq 10^{12}\mathrm{cm}^{-2}$.

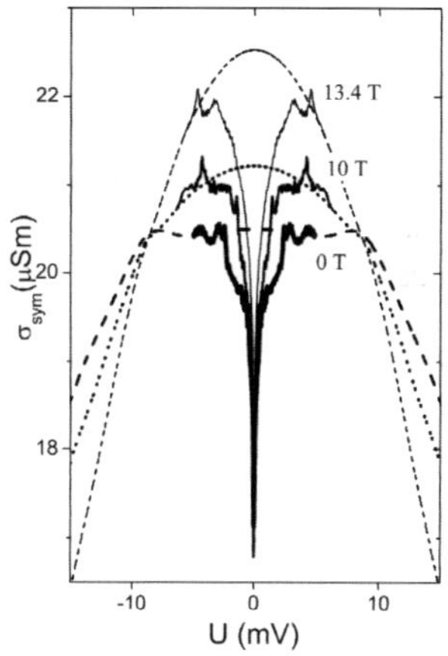

**Figure 5.** Tunnel conductivity $\sigma_{sym}$ in magnetic field B perpendicular to $\delta$-layer plane for sample "e" at 0.03 K. Dashed lines - $\sigma_{bkg}$ for different B.

Firstly, we analyse the $\sigma_{sym}$(U) dependence near the zero bias at different temperatures (see Fig.3). $\sigma_{sym}$(U) can be presented as $\sigma_{sym}$(U)=($\Delta\sigma_{sym} + \sigma_{bkg}$), where $\sigma_{bkg}$ is shown in Fig.2 at T=4.2 K.$\Delta$ $\sigma_{sym}$(U) proportional to the change$\Delta$ $\rho$ of TDOS near Fermi level and $\sigma_{bkg}(U) \simeq const$ for the bias range $0 - 4\,\mathrm{mV}$ for the $T \leq 4$ K. Consequently the dependencies in Fig.3 demonstrate$\Delta$ $\rho$ as a function of U or energy $\varepsilon$. Aronov and Altshuler [3] have shown that$\Delta$ $\rho/\rho \sim \ln(\varepsilon\tau/\hbar)$ at kT<$|\varepsilon| < \hbar/\tau$ and phase relaxation due to electron-electron collision. This dependence is in a good agreement with the data in Fig.3. The low limits of energy region of logarithmic dependence$\Delta$ $\rho$(U) are shown with arrows in Fig.3 for each temperature. The upper limit is $\hbar/\tau$ ~10 meV as follows from mobility value for our samples. The dependence of$\Delta$ $\sigma_{sym}(U = 0)/\sigma_{bkg} \propto \Delta\rho(\varepsilon = 0)/\rho$ as a function of T is presented in Fig.4. One can see that the data correspond to expected dependence$\Delta$ $\rho(\varepsilon = 0)/\rho \propto \ln(kT\tau/\hbar)$ [3]. At the same time the width of the dip near the Fermi level is linear function of T. This dependence is found at the range of temperatures $0.1 - 20\,\mathrm{K}$. Note that the dip of TDOS$\Delta$ $\rho(\varepsilon = 0)$ is equal to 20% of $\rho$ at temperature 0.03 K in our experiment (Fig.4). Such result points out to possibility of localization of 2D carriers in our $\delta$-layer structure at lower temperatures. Data for another sample with $n_{2D} \simeq 10^{12}\mathrm{cm}^{-2}$ (see point curve in Fig.4) show that the strong localization can be achived even at 0.1 K. The investigation of such samples are in progress. Thus our results demonstrate that the phase relaxation time $\tau_\varphi \sim 1/T$ , i.e. $\tau_\varphi$ is associated with electron-electron scattering in diffusion channel of our high density 2DES.

Unexpectedly the tunnel conductivity at low bias strongly depends on magnetic field B perpendicular to $\delta$-layer, as can be seen in Fig.5. The positive magnetoconductivity (for $\sigma_{bkg}$) was observed at low bias voltages and negative one was found at $\mid U \mid > 10\,\mathrm{mV}$. At the same time position of the dip in TDOS does not depend on magnetic field. Additional experiments are needed for the explanation of this effect.

## 4. Conclusion

We performed the first direct spectroscopic measurements of TDOS near the Fermi level of the high density 2DES in GaAs structure. Logarithmic dependencies of TDOS were obtained in wide ranges of energies, $\varepsilon = 0.02 - 5$ meV, and temperatures T, T=$0.1 - 20$. The data are in a good agreement with the Aronov-Altshuler theory of the electron-electron interaction in disordered 2D conductors at $k_F l \gg 1$. We found also the non-monotonic tunneling magnetoconductivity for the magnetic fields perpendicular to the 2DES plane.

## 5. Acknowledgments

The work was supported by RFBR and RAS. We gratefully acknowledge A.S Bugaev and Yu. V. Fedorov for the sample preparation, and M.N. Feiginov for self-consistent calculations of the tunnel spectrum.

## References

[1] Gantmaher V.F., Dolgopolov V.T., Physics-Uspehi, **51**, 3 (2008).

[2] Shashkin A.A., Physics-Uspehi, **48**, 129 (2005).

[3] Altshuler B.I., Aronov A.G. "Electron-electron interaction in disordered conductors", in *Electron-electron interaction in disordered systems*, (Eds.:Efros A.L., Pollak M.), North-Holland (1985).

[4] White A.E., Dynes R.C., Garno J.P., Phys. Rev. B, **31**, 1174 (1985).

[5] Butko V.Yu., DiTusa J.F., Adams P.W., Phys. Rev. Lett., **84**, 1543 (2000).

[6] Uren M.J., Davies R.A., Pepper M., J.Phys. C.: Solid St. Phys., **13**, L985 (1980).

[7] Van Keuls F.W., Marthur H., Jiang H.W., Dahm A.J., Phys. Rev. B, **56**, 13263 (1997).

[8] Khavin Yu., Gershenzon M., Bogdanov A., Phys.Rev B, **58**, 8009 (1998).

[9] Dizhur E.M., Voroovskii A.N., Fedorov A.V. at al., ZETPh Letters, **80**, 489 (2004).

[10] Kotel'nikov I.N. et al, Abstracts of Advance Research Workshop "Fundamentals of electronic nanosystems", S.-Petersburg, Russia, June 28 - July 4, 2008, p.46.

[11] Feiginov M.N. and Kotel'nikov I.N., Appl. Phys. Lett. **91**, 083510 (2007).

Journal of Physics: Conference Series **193** (2009) 012128 doi:10.1088/1742-6596/193/1/012128

# The characterization of the hole transport in Sb based strained quantum wells

**Zbynek Soban[1,2] Jan Voves[1], Karol Kalna[3]**

[1]Department of Microelectronics, Faculty of Electrical Engineering, The Czech Technical University in Prague, Technick 2, CZ-166 27 Praha 6 - Czech Republic
[2]Institute of Physics, Academy of Sciences of the Czech Republic, Cukrovarnicka 10, 16253 Praha 6, Czech Republic
[3]Device Modelling Group, Department of Electronics & Electrical Engineering, University of Glasgow, Oakfield Avenue, Glasgow G12 8LT - U.K.

E-mail: `sobanz1@fel.cvut.cz`

**Abstract.** Compressively strained InSb structures aimed for *p*-channel heterostructure field effect transistors (HFETs) are analysed in order to maximise their hole mobility. We optimise the heterostructure by the change of the material composition, thickness of the layers and position of $\delta$-doping. The splitting of light and heavy hole bands in compressively strained channels are calculated by *nextnano*$^3$ software as a function of material mole fraction and quantum well thickness. The $8 \times 8$ **k.p** method is used for the determination of corresponding hole eigenvalues. The strained layer bandstructure calculation is carried out with *Quantumwise ATK* for the Kohn-Sham DFT using double-zeta polarised basis set orbitals as well.

## 1. Introduction and Motivation

III-V Heterostructure Field Effect Transistors (HFETs) are very promising candidates for future high-performance, low-power digital applications [1, 2] possibly replacing a strained Si transistors. The development of appropriate *p*-type MOSFET transistor compatible with III-V technology would be advantageous for a possible co-integration of high mobility III-V channels into the CMOS Si architecture even alternative Ge solutions exist [3].

While the electron mobility in III-Vs is larger than in Si, the increase in the hole mobility by strain is crucial for the development of the *p*-type channel HFETs. A small band gap and high mobility of InSb seems to be right candidate for this purpose. Successful realisation of HFETs with InSb resp. GaSb ultrathin strained layers has been reported recently [4], [5].

## 2. Structure

In Fig. 1 we show a structure of QWFET under study. We assume a bottom barrier of $Al_xIn_{1-x}Sb$, then a narrow bandgap *p*-channel of InSb and a top barrier of $Al_xIn_{1-x}Sb$ with the same mole fraction of $x_{Al}$ as at the bottom layer. In the top barrier, we consider a $\delta$-doped layer with a doping concentration of $2\text{x}10^{12}$ cm$^{-2}$ and assume a Fermi-level pinning of 0.55 eV on the top barrier which will induce a band bending.

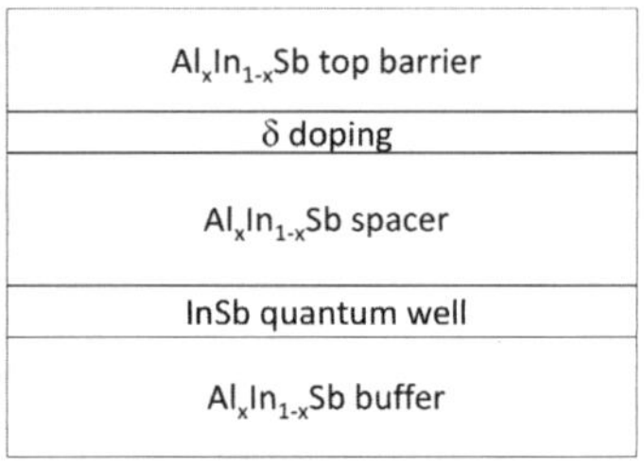


**Figure 1.** Layer structure of the InSb QWFET: Thickness of top barrier is 13 nm. $\delta$-doping is placed with a nominal $2\text{x}10^{12}$ ($\text{cm}^{-2}$). A spacer between $\delta$-doping and quantum well is 5 nm thick. Quantum well thickness was changed in the interval from 5 to 15 nm. A thickness of the bottom barrier is 25 nm.

## 3. Simulations

We employ a single band and $8 \times 8$ **k.p** method to calculate a band profile and carrier density using *nextnano*$^3$ [6] simulation tool within a self-consistent solution. A hole density along the InSb QWFET is shown in Fig. 2.

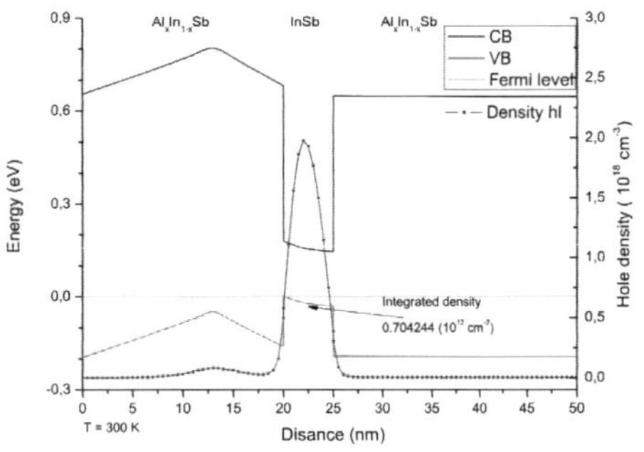


**Figure 2.** The band gap of calculated structure.

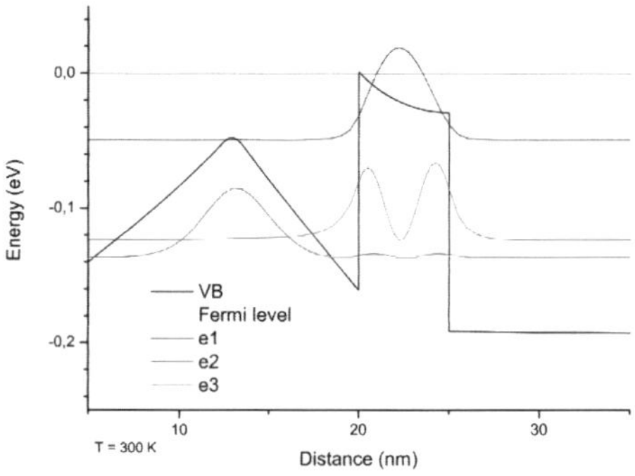


**Figure 3.** Detail of wavefunctions calculated by **k.p** method.

Wavefunctions in the quantum well of the QWFET transistor are shown in Fig. 3 in a simplified way. We have changed the hole sheet density in the quantum well by varying the $\delta$-doping concentration assuming the carbon dopant with an activation energy of 27 meV.

### *3.1. Bulk dispersion*

We have considered various aluminium contents for $Al_xIn_{1-x}Sb$ substrate is order to find the optimal strain conditions. Fig. 4 shows the calculated energy dispersion in two **k**-space directions: [001] and [011] respectively. Three bands are depicted: heavy holes (HH), light holes (LH) and spin-orbit split-off holes (SO). In this figure, we compare an unstrained 5 nm thick layer of InSb (bold curves) with a strained layer of InSb grow on $Al_xIn_{1-x}Sb$ substrate assuming $x = 0.3$.

The valence band splits off in the $Al_{0.3}In_{0.7}Sb$ barriers due to the strain. The difference of HH and LH at **k**=0 is 0.121 eV. The difference will reduce to only 0.063eV when x=0.15. As we increase the strain in InSb layer, the effective mass of holes decreases which is very beneficial for the improvement in the hole mobility.

### *3.2. Effective mass calculation*

The InSb strained layer bandstructure calculation is carried out using *Quantumwise ATK* simulation software. We have employed the Kohn-Sham Density Function Theory (DFT)

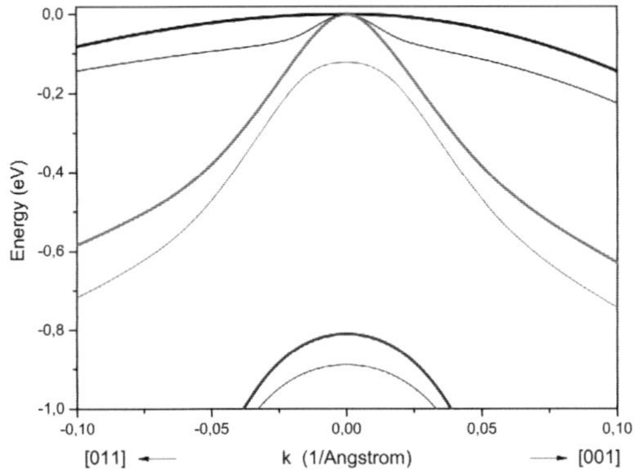


**Figure 4.** Dispersion $E(k_{||})$ of 5 nm thick layer of InSb, bold line - unstrained case, thin line - strained on $Al_xIn_{1-x}Sb$, x=0.15 on-line colour: black - HH, red - LH, blue - SO.

approach using double-zeta polarised basis set orbitals. We have used the both DFT models: Local Density Approximation (LDA) and Generalized Gradient Approximation (GGA). LDA underestimates the bandgap of the semiconductors and narrow bandgap materials as InSb are found to be metallic within LDA. On the other hand, GGA gives higher bandgap values for InSb then those tabulated. We have used a bulk model with the compressive strain in the $x$-$y$ plane. Two lattice vectors in this plane were reduced thus a face-centered cubic lattice was transformed to face-centered orthorombic. A Fig. 5 illustrates the change of the carrier effective masses in InSb under applied strain for LDA (a) and GGA (b) approximations.

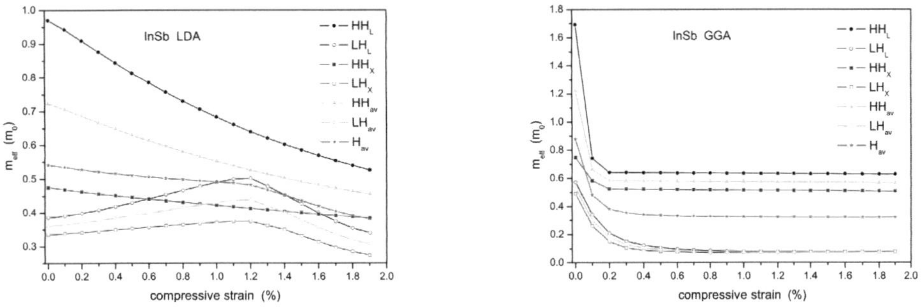


(a) (b)

**Figure 5.** Effective mass calculation based on LDA (a) and GGA (b) approximations.

$HH_X$ and $HH_L$ are effective masses of heavy holes calculated from the $E(\mathbf{k})$ dispersion along the [100] and [111] directions, respectively. Using an analytical approach, the effective mass can be calculated as: $m^* = \hbar^2 \cdot \left[\frac{d^2\varepsilon}{dk^2}\right]^{-1}$. The analogous values $LH_X$ and $LH_L$ are calculated for the light hole band. The average values for LH and HH effective masses are indicated as $HH_{av}$ and $LH_{av}$. $H_{av}$ is an average value of $HH_{av}$ and $LH_{av}$.

### *3.3. Hole mobility calculation*

Finally, we have calculated a carrier mobility $\mu$ in a quantum well of $p$-type FET transistor. We can assume that a two dimensional hole gas (2DHG) occurs in the quantum well since the layer thickness of InSb is 5 nm. We employ an simple approach of the calculation of hole mobility that includes different scattering mechanisms [7]. We have taken into account the acoustic phonon scattering, the polar optical phonon scattering and the ionised impurity scattering. The carrier mobility is calculated from the expression $\mu = \frac{e\tau}{m^*}$ where $\tau$ is the scattering time associated with the included scattering process. A total carrier mobility is calculated according to Mathieson's rule. A detailed description of single scattering mechanisms can be found in [7].

### *3.4. Simulation parameters*

Parameters that we used in simulation are summarised in Tab. 1. Even the most of these material parameters are well known and described in literature, some of them are less accurate, e.g., an acoustic deformation potential. We have used a value of 7.2 eV instead of a value 30 eV that have been quoted in past. The value of 7.2 eV is in a very good agreement with the experimental data. The problem with the acoustic deformation potential is discussed in [7]. We

**Table 1.** Material parameters

| Name | Value |
|---|---|
| bandgap at $\Gamma$ (eV) | 0.1737 |
| lattice constant (nm) | 0.64794 |
| LO phonon energy (eV) | 22.0 |
| static dielectric const ( - ) | 17.5 |
| acoustic deformation potential (eV) | 7.2 |
| optical dielectric const ( - ) | 15.68 |
| sound velocity (m.s$^{-1}$) | 3.7 |
| hole effective mass HH ($m_0$) | 0.405 |
| hole effective mass LH ($m_0$) | 0.016 |

carried out an extensive study of an effective mass of the hole in dependence on the strain. The results of this study are depicted on Fig. 5. In Tab. 1 we quote a unstrained bulk case for InSb.

## 4. Conclusion

We have simulated effective masses and mobilities for the case of thin, 5 nm InSb quantum well embedded into $Al_xIn_{1-x}Sb$ layer which is incorporated into a heterostructure *p*-FET. When we have increased a doping concentration from 1 to 3x10$^{12}$cm$^{-2}$, we have achieved a different integrated hole sheet density in the quantum well. The sheet density increased from 0.2x10$^{12}$ to 1.4x10$^{12}$ cm$^{-2}$, respectively. We have also carried intensive study by changing material and structure parameters to gain a higher mobility in the quantum well. For example, we have increased the spacer between the $\delta$-doping position and the quantum well and we broadened the thickness of the quantum well. The carrier mobility has increased more than by factor of two reaching 1085 cm$^2$/Vs which is close to measured results from literature [4].
**Acknowledgements.** This work was supported by The Ministry of Education, Czech Republic, (KAN400100652 GA CAS and No. MSM 6840770014), and by The Royal Society of Edinburgh (52026).

## References

[1] Chau R et al. 2005 *IEEE Trans. Nanotechnol.* **4** 153
[2] Kalna K et al. 2008 *IEEE Trans. Electron Devices* **5**, 2297-2306.
[3] Skotnicki T 2007, *Microelectron. Eng.* **84**, 1845-1852.
[4] Radosavljevic M et al. 2008 *IEDM Tech. Dig.* 727.
[5] Bennett B R, Ancona M G, Boos J B, Canedy C B and Khan S A 2008 *J. Cryst. Growth* **311** 47
[6] Birner S et al. 2007 *IEEE Transactions on Electron Devices* **54** 2137
[7] Shao Y, Solin S A and Ram-Mohan L R 2006 Carrier Mobilities in $\delta$-doped Heterostructures arXiv:cond-mat/0602140

EDISON 16 IOP Publishing
Journal of Physics: Conference Series **193** (2009) 012129 doi:10.1088/1742-6596/193/1/012129

# Theory of relaxation and decoherence of intersublevel transitions in semiconductor quantum dots

**T. Grange[1,2], R. Ferreira[2] and G. Bastard[2]**

[1] Walter Schottky Institut, Technische Universität München, Am Coulombwall 3, 85748 Garching, Germany

[2] Laboratoire Pierre Aigrain, Ecole Normale Supérieure, CNRS, 24 rue Lhomond, 75005 Paris, France

E-mail: thomas.grange@wsi.tum.de

**Abstract.** We present calculations of intersublevel population and phase relaxation in semiconductor quantum dots. Anharmonic polaron decay is shown to be responsible for population relaxation over a large spectral range. Additional pure dephasing processes are caused by acoustic phonons, which cause the appearance of peaked sidebands and induce broadening through real and virtual transitions between the excited states.

## 1. Introduction

Energy relaxation in quasi zero-dimensional systems has been the subject of debates for nearly two decades. It was first pointed out that a phonon bottleneck should occur in these nanostructures [1, 2, 3]: the emission of acoustic phonon should be strongly inhibited as soon as the energy separation between the levels is higher than a few meV, while LO-phonon emission should not be possible unless the intersublevel energy separation match the LO-phonon energy.

Later, it was found that a strong coupling regime occurs between confined electronic states and LO-phonon [4, 5, 6]. This is due to the fact that the Fröhlich matrix elements (of the order of a few meV) are much larger than both the electronic linewidth (discussed below) and the dispersion of long-wavelength LO-phonon involved (typically 0.1meV). A striking consequence is the observation of anticrossing when two confined states (electron, hole, or exciton) have energies differing by a multiple of LO-phonon energy [5, 6, 7, 8, 9, 10]. The eigenstates are quasiparticles called polarons. This polaronic nature of the intersublevel transitions have important consequences for energy relaxation.

Recently we have proposed a model for energy relaxation in semiconductor quantum dots (QDs) that rely on a microscopic description of polaron anharmonicity [11]. This approach results in dramatic differences compared to previous established models [12, 13]. We show that this mechanism is responsible for energy relaxation in QDs in a large spectral range, in good agreement with pump-probe measurements [14, 15, 16, 17].

We also study the coherence of such QDs intersublevel transitions. At low temperature, dephasing is limited by the anharmonic polaron decay to the ground state ($T_2 = 2T_1$), while at higher temperature additional mechanisms of dephasing appear. We show below that acoustic phonons induce (i) peaked acoustic phonon sidebands and (ii) homogeneous broadening of the intersublevel transition due to both real and virtual transitions towards the second excited $p$-state.

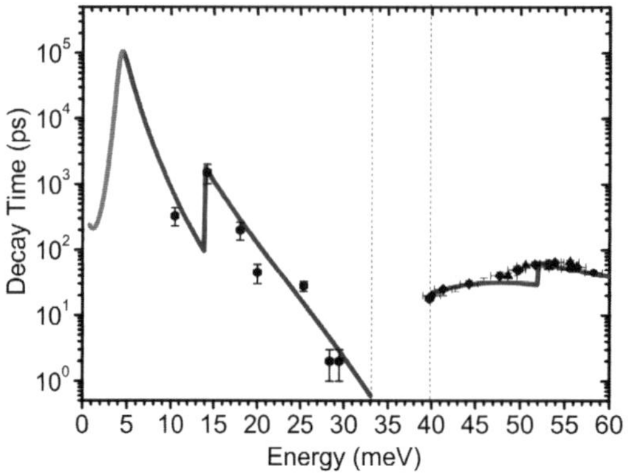


**Figure 1.** Intersublevel relaxation time as a function of the transition energy: Calculation in solid line and measurements from the references [15] and [17] in symbols. The transition energy on the abscissa corresponds to the energy difference between the ground and excited polaron state with dominant electronic component. The blue line corresponds to the domain where the relaxation is dominated by anharmonic relaxation decay while the red one corresponds to acoustic phonon emission. The dashed lines indicate the polaron splitting at the LO-phonon resonance.

## 2. Energy relaxation by anharmonic polaron decay

It was first pointed out by Li and co-workers [12] that Fröhlich couplings combined to LO-phonon instability allow electronic relaxation. In their model, a constant phonon damping was added phenomenologically in the coupled equation of the electron-phonon system. This semi-classical model was used in order to explain the lifetime increases for energies comprise between 40 and 50 meV [14, 15] but failed in explaining behaviour at higher energies [15]. More recently, in [11], we have presented a microscopic model of anharmonicity induced relaxation. We have shown that this approach is mandatory: the efficiency of anharmonic processes varies strongly with the energy to dissipate. These relative variations are of the same order, or even larger that the relative variations of the LO-phonon component with energy. As a consequence, the semi-classical model fails in predicting, even qualitatively, the variations of the intersublevel lifetimes with energy.

The microscopic theory for anharmonic polaron relaxation was presented in [11], and has been recently extended to take into account electronic states far detuned from the LO-phonon resonance, in particular in the THz range [17]. The polaron relaxation rate between two QDs levels reads:

$$\Gamma_i = \frac{4\Delta^2}{(\hbar\omega_{\rm LO} + \Delta)^2} \frac{V_F^2}{(\Delta - \hbar\omega_{\rm LO})^2 + V_F^2} \Gamma^{ph}(\Delta) \tag{1}$$

where $V_F$ is the Fröhlich matrix element involved, $\omega_{\rm LO}$ is the bulk LO-phonon frequency, $\Delta$ is the transition energy between polaron states, and the function $\Gamma^{ph}$ was defined in [11]. $\Gamma^{ph}(\Delta)$ can be seen as the decay rate of a LO-phonon vibration that would be driven at the polaron frequency $\Delta/\hbar$. The first term in the above formula is needed only for transitions well detuned from the resonance $\hbar\omega_{\rm LO}$, and accounts for interferences between resonant and non-resonant Fröhlich interactions. The second term corresponds to the weight of the LO-phonon component in the excited polaron (if the polaron wavefunction is of the form $|\phi_i\rangle = \alpha_i|p,0\rangle + \beta_i|s,1\rangle$, where 0 and 1 refer to LO-phonon occupation numbers, this quantity reads $|\beta_i|^2$).

The above formula can be split into two contributions: the first two terms describe the variations of the polaronic nature of the excitation while the term $\Gamma^{ph}$ account for the variations in the efficiency of anharmonic mechanism. As shown in [11], $\Gamma^{ph}(E)$ is made of a sum of different anharmonic channels,

Journal of Physics: Conference Series **193** (2009) 012129 doi:10.1088/1742-6596/193/1/012129

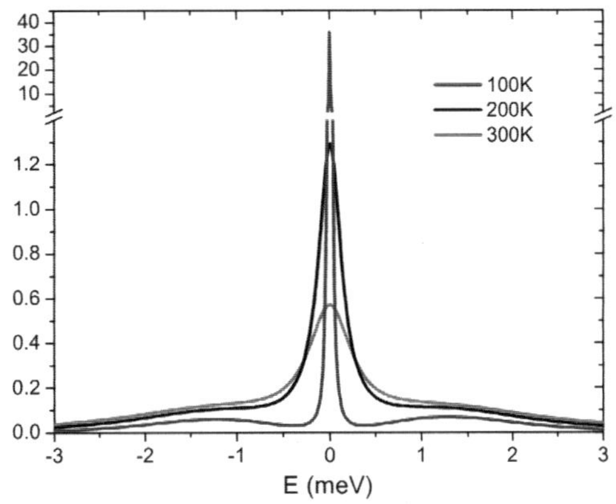


**Figure 2.** Intersublevel absorption from $s$ to first $p$ excited state of a QD for different temperatures (100K, 200K, 300K) assuming a splitting of 5 meV between the $p$-states. The absorption exhibits zero-phonon lines (ZPL) and acoustic phonon sidebands on both sides of the ZPL.

which consist of two-phonon combinations in the cubic approximation for crystal anharmonicity. Each of these channels efficiency shows dramatic variations with energy. As a consequence, one cannot approximate $\Gamma^{ph}$ by a constant $\Gamma_{LO}$ corresponding to LO-phonon relaxation in bulk. In the semiclassical model, the polaron relaxation rate was governed by the sole LO-phonon component $|\beta_i|^2$ and thus its behaviour was expected to be symmetric with respect to the LO-phonon resonance. Here, in strong contrast, a dramatic increase of intersublevel times is found below the LO-phonon energy when reducing the transition energy [17].

## 3. Acoustic phonon induced decoherence

Beyond the knowledge of population relaxation mechanism, it is important to know if there is additional dephasing processes. For example, Sherwin and co-workers proposed that intersublevel transitions in QDs in the terahertz range could be used a quantum bit [18]. Homogeneous linewidth are also important for optoelectronic applications such as polaron laser proposed by Sauvage and Boucaud [19].

Acoustic phonon plays a role not only by providing a population relaxation mechanism at low transition energy, but also by inducing pure dephasing processes. As shown in [20] and [21], acoustic phonon brings two kinds of contributions to the dephasing of intersublevel transitions. The diagonal part of the acoustic phonon-electron interaction is responsible for acoustic phonon sidebands. On the other hand, non-diagonal elements can induce real or virtual transitions between the excited levels.

The calculated absorption of a single quantum dot is presented on figure 2. In this calculation, the transition energy is 55 meV, and a splitting of 5 meV is assumed between the $p$-states. In the lineshape calculation, three mechanisms are taken into account: (i) anharmonic polaron decay, (ii) acoustic phonon sidebands, (iii) acoustic phonon induced real and virtual transitions towards the higher $p$-states.

For interband transitions, it was already shown that acoustic phonon causes the appearance of sidebands. However, in the latter case, these sidebands consist at non-zero temperature of a broad pedestal centered near the zero-phonon line (ZPL) [22, 23]. In the time-domain, these sidebands are responsible for a partial decay of coherence on the ps time scale [24, 25]. In contrast, for intersublevel transitions, these sidebands are instead peaked on both sides of the transition (see figure 2), which results in an oscillatory behaviour in photon echo experiments [20].

The ZPL broadening with temperature is essentially attributed to real and virtual transitions between the $p$-states. Acoustic phonons can induce efficient one-phonon absorption or emission only if the energy separations between the $p$-levels is of the order of a few meV (with a maximum of efficiency for typically

1-2 meV). Hence it is crucial to take also into account two phonon processes. Note that virtual transitions have also been proposed in order to explain broadening of the ZPL for interband transitions [26, 27]. In [20] and [21], we have calculated real and virtual transitions, taking into account up to two-phonon processes in a non-perturbative manner. A very good agreement between calculations and experiments is obtained. In particular, virtual transitions plays a dominant role in decoherence at high temperature.

## 4. Conclusion

We have presented theoretical investigations of intersublevel population and phase relaxation in QDs. Population relaxation is due to anharmonic polaron decay in a large spectral range, and a microscopic model of crystal anharmonicity is mandatory in order to calculate relaxation times. Additional dephasing processes are caused by interaction with acoustic phonon, resulting in the appearance of acoustic phonon sidebands and broadening of the ZPL due to real and virtual transitions.

## Acknowledgments

T. Grange acknowledges support from the Alexander von Humboldt foundation.

## References

[1] Bockelmann U and Bastard G 1990 *Phys. Rev. B* **42** 8947–8951
[2] Benisty H, Sotomayor-Torrès C M and Weisbuch C 1991 *Phys. Rev. B* **44** 10945–10948
[3] Inoshita T and Sakaki H 1992 *Phys. Rev. B* **46** 7260–7263
[4] Inoshita T and Sakaki H 1997 *Phys. Rev. B* **56** 4355–4358(R)
[5] Hameau S, Guldner Y, Verzelen O, Ferreira R, Bastard G, Zeman J, Lemaître A and Gérard J M 1999 *Phys. Rev. Lett.* **83** 4152–4155
[6] Hameau S, Isaia J N, Guldner Y, Deleporte E, Verzelen O, Ferreira R, Bastard G, Zeman J and Gérard J M 2002 *Phys. Rev. B* **65** 085316
[7] Preisler V, Ferreira R, Hameau S, de Vaulchier L, Guldner Y, Sadowski M and Lemaitre A 2005 *Phys. Rev. B* **72** 115309
[8] Preisler V, Grange T, Ferreira R, de Vaulchier L, Guldner Y, Teran F, Potemski M and Lemaître A 2006 *Phys. Rev. B* **73** 75320
[9] Carpenter B, Zibik E, Sadowski M, Wilson L, Whittaker D, Cockburn J, Skolnick M, Potemski M, Steer M and Hopkinson M 2006 *Phys. Rev. B* **74** 161302
[10] Preisler V, Grange T, Ferreira R, de Vaulchier L, Guldner Y, Teran F, Potemski M and Lemaître A 2008 *The European Physical Journal B* **67** 51–56
[11] Grange T, Ferreira R and Bastard G 2007 *Phys. Rev. B* **76** 241304(R)
[12] Li X Q, Nakayama H and Arakawa Y 1999 *Phys. Rev. B* **59** 5069–5073
[13] Verzelen O, Ferreira R and Bastard G 2000 *Phys. Rev. B* **62** 4809–4812(R)
[14] Sauvage S, Boucaud P, Lobo R P S M, Bras F, Fishman G, Prazeres R, Glotin F, Ortega J M and Gérard J M 2002 *Phys. Rev. Lett.* **88** 177402
[15] Zibik E A, Wilson L R, Green R P, Bastard G, Ferreira R, Phillips P J, Carder D A, Wells J P R, Cockburn J W, Skolnick M S, Steer M J and Hopkinson M 2004 *Physical Review B* **70** 161305
[16] Grange T, Zibik E, Ferreira R, Bastard G, Carpenter B, Phillips P, Stehr D, Winnerl S, Helm M, Steer M, Hopkinson M, Cockburn J, Skolnick M and Wilson L 2007 *New J. Phys.* **9** 259
[17] Zibik E A, Grange T, Carpenter B A, Porter N, Ferreira R, Bastard G, Stehr D, Winnerl S, Helm M, Liu H Y, Skolnick M S and Wilson L R *Nature Materials* advance online publication, 16 August 2009 (doi:10.1038/nmat2511)
[18] Sherwin M S, Imamoglu A and Montroy T 1999 *Phys. Rev. A* **60** 3508–3514
[19] Sauvage S and Boucaud P 2006 *Appl. Phys. Lett.* **88** 063106
[20] Zibik E A, Grange T, Carpenter B A, Ferreira R, Bastard G, Vinh N Q, Phillips P J, Steer M J, Hopkinson M, Cockburn J W, Skolnick M S and Wilson L R 2008 *Phys. Rev. B* **77** 041307(R)
[21] Grange T 2009 *arXiv:0907.0369v1*
[22] Besombes L, Kheng K, Marsal L and Mariette H 2001 *Phys. Rev. B* **63** 155307
[23] Favero I, Cassabois G, Ferreira R, Darson D, Voisin C, Tignon J, Delalande C, Bastard G, Roussignol P and Gérard J M 2003 *Phys. Rev. B* **68** 233301
[24] Borri P, Langbein W, Schneider S, Woggon U, Sellin R L, Ouyang D and Bimberg D 2001 *Phys. Rev. Lett.* **87** 157401
[25] Vagov A, Axt V M, Kuhn T, Langbein W, Borri P and Woggon U 2004 *Phys. Rev. B* **70** 201305
[26] Muljarov E A and Zimmermann R 2004 *Phys. Rev. Lett.* **93** 237401
[27] Muljarov E A, Takagahara T and Zimmermann R 2005 *Phys. Rev. Lett.* **95** 177405

# Wigner distribution function description of a multilayered nanostructure with magnetic impurities

**B.J. Spisak[1,2], M. Wołoszyn[2], P. Wójcik[2] and G.J. Morgan[1]**

[1] School of Physics and Astronomy, University of Leeds, Leeds LS2 9JT, United Kingdom
[2] Faculty of Physics and Applied Computer Science, AGH University of Science and Technology, Al. Mickiewicza 30, 30-059 Kraków, Poland

E-mail: woloszyn@agh.edu.pl

**Abstract.** The Wigner distribution function formalism is applied to the description of transport properties of spintronic multilayer nanodevice. The central layer of the nanodevice is doped with magnetic impurities. The current-voltage characteristics and the spin polarisation of current are calculated.

## 1. Introduction

The high precision of techniques based on semiconductor technology enables the construction of semiconductor heterostructure on a nanometre scale. On this length scale, quantum effects determine the basic characteristics of the nanodevice, and the nonclassical behaviour is utilised to design a new generation of digital elements and circuits [1]. The current trends in designing nanodevices are focused on the use and manipulation of the electron charge as well as the spin degree of freedom to enhance the performance of nanodevices. One of the simplest examples of spintronic nanodevices based on the hybrid semiconductor technology is a trilayer nanostructure where two dilute ferromagnetic semiconductors (DMS) are separated by a nonmagnetic semiconductor (NMS) layer [1].

In this report we analyse the transport properties of a simple model of the spintronic nanostructure DMS/NMS/DMS assuming that the NMS region of nanodevice contains magnetic impurities. Our analysis is based on the quantum kinetic equation for the Wigner distribution function because of its intrinsic interesting description of quantum mechanics and its potential usefulness in the simulation of nanodevice.

## 2. The Model nanodevice and method of calculation

The structure and the potential profile of the simulated spintronic nanostructure is shown in Fig. 2. The total length of the nanodevice is 16 nm with a 4 nm nonmagnetic semiconductor layer of GaN that is separated from the dilute magnetic semiconductor layers of GaMnN by spin-dependent potential barriers made from AlGaN. The existence of the spin-dependent barriers results from the difference between majority and minority spin electron states at the Fermi level in GaMnN [2].

We assume the same magnetisation axis for both GaMnN layers and the exchange splitting energy, $\Delta_{\sigma}^{DMS}$. The magnetic elements (Mn) in the clean layer of GaN are a consequence of the fabrication process. The concentration of impurities is low, and assumed to be smeared out.

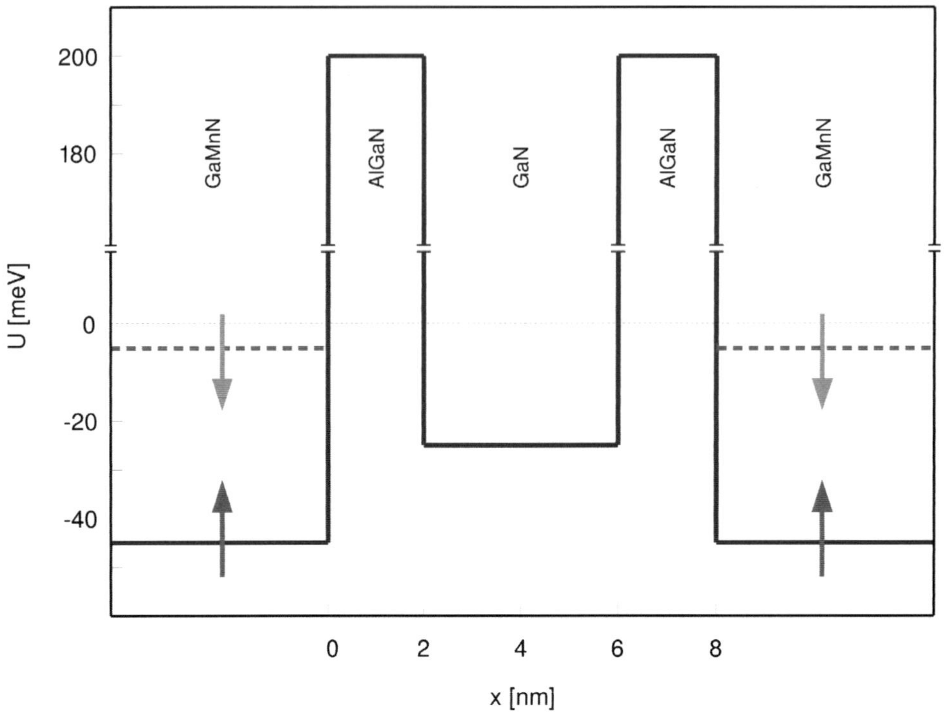


**Figure 1.** Potential energy profile in the spintronic nanodevice for zero bias voltage $V_b$ between contacts and parallel magnetisation in the DMS layers. Energy scale relative to the Fermi level.

The impurities can be characterised by a splitting parameter$\Delta\ _{\sigma}^{NMS}$, according to [3] such that $\Delta_{\sigma}^{NMS} < \Delta_{\sigma}^{DMS}$. The impurities in the NMS lead to loss of phase coherence after a time $\tau_{\varphi}$ [4].

When a small bias voltage is applied the spin-polarised electrons are in a nonequilibrium state and the spin-polarised current density can be evaluated by the formula

$$j_{\sigma}(x) = \frac{e}{2\pi} \int dk \frac{\hbar k}{m} \rho_{\sigma}(x, k), \tag{1}$$

where $m$ is the effective mass of conduction electrons, and $\rho_{\sigma}(x, k)$ is the Wigner distribution function for electrons with spin $\sigma$ for a 1-D system. We model the nanostructure in this fashion on the grounds that the system is translationally invariant in the $y$ and $z$ direction.

The Wigner distribution function, $\rho_{\sigma}(x, k)$ can be found by the solution of the simplified version of the quantum kinetic equation [5, 6], namely

$$\frac{\partial \rho_{\sigma}(x,k,t)}{\partial t} + \frac{\hbar k}{m} \frac{\partial \rho_{\sigma}(x,k,t)}{\partial x} + \frac{1}{2\pi i \hbar} \int dk' \, \mathcal{U}(x, k-k') \rho_{\sigma}(x, k', t) = -\frac{\rho_{\sigma}(x,k,t) - \rho_{\sigma}^{0}(x,k,t)}{\tau_{\varphi}}, \tag{2}$$

where $\rho_{\sigma}^{0}(x, k, t)$ is the equilibrium distribution function for electrons with spin $\sigma$, and the integral kernel $\mathcal{U}(x, k-k')$ represents the non-local potential energy. The form of the integral kernel can be found in [7]. The non-local potential energy includes the band offset, and the smeared impurity potential.

We solve the quantum kinetic equation for the Wigner distribution function (3) using the open boundary conditions in the form [8]

$$\rho_\sigma(0,k)\Big|_{k>0} = f_\sigma^L(k)\ , \qquad (3)$$

$$\rho_\sigma(L,k)\Big|_{k<0} = f_\sigma^R(k)\ ,$$

where $f_\sigma^{L(R)}(k)$ is chosen to be the semiclassical form

$$f_\sigma^{L(R)}(k) = \frac{mk_BT}{\pi\hbar^2}\ln\left\{1+\exp\left[-\frac{1}{k_BT}\left(\frac{\hbar^2k^2}{2m}-\mu_\sigma^{L(R)}\right)\right]\right\}, \qquad (4)$$

where $T$ is the temperature and $\mu_\sigma^{L(R)} = E_F \pm \Delta_\sigma^{DMS}$ is the electrochemical potential of the left (right) DMS.

The numerical calculations were carried out using the computational grid with $N_x = 116$ mesh points for the position $x$ and $N_k = 314$ for the wave vector $k$. We assume that the conduction electrons are described by the conduction band effective mass of GaN, i.e. $m = 0.228\,m_0$.

## 3. Results and discussion

When the amount of dopants inside the NMS increases, the maximum current density is decreasing in case of the majority carriers by about 15 %. At the same time the maximum value is slightly shifted towards higher voltages for positive values of$\Delta_\sigma^{NMS}$, and toward lower voltages for negative values of$\Delta_\sigma^{NMS}$. Similar shifts are observed for the minority carriers, but with rather insignificant changes of the current density values (less than 2 %).

One of the most important transport characteristics of spintronic nanodevice is the polarisation of the current density that is defined as follows [9]

$$P_j(V_b) = \frac{j^\uparrow(V_b) - j^\downarrow(V_b)}{j^\uparrow(V_b) + j^\downarrow(V_b)}. \qquad (5)$$

Figure 2 presents the polarisation of the current density as a function of the parameter$\Delta_\sigma^{NMS}$ for several different values of the bias voltage $V_b$. Degree of the polarisation depends on the applied bias voltage. For example at 0.18 V the current density polarisation slightly decreases from 98 % at$\Delta_\sigma^{NMS} = -4$ meV to 90 % at$\Delta_\sigma^{NMS} = 4$ meV. When the bias voltage is increased to 0.20 V the polarisation virtually does not depend on$\Delta_\sigma^{NMS}$ and is about 85 %. After a further increase of the bias voltage polarisation is decreasing. At $V_b = 0.22$ V it is an increasing function of$\Delta_\sigma^{NMS}$, from zero at -4 meV to about 65 % at 4 meV.

## 4. Concluding remarks

Using the Wigner distribution function formalism, we investigated the influence of magnetic impurities on the transport characteristics of the spintronic nanodevice DMS/NMS/DMS. We find that the polarisation of the current density is very sensitive to the magnetic impurities which leads to the conclusion that the purity of the non magnetic layer influences the basic transport characteristics of nanodevice.

## Acknowledgments

This paper has been supported by Polish Ministry of Science and Higher Education Programme "International scientists mobility support", AGH-UST under the project no. 11.11.220.01 "Basic and applied research in nuclear and solid state physics" and the Foundation for Polish Science MPD Programme co-financed by the EU European Regional Development Fund.

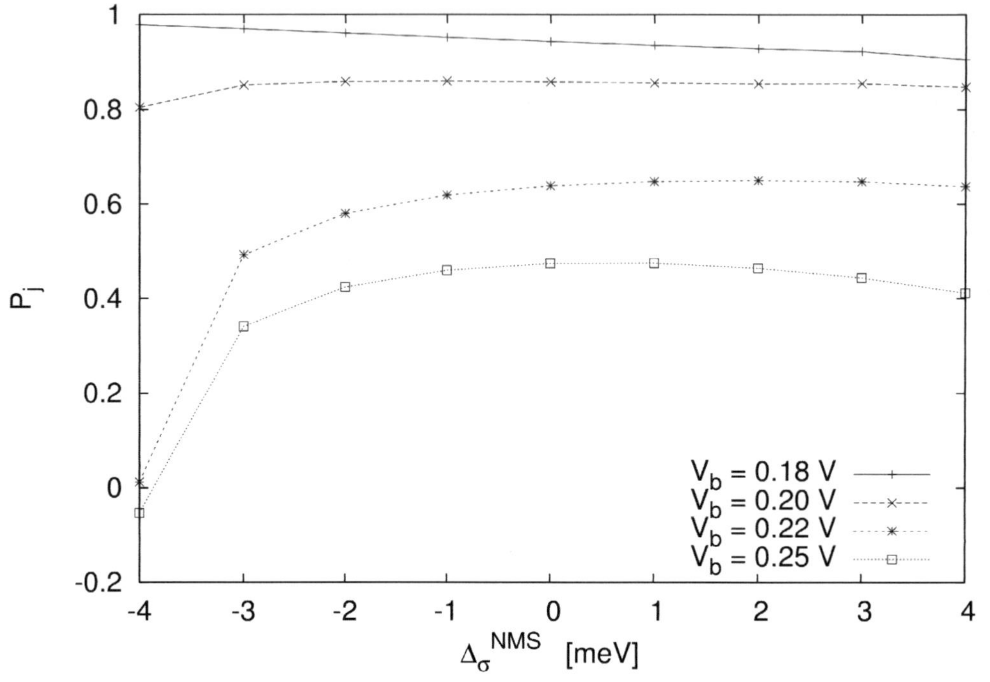


**Figure 2.** Polarisation of the current density for various values of the bias voltage. The dephasing time $\tau_\varphi$ is 100 fs.

## References


[1] Žutić I, Fabian J and Das Sarma S, 2004 *Rev. Mod. Phys.*, **76** 323
[2] Ertler Ch and Fabian J, 2007 *Phys. Rev. B* **75** 195323
[3] Lee B, Jungwirth T, and MacDonald A H, 2000 *Phys. Rev. B*, **61** 15606
[4] Zaránd G, Borda L, von Delft J and Andrei N, 2004 *Phys. Rev. Lett.* **93** 107204
[5] Yamauchi T and Mizushima K, 2000 *Phys. Rev. B*, **61** 8242
[6] Weng M Q and We M W, 2003 *J. Appl. Phys.*, **93** 410
[7] Wójcik P, Spisak B J, Wołoszyn M and Adamowski J 2008 *Acta Phys. Pol. A* **114** 1431
[8] Frensley W R 1990 *Rev. Mod. Phys.* **62** 745
[9] Maekawa S *Concepts in Spin Electronics* Oxford University Press 2006

Journal of Physics: Conference Series **193** (2009) 012131 doi:10.1088/1742-6596/193/1/012131

# Tunneling accross mesoscopic Hall bars

**C. Chaubet[1], O. Couturaud[1], M. El Khalifi[1], S. Bonifacie[1] and D. Mailly[2]**

[1] Groupe d'Etude des Semiconducteurs, Université Montpellier 2, PL. E. Bataillon, 34095 Montpellier Cedex 05, France

[2] Laboratoire de Photonique et de Nanostructures, CNRS, route de Nozay, 91460 Marcoussis, France

E-mail: christophe.chaubet@univ-montp2.fr

**Abstract.** At mesoscopic scale, the variable range hopping can be described by a tunnel effect through a single barrier. We have placed a GaALAs/GaAs HEMT in the quantum Hall regime (high magnetic field and very low temperature) and measured the components of the conductance tensor $\sigma_{xx}$ and $\sigma_{xy}$. We study the regime of the Shubnikov de Haas peaks and compare it with the same regime in large sample, commonly understood as a VRH regime. We analyse the variation in temperature in the range [0.1K-1K] for $kT$ to scale the fundamental energies, and demonstrate that in the mesoscopic scale, the measured longitudinal conductivity corresponds exactly to the tunneling through a single barrier. We deduce the barrier characteristics by fitting our data with the conductivity of a square tunnel barrier, and we retrieve the characteristic height and length of the bare disorder potential.

## 1. Introduction

In large Hall bar placed in the quantum Hall regime at very low temperature and high magnetic field, the absence of backscattering [1] guaranties a perfect quantization of the Hall resistance which allows to define the Ohm's etalon [2]. In contrast, in nanoHall bars, backscattering between opposite edges of the sample gives rise to sharp peaks on the Shubnikov de Haas maxima and on the quantized Hall plateaus, when cooled to a very low temperature [3, 4, 5]. We have processed gated Hall bars (see fig.1) to make a spectroscopy of Landau levels by biaising the top-gate for fixed values of the magnetic field. Indeed the gate voltage allows to tune the electronic density and to modify the Fermi level position with respect to the long range impurity potential. We measured the longitudinal conductivity (SdH peak) as a function of the gate voltage for a set of temperature values in the range $[0.1K, 1K]$ (see fig.2).

It is remarkable that unlike the case of wide samples, the longitudinal conductivity does not vanish when the temperature tends to zero [6, 7] as shown in fig.2. At mesoscopic scale, the conductivity tends to a skeleton function, which is a noisy image of the Landau level density of states [9, 10]. In a precedent work, we have analysed the temperature dependance of the longitudinal conductivity for fixed values of the gate voltage. On the high-$\nu$ side of the transition, the temperature dependence of the sharp peaks allows to identify this process with a double barrier tunnelling [5, 11]. On the low-$\nu$ side of the transition, the temperature dependence of the conductivity allows to identify this process with the variable range hopping [9]. We have

Journal of Physics: Conference Series **193** (2009) 012131 doi:10.1088/1742-6596/193/1/012131

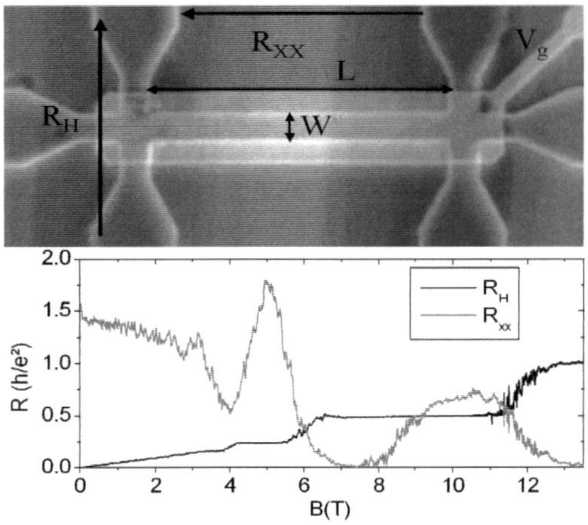


**Figure 1.** The magnetotransport coefficients are measured as a function of the magnetic field. Fluctuations are both on $R_{xx}$ and $R_{xy}$.

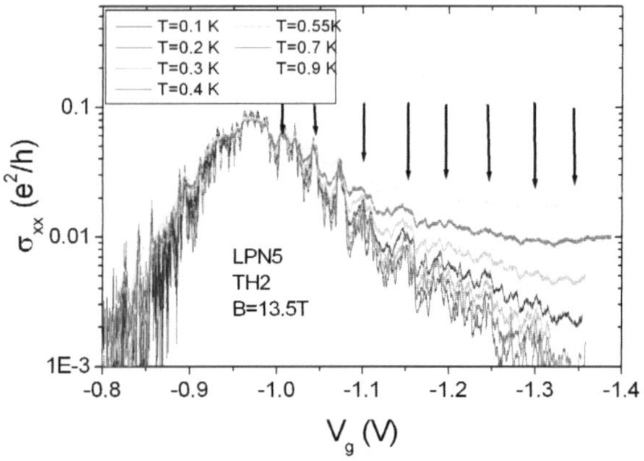


**Figure 2.** The longitudinal conductivity $\sigma_{xx}$ is deduced from the measurements of $R_{xx}$. At a fixed value of $B$, we have scanned the density of states by tunning the gate voltage. Arrows indicate the values of $V_g$ for which we analysed the temperature dependance of $\sigma_{xx}$.

demonstrated in ref [9] that the conductivity is well described by the relation [12]:

$$\sigma_{xx}(\nu, T) = \sigma_S(\nu) + \sigma_H(T) exp\left(-\sqrt{T_0/T}\right) \tag{1}$$

At vanishing temperature the hopping conductivity vanishes as well. Thus, the observed residual conductivity is obviously due to the recovering of waves functions between edges.

The purpose of this work is to demonstrate that the basic tunnel effects, which result in this residual conductivity, evolve in temperature in such a way that it reproduces the experimental variations. We will then have a description of the variable range hopping in terms of pure tunnel barrier transmission. We first give the description of the transmission through a tunnel barrier and then present our experimental results. We obtain the characteristics of the tunnel barriers for several values of $V_g$, and we retrieve the expected height and width of the barriers.

## 2. The conductance of a square tunnel barrier

Our two dimensional electrons are in the long range disorder potential created by the Si donor-impurities located in the GaAlAs barrier. Depending on the position of the Fermi energy with respect to the disorder potential, electrons will tunnel through different barriers located at different places in the sample. Of course, the shapes of tunnel barriers are not known, and our work is based on the idea that a square barrier would allow a reasonable fit of our data (as depicted in fig.3), an hypothesis which is verified a-posteriori. In the quantum Hall regime, the longitudinal conductivity corresponds to the conductivity accross the sample, i.e. to the backscattering. Applying the Landauer formula [13]:

$$\sigma_{xx}(\nu, T) = \frac{2e^2}{h} \times t \tag{2}$$

where $t$ is the transmission between sample edges through tunnel barriers, and $e^2/h$ the quantum of conductance. Let us make the hypothesis of a single square barrier. In the limit $t << 1$, the transmission of an electron whose energy is $E$ through a square barrier of height $V_0$ and width $a$, is given by the well known formula where $m$ is the electron effective mass [14]:

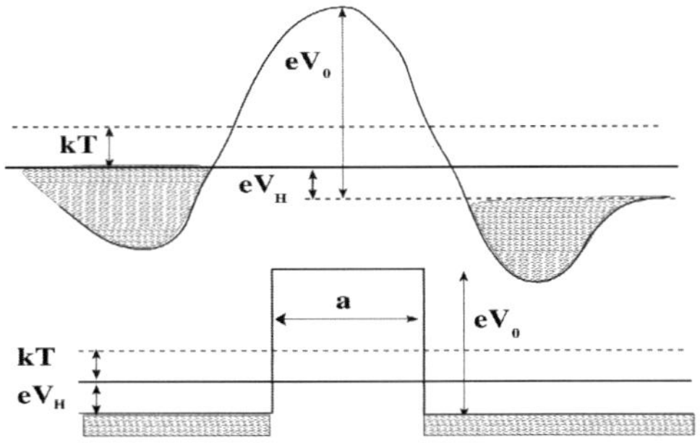


**Figure 3.** Representation of the real tunnel barrier due to the disorder potential and the corresponding square barrier model. Relevant energies are the barrier height $V_0$, the thermal energy $kT$ and the electric potential $eV_H$. a is the barrier width. The energy levels occupied by electrons are depicted by the gray region.

$$t = \frac{16E(V_0 - E)}{V_0^2} \times exp\left(-2a\sqrt{\frac{2m(V_0 - E)}{\hbar^2}}\right) \tag{3}$$

At the temperature $T$, the energy of the electron on one side of the tunnel barrier is $E = eV_H + kT$ as represented on fig.3 where $V_H$ is the Hall potential accross the channel. Introducing this expression of $E$ into eq.(3), and making the approximation $eV_H << V_0$ and $kT << V_0$, leads to the following simplified equation (see Appendix):

$$\frac{\sigma_{xx}(\nu, T)}{\frac{2e^2}{h}} = 32\left(\frac{V_H + \beta T}{v_0}\right) exp\left(-2a\gamma\sqrt{v_0 - \beta T}\right) \tag{4}$$

where the Hall potential is known from the experimental conditions $V_H = R_H I$ where $R_H$ is the Hall resistance and $I$ the current through the sample; $\beta = k/e \approx 0.861 \times 10^{-4}$; $\gamma = \sqrt{2me/\hbar^2} \approx 10^{-9}$ an $v_0 = V_0/e$ the barrier height expressed in eV.

## 3. The discrete nature of variable range hopping

The transition $\nu = 2 \rightarrow \nu = 1$ is shown in fig.2 for various temperatures: we have varied the gate voltage to tune the electron density, and consequently to lower the Fermi level with respect to the long range impurity potential. So doing, we changed the paths followed by electrons through the tunnel barriers. We concentrated here on the low-$\nu$ side of the transition, where the conductivity obviously increases with the temperature.

For several values of the gate voltage, indicated by arrows on fig.2, we have reported on a graph (see fig.4), the conductivity versus the temperature. In Fig.4, all data were fitted successfully with the square barrier transmission (eq.(4)). We thus obtained both the residual conductivity at $T = 0$, and the temperature dependance.

Once we obtained the parameters of the tunnel barrier, we report them on a graph as a function of the gate voltage. In fig.5, we observed that all values concentrate around a mean value of $a = 250$nm for the width, and $V_0 = 0.25$eV for the barrier height. Those values are completely coherent with all parameters in our experiments. Indeed, taking into account the depletion length, the width of the channel is around 1$\mu$m, and 250nm of tunnel barriers represents 1/4 of the path length followed by electrons, which is reasonable. Also, 250$\mu$eV is 2.5 times the Hall potential, a reasonable value for the barrier height.

We conclude from our results that the discrete nature of variable range hopping (VRH) has been revealed in these experiments. We have demonstrated that the VRH [8, 12, 15] corresponds in fact to a transport through tunnel barriers that we could characterize as a function of the filling

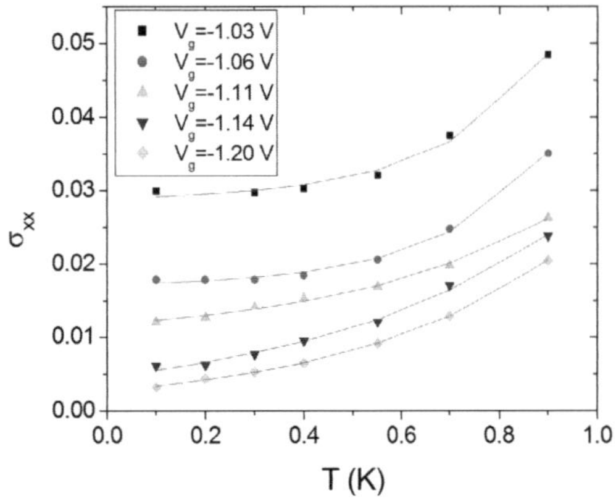


**Figure 4.** For several values of the gate voltage, we reported the conductivity versus the temperature, and fitted the data with eq.(4).

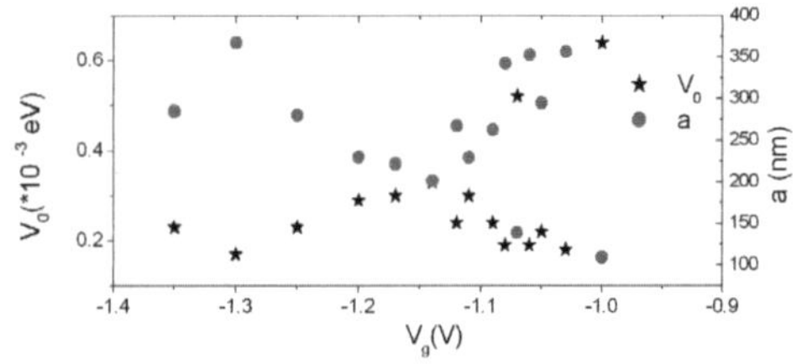


**Figure 5.** Barrier width $a$ (stars) and height $V_0$ (circles) obtained from the fit, as a function of the gate voltage.

factor. These experiments can therefore be used to characterize precisely the bare disorder potential which is "seen" by two dimensional electrons, and which create the tunnel barriers in the VRH regime.

## Appendix

Reporting the expression of the energy $E$ into eq.(3), using eq.(2) and assuming $eV_H << V_0$ and $kT << V_0$, leads to the equation:

$$\sigma_{xx} = \frac{32e^2}{hV_0^2}\left(eV_H V_0 + kT(V_0 - 2eV_H)\right) e^{-2(a/h)\sqrt{2m(V_0 - eV_H - kT)}} \tag{A.1}$$

We define $v_0 = V_0/e$, the value of the barrier height in $eV$.

$$\sigma_{xx} = \frac{32e^2}{h}\frac{1}{V_0^2}\left(e^2 V_H v_0 + ekT(v_0 - 2V_H)\right) e^{-2(a/h)\sqrt{2me(v_0 - V_H - (k/e)T)}} \tag{A.2}$$

when $kT >> eV_H$ which is our case because we heat the cristal and not the electron gaz, the above equation reduces to eq.(4).

## References

[1] M. Buttiker, Phys. Rev. B **38**, 9375 (1988).
[2] R. J. Haug, R. R. Gerhardts, K. v. Klitzing and K. Ploog, Phys. Rev. Lett. **59**, 1349 (1987).
[3] D. H. Cobden, C. H. W. Barnes and C. J. B. Ford, Phys. Rev. Lett. **82**, 4695 (1999).
[4] T. Machida, S. Ishizuka, S. Komiyama, K. Muraki and Y. Hirayama, Phys. Rev. B **63**, 45318 (2001).
[5] B. Jouault, O. Couturaud, C. Chaubet, Phys. Rev. B **76**, 161302(R) (2007).
[6] M. Furlan, Phys. Rev. B **57**, 14818 (1998).
[7] P.T. Coleridge, P. Zawadzki, and S. Sachrajda, Phys. Rev. B **49**, 10798 (1994).
[8] The width$\Delta\ \nu$ of the SdH peaks scales like$\Delta\ \nu \propto T^{0.4}$, as shown by: D.G. Polyakov and B.I. Shklovskii, Phys. Rev. Lett.**70**, 3796 (1993).
[9] O. Couturaud, S. Bonifacie, B. Jouault, A. Raymond, D. Mailly, C. Chaubet, Phys. Rev. B **80**, 033304 (2009).
[10] S. Bonifacie, C. Chaubet, B. Jouault and A. Raymond, Phys. Rev. B**74**, 245303 (2006).
[11] J. K. Jain and S. A. Kivelson, Phys. Rev. Lett. **60**, 1542 (1988).
[12] Y. Ono, J. Phys. Soc. Jap. **51**, 237 (1982).
[13] S. Datta, *Electronic Transport in Mesoscopic System*, Cambridge University Press, Cambridge U.K., 1997.
[14] C. Cohen Tanoudji, B. Diu, F. Laloe, *Mecanique Quantique*, p75, Paris, Hermann (1977).
[15] F. Hohls, U. Zeitler and R.J. Haug, Phys. Rev. Lett.**88**, 036802 (2002).

# Exciton coherence in clean single InP/InAsP/InP nanowire quantum dots emitting in infra-red measured by Fourier spectroscopy

**H Sasakura[1], H Kumano[1,5], I Suemune[1,5], J Motohisa[2], Y Kobayashi[2], M van Kouwen[2,4], K Tomioka[2,3], T Fukui[2,3], N Akopian[4], and V. Zwiller[4]**

[1]Research Institute for Electron Science (RIES), Hokkaido University, Sapporo 001-0021, Japan
[2]Graduate School of Information Science Technology, Hokkaido University, Sapporo 060-0814, Japan
[3]Research Center for Integrated Quantum Electronics (RCIQE), Hokkaido University, Sapporo 060-8628, Japan
[4]Quantum Transport, Kavli Institute of Nanoscience, Delf University of Technology, The Netherlands
[5]CREST, Japan Science and Technology Agency, Kawaguchi 332-0012

E-mail: hirotaka@eng.hokudai.ac.jp

**Abstract.** We report optical properties of InP/InAsP/InP nanowire quantum dots and single-photon Fourier spectroscopy of an exciton in a single InAsP quantum dot embedded in an InP nanowire. The coherent length of the time-averaged emission originating from the single InAsP QD was measured by a Mach-Zehnder interferometer inserted in the photoluminescence path. Effects of fluctuations in surrounding excess charges trapped in the InP nanowire were investigated by excitation power and energy dependencies.

Single-photon and/or entangled-photon pair sources based on solid-state device have drawn increasing attention, which are key to quantum information networks [1]. Emissions from 0D-exciton complexes localized in semiconductor quantum dots (QDs) are one of the most promising candidates for single-photon and/or entangled-photon pair sources [2, 3, 4, 5, 6, 7, 8, 9, 10, 11, 12]. Recently, research on semiconductor QD embedded in epitaxially grown nanowire (NW) structures has been a subject of interest in practical nano-scale device applications, such as photon sources, single-electron storage, and quantum logic circuit. This 0D+1D combined structures have been expected to improve the controllability of emission energy corresponding to size, shape anisotropy, and position of QD. In addition, epitaxially grown NWs can remediate fast dephasing of carriers in NW structures fabricated by top-down techniques due to the reduction of surface traps induced by defects and/or disorders. Besides, 0D excitons confined in a semiconductor QD can interact with its environment through carrier-phonon, carrier-carrier, and electron-nuclei interactions. The coherence length is an important parameter for the characterization of the single-photon source because decoherence (dephasing) of single-photon source sets up the bit error in the demodulation at the receiver [13]. In this study we report the optical properties of InAsP QDs embedded in InP NWs (hereafter called NW-QDs) and discuss dephasing of exciton confined in the InAs QD. Excitation power and energy dependencies of coherence times of a single NW-QD emitting in infra-red are measured by single self-photon Fourier spectroscopy.

The sample used here is InP/InAsP/InP NW-QDs grown by selective-area metalorganic vapor phase epitaxy. The lateral diameter of QD was ~100 nm corresponding to that of NW (Fig. 1(a)). The detailed growth conditions and the characteristics are seen in Ref. [14]. The single NW-QD photoluminescence

(PL) spectroscopy was performed by standard micro-PL measurements in the far field as shown in Fig. 1(b). Excitation laser beam travelling along NW growth direction was focused on the sample surface by a microscope objective. The QD and NW emissions collected by the same objective lens were dispersed by a double grating spectrometer ($f$ = 1.0 m) and were detected with a liquid-nitrogen-cooled InGaAs photodiode array. The typical exposure time was 1 s to obtain one spectrum with a high signal-to-noise ratio.

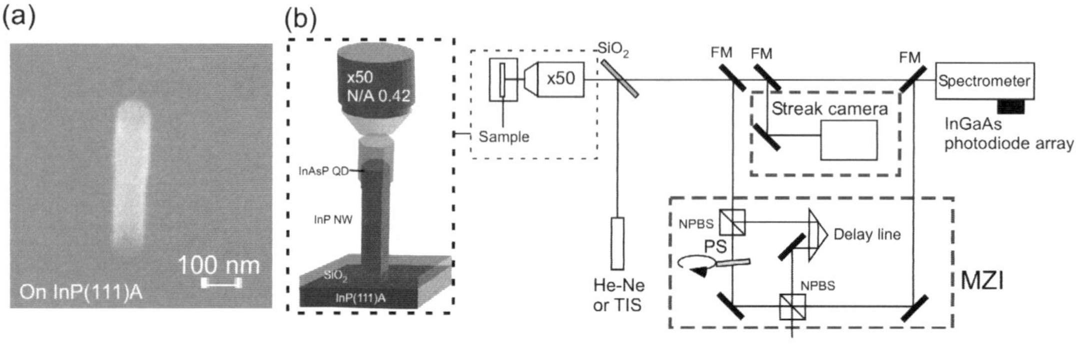


**Figure 1.** (a) SEM image of NW-QD. (b) Schematic of the experimental setup: single NW-QD-PL passing through a Mach-Zehnder interferometer (MZI). The radiative recombination decay was obtained by streak camera. FM: flipper-mounted mirror. PS: phase shifter. NPBS: non-polarized beam splitter cube.

Figure 2(a) shows a time-integrated PL spectrum of a single NW-QD at 5 K. The excitation has been carried out with He-Ne laser above InP NW bandgap energy. At low excitation intensity, the sharp peak centered around 1.023282 eV (FWHM of 46 $\mu$ eV) for QD is observed. This sharp emission shows no significant fine structure splitting induced by an anisotropic exchange interaction [15, 16, 17]. Figure 2(b) shows excitation power dependence of PL intensity (red solid circles) and emission energy (blue solid triangles) of QD. With increasing excitation power density, the PL intensity linearly increased and saturated. Before the saturation of PL intensity, the emission energy is slightly shifted (~200 $\mu$eV). This energy-shift is attributed to the Stark effect due to the internal electric field induced by the dipoles at the wurtzite and zinc-blende heterointerfaces in the InP NW [18].

We have performed a time resolved PL (TR-PL) measurement using pulsed-Ti:sapphire laser excitation at a repetition rate of 76 MHz. Streak cameras (Hamamatsu: C9510-NIR and C5683) were used to obtain TR-PLs of both structures. Figure 2(c) shows transient PL signals of single InP NW (blue solid triangles) and InAsP QD (red solid circles). The observed TR-PL of InP NW proves to consist of at least two decay components: InP substrate (fast decay components) and InP NW (slow decay components). From the streak camera image overlapping of NW and substrate emissions (inset of Fig. 2(c)), the remaining PL signals of InP NW were observed before zero delay time (marked by circle), corresponding to type-II recombination [18]. In what follows we focus on the QD emission with recombination lifetime of ~2 ns obtained by fitting to a single exponential function (red solid curve).

A Mach-Zehnder interferometer inserted in the PL path was used to apply a first-order correlation measurement on single NW-QD exciton emissions; this is a type of time-domain spectroscopy called single-photon Fourier spectroscopy and first demonstrated by Kammerer *et al* [19]. Fourier spectroscopy is an interesting method to explore the spectral dynamics of a single transition by both high temporal and high spectral resolutions with very low photon losses [19, 20, 21, 22, 23, 24]. The PL passing through the interferometer was sent to the double-grating spectrometer equipped with an InGaAs photodiode array. Here two types of excitation sources were used: a cw-Ti:sapphire laser (925 nm) below nanowire bandgap excitation and He-Ne laser for barrier excitation. Rotating a thin glass plate set in one of the interferometer arm gave a fine tuning of the relative phase (PS in Fig. 1(b)), $E_0\tau_1/\hbar$,

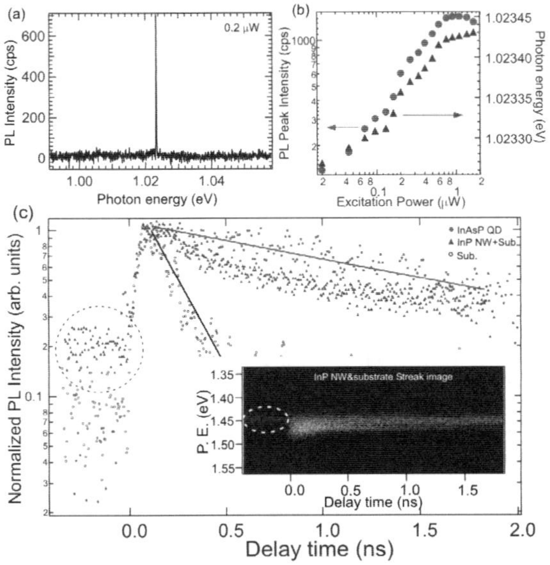


**Figure 2.** (a) PL spectrum of a single QD under He-Ne laser excitation (0.2 $\mu$W) at 5 K. (b) Solid (red) circles and solid (blue) triangles indicate the PL peak intensity and PL peak emission energy as a function of excitation power, respectively. (c) Time-resolved PL signals: single InAsP QD (solid (red) circles), InP nanowire (solid (blue) triangles), and InP substrate (open circles). Inset: streak camera image of a single NW exciton emission.

between the two arms. Short-period fringe evolution of single NW-QD exciton PL emission is shown in inset of Fig. 2(a) at $\tau$ = 10 ps. The fringe evolution as a function of the delay $\tau$ are given by $I(\tau) = I_0[1+V(\tau)\cos(E_0\tau_1/\hbar+\theta)]$, where $E_0$, $I_0$, $V(\tau)$, and $\theta$ are the central detection energy (~1.023282 eV), the averaged signal intensity, the visibility contrast given by the modulus of the Fourier transform of the intensity spectrum, and the phase, respectively. By varying the time delay $\tau$, we recorded interference fringes of single-photon events. Since the visibility contrast decays with increasing delay time $\tau$ between two arms, we can obtain the coherence time by measuring the function $V(\tau)$. Figure 3(a) shows the visibility plot of the single QD-PL interferogram as a function of the path-length difference between the two arms of the interferometer. Visibilities of interference fringes decays with an almost simple exponential function. With decreasing excitation power, the coherence time increases up to ~30 ps as shown in Fig. 3(b). This suggests that the dephasing induced by the environmental charge fluctuations were sufficiently suppressed under weak excitation. The prolonged dephasing time corresponds to the energy-shift, $\Delta E$, of the QD exciton emission as shown in the right axis of Fig. 3(b), which is corresponding to the number of surrounding excess charges in NW. In fact, we obtained the same coherence time in the excitation power of 80 $\mu$W with below InP nanowire bandgap energy (using cw-Ti:sapphire laser) and FWHM of PL peak deduced by coherence time, $2\hbar/T_2 = 44$ $\mu$eV.

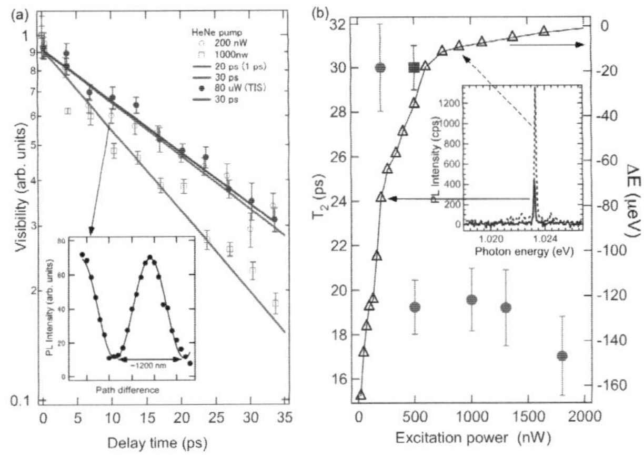


**Figure 3.** (a) Visibility plot of the single-dot PL. Inset: Short-period fringe evolution of a single-dot PL at delay time of 10 ns. (b) Excitation power ($P$) dependence of $T_2$ (solid (red) circles) with He-Ne laser excitation. Solid (blue) circle indicates $T_2$ with cw-TIS (925 nm) excitation at $P$=80 $\mu$W. PL peak energy, $\Delta E = E(P) - E(2\ \mu\text{W})$, (open triangles) are replotted. Inset: PL spectra with $P$ of 0.25 $\mu$W (solid line) and 0.9 $\mu$W (dashed line).

We report the single-photon Fourier spectroscopy of exciton in a single NW-QD. With decreasing excitation power and energy, the effects of fluctuations induced by the environmental excess charges are suppressed. These experimental facts suggest that the emissions originating from InAsP QDs embedded in the standing InP nanowires are promising candidates for single-photon source at a 1.2 micron-meter wavelength.

## References

[1] D. D. Awschalom, N. Samarth, and D. Loss, *Semiconductor Spintronics and Quantum Computation* (Springer, Berlin, 2002).
[2] Z. Yuan, B. E. Kardynal, R. M. Stevenson, A. J. Shields, C. J. Lobo, K. Cooper, N. S. Beattie, D. A. Ritchie, and M. Pepper, Science **295**, 102 (2002).
[3] A. J. Benetto, D. C. Unitt, P. See, A. J. Shields, P. Atkinson, K. Cooper, and D. A. Ritchie, Appl. Phys. Lett. **86**, 181102 (2005).
[4] C. Zinoni, B. Alloing, C. Paranthoen, and A. Fiore, Appl. Phys. Lett. **85**, 2178 (2004).
[5] M. B. Ward, T. Farrow, P. See, Z. L. Yuan. O. Z. Karimov, A. J. Benett, A. J. Shields, P. Atkinson, K. Cooper, and A. Ritchie, Appl. Phys. Lett. **90**, 063512 (2007).
[6] T. Miyazawa, J. Tatebayashi, S. Hirose, T. Nakaoka, S. Ishida, S. Iwamoto, K. Takemoto, T. Usuki, N. Yokoyama, M. Takatsu, and Y. Arakawa, Jpn. J. Appl. Phys., Part 1 **45**, 3621 (2006).
[7] R. Schmidt, U. Scholz, M. Vitzethum, R. Fix, C. Metzner, P. Kailuweit, D. Reuter, A. Wieck, M. C. Hubner, S. Stufler, A. Zrenner, S. Malzer, and G. H. Dohler, Appl. Phys. Lett. **88**, 121115 (2006).
[8] D. J. P. Elli, A. J. Bennett, A. J. Schields, P. Atkinson, and D. A. Ritchie, Appl. Phys. Lett. **88**, 131102 (2006).
[9] B. Lounis and W. E. Moerner, Nature (London) **407**, 491 (2000).
[10] V. Zwiller, H. Blom, P. Jonsson, N. Panev, S. Jeppesen, T. Tsegaya, E. Goobar, M.-E. Pistol, L. Samuelson, and G. Bjork, Appl. Phys. Lett. **78**, 2476 (2001).
[11] K. Takemoto, Y. Sakuma, S. Hirose, T. Usuki, N. Yokoyama, T. Miyazawa, M. Takatsu, and Y. Arakawa, Jpn. J. Appl. Phys., **43**, L993 (2004).
[12] S. Kimura, H. Kumano, M. Endo, I. Suemune, T. Yokoi, H. Sasakura, S. Adachi, S. Muto, H. Z. Song, S. Hirose, and T. Usuki, Jpn. J. Appl. Phys., **44**, L793 (2005).
[13] K. Inoue, T. Ohashi, T. Kukita, K. Watanabe, S. Hayashi, T. Honjo, H. Takesue, Opt. Express, **16**, 15469 (2008).
[14] P. Mohan, J. Motohisa, and T. Fukui, Nanotechnology, **16**, 2903 (2005).
[15] M. Bayer, G. Ortner, O. Stern, A. Kuther, A. A. Gorbunov, A. Forchel, P. Haweylak, S. Fafard, K. Hinzer, T. L. Reinecke, S. N. Walck, J. P. Reithmaier, F. Klopf, and F. Schafer, Phys. Rev. B **65**, 195315 (2002).
[16] I. A. Akimov, K. V. Kavokin, A. Hundt, and F. Henneberger, Phys. Rev. B **71**, 075326 (2005).
[17] R. Singh and G. Bester, Phys. Rev. Lett. **103**, 063601 (2009).
[18] B. Pal, K. Goto, M. Ikezawa, Y. Masumoto, P. Mohan, J. Motohisa, and T. Fukui, Appl. Phys. Lett. **93**, 073105 (2008).
[19] C. Kammerer, G. Cassabois, C. Voisin, M. Perrin, C. Delalande, Ph. Roussignol, and J. M. Gerard, Appl. Phys. Lett. **81**, 2737 (2002).
[20] V. Zwiller, T. Aichele, and O. Benson, Phys. Rev. B **69**, 165307 (2004).
[21] T. Kuroda, K. Sakoda, K. Watanabe, N. Koguchi, and G. Kido, Appl. Phys. Lett. **88**, 124101 (2006).
[22] C. Santori, D. Fattal, J. Vu.kovi., G. S. Solomon, and Y. Yamamoto, Nature (London) **419**, 594 (2002).
[23] S. Adachi, N. Yatsu, R. Kaji, S. Muto, and H. Sasakura, Appl. Phys. Lett. **91**, 161910 (2007).
[24] H. Kamada and T. Kutsuwa, Phys. Rev. B **78**, 155324 (2008).

Journal of Physics: Conference Series **193** (2009) 012133 doi:10.1088/1742-6596/193/1/012133

# Coulomb correlations of a few body system of spatially separated charges

**Ian Mondragon-Shem and Boris A. Rodriguez***

Instituto de Física, Universidad de Antioquia, AA 1226, Medellín, Colombia

E-mail: *`banghelo@fisica.udea.edu.co`

**Francisco E. Lopez**

Centro de Investigaciones, Instituto Tecnológico Metropolitano, Calle 73 No 76A -354, Medellín, Colombia

**Abstract.** A Hartree-Fock and Hartree-Fock-Bogoliubov study of a few body system of spatially separated charge carriers was carried out. Using these variational states, we compute an approximation to the correlation energy of a finite system of electron-hole pairs. This energy is shown as a function of the Coulomb coupling and the interplane distance. We discuss how the correlation energy can be used to theoretically determine the formation of indirect excitons in semiconductors which is relevant for collective phenomena such as Bose-Einstein condensation (BEC).

## 1. Introduction

The physical properties displayed by charge carriers in spatially confining settings could have a diverse range of potential applications relevant to technological developments in the near future. Systems with confined charge carriers such as semiconductor quantum wells, quantum wires and quantum dots, will likely have an important impact in the development of quantum computation and quantum information [1] and in producing new technological devices in electronics, spintronics and optoelectronics [2]. In order to take advantage of this important practical potential, it is necessary to have a thorough theoretical understanding of the physics of such systems.

In particular, it is well known that the physics of electron-hole pairs in quantum systems, such as quantum wells and quantum dots, is fundamentally affected by Coulomb correlations (see, for example, [3] and [4]). As a consequence, a complete theoretical study of the physical properties of charge carriers in confining settings must take into account the very important long-ranged Coulomb interaction.

With this in sight, we focus our study on a few-body system of electron-hole pairs confined to vertically stacked parabolic quantum dots. This type of system has been studied recently in the theory of quantum dot molecules (e.g. [5], [6]). Our calculations will involve twelve particles: six electrons and six holes, for a total of six electron-hole pairs. It is clear that such a system is a long way from the $O(10^2)$ pairs of real systems. However, the correlations that arise here should give us a first reasonable insight into the real physical system.

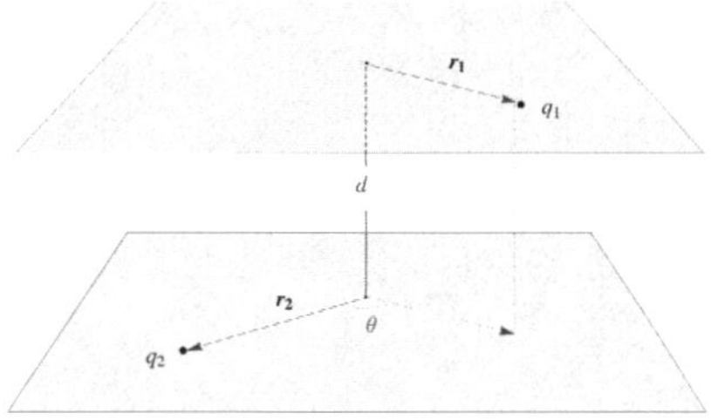


**Fig 1:** Schematic of our system with two charge carriers $q_1$ and $q_2$. They are in two different planes, spatially separated by a vertical distance $d$. The angle $\theta$ is the angle between the position vectors $\mathbf{r_1}$ and $\mathbf{r_2}$.

## 2. Theoretical framework

We compute the ground state energy of the system of electron-hole pairs using two variational approximations used in finite system studies: the Hartree-Fock (HF) state and a Hartree-Fock-Bogoliubov-type (HFB) state. These states differ in that the HF state fails to provide the so-called correlation energy of the few-body system, whereas the HFB state is able to yield a first approximation this energy. With this information, we will be able to assess under which conditions the correlation energy due to electron-hole pair formation is the greatest i.e. to assess when the formation of spatially indirect excitons is favoured.

To achive our goal, it is necessary to have an efficient way of computing the representation of the Coulomb interaction in the $2D$ harmonic oscillator basis. The Coulomb matrix elements read: $\langle ij|\hat{V}(d)|kl\rangle = \langle ij|\frac{1}{\sqrt{\mathbf{r}_{12}^2+d^2}}|kl\rangle$, where $d$ denotes the separation distance (see Fig.(1)) and the indices in the bras and kets label the 2D harmonic oscillator eigenstates . We have been able to compute the matrix elements [7] by expanding the integrals in terms of a single infinite series followed by a series acceleration algorithm to achieve rapid convergence.

The hamiltonian of the electron-hole pairs in this system is given by

$$\begin{aligned} H &= \sum_i \{\epsilon_i^{(e)} e_i^\dagger e_i + \epsilon_i^{(h)} h_i^\dagger h_i\} + \beta_{ee} \sum_{ijkl} \langle ij|V(0)|kl\rangle e_i^\dagger e_j^\dagger e_l e_k + \\ &+ \beta_{hh} \sum_{ijkl} \langle ij|V(0)|kl\rangle h_i^\dagger h_j^\dagger h_l h_k - \beta_{eh} \sum_{i\bar{j}k\bar{l}} \langle ij|V(d)|kl\rangle e_i^\dagger h_{\bar{j}}^\dagger h_{\bar{l}} e_k. \end{aligned} \quad (1)$$

The $\beta_{ee}$, $\beta_{hh}$ and $\beta_{eh}$ parameters are dimensionless quantities that measure the relative strength of the Coulomb interaction and the harmonic potential: $\beta_{xx} = \left(\frac{me^4}{\epsilon_{xx}^2 \hbar^2}/(\hbar\omega_o)\right)^{1/2}$. In this work, we take $\beta_{ee} = \beta_{hh} = \beta_{eh} = \beta$, because we only seek to understand the immediate physical effect of having spatial separation between the charge carriers. A more realistic calculation will need to take into account the fact that the $\beta$ factor dependends on the dielectric constant of the materials involved in the semiconductor and, thus, their values are going to be different in general. Through out this work, we take the energy scale to be the harmonic potential energy $\hbar\omega$ and the length scale to be the harmonic oscillator length $l = \sqrt{\frac{\hbar}{m\omega}}$.

### 2.1. The Hartree-Fock ground state

In the first step of our calculation, we minimized the average energy using a slater determinant in order to compute the ground state energy of the few body system. The HF state is given by: $|\phi_{HF}\rangle = \prod_{\mu=1} e_{\alpha_\mu}^\dagger \prod_{\nu=1} h_{\beta_\nu}^\dagger |0\rangle$. As is well known, the Hartree-Fock energy provides the best energy for uncorrelated single particle wave functions. Deviations from this result will yield information of the correlations beyond a single particle picture of the system. This transition to a correlated state will involve, in a first approach, two-particle correlations between electrons and holes, thus generating a picture of bound electron-hole pairs.

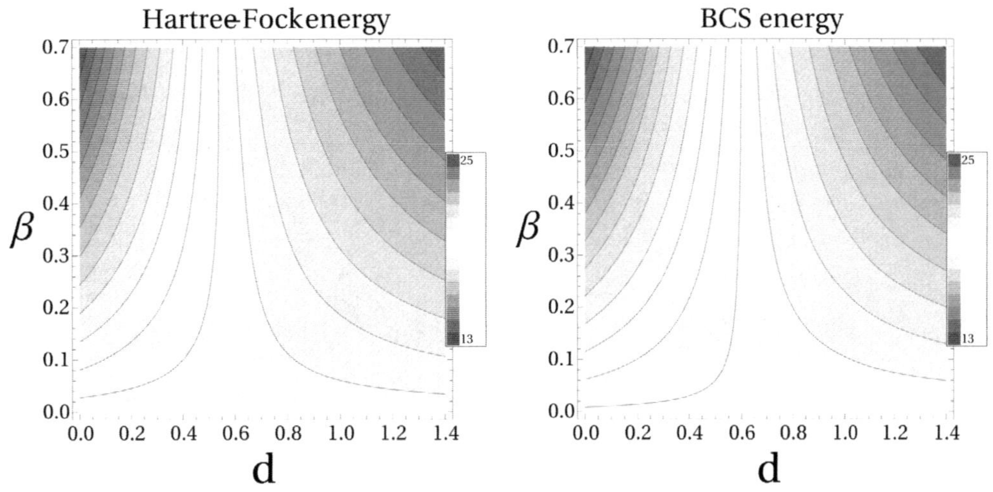


**Figure 1.** Contour plots of the HF and the BCS ground states in the $(\beta, d)$ parameter space.

### *2.2. The BCS ground state*

In order to include correlation effects, in the second step of our calculations we made use of the BCS (Bardeen-Cooper-Schrieffer) state, given by: $|\phi_{BCS}\rangle = \prod_i (u_i + v_i e_i^\dagger h_{\bar{i}}^\dagger)|0\rangle$ which is one of a various types of Hartree-Fock-Bogoliubov correlated variational ground state approximations. Such a state introduces pair correlations between electrons and holes from the outset. The coefficients $v_i$ are the variational parameters in the system and are a measure of the likelihood for an $i^{th}$ correlated pair to appear in the system.

## 3. Results and discussion

In Fig.(1), we show the HF and BCS energies in the parameter space $(\beta, d)$. Both of these plots tell us how, by a suitable combination of the Coulomb coupling $\beta$ and the interplane distance, we can go from energies below the energy of the noninteracting particles to energies well above this value. The appearance of the blue region shows that the attractive interaction between electrons and holes is dominant over that of the repulsive electron-electron and hole-hole ones, when $\beta$ increases or when $d$ decreases. However, as we separate the planes, the electron-hole energy gradually loses its predominance until we achieve a state in which the contribution from repulsive and attractive interactions cancels out, and we are left with the energy value of the noninteracting system.

Note that the contour level corresponding to the particular situation of equal contributions from attractive and repulsive energies is given approximately by a small region of constant distances, *independent of the Coulomb coupling.* This region is shifted to the right in the BCS plot, with respect to that in the Hartree-Fock plot.

The correlation energy, which is approximately given by $E_C(d,\beta) \approx E_{BCS}(d,\beta) - E_{HF}(d,\beta)$, is shown in Fig.(2); both plots in this figure provide different views of the correlation energy. We arrive at two important conclusions from the behaviour of $E_C(d,\beta)$.

First, the system electron-hole pairs show the strongest excitonic behaviour at a non-trivial value of interplane distance. This can be understood as follows: when $d = 0$, the interactions between the particles are effectively screened since they all exist in the same space i.e. in

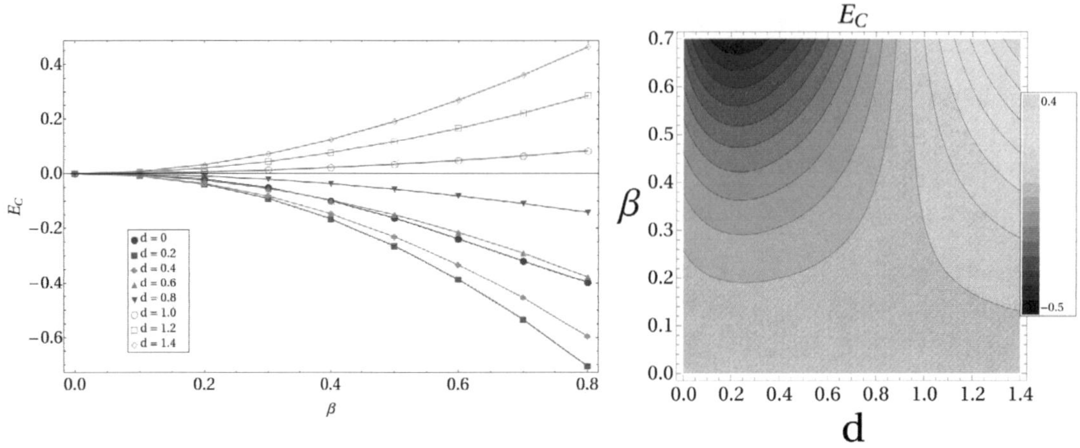


**Figure 2.** (**Left**) Correlation energy $E_C$ for various values of the distance $d$. (**Right**) Contour plot of the correlation energy in the ($d$,$\beta$ ) plane.

the two-dimensional plane. However, when there is a finite distance between the planes, the screening between electrons and holes is reduced and, thus, the electron-hole attractive Coulomb interaction is reinforced. This enhanced attraction effect competes with the electron-electron and hole-hole repulsive interactions and with the suppression of the Coulomb interaction due to the finite distance between the electrons and holes. The attractive interactions eventually lose this competition when the repulsive interaction dominates the ground state of the system, which happens at a critical distance, as can be noted in the plot of the correlation energy.

This last physical turning point brings us to the second important feature of the correlation energy: at the aforementioned critical distance, the energy of the BCS state looses its superiority over the Hartree-Fock energy, since the BCS state energy becomes larger than that of the Hartree-Fock state. This means that the system is no longer suitably described through correlated pairs i.e. the system has suffered a structural change of quantum state, which is reminiscent of the quantum phase transitions that occur in large many-body systems. This transition point, together with the minimum of the correlation energy for finite $d$, are important pieces of information for understanding the BEC state of indirect excitons in semiconductors.

## Acknowledgments

The authors acknowledge partial financial support from the CODI- Universidad de Antioquia and the Centro de Investigaciones- Instituto Tecnológico Metropolitano de Medellín. I.M.S. is grateful for useful discussions with J. L. Sanz on the Hartree-Fock method.

## References

[1] M. A. Nielsen and I. L. Chuang. Quantum Computation and Quantum Information, Cambridge University Press 2000.
[2] I. Zutic, J. Fabian and S. Das Sarma, 2004 *Rev. Mod. Phys.* **76** 323.
[3] M. Alatalo, M.A. Salmi, P. Pietilainen and Tapash Chakraborty, 1995 *Phys. Rev. B* **52** 7845.
[4] J. P. Eisenstein, G.S. Boebinger, L.N. Pfeiffer, K.W. West and Song He, 1992 *Phys. Rev. Lett.* **68** 1383.
[5] S. K. Saikin, C. Emary, D. G. Steel and L. J. Sham, 2008 *Phys. Rev. B* **78** 235314.
[6] M. F. Doty, M. Scheibner, A. S. Bracker, I. V. Ponomarev, T. L. Reinecke, and D. Gammon, 2008 *Phys. Rev. B* **78** 115316.
[7] I. Mondragon-Shem, F. E. Lopez and B.A. Rodriguez. Submitted.

Journal of Physics: Conference Series **193** (2009) 012134 doi:10.1088/1742-6596/193/1/012134

# Dicke model: entanglement as a finite size effect

**Oleksandr Tsyplyatyev and Daniel Loss**

Department of Physics, University of Basel, Klingelbergstrasse 82, CH-4056 Basel, Switzerland

E-mail: o.tsyplyatyev@unibas.ch

**Abstract.** We analyze the Dicke model at zero temperature by matrix diagonalization to determine the entanglement in the ground state. In the infinite system limit the mean field approximation predicts a quantum phase transition from a non-interacting state to a Bose-Einstein condensate at a threshold coupling. We show that in a finite system the spin part of the ground state is a bipartite entangled state, which can be tested by probing two parts of the spin system separately, but only in a narrow regime around the threshold coupling. Around the resonance, the size of this regime is inversely proportional to the number of spins and shrinks down to zero for infinite systems. This spin entanglement is a non-perturbative effect and is also missed by the mean-field approximation.

Coherent interaction between electromagnetic photon fields and matter attracted interest a long time ago with renewed attention gained in the last decade due to significant developments in the experimental techniques in various areas of physics. Achievement of Bose Einstein condensation of cold atomic gases in electromagnetic traps enabled the coherent coupling of hyperfine states of $10^5$ atoms to a single photon mode of an optical resonator [1]. Advances in the semiconductor technology allowed to obtain optical microcavities where electron-hole excitations inside the semiconductor quantum well are strongly coupled to an eigenmode of the optical resonator [2]. Strong coupling of a single mode of a transmission line resonator to a Cooper pair box and a quantized mode of an optical crystal cavity to several semiconductor quantum dots [3] have been demonstrated as a possible way to a quantum computing device [4].

Theoretical understanding of all these systems is based on a model proposed by Dicke [5] which describes $N$ spins $1/2$ (identical two-level systems) with splitting energy $2\epsilon$ coupled to a single mode of electromagnetic field $\omega$. It was shown that this model is exactly diagaonalisable [6]. At zero temperature it undergoes a quantum phase transition from a non interacting state with unpopulated bosonic mode to a condensed state with with a highly populated bosonic mode [7] if coupling between the boson and a single spin $g$ is greater than a threshold value. In the thermodynamic limit a phase transition occurs in the region of strong coupling if temperature is less than a critical temperature which can be described by a Bogolyubov Hamiltonian similarly to the pairing model of superconductivity [8]. Recently, a variational wave function approach to the generalised Dicke model was used [9] to describe Bose Einstein condensation of exciton-polaritons in a semiconductor optical cavity.

In this paper we analyse the Dicke model at zero temperature for a finite $N$ using matrix diagonalisation methods. We find that for the particular coupling strength,

$$\omega\epsilon < g^2 \lesssim \omega\epsilon\left(1+\frac{1}{N}\right), \tag{1}$$

Journal of Physics: Conference Series **193** (2009) 012134 doi:10.1088/1742-6596/193/1/012134

the ground state of the spin subsystem is a bipartite entangled state and it is not entangled outside of this region. The lower bound of the inequality is onset of the quantum transition described by the mean field theory. The upper bound is the condition to have only singly populated bosonic mode in the ground state. The approximated value is the result of $1/N$ expansion around the resonance.

In the thermodynamic limit the ground state projected onto the subspace of N spins is not entangled as it is a product state. For a weak coupling below the quantum transition the ground state is a product of all unexcited single spin states. For a strong coupling above the transition threshold the ground state is also a product state of all single spin states [9] as a result of the mean-field approximation. In a finite system the mean-field approximation is not applicable in a small region above the transition threshold where the expectation value of the boson is of the order of 1 and its fluctuations are also of the order of 1. The ground state in this region is a superposition of the unexcited state of all spins and a spin state with only one single spin-flip excitation, the $N$ spin W-state, $|W\rangle = (|\uparrow\downarrow\downarrow\ldots\rangle + |\downarrow\uparrow\downarrow\ldots\rangle + |\downarrow\downarrow\uparrow\ldots\rangle + \ldots)/\sqrt{N}$. The W-state can also be interpreted as 'magnon state' at vanishing wave vector [10].

Furthermore, the W-state can be considered as a bipartite entangled state in the following sense. Dividing all spins into two groups the W-state is a Bell state in the subspace restricted by only one spin-flip excitation above the unexcited states of each group of the spins. In the course of a bipartite measurement if the first group is found in the excited state then the second group is projected onto the unexcited state and if the first group is found in the unexcited state then the second group is projected onto the excited state.

We diagonalize the Dicke model for $N$ spins $1/2$ coupled to a single bosonic mode [11]

$$H = \omega b^\dagger b + \epsilon \sum_j S_j^z + \frac{g}{\sqrt{N}} \sum_j \left( S_j^+ b + S_j^- b^\dagger \right), \tag{2}$$

where the sum runs over $N$ spin-1/2 operators $\mathbf{S}_j$ that obey the commutation relations $[S_i^\alpha, S_j^\beta] = \epsilon_{\alpha\beta\gamma}\delta_{ij}S_i^\gamma$, and $b$ ($b^\dagger$) is standard bosonic annihilation (creation) operator.

The Dicke model possesses the following conserved quantities. One is the number of excitations of the coupled spin-boson system, $L = n + J_z$, expressed in terms of the z-component of total spin operator $J_\alpha = \sum_j S_j^\alpha, \alpha = x, y, z$, and the occupation number operator $n = b^\dagger b$ of the boson mode. Note that $J_\alpha$ and $n$ are not conserved separately. The eigenvalues of $L$ are the so-called cooperation numbers $c$, given by the sum of expectation values of $n$ and $J_z$. A second conserved quantity is the total spin, $J^2 = J_z^2 + (J_+J_- + J_-J_+)/2$ with eigenvalues $j(j+1)$.

We represent the Dicke Hamiltonian (2) in a basis where $J^2$ and $L$ are block-diagonal. Within a block of given $c$ and $j$ the remaining degrees of freedom can be labeled by the eigenvalues $m$ of $J_z$. In the representation $|c, j, m\rangle$ each block has a tridiagonal form. The diagonal matrix elements represent the energies of states containing $c - m$ bosons and $m$ excited spins, $\langle m|H|m\rangle = \omega(c-m) + \epsilon m$.The first above- and below-diagonal matrix elements are transition amplitudes connecting all pairs of states which differ by just one flipped spin, $\langle m|H|m+1\rangle = \frac{g}{\sqrt{N}}\sqrt{(c-m)(j(j+1) - m(m+1))}$. The size of each block is limited by the fact that $-N/2 \leq m \leq N/2$ for $j = N/2$. However,the upper bound on $m$ is further constrained by the cooperation number $c$ which has the lower bound $-N/2$.

The ground state of Eq. (2) is the lowest energy state of all blocks. For different values of the parameters $\omega$, $\epsilon$, and $g$ the ground state can have different $c's$ and $j's$ (see below). We consider now the case with the bosonic ($\omega$) and spin excitation energies ($\epsilon$) close to each other. For the uncoupled case, $g = 0$, the ground state contains no bosons and no excited spins (i.e. all spins are, say, down). From the definition of $L$ and the diagonal matrix elements we see that the cooperation number of this state is then $c = -N/2$ with corresponding ground state energy $E_0 = -N\epsilon/2$. For finite but still small coupling, $g \ll 1$, we can use a perturbative approach to remove the boson mode via a Schrieffer-Wolff transformation to obtain an effective spin-Hamiltonian [13],

$$H = \omega b^\dagger b + \left(\epsilon + \frac{2g^2 b^\dagger b}{N(\epsilon - \omega)}\right) \sum_j S_j^z + \frac{g^2}{N(\epsilon - \omega)} \sum_{i,j} S_i^+ S_j^- , \tag{3}$$

where the spin-boson mixing is eliminated up to the second order in $g$, introducing an effective XY coupling between the spins within a subband of given boson occupation number $n$. Such a perturbative approach gives a qualitatively correct description of the energies and wave functions until $g$ crosses a threshold value $g_c$ where a transition to a strongly correlated non-perturbative regime takes place (see below).

Increasing $g$ further above the transition threshold a sequence of states with $c = 1 - N/2$, $c = 2 - N/2$, and so on becomes subsequently the ground state due to the interaction energy. The ground state with $c$ greater but close to $-N/2$ can not already be approximated with help from the perturbation theory as coupling is too strong and can not yet be approximated using the mean field approach as the fluctuations of the order parameter are too large compared to its mean. Here we diagonalize the Hamiltonian Eq. (2) using the matrix representation. Sizes of the matrices in this regime are limited by the cooperation number $c$. In the presentation $|c, j, m\rangle$ the matrices are not larger then $2 \times 2$ and $3 \times 3$ for $c = 1 - N/2$ and $c = 2 - N/2$ respectively. The lowest eigenenergies of these matrices are $E_1 = -\frac{(N-1)\epsilon}{2} + \frac{\omega}{2} - \sqrt{g^2 + \left(\frac{\omega-\epsilon}{2}\right)^2}$ for $c = 1 - N/2$ and $E_2 = -\frac{(N-2)\epsilon}{2} + \omega - 2g\sqrt{1 - \frac{1}{2N}} + \frac{\omega-\epsilon}{2(2N-1)}$ for $c = 2 - N/2$. The last expression is a result of expansion at the resonance in powers of $\omega - \epsilon$. Both $E_1$ and $E_2$ belong to subblocks with $j = N/2$ in accordance with a theorem from [8]. Comparing $E_0$ and $E_1$ we find $g_c = \sqrt{\omega\epsilon}$. The same condition to have a non zero population of the bosonic mode in the ground state was established in [7].

Comparing $E_1$ and $E_2$ near the resonance $\omega = \epsilon$ we find that a state with $c = 2 - N/2$ becomes the ground state when $g$ exceeds $g_2 = (\epsilon + \omega + (\omega - \epsilon)/(2N-1)) / \left(4\sqrt{1 - 1/2N} - 1\right)$. Thus, $g < g_2$ together with $g > g_c$ define the regime of the model parameters where matrix diagonalization is the only way to study the Dicke model. The upper bound of Eq. (1) coincides with $g_2$ for $\epsilon \approx \omega$ and $N \gg 1$. Increasing $g$ further, we can determine the boundaries between the ground states with different values of $c$ ( i.e. $c = 3 - N/2, 4 - N/2, \ldots$) by numerical diagonalization. A result similar to $E_1$, $E_2$, and $g_2$ was obtained in [14] but the descreete jumps in the cooperation number were interpreted as an infinite sequence of instabilities.

In the strong coupling regime, $g \gg g_c$, the mean field approach provides a good approximation to the exact ground state. Indeed, introducing the expectation value of the bosonic operator $\mathcal{B} = \langle b \rangle$ and neglecting quantum fluctuations around it, the Dicke Hamiltonian Eq. (2) becomes

$$H = \omega |\mathcal{B}|^2 + \sum_j \left(\epsilon S_j^z + \frac{g}{\sqrt{N}} \left(S_j^+ \mathcal{B} + S_j^- \mathcal{B}^*\right)\right). \tag{4}$$

Eigenstates of this Hamiltonian are product states of the N spins, and thus are manifestly not entangled. Diagonalization of the $2 \times 2$ - matrices for each spin and subsequent minimization of the sum of the lowest eigenenergies over $|\mathcal{B}|$ gives the following self-consistency (mean-field) equation $\omega = g^2/\sqrt{4|\mathcal{B}|^2 g^2/N + \epsilon^2}$, which describes a quantum phase transition at a threshold value of the coupling strength $g_c$ that we have already found from the matrix diagonalization.

The mean field approximation is satisfactory even for a system of only a few spins. To see this, we compare numerically for $N = 3$ spins the expectation value of $n$ taken with respect to the exact ground state with $|\mathcal{B}|^2$ at the resonance $\omega = \epsilon$. There we see that the largest deviation is in the region of intermediate coupling strength where the cooperation number of the exact ground state is $c = 1 - N/2$. For weak coupling $g < g_c$ the ground state coincides with the non-interacting one and $\langle n \rangle = |\mathcal{B}|^2 = 0$. The quantum fluctuations of $n$ compared to its mean value are already small for the coupling corresponding to the exact ground state with $c = 4 - N/2$, $\left(\langle n^2 \rangle - \langle n \rangle^2\right) / \langle n \rangle^2 \simeq 0.2$. Thus, in the strong coupling regime the approximation of neglecting these fluctuations in Eq. (4) is already good for $g \gtrsim g_2$ irrespective of $N$.

The ground state can be characterized in terms of entanglement between different parts of the spin-subsystem. In the weak coupling regime, the ground state, being a direct product of (unexcited) spin states, is not entangled. In the regime of strong coupling, where the approximate Hamiltonian Eq. (4) is valid, the ground state is also a product of the individual spin states which, thus, also has no entanglement of any pair of spins. The ground state in the intermediate region Eq. (1) has to be found by matrix diagonalization and will be analyzed below.

Diagonalization of the matrix for $c = 1 - N/2$ and $g = \sqrt{\omega\epsilon}$ gives a ground state in the regime defined by Eq. (1). We change the mixed spin-boson representation $|c, l, m\rangle$ to a separate representation of spins and the boson $|\mathfrak{n}, l, m\rangle$, where $\mathfrak{n}$ is the bosonic occupation number. Then, tracing out the bosonic degree of freedom $\mathfrak{n}$ we obtain the reduced density matrix of the spins only, which in the representation of $J_z$ eigenstates $|m\rangle$ is given by $\hat{\rho} = \frac{\epsilon}{\epsilon+\omega} |-N/2\rangle \langle -N/2| + \frac{\omega}{\epsilon+\omega} |-N/2+1\rangle \langle -N/2+1|$.There is a finite probability to find either all spins in the completely polarized state or in the W-state $|-N/2+1\rangle$.

The W-state is a Bell state for $N = 2$. For $N > 2$ it is a bipartite entangled state. The set of spins can be divided into two equal groups consisting of $N/2$ spins each (assuming N even). In the basis $|m_1\rangle |m_2\rangle$, $m_1$ and $m_2$ are the eigenvalues of the operators $J_z^{1(2)}$ belonging to the first (second) group, the state is $|-N/2+1\rangle$ is $\left(|-N/4\rangle|-N/4+1\rangle + |-N/4+1\rangle|-N/4\rangle\right)/\sqrt{2}$, i.e. a measurement of one group projects another group onto the definite state. Therefore in the intermediate region Eq.(1) the ground state of the Hamiltonian Eq. (2) is a bipartite entangled state. Note that $N$ spin W-state does not belong to the class of $N$ spin entangled states prepared by squeezing [15]. The squeezing parameter is not defined for the $N = 2$ W-state and is greater than one for $N > 2$ W-states.

Let the spins have different splitting energies $\epsilon_j$ in Eq.(2) instead of the case $\epsilon_j \equiv \epsilon$ which we have considered in the paper. A small variation $\langle \epsilon_j \rangle - \epsilon_j \ll \epsilon_j$, $\langle \epsilon_j \rangle$ being the average over splitting energies of all spins, can be treated using perturbation theory and will not affect our results, Eqs.(1,13) much. For instance, the $N = 2$ W-state in Eq.(13) becomes $\left((1 - (\epsilon_1 - \epsilon_2)/\sqrt{2}g) |\uparrow\downarrow\rangle + |\downarrow\uparrow\rangle\right)/\sqrt{2}$ at the resonance $\epsilon = \omega$. The amplitudes of different components in the singlet will alter slightly while the W-state will not change qualitatively. For a more detailed analysis of the inhomogeneous Dicke model see [16].

We acknowledge support from the Swiss NF, NCCR Nanoscience Basel, and JST ICORP.

[1] F. Brennecke, T. Donner, S. Ritter, T. Bourdel, M. Köhl, and T. Esslinger, Nature **450**, 268 (2007).
[2] J. Kasprzak *et al.*, Nature **443**, 409 (2006).
[3] A. Wallraff *et al.*, Nature **431**, 162 (2004); K. Hennessy *et al.*, Nature **445**, 896 (2007).
[4] A. İmamoğlu *et al.*, Phys. Rev. Lett. **83**, 4204 (1999).
[5] R. H. Dicke, Phys. Rev. **93**, 99 (1954).
[6] M. Tavis and F. W. Cummings, Phys. Rev. **170**, 379 (1968).
[7] G. Scharf, Helv. Phys. Acta **43**, 806(1970).
[8] K. Hepp and E. H. Lieb, Ann. Phys. (N.Y.) **76**, 360 (1973).
[9] P. R. Eastham and P. B. Littlewood, Phys. Rev. B **64**, 235101 (2001).
[10] J. S. Pratt, Phys. Rev. B **73**, 184413 (2006).
[11] Dicke model implies the rotating wave approximation. For analysis without this approximation see [12].
[12] N. Lambert *et al.*, Phys. Rev. Lett. **92**, 073602 (2004); J. Vidal *et al.*, Europhys. Lett. **74**, 817 (2006).
[13] M. Trif, V. N. Golovach, and D. Loss, Phys. Rev. B **77**, 045434 (2008).
[14] Vladimir Buzek, Miguel Orszag, and Marian Rosko, Phys. Rev. Lett. **94**, 163601 (2005).
[15] A. Sørensen, L.-M. Duan, J. I. Cirac, and P. Zoller, Nature **409**, 63 (2001).
[16] O. Tsyplyatyev and D. Loss, Phys. Rev. A **80**, 023803 (2009).

Journal of Physics: Conference Series **193** (2009) 012135 doi:10.1088/1742-6596/193/1/012135

# Microscopic theory of energy dissipation and decoherence in open systems: a quantum Fermi's golden rule

**D Taj[1,2,3], R C Iotti[1] and F Rossi[1]**

[1] Dip. Fisica, Politecnico di Torino, C.so Duca degli Abruzzi 24, 10129 Torino, Italy
[2] Laboratoire LPA, ENS, UMR 8551 CNRS, UPMC Univ. Paris 6 75005 Paris, France
[3] Physics Department, University of Fribourg, Chemin du Musee 3, 1700 Fribourg, Switzerland

E-mail: david.taj@unifr.ch

**Abstract.** We shall revisit the conventional adiabatic or Markov approximation, which — contrary to the semiclassical case— does not preserve the positive-definite character of the corresponding density matrix, thus leading to highly non-physical results. To overcome this serious limitation, originally addressed by Davies and co-workers almost three decades ago, we shall propose an alternative more general adiabatic procedure, able to provide a reliable/robust treatment of energy-dissipation and dephasing processes in electronic quantum devices. Unlike standard master-equation formulations, our procedure guarantees a positive evolution for a variety of physical subsystem (including the common partial trace), and quantum scattering rates are well defined even for subsystems with internal structure/ continuous energy spectrum. We shall compare the proposed Markov dissipation model with the conventional one also through basic simulations of energy-relaxation versus decoherence channels in prototypical semiconductor nanodevices.

Present-day technology pushes device dimensions toward limits where the traditional semiclassical or Boltzmann theory [1] can no longer be applied, and more rigorous quantum-kinetic approaches are imperative [2, 3]. However, in spite of the quantum-mechanical nature of electron and photon dynamics in the core region of typical solid-state nanodevices [4, 5], the overall behavior of such quantum systems is often governed by a complex interplay between phase coherence and energy relaxation/dephasing [6], the latter being also due to the presence of spatial boundaries [7, 8]. Therefore, a proper treatment of such novel nanoscale devices requires a theoretical modeling able to properly account for both coherent and incoherent —i.e., phase-breaking— processes on the same footing.

To this aim, a quantum generalization of the standard Boltzmann collision operator [1] has been proposed [6], which however does not preserve the positive-definite character of the density-matrix [9]. Actually, the physical extent to which positivity is lost was not at all clear in [9], and thus the standard Boltzmann operator has survived in the semiconductor community, up to now: we shall clearly show, through some basic simulation for prototypical semiconductor nanodevices, that the problem is indeed far more serious than one may suspect.

The only available alternative that guarantees positivity was first formalized by Davies [10, 11], but its range of application is far narrower than the conventional markovian procedure cited above, as it only applies to quantum devices without internal structure (with discrete spectrum, that is, quantum dots), while it fails to describe relaxation/ decoherence in infinitely extended/continuous devices (as it is the case of quantum transport [8]).

Aim of the present Letter is to propose an alternative and more general adiabatic procedure which (i) in the discrete-spectrum case reduces to Davies' model [10], (ii) for diagonal states gives the well known Fermi's golden rule [12, 13], and (iii) always describes a Lindblad (and thus positive) evolution, even for continuous energy spectra, thus providing a reliable/robust treatment of energy-dissipation and dephasing processes in semiconductor quantum devices.

Our approximation scheme holds true under the same validity regime of the conventional Markov approach: the so called weak-coupling limit [14], where the subsystem density matrix in the interaction frame moves slowly with respect of perturbative effects.

In order to recall the main features of the problem, let us consider its general formulation based on the fully operatorial approach proposed in [15]. Given a generic physical observable $A$ —described by the operator $\hat{A}$— its quantum plus statistical average value is given by $A = \mathrm{tr}\left\{\hat{A}\hat{\rho}\right\}$, where $\hat{\rho}$ is the so-called density-matrix operator. Its time evolution is dictated by the total (system plus environment) Hamiltonian, that can be regarded as the sum of a noninteracting (system plus environment) contribution plus a system-environment coupling term: $\hat{H} = \hat{H}_\circ + \hat{H}'$; the corresponding equation of motion for the density-matrix operator —also known as Liouville-von Neumann equation— in the interaction picture is given by:

$$\frac{d\hat{\rho}^i}{dt}(t) = -i\left[\hat{\mathcal{H}}^i(t), \hat{\rho}^i\right] , \tag{1}$$

where $\hat{\mathcal{H}}^i$ denotes the interaction Hamiltonian $\hat{H}'$ written in units of $\hbar$.

The key idea beyond any perturbation approach is that the effect of the interaction Hamiltonian $\hat{H}'$ is "small" compared to the free evolution dictated by the noninteracting Hamiltonian $\hat{H}_\circ$. Following this spirit, by formally integrating Eq. (1) from $t_\circ$ to the current time $t$, and inserting such formal solution for $\hat{\rho}^i(t)$ on the right-hand side of Eq. (1), we obtain an integro-differential equation of the form:

$$\frac{d\hat{\rho}^i}{dt}(t) = -i\left[\hat{\mathcal{H}}^i(t), \hat{\rho}^i(t_\circ)\right] - \int_{t_\circ}^{t} dt' \left[\hat{\mathcal{H}}^i(t), \left[\hat{\mathcal{H}}^i(t'), \hat{\rho}^i(t')\right]\right] . \tag{2}$$

We stress that so far no approximation has been introduced: Eq.s (1) and (2) are fully equivalent, we have just isolated the first-order contribution from the full time evolution in Eq. (1).

Let us now focus on the time integral in Eq. (2). Here, the two quantities to be integrated over $t'$ are the interaction Hamiltonian $\hat{\mathcal{H}}^i$ and the density-matrix operator $\hat{\rho}^i$. In the spirit of the perturbation approach previously recalled, the time variation of $\hat{\rho}^i$ can be considered adiabatically slow compared to that of the Hamiltonian $\hat{\mathcal{H}}$ written in the interaction picture, i.e., $\hat{\mathcal{H}}^i(t') = \hat{U}_\circ^\dagger(t')\hat{\mathcal{H}}\hat{U}_\circ(t')$; indeed, the latter exhibits rapid oscillations due to the noninteracting evolution operator $\hat{U}_\circ(t) = e^{-\frac{i\hat{H}_\circ t}{\hbar}}$. Therefore, in the standard (and problematic) Markov approximation the density-matrix operator $\hat{\rho}^i$ in interaction frame is simply taken out of the time integral and evaluated at the current time $t$, obtaining an effective equation that *i)* has still the double-commutator structure in (2) (and thus conserves the trace, i.e. the total probability can always be normalized to unity); *ii)* is local in time (i.e. Markovian); *iii)* does not guarantee positivity in general.

Although the last point is of fundamental importance from a physical perspective, it is a common attitude in solid state physics to gauge the problem away, and keep on using the common Markov approximation, by looking at the possible lack of positivity as a transient and negligible effect. As opposed to this, we have performed a simulation of a semiconductor prototypical model, as simple as a two-level quantum dot, interacting with just one resonant phonon at thermal equilibrium. To be more specific, we chose the quantum dot unperturbed hamiltonian $H_{\mathrm{qd}} = \sum_{\alpha=1}^{2} \omega_\alpha c_\alpha^\dagger c_\alpha$, written in units of $\hbar$, where$\Delta\ \omega = \omega_1 - \omega_2 = 7.25$ THz and $c_\alpha$ is the annihilation operator for the level $\alpha$. Then the perturbation was taken to be $H' = \sum_{\alpha\beta} a_{\alpha\beta}\, c_\alpha^\dagger c_\beta\, b + \mathrm{h.c.}$, where $a_{\alpha\beta}$ are generic coupling parameters, and $b$ is the annihilation operator for the resonant phonon mode with bosonic population $N = (\exp\{\hbar\Delta\omega/kT\} - 1)^{-1}$, $k$ being the Boltzmann constant and $T$ the absolute temperature. Using the common Markov approximation, we have indeed found the thermal state of the quantum dot as the steady state of the Markov dynamics. But astonishingly enough, we were able to prove that, for quite general choices of the coupling parameters, even a perturbation of the thermal population probabilities as small as $10^{-5}$ could induce exponential divergence of the density matrix eigenvalues: non-negligible lack of positivity is reached within 10 ps (see Fig.(1). It is important to stress that these divergences cannot possibly come from numerical rounding errors, as the model has been solved analytically, and only at that point the numerical value of the parameters has been substituted. Moreover, divergences show up only *above* a critical temperature (computed from

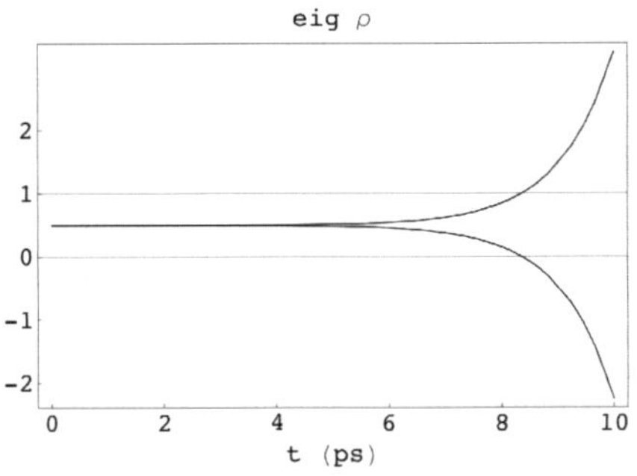


**Figure 1.** Time evolution of the density matrix eigenvalues with initial conditions $\rho_{11}(0){=}(1 + 10^{-5})\,\frac{N+1}{2N+1}$, $\rho_{22}(0) = 1 - \rho_{11}(0)$ and $\rho_{12}(0) = \rho_{21}(0)^* = 0$ (a very small perturbation of the thermal state). Parameters are$\Delta\ \omega{=}7.25$ THz, $a_{11} = 1/20$, $a_{22} = -1/20$, $a_{21} = 1/20\sqrt{2}$, $a_{12} = \sqrt{1/2 - 100a_{21}^2/10}$, $N{=}100$. Horizontal gridlines define the physical region.

the coupling parameters). Both the eigenvalue divergence *per se*, and its onset at very high temperatures, are clear hallmarks of a non-physical behavior, which in turn forces us to abandon the conventional Markovian approximation for large time modeling of nowadays semiconductor quantum devices.

In order to introduce our alternative formulation of the problem, let us go back to the integro-differential equation (2), and let us consider the following time symmetrization: given the two times $t'$ and $t$, we shall introduce the "average" or "macroscopic" time $T = \frac{t+t'}{2}$ and the "relative" time $\tau = t - t'$. This change of variable has very solid bases, as it is common and well established in a wide variety of contests, such as Wigner's phase-space formulation of quantum mechanics [16], standard quantum kinetics Green functions (see e.g. [17]), and even classical radiation theory (e.g., in the treatment of Bremsstrahlung): the basic idea is that the relevant time characterizing/describing our effective system evolution is the macroscopic time $T$. It is now easy to rewrite the second-order contribution in Eq. (2) in terms of these new time variables:

$$\frac{d}{dT}\hat{\rho}^i(T) = -\int_0^{t-t_\circ} d\tau \left[\hat{\mathcal{H}}^i\left(T + \frac{1}{2}\tau\right), \left[\hat{\mathcal{H}}^i\left(T - \frac{1}{2}\tau\right), \hat{\rho}^i\left(T - \frac{1}{2}\tau\right)\right]\right] . \qquad (3)$$

In the spirit of the adiabatic approximation previously recalled, the density-matrix operator $\hat{\rho}^i$ can be taken out of the time integral and evaluated at the current time $T$, so that $\hat{\rho}^i\left(T - \frac{1}{2}\tau\right) \rightarrow \hat{\rho}^i(T)$. As already stressed, this important approximation is valid when the density matrix varies slowly in interaction picture, that is, in the limit of weak-coupling. It is now convenient to replace the finite-domain time integration over $\tau$ by introducing a corresponding Gaussian correlation function $\exp\{-\frac{\tau^2}{2\bar{t}^2}\}$ whose width $\bar{t}$ may be regarded as a safe overestimation of the so-called "correlation time", which is shorter than the period of the system evolution; indeed, for $t - t_\circ$ greater than the correlation time, the time integration may be safely extended up to infinity. To compare with the exact dynamics in the weak-coupling limit, the correlation time has to finally be brought back to infinity: this in turn may be accomplished by a scaling property of the form $\bar{t}(g) \sim g^{-\xi}\,\overline{T}$, for $g \sim 0$, where $g$ is the coupling constant, $\overline{T}$ is a fixed reference time, and $\xi > 0$. Thus in brief, $i$) we assume that the density matrix moves slowly with respect to the correlation time $\bar{t}$; $ii$) we insert the gaussian weight $\exp\{-\frac{\tau^2}{2\bar{t}^2}\}$ inside the time integral in Eq.(3); $iii$) we put $t - t_\circ \to +\infty$ for the upper boundary of the latter time integral.

The second crucial step in order to get a genuine Lindblad superoperator for the global dynamics is to exploit once again the slowly-varying character of the density-matrix operator $\hat{\rho}^i$. The key idea is to perform on both sides of Eq. (3), once steps $i$)-$iii$) above are undertaken, a further "temporal coarse graining", i.e., a weighted time average on a so-called microscopic scale, a scale over which the variation of $\hat{\rho}^i(T)$ is negligible. Since in the small and intermediate coupling regime such time-scale is fully compatible with the correlation-time-scale $\bar{t}$, we shall perform such time average employing once again the Gaussian correlation function. Focussing on the skew-adjoint part of Eq. (3), i.e., the so-called scattering part, which is clearly our main interest (the self-adjoint part is just an energy-renormalization term and does not threaten positivity), we get:

$$\frac{d\hat{\rho}^i}{dT}(T) = -\frac{1}{\sqrt{2\pi}\bar{t}}\int dT'\, e^{-\frac{2T'^2}{\bar{t}^2}} \int d\tau e^{-\frac{\tau^2}{2\bar{t}^2}} \left[\hat{\mathcal{H}}^i\left(T+T'+\frac{\tau}{2}\right), \left[\hat{\mathcal{H}}^i\left(T+T'-\frac{\tau}{2}\right), \hat{\rho}^i(T)\right]\right] . \quad (4)$$

Moving back to the original Schrödinger picture and combining the two Gaussian distributions, the above equation can be rewritten in the following compact form:

$$\frac{d\hat{\rho}}{dT} = -\frac{1}{2}\left[\hat{\mathcal{L}}, \left[\hat{\mathcal{L}}, \hat{\rho}\right]\right] \qquad \text{with} \qquad \hat{\mathcal{L}} = \left(\frac{2}{\pi\bar{t}^2}\right)^{\frac{1}{4}} \int_{-\infty}^{\infty} dt'\, \hat{\mathcal{H}}^i(t')\, e^{-\frac{t'^2}{\bar{t}^2}} . \quad (5)$$

This is the genuine Lindblad-like superoperator we were looking for; indeed, the operators $\mathcal{L}$ are always Hermitian, and such effective dynamics is positive-definite. We stress how the proposed time symmetrization gives rise to a fully symmetric superoperator, compared to the strongly asymmetric Markov superoperator in [14].

Let us finally rewrite the new Markov superoperator (5) in our noninteracting basis $\lambda$, defined by the (possibly generalized) eigenvectors of $H_\circ$: we obtain an effective equation of motion of the form

$$\frac{d\rho_{\lambda_1\lambda_2}}{dt} = \frac{1}{2}\sum_{\lambda'_1\lambda'_2}\left[\mathcal{P}_{\lambda_1\lambda_2,\lambda'_1\lambda'_2}\rho_{\lambda'_1\lambda'_2} - \mathcal{P}_{\lambda_1\lambda'_2,\lambda'_1\lambda'_1}\rho_{\lambda'_2\lambda_2}\right] + \text{H.c.} \quad (6)$$

with symmetrized quantum scattering rates

$$\mathcal{P}_{\lambda_1\lambda_2,\lambda'_1\lambda'_2} = \frac{2\pi}{\hbar} H'_{\lambda_1\lambda'_1} H'^{*}_{\lambda_2\lambda'_2} \frac{1}{\sqrt{2\pi\bar{\epsilon}}} \exp\left\{-\frac{\left(\epsilon_{\lambda_1} - \epsilon_{\lambda'_1}\right)^2 + \left(\epsilon_{\lambda_2} - \epsilon_{\lambda'_2}\right)^2}{4\bar{\epsilon}^2}\right\} \quad (7)$$

substituting the strongly asymmetrical scattering superoperator given by the conventional Markov approximation [18]. Here, $\hbar$ has been shown explicitly, and $\bar{\epsilon} = \frac{\hbar}{\bar{t}}$ is a measure of the energy uncertainty in the interaction process induced by our temporal coarse graining. The above scattering superoperator can be regarded as a generalization of the conventional Fermi's golden rule to the density matrix formalism; indeed, in the semiclassical diagonal case ($\lambda_1 = \lambda_2, \lambda'_1 = \lambda'_2$) the above scattering superoperator boils down to what could be considered a dressed vertex-smoothed version of the Fermi's Golden Rule

$$P_{\lambda\lambda'} = \mathcal{P}_{\lambda\lambda,\lambda'\lambda'} = \frac{2\pi}{\hbar}|H'_{\lambda\lambda'}|^2 \frac{1}{\sqrt{2\pi\bar{\epsilon}}} \exp\left\{-\frac{(\epsilon_\lambda - \epsilon_{\lambda'})^2}{2\bar{\epsilon}^2}\right\}. \quad (8)$$

and in the limit of infinite correlation-time ($\bar{\epsilon} \to 0$) the standard Fermi's Golden Rule [19] is readily recovered.

In passing, we note that the transition rates (7) could be regarded as a "Quantum" version of the celebrated Fermi's Golden rule. This should not generate confusion: of course the Fermi's Golden Rule transition rates are computed according to quantum mechanical calculations (see [19] among thousands of textbooks), but, once computed, they give rise to the Boltzmann equation, which describes a *classical* Markov process [20] for *classical* probabilities. Instead, the transition rates (7) do *not* describe a classical Markov process, but rather its *quantum* analog: a so called Quantum Dynamical Semigroup [21] for the full density matrix.

Now, the whole theoretical scheme becomes meaningful and applicable only when a well-defined subsystem of interest is identified (together with a corresponding infinite-dimensional environment), so that its correlation time $\bar{t}$ can be estimated, and our (irreversible) semigroup dynamics can correctly describe the subsystem-projected (but fully reversible [22]) exact Hamiltonian dynamics. As final crucial step, we shall show that our conclusions about positivity remain valid no matter how the subsystem is chosen.

To this end, let us now consider a generic projection $P_0$/subsystem [23] on a subalgebra $\mathcal{X}$ of the observable algebra/global system in Heisenberg picture, which is also a completely positive map [24], and let $\{\hat{V}_\alpha\}$ be its Kraus decomposition [25], so that $P_0\hat{A} = \sum_\alpha \hat{V}_\alpha^\dagger \hat{A}\hat{V}_\alpha$. Then one could easily verify that the partial trace over a bath is embodied above as a particular case (see for example the formal approaches in [26]). We now observe that, due to its symmetry, our generator keeps the same form of Eq.(5) also in Heisenberg picture: by projecting the latter with $P_0$, and using the completeness relation $\sum_\alpha \hat{V}_\alpha^\dagger \hat{V}_\alpha = 1$, one can easily write the form for the subsystem's generator in Schrödinger picture, dual to the projected dynamics on the subalgebra $\mathcal{X}$, obtaining $\frac{d}{dT}\hat{\rho} = -\frac{1}{2}\sum_{\alpha\beta}\{\hat{D}_{\alpha\beta}^\dagger \hat{D}_{\alpha\beta}, \hat{\rho}\} + \sum_{\alpha\beta}\hat{D}_{\alpha\beta}\hat{\rho}\hat{D}_{\alpha\beta}^\dagger$. Here the "quantum transition amplitude" operators are given, according to (5), by $\hat{D}_{\alpha\beta} = \hat{V}_\alpha\hat{\mathcal{L}}\hat{V}_\beta$ (note that these operators are indeed $\bar{t}$-dependent, as can be seen from (5)). This Lindblad form [21] shows indeed that we have obtained the generator of the completely positive Quantum Dynamical Semigroup we were looking for. Moreover, the effort we made to consider this rather abstract class of projection is completely justified, as, for example, it paves the way for a new formalism for Quantum Transport: suppose that $\hat{P}$ is a projection in our Hilbert space that identifies, say, a one-dimensional nano-device, and let $\hat{Q}_l$ and $\hat{Q}_r$ project on the left and right contact respectively (obviously $\hat{P} + \hat{Q}_l + \hat{Q}_r = 1$). Then $P_0\hat{A} = \hat{P}\hat{A}\hat{P} + \hat{Q}_l\hat{A}\hat{Q}_l + \hat{Q}_r\hat{A}\hat{Q}_r$ does belong to the class of projections we have just studied, but the chosen subsystem does not come from a partial trace, nor does it have finite dimensions or discrete spectral properties: its weak-coupling dynamics needs the full power of our theory, in contrast with the previous ones [10, 14].

At this point a few comments are in order. As discussed extensively in [15], also for the simplest case of a standard two-level system —i.e., a generic quantum bit— the standard Markov superoperator predicts a non-trivial coupling between level population and polarization described by the so-called $T_3$ contributions. In contrast, for a two-level system coupled to its environment, the proposed quantum Fermi's golden rule does not predict any $T_3$ coupling term (they vanish in the infinite correlation time limit $\bar{\epsilon} \to 0$), thus providing a rigorous derivation of the well-known and successfully employed $T_1T_2$ dephasing model [27]. In general in fact, one could show that for subsystems with discrete spectra, the limit $\bar{\epsilon} \to 0$ reduces to Davies' theory [10]. However, for subsystems with continuous spectra, such limit is not defined, but for all finite collision times $\bar{t} > 0$ the proposed approach gives $\bar{t} \sim T_3$, so $T_3$ contributions are indeed present, but they become less and less important as the collision time $\bar{t}$ is raised, as it must be in the weak-coupling limit, to compare with the exact hamiltonian dynamics (see before).

It is also important to say that the pathological eigenvalue divergences we have outlined in our example, for a two-level quantum dot, are completely cured. This is so because the steady state in our approach can be always be shown to be thermal, in the infinite collision time limit, for discrete energy spectra and under suitable ergodic coupling hypotheses [10]. Finite collision times would then induce a shift in the thermal steady distribution, which in turn would always remain physically meaningful (due to the Lindblad structure of our scattering operator). A detailed study of the foretold shift goes however beyond the scope of the present paper.

To summarize, we have proposed a new approach to modeling nowadays semiconductor nanodevices, by critically reviewing the standard Markov procedure. Indeed, the latter does not preserve the positive-definite character of the density-matrix operator, thus leading to highly non-physical results. To overcome this serious limitation, we have identified an alternative and more general adiabatic procedure which (i) is physically justified under the same validity restrictions of the conventional Markov approach, (ii) in the semiclassical limit reduces to the standard Fermi's golden rule, and (iii) describes a genuine Lindblad evolution, thus providing a reliable/robust treatment of energy-dissipation and dephasing in state-of-the-art quantum devices. We stress that our formulation generalizes preexisting theories significantly, as it gives a positive dynamics for a considerably large class of projections, i.e. ways to chose the subsystem, and it is well defined even for infinitely extended subsystems (i.e. with continuous spectrum). In turn, this allows to investigate subsystems with both discrete and continuous spectra, a feature largely shared by mesoscale electronic and opto-electronic quantum devices; on the other side, it suggests a new way to treat electrical contacts for infinitely extended quantum devices, thus opening up the exciting possibility of a new formalism for Quantum Transport.

## References

[1] Jacoboni C & Lugli P 1989 *The Monte Carlo Method for Semiconductor Device Simulations* (Springer, Wien)
[2] Shah J 1992 *Hot Carriers in Semiconductor Nanostructures* (Academic Press inc., Boston)
[3] Schöll E 1998 *Theory of Transport Properties of Semiconductor Nanostructures* (Chapman and Hall, London)
[4] Rossi F 2005 *Semiconductor Macroatoms* (Imperial College Press, London)
[5] Fazio R 2005 *Solid State Quantum Information, New J. Phys.* **7**
[6] Rossi F and Kuhn T 2002 *Rev. Mod. Phys.* **74** 895
[7] Frensley W 1990 *Rev. Mod. Phys.* **62** 3
[8] Taj D, Genovese L and Rossi F 2006 *Europhys. Lett.* **74** 1060
[9] Dümcke R and Spohn H 1979 *Z. Phys. B* **34** 419
[10] Davies E B 1974 *Comm. Math. Phys.* **39** 91
[11] Sphon H 1980 *Rev. Mod. Phys.* **53** 569
[12] Alicki R 1977 *International Journal of Theoretical Physics* bf 16 351
[13] Fermi E 2005 *Nuclear Physics* (University of Chicago Press)
[14] Davies E B 1976 *Math. Ann.* **219** 147–158
[15] Iotti R C, Ciancio E and Rossi F 2005 *Phys. Rev. B* **72** 125347
[16] Kim Y S and Noz M E 1991 *Phase Space Picture of Quantum Mechanics*, World Scientific LNP, **40**
[17] Haug H and Jahuo A P 1996 *Quantum Kinetics in Transport and Optics of Semiconductors*, (Springer, Berlin)
[18] Taj D, Iotti R C and Rossi F 2009 *Semiconductor Science and Technology* **24** 065004.
[19] Cohen Tannoudji C. *et al.* 1977 *Quantum Mechanics*, Vol. **II**, John Wiley & Sons, New York.
[20] Ikeda N and Watanabe S 1981 *Stochastic differential equations and diffusion processes*, (North-Holland Publ.)
[21] Lindblad G 1976 *Comm. Math. Phys.* **48** 119
[22] Davies E B 1976 *Quantum Theory of Open Systems* (Academic Press, London)
[23] Taj D 2009 Weak-Coupling Limit. I *Preprint* arXiv:0905.1012v1
[24] Taj D 2009 Weak-Coupling Limit. II *Preprint* arXiv:0905.1019v1
[25] Kraus 1970 *Ann. Phys.* **64** 311
[26] Taj D and Rossi F 2008 *Phys. Rev.* A **78** 052113
[27] Shah J 1996 *Ultrafast Spectroscopy of Semiconductors and Semiconductor Nanostructures* (Springer, Berlin)

# Radiative and phonon-induced dephasing in double quantum dots

**Paweł Machnikowski**

Institute of Physics, Wrocław University of Technology, 50-370 Wrocław, Poland

E-mail: Pawel.Machnikowski@pwr.wroc.pl

**Abstract.** A simple method for describing the evolution of a quantum state of a double quantum dot system interacting simultaneously with the electromagnetic environment and with the lattice modes is developed. It is shown that the joint action of the two reservoirs leads to nontrivial effects in the system dephasing. As an example, the impact of phonon-induced initial dephasing on the radiative decay of delocalized exciton states is discussed.

## 1. Introduction

Double quantum dots (DQDs) are systems composed of two quantum dots placed close enough to each other for new coherent and collective phenomena to occur which cannot be reduced to the known properties of individual dots. Such new phenomena related to the optical properties of the system include modification of the optical response due to inter-dot coupling [1; 2] or collective effects in the spontaneous emission [3; 4]. Another class of effects is related with carrier-phonon interaction. Examples include phonon-assisted exciton transfer between the dots [5–7], modification of optical spectra due to phonon packets traveling between the dots [8], or phonon-induced decay of entanglement [9]. The special properties of DQDs may open the way to new applications, like long-time storage of quantum information [10], conditional optical control [1] that may lead to an implementation of a two-qubit quantum gate [11], generation of entangled photons [12] or coherent optical spin control and entangling [13–15]. Although both spontaneous emission and phonon-related phenomena have been studied to some extent for DQD systems, the interplay between these two classes of effects has not been investigated.

In the present contribution, we formulate a theoretical description of the evolution of a double quantum dot coupled simultaneously to the two reservoirs: the quantum electromagnetic field (photon vacuum) and lattice vibrational modes (phonons). After defining the model (Sec. 2), the approach to the simulation of the quantum open system is presented (Sec. 3), followed by an example of an interplay between the two couplings (Sec. 4).

## 2. The system

The system under study is composed of two stacked self-assembled QDs with transition energies $\epsilon_1$ and $\epsilon_2$, interacting with their phonon and photon (radiative) environments. We restrict the discussion to the ground states of excitons in each dot and assume that the spin polarizations of the excitons are fixed. As the exciton dissociation energy in absence of external electric fields is rather large (several to a few tens of meV), we consider only spatially direct exciton states, i.e., such that the electron-hole pairs reside in one and the same dot.

Journal of Physics: Conference Series **193** (2009) 012136 doi:10.1088/1742-6596/193/1/012136

Thus, the model includes four basis states: the state without excitons, $|0\rangle$, the states $|1\rangle$ and $|2\rangle$ with an exciton in the first and second dot, respectively, and the 'molecular biexciton' state $|3\rangle$ with excitons in both dots. The system evolution will be described in a 'rotating basis' defined by the unitary transformation

$$U = e^{iEt(|1\rangle\langle 1|+|2\rangle\langle 2|+2|3\rangle\langle 3|)/\hbar},$$

where $E = (\epsilon_1 + \epsilon_2)/2$. Note that this is not equivalent to interaction picture with respect to the Hamiltonian of uncoupled dots. Upon the standard weak-coupling derivation of the Master equation, the latter would always lead to uncorrelated emission for non-identical dots. On the contrary, the present transformation allows us to explicitly keep the energy difference $\Delta = (\epsilon_1 - \epsilon_2)/2$ which yields unitary evolution superposed on the dissipative one, consistent with the result of the Weisskopf–Wigner approach [3], and allows one to correctly account for the transition from "identical dots" to "different dots". The Hamiltonian is then

$$H = H_{\rm DQD} + H_{\rm ph} + H_{\rm rad} + H_{\rm c-ph} + H_{\rm c-rad}.$$

The first term describes exciton states in the DQD structure,

$$H_{\rm DQD} = \Delta(|1\rangle\langle 1| - |2\rangle\langle 2|) + V(|1\rangle\langle 2| + |2\rangle\langle 1|), \qquad (1)$$

where $V$ is the coupling between the dots, which may originate either from the Coulomb (Förster) interaction or from tunnel coupling (it can be assumed real). It is convenient to introduce the parametrization $\Delta = \mathcal{E}\cos 2\theta$, $V = \mathcal{E}\sin 2\theta$, where $2\mathcal{E}$ is the energy splitting between the single-exciton eigenstates of $H_{\rm DQD}$ and $\theta$ is the mixing angle of the single-exciton states. Electron and hole wave functions are modelled by identical anisotropic Gaussians with identical extensions $l$ in the $xy$ plane and $l_z$ along $z$ for both particles,

$$\psi_{1,2}(\boldsymbol{r}) \sim \exp\left[-\frac{1}{2}\frac{x^2+y^2}{l^2} - \frac{1}{2}\frac{(z \pm D/2)^2}{l_z^2}\right],$$

where $D$ is the distance between the dots.

The phonon modes are described by the free phonon Hamiltonian $H_{\rm ph} = \sum_{\boldsymbol{k}} \hbar\omega_{\boldsymbol{k}} b_{\boldsymbol{k}}^\dagger b_{\boldsymbol{k}}$, where $b_{\boldsymbol{k}}, b_{\boldsymbol{k}}^\dagger$ are bosonic operators of the phonon modes and $\omega_{\boldsymbol{k}}$ are the corresponding frequencies. Interaction of carriers confined in the DQD with phonons is modelled by the independent boson Hamiltonian

$$H_{\rm c-ph} = (|1\rangle\langle 1| + |3\rangle\langle 3|)\sum_{\boldsymbol{k}} f_{\boldsymbol{k}}^{(1)}(b_{\boldsymbol{k}}^\dagger + b_{-\boldsymbol{k}}) + (|2\rangle\langle 2| + |3\rangle\langle 3|)\sum_{\boldsymbol{k}} f_{\boldsymbol{k}}^{(2)}(b_{\boldsymbol{k}}^\dagger + b_{-\boldsymbol{k}}), \qquad (2)$$

where $f_{\boldsymbol{k}}^{(1,2)}$ are system-reservoir coupling constants. For Gaussian wave functions, the coupling constants for the deformation potential coupling between confined charges and longitudinal phonon modes have the form $f_{\boldsymbol{k}}^{(1,2)} = f_{\boldsymbol{k}} e^{\pm ik_z D/2}$, where

$$f_{\boldsymbol{k}} = (\sigma_{\rm e} - \sigma_{\rm h})\sqrt{\frac{k}{2\varrho v c_{\rm l}}}\exp\left[-\frac{l_z^2 k_z^2 + l^2 k_\perp^2}{4}\right].$$

Here $v$ is the normalization volume, $k_{\perp/z}$ are momentum components in the $xy$ plane and along the $z$ axis, $\sigma_{\rm e/h}$ are deformation potential constants for electrons/holes, $c_{\rm l}$ is the speed of longitudinal sound, and $\varrho$ is the crystal density. We assume that off-diagonal carrier–phonon couplings are negligible due to small overlap of the wave functions confined in different dots.

The third component in our modeling is the radiative reservoir (modes of the electromagnetic field), described by the Hamiltonian $H_{\rm rad} = \sum_{\boldsymbol{k},\lambda} \hbar w_{\boldsymbol{k}} c^{\dagger}_{\boldsymbol{k},\lambda} c_{\boldsymbol{k},\lambda}$, where $c_{\boldsymbol{k},\lambda}, c^{\dagger}_{\boldsymbol{k},\lambda}$ are photon creation and annihilation operators and $w_{\boldsymbol{k}}$ are the corresponding frequencies ($\lambda$ denotes polarizations). The QDs are separated by a distance much smaller than the relevant photon wavelength $\lambda = 2\pi\hbar c/E$, so that the spatial dependence of the EM field may be neglected (the Dicke limit). The Hamiltonian describing the interaction of carriers with the EM modes in the dipole and rotating wave approximations is

$$H_{\rm c-rad} = \Sigma_- \sum_{\boldsymbol{k},\lambda} e^{-iEt/\hbar} g_{\boldsymbol{k}\lambda} c^{\dagger}_{\boldsymbol{k},\lambda} + \text{H.c.}, \tag{3}$$

with $\Sigma_- = |0\rangle\langle 1| + |2\rangle\langle 3| + |0\rangle\langle 2| + |1\rangle\langle 3|$ and $g_{\boldsymbol{k}\lambda} = i\boldsymbol{d}\cdot\hat{e}_\lambda(\boldsymbol{k})\sqrt{\frac{\hbar w_{\boldsymbol{k}}}{2\varepsilon_0\varepsilon_{\rm r} v}}$, where $\boldsymbol{d}$ is the interband dipole moment, $\varepsilon_0$ is the vacuum permittivity, $\varepsilon_{\rm r}$ is the dielectric constant of the semiconductor and $\hat{e}_\lambda(\boldsymbol{k})$ is the unit polarization vector of the photon mode with the wave vector $\boldsymbol{k}$ and polarization $\lambda$. For wide-gap semiconductors with $E \sim 1$ eV, zero-temperature approximation may be used for the radiation reservoir at any reasonable temperature.

## 3. The system evolution

In certain limiting cases, analytical formulas for the evolution of the DQD system may be found. For uncoupled dots ($V = 0$) interacting only with lattice modes (phonons), an exact solution is available [9]. If only the radiative decay is included, a solution in the Markov limit can be obtained [3]. Here, we propose a general description which allows one to deal with the simultaneous action of both these environments. We describe the evolution of the reduced density matrix of the DQD system in the interaction picture with respect to $H_{\rm DQD}$ by the equation

$$\dot{\rho} = \mathcal{L}_{\rm rad}[\rho] + \mathcal{L}_{\rm ph}[\rho].$$

Here the first term describes the effect of the radiative decoherence in the Markovian limit in terms of the Lindblad dissipator

$$\mathcal{L}_{\rm rad}[\rho] = \Gamma_{\rm rad}\left[\Sigma_-(t)\rho\Sigma_+(t) - \frac{1}{2}\{\Sigma_+(t)\Sigma_-(t), \rho\}_+\right],$$

where $\Sigma_-(t) = \Sigma_+^{\dagger}(t) = e^{iH_{\rm DQD}t/\hbar}\Sigma_- e^{-iH_{\rm DQD}t/\hbar}$ and $\Gamma_{\rm rad} = E^3|\boldsymbol{d}|^2\sqrt{\epsilon_{\rm r}}/(3\pi\epsilon_0 c^3\hbar^4)$ is the spontaneous decay rate for a single dot. The second term accounts for the interaction with the non-Markovian phonon reservoir. We use the time-convolutionless equation

$$\mathcal{L}_{\rm ph}[\rho] = -\int_0^t d\tau\, \mathrm{tr}_{\rm ph}\left[H_{\rm c-ph}(t), [H_{\rm c-ph}(\tau), \rho(t)\otimes\rho_{\rm ph}]\right], \tag{4}$$

where $H_{\rm c-ph}(t) = e^{i(H_{\rm DQD}+H_{\rm ph})t/\hbar} H_{\rm c-ph} e^{-i(H_{\rm DQD}+H_{\rm ph})t/\hbar}$ is the carrier-phonon interaction Hamiltonian in the interaction picture, $\rho_{\rm ph}$ is the phonon density matrix at the thermal equilibrium, and $\mathrm{tr}_{\rm ph}$ denotes partial trace with respect to phonon degrees of freedom. While some phonon–assisted processes allow a Markovian limit, there are many effects that can only be reproduced if the reservoir memory is included (such as the initial phonon–assisted dephasing). Moreover, there seems to be no universal way to extract the Markov limit in various physical situations. By using the general Eq. (4), we obtain an all-purpose equation of motion at a modest computational cost.

In the evolution generated by the $H_{\rm DQD}$, the states $|0\rangle$ and $|3\rangle$ are invariant and nontrivial evolution takes place only in the subspace spanned by $|1\rangle, |2\rangle$. The corresponding evolution operator can therefore be found easily. Transforming Eq. (2) to the interaction picture and substituting to Eq. (4) we find the phonon-related contribution to the system evolution in the explicit form

$$\mathcal{L}_{\rm ph}[\rho]=\sum_{i=1,2}\left[T_i(t)\rho(t)S_i(t)-S_i(t)T_i(t)\rho(t)\right]+\text{H.c.}$$

Here the nonzero elements of the matrices $S_i$ are

$$\begin{aligned}\langle 1|S_1|1\rangle=1-\langle 2|S_1|2\rangle=1-\langle 1|S_2|1\rangle=\langle 2|S_2|2\rangle &= \frac{1}{2}\sin^2 2\theta\left(1-\cos\frac{2\mathcal{E}t}{\hbar}\right),\\ \langle 1|S_1|2\rangle=\langle 2|S_1|1\rangle^*=-\langle 1|S_2|2\rangle=-\langle 2|S_2|1\rangle^* &= \frac{i}{2}\sin 2\theta\sin\frac{2\mathcal{E}t}{\hbar}-\frac{1}{4}\sin 4\theta\left(1-\cos\frac{2\mathcal{E}t}{\hbar}\right),\\ \langle 3|S_1|3\rangle=\langle 3|S_2|3\rangle &= 1,\end{aligned}$$

and the matrices $T_i(t)$ are defined by

$$T_i(t)=\sum_{j=1,2}\int_0^t d\tau S_j(\tau)R_{ij}(t-\tau).$$

Here

$$\begin{aligned}R_{11}(t)=R_{22}(t) &= \frac{1}{\hbar^2}\sum_{\boldsymbol{k}}|f_{\boldsymbol{k}}|^2\left[n_{\boldsymbol{k}}e^{i\omega_{\boldsymbol{k}}t}+(n_{\boldsymbol{k}}+1)e^{-i\omega_{\boldsymbol{k}}t}\right],\\ R_{12}(t)=R_{21}(t) &= \frac{1}{\hbar^2}\sum_{\boldsymbol{k}}|f_{\boldsymbol{k}}|^2\cos k_z D\left[n_{\boldsymbol{k}}e^{i\omega_{\boldsymbol{k}}t}+(n_{\boldsymbol{k}}+1)e^{-i\omega_{\boldsymbol{k}}t}\right]\end{aligned}$$

are memory functions with $n_{\boldsymbol{k}}$ denoting the Bose distribution for phonon modes. The above equation of motion for the reduced density matrix strictly reproduces the results in the limiting cases mentioned above. Moreover, for the case of a DQD coupled to phonons with non-vanishing inter-dot coupling $V$, it yields results reasonably close to those obtained by a much more complex correlation expansion technique [5].

In numerical simulations, we take the parameters corresponding to a self-assembled InAs/GaAs system: $\sigma_{\rm e}-\sigma_{\rm h}=9$ eV, $\rho=5350$ kg/m$^3$, $c_{\rm l}=5150$ m/s, the wave function parameters $l=4.5$ nm, $l_z=1$ nm, $D=6$ nm, and the radiative recombination time (for a single dot) $1/\Gamma_{\rm rad}=400$ ps.

## 4. Example: Impact of phonon-induced pure dephasing on spontaneous emission

As an example of an application of the formalism developed above, let us study a simple example of interplay between phonon-related and radiative phenomena. First, we consider identical, uncoupled dots, that is, $V=0$, $\Delta=0$. In such case of two identical emitters, the spontaneous emission has a collective character which leads to a strong modification of the exciton recombination process [3]. In particular, out of the two delocalized single exciton states $|\psi_\pm\rangle=(|1\rangle\pm|2\rangle)/\sqrt{2}$, $|\psi_+\rangle$ has a "superradiant" character and decays twice faster than a single-dot excitation, while $|\psi_-\rangle$ is stable against recombination ("subradiant"). In Fig. 1(a), the decay of the state $|\psi_+\rangle$ in the absence of phonon perturbation is shown by the black dotted line.

For a single dot, phonon-induced dephasing (within an independent boson model with a harmonic reservoir) leads only to coherence decay on picosecond time scales and does not affect the radiative recombination on long time scales. This situation changes in quite an interesting manner in the two-dot case. As one can see in Fig. 1(a), in the presence of phonon-induced

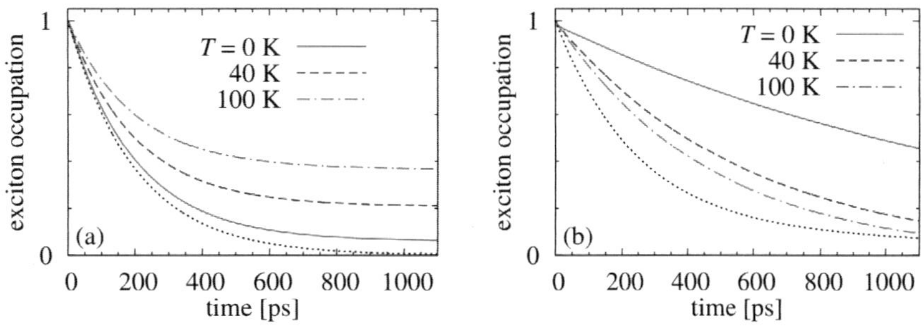


**Figure 1.** Decay of the exciton occupation for the superradiant initial state $|\psi_+\rangle$ in the absence of phonon-induced dephasing (black dotted lines) and in the presence of phonon-induced dephasing at various temperatures as shown. (a) identical, uncoupled dots, $\Delta = V = 0$; (b) non-identical, coupled dots, $\Delta = V = 1$ meV.

dephasing, the exciton recombination slows down and a long-lived occupation appears in the system. Moreover, the final exciton occupation grows as the temperature increases. Obviously, this effect must be due to some kind of coherence leading to occupation trapping in a subradiant state which is only possible in such a clear form because the model does not contain dephasing channels that would destroy this coherence (like anharmonicity-induced scattering [16]). Still, such a noise-induced coherence is remarkable.

The reason for this behavior is that during the first few picoseconds of the open system evolution the coherence between the single exciton states decays due to carrier-phonon dynamics according to $\langle 1|\rho|2\rangle = \frac{1}{2}e^{-h(t)}$, where $\rho$ is the density matrix of the exciton subsystem and

$$h(t) = 4\sum_{\boldsymbol{k}} \left|\frac{f_{\boldsymbol{k}}}{\omega_{\boldsymbol{k}}}\right|^2 \sin^2\frac{k_z D}{2}(1-\cos\omega_{\boldsymbol{k}} t) \xrightarrow{t\gg t_{\mathrm{ph}}} h_\infty = 4\sum_{\boldsymbol{k}} \left|\frac{f_{\boldsymbol{k}}}{\omega_{\boldsymbol{k}}}\right|^2 \sin^2\frac{k_z D}{2},$$

where $t_{\mathrm{ph}}$ is the characteristic time of the phonon-induced initial dephasing which is of the order of a few picoseconds [9]. Actually, the above formulas are valid in the absence of spontaneous emission. However, on the time scales $t \ll \Gamma$, when the dephasing takes place, radiative contribution can be approximately neglected. As a result, the initial density matrix $\rho = |\psi_+\rangle\langle\psi_+|$ is transformed into

$$\rho' = \frac{1+e^{-h_\infty}}{2}|\psi_+\rangle\langle\psi_+| + \frac{1-e^{-h_\infty}}{2}|\psi_-\rangle\langle\psi_-|.$$

This dephased state is a mixture of a superradiant and a subradiant component. The latter does not decay radiatively and leads to the persistent occupation tail visible in Fig. 1(a).

In the more realistic case of dots that differ in their transition energies and are coupled [Fig. 1(b)], the radiative recombination process still depends on temperature but this dependence is reversed: Now, the emission speeds up as the temperature is increased. In any case, however, the decay in the presence of carrier-phonon coupling is slower than in the absence of phonons. In order to explain this behavior, we note that although the eigenstates of $H_{\mathrm{DQD}}$ in the presence of energy mismatch are not purely sub- and superradiant one of them still decays faster than the other [3]. For $V > 0$, the lower eigenstate has a partly subradiant character. Moreover, the coupling between the dots enables excitation transfer between these eigenstates [5]. Coupling to phonons provides a thermalization mechanism for the occupations of the two eigenstates. At $T = 0$ this means a transfer to the lower-energy, subradiant state, which strongly suppresses emission. At higher temperatures, however, the contribution from the higher-energy, superradiant state increases and the radiative decay speeds up.

## 5. Conclusion

Double quantum dots not only show a richer structure of exciton states than a single dot but also admit much more complicated dephasing channels. The approach to numerical simulation of the quantum open system dynamics of a single DQD structure presented in this contribution opens the way to a systematic study of these effects. Although much more accurate (sometimes even exact) methods can be used in the case of two dots interacting with a single reservoir [5; 7; 17] treating the evolution under joint action of both reservoirs at the same level is either impossible or very demanding numerically. The method presented here, based on the TCL equation for the non-Markovian phonon-related effects and the Lindblad equation for spontaneous emission, yields simple equations for the complete biexciton density matrix of the DQD system and allows one to calculate arbitrary exciton-related properties of an undriven system (or a system excited with a single ultrashort pulse).

## Acknowledgments

This work was supported by the Polish MNiSW under Grant No. N N202 1336 33.

## References

[1] Unold T, Mueller K, Lienau C, Elsaesser T and Wieck A D 2005 *Phys. Rev. Lett.* **94** 137404
[2] Danckwerts J, Ahn K J, Förstner J and Knorr A 2006 *Phys. Rev. B* **73** 165318
[3] Sitek A and Machnikowski P 2007 *Phys. Rev. B* **75** 035328
[4] Sitek A and Machnikowski P 2009 *Phys. Rev. B* **80** 115301
[5] Rozbicki E and Machnikowski P 2008 *Phys. Rev. Lett.* **100** 027401
[6] Govorov A O 2003 *Phys. Rev. B* **68** 075315
[7] Richter M, Ahn K J, Knorr A, Schliwa A, Bimberg D, Madjet M E A and Renger T 2006 *Phys. Stat. Sol. (b)* **243** 2302
[8] Huneke J, Krügel A, Kuhn T, Vagov A and Axt V M 2008 *Phys. Rev. B* **78** 085316
[9] Roszak K and Machnikowski P 2006 *Phys. Rev. A* **73** 022313
[10] Pazy E, D'Amico I, Zanardi P and Rossi F 2001 *Phys. Rev. B* **64** 195320
[11] Biolatti E, Iotti R C, Zanardi P and Rossi F 2000 *Phys. Rev. Lett.* **85** 5647
[12] Gywat O, Burkard G and Loss D 2002 *Phys. Rev. B* **65** 205329
[13] Troiani F, Molinari E and Hohenester U 2003 *Phys. Rev. Lett.* **90** 206802
[14] Nazir A, Lovett B W, Barrett S D, Spiller T P and Briggs G A D 2004 *Phys. Rev. Lett.* **93** 150502
[15] Gauger E M, Nazir A, Benjamin S C, Stace T M and Lovett B W 2008 *New J. Phys.* **10** 073016
[16] Machnikowski P 2006 *Phys. Rev. Lett.* **96** 140405
[17] Krügel A, Axt V M and Kuhn T 2006 *Phys. Rev. B* **73** 035302

# AUTHOR INDEX

# AUTHOR INDEX

# AUTHOR INDEX

# AUTHOR INDEX

# AUTHOR INDEX